Lecture Notes in Computer Science 8560

Advanced Research in Computing and Software Science
Subline of Lectures Notes in Computer Science

Gilles Dowek (Ed.)

Rewriting and Typed Lambda Calculi

Joint International Conference, RTA-TLCA 2014
Held as Part of the Vienna Summer of Logic, VSL 2014
Vienna, Austria, July 14-17, 2014
Proceedings

Springer

Volume Editor

Gilles Dowek
Inria
23 avenue d'Italie, CS 81321
75214 Paris Cedex 13, France
E-mail: gilles.dowek@inria.fr

ISSN 0302-9743 e-ISSN 1611-3349
ISBN 978-3-319-08917-1 e-ISBN 978-3-319-08918-8
DOI 10.1007/978-3-319-08918-8
Springer Cham Heidelberg New York Dordrecht London

Library of Congress Control Number: 2014942549

LNCS Sublibrary: SL 1 – Theoretical Computer Science and General Issues

Typesetting: Camera-ready by author, data conversion by Scientific Publishing Services, Chennai, India

Printed on acid-free paper

Springer is part of Springer Science+Business Media (www.springer.com)

logic **n. 1** the science of reasoning.
– ORIGIN from Greek *logikē tekhnē* 'art of reason'.

Foreword

In the summer of 2014, Vienna hosted the largest scientific conference in the history of logic. The Vienna Summer of Logic (VSL, http://vsl2014.at) consisted of twelve large conferences and 82 workshops, attracting more than 2000 researchers from all over the world. This unique event was organized by the Kurt Gödel Society and took place at Vienna University of Technology during July 9 to 24, 2014, under the auspices of the Federal President of the Republic of Austria, Dr. Heinz Fischer.

The conferences and workshops dealt with the main theme, logic, from three important angles: logic in computer science, mathematical logic, and logic in artificial intelligence. They naturally gave rise to respective streams gathering the following meetings:

Logic in Computer Science / Federated Logic Conference (FLoC)

- 26th International Conference on Computer Aided Verification (CAV)
- 27th IEEE Computer Security Foundations Symposium (CSF)
- 30th International Conference on Logic Programming (ICLP)
- 7th International Joint Conference on Automated Reasoning (IJCAR)
- 5th Conference on Interactive Theorem Proving (ITP)
- Joint meeting of the 23rd EACSL Annual Conference on Computer Science Logic (CSL) and the 29th ACM/IEEE Symposium on Logic in Computer Science (LICS)
- 25th International Conference on Rewriting Techniques and Applications (RTA) joint with the 12th International Conference on Typed Lambda Calculi and Applications (TLCA)
- 17th International Conference on Theory and Applications of Satisfiability Testing (SAT)
- 76 FLoC Workshops
- FLoC Olympic Games (System Competitions)

Mathematical Logic

- Logic Colloquium 2014 (LC)
- Logic, Algebra and Truth Degrees 2014 (LATD)
- Compositional Meaning in Logic (GeTFun 2.0)
- The Infinity Workshop (INFINITY)
- Workshop on Logic and Games (LG)
- Kurt Gödel Fellowship Competition

Logic in Artificial Intelligence

- 14th International Conference on Principles of Knowledge Representation and Reasoning (KR)
- 27th International Workshop on Description Logics (DL)
- 15th International Workshop on Non-Monotonic Reasoning (NMR)
- 6th International Workshop on Knowledge Representation for Health Care 2014 (KR4HC)

The VSL keynote talks which were directed to all participants were given by Franz Baader (Technische Universität Dresden), Edmund Clarke (Carnegie Mellon University), Christos Papadimitriou (University of California, Berkeley) and Alex Wilkie (University of Manchester); Dana Scott (Carnegie Mellon University) spoke in the opening session. Since the Vienna Summer of Logic contained more than a hundred invited talks, it would not be feasible to list them here.

The program of the Vienna Summer of Logic was very rich, including not only scientific talks, poster sessions and panels, but also two distinctive events. One was the award ceremony of the Kurt Gödel Research Prize Fellowship Competition, in which the Kurt Gödel Society awarded three research fellowship prizes endowed with 100.000 Euro each to the winners. This was the third edition of the competition, themed Logical Mind: Connecting Foundations and Technology this year.

The 1st FLoC Olympic Games formed the other distinctive event and were hosted by the Federated Logic Conference (FLoC) 2014. Intended as a new FLoC element, the Games brought together 12 established logic solver competitions by different research communities. In addition to the competitions, the Olympic Games facilitated the exchange of expertise between communities, and increased the visibility and impact of state-of-the-art solver technology. The winners in the competition categories were honored with Kurt Gödel medals at the FLoC Olympic Games award ceremonies.

Organizing an event like the Vienna Summer of Logic was a challenge. We are indebted to numerous people whose enormous efforts were essential in making this vision become reality. With so many colleagues and friends working with us, we are unable to list them individually here. Nevertheless, as representatives of the three streams of VSL, we would like to particularly express our gratitude to all people who helped to make this event a success: the sponsors and the Honorary Committee; the Organization Committee and the local organizers; the conference and workshop chairs and Program Committee members; the reviewers and authors; and of course all speakers and participants of the many conferences, workshops and competitions.

The Vienna Summer of Logic continues a great legacy of scientific thought that started in Ancient Greece and flourished in the city of Gödel, Wittgenstein and the Vienna Circle. The heroes of our intellectual past shaped the scientific world-view and changed our understanding of science. Owing to their achievements, logic has permeated a wide range of disciplines, including computer science, mathematics, artificial intelligence, philosophy, linguistics, and many more. Logic is everywhere – or in the language of Aristotle, πάντα πλήρη λογικῆς τέχνης.

July 2014

Matthias Baaz
Thomas Eiter
Helmut Veith

Preface

This volume contains the proceedings of the 25th International Conference on Rewriting Techniques and Applications (RTA 2014) and 12th International Conference on Typed Lambda Calculus and Applications (TLCA 2014), which were held jointly during July 14–17 2014, in Vienna, Austria.

RTA is the major forum for the presentation of research on all aspects of rewriting. Previous RTA conferences were held in Dijon (1985), Bordeaux (1987), Chapel Hill (1989), Como (1991), Montreal (1993), Kaiserslautern (1995), New Brunswick (1996), Sitges (1997), Tsukuba (1998), Trento (1999), Norwich (2000), Utrecht (2001), Copenhagen (2002), Valencia (2003), Aachen (2004), Nara (2005), Seattle (2006), Paris (2007), Hagenberg/Linz (2008), Brasilia (2009), Edinburgh (2010), Novi Sad (2011), Nagoya (2012), and Eindhoven (2013).

TLCA is the major forum for the presentation of research on all aspects of typed lambda calclus. Previous TLCA conferences were held in Utrecht (1993), Edinburgh (1995), Nancy (1997), L'Aquila (1999), Kraków (2001), Valencia (2003), Nara (2005), Paris (2007), Brasilia (2009), Novi Sad (2011), and Eindhoven (2013).

Of the 87 papers submitted to RTA-TLCA 2014, the Programme Committee selected 31 for presentation at the conference. Moreover, it nominated the paper "Implicational Relevance Logic is 2-ExpTime-Complete" by Sylvain Schmitz for the Best Paper Award and "A Coinductive Confluence Proof for Infinitary Lambda-Calculus" by Łukasz Czajka for the Best Student Paper Award.

We are grateful to Nicola Gambino, Manfred Schmidt-Schauss, and Nobuko Yoshida for accepting our invitation to present selected topics of their choice and for enriching the conference by their invited talks, their participation, and their contributions to the present proceedings.

We used the EasyChair system for many aspects of the reviewing process. We wish to thank Andrei Voronkov and all others of the EasyChair team for this invaluable tool. Moreover, we wish to thank the editorial board of ARCoSS, in particular Vladimiro Sassone, for agreeing to publish these proceedings as a volume in the ARCoSS series.

RTA-TLCA 2014 was organized as part of the Federated Logic Conference (FLoC 2014), itself part of the Vienna Summer of Logic (VSL 2014). We would like to thank Gernot Salzer, the conference chair of RTA-TLCA 2014, as well as the other members the Organizing Committees of FLOC 2014 and VSL 2014 for their indispensable contribution to the success of this conference.

Last but not least we want to thank all members of the RTA-TLCA 2014 Program Committee for their reviews, their constructive comments, and the pleasant collaboration.

July 2014 Gilles Dowek

Organization

Program Committee

Andreas Abel	Gothenburg University, Sweden
Beniamino Accattoli	Università di Bologna, Italy
Zena Ariola	University of Oregon, USA
Pierre Clairambault	CNRS, France
Ugo Dal Lago	Università di Bologna, Italy
Fer-Jan de Vries	University of Leicester, UK
Gilles Dowek	Inria, France
Santiago Escobar	Universitat Politècnica de València, Spain
Amy Felty	University of Ottawa, Canada
Maribel Fernández	King's College London, UK
Adrià Gascón	SRI International, USA
Masahito Hasegawa	Kyoto University, Japan
Olivier Hermant	MINES ParisTech, France
Jeroen Ketema	Imperial College London, UK
Christopher Lynch	Clarkson University, USA
Paul-André Melliès	CNRS, France
Alexandre Miquel	Universidad de la República Uruguay, Uruguay
César Muñoz	NASA, USA
Vivek Nigam	Universidade Federal da Paraíba, Brazil
Luke Ong	University of Oxford, UK
Brigitte Pientka	McGill University, Canada
Jakob Rehof	Technical University of Dortmund, Germany
David Sabel	Goethe-University Frankfurt am Main, Germany
Sylvain Salvati	Inria, France
Gernot Salzer	Technische Universität Wien, Austria
Aleksy Schubert	University of Warsaw, Poland
Peter Selinger	Dalhousie University, Canada
Paula Severi	University of Leicester, UK
Christian Sternagel	University of Innsbruck, Austria
Kazushige Terui	RIMS, Kyoto University, Japan
Steffen van Bakel	Imperial College London, UK
Femke van Raamsdonk	VU University Amsterdam, The Netherlands
Mateu Villaret	Universitat de Girona, Spain
Harald Zankl	University of Innsbruck, Austria

Invited Talk
(Abstract)

A Unified Approach to Univalent Foundations and Homotopical Algebra

Nicola Gambino

Voevodsky's Univalent Foundations of Mathematics programme seeks to develop a new approach to the foundations of mathematics, based on dependent type theories extended with axioms inspired by homotopy theory. The most remarkable of these new axioms is the so-called Univalence Axiom, which allows us to treat isomorphic types as if there were equal.

Quillen's homotopical algebra, instead, provides a category-theoretic framework in which it is possible to develop an abstract version of homotopy theory, giving a homogeneous account of several situations where objects are identified up to a suitable notion of 'weak equivalence'. The central notion here is that of a model category, examples of which arise naturally in several different areas of mathematics.

The aim of this talk is to explain how the type theories considered in Univalent Foundations and the categorical structures considered in homotopical algebra are related, but different, and to describe categorically the common core of Univalent Foundations and homotopical algebra, which allows a simoultaneous development of the two subjects. The axiomatisation will be based on work of several authors, including Awodey, van den Berg, Garner, Joyal, Lumsdaine, Shulman and Warren.

Acknowledgement. This material is based on research sponsored by the Air Force Research Laboratory, under agreement number FA8655-13-1-3038. The U.S. Government is authorized to reproduce and distribute reprints for Governmental purposes notwithstanding any copyright notation thereon. The views and conclusions contained herein are those of the authors and should not be interpreted as necessarily representing the official policies or endorsements, either expressed or implied, of the Air Force Research Laboratory or the U.S. Government.

Table of Contents

Process Types as a Descriptive Tool for Interaction*
Control and the Pi-Calculus

Kohei Honda[1,**], Nobuko Yoshida[2], and Martin Berger[3]

[1] Queen Mary, University of London, London, UK
[2] Imperial College London, London, UK
[3] University of Sussex, Sussex, UK

Abstract. We demonstrate a tight relationship between linearly typed π-calculi and typed λ-calculi by giving a type-preserving translation from the call-by-value $\lambda\mu$-calculus into a typed π-calculus. The $\lambda\mu$-calculus has a particularly simple representation as typed mobile processes. The target calculus is a simple variant of the linear π-calculus. We establish full abstraction up to maximally consistent observational congruences in source and target calculi using techniques from games semantics and process calculi.

1 Introduction

At TLCA 2001 [4] the authors started a research programme relating typed λ-calculi and typed π-calculi [5, 17, 18, 39, 40]. The rationale behind the programme has been twofold: first to demonstrate that functional computation can be decomposed into name-passing and thus be fruitfully understood as a constrained and well-behaved form of interaction. Secondly, the authors wanted their nascent investigations of the vast space of typing systems for interacting processes be guided by the λ-calculus community's insights into the nature of typing, one of the great contributions to computer science. A key aspect of our methodology for generalising λ-calculus types to interaction has been to study full-abstraction: Milner's encoding of untyped λ-calculus [24] is sound but not complete in the absence of types. This is because atomic β-reduction steps are decomposed into multiple name-passing interactions where intermediate steps can be observed. To achieve completeness, the ability to observe such intermediate steps must be ruled out by typing – in other words, only processes that interact 'functionally' should be well-typed. A key insight has been the usefulness of *linearly* typed π-calculi for the understanding of typed λ-calculi. Linearity and it's close cousin *affinity* yield full abstraction using proof techniques coming from games semantics, linear logic and process calculi. This paper continues our programme by studying the call-by-value $\lambda\mu$-calculus [26], henceforth $\lambda\mu$v-calculus, a λ-calculus with non-local control-flow manipulation (often referred to as "control" or "jumping"). The $\lambda\mu$v-calculus is a proof-calculus for classical propositional logic. Non-local control can be encoded in pure λ-calculus using continuation-passing style, but a direct representation of jumping is of interest, not just because of the connection with classical logic, and because

* Partially supported by EPSRC EP/K011715/1, EP/L00058X/1 and EP/K034413/1.

** http://mrg.doc.ic.ac.uk/kohei.html [29,33]

G. Dowek (ed.): RTA-TLCA 2014, LNCS 8560, pp. 1–20, 2014.

it leads to more readable programs in comparison with continuation-passing style, but also because jumping can be seen as a form of interaction that is less restrictive than the last-in/first-out calling discipline of functional computation.

Technically, we present a type-preserving translation from the $\lambda\mu$v-calculus into a typed π-calculus, and show that it is fully abstract up to maximally consistent observational congruences in source and target calculi. Full abstraction is proved via an inverse transformation from the typed π-terms which inhabit the $\lambda\mu$v-types into the $\lambda\mu$v-calculus [26].

There are different notions of control. In the π-calculus they can be represented as distinct forms of typed interaction. The $\lambda\mu$-calculus introduced by Parigot [27] provides unrestricted control, and was later studied in call-by-value form by Ong and Stewart [26]. Surprisingly, as this paper demonstrates, unrestricted control has a particularly simple representation as typed name passing processes; processes used for the embedding are exactly characterised as a proper subset of the linear π-calculus introduced in [39], with a clean characterisation in types and behaviour. The linear π-calculus can embed, for example, the simply typed λ-calculus full abstractly. We call the subset of the linear π-calculus corresponding to full control the π^{c}-calculus ("c" indicates control). The π^{c}-calculus is restricted in that each channel is used only for a unique stateless replicated input and for zero or more dual outputs. The π^{c}-calculus also precludes circular dependency between channels. In spite of its simplicity, both, call-by-value and call-by-name full control, are precisely embeddable into the π^{c}-calculus by changing the translation of types. Because the π^{c}-calculus is a proper subset of the linear π-calculus, many known results about the linear π-calculus can be carried over to the π^{c}-calculus. This can be used for establishing properties of the $\lambda\mu$v-calculus. For example, strong normalisability of the π^{c}-calculus is an immediate consequence of the same property of the linear π-calculus, and that can be used for showing strong normalisability of the $\lambda\mu$v-calculus [24]. The tight operational and equational correspondence enables the use of typed π-calculi for investigating and analysing control operators and calculi in a uniform setting, possibly integrated with other language primitives and operational structures. After studying the call-by-value $\lambda\mu$-calculus, we also demonstrate applicability of our framework by an embedding of the call-by-name $\lambda\mu$-calculus into the same π^{c}-calculus by changing translation of types.

Section 2 summarises the π^{c}-calculus. Section 3 gives our encoding of the $\lambda\mu$v-calculus. Section 4 establishes full abstraction. The paper concludes with an outline of the call-by-name encoding, and discusses related work as well as open issues. Full proofs and additional discussions are delegated to [1].

2 Processes and Types

Processes. Types for processes prescribe usage of names. To be able to do this with precision, it is important to control dynamic sharing of names. For this purpose, it is useful to restrict name passing to *bound (private) name passing*, where only bound names are passed in interaction. This allows tighter control of sharing without losing essential expressiveness, making it easier to administer name usage in more stringent ways. The resulting calculus is an asynchronous version of the πI-calculus [31] and has

(Structural Rules)

(S0) $P \equiv Q \quad \text{if } P \equiv_\alpha Q$

(S1) $P|\mathbf{0} \equiv P$ (S2) $P|Q \equiv Q|P$

(S3) $P|(Q|R) \equiv (P|Q)|R$

(S4) $(\nu\, x)\mathbf{0} \equiv \mathbf{0}$

(S5) $(\nu\, x)(\nu\, y)P \equiv (\nu\, y)(\nu\, x)P$

(S6) $(\nu\, x)(P|Q) \equiv ((\nu\, x)P)|Q \quad (x \notin \mathsf{fn}(Q))$

(S7) $\overline{x}(\vec{y})\overline{z}(\vec{w})P \equiv \overline{z}(\vec{w})\overline{x}(\vec{y})P \quad (x, z \notin \{\vec{w}\vec{y}\})$

(S8) $(\nu\, z)\overline{x}(\vec{y})P \equiv \overline{x}(\vec{y})(\nu\, z)P \quad (z \notin \{x\vec{y}\})$

(S9) $\overline{x}(\vec{y})(P|Q) \equiv (\overline{x}(\vec{y})P)|Q \quad (\{\vec{y}\} \cap \mathsf{fn}(Q) = \emptyset)$

(Reduction)

(Com$_!$)
$! \, x(\vec{y}).P \,|\, \overline{x}(\vec{y})Q \longrightarrow ! \, x(\vec{y}).P|(\nu\, \vec{y})(P|Q)$

(Res)
$P \longrightarrow Q \implies (\nu\, x)P \longrightarrow (\nu\, x)Q$

(Par)
$P \longrightarrow P' \implies P|Q \longrightarrow P'|Q$

(Out)
$P \longrightarrow Q \implies \overline{x}(\vec{y})P \longrightarrow \overline{x}(\vec{y})Q$

(Cong)
$P \equiv P' \longrightarrow Q' \equiv Q \implies P \longrightarrow Q$

Fig. 1. Reduction and Structural Rules

expressive power equivalent to the calculus with free name passing (for the result in the typed setting, see [39]). In the present study, the restriction to bound name passing leads to, among others, a clean inverse transformation from the π-calculus into the $\lambda\mu$-calculus. The grammar of the calculus is given below.

$$P \;::=\; !\,x(\vec{y}).P \;\mid\; \overline{x}(\vec{y})P \;\mid\; P|Q \;\mid\; (\nu\, x)P \;\mid\; \mathbf{0}$$

The initial x in $!\,x(\vec{y}).P$ and $\overline{x}(\vec{y})P$ is called *subject*. We write $!x.P$ for $!x(\epsilon).P$ and $\overline{x}P$ for $\overline{x}(\epsilon)P$, where ϵ denotes the empty vector. $|$ is parallel composition, $!\,x(\vec{y}).P$ is replicated input, and $\overline{x}(\vec{y})P$ is asynchronous bound output, $(\nu\, x)P$ is name hiding and $\mathbf{0}$ denotes nil. The full definition of the reduction rules and the structure rules is found in Figure 1 ($\longrightarrow$ is generated from the given rules; $\equiv$ is generated from the given rules together with the closure under all contexts). We write $\twoheadrightarrow$ for $\longrightarrow^* \cup \equiv$.

Types. First we introduce *channel types*. They indicate possible usage of channels.

$$\tau \;::=\; (\vec{\tau})^p \qquad\qquad p ::= \,! \mid\, ?$$

$\tau, \tau', \ldots$ (resp. $p, p', \ldots$) range over types (resp. modes). $!$ and $?$ are called *server* mode and *client* mode, respectively, and they are *dual* to each other. Here, by server we mean that the process is waiting with an input to be invoked. Conversely, a process is a client, if its next action is sending a message to a server. We write $\mathsf{md}(\tau)$ for the outermost mode of τ. For example, $\mathsf{md}((\tau_1\,\tau_2)^!) = !$. We write $()^p$ for $(\epsilon)^p$, which stands for a channel that carries no names. We further demand the following condition to hold for channel types. A channel type τ is *IO-alternating* if, for each of its subexpression $(\tau_1..\tau_n)^p$, if $p = !$ (resp. $p = ?$) then each $\mathsf{md}(\tau_i) = ?$ (resp. $\mathsf{md}(\tau_i) = !$). *Hereafter we assume all channel types we use are IO-alternating*. The *dual of* τ, written $\overline{\tau}$, is defined as the result of dualising all modes in τ_i. For example, $(\overline{\tau_1}\,\overline{\tau_2})^?$ is the dual of the above type. To guarantee the uniqueness of a server (replicated) process, we introduce the partial operation $\odot$ on types, which is generated from: $\tau \odot \overline{\tau} = \overline{\tau} \odot \tau = \overline{\tau}$ and $\tau \odot \tau = \tau$ with $(\mathsf{md}(\tau) = ?)$. Note $\odot$ is indeed partial since it is not defined in other cases. This operation means that a server should be unique, but an arbitrary number of clients can request interactions. To guarantee the second condition, we introduce an *action type* ranged over by $A, B, C\ldots$. The syntax is given as follows:

$$A \ ::= \ \emptyset \mid x{:}\tau \mid x{:}(\vec{\tau_1})^! \rightarrow y{:}(\vec{\tau_2})^? \mid A, B$$

The idea behind this definition is that action types are graphs where nodes are of the form $x : \tau$, provided names like x occur at most once.

Typing. The typing judgement are of the form $\vdash_\phi P \triangleright A$ which is read: *P has type A with mode* ϕ where *IO-modes*, $\phi \in \{\text{I}, \text{O}\}$, guarantees a restriction to a single thread. We present the typing system in Appendix A. The rules are obtained just by restricting the typing system in [39] to the replicated fragment of the syntax we are now using. The resulting typed calculus is called π^{C}. The subject reduction of π^{C} is an immediate consequence of that in [39], since both the action types and the reduction of the present calculus are projection of those of the sequential linear π-calculus in [39, §5.3].

Proposition 2.1. (Subject Reduction) *If* $\vdash_\phi P \triangleright A$ *and* $P \longrightarrow Q$ *then* $\vdash_\phi Q \triangleright A$.

In addition to the standard reduction, we define an extended notion of reduction, called *the extended reduction*, written $\searrow$, again precisely following [39]. We shall use this reduction extensively in the present study. While $\longrightarrow$ gives a natural notion of dynamics which makes sense in both sequential and concurrent computation, $\searrow$ extends $\longrightarrow$ by exploiting the stateless nature of π^{C}-processes. It offers a close correspondence with the reduction in the $\lambda\mu$v-calculus through the encoding. For that reason $\searrow$ is useful for studying the correspondence between two calculi. Formally $\searrow$ is the least compatible relation, i.e. closed under typed context, taken modulo $\equiv$, that includes:

$$C[\overline{x}(\vec{y})P] | !x(\vec{y}).Q \searrow_r C[(\boldsymbol{\nu}\, \vec{y})(P|Q)] \mid !x(\vec{y}).Q \qquad (\boldsymbol{\nu}\, x)!x(\vec{y}).Q \searrow_g \mathbf{0}$$

where $C[\,\cdot\,]$ is an arbitrary (typed) context. We can immediately see that $\longrightarrow \subset \searrow$. Note $\searrow$ calculates *under* prefixes, which is unusual in process calculi. Another observation is that a given typed process in the π^{C}-calculus can have at most one redex for the standard reduction $\longrightarrow$ while it may have more than one redex for $\searrow$. The extended reduction $\searrow$ is the exact image of extended reduction in [39] onto the present subcalculus, so that we immediately conclude, from the results in [39]:

Proposition 2.2. *1.* (Subject Reduction) *If* $\vdash_\phi P \triangleright A$ *and* $P \searrow Q$ *then* $\vdash_\phi Q \triangleright A$.
2. (CR) *If P is typable and* $P \searrow Q_i$ $(i = 1, 2)$ *with* $Q_1 \not\equiv Q_2$*, we have* $Q_i \searrow^+ R$ $(i = 1, 2)$ *for some R.*
3. (SN) *If P is typable then P does not have infinite* $\searrow$*-reductions.*

It may be useful to state at this point that, possibly contrary to what is suggested by the asymmetric notation, $\searrow$ does *not* introduce a new form of computation step, a new form of interaction. Instead, $P \searrow Q$ says that P and Q cannot be distinguished by *well-typed* observers. This indistinguishability is an artefact of our restrictive typing discipline and does not hold in the untyped calculus. The notation was chosen to emphasise that Q in $P \searrow Q$ is 'smaller' or more reduced than P in a sense that can be made precise.

There are three further observations on the extended reduction. First, while we do not use the property directly in the present work, the convertibility induced by $\searrow$ (i.e. the typed congruent closure of $\searrow$) coincides with the weak bisimilarity $\approx$ [39, Theorem 4.1], because the transition relation is the faithful image of that of the pure sequential linear π-calculus in [39]. Second, Proposition 2.2 (3) indicates all π^{c}-processes are represented by their $\searrow$-normal forms, i.e. those π^{c}-processes which do not have a $\searrow$-redex, which own a very simple syntactic structure characterised inductively. Finally, in the definition of $\searrow$ it is not necessary to cater for replicated inputs occurring freely under other input prefixes as that is impossible by typing. Similarly, any replicated input with free subject under an output can be put into parallel with that output by the structural rules in the typed setting.

Definition 2.1. *Let the set* NF_e *of* π^{c}*-processes be generated by the following induction, assuming typability in each clause. (1)* $\mathbf{0} \in \mathsf{NF}_e$*; (2) if* $P, Q \in \mathsf{NF}_e$ *and* P *and* Q *do not share a common free name of different polarities, then* $P|Q \in \mathsf{NF}_e$*; (3)* $P \in \mathsf{NF}_e$ *then* $!x(\vec{y}).P \in \mathsf{NF}_e$*; (4)* $\overline{x}(\vec{y})P \in \mathsf{NF}_e$ *if* $P \in \mathsf{NF}_e$ *and* $\overline{x}(\vec{y})P$ *is a prime output, where we call* $\overline{x}(\vec{y})P$ prime *if the initial* x *is its only free name not under input prefix; and (5) If* $P \in \mathsf{NF}_e$ *and* $P \equiv Q$ *then* $Q \in \mathsf{NF}_e$*. Clearly if* P *is typable and* $P \not\searrow$ *then* $P \in \mathsf{NF}_e$*.*

Contextual Congruence for π^{c}. The Church-Rosser property of typed processes, as stated in Proposition 2.2, suggests that non-deterministic state change (which plays a basic role in e.g. bisimilarity and testing/failure equivalence) may safely be ignored in typed equality, so that a Morris-like contextual equivalence suffices as a basic equality over processes. Let us define:

$$P \Downarrow_x \text{ iff } P \twoheadrightarrow \overline{x}(\vec{y})Q \text{ for some } Q$$

We can now define a basic typed congruence. Below, a relation over typed processes is *typed* if it relates only processes with identical action type and IO-mode. If $\mathcal{R}$ is a typed relation and $\vdash_\phi P_{1,2} \triangleright A$ are related by $\mathcal{R}$ then we write $\vdash_\phi P_1 \mathcal{R} P_2 \triangleright A$ or, when no confusion arises, $P_1 \mathcal{R} P_2$. A relation is a *typed congruence* when (1) $\mathcal{R} \supseteq \equiv$, and (2) $\mathcal{R}$ is a typed equivalence relation closed under typed contexts (note we are taking $\equiv$ as if it were the α-equality: this is essentially because the notion of reduction depends on this relation, just as reduction in the λ-calculus depends on the α-equality).

Definition 2.2. $\cong_\pi$ is the maximum typed congruence satisfying: if $\vdash_{\mathrm{O}} P \cong_\pi Q \triangleright x : ()^?$, then $P \Downarrow_x$ iff $Q \Downarrow_x$.

Below a typed congruence is *maximally consistent* [15] if adding any additional equation to it leads to inconsistency, i.e. equations on all processes with identical typing.

Proposition 2.3. (1) $\searrow \subset \cong_\pi$; (2) $\cong_\pi$ *is a maximally consistent typed congruence;* (3) $\cong_\pi$ *is the unique maximally consistent congruence containing* $\searrow$*.*

Our choice of observable in π^{c}corresponds to the usual output-barbed congruence one considers in the untyped calculus. It is also *the* canonical choice for the calculus fragment under discussion, for the following reasons. $?$-actions are not considered as observables in linear/affine π-calculi [4, 39] since, intuitively, invoking replicated processes do not affect them. Proposition 2.3 suggests that the existence/non-existence of $?$-actions may be the only sensible way to obtain a non-trivial large equality in π^{c}, equationally justifying the use of $?$-actions as observables.

3 Encoding

Call-by-value $\lambda\mu$-calculus. This section presents a type-preserving embedding of the call-by-value $\lambda\mu$-calculus by Ong and Stewart [26] in π^{c}. Apart from tractable syntactic properties of the calculus in comparison with its call-by-name counterpart, [26] showed how various control primitives of call-by-value languages (such as call/cc in ML) can be encoded in this calculus and its extension with recursion [26]. The calculus represents full control in a call-by-value setting, just like the call-by-value λ-calculus with Felleisen's $\mathcal{C}$ operator.

Types $(\alpha, \beta, \ldots)$ are those of simply typed λ-calculus with the atomic type $\bot$ (we can add other atomic types with appropriate values and operations on them). We use *variables* $(x, y, \ldots)$ and *control variables* (or *names*) $(a, b, \ldots)$. *Preterms* $(M, N, \ldots)$ and *values* $(V, W, \ldots)$ are generated from the grammar:

$$M, N ::= x \mid \lambda x^{\alpha}.N \mid MN \mid \mu a^{\alpha}.M \mid [a]M \qquad V, W ::= x \mid \lambda x^{\alpha}.N$$

Apart from variables, abstraction and application, we have a *named term* $[a]M$ and a *μ-abstraction* $\mu a.M$, both of which use names. The typing judgement has the form $\Gamma \vdash M : \alpha; \Delta$ where Γ is a finite map from variables to types, M is a preterm given above, and Δ is a finite map from names to non-$\bot$-types. The typing rules are given below:

(Id)
$$\frac{-}{\Gamma \cdot x{:}\alpha \vdash x{:}\alpha\,;\Delta}$$

(C-var)
$$\frac{\Gamma \cdot x{:}\alpha \cdot y{:}\alpha \vdash M{:}\beta\,;\Delta}{\Gamma \cdot z{:}\alpha \vdash M\{z/xy\}{:}\beta\,;\Delta}$$

(C-name)
$$\frac{\Gamma \vdash M{:}\beta\,;\Delta \cdot a{:}\alpha \cdot b{:}\alpha}{\Gamma \vdash M\{c/ab\}{:}\beta\,;\Delta \cdot c{:}\alpha}$$

($\Rightarrow$-I)
$$\frac{\Gamma \cdot x{:}\alpha \vdash M{:}\beta\,;\Delta}{\Gamma \vdash \lambda x^{\alpha}.M{:}\alpha \Rightarrow \beta\,;\Delta}$$

($\Rightarrow$-E)
$$\frac{\Gamma \vdash M{:}\alpha \Rightarrow \beta\,;\Delta \quad \Gamma \vdash N{:}\alpha\,;\Delta}{\Gamma \vdash MN{:}\beta\,;\Delta}$$

($\bot$-I)
$$\frac{\Gamma \vdash M{:}\alpha\,;\Delta\alpha \neq \bot}{\Gamma \vdash [a]M{:}\bot\,;\Delta \cdot a{:}\alpha}$$

($\bot$-E)
$$\frac{\Gamma \vdash M{:}\bot\,;\Delta \cdot a{:}\alpha}{\Gamma \vdash \mu a^{\alpha}.M : \alpha\,;\Delta}$$

In the rules, we assume newly introduced names/variables in the conclusion are always fresh. The notation $\Gamma \cdot x : \tau$ indicates x is not in the domain of Γ. $M\{z/xy\}$ denotes the result of substituting z in M for both x and y, similarly for $M\{c/ab\}$. A typable preterm is called a *$\lambda\mu v$-term*. The reduction rules for the $\lambda\mu$v-calculus is given next:

(β_v) $(\lambda x.M)V \longrightarrow M\{V/x\}$

(ζ_{arg}) $V^{\alpha \Rightarrow \beta}(\mu a^{\alpha}.M) \longrightarrow \mu b.(M\{\,[b](V[\cdot])\,/\,[a][\cdot]\,\})$

(η_v) $\lambda x.(Vx) \longrightarrow V \quad (x \notin \mathsf{fv}(V))$

$(\zeta_{\mathrm{fun},\bot})$ $(\mu a^{\alpha \Rightarrow \bot}.M)N \longrightarrow M\{\,[\cdot]N\,/\,[a][\cdot]\,\}$

$(\mu\text{-}\beta)$ $[b]\mu a.M \longrightarrow M\{b/a\}$

$(\zeta_{\mathrm{arg},\bot})$ $V^{\alpha \Rightarrow \bot}(\mu a^{\alpha}.M) \longrightarrow M\{\,V[\cdot]\,/\,[a][\cdot]\,\}$

$(\mu\text{-}\eta)$ $\mu a.[a]M \longrightarrow M \quad (a \notin \mathsf{fn}(M))$

$(\bot)$ $V^{\bot \Rightarrow \beta}M \longrightarrow \mu b^{\beta}.M \quad (b \text{ fresh})$

(ζ_{fun}) $(\mu a^{\alpha \Rightarrow \beta}.M)N \longrightarrow \mu b.M\{\,[b]([\cdot]N)\,/\,[a][\cdot]\,\}$

$(\bot_{\bot})$ $V^{\bot \Rightarrow \bot}M \longrightarrow M$

We let $\beta \neq \bot$. In $(\zeta_{\mathrm{arg}}, \zeta_{\mathrm{arg},\bot})$, $\alpha \neq \bot$. In the rules we include η_v-reduction, unlike [26]. Inclusion or non-inclusion does not affect the subsequent technical development. Some rules use substitution $M\{\,C[\cdot]\,/\,[a][\cdot]\,\}$ defined next, assuming the bound name convention. Note that substitution is applied in a nested fashion in the last line.

$$
\begin{aligned}
x\{C[\cdot]/[a][\cdot]\} &\stackrel{\text{def}}{=} x\\
(\lambda x.M)\{C[\cdot]/[a][\cdot]\} &\stackrel{\text{def}}{=} \lambda x.(M\{C[\cdot]/[a][\cdot]\})\\
MN\{C[\cdot]/[a][\cdot]\} &\stackrel{\text{def}}{=} (M\{C[\cdot]/[a][\cdot]\})(N\{C[\cdot]/[a][\cdot]\})\\
(\mu b^{\alpha}.M)\{C[\cdot]/[a][\cdot]\} &\stackrel{\text{def}}{=} \mu b^{\alpha}.(M\{C[\cdot]/[a][\cdot]\})\\
([a']M)\{C[\cdot]/[a][\cdot]\} &\stackrel{\text{def}}{=} \begin{cases}[a'](M\{C[\cdot]/[a][\cdot]\}) & (a\neq a')\\ [a'](C[M\{C[\cdot]/[a][\cdot]\}]) & (a=a')\end{cases}
\end{aligned}
$$

Encoding (1): Types. The general idea of the encoding is simple, and closely follows the standard call-by-value encoding of the λ-calculus, due to Milner [24]. The reading is strongly operational, elucidating the dynamics of $\lambda\mu$-terms up to a certain level of abstraction. Given a $\lambda\mu$-term, $\Gamma \vdash M : \alpha; \Delta$, its encoding considers Γ as the interaction points of the program/process where it queries the environment and gets information; while either at its main port, typed as α, or at one of the control variables given as Δ, the program/process would return a value: at which port it would return depends on how its sequential thread of control will proceed during execution. If Δ is empty, this reading precisely coincides with Milner's original one [24]. One of the distinguishing features of the π-calculus encodings of programming languages in general (including those for untyped calculi) and that of the present encoding in particular, is that the operational interpretation of this sort in fact obeys a clean and rigid type structure.

We start with the encoding of types, using two maps, $\alpha^{\bullet}$ and α°. Intuitively α° maps α as a type for values; while $\alpha^{\bullet}$ maps α as a type for threads which may converge to values of type α or which may diverge, or "computation" in Moggi's terminology [25].

$$
\alpha^{\bullet} \stackrel{\text{def}}{=} \begin{cases}\epsilon & (\alpha=\bot)\\ (\alpha^{\circ})^{?} & (\alpha\neq\bot)\end{cases}
\qquad
(\alpha\Rightarrow\beta)^{\circ} \stackrel{\text{def}}{=} \begin{cases}(\beta^{\bullet})^{!} & (\alpha=\bot)\\ (\overline{\alpha^{\circ}}\beta^{\bullet})^{!} & (\alpha\neq\bot)\end{cases}
$$

Note a type for computation is the lifting of a type for values. The encoding of $\bot$ indicates that we assume there is no (closed) value, or a proof without assumptions, inhabiting $\bot$. This leads to the degenerate treatment of $(\bot \Rightarrow \alpha)^{\bullet}$ since "asking at the assumed absurdity" does not make sense. By "degenerate" we mean that the argument in $(\bot \Rightarrow \alpha)$ is simply ignored.

Example 3.1 As simple examples, consider: $(\bot \Rightarrow \bot)^{\circ} \stackrel{\text{def}}{=} ()^{!}$ and $((\bot \Rightarrow \bot) \Rightarrow \bot)^{\circ} \stackrel{\text{def}}{=} (()^{?})^{!}$. Note if $\alpha \neq \bot$ we always have $(\alpha \Rightarrow \bot)^{\circ} = (\overline{\alpha^{\circ}})^{!}$ which corresponds to the standard translation, $\neg A \stackrel{\text{def}}{=} A \supset \bot$.

Following the mappings of types, the environments for variables and names are mapped as follows, starting from $\emptyset^{\bullet} \stackrel{\text{def}}{=} \emptyset$ and $\emptyset^{\circ} \stackrel{\text{def}}{=} \emptyset$.

$$
(a{:}\alpha\cdot\Delta)^{\bullet} \stackrel{\text{def}}{=} \begin{cases}a{:}\alpha^{\bullet}\cdot\Delta^{\bullet} & (\alpha\neq\bot)\\ \Delta^{\bullet} & (\alpha=\bot)\end{cases}
\qquad
(x{:}\alpha\cdot\Gamma)^{\circ} \stackrel{\text{def}}{=} \begin{cases}x{:}\overline{\alpha^{\circ}}\cdot\Gamma^{\circ} & (\alpha\neq\bot)\\ \Gamma^{\circ} & (\alpha=\bot)\end{cases}
$$

The special treatment of $\bot$ follows the encoding of types above and reflects its special role in classical natural deduction. Simply put, if we have a proof whose conclusion

$$[\![x:\alpha]\!]_u \stackrel{\text{def}}{=} \begin{cases} \overline{u}\langle x^{\overline{\alpha^\circ}}\rangle & (\alpha \neq \bot) \\ \mathbf{0} & (\alpha = \bot) \end{cases} \quad [\![\lambda x^\alpha.M : \alpha \Rightarrow \beta]\!]_u \stackrel{\text{def}}{=} \begin{cases} \overline{u}(c)!c(xz).[\![M:\beta]\!]_z & (\alpha \neq \bot, \beta \neq \bot) \\ \overline{u}(c)!c(z).[\![M:\beta]\!]_z & (\alpha = \bot, \beta \neq \bot) \\ \overline{u}(c)!c(x).[\![M:\bot]\!]_z & (\alpha \neq \bot, \beta = \bot) \\ \overline{u}(c)!c.[\![M:\bot]\!]_z & (\alpha = \bot, \beta = \bot) \end{cases}$$

$$[\![MN:\beta]\!]_u \stackrel{\text{def}}{=} \begin{cases} [\![M:\alpha\Rightarrow\beta]\!]_m\{m(c) = ([\![N:\alpha]\!]_n\{n(e) = \overline{c}\langle eu^{\overline{\alpha^\circ\beta^\circ}}\rangle\})\} & (\alpha\neq\bot, \beta\neq\bot) \\ [\![M:\alpha\Rightarrow\beta]\!]_m\{m(c) = \overline{c}\langle u^{\overline{\beta^\circ}}\rangle\} & (\alpha=\bot, \beta\neq\bot) \\ [\![M:\alpha\Rightarrow\beta]\!]_m\{m(c) = [\![N:\alpha]\!]_u\} & (\alpha=\bot) \end{cases}$$

$$[\![[a]M:\bot]\!]_u \stackrel{\text{def}}{=} [\![M:\alpha]\!]_m\{a/m\} \quad [\![\mu a^\alpha.M:\alpha]\!]_u \stackrel{\text{def}}{=} [\![M:\bot]\!]_m\{u/a\}$$

Fig. 2. Encoding of $\lambda\mu$-terms

is the falsity $\bot$, then it is given there, for its all usefulness, for the purpose of having a contradiction and negating a stipulated assumption. Operationally this suggests the proof whose (conclusion's) type is $\bot$ has nothing positive to communicate to the outside, which explains why the map for computation $(\cdot)^\bullet$ ignores the control channel of type $\bot$. Dually you get no information from the proof of type $\bot$, so querying at that environment port is insignificant, hence we ignore $\bot$-types in the negative positions.

Encoding (2): Terms. For the encoding of terms, we introduce the following notations, which we shall use throughout the paper. Below in (3) we use the notation from [12, Remark 15] in the context of CPS calculus (cf. Section 5).

1. (copycat) Let τ be an input type. Then $[x \to y]^\tau$, *copy-cat of type* τ, is inductively defined by the following clause.

$$[x \to x']^{(\tau_1..\tau_n)^!} \stackrel{\text{def}}{=} !x(\vec{y}).\overline{x'}(\vec{y'})\Pi_{1\leq i\leq n}[y'_i \to y_i]^{\overline{\tau_i}}$$

   where $\prod_{1\leq i\leq n} P_i$ (or $\prod_i P_i$) stands for the n-fold parallel composition $P_1|\cdots|P_n$.
2. (free output) $\overline{x}\langle\vec{y}^{\vec{\tau}}\rangle \stackrel{\text{def}}{=} \overline{x}(\vec{z})\Pi[z_i \to y_i]^{\overline{\tau_i}}$ with each τ_i having an output mode.
3. (substitution environment) $P\{x(\vec{y}) = R\} \stackrel{\text{def}}{=} (\boldsymbol{\nu}\, x)(P \mid !x(\vec{y}).R)$.

Figure 2 presents the encoding of terms. The encoding closely follows that of types, mapping a typing judgement $\Gamma \vdash M:\alpha\,;\Delta$ and a fresh name (called *anchor*) to a process. We omit the type environment from the source term in Figure 2. In each rule, we assume newly introduced names (among others an anchor) are always fresh. The anchor u in $[\![M:\alpha]\!]_u$ represents the point of interaction which M may have as a process [24] or, more concretely, the channel through which the process returns the resulting value to the environment. The process $[\![M:\alpha]\!]_u$ may also have interactions at its free variables (for querying information) and at its free control variables (for returning values). Note both of them are now channel names.

Proposition 3.1. *$\Gamma \vdash M:\alpha\,;\Delta$ implies* $\vdash_{\mathsf{O}} [\![M:\alpha]\!]_u \triangleright (u:\alpha\cdot\Delta)^\bullet,\ \Gamma^\circ$.

In Proposition 3.1, the type of the term and the types of control names are both mapped with $(\ )^\bullet$, conforming to the shape of the sequent $\Gamma \vdash M : \alpha; \Delta$. In particular, there is no causality arrow in the types for translations of $\lambda\mu$-terms. This is because all types (including environments and types for names) are mapped to output types, and causality can only from **!** to **?**.

Example 3.2 (variable) As a simplest example, consider $[\![x : \bot]\!]_u \stackrel{\text{def}}{=} \mathbf{0}$. Since $(x : \bot)^\circ = (u : \bot)^\bullet = \emptyset$, we have $\vdash_{\mathsf{O}} [\![x : \bot]\!]_u \triangleright (u : \bot)^\bullet, (x : \bot)^\circ$ This encoding intuitively represents a trivial proof which assumes $\bot$ and concludes $\bot$, or, in the terminology of Linear Logic, the axiom link of the empty type.

Example 3.3 (identity, 1) By closing x in Example 3.2, $[\![\lambda x^\bot.x : \bot \Rightarrow \bot]\!]_u \stackrel{\text{def}}{=} \overline{u}(c)!c.\mathbf{0}$. Since $(\bot \Rightarrow \bot)^\bullet = (()^!)^?$, we have $\vdash_{\mathsf{O}} [\![\lambda x^\bot.x : \bot \Rightarrow \bot]\!]_u \triangleright u{:}(\bot \Rightarrow \bot)^\bullet$.

Example 3.4 (identity, 2) If $\alpha \neq \bot$, then $[\![\lambda x^\alpha.x : \alpha \Rightarrow \alpha]\!]_u \stackrel{\text{def}}{=} \overline{u}(c)!c(xz).\overline{z}\langle x^{\overline{\alpha^\circ}}\rangle$.

Example 3.5 (control operator, 1) The following term essentially corresponds to $\mathcal{C}$ in $\lambda\mathcal{C}v$ introduced by Felleisen and his colleagues [10,11]. Logically it is a shortest proof of $\neg\neg A \supset A$. Below we let $\neg\alpha \stackrel{\text{def}}{=} \alpha \Rightarrow \bot$: $\aleph \stackrel{\text{def}}{=} \lambda z^{\neg\neg\alpha}.\mu a^\alpha.z(\lambda x^\alpha.[a]x)$. Its direct encoding is, assuming $\alpha \neq \bot$:

$$[\![\aleph]\!]_u \stackrel{\text{def}}{=} \overline{u}(c)!c(za).(\boldsymbol{\nu}\, m)(\overline{m}\langle z\rangle \mid !m(z).(\boldsymbol{\nu}\, n)(\overline{n}(f)!f(x).\overline{a}\langle x\rangle \mid !n(f).\overline{z}\langle f\rangle))$$

which, through a couple of $\searrow$ uses, can be simplified into $\overline{u}(c)!c(za).\overline{z}(f)!f(x).\overline{a}\langle x\rangle$. This agent first signals itself: then it is invoked with a function in the environment (of type $\neg\neg\alpha$) as an argument and a continuation a (of type α), invokes the former with the identity agent (whose continuation is a) and a continuation a. Then if that function asks back at the identity with an argument, say x, then this x is returned to a as the answer to the initial invocation. Note how the π^{c}-translation makes explicit the operational content of the agent, especially when simplified using $\searrow$.

Example 3.6 (control operator, 3) The following is the well-known witness of Peirce's law, $((A \supset B) \supset A) \supset A$, and corresponds to `call/cc` in Scheme.

$$\kappa \stackrel{\text{def}}{=} \lambda y^{(\alpha \Rightarrow \beta) \Rightarrow \alpha}.\mu a^\alpha.[a](y(\lambda x^\alpha.\mu b^\beta.[a]x)).$$

The direct encoding becomes:

$$[\![\kappa]\!]_u \stackrel{\text{def}}{=} \overline{u}(c)!c(za).(\boldsymbol{\nu}\, m)(\overline{m}\langle z\rangle | !m(z).(\boldsymbol{\nu}\, n)(\overline{n}(f)!f(xb).\overline{a}\langle x\rangle | !n(f).\overline{z}\langle fa\rangle))$$

which is simplified with $\searrow$ into: $\overline{u}(c)!c(ya).\overline{y}(fa')(!f(xb).\overline{a}\langle x\rangle \mid [a' \to a])$. The process first signals itself at u, then, when invoked with an argument y and a return point a, asks at y with an argument f and a new return point a'. Then whichever is invoked, it would return with the received value to the initial return point a. Note that the only difference from the encoding of $\aleph$ is whether, in addition to the invocation of the identity function at f, there is the possibility that the direct return comes from the environment: the difference, thus, is, in the standard execution, whether it preserves a current stack to forward the value from the environment or not.

Correspondence in Dynamics. The dynamics of $\lambda\mu$-calculi, including its call-by-name and call-by-value versions, has additional complexity due to the involvement of μ-abstraction. Among others it becomes necessary to use a nested context substitution

$M\{C[\cdot] / [a][\cdot]\}$ when μ-abstraction and application interact. In the following we analyse the dynamics of the $\lambda\mu$v-calculus through the embedding, using the interaction-oriented dynamics of π^{c}. The strong normalisability of $\lambda\mu$v-reduction is an immediate consequence of this analysis.

We need some preparations. First, for a $\lambda\mu$v-term which is also a value, the following construction is useful.

Definition 3.1. Let $\Gamma \vdash V : \alpha; \Delta$. Then we set $[\![V]\!]^{\mathsf{val}}_m \stackrel{\text{def}}{=} P$ iff $[\![V]\!]_u \stackrel{\text{def}}{=} \overline{u}(m)P$.

Note $\overline{u}(m)[\![V]\!]^{\mathsf{val}}_m$ is identical with $[\![V]\!]_u$ up to alpha-equality. Note further, by typing, $[\![V]\!]^{\mathsf{val}}_m$ always has the form $!m(\vec{y}).P$. These observations are useful when we think about the encodings, especially when we apply extended reduction on them.

We are now ready to embark on the analysis of $\lambda\mu$v-reduction through its encoding into π^{c}. Suppose we have reduction $M \longrightarrow M'$ for a $\lambda\mu$v-term M. Using the definitions above, the generation of reduction can be attributed to one of the following cases: (1) (β_v)-rule or (η_v)-rule; (2) one of the μ-reduction rules; or (3) one of the ζ-reduction rules. Of those, ζ-reductions require the most attention. Instead of considering the general case (which we shall treat later), let us first take a look at the following concrete $\lambda\mu$v-reduction. Below f and g are typed as $\alpha \Rightarrow \gamma$ and α.

$$M \stackrel{\text{def}}{=} (\mu a^{\alpha\Rightarrow\beta}.[a]\lambda y^{\alpha}.\mu e^{\beta}.[a]f)g \longrightarrow \mu b^{\beta}.[b](\lambda y.\mu e.[b](fg))g \stackrel{\text{def}}{=} M' \quad (1)$$

The encoding into π^{c}elucidates ζ-reductions on the uniform basis of name passing interaction. Let us first encode M, writing $\overline{c}\langle\!\langle xu\rangle\!\rangle$ for $(\boldsymbol{\nu}\, n)(!n(y).\overline{c}\langle yu\rangle|\overline{n}\langle x\rangle)$:

$$[\![M : \alpha \Rightarrow \beta]\!]_u \stackrel{\text{def}}{=} (\boldsymbol{\nu}\, a)(\overline{a}(c)!c(ye).\overline{a}\langle f\rangle \mid !a(c).\overline{c}\langle\!\langle gu\rangle\!\rangle) \quad (2)$$

On the right of (2), we find two $\searrow_r$-redexes (apart from in $\overline{c}\langle\!\langle gu\rangle\!\rangle$), two outputs and a shared input at a, which are ready to interact. Redexes for the ζ-reduction now arise explicitly as redexes for interactions. Note also these redexes do not depend on whether the argument (g above) is a value or not, explaining the shape of (ζ_{fun}).

To see how M' in (1) results from M in the encoding, we "copy" replications to make these two redexes contiguous, obtaining:

$$(\boldsymbol{\nu}\, a)(\overline{a}(c)!c(ye).(\boldsymbol{\nu}\, a)(\overline{a}\langle f\rangle \mid !a(c).\overline{c}\langle\!\langle gu\rangle\!\rangle) \mid !a(c).\overline{c}\langle\!\langle gu\rangle\!\rangle) \quad (3)$$

This term is an intermediate form before reducing the mentioned two redexes in (2) and is behaviourally equivalent to (2) (even in the untyped weak bisimilarity). We observe:

$$[\![M' {:} \alpha{\Rightarrow}\beta]\!]_u \stackrel{\text{def}}{=} (\boldsymbol{\nu}\, a)(\overline{a}(c)!c(ye).(\boldsymbol{\nu}\, a')(\overline{a'}\langle f\rangle \mid !a'(c).\overline{c}\langle\!\langle gu\rangle\!\rangle) \mid !a(c).\overline{c}\langle\!\langle gu\rangle\!\rangle)$$

so the intermediate form (3) is nothing but the encoding of M'. This also shows if we really reduce the two $\searrow$-redexes from (2), the result goes past (3). In general, $M \longrightarrow M'$ does not imply $[\![M]\!]_u \searrow^{+} [\![M']\!]_u$ since $[\![M]\!]_u$ reduces a little further than $[\![M']\!]_u$. However $[\![M']\!]_u$ can catch up with the result by reducing the mentioned two redexes in (3).Based on this observation, we formally state the main result. Below size(M) is the size of M, which is inductively defined as: size$(x) = 1$, size$([a]M) = 1 + $ size(M), size$(\lambda x.M) = $ size$(M) + 1$, size$(\mu a.M) = 1 + $ size(M), size$(MN) = $ size$(M) +$ size(N). We use this index for maintaining the well-ordering on reduction. Below $\rightarrow_{\lambda\mu\mathsf{v}}$ is the reduction relation on $\lambda\mu$-terms presented in [26].

Proposition 3.2. *$M \to_{\lambda\mu\mathsf{v}} M'$ with M and M' typed implies either $[\![M:\alpha]\!] \equiv [\![M':\alpha]\!]$ such that $size(M) \gneq size(M')$, or $[\![M:\alpha]\!]_u \searrow^+ P$ such that $[\![M':\alpha]\!]_u \searrow^* P$. In particular, $\to_{\lambda\mu\mathsf{v}}$ on $\lambda\mu$-terms is strongly normalising.*

4 Decoding and Full Abstraction

Canonical Normal Forms. In the previous section we have shown that types and dynamics of $\lambda\mu\mathsf{v}$-terms are faithfully embeddable into π^{c}. In this section we show this embedding is as faithful as possible — if a process lives in the encoding of a $\lambda\mu\mathsf{v}$-type, then it is indeed the image of a $\lambda\mu\mathsf{v}$-term of that type. This result corresponds to the standard definability result in denotational semantics, and immediately leads to full abstraction for a suitably defined observational congruence for $\lambda\mu\mathsf{v}$.

A key observation towards definability is that we can algorithmically translate back processes having the encoded $\lambda\mu\mathsf{v}$-types into the original $\lambda\mu\mathsf{v}$-terms. To study the decoding, it is convenient to introduce *canonical normal forms* (CNFs) [2, 4, 19], which are essentially a subset of $\lambda\mu\mathsf{v}$-terms whose syntactic structures precisely correspond to their process representation.

First, *CNF preterms* ($\mathtt{N}, \ldots$) and *CNF value preterms* ($\mathtt{U}, \ldots$) are given by:

$$\mathtt{N} ::= \mathtt{c} \mid \lambda x^\alpha.\mathtt{N} \mid \mathtt{let}\ x = y\mathtt{U}\ \mathtt{in}\ \mathtt{N} \mid \mathtt{let}\ \_ = y\mathtt{U} \mid [a]\mathtt{U} \mid \mu a^\alpha.\mathtt{N}$$
$$\mathtt{U} ::= \mathtt{c} \mid \lambda x^\alpha.\mathtt{N} \mid \mu a^\alpha.[a]\mathtt{U}$$

We further assume the following conditions on CNF preterms: (1) In $[a]\mathtt{N}$, $\mathtt{N}$ does not have form $\mu b^\beta.\mathtt{N}'$. (2) In $\mu a^\alpha.\mathtt{N}$, (a) if $\mathtt{N}$ is $[a]\mathtt{U}$ then $a \in \mathsf{fn}(\mathtt{U})$; and (b) if $\mathtt{N}$ is $\mathtt{let}\ x = y\mathtt{U}'\ \mathtt{in}\ \mathtt{N}'$ then $a \in \mathsf{fn}(\mathtt{U}')$. (3) In $\mu a^\alpha.[a]\mathtt{U}$, $a \in \mathsf{fn}(\mathtt{U})$. The conditions 1, 2-a and 3 are to avoid a μ-redex. The condition 2-b is to determine the shape of a normal form, since without this condition $\mu a.\mathtt{let}\ x = y\mathtt{U}'\ \mathtt{in}\ \mathtt{N}'$ can be written $\mathtt{let}\ x = y\mathtt{U}'\ \mathtt{in}\ \mu a.\mathtt{N}'$.

Under these conditions, the set of CNFs are those which are typable by the following typing rules combined with those for the $\lambda\mu$-calculus except the rule for application.

(⊥-const)
$$\frac{-}{\Gamma\cdot x:\bot \vdash \mathtt{c}:\bot\,;\Delta}$$

(let)
$$\frac{\Gamma\cdot x:\beta \vdash \mathtt{N}:\gamma\,;\Delta \qquad \Gamma \vdash y\mathtt{U}:\beta\,;\Delta \quad (\beta \neq \bot)}{\Gamma \vdash \mathtt{let}\ x^\beta = y\mathtt{U}\ \mathtt{in}\ \mathtt{N}:\gamma\,;\Delta}$$

(let-⊥)
$$\frac{\Gamma \vdash y:\alpha \Rightarrow \bot\,;\emptyset \qquad \Gamma \vdash \mathtt{U}:\alpha\,;\Delta}{\Gamma \vdash \mathtt{let}\ \_ = y\mathtt{U}:\bot\,;\Delta}$$

In (⊥-const), c, which witnesses absurdity, is introduced only when $\bot$ is assumed in the environment (logically this says that we can say an absurd thing only when the environment is absurd). CNFs which are also CNF value preterms are called *CNF values*. Note a CNF value is either c (which is the sole case when it has a type $\bot$), a λ-abstraction, or a μ-abstraction followed by a λ-abstraction.

CNFs correspond to $\lambda\mu\mathsf{v}$-terms as follows. In the first rule we assume x is chosen arbitrarily from variables assigned to $\bot$. Below in the first line, it is semantically (and logically) irrelevant which $\bot$-typed variable we choose: for example, we may assume there is a total order on names and choose the least one from the given environment.

$$(\Gamma\cdot x:\bot \vdash \mathtt{c}:\bot\,;\Delta)^* \stackrel{\text{def}}{=} \Gamma\cdot x:\bot \vdash x:\bot\,;\Delta \quad (\Gamma \vdash \mathtt{let}\ \_ = y\mathtt{U}:\bot\,;\Delta)^* \stackrel{\text{def}}{=} \Gamma \vdash y\mathtt{U}^*:\bot\,;\Delta$$
$$(\Gamma \vdash \mathtt{let}\ x^\beta = y\mathtt{U}\ \mathtt{in}\ \mathtt{N}:\gamma\,;\Delta)^* \stackrel{\text{def}}{=} \Gamma \vdash (\lambda x.\mathtt{N}^*)(y\mathtt{U}^*):\gamma\,;\Delta$$

$$[\mathbf{0}]_u^{\Gamma;\Delta} \stackrel{\text{def}}{=} \mathtt{c}:\bot \qquad [\overline{u}(c)!c(xz).R]_u^{\Gamma;\Delta\cdot u:(\alpha\Rightarrow\beta)} \stackrel{\text{def}}{=} \lambda x^\alpha.[R]_z^{\Gamma\cdot x:\overline{\alpha};\Delta\cdot z:\beta}$$

$$[\overline{y}]_u^{\Gamma\cdot y:\alpha\Rightarrow\beta;\Delta} \stackrel{\text{def}}{=} \mathtt{let}\ \_ = y\mathtt{c} \qquad [\overline{u}(c)!c(x).R]_u^{\Gamma;\Delta\cdot u:(\alpha\Rightarrow\bot)} \stackrel{\text{def}}{=} \lambda x^\alpha.[R]_m^{\Gamma\cdot x:\overline{\alpha};\Delta}$$

$$[\overline{u}(c)!c.R]_u^{\Gamma;\Delta\cdot u:(\bot\Rightarrow\bot)} \stackrel{\text{def}}{=} \lambda x^\bot.[R]_m^{\Gamma;\Delta} \quad [\overline{u}(c)!c(z).R]_u^{\Gamma;\Delta\cdot u:(\bot\Rightarrow\beta)} \stackrel{\text{def}}{=} \lambda x^\bot.[R]_z^{\Gamma;\Delta\cdot z:\beta}$$

$$[P_{\langle a\rangle}]_u^{\Gamma;\Delta\cdot a:\alpha} \stackrel{\text{def}}{=} [a][P_{\langle m/a\rangle}]_m^{\Gamma;\Delta\cdot a:\alpha\cdot m:\alpha} \qquad [\overline{y}(w)R]_u^{\Gamma\cdot y:\alpha\Rightarrow\beta;\Delta} \stackrel{\text{def}}{=} \mathtt{let}\ \_ = y[\overline{c}(w)P]_c^{\Gamma;\Delta}$$

$$[\overline{y}(wz)(R\,|\,!z(x).Q)]_u^{\Gamma\cdot y:\alpha\Rightarrow\beta;\Delta} \stackrel{\text{def}}{=} \mathtt{let}\ x^\beta = y[\overline{c}(w)R]_c^{\Gamma\cdot y:\alpha\Rightarrow\beta;\Delta}\ \mathtt{in}\ [Q]_u^{(\Gamma\cdot x:\beta);\Delta}$$

$$[\overline{y}(z)!z(x).Q]_u^{\Gamma\cdot y:\alpha\Rightarrow\beta;\Delta} \stackrel{\text{def}}{=} \mathtt{let}\ x^\beta = y\mathtt{c}\ \mathtt{in}\ [Q]_u^{\Gamma\cdot y:\alpha\Rightarrow\beta\cdot x:\beta;\Delta}$$

$$[P]_u^{\Gamma;\Delta\cdot u:\alpha} \stackrel{\text{def}}{=} \mu u^\alpha.[P]_m^{\Gamma;\Delta\cdot u:\alpha} \text{ other cases}$$

Fig. 3. Decoding of $\lambda\mu$-typed processes ($\alpha, \beta \neq \bot$ and m, u are fresh)

For CNFs which are λ-abstraction, μ-abstraction and named terms, the mapping uses the same clauses as in Figure 2, replacing $[\![\cdot]\!]$ in the defining clauses with $(\cdot)^*$.

Via $(\ )^*$ we can encode CNFs to processes:

$$\Gamma \vdash \mathtt{N} : \alpha; \Delta \quad \mapsto \quad \Gamma \vdash \mathtt{N}^* : \alpha; \Delta \quad \mapsto \quad \vdash_{\mathsf{O}} [\![(\mathtt{N} : \alpha)^*]\!]_u \triangleright (u{:}\alpha, \Delta)^\bullet, \Delta^\circ$$

CNFs can also be directly encoded into π^{c}-processes, using the following rules combined with those for abstraction, naming and μ-abstraction given in Figure 2 (replacing $[\![\cdot]\!]$ with $\langle\cdot\rangle$ in each clause).

$$\langle \mathtt{let}\ x = y\mathtt{U}\ \mathtt{in}\ \mathtt{N}{:}\gamma\rangle_u \stackrel{\text{def}}{=} \begin{cases} \overline{y}(wz)(P|!z(x).\langle \mathtt{N}{:}\gamma\rangle_u) & (\mathtt{U} \neq \mathtt{c},\ \langle \mathtt{U}\rangle_c \stackrel{\text{def}}{=} \overline{c}(w)P) \\ \overline{y}(z)!z(x).\langle \mathtt{N}{:}\gamma\rangle_u & (\mathtt{U} = \mathtt{c}) \end{cases}$$

$$\langle \mathtt{let}\ \_ = y\mathtt{U}{:}\bot\rangle_u \stackrel{\text{def}}{=} \begin{cases} \langle \mathtt{U}\rangle_y & (\mathtt{U} \neq \mathtt{c}) \\ \overline{y} & (\mathtt{U} = \mathtt{c}) \end{cases} \qquad \langle \mathtt{c}{:}\bot\rangle_u \stackrel{\text{def}}{=} \mathbf{0}$$

Two process encodings of CNFs coincide up to $\searrow$.

Proposition 4.1. *1. Let $\Gamma \vdash \mathtt{N}{:}\alpha\,;\Delta$. Then $\vdash_{\mathsf{O}} \langle \mathtt{N}\rangle_u \triangleright (u{:}\alpha, \Delta)^\bullet, \Gamma^\circ$.*
2. Let $\Gamma \vdash \mathtt{N}{:}\alpha\,;\Delta$. Then $[\![\mathtt{N}^* : \alpha]\!]_u \searrow^* \langle \mathtt{N}\rangle_u \not\searrow$.

Definability. The decoding of π^{c}-processes (of encoded $\lambda\mu$v-types) to $\lambda\mu$v-preterms is written $[P]_u^{\Gamma;\Delta}$, which translates $P \in \mathsf{NF}_e$ such that $\vdash_{\mathsf{O}} P \triangleright \Gamma^\circ, \Delta^\bullet$ with $u \notin \mathrm{dom}(\Gamma)$ to a $\lambda\mu$v-preterm M. Without loss of generality, we assume P does not contain redundant $\mathbf{0}$ or hiding. The mapping is defined inductively by the rules given in Figure 3. In the second last line, $P_{\langle a\rangle}$ indicates P is a prime output with subject a, whereas $P_{\langle m/a\rangle}$ is the result of replacing the subject a in $P_{\langle a\rangle}$ with m.

Proposition 4.2. *Let $\bot \notin image(\Gamma)$, $u \notin dom(\Gamma)$ and $P \in \mathsf{NF}_e$. Then $\vdash P \triangleright \Gamma^\circ \cdot \Delta^\bullet$ implies, with x fresh:* (1) *if $\Delta = \Delta_0 \cdot u{:}\alpha$ then $\Gamma \cdot x{:}\bot \vdash [P]_u^{\Gamma^\circ;\Delta^\bullet} : \alpha\,; \Delta_0$ and;* (2) *if $u \notin dom(\Delta)$ then $\Gamma \cdot x{:}\bot \vdash [P]_u^{\Gamma^\circ;\Delta^\bullet} : \bot\,; \Delta_0$.*

Let us say $\Gamma \vdash M : \beta\,; \Delta$ with $u \notin \mathrm{dom}(\Gamma)$ *defines* $\vdash P \triangleright \Gamma^\circ \cdot \Delta^\bullet \in \mathsf{NF}_e$ *at* u iff $[\![M : \beta]\!]_u \searrow^* P$. A $\lambda\mu$v-term is *closed* if it contains neither free names nor free variables. We can now establish the definability.

Theorem 4.1. (definability) *Let* $\vdash P \triangleright \Gamma^{\circ}{\cdot}\Delta^{\bullet}{\cdot}u{:}\alpha^{\bullet} \in \mathsf{NF}_e$ *such that* $\bot \notin image(\Gamma)$. *Then* $\Gamma \cdot x{:}\bot \vdash [P]_u : \alpha\, ;\, \Delta$ *defines* P. *Further if* $\Gamma = \Delta = \emptyset$ *and* $P \not\equiv \mathbf{0}$, *then there is a closed* $\lambda\mu\nu$*-term which defines* P.

Full Abstraction. To prove full abstraction, our first task is to define a suitable observational congruence in the $\lambda\mu\nu$-calculus. There can be different notions of observational congruences for the calculus; here we choose a large, but consistent congruence. This equality is defined solely using the terms and dynamics of the calculus; yet, as we shall illustrate later, its construction comes from an analysis of $\lambda\mu\nu$-terms' behaviour through their encoding into π^{c}-processes and the process equivalence $\cong_\pi$. The analysis is useful since the notion of observation in pure $\lambda\mu\nu$-calculus may not be too obvious, while $\cong_\pi$ is based on a clear and simple idea of observables. Two further observations on the induced congruence: (1) The congruence is closely related with (and possibly coincide with some of) the notions of equality over full controls, as studied by Laird [20], Selinger [34] and others; and (2) If we extend $\lambda\mu\nu$ with sums or non-trivial atomic types, and define the congruence based on the convergence to distinct normal forms of these types, then the resulting congruence restricted to the pure $\lambda\mu\nu$-calculus is precisely what we obtain by the present congruence.

Definition 4.1. $\equiv_\bot$ *is the smallest typed congruence on* $\lambda\mu\nu$*-terms which includes:*

1. $\Gamma \vdash M \equiv_\bot N : \beta; \Delta$ *when* $M \equiv_\alpha N$.
2. $\Gamma \vdash M \equiv_\bot N : \beta; \Delta$ *when* $N \stackrel{\text{def}}{=} M\{y/x\}$ *where* $\Gamma(x) = \Gamma(y) = \bot$.

For example, we have, under the environment $x : \bot, y : \bot$: $x \equiv_\bot y$. We also have: $\lambda x^\bot.\lambda y^\bot x \equiv_\bot \lambda x^\bot.\lambda y^\bot y$. We can easily check that, in the encoding, $\equiv_\bot$-related terms are always mapped to an identical process.

Convention 1. *Henceforth we always consider* $\lambda\mu\nu$*-terms and CNFs up to* $\equiv_\bot$.

We can now define observables, which is an infinite series of closed terms of the type $\bot \Rightarrow \bot \Rightarrow \bot$.

Definition 4.2. Define $\{W_i\}_{i\in\omega}$ by the following induction: $W_0 \stackrel{\text{def}}{=} \lambda z^\bot.\mu u^{\bot\Rightarrow\bot}.z$ and $W_{n+1} \stackrel{\text{def}}{=} \lambda z^\bot.\mu u^{\bot\Rightarrow\bot}.[w]W_n$. Let $\gamma = \bot \Rightarrow \bot \Rightarrow \bot$. We then define: *Obs* $\stackrel{\text{def}}{=}$ $\{W_0\} \cup \{\mu\, w^\gamma.[w]W_{n+1},\ n \in \mathbb{N}\}$ where we take terms up $\equiv_\bot$.

All terms in *Obs* are closed $\rightarrow_{\lambda\mu\nu}$-normal forms of type γ (W_0 can also be written as $\mu w.[w]W_0$, but is treated separately since $\mu w.[w]W_0$ is not a normal form).

To illustrate the choice of *Obs*, we show below the π-calculus representation of W_0, $\mu\, w.[w]W_1, \mu\, w.[w]W_2, \ldots$ through $[\![\cdot]\!]_u$, which is in fact the origin of *Obs*.

Definition 4.3. Define $\{P_i\}_{i\in\omega}$ as follows (below we use the same names for bound names for simplicity). $P_0 \stackrel{\text{def}}{=} \overline{w}(c)!c(u).\mathbf{0}$ and $P_{n+1} \stackrel{\text{def}}{=} \overline{w}(c)!c(u).P_n$. We set $Obs_\pi \stackrel{\text{def}}{=}$ $\{\vdash_{\mathrm{O}} P_i \triangleright w : \gamma^\bullet\}_{i\in\omega}$, taking processes modulo $\equiv$.

Note each P_i only outputs at w (if ever) at any subsequent invocation, even though an output at any one of the bound names (u above) is well-typed. For example,

$P_1' \stackrel{\text{def}}{=} \overline{w}(c)!c(u).\overline{u}(c)!c(u).\mathbf{0}$ has type $w : \gamma^\bullet$ but differs from P_1 by outputting at the bound u when it is invoked the second time. One can check P_0 is the smallest (w.r.t. process size, i.e. number of constructors) non-trivial inhabitant of this type: in particular it is smaller than $[\![\lambda z^\perp.\lambda x^\perp.x]\!]_w \stackrel{\text{def}}{=} P_1'$.

Proposition 4.3. *1.* $\vdash_{\mathsf{O}} P_i \cong_\pi P_j \triangleright w : \gamma^\bullet$ *for arbitrary* i *and* j.
2. If $\vdash_{\mathsf{O}} Q \triangleright w : \gamma^\bullet$, $Q \in \mathsf{NF}_e$ *and* $Q \cong_\pi P_i$ *then* $Q \in \mathit{Obs}_\pi$.

Processes in Obs_π have uniform behaviours: indeed they are closed under $\cong_\pi$. These observations motivate the following definition. Below $C[\,\cdot\,]^\beta_{\Gamma;\alpha;\Delta}$ is a typed context whose hole takes a term typed as $\alpha;\Delta$ under the base Γ and which returns a closed term of type β.

Definition 4.4. We write $\Gamma \vdash M \cong_{\lambda\mu} N \ : \ \alpha;\Delta$ when, for each typed context $C[\,]^{\perp\Rightarrow\perp\Rightarrow\perp}_{\Gamma;\alpha;\Delta}$, we have: $\exists L.(C[M] \Downarrow L \in \mathit{Obs})$ iff $\exists L'.(C[N] \Downarrow L' \in \mathit{Obs})$.

Note that we treat all values in *Obs* as an identical observable. Immediately $\rightarrow_{\lambda\mu\mathrm{v}} \subset \cong_{\lambda\mu}$. We can now establish the full abstraction, following the standard routine. We start with the computational adequacy. We write $M \Downarrow L$ when $M \rightarrow^*_{\lambda\mu\mathrm{v}} L \not\rightarrow_{\lambda\mu\mathrm{v}}$.

Proposition 4.4. (computational adequacy) *Let* $M : \perp \Rightarrow \perp \Rightarrow \perp$ *be closed. Then* $\exists L.(M \Downarrow L \in \mathit{Obs})$ *iff* $\exists P.([\![M]\!]_u \searrow^* P \in \mathit{Obs}_\pi)$.

Corollary 4.1. (soundness) $[\![M]\!]_u \cong_\pi [\![N]\!]_u$ *implies* $M \cong_{\lambda\mu} N$.

Proof. Assume $[\![M]\!]_u \cong_\pi [\![N]\!]_u$. We show, for each well-typed $C[\cdot]$, $\exists L.(C[M] \Downarrow L \in \mathit{Obs})$ iff $\exists L'.(C[M] \Downarrow L' \in \mathit{Obs})$. Let $C[\,\cdot\,]$ be well-typed. Now we reason:

$$\begin{array}{llll} C[M] \Downarrow L \in \mathit{Obs} & \Rightarrow & [\![C[M]]\!]_v \Downarrow [\![L]\!]_v \in \mathit{Obs}_\pi & \text{(Proposition 4.4)} \\ & \Rightarrow & \exists O.[\![C[N]]\!]_v \searrow^* O \in \mathit{Obs}_\pi & ([\![C[M]]\!]_v \cong_\pi [\![C[N]]\!]_v) \\ & \Rightarrow & C[N]_v \searrow^* L' \in \mathit{Obs} & \text{(Proposition 4.4)} \end{array}$$

Theorem 4.2. (full abstraction) *Let* $\Gamma \vdash M_i : \alpha;\Delta$ $(i = 1,2)$*. Then* $M_1 \cong_{\lambda\mu} M_2$ *if and only if* $[\![M_1]\!]_u \cong_\pi [\![M_2]\!]_u$.

Proof. Suppose $\emptyset \vdash M_1 \cong_{\lambda\mu} M_2 : \alpha;\emptyset$ but $\vdash_{\mathsf{O}} [\![M_1]\!]_u \not\cong_\pi [\![M_2]\!]_u \triangleright u : \alpha^\bullet$. By this, converting the observable $()^?$ to the convergence to Obs_π in $\gamma^\bullet$ with $\gamma = \perp \Rightarrow \perp \Rightarrow \perp$, there exists $\vdash_{\mathtt{I}} R \triangleright u : \overline{\alpha^\bullet}, v : \gamma^\bullet$ such that $R \in \mathsf{NF}_e$ and (say) $\exists P.\ (\boldsymbol{\nu}\, u)([\![M_1]\!]|R) \searrow^* P \in \mathit{Obs}_\pi$ and $\neg\exists P.\ (\boldsymbol{\nu}\, u)([\![M_2]\!]|R) \searrow^* P \in \mathit{Obs}_\pi$. Since $R \in \mathsf{NF}_e$, we can safely set $R \stackrel{\text{def}}{=} !u(c).R'$. Now take $\vdash_{\mathtt{I}} !u(cv).R' \triangleright u : (\alpha \Rightarrow \gamma)^\bullet$. By Theorem 4.1 (definability), we can find L such that $\vdash L : \alpha \Rightarrow \gamma$ where $[\![L]\!]_u \cong_\pi !u(cv).R'$. Since $[\![LM_i]\!]_u \searrow^+ (\boldsymbol{\nu}\, u)([\![M_i]\!]|R)$, we conclude $\exists L'.\ LM_1 \Downarrow L' \in \mathit{Obs}$ and $\neg\exists L'.\ LM_2 \Downarrow L' \in \mathit{Obs}$, which contradicts the assumption. Since precisely the same argument holds when Γ and Δ are possibly non-empty in $\Gamma \vdash M_{1,2} : \alpha;\Delta$ by closing them by λ/μ-abstractions, we have now established the full abstraction.

5 Discussion

This paper explored the connection between control and the π-calculus, first pointed out by Thielecke [35] who showed that the target of CPS-transform can be written down as name passing processes. This paper presented the typed π-calculus for full control, which arises as a subcalculus of the linear π-calculus [39] where all inputs are replicated. The main contribution of the present work is the use of a duality-based type structure in the π-calculus by which the embedding of control constructs in processes becomes semantically exact.

Control and Name Passing (1). The notion of full control arises in several related contexts. Historical survey of studies of controls and continuations can be found in [28,36]. Here we pick up three strands of research to position the present work in a historical context. In one strand, notions of control operators have been formulated and studied as a way to represent jumps and other non-trivial manipulation of control flows as an extension of the λ-calculus and related languages. Among many works, Felleisen and others [10, 11] studied syntactic and equational properties of control operators in the context of the call-by-value λ-calculus, clarifying their status. Griffin [13] shows a correspondence between the λ-calculus with control operators, classical proofs and the CPS transform. Finally Parigot [27] introduced the $\lambda\mu$-calculus, the calculus without control operators but which manipulates names, as term-representation of classical natural deduction proofs. The control-operator-based presentation and name-based presentation, which are shown to be equivalent by de Groote [9], elucidate statics and dynamics of full control in different ways: the latter gives a more fine-grained picture while the former often offers a more condensed representation. In this context, the present work shows a further decomposition (and arguably simpler presentation) of the dynamics of full control on the uniform basis of name passing interaction.

Control and Name Passing (2). Another closely related context is the CPS transform [8,12,30]. In this line of studies, the main idea is to represent the dynamics of the λ-calculus, or procedural calls, in a way close to implementations. Consider for example the reduction: $(\lambda x.x)1 \longrightarrow_\beta 1$. To model implemented execution of this reduction, we elaborate each term with a continuation to which the resulting value should be returned. We write this transformation $\langle\!\langle M\rangle\!\rangle$. In the above example, $\langle\!\langle \lambda x.x\rangle\!\rangle \stackrel{\text{def}}{=} \lambda h.h(\lambda x.\langle\!\langle x\rangle\!\rangle)$ (which receives a next continuation and "sends out" its resulting value to that continuation, with $\langle\!\langle x\rangle\!\rangle = \lambda k.kx$); whereas $\langle\!\langle 1\rangle\!\rangle \stackrel{\text{def}}{=} \lambda h'.h'1$. The term $(\lambda x.x)1$ as a whole is transformed as follows:

$$\lambda k.(\lambda h.h\lambda x.\langle\!\langle x\rangle\!\rangle)(\lambda m.\langle\!\langle 1\rangle\!\rangle(\lambda n.mnk)) \qquad (4)$$

This transformation may need some illustration. Assume first we apply to the above abstraction the ultimate continuation k (to which the result of evaluating the whole term should jump), marking the start of computation. Write M for $\lambda x.x$ and N for 1. After the continuation k is fed to the left-hand side, we first give $\langle\!\langle M\rangle\!\rangle$ its next continuation $(\lambda m.\langle\!\langle N\rangle\!\rangle(\lambda n.mnk))$, to which the result of evaluating M, say V, is fed, replacing m, then we send $\langle\!\langle N\rangle\!\rangle$ its continuation $\lambda n.Vnk$, to which the result of evaluating $\langle\!\langle N\rangle\!\rangle$ is fed, replacing n, so that finally the "real" computation VW can be performed, to whose result the ultimate continuation k is applied.

As may be seen from the example above, the CPS transform can be seen as a way to mimic the operational idea of "jumping to an address with a value" solely using function application and abstraction. This representation is useful to connect the procedural calls in high-level languages to their representation at an execution level. The representation is somewhat shy about the use of "names" by abstracting them immediately after their introduction, partly because this is the only way to use the notion in the world of pure functions (note in $(\lambda h.hM)V$, the bound h in fact names V). This however does not prevent us from observing (4) is isomorphic to its process encoding via $[\![\cdot]\!]$ of Section 3, given as:

$$(\boldsymbol{\nu}\, h)([\![\lambda x.x]\!]_h \mid !h(n).(\boldsymbol{\nu}\, h')([\![1]\!]_{h'} \mid !h'(m).\overline{n}\langle mk\rangle)) \qquad (5)$$

In (5), k, h, h' are all channel names at which processes interact: the input/output polarities make it clear what is named (used as replicated inputs) and to which it is jumping (used as outputs, i.e. subscripts of the encoding). The "book-keeping" abstractions of h and h' in (4) are replaced by hiding. Setting $[\![1]\!]_{h'}$ to be $\overline{h'}\langle 1\rangle$ (regarding 1 as a specific name), we can see how (5) reduces precisely as (4) reduces modulo the book-keeping reductions. Sangiorgi [32] observed that we can regard (4) as terms in the applicable part of the higher-order π-calculus (a variant of π-calculus which processes communicate not only names but also terms) and that the translation from a λ-term to its process representation can be factored into the former's CPS transformation and its encoding into the π-calculus.

In the context of these studies, where the control is studied purely in the context of the λ-calculi, the main contribution of our work may lie in identifying the precise realm of typed processes which, when it is used for the encoding of λ-terms, gives exactly the same equational effect as the standard CPS transform embedded in the λ-calculus. As we have shown in [4, 39], the encoding of the λ-calculi into the linear/affine-π-calculi [4] results in full abstraction. π^{c} offers a refined understanding on CPS-transform, with precisely the same induced equivalence. As related points, we have suggested possible relationship between existing CPS transformations/inversions [8, 12, 30], on the one hand, and the encoding/decoding in Sections 3 and 4 in this paper on the other.

Control and Name Passing (3). There are many studies of semantics and equalities in calculi with full control, notably those which aim to investigate appropriate algebraic structures of suitable categories (for example those by Thielecke [35], Laird [21] and Selinger [34]). The present work may have two interests in this context.

The basis of the observational equivalence for $\lambda\mu$v-terms, the behavioural equivalence over π^{c}-processes, has very simple operational content, while inducing the equality closely related with those studied in the past. Among others we believe that $\cong_{\lambda\mu}$ coincides with the call-by-value, total and extensional version of Laird's games for control [16,20] (it is easy to check all terms in *Obs* are equated in such a universe). We also suspect it is very close to the equality induced by the call-by-value part of Selinger's dualised universe [34] (for the same reason), though details are to be checked. The combination of clear observational scenario and correspondence with good denotational universes is one of the notable aspects of the use of the π-calculus.

In another and related view, we may consider processes in π^{c} as name passing transition systems (or name passing synchronisation trees). As such, a process identifies meaning of a denoted program as an abstract entity. The rich repertoire of powerful reasoning techniques developed for π-calculi is now freely available; further this representation has enriching connection with studies on game-based semantics, most notably games for control studied by Laird [21]. Indeed, Laird's work may be regarded as a characterisation of dynamic interaction structure of π^{c} (or, to be precise, its affine extensions), where the lack of well-bracketing corresponds to the coalescing of linear actions into replicated actions. Another intensional structure in close connection is *Abstract Böhm Trees* studied by Curien and Herbelin [6]. We expect the variant of these structures for full control to have a close connection to name passing transition of π^{c}.

It is also notable that representation of programs and other algorithmic entities as name passing transition, together with basic operators such as parallel composition, hiding and prefixing, is not limited to the control nor to sequential computation.

Control as Proofs and Control as Processes. The present work has a close connection with recent studies on control from a proof-theoretic viewpoint, notably Polarised Linear Logic by Laurent [22, 23] and $\overline{\lambda}\mu\tilde{\mu}$-calculus by Curien and Herbelin [7]. The type structures for the linear/affine π-calculi are based on duality, here arising in a simplest possible way, as mutually dual input and output modes of channel types. This duality has a direct applicability for analysis of processes and programs, as may be seen in the new flow analysis we have recently developed for typed π-calculi [17]. This duality allows a clean decomposition of behaviours in programming languages into name passing interaction, and is in close correspondence with polarity in Polarised Linear Logic by Laurent [22, 23]. Laurent and the first author recently obtained a basic result on the relationship between π^{c} and Polarised Linear Logic, as discussed in [14].

In a different context, Curien and Herbelin [7] presents $\overline{\lambda}\mu\tilde{\mu}$, a calculus for control, based on Gentzen's LK, in which a strong notion of duality elucidates the distinction between the call-by-name and call-by-value evaluations in the setting of full control. One interesting aspect is the way non-determinism arises in their calculus, which suggests an intriguing connection between the dynamics of their calculus and name passing processes. From the same viewpoint, the connection with a recent work by Wadler [38] on duality and λ-calculi is an interesting subject for further studies.

The present study concentrates on the call-by-value encoding of the $\lambda\mu$-calculus. As in the λ-calculus [4], we can similarly embed the call-by-name $\lambda\mu$-calculus into π^{c} by changing the encoding of types (hence terms). The mapping $[\alpha_1, ..., \alpha_n, \bot]^\circ \stackrel{\text{def}}{=} (\overline{\alpha_1^\circ}, ..., \overline{\alpha_n^\circ})^!$ is the standard Hyland-Ong encoding of call-by-name types [4] assuming the only atomic type is $\bot$. In the presence of control, we can simply augment this map with $\alpha^\bullet \stackrel{\text{def}}{=} (\alpha^\circ)^?$ which says: "a program may jump to a continuation" (this corresponds to the "player first" in Laurent's games [22]). This determines, together with the one given in [4], the encoding of programs. We strongly believe the embedding is fully abstract, though details are to be checked.

Van Bakel and Vigliotti [37] present a different approach towards representing control as interaction. Milner's encoding of λ-calculus does not model reductions under λ-binders by matching reductions in the π-calculus. They consider a variant of the $\lambda\mu$-calculus with explicit substitution that preserves single-step explicit head reduction.

Finally, the idea of viewing non-local control-flow manipulation in λ-calculi as typed interacting processes in π^{c} has been fruitful for finding Hoare logics for languages with call/cc: [3] produces the first such logic by designing a Hennessy-Milner logic for a variant of π^{c} and then pushes that logic back to call-by-value PCF extended with with call/cc to obtain a conventional Hoare logic.

References

1. A full version of this paper. Doc Technical Report DTR 2014/2, Department of Computing, Imperial College London (April 2014)
2. Abramsky, S., Jagadeesan, R., Malacaria, P.: Full Abstraction for PCF. Info. & Comp. 163, 409–470 (2000)
3. Berger, M.: Program Logics for Sequential Higher-Order Control. In: Arbab, F., Sirjani, M. (eds.) FSEN 2009. LNCS, vol. 5961, pp. 194–211. Springer, Heidelberg (2010)
4. Berger, M., Honda, K., Yoshida, N.: Sequentiality and the π-calculus. In: Abramsky, S. (ed.) TLCA 2001. LNCS, vol. 2044, pp. 29–45. Springer, Heidelberg (2001)
5. Berger, M., Honda, K., Yoshida, N.: Genericity and the pi-calculus. Acta Inf. 42(2-3), 83–141 (2005)
6. Curien, P.-L., Herbelin, H.: Computing with Abstract Böhm Trees. In: Functional and Logic Programming, pp. 20–39. World Scientific (1998)
7. Curien, P.-L., Herbelin, H.: The duality of computation. In: Proc. ICFP, pp. 233–243 (2000)
8. Danvy, O., Filinski, A.: Representing control: A study of the CPS transformation. MSCS 2(4), 361–391 (1992)
9. de Groote, P.: On the Relation between the lambda-mu-calculus and the Syntactic Theory of Sequential Control. In: Pfenning, F. (ed.) LPAR 1994. LNCS, vol. 822, pp. 31–43. Springer, Heidelberg (1994)
10. Felleisen, M., Friedman, D.P., Kohlbecker, E., Duba, B.: Syntactic theories of sequential control. TCS 52, 205–237 (1987)
11. Felleisen, M., Hieb, R.: The revised report on the syntactic theories of sequential control and state. TCS 103(2), 235–271 (1992)
12. Führmann, C., Thielecke, H.: On the call-by-value CPS transform and its semantics. Inf. Comput. 188(2), 241–283 (2004)
13. Griffin, T.G.: A formulae-as-type notion of control. In: Proc. POPL, pp. 47–58 (1990)
14. Honda, K., Laurent, O.: An exact correspondence between a typed pi-calculus and polarised proof-nets. Theor. Comput. Sci. 411(22-24), 2223–2238 (2010)
15. Honda, K., Yoshida, N.: On reduction-based process semantics. TCS 151, 393–456 (1995)
16. Honda, K., Yoshida, N.: Game-theoretic analysis of call-by-value computation. TCS 221, 393–456 (1999)
17. Honda, K., Yoshida, N.: Noninterference through flow analysis. JFP 15(2), 293–349 (2005)
18. Honda, K., Yoshida, N.: A uniform type structure for secure information flow. TOPLAS 29(6) (2007)
19. Hyland, J.M.E., Ong, C.H.L.: On full abstraction for PCF. Inf. & Comp. 163, 285–408 (2000)
20. Laird, J.: A semantic analysis of control. PhD thesis, University of Edinburgh (1998)
21. Laird, J.: A game semantics of linearly used continuations. In: Gordon, A.D. (ed.) FOSSACS 2003. LNCS, vol. 2620, pp. 313–327. Springer, Heidelberg (2003)
22. Laurent, O.: Polarized games. In: Proc. LICS, pp. 265–274 (2002)
23. Laurent, O.: Polarized proof-nets and $\lambda\mu$-calculus. TCS 290(1), 161–188 (2003)
24. Milner, R.: Functions as Processes. MSCS 2(2), 119–141 (1992)
25. Moggi, E.: Notions of computation and monads. Inf. & Comp. 93(1), 55–92 (1991)
26. Ong, C.-H.L., Stewart, C.A.: A Curry-Howard foundation for functional computation with control. In: Proc. POPL, pp. 215–227 (1997)

27. Parigot, M.: λ-μ-Calculus: An Algorithmic Interpretation of Classical Natural Deduction. In: Voronkov, A. (ed.) LPAR 1992. LNCS, vol. 624, pp. 190–201. Springer, Heidelberg (1992)
28. Reynolds, J.: The discovers of continuations. Lisp and Symbolic Computation 6, 233–247 (1993)
29. Robinson, E.: Kohei Honda. Bulletin of the EATCS 112 (February 2014)
30. Sabry, A., Felleisen, M.: Reasoning about programs in continuation-passing style. In: Proc. LFP, pp. 288–298 (1992)
31. Sangiorgi, D.: π-calculus, internal mobility and agent-passing calculi. TCS 167(2), 235–274 (1996)
32. Sangiorgi, D.: From λ to π: or, rediscovering continuations. MSCS 9, 367–401 (1999)
33. Sassone, V.: ETAPS Award: Laudatio for Kohei Honda. Bulletin of the EATCS 112 (February 2014)
34. Selinger, P.: Control categories and duality: on the categorical semantics of the lambda-mu calculus. MSCS 11(2), 207–260 (2001)
35. Thielecke, H.: Categorical Structure of Continuation Passing Style. PhD thesis, University of Edinburgh (1997)
36. Thielecke, H.: Continuations, functions and jumps. SIGACT News 30(2), 33–42 (1999)
37. van Bakel, S., Vigliotti, M.G.: An Output-Based Semantics of $\Lambda\mu$ with Explicit Substitution in the π-Calculus. In: Baeten, J.C.M., Ball, T., de Boer, F.S. (eds.) TCS 2012. LNCS, vol. 7604, pp. 372–387. Springer, Heidelberg (2012)
38. Wadler, P.: Call-by-value is dual to call-by-name. In: Proc. ICFP, pp. 189–201 (2003)
39. Yoshida, N., Berger, M., Honda, K.: Strong normalisation in the π-calculus. Inf. Comput. 191(2), 145–202 (2004)
40. Yoshida, N., Honda, K., Berger, M.: Linearity and bisimulation. J. Log. Algebr. Program. 72(2), 207–238 (2007)

A Typing

We write $|A|$ for the set of A's nodes. Edges, which are always from input-moded nodes to output-moded nodes, denote dependency between channels and are used to prevent vicious cycles between names. If A is such a graph and $x : \tau$ is one of its nodes, we also write $A(x) = \tau$. By $\mathsf{fn}(A)$ we denote the set of all names x such that $A(x) = \tau$ for some τ. Sometimes we also write $x : \tau \in A$ to indicate that $A(x) = \tau$. We write $\mathsf{md}(A) = p$ to indicate that $\mathsf{md}(A(x)) = p$ for all $x \in \mathsf{fn}(A)$. We write $x \to y$ if $x : \tau \to y : \tau'$ for some τ and τ', in a given action type. We compose two processes typed by A and B iff: (1) $A(a) \odot B(a)$ is defined for all $a \in \mathsf{fn}(A) \cap \mathsf{fn}(B)$; and (2) the composition creates no circularity between names. We define $A \asymp B$ iff: (1) whenever $x : \tau \in A$ and $x : \tau' \in B$, $\tau \odot \tau'$ is defined; and (2) whenever $x_1 \to x_2$, $x_2 \to x_3$, $\dots$, $x_{n-1} \to x_n$ alternately in A and B ($n \geq 2$), we have $x_1 \neq x_n$.

Then $A \odot B$, defined iff $A \asymp B$, is the following action type.

- $x : \tau \in |A \odot B|$ iff either (1) $x : \tau$ occurs in either A or B, but not both ; or (2) $x : \tau' \in A$ and $x : \tau'' \in B$ and $\tau = \tau' \odot \tau''$.
- $x \to y$ in $A \odot B$ iff $x : \tau, y : \tau' \in |A \odot B|$ and $x = z_1 \to z_2$, $z_2 \to z_3, \dots, z_{n-1} \to z_n = y$ ($n \geq 2$) alternately in A and B.

Finally, the third condition, the restriction to a single thread, is guaranteed by using *IO-modes*, $\phi \in \{\mathrm{I}, \mathrm{O}\}$, in the typing judgement. These IO-modes are given the following partial algebra, using the overloaded notation $\odot$: $\mathrm{I}\odot\mathrm{I}=\mathrm{I}$ and $\mathrm{I}\odot\mathrm{O}=\mathrm{O}\odot\mathrm{I}=\mathrm{O}$. Among the two IO-modes, O indicates a unique active output: thus $\mathrm{O}\odot\mathrm{O}$ is undefined, which means that we do not want more than one active thread at the same time. We write $\phi_1 \asymp \phi_2$ if $\phi_1 \odot \phi_2$ is defined. IO-modes sequentialise the computation in our typed calculus. This makes reductions deterministic which in turn simplifies reasoning.

$$\text{(Zero)}\ \frac{-}{\vdash_{\mathrm{I}} \mathbf{0} \triangleright \emptyset} \qquad \text{(Par)}\ \frac{\vdash_{\phi_i} P_i \triangleright A_i \quad (i=1,2) \qquad A_1 \asymp A_2 \quad \phi_1 \asymp \phi_2}{\vdash_{\phi_1\odot\phi_2} P_1|P_2 \triangleright A_1 \odot A_2} \qquad \text{(Res)}\ \frac{\vdash_{\phi} P \triangleright A \qquad \mathsf{md}(A(x)) = !}{\vdash_{\phi} (\boldsymbol{\nu}\, x)P \triangleright A/x} \qquad \text{(Weak)}\ x \notin \mathsf{fn}(A)\ \frac{\vdash_{\phi} P \triangleright A \qquad \mathsf{md}(\tau) = ?}{\vdash_{\phi} P \triangleright A,\ x{:}\tau}$$

$$\text{(Weak-}io\text{)}\ \frac{\vdash_{\mathrm{I}} P \triangleright A}{\vdash_{\mathrm{O}} P \triangleright A} \qquad (\mathsf{In}^!)\ x \notin \mathsf{fn}(A),\ \mathsf{md}(A) = ?\ \frac{\vdash_{\mathrm{O}} P \triangleright \vec{y}{:}\vec{\tau},\ A}{\vdash_{\mathrm{I}} !x(\vec{y}).P \triangleright x{:}(\vec{\tau})^! {\to} A} \qquad (\mathsf{Out}^?)\ y_i{:}\tau_i \in A\ \frac{\vdash_{\mathrm{I}} P \triangleright A \asymp x{:}(\vec{\tau})^?}{\vdash_{\mathrm{O}} \overline{x}(\vec{y})P \triangleright A/\vec{y} \odot x{:}(\vec{\tau})^?}$$

In the following, we briefly illustrate each typing rule. In (Zero), we start in I-mode with empty type since there is no active output. In (Par), "$\asymp$" controls composability, ensuring that at most one thread is active in a given term (by $\phi_1 \asymp \phi_2$) and uniqueness of replicated inputs and non-circularity (by $A_1 \asymp A_2$). The resulting type is given by merging two types. In (Res), we do not allow ? to be restricted since this action expects its dual server always exists in the environment. $A/\vec{y}$ means the result of deleting the nodes $\vec{x} : \vec{\tau}$ in A (and edges from/to deleted nodes). In (Weak), we weaken a ?-moded channel since this mode means zero or more output actions at a given channel. In (Weak-io), we turn the input mode into the output mode. ($\mathsf{In}^!$) ensures non-circularity at x (by $x \notin \mathsf{fn}(A)$) and no free input occurrence under input (by $\mathsf{md}(A) = ?$). Then it records the causality from input to free outputs. If A is empty, $x : (\vec{\tau})^! \to A$ simply stands for $x : (\vec{\tau})^!$. ($\mathsf{Out}^?$) essentially the rule composes the output prefix and the body in parallel. In the condition, $y_i : \tau_i \in A$ means each $y_i : \tau_i$ appears in A. This ensures bound input channels $\vec{y}$ become always active after the message received. It also changes the mode to output, to indicate an active thread or server. Note that this rule does not suppress the body by prefix since output is asynchronous.

Concurrent Programming Languages and Methods for Semantic Analyses (Extended Abstract of Invited Talk)

Manfred Schmidt-Schauß

Goethe University, Frankfurt, Germany
schauss@ki.informatik.uni-frankfurt.de

Abstract. The focus will be a presentation of new results and successes of semantic analyses of concurrent programs. These are accomplished by contextual equivalence, observing may- and should-convergence, and by adapting known techniques from deterministic programs to non-determinism and concurrency. The techniques are context lemmata, diagram techniques, applicative similarities, infinite tree reductions, and translations. The results are equivalences, correctness of program transformations, correctness of implementations and translations.

1 Motivation

The central issue is the semantics of realistic concurrent programming languages. Several other questions are related to this. From a compiler writers perspective the questions are which optimizations or program transformations are permitted, which implementations in low level languages represent the intention of the high level program, and which evaluation strategy or scheduling of processes can be used and how (un)fair can the evaluation be. A programmer may ask how to model a given problem, or how to write efficient and reliable code, and how to debug. A programming language designer may also ask which constructs are useful and which are harmful.

For standard deterministic programming languages there are different approaches to semantics, which in general also lead to the same equivalences of programs. The situation is different for languages with nondeterministic or concurrent constructs.

In collaboration with David Sabel and building on experience with contextual semantics for smaller languages, we applied, thereby adapting and extending, the contextual semantics to an interesting language model: a call-by-need, functional programming language with a separation of pure and side-effecting computations where side effects are encapsulated by monadic programming. A real programming language embodying these concepts is Concurrent Haskell, and in particular, the current implementation in the Glasgow Haskell Compiler (GHC).

There are several challenges: A first one is the construction of a model calculus (called CHF) that captures the essence of the operational behavior. A second challenge is the design of the semantics, which we selected as a combination

G. Dowek (ed.): RTA-TLCA 2014, LNCS 8560, pp. 21–30, 2014.

of the observations of may- and should-convergence. The third challenge is to demonstrate that the known tools can be adapted and permit to prove useful results for the calculus.

This talk will mainly present two results: A conservativity result that can be seen as a theoretical justification of the separation into pure and monadic programming, and a correctness result for an implementation of a software transactional memory specification into a highly concurrent (call-by-need, functional) implementation.

The structure will be to first go into known details and ingredients of call-by-need functional calculi, the syntactic and operational details. Then we detail the contextual semantics, and may- and should-convergence and the relation to must-convergence. Then a short overview of the used techniques, tools and methods is given, where of course, rewriting techniques play an important role at several places. A short overview of the proof details for the conservativity and also for the concurrent implementation is given.

As a disclaimer, the talk is subjective and so will only mention a small selection of related results and approaches.

2 Start: Lambda Calculus

A program calculus modelling functions and the application of functions to arguments is the untyped lambda calculus with the syntax $s := V \mid (s_1\ s_2) \mid \lambda V.s$. The only operation is the beta-reduction $(\lambda x.s)\ t\ \rightarrow s[t/x]$, which is a (higher-order) rewrite rule. Evaluation in programming languages usually follows a fixed (rewriting) strategy. An early study is [1], where the call-by-name outermost strategy is selected, and where the values are selected as the abstractions, with the motivation to properly model the evaluation strategy of lazy functional languages. An expression s is terminating (or *converging*) if s reduces after finitely many reduction steps to $\lambda x.s'$, which is denoted as $s\downarrow$.

Following [17], the *contextual equivalence* is defined as $s \sim_c t$ iff for all contexts C: $C[s]\downarrow \iff C[t]\downarrow$. This is a congruence on expressions. An *applicative similarity* characterizes contextual equivalence in [1]: $\sim_b = \sim_c$, where $\sim_b$ is the greatest fixpoint satisfying the following condition: $s \sim_b t \iff \forall$ closed r : $s\ r \sim_b s\ r$. This implies that beta-reduction is a correct program transformation, but η-reduction is in general not correct, since $\Omega \not\sim \lambda x.(\Omega\ x)$. Other strategies, like call-by-value, or innermost, with a slightly specialized beta-reduction $(\lambda x.s)\ (\lambda y.t)\ \rightarrow s[(\lambda y.t)/x]$ are incomparable with call-by-need [20].

A related early work is by Robin Milner [14] who constructed a (fully abstract) term model for the lambda calculus, which is exactly the quotient by contextual equivalence.

3 Increasingly More Expressive Calculi

3.1 Deterministic Constructs

The lambda calculus is an example of a deterministic programming language. Let us look at the possible further language constructs and concepts:

Sharing: An explicit syntactic treatment is by (`let` $x = s$ `in` t), where s may be shared in t and the occurrences of x in t represent s; similar to explicit substitutions (in rewriting terms). A more intricate construct is the `let(rec)` in Haskell [6], also called cyclic let, which has syntax (`let` $x_1 = s_1, \ldots, x_n = s_n$ `in` s) and a recursive scope (every x_i is bound in every s_j and s).

Evaluation strategy: Known strategies are call-by-name and call-by-value. The call-by-need (or lazy) strategy is a combination of call-by-name with a proper treatment of sharing.

Data can be constructed by data constructors and analyzed by case-constructs (`case` s`of` $(patt_1 \to s_1)\ldots$). An if-then-else is a special case.

The operator `seq` in Haskell influences the evaluation strategy, i.e. (`seq` s t) has the same value as the value v_t of t, but s must also be evaluated before the value v_t is returned. Though looking innocent, it has a serious influence on the semantics. For example in a polymorphically typed language, the parametricity theorems do no longer hold, see [9]. Operationally, `seq` can also be interpreted as permitting the start of a parallel evaluation.

Typing is ubiquitous in programming languages. We will mainly take into account monomorphic or polymorphic, predicative typing.

3.2 Non-Deterministic Constructs

The operator `choice` is used as (`choice` s t) with the idea that at runtime, on evaluation, either s or t is returned. There are several variants of behaviors, the most prominent one is `amb` [13], which permits the selection of one of its arguments only after it is evaluated to a value. It is also called *bottom-avoiding* `amb`, and often analyzed in connection with fair evaluation.

Concurrent non-determinism is available if several threads are running at different and unpredictable speed and communicate (by message passing as in the pi-calculus [15,16] or by accessing the shared-memory). This leads to a non-deterministic sequence of communications and of side-effects. There may be a synchronizing construct (e.g.semaphores or synchronizing variables, like MVars in Haskell) or a selection of one of several possible interactions on a channel. In Haskell, the monadic programming separates the sequential side-effecting evaluation from the pure functional evaluation [6].

4 Contextual Equality; May and Should

In a programming language L with a notion of successful evaluation (or may-convergence) $(e{\downarrow}e')$, and a notion of contexts (programs with hole) $C[\cdot]$, **contextual equivalence** is defined as $s \leq_L t$ iff $\forall C[\cdot]$ s.t. $C[s], C[t]$ are closed: $C[s]{\downarrow} \implies C[t]{\downarrow}$. Equivalence $s \sim_L t$ holds iff $s \leq_L t \wedge t \leq_L s$. This is based on the operational semantics, which occurs here as the $\downarrow$-symbol.

Small-step vs. big-step operational semantics: Big-step operational semantics is like a specification (or logical deduction system for the obtained value(s)). A *small-step operational semantics* is the same as a (higher-order) rewriting system for programs which is restricted by a strategy. In a non-deterministic/concurrent language, the small-step operational semantics is in my opinion a more adequate notion of evaluation, since it naturally describes the infinite evaluation graph and also naturally specifies the granularity of the atomic single reduction steps.

Let us assume small-step operational semantics from now on.

In non-deterministic and concurrent programming languages, the may-contextual equivalence does not fully capture the idea of equivalent programs, since there may be different evaluation possibilities, and non-terminating or failing evaluation paths. So there are other, additional conditions for equivalence. One is *must-convergence*: $P{\Downarrow_{must}}$, iff every reduction possibility of P is terminating and successful. There is also a third one: *should-convergence*: $P{\Downarrow_{should}}$, if every program reachable by reduction is may-convergent. This leads to variants of contextual equivalence $s \sim_{L,must} t$ iff $\forall C[\cdot] : C[s]{\Downarrow_{must}} \iff C[t]{\Downarrow_{must}}$ and $s \sim_{L,should} t$ iff $\forall C[\cdot] : C[s]{\Downarrow_{should}} \iff C[t]{\Downarrow_{should}}$, which usually are combined as $s \sim_L t$ iff $s \sim_{L,may} t \wedge s \sim_{L,must}$ or alternatively, $s \sim_L t$ iff $s \sim_{L,may} t \wedge s \sim_{L,should}$.

Instead of arguing which of the combinations, "may and must" (M&M), or "may and should" (M&S), is the right one, we provide some observations

1. The equivalences are different, and so have different invariants;
2. After combining it with fairness (redexes are not indefinitely delayed), the (M&S) contextual equivalence is unchanged, whereas the (M&M) - equivalence changes;
3. There are more papers using (M&M);
4. The invariant under (M&M)-$\sim$ is guarantee of a success;
5. The invariant under (M&S)-$\sim$ is a guarantee of error-freeness;
6. The (M&S)-equivalence is closed w.r.t. certain extensions (see [30]);
7. (M&S): Busy-waiting-implementations of concurrency constructs are valid and can be proved equivalent to other implementations. (see e.g. [32]) in contrast to M&M without fairness restrictions.

Let us assume may- and should-convergence from now on.

5 Methods for Proving Contextual Equality

Since direct proofs of contextual equivalences would require an equivalence reasoning over all contexts, it is very helpful if tools are provided that either restrict this quantification to a smaller set of contexts that are more friendly to reasoning, or tools providing easier ways of reasoning.

Context Lemmas: These are usually of the form: if for all contexts R of a class $\mathcal{R}$ of contexts: $R[s]{\downarrow} \iff R[t]{\downarrow}$; then $s \sim_{may} t$. Similarly for should- and must-convergence (see [27]). A CIU-Lemma is a variant: if for all closed contexts R of a class $\mathcal{R}$ of contexts and for all closed (special) substitutions σ:

$R[\sigma(s)]\downarrow \iff R[\sigma(t)]\downarrow$; then $s \sim_{may} t$. For example, a context lemma appears in [14], and also in many other papers, and for ciu-theorems (see e.g. [12]).

Applicative Similarities or Applicative Bisimilarities: Testing equivalence of s, t is reduced to checking the behavior of s and t as functions on all (closed) arguments r. The relation is usually defined as a greatest fixpoint and thus also covers for example infinite lists as results, but requires co-induction as a proof principle. A basic technique for proving soundness of applicative similarities, often modified and adapted to different calculi, is Howe's technique [7,8]. The core of the technique is the definition of a specific congruence-closure relation of the applicative similarity relation, which is then used to show that applicative similarity (in the current calculus) is a pre-congruence.

Diagram-Technique: This is reminiscent of the overlapping and reduction completion technique in Knuth-Bendix completion. However, it is different, since reduction is in general non-terminating, a rewriting strategy has to be obeyed, and instead of an equality, a converging reduction is looked for.

Examples for a diagram technique application is the proof of strong confluence from local strong confluence, the use of the so-called (parallel) 1-reduction to show confluence in various lambda-calculi. [31] develops and applies the diagram method in a call-by-need lambda-calculus. Overlap diagrams between standard reductions and internal reductions are computed, in the form of so-called forking and commuting diagrams. An automatic method using nominal letrec-unification is described in [21]. Often there are finite sets of them (they even represent infinitely many possibilities) and so they can be used to show may-convergencies by constructing converging standard reduction sequences: Given s and a successful standard reduction sequence $s \xrightarrow{*} s'$ where s' is a value (WHNF), and $s \to t$ by a single transformation step, a converging standard reduction $t \xrightarrow{*} t'$ can be constructed using the diagrams. The main idea can be illustrated as follows:

Diagram	Given	After completion
$\begin{array}{ccc} \cdot & \xrightarrow{T} & \cdot \\ {\scriptstyle sr}\downarrow & & \downarrow {\scriptstyle sr} \\ \cdot & \xrightarrow{T,*} & \cdot \end{array}$	$\begin{array}{ccc} s & \xrightarrow{T} & t \\ {\scriptstyle sr}\downarrow & & \\ s_1 & & \\ {\scriptstyle sr,*}\downarrow & & \\ s_n & & \end{array}$	$\begin{array}{ccc} s & \xrightarrow{T} & t \\ {\scriptstyle sr}\downarrow & & \downarrow {\scriptstyle sr} \\ s_1 & \xrightarrow{T,*} & t_1 \\ {\scriptstyle sr,*}\downarrow & & \downarrow {\scriptstyle sr,*} \\ s_n & \xrightarrow{T,*} & t_n \end{array}$

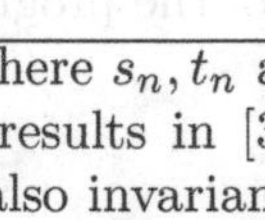

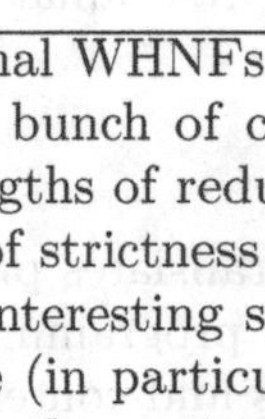

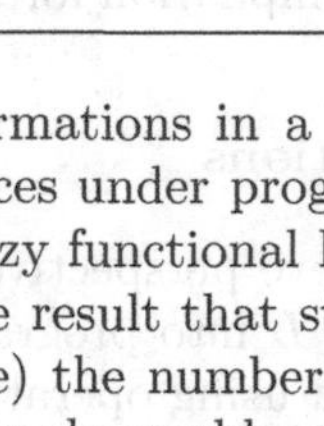

Where s_n, t_n are the final WHNFs.

The results in [31] are a bunch of correct transformations in a core-calculus, but also invariants on lengths of reductions sequences under program transformations. A certain form of strictness analysis for lazy functional languages was proved to be sound. An interesting spin-off was the result that strictness optimization does not change (in particular not reduce) the number of (essential) reduction steps. However, the optimization effect is observable, and has other reasons located in the abstract machine program.

Unfortunately, the obtained diagram sets are in general a bit more complicated than expected, for example, closing them may require $\xrightarrow{sr,*}$, and moreover, the diagram sets have an inherent non-determinism.

Abstracting a bit, a diagram set is a rewriting system on reduction sequences (a mix of standard reductions and transformations), where the desired results are standard reduction sequences. But due to the $\xrightarrow{sr,*}$ the diagrams can also be seen as an infinite rewrite system, finitely represented. The question of existence of a standard reduction sequence can be translated into a termination problem. This was explored in [22], where the problem of correctness of a program transformation was (after a massage) a termination problem, which could be solved by an automatic termination prover (like AProVE [2]).

Infinite Expressions are used in different deterministic call-by-need (letrec)-calculi to bridge the gap between call-by-need and call-by-name variants. The infinite expressions together with call-by-name reductions are used as an intermediate calculus where it can be shown that the translations are convergence equivalent. This method is for example used in [29,25].

5.1 Program Transformations in a Programming Language

In programming languages and compilers of programming languages, there are a class of optimizations that are source-to-source, also on intermediate levels, which can all be subsumed under the notion of program transformations.

Haskell [6] and the GHC make use of such program transformations for optimizations. See [18,19] for small transformations like let-shifting. Haskell also has a rewriting-feature to permit a set of source-to-source transformations to modify and optimize programs, where the correctness proof of the transformations is due to the programmer.

Another form of correct transformations is deforestation [5] which avoids the construction of intermediate data structures (see also [33]). In a similar direction are the transformation ideas in the work of Bird and Moor [4] and the design principles and transformations in [3].

A technique that is used in compilers of programming languages is pre-evaluation of expressions, which is like partial evaluation [10,11]. Programming languages based on calculi where reduction rules are correct transformations can make use of this method. This exactly corresponds to proving the correctness of reduction rules in higher-order program calculi, and the use of these reduction rules during compilation for simplifying or pre-evaluating parts of the program.

5.2 Translations

From a bird's eye perspective, a compiler translates programs from a programming language L into programs of another programming language L'. Making this formal and using operational semantics and contextual preorder (or equivalence) as semantics leads to the notion of translations ϕ from one calculus L into another L', mapping expressions, contexts, convergencies, and leads to two useful correctness notions: **adequacy** holds if $\phi(s) \leq \phi(t) \implies s \leq t$, and **full abstractness** holds if $\phi(s) \leq \phi(t) \iff s \leq t$ (see e.g. [26]). An adequate translation can be interpreted as a correct result of a compilation from a high-level language into a low-level language, whereas a fully abstract translation is a translation from a language into another one of equal level. A fully abstract translation can also be viewed as an embedding of language L in L' (modulo the contextual equivalence).

Adequacy holds if two criteria (convergence equivalence and compositional) hold of a translation. Convergence equivalence means that $s\downarrow \iff \phi(s)\downarrow'$, and compositional means that $\phi(C[e]) \sim' \phi(C)[\phi(e)])$. Since the definition of contextual equivalence can be applied to almost every programming language, also the notions above are useful for all these languages.

For concurrent languages, the convergence equivalence criteria consist usually of two convergencies. The methods are applied and turned out to be crucial for several analyses in concurrent languages: The theoretical essence of [28] is the proof of adequacy of a translation. Translations are used several times in [25].

6 Concurrent Programming Languages

Modelling concurrency and non-determinism in lazy (call-by-need) functional languages has two sensible alternatives: (1) permitting calls anywhere in the program; and (2) using monadic programming in connection with actions that manipulate external objects.

(1) Permitting calls anywhere in the program: Call-by-need extended lambda calculi with a choice-operator are appropriate for modelling this method. Among the possibilities are `choice` or `amb`, where `choice` can model interactions with the environment, whereas `amb` is a bit more ambitious and models concurrent threads, where threads can also be killed or be garbage collected. In [23] an expressive calculus with `amb` and `letrec` is analyzed with may- and should convergence also under fair evaluation of threads. The results are context lemmas for may- and should-preorders, and a large set of correct program transformations and some equalities for `amb` and relations between the two contextual preorders.

(2) Using monadic programming in connection with actions that manipulate external objects.: Concurrent Haskell is such a language, but it is too complex to be used for theoretical analyses, so we approached it with the calculus CHF (concurrent Haskell with futures) [24,25], which is rather close to the realistic language "concurrent Haskell":

Threads are generalized to so-called futures, which are named threads that can deliver a result. Monomorphic typing is necessary to keep monadic programming and the pure functional part separate. Using contextual equivalence, the calculus can indeed be semantically analyzed and again all deterministic reduction rules are correct. The non-deterministic ones are `putMVar` and `takeMVar` which empty or fill a synchronizing variable MVar which cannot be used as program transformations for obvious reasons.

One result of [24] are context lemmas, and a further result is that after a minor change w.r.t. Haskell (restricting `seq` in the first argument to pure functional expressions), the monadic axioms hold, which can be transferred to concurrent Haskell. Without this restriction these axioms do not hold, thus these are false in concurrent Haskell.

The work in [25] is a deeper analysis of CHF, where the goal was to find out about the equivalences that hold in the pure functional part (say Haskell) and whether these equivalences still hold in full CHF. Formally, this is the question whether the pure part of CHF with its local contextual equivalence is conservatively embedded in CHF. Indeed, this is the case, which is an important result with the practical consequence that the optimizations of pure Haskell can still be used in concurrent Haskell.

It is also shown that having a too restricted scheduling of futures (threads), then conservativity fails. In particular, if lazy futures (evaluation starts only if requested from another thread) are added, then sequencing is observable: a particular witness is: $\texttt{seq}\, y\, (\texttt{seq}\, x\; y) \sim_{CHF,pure} (\texttt{seq}\; x\;\; y)$, but $\texttt{seq}\, y\, (\texttt{seq}\, x\; y) \not\sim_{CHF,lazyfuture} (\texttt{seq}\, x\; y)$.

Proof methods are again: translating the call-by-need calculus into a call-by-name infinite calculus, applying Howe's technique for the pure part and the infinite expression calculus, and then showing correctness of embeddings.

7 Software Transactional Memory

Looking at a real application: software transactional memory in a CHF-like calculus of a variant of concurrent Haskell [28].

The main idea is programming of transactions on main memory, where updates made by a transaction are visible to other transactions only if the transaction successfully ended, and which can be stopped and retried in case of a conflicting access. Additionally, an `Orelse` allows a programmed alternative instead of a complete `Retry`. The specification of the operational semantics of a transaction says: execute it, but only if it can be executed successfully as one atomic and isolated step.

The result of our work is that the operational specification with undecidable evaluation conditions can be implemented, where every rule application has decidable execution conditions, and where the target calculus is much more explicit about the variables, locking variables of the transactional memory, running threads in parallel and killing other threads.

The main result is that this implementation is semantically correct (i.e. adequate) w.r.t. contextual equivalence with may- and should-convergence. The core of the proof is to show convergence-equivalence of a translation for may- and should-convergence. It has four parts, two for may- and two for should-convergence, and then one for every implication.

That a theoretical result is possible at all is surprising, since the implementation, using more than 20 compound rules, is rather complex.

8 Conclusion, Further Work

Contextual equivalence as semantics for concurrent languages and using may- and should-convergence have a good coverage of invariants and a potential as a foundation for their correctness.

Further work may be to strengthen the methods and apply it to other realistic programming languages e.g. one may include exception handling in the analysis.

Programming languages where the operational semantics is not really part of the specification or subject to permanent changes are prohibitive for such analyses.

References

1. Abramsky, S.: The lazy lambda calculus. In: Turner, D.A. (ed.) Research Topics in Functional Programming, pp. 65–116. Addison-Wesley (1990)
2. Aprove-team: AProVE website (2014), `http://aprove.informatik.rwth-aachen.de`
3. Bird, R.: Pearls of Functional Algorithm Design. Cambridge University Press, Cambridge (2010)
4. Bird, R.S., De Moor, O.: Algebra of Programming. International Series in Computer Science, vol. 100. Prentice Hall (1997)
5. Gill, A.: Cheap Deforestation for Non-strict Functional Languages. PhD thesis, Glasgow University, Department of Computing Science (1996)
6. Haskell-community: The Haskell Programming Language (2014), `http://www.haskell.org/haskellwiki/Haskell`
7. Howe, D.: Equality in lazy computation systems. In: 4th IEEE Symp. on Logic in Computer Science, pp. 198–203 (1989)
8. Howe, D.: Proving congruence of bisimulation in functional programming languages. Inform. and Comput. 124(2), 103–112 (1996)
9. Johann, P., Voigtländer, J.: The impact of seq on free theorems-based program transformations. Fundamenta Informaticae 69(1-2), 63–102 (2006)
10. Jones, M.P.: Dictionary-free Overloading by PaRTIAL Evaluation. In: Partial Evaluation and Semantics-Based Program Manipulation, Orlando, Florida, pp. 107–117. Technical Report 94/9, Department of Computer Science, University of Melbourne (June 1994)
11. Jones, N.D.: An introduction to partial evaluation. ACM Comput. Surv. 28(3), 480–503 (1996)
12. Mason, I., Smith, S.F., Talcott, C.L.: From operational semantics to domain theory. Inform. and Comput. 128, 26–47 (1996)
13. McCarthy, J.: A Basis for a Mathematical Theory of Computation. In: Braffort, P., Hirschberg, D. (eds.) Computer Programming and Formal Systems, pp. 33–70. North-Holland, Amsterdam (1963)
14. Milner, R.: Fully abstract models of typed λ-calculi. Theoret. Comput. Sci. 4, 1–22 (1977)
15. Milner, R.: Functions as processes. Mathematical Structures in Computer Science 2(2), 119–141 (1992)
16. Milner, R., Parrow, J., Walker, D.: A calculus of mobile processes, i & ii. Inform. and Comput. 100(1), 1–77 (1992)

17. Morris, J.: Lambda-Calculus Models of Programming Languages. PhD thesis, MIT (1968)
18. Peyton Jones, S., Partain, W., Santos, A.: Let-floating: moving bindings to give faster programs. In: Harper, R., Wexelblat, R.L. (eds.) Proceedings of the first ACM SIGPLAN International Conference on Functional Programming, pp. 1–12. ACM Press (1996)
19. Peyton Jones, S.L., Santos, A.L.M.: A transformation-based optimiser for Haskell. Science of Computer Programming 32(1-3), 3–47 (1998)
20. Plotkin, G.D.: Call-by-name, call-by-value, and the lambda-calculus. Theoret. Comput. Sci. 1, 125–159 (1975)
21. Rau, C.: Automatic correctness proofs of program transformations (2014) (thesis in preparation)
22. Rau, C., Sabel, D., Schmidt-Schauß, M.: Correctness of program transformations as a termination problem. In: Gramlich, B., Miller, D., Sattler, U. (eds.) IJCAR 2012. LNCS, vol. 7364, pp. 462–476. Springer, Heidelberg (2012)
23. Sabel, D., Schmidt-Schauß, M.: A call-by-need lambda-calculus with locally bottom-avoiding choice: Context lemma and correctness of transformations. Math. Structures Comput. Sci. 18(3), 501–553 (2008)
24. Sabel, D., Schmidt-Schauß, M.: A contextual semantics for Concurrent Haskell with futures. In: Schneider-Kamp, P., Hanus, M. (eds.) PPDP 2011, pp. 101–112. ACM, New York (2011)
25. Sabel, D., Schmidt-Schauß, M.: Conservative concurrency in Haskell. In: Dershowitz, N. (ed.) LICS, pp. 561–570. IEEE (2012)
26. Schmidt-Schauß, M., Niehren, J., Schwinghammer, J., Sabel, D.: Adequacy of compositional translations for observational semantics. In: Ausiello, G., Karhumäki, J., Mauri, G., Ong, C.H.L. (eds.) 5th IFIP TCS 2008. IFIP, vol. 273, pp. 521–535. Springer, Heidelberg (2008)
27. Schmidt-Schauß, M., Sabel, D.: On generic context lemmas for higher-order calculi with sharing. Theoret. Comput. Sci. 411(11-13), 1521–1541 (2010)
28. Schmidt-Schauß, M., Sabel, D.: Correctness of an STM Haskell implementation. In: Morrisett, G., Uustalu, T. (eds.) Proceedings of the 18th ACM SIGPLAN International Conference on Functional Programming, ICFP 2013, pp. 161–172. ACM, New York (2013)
29. Schmidt-Schauß, M., Sabel, D., Machkasova, E.: Simulation in the call-by-need lambda-calculus with letrec. In: Lynch, C. (ed.) Proc. of 21st RTA 2010. LIPIcs, vol. 6, pp. 295–310. Schloss Dagstuhl - Leibniz-Zentrum für Informatik (2010)
30. Schmidt-Schauß, M., Sabel, D., Schütz, M.: Deciding inclusion of set constants over infinite non-strict data structures. RAIRO-Theoretical Informatics and Applications 41(2), 225–241 (2007)
31. Schmidt-Schauß, M., Schütz, M., Sabel, D.: Safety of Nöcker's strictness analysis. J. Funct. Programming 18(4), 503–551 (2008)
32. Schwinghammer, J., Sabel, D., Schmidt-Schauß, M., Niehren, J.: Correctly translating concurrency primitives. In: Rossberg, A. (ed.) ML 2009: Proceedings of the 2009 ACM SIGPLAN Workshop on ML, pp. 27–38. ACM, New York (2009)
33. Sculthorpe, N., Farmer, A., Gill, A.: The HERMIT in the tree - mechanizing program transformations in the GHC core language. In: Hinze, R. (ed.) IFL 2012. LNCS, vol. 8241, pp. 86–103. Springer, Heidelberg (2013)

Unnesting of Copatterns

Anton Setzer[1], Andreas Abel[2], Brigitte Pientka[3], and David Thibodeau[3]

[1] Dept. of Computer Science, Swansea University, Swansea SA2 8PP, UK
a.g.setzer@swan.ac.uk
[2] Computer Science and Engineering, Chalmers and Gothenburg University, Rännvägen 6, 41296 Göteborg, Sweden
andreas.abel@gu.se
[3] School of Computer Science, McGill University, Montreal, Canada
{bpientka,dthibo1}@cs.mcgill.ca

Abstract. Inductive data such as finite lists and trees can elegantly be defined by constructors which allow programmers to analyze and manipulate finite data via pattern matching. Dually, coinductive data such as streams can be defined by observations such as head and tail and programmers can synthesize infinite data via copattern matching. This leads to a symmetric language where finite and infinite data can be nested. In this paper, we compile nested pattern and copattern matching into a core language which only supports simple non-nested (co)pattern matching. This core language may serve as an intermediate language of a compiler. We show that this translation is conservative, i.e. the multi-step reduction relation in both languages coincides for terms of the original language. Furthermore, we show that the translation preserves strong and weak normalisation: a term of the original language is strongly/weakly normalising in one language if and only if it is so in the other. In the proof we develop more general criteria which guarantee that extensions of abstract reduction systems are conservative and preserve strong or weak normalisation.

Keywords: Pattern matching, copattern matching, algebraic data types, codata, coalgebras, conservative extension, strong normalisation, weak normalisation,abstract reduction system, ARS.

1 Introduction

Finite inductive data such as lists and trees can be elegantly defined via constructors, and programmers are able to case-analyze and manipulate finite data in functional languages using pattern matching. To compile functional languages supporting pattern matching, we typically elaborate complex and nested pattern matches into a series of simple patterns which can be easily compiled into efficient code (see for example [3]). This is typically the first step in translating the source language to a low-level target language which can be efficiently executed. It is also an important step towards developing a core calculus supporting well-founded recursive functions.

G. Dowek (ed.): RTA-TLCA 2014, LNCS 8560, pp. 31–45, 2014.

Dually to finite data, coinductive data such as streams can be defined by observations such as head and tail. This view was pioneered by Hagino [7] who modelled finite objects via initial algebras and infinite objects via final coalgebras in category theory. This led to the design of symML, a dialect of ML where we can for example define the codata-type of streams via the destructors head and tail which describe the observations we can make about streams [8]. Cockett and Fukushima [6] continued this line of work and designed a language *Charity* where one programs directly with the morphisms of category theory. Our recent work [2] extends these ideas and introduces *copattern* matching for analyzing infinite data. This novel perspective on defining infinite structures via their observations leads to a new symmetric foundation for functional languages where inductive and coinductive data types can be mixed.

In this paper, we elaborate our high-level functional language which supports nested patterns and copatterns into a language of simple patterns and copatterns. Similar to pattern compilation in Idris or Agda, our translation into simple patterns is guided by the coverage algorithm. We show that the translation into our core language of simple patterns is conservative, i.e. the multi-step reduction relations of both languages coincide for terms of the original language. Furthermore, we show that the translation preserves strong normalisation (*SN*) and weak normalisation (*WN*): a term of the original language is SN or WN in one language if and only if it has this property in the other.

The paper is organized as follows: We describe the core language including pattern and copattern matching in Sect. 2. In Sect. 3, we explain the translation into simple patterns. In Sect. 4 we develop criteria which guarantee that extensions of abstract reduction systems are conservative and preserve SN or WN. We use this these criteria in Section 5 to show that the translation of patterns into simple patterns is a conservative extension preserving SN and WN.

2 A Core Language for Copattern Matching

In this section, we summarize the basic core language with (co)recursive data types and support for (co)pattern described in previous work [2].

2.1 Types and Terms

A language $\mathcal{L} = (\mathcal{F}, \mathcal{C}, \mathcal{D})$ consists of a finite set $\mathcal{F}$ of *constants* (function symbols), a finite set $\mathcal{C}$ of *constructors*, and a finite set $\mathcal{D}$ of *destructors*. We will in the following assume one fixed language $\mathcal{L}$, with pairwise disjoint $\mathcal{F}$, $\mathcal{C}$, and $\mathcal{D}$. We write f, c, d for elements of $\mathcal{F}, \mathcal{C}, \mathcal{D}$, respectively.

Our type language includes $\mathbf{1}$ (unit), $A \times B$ (products), $A \to B$ (functions), disjoint unions D (labelled sums, "data"), records R (labelled products), least fixed points $\mu X.D$, and greatest fixed points $\nu X.R$.

$$
\begin{array}{lll}
\text{Types } A, B, C & ::= & X \mid \mathbf{1} \mid A \times B \mid A \to B \mid \mu X.D \mid \nu X.R \\
\text{Variants } D & ::= & \langle c_1\ A_1 \mid \ldots \mid c_n\ A_n \rangle \\
\text{Records } R & ::= & \{d_1 : A_1, \ldots, d_n : A_n\}
\end{array}
$$

In the above let c_i be different, and d_i be different. Variant types $\langle c_1\, A_1 \mid \ldots \mid c_n\, A_n \rangle$, finite maps from constructors to types, appear only in possibly recursive data types $\mu X.D$. Records $\{d_1 : A_1, \ldots, d_n : A_n\}$, finite maps from destructors to types, list the fields d_i of a possibly recursive record type $\nu X.R$. To illustrate, we define natural numbers Nat, lists and Nat-streams:

$$\begin{array}{ll} \mathsf{Nat} & := \mu X.\langle \mathsf{zero}\ \mathbf{1} \mid \mathsf{suc}\ X \rangle \\ \mathsf{List}\ A & := \mu X.\langle \mathsf{nil}\ \mathbf{1} \mid \mathsf{cons}\ (A \times X) \rangle \\ \mathsf{StrN} & := \nu X.\{\mathsf{head} : \mathsf{Nat}, \mathsf{tail} : X\} \end{array}$$

In our non-polymorphic calculus, type variables X only serve to construct recursive data types and recursive record types. As usual, $\mu X.D$ ($\nu X.R$, resp.) binds type variable X in D (R, resp.). Capture-avoiding substitution of type C for variable X in type A is denoted by $A[X := C]$. A type is *well-formed* if it has no free type variables; in the following, we assume that all types are well-formed.

We write $c \in D$ for $c\, A$ for some A being part of variant D and define the type of constructor c as $(\mu X.D)_c := A[X := \mu X.D]$. Analogously, we write $d \in R$ for $d : A$ for some A being part of the record R and define the type of the destructor d as $(\nu X.R)_d := A[X := \nu X.R]$.

A signature for $\mathcal{L}$ is a map Σ from $\mathcal{F}$ into the set of types. Unless stated differently, we assume one fixed signature Σ. A typed language is a pair $(\mathcal{L}, \Sigma)$ where $\mathcal{L}$ is a language and Σ is a signature for $\mathcal{L}$. We sometimes write Σ instead of $(\mathcal{L}, \Sigma)$. We write $f \in \Sigma$ if $\Sigma(f)$ is defined, i.e. $f \in \mathcal{F}$. Next, we define the grammar of *terms* of a language $\mathcal{L} = (\mathcal{F}, \mathcal{C}, \mathcal{D})$. Herein, $f \in \mathcal{F}$, $c \in \mathcal{C}$, and $d \in \mathcal{D}$.

$$\begin{array}{rlllll} e, r, s, t, u ::= & f & \text{Defined constant (function)} & \mid x & \text{Variable} \\ & \mid () & \text{Unit (empty tuple)} & \mid (t_1, t_2) & \text{Pair} \\ & \mid c\, t & \text{Constructor application} & \mid t_1\, t_2 & \text{Application} \\ & \mid t\,.d & \text{Destructor application} & & \end{array}$$

Terms include identifiers (variables x and defined constants f) and introduction forms: pairs (t_1, t_2), unit $()$, and constructed terms $c\, t$, for the *positive* types $A \times B$, $\mathbf{1}$, and $\mu X.D$. There are however no elimination forms for positive types, since we define programs via rewrite rules and employ *pattern matching*. On the other hand we have *eliminations*, application $t_1\, t_2$ and projection $t\,.d$, of *negative* types $A \to B$ and $\nu X.R$ respectively, but omit introductions for these types, since this will be handled by *copattern matching*.

We write term substitutions as $s[x_1 := t_1, \ldots, x_n := t_n]$ or short $s[\vec{x} := \vec{t}]$. Contexts Δ are finite maps from variable to types, written as lists of pairs $x_1 : A_1, \ldots, x_n : A_n$, or short $\vec{x} : \vec{A}$, with $\cdot$ denoting the empty context. We write $\Delta \to A$ or $\vec{A} \to A$ for n-ary curried function types $A_1 \to \cdots \to A_n \to A$ (but A may still be a function type), and $s\, \vec{t}$ for n-ary curried application $s\, t_1\, \cdots\, t_n$.

The typing rules for terms (relative to a typed language Σ) are defined in Figure 1. If we want to explicitly refer to a given typed language $(\mathcal{L}, \Sigma)$ or Σ we write $\Delta \vdash_{\mathcal{L},\Sigma} A$ or $\Delta \vdash_{\Sigma} A$, similarly for later notions of $\vdash$.

$$\frac{\Delta(x) = A}{\Delta \vdash x : A} \qquad \frac{}{\Delta \vdash () : \mathbf{1}} \qquad \frac{\Delta \vdash t : (\mu X.D)_c}{\Delta \vdash c\,t : \mu X.D} \qquad \frac{\Delta \vdash t_1 : A_1 \quad \Delta \vdash t_2 : A_2}{\Delta \vdash (t_1, t_2) : A_1 \times A_2}$$

$$\frac{}{\Delta \vdash f : \Sigma(f)} \qquad \frac{\Delta \vdash t : A \to B \quad \Delta \vdash t' : A}{\Delta \vdash t\,t' : B} \qquad \frac{\Delta \vdash t : \nu X.R}{\Delta \vdash t\,.d : (\nu X.R)_d}$$

Fig. 1. Typing rules

2.2 Patterns and Copatterns

For each $f \in \mathcal{F}$, we will determine the rewrite rules for f as a set of pairs $(q \longrightarrow r)$ where q is a copattern sometimes referred to as *left hand side*, and r a term, sometimes referred to as *right hand side*. Patterns p and copatterns q are special terms given by the grammar below, where $c \in \mathcal{C}$ and $d \in \mathcal{D}$.

$p ::= x$	Variable pattern	$q ::= f$	Head (constant)
$\mid ()$	Unit pattern	$\mid q\,p$	Application copattern
$\mid (p_1, p_2)$	Pair pattern	$\mid q\,.d$	Destructor copattern
$\mid c\,p$	Constructor pattern		

In addition we require p and q to be *linear*, i.e. each variable occurs at most once in p or q. When later defining typed patterns $\Delta \vdash q : A$ as part of a coverage complete pattern set for a constant f, we will have that, if this judgement is provable as a typing judgement for terms, the variables in q are exactly the variables in Δ, and f is the head of q.

The distinction between patterns and copatterns is in this article only relevant in this grammar, therefore we often write simply "pattern" for both.

Example 1 (Cycling numbers). Function cyc of type Nat → StrN, when passed an integer n, produces a stream $n, n-1, \ldots, 1, 0, N, N-1, \ldots, 1, 0, N, N-1, \ldots$ for some fixed N. To define this function we match on the input n and also observe the resulting stream, highlighting the mix of pattern and copattern matching. The rules for cyc are the following:

$$\begin{array}{lll} \mathsf{cyc}\ x & .\mathsf{head} & \longrightarrow x \\ \mathsf{cyc}\ (\mathsf{zero}\ ()) & .\mathsf{tail} & \longrightarrow \mathsf{cyc}\ N \\ \mathsf{cyc}\ (\mathsf{suc}\ x) & .\mathsf{tail} & \longrightarrow \mathsf{cyc}\ x \end{array}$$

Example 2 (Fibonacci Stream). Nested destructor copatterns appear in the following definition of the stream of Fibonacci numbers. It uses zipWith _+_ which is the pointwise addition of two streams.

$$\begin{array}{ll} \mathsf{fib}\ .\mathsf{head} & \longrightarrow 0 \\ \mathsf{fib}\ .\mathsf{tail}\ .\mathsf{head} & \longrightarrow 1 \\ \mathsf{fib}\ .\mathsf{tail}\ .\mathsf{tail} & \longrightarrow \mathsf{zipWith}\ \_{+}\_\ \mathsf{fib}\ (\mathsf{fib}\ .\mathsf{tail}) \end{array}$$

2.3 Coverage

For our purposes, the rules for a constant f are *complete*, if every closed, well-typed term t of positive type can be reduced with exactly one of the rules of f. Alternatively, we could say that all cases for f are uniquely *covered* by the reduction rules. Coverage implies that the execution of a program progresses, i.e. does not get stuck, and is deterministic. Note that by restricting to positive types, which play the role of ground types, we ensure that t is not stuck because f is underapplied. Progress has been proven in previous work [2]; in this work, we extend coverage checking to an algorithm for pattern compilation.

We introduce the judgement $f : A \lhd\!\mid Q$, called a *coverage complete pattern set for f (cc-pattern-set for f)*. Here Q is a *set* $Q = (\Delta_i \vdash q_i : C_i)_{i=1,\ldots,n}$. If $f : A \lhd\!\mid Q$ then constant f of type A can be defined by the coverage complete patterns q_i (depending on variables in Δ_i) together with rewrite rules $q_i \longrightarrow t_i$ for some $\Delta_i \vdash t_i : C_i$.

The rules for deriving cc-pattern-sets are presented in Figure 2. In the variable splitting rules, the split variable is written as the last element of the context. Because contexts are finite maps they have no order—any variable can be split. Note as well that patterns and copatterns are by definition required to be linear.

Result splitting:

$$\frac{}{f : A \lhd\!\mid (\cdot \vdash f : A)}\;\mathrm{C_{Head}} \qquad \frac{f : A \lhd\!\mid Q\,(\Delta \vdash q : B \to C)}{f : A \lhd\!\mid Q\,(\Delta, x : B \vdash q\,x : C)}\;\mathrm{C_{App}}$$

$$\frac{f : A \lhd\!\mid Q\,(\Delta \vdash q : \nu X.R)}{f : A \lhd\!\mid Q\,(\Delta \vdash q\,.d : (\nu X.R)_d)_{d\in R}}\;\mathrm{C_{Dest}}$$

Variable splitting:

$$\frac{f : A \lhd\!\mid Q\,(\Delta, x : \mathbf{1} \vdash q : C)}{f : A \lhd\!\mid Q\,(\Delta \vdash q[x := ()] : C)}\;\mathrm{C_{Unit}}$$

$$\frac{f : A \lhd\!\mid Q\,(\Delta, x : A_1 \times A_2 \vdash q : C)}{f : A \lhd\!\mid Q\,(\Delta, x_1 : A_1, x_2 : A_2 \vdash q[x := (x_1, x_2)] : C)}\;\mathrm{C_{Pair}}$$

$$\frac{f : A \lhd\!\mid Q\,(\Delta, x : \mu X.D \vdash q : C)}{f : A \lhd\!\mid Q\,(\Delta, x' : (\mu X.D)_c \vdash q[x := c\,x'] : C)_{c\in D}}\;\mathrm{C_{Const}}$$

Fig. 2. Coverage rules

The judgement $f : \Sigma(f) \lhd\!\mid (\Delta_i \vdash q_i \longrightarrow t_i : C_i)_{i=1,\ldots,n}$ called a *coverage complete set of rules for f (cc-rule-set for f)* has the following derivation rule

$$\frac{f : \Sigma(f) \lhd\!\mid (\Delta_i \vdash q_i : C_i)_{i=1,\ldots,n} \qquad \Delta_i \vdash t_i : C_i \; (i = 1,\ldots,n)}{f : \Sigma(f) \lhd\!\mid (\Delta_i \vdash q_i \longrightarrow t_i : C_i)_{i=1,\ldots,n}}$$

Then $f : \Sigma(f) \lhd\!\mid (\Delta_i \vdash q_i : C_i)_{i=1,\ldots,n}$ is called the *underlying cc-pattern-set of the cc-rule set*. The corresponding *term rewriting rules for f* are $q_i \longrightarrow t_i$.

A program $\mathcal{P}$ over the typed language Σ is a function mapping each constant f to a cc-rule-set $\mathcal{P}_f$ for f. We write $t \longrightarrow_{\mathcal{P}} t'$ for one-step reduction of term t to t' using the compatible closure[1] of the term rewriting rules in $\mathcal{P}$, and drop index $\mathcal{P}$ if clear from the context of discourse. We further write $\longrightarrow^*_{\mathcal{P}}$ for its transitive and reflexive closure and $\longrightarrow^{\geq 1}_{\mathcal{P}}$ for its transitive closure.

Example (Deriving a cc-pattern-set for cyc*)* We start with $\mathrm{C}_{\mathrm{Head}}$

$$\mathsf{cyc} : \mathsf{Nat} \to \mathsf{StrN} \ \lhd| \ (\cdot \vdash \mathsf{cyc} : \mathsf{Nat} \to \mathsf{StrN})$$

We apply x to the head by $\mathrm{C}_{\mathrm{App}}$.

$$\mathsf{cyc} : \mathsf{Nat} \to \mathsf{StrN} \ \lhd| \ (x : \mathsf{Nat} \vdash \mathsf{cyc}\ x : \mathsf{StrN})$$

Then we split the result by $\mathrm{C}_{\mathrm{Dest}}$.

$$\mathsf{cyc} : \mathsf{Nat} \to \mathsf{StrN} \ \lhd| \ \begin{array}{l} (x : \mathsf{Nat} \vdash \mathsf{cyc}\ x\ .\mathsf{head} : \mathsf{Nat}) \\ (x : \mathsf{Nat} \vdash \mathsf{cyc}\ x\ .\mathsf{tail} \ \ : \mathsf{StrN}) \end{array}$$

In the second copattern, we split x using $\mathrm{C}_{\mathrm{Const}}$.

$$\mathsf{cyc} : \mathsf{Nat} \to \mathsf{StrN} \ \lhd| \ \begin{array}{llll} (x : \mathsf{Nat} & \vdash \mathsf{cyc}\ x & .\mathsf{head} & : \mathsf{Nat}) \\ (x : \mathbf{1} & \vdash \mathsf{cyc}\ (\mathsf{zero}\ x) & .\mathsf{tail} & : \mathsf{StrN}) \\ (x : \mathsf{Nat} & \vdash \mathsf{cyc}\ (\mathsf{suc}\ x) & .\mathsf{tail} & : \mathsf{StrN}) \end{array}$$

We finish by applying $\mathrm{C}_{\mathrm{Unit}}$ which replaces x by $()$ in the second clause.

$$\mathsf{cyc} : \mathsf{Nat} \to \mathsf{StrN} \ \lhd| \ \begin{array}{llll} (x : \mathsf{Nat} & \vdash \mathsf{cyc}\ x & .\mathsf{head} & : \mathsf{Nat}) \\ (\cdot & \vdash \mathsf{cyc}\ (\mathsf{zero}\ ()) & .\mathsf{tail} & : \mathsf{StrN}) \\ (x : \mathsf{Nat} & \vdash \mathsf{cyc}\ (\mathsf{suc}\ x) & .\mathsf{tail} & : \mathsf{StrN}) \end{array}$$

This concludes the derivation of the cc-pattern-set for the cyc function.

3 Reduction of Nested to Simple Pattern Matching

In the following, we describe a translation of deep (aka nested) (co)pattern matching (i.e. pattern matching as defined before) into shallow (aka non-nested) pattern matching, which we call *simple* pattern matching, as defined below. We are certainly not the first to describe such a translation, except maybe for copatterns, but we have special requirements for our translation. The obvious thing to ask for is *simulation, i.e.* each reduction step in the original program should correspond to one or more reduction steps in the translated program. However, we want the translation also to *preserve and reflect normalization*: A term in the original program terminates, if and only if it terminates in the translated program. Preservation of normalization is important for instance in dependently typed languages such as Agda, where the translated programs are run during

[1] See e.g. Def. 2.2.4 of [12].

type checking and need to behave exactly like the original, user-written programs.

The strong normalization property is lost by some of the popular translations. For instance, translating rewrite rules to fixed-point and case combinators breaks normalization, simply because fixed-point combinators reduce by themselves, allowing infinite reduction sequences immediately. But also special fixed-point combinators that only unfold, if their principal argument is a constructor term, or dually co-fixed-point combinators that only unfold, if their result is observed[2], have such problems. Consider the following translation of a function f with deep matching into such a fixed-point combinator:

$$\begin{array}{lll} f\ (\mathsf{zero}\ ()) & \longrightarrow \mathsf{zero}\ () \\ f\ (\mathsf{suc}\ (\mathsf{zero}\ ())) & \longrightarrow \mathsf{zero}\ () \\ f\ (\mathsf{suc}\ (\mathsf{suc}\ x)) & \longrightarrow f\ (\mathsf{suc}\ x)) \end{array} \rightsquigarrow \mathsf{fix}\ f\ (x).\mathsf{case}\ x\ \mathsf{of} \begin{cases} \mathsf{zero}\ () & \longrightarrow \mathsf{zero}\ () \\ \mathsf{suc}\ (\mathsf{zero}\ ()) & \longrightarrow \mathsf{zero}\ () \\ \mathsf{suc}\ (\mathsf{suc}\ x) & \longrightarrow f(\mathsf{suc}\ x)) \end{cases}$$

While the term $f\ (\mathsf{suc}\, x)$ terminates for the original program simply because no pattern matches (i.e. no rewrite rule applies), it diverges for the translated program since the fixed-point applied to a constructor unfolds to a term containing the original term as a subterm. A closer look reveals that this special fixed-point combinator preserves normalization for simple pattern matching only.

3.1 Simple Patterns

A simple copattern q_{s} is of one of the forms $f\ \vec{x}$ (no matching), $f\ \vec{x}\ .d$ (shallow result matching) or $f\ \vec{x}\ p_{\mathsf{s}}$ (shallow argument matching) where $p_{\mathsf{s}} ::= () \mid (x_1, x_2) \mid c\, x$ is a simple pattern.

Definition 1 (Simple coverage-complete pattern sets)

(a) Simple cc-pattern-sets $f : A \lhd|_{\mathsf{s}} Q$ *are defined as follows* ($\Delta = \vec{x} : \vec{A}$)*:*

$$\begin{array}{rl} f : \Delta \to A & \lhd|_{\mathsf{s}}\ (\Delta \vdash f\ \vec{x} : A) \\ f : \Delta \to \nu X.R & \lhd|_{\mathsf{s}}\ (\Delta \vdash f\ \vec{x}\ .d : (\nu X.R)_d)_{d \in R} \\ f : \Delta \to \mathbf{1} \to A & \lhd|_{\mathsf{s}}\ (\Delta \vdash f\ \vec{x}\ () : A) \\ f : \Delta \to (B_1 \times B_2) \to A & \lhd|_{\mathsf{s}}\ (\Delta, y_1 : B_1, y_2 : B_2 \vdash f\ \vec{x}\ (y_1, y_2) : A) \\ f : \Delta \to (\mu X.D) \to A & \lhd|_{\mathsf{s}}\ (\Delta, x' : (\mu X.D)_c \vdash f\ \vec{x}\ (c\ x') : A)_{c \in D} \end{array}$$

(b) A cc-rule-set is simple *if the underlying cc-pattern-set is simple. A constant in a program is* simple, *if its cc-rule-set is simple. A program is* simple *if all its constants are simple.*

Remark 2. If $f : A \lhd|_{\mathsf{s}} Q$ then $f : A \lhd| Q$.

[2] Such fixed-point combinators are used in the *Calculus of Inductive Constructions*, the core language of Coq [9], but have also been studied for sized types [4,1].

3.2 The Translation Algorithm by Example

Neither the cyc function, nor the Fibonacci stream are simple. The translation into simple patterns introduces auxiliary function symbols, which are obtained as follows: We start from the bottom of the derivation tree of a non simple cc-pattern-set, remove the last derivation step, and create a new function symbol. This function takes as arguments the variables we have not split on from the original function and (co)pattern matches just as the last derivation set of the original derivation did. Let us walk through the algorithm of transforming patterns into simple patterns for the cyc function. The original program is

$$\mathsf{cyc} : \mathsf{Nat} \to \mathsf{StrN} \vartriangleleft \left| \begin{array}{llll} (x : \mathsf{Nat} \vdash \mathsf{cyc}\ x & .\mathsf{head} & \longrightarrow x & : \mathsf{Nat}) \\ (& \vdash \mathsf{cyc}\ (\mathsf{zero}\ ()) & .\mathsf{tail} & \longrightarrow \mathsf{cyc}\ N : \mathsf{StrN}) \\ (x : \mathsf{Nat} \vdash \mathsf{cyc}\ (\mathsf{suc}\ x) & .\mathsf{tail} & \longrightarrow \mathsf{cyc}\ x & : \mathsf{StrN}) \end{array}\right.$$

In the derivation of the underlying cc-pattern-set, the last step was $\mathrm{C}_{\mathrm{Unit}}$ replacing pattern variable $x : \mathbf{1}$ by pattern (). We introduce a new constant g_2 with simple cc-rule-set and replace the right hand side of the split clause with a call to g_2 in the cc-rule-set of cyc. We obtain the following program:

$$\begin{array}{lll} \mathsf{cyc} : \mathsf{Nat} \to \mathsf{StrN} & \vartriangleleft | & \begin{array}{llll} (x : \mathsf{Nat} \vdash \mathsf{cyc}\ x & .\mathsf{head} & \longrightarrow x & : \mathsf{Nat}) \\ (x : \mathbf{1} \ \ \vdash \mathsf{cyc}\ (\mathsf{zero}\ x) & .\mathsf{tail} & \longrightarrow g_2\ x & : \mathsf{StrN}) \\ (x : \mathsf{Nat} \vdash \mathsf{cyc}\ (\mathsf{suc}\ x) & .\mathsf{tail} & \longrightarrow \mathsf{cyc}\ x & : \mathsf{StrN}) \end{array} \\ g_2 \ : \mathbf{1} \to \mathsf{StrN} & \vartriangleleft |_{\mathsf{s}} & (\cdot \ \ \vdash g_2\ () \ \ \longrightarrow \mathsf{cyc}\ N : \mathsf{StrN}) \end{array}$$

Let a term in the new language be *good*, if all occurrences of g_2 are applied at least once. We can define a back-translation int of good terms into the original language by recursively replacing $g_2\ s$ by cyc (zero s) .tail.

The second last step in the derivation of the cc-pattern-set was a split of pattern variable x : Nat into zero x and suc x using $\mathrm{C}_{\mathrm{Const}}$. Again, we introduce a simple auxiliary function g_1, which performs just this split and obtain a simple program with mutually recursive functions cyc, g_1, and g_2:

$$\begin{array}{lll} \mathsf{cyc} : \mathsf{Nat} \to \mathsf{StrN} & \vartriangleleft |_{\mathsf{s}} & \begin{array}{lll} (x : \mathsf{Nat} \vdash \mathsf{cyc}\ x\ .\mathsf{head} & \longrightarrow x & : \mathsf{Nat}) \\ (x : \mathsf{Nat} \vdash \mathsf{cyc}\ x\ .\mathsf{tail} & \longrightarrow g_1\ x & : \mathsf{StrN}) \end{array} \\ g_1 \ : \mathsf{Nat} \to \mathsf{StrN} & \vartriangleleft |_{\mathsf{s}} & \begin{array}{lll} (x : \mathbf{1} \ \ \vdash g_1\ (\mathsf{zero}\ x) & \longrightarrow g_2\ x & : \mathsf{StrN}) \\ (x : \mathsf{Nat} \vdash g_1\ (\mathsf{suc}\ x) & \longrightarrow \mathsf{cyc}\ x & : \mathsf{StrN}) \end{array} \\ g_2 \ : \mathbf{1} \to \mathsf{StrN} & \vartriangleleft |_{\mathsf{s}} & (\cdot \ \ \vdash g_2\ () \ \ \longrightarrow \mathsf{cyc}\ N : \mathsf{StrN}) \end{array}$$

The back-interpretation of g_1 for good terms of the new program replaces recursively $g_1\ s$ by cyc s .tail. We note the following:

(a) The translation can be performed by induction on the derivation of coverage; or, one can do the translation while checking coverage.[3]

[3] This is actually happening in the language Idris [5]; Agda [10] has separate phases, but uses the split tree generated by the coverage checker to translate pattern matching into case trees.

(b) The generated functions are simple upon creation and need not be processed recursively. The right hand sides of these functions are either right hand sides of the original program or calls to earlier generated functions applied to exactly the pattern variables in context.
(c) When generating a function, it is invoked on the pattern variables in context. We can define a function int which interprets this generated function back into terms of the original program (if applied to good terms).
(d) Since we gave earlier created functions (here: g_2) a higher index than later created functions (here: g_1), calls between generated functions increase the index. There can only be finitely many calls between generated functions before executing an original right hand side again. This fact ensures preservation of normalization (see later).
(e) Calls between generated functions are undone by the back translation int, thus the corresponding reduction steps vanish under int.

In the case of the Fibonacci stream, the translated simple program is as follows:

$$\begin{array}{lll} \mathsf{fib}\ .\mathsf{head} \longrightarrow 0 & \quad & g\ .\mathsf{head} \longrightarrow 1 \\ \mathsf{fib}\ .\mathsf{tail} \longrightarrow g & \quad & g\ .\mathsf{tail} \longrightarrow \mathsf{zipWith}\ \_+\_\ \mathsf{fib}\ (\mathsf{fib}\ .\mathsf{tail}) \end{array}$$

3.3 The Translation Algorithm

Let $\mathcal{P}$ be the input program for typed language Σ. Let $\mathcal{P}_f$ be a non-simple cc-rule-set of $\mathcal{P}$. Consider the last step in the derivation of the underlying cc-pattern-set. Since $\mathcal{P}_f$ is non-simple, this step cannot be $\mathsf{C}_{\mathbf{Head}}$. Assume

$$\mathcal{P}_f = f : \Sigma(f) \ \lhd |\ Q\ (\Delta_i \vdash q_i \longrightarrow t_i : C_i)_{i \in I}.$$

where in some cases $I = \{0\}$. Let the last step in the derivation of the underlying cc-pattern-set be

$$\frac{f : \Sigma(f) \ \lhd |\ Q\ (\Delta' \vdash q : A)}{f : \Sigma(f) \ \lhd |\ Q\ (\Delta_i \vdash q_i : C_i)_{i \in I}}\ \mathsf{C}$$

We extend Σ to Σ' by adding one fresh constant $g : \Delta' \to A$. Let $\Delta' = \vec{y} : \vec{A}$. Depending on C we introduce below a simple q_i' and define the program $\mathcal{P}'$ for the typed language Σ' by

$$\begin{array}{lll} \mathcal{P}_f' = f : \Sigma(f) & \lhd| & Q\ (\Delta' \vdash q \longrightarrow g\ \vec{y} : A) \\ \mathcal{P}_g' = g : \Delta' \to A & \lhd|_{\mathsf{s}} & (\Delta_i \vdash q_i' \longrightarrow t_i \ : C_i)_{i \in I} \\ \mathcal{P}_h' = \mathcal{P}_h \quad \text{otherwise} & & \end{array}$$

Note that the underlying cc-pattern-set for f is as in the premise of C, $\mathcal{P}_g'$ is simple, and all other constants are left unchanged. Therefore the height of the derivation for the cc-pattern-set for f is reduced by 1. We then recursively apply the algorithm on $\mathcal{P}'$. Since each step of the algorithm makes the coverage derivation of one non-simple function shorter, and new constants are simple, the algorithm terminates, returning only simple constants.

In case of variable splitting, we always reorder Δ' such that the variable we split on appears last. When referring to a context Δ, assume $\Delta = \vec{x} : \vec{A}$.
Case $q\ x \longrightarrow t$ and C is

$$\frac{f : \Sigma(f)\ \lhd|\ Q\ (\Delta \vdash q : B \to C)}{f : \Sigma(f)\ \lhd|\ Q\ (\Delta, x : B \vdash q\ x : C)}\ \mathrm{C_{App}}$$

Define $q'_0 = g\ \vec{x}\ x$. Therefore,

$$\begin{array}{ll} \mathcal{P}'_f = f : \Sigma(f) & \lhd|\ Q(\Delta \vdash q \longrightarrow g\ \vec{x} : B \to C) \\ \mathcal{P}'_g = g : \Delta \to B \to C & \lhd|_{\mathsf{s}}\ (\Delta, x : B \vdash g\ \vec{x}\ x \longrightarrow t : C) \end{array}$$

Case $q\ .d \longrightarrow t_d$ for all $d \in R$ and C is

$$\frac{f : \Sigma(f)\ \lhd|\ Q\ (\Delta \vdash q : \nu X.R)}{f : \Sigma(f)\ \lhd|\ Q\ (\Delta \vdash q\ .d : (\nu X.R)_d)_{d \in R}}\ \mathrm{C_{Dest}}$$

Define $q'_d = g\ \vec{x}\ .d$. Therefore,

$$\begin{array}{ll} \mathcal{P}'_f = f : \Sigma(f) & \lhd|\ Q\ (\Delta \vdash q \longrightarrow g\ \vec{x} : \nu X.R) \\ \mathcal{P}'_g = g : \Delta \to \nu X.R & \lhd|_{\mathsf{s}}\ (\Delta \vdash g\ \vec{x}\ .d \longrightarrow t_d : (\nu X.R)_d)_{d \in R} \end{array}$$

Case $q[x' := ()] \longrightarrow t$ and C is

$$\frac{f : \Sigma(f)\ \lhd|\ Q\ (\Delta, x' : \mathbf{1} \vdash q : C)}{f : \Sigma(f)\ \lhd|\ Q\ (\Delta \vdash q[x' := ()] : C)}\ \mathrm{C_{Unit}}$$

Define $q'_0 := g\ \vec{x}\ ()$. Therefore,

$$\begin{array}{ll} \mathcal{P}'_f = f : \Sigma(f) & \lhd|\ Q\ (\Delta, x' : \mathbf{1} \vdash q \longrightarrow g\ \vec{x}\ x' : C) \\ \mathcal{P}'_g := g : \Delta \to \mathbf{1} \to C & \lhd|_{\mathsf{s}}\ (\Delta \vdash g\ \vec{x}\ () \longrightarrow t : C) \end{array}$$

Case $q[x' := (x_1, x_2)] \longrightarrow t$ and C is

$$\frac{f : \Sigma(f)\ \lhd|\ Q\ (\Delta, x' : A_1 \times A_2 \vdash q : C)}{f : \Sigma(f)\ \lhd|\ Q\ (\Delta, x_1 : A_1, x_2 : A_2 \vdash q[x' := (x_1, x_2)] : C)}\ \mathrm{C_{Pair}}$$

Define $q'_0 = g\ \vec{x}\ (x_1, x_2)$. Therefore,

$$\begin{array}{ll} \mathcal{P}'_f = f : \Sigma(f) & \lhd|\ Q\ (\Delta, x' : A_1 \times A_2 \vdash q \longrightarrow g\ \vec{x}\ x' : C) \\ \mathcal{P}'_g := g : \Delta \to (A_1 \times A_2) \to C & \lhd|_{\mathsf{s}}\ (\Delta, x_1 : A_1, x_2 : A_2 \vdash g\ \vec{x}\ (x_1, x_2) \longrightarrow t : C) \end{array}$$

Case $q[x' := c\ x'] \longrightarrow t_c$ for all $c \in D$ and C is

$$\frac{f : \Sigma(f)\ \lhd|\ Q\ (\Delta, x' : \mu X.D \vdash q : C)}{f : \Sigma(f)\ \lhd|\ Q\ (\Delta, x' : (\mu X.D)_c \vdash q[x' := c\ x'] : C)_{c \in D}}\ \mathrm{C_{Const}}$$

Define $q'_c := g\ \vec{x}\ (c\ x')$. Therefore,

$$\begin{array}{ll} \mathcal{P}'_f = f : \Sigma(f) & \lhd|\ Q\ (\Delta, x' : \mu X.D \vdash q \longrightarrow g\ \vec{x}\ x' : C) \\ \mathcal{P}'_g = g : \Delta \to \mu X.D \to C & \lhd|_{\mathsf{s}}\ (\Delta, x' : (\mu X.D)_c \vdash g\ \vec{x}\ (c\ x') \longrightarrow t_c : C)_{c \in D} \end{array}$$

4 Extensions of Abstract Reduction Systems

It is easy to see that a reduction in the original program $\mathcal{P}$ (over Σ) corresponds to possibly multiple reductions in the translated language $\mathcal{P}'$ (over Σ'). What is more difficult to prove is that we do not get *additional* reductions, i.e. if $t \not\longrightarrow^*_{\mathcal{P}} t'$ then it is impossible to reduce t to t' using reductions and intermediate terms in $\mathcal{P}'$. We call this notion conservative extension. Even this will not be sufficient as pointed out in Sect. 3, we need in addition preservation of normalisation. We will define and explore the corresponding notions more generally for *abstract reduction systems (ARS)*.

An *ARS* is a pair $(\mathcal{A}, \longrightarrow)$, often just written $\mathcal{A}$, such that $\mathcal{A}$ is a set and $\longrightarrow$ is a binary relation on $\mathcal{A}$ written infix. Let $\longrightarrow^*$ be the transitive-reflexive and $\longrightarrow^{\geq 1}$ be the transitive closure of $\longrightarrow$. An element $a \in \mathcal{A}$ is in *normal form (NF)* if there is no $a' \in \mathcal{A}$ such that $a \longrightarrow a'$. It is *weakly normalising (WN)* if there exists an $a' \in \mathcal{A}$ in NF such that $a \longrightarrow^* a'$. a is *strongly normalising (SN)* if there exist no infinite reduction sequence $a = a_0 \longrightarrow a_1 \longrightarrow a_2 \longrightarrow \cdots$. Let SN, WN, NF be the set of elements in $\mathcal{A}$ which are SN, WN, NF respectively. For a reduction system $(\mathcal{A}', \longrightarrow')$, let SN', WN', NF' be the elements of $\mathcal{A}'$ which are $\longrightarrow'$-SN, -WN, -NF.

Let $(\mathcal{A}, \longrightarrow)$, $(\mathcal{A}', \longrightarrow')$ be ARS such that $\mathcal{A} \subseteq \mathcal{A}'$. Then,

$\mathcal{A}'$ *is a conservative extension of* $\mathcal{A}$ iff $\forall a, a' \in \mathcal{A}.\ a \longrightarrow^* a' \Leftrightarrow a \longrightarrow'^* a'$
$\mathcal{A}'$ *is an* SN*-preserving extension of* $\mathcal{A}$ iff $\forall a \in \mathcal{A}.\ a \in \mathsf{SN} \Leftrightarrow a \in \mathsf{SN}'$
$\mathcal{A}'$ *is a* WN*-preserving extension of* $\mathcal{A}$ iff $\forall a \in \mathcal{A}.\ a \in \mathsf{WN} \Leftrightarrow a \in \mathsf{WN}'$

Lemma 3 (Transitivity of conservative/SN/WN-preserving extensions). Let $\mathcal{A}, \mathcal{A}', \mathcal{A}''$ be ARSs, $\mathcal{A}'$ be an extension of $\mathcal{A}$ and $\mathcal{A}''$ an extension of $\mathcal{A}'$, both of which are conservative, SN-preserving, or WN-preserving extensions. Then $\mathcal{A}''$ is a conservative, SN-preserving, or WN-preserving extension, respectively, of $\mathcal{A}$.

In order to show the above properties, we use the notion of a back-translation from the extended ARS into the original one:

Let $(\mathcal{A}, \longrightarrow)$, $(\mathcal{A}', \longrightarrow')$ be ARSs such that $\mathcal{A} \subseteq \mathcal{A}'$. Then a *back-interpretation of* $\mathcal{A}'$ *into* $\mathcal{A}$ is given by

– a set Good such that $\mathcal{A} \subseteq \mathsf{Good} \subseteq \mathcal{A}'$; we say a *is good* if $a \in \mathsf{Good}$;
– a function $\mathsf{int} : \mathsf{Good} \to \mathcal{A}$ such that $\forall a \in \mathcal{A}. \mathsf{int}(a) = a$.

We define 3 conditions for a back-interpretation $(\mathsf{Good}, \mathsf{int})$ where condition (SN 2) refers to a measure $\mathsf{m} : \mathsf{Good} \to \mathbb{N}$:

(SN 1) $\forall a, a' \in \mathcal{A}. a \longrightarrow a' \Rightarrow a \longrightarrow'^{\geq 1} a'$.
(SN 2) If $a \in \mathsf{Good}$, $a' \in \mathcal{A}'$ and $a \longrightarrow' a'$ then $a' \in \mathsf{Good}$ and we have $\mathsf{int}(a) \longrightarrow^{\geq 1} \mathsf{int}(a')$ or $\mathsf{int}(a) = \mathsf{int}(a') \wedge \mathsf{m}(a) > \mathsf{m}(a')$.
(WN) If $a \in \mathsf{Good} \cap \mathsf{NF}'$ then $\mathsf{int}(a) \in \mathsf{NF}$.

The following theorem substantially extends Lem. 1.1.27 of [13] and Lem. 2.2.5 of [11]:

Theorem 4 (Backinterpretations for ARSs and conservativity, SN, WN). Let $(\mathcal{A}, \longrightarrow)$, $(\mathcal{A}', \longrightarrow')$ be ARSs such that $\mathcal{A} \subseteq \mathcal{A}'$. Let $(\mathsf{Good}, \mathsf{int})$ be a back-interpretation from $\mathcal{A}'$ into $\mathcal{A}$, $\mathsf{m} : \mathsf{Good} \to \mathbb{N}$. Then the following holds:

(a) (SN 1), (SN 2) imply that $\mathcal{A}'$ is a conservative extension of $\mathcal{A}$ preserving SN.
(b) (SN 1), (SN 2), (WN) imply that $\mathcal{A}'$ is an extension of $\mathcal{A}$ preserving WN.

Proof: (a): Proof of Conservativity: $a \longrightarrow^* a'$ implies by (SN 1) $a \longrightarrow'^* a'$. If $a, a' \in \mathcal{A}$, $a \longrightarrow'^* a'$ then by (SN 2) $a = \mathsf{int}(a) \longrightarrow^* \mathsf{int}(a') = a'$.
Proof of preservation of SN: We show the classically equivalent statement $\forall a \in \mathcal{A}.\neg(a \text{ is } \longrightarrow\text{-SN}) \Leftrightarrow \neg(a \text{ is } \longrightarrow'\text{-SN})$.
For "$\Rightarrow$" assume $a = a_0 \longrightarrow a_1 \longrightarrow a_2 \longrightarrow \cdots$ is an infinite $\longrightarrow$-reduction sequence starting with a. Then by (SN 1) $a = a_0 \longrightarrow'^{\geq 1} a_1 \longrightarrow'^{\geq 1} a_2 \longrightarrow'^{\geq 1} \cdots$ is an infinite $\longrightarrow'$-reduction sequence.
For "$\Leftarrow$" assume $a = a'_0 \longrightarrow' a'_1 \longrightarrow_{\mathcal{P}'} a'_2 \longrightarrow_{\mathcal{P}'} \cdots$.
Then by (SN 2) $a = \mathsf{int}(a_0) = \mathsf{int}(a'_0) \longrightarrow^* \mathsf{int}(a'_1) \longrightarrow^* \mathsf{int}(a'_2) \longrightarrow'^* \cdots$. If $\mathsf{int}(a'_i) = \mathsf{int}(a'_{i+1})$ then $\mathsf{m}(a'_i) > \mathsf{m}(a'_{i+1})$, so by (SN 2) after finitely many steps, where $\mathsf{int}(a'_i) = \mathsf{int}(a'_{i+1})$, we must have one step $\mathsf{int}(a'_j) \longrightarrow^{\geq 1} \mathsf{int}(a'_{j+1})$. Thus, we obtain an infinite reduction sequence starting with a in $\mathcal{A}$.
(b) Assume $a \in \mathcal{A}$, $a \in \mathsf{WN}$. Then $a \longrightarrow^* a' \in \mathsf{NF}$ for some a', therefore $a' \in \mathsf{SN}$, by (a) $a' \in \mathsf{SN}'$, $a' \longrightarrow'^* a''$ for some $a'' \in \mathsf{NF}'$, therefore $a \longrightarrow'^* a' \longrightarrow'^* a'' \in \mathsf{NF}'$, $a \in \mathsf{WN}'$. For the other direction, assume $a \in \mathcal{A}$, $a \in \mathsf{WN}'$. Then $a \longrightarrow'^* a' \in \mathsf{NF}'$ for some a', by (SN 2), (WN) $a = \mathsf{int}(a) \longrightarrow^* \mathsf{int}(a') \in \mathsf{NF}$, $a \in \mathsf{WN}$.

5 Proof of Correctness of the Translation

In our translation we extend our language by new auxiliary constants while keeping the old ones, including their types. More formally, we define $\Sigma \subseteq_{\mathsf{F}} \Sigma'$, pronounced Σ' *extends* Σ *by constants*, if (1) Σ' and Σ have the same constructor and destructor symbols $\mathcal{C}, \mathcal{D}$, (2) the constants $\mathcal{F}$ of $\mathcal{L}$ form a subset of the constants of $\mathcal{L}'$, and (3) Σ and Σ' assign the same types to $\mathcal{F}$.

Let $\mathcal{P}$ be a program for Σ, $\mathsf{Term}_\Sigma = \{t \mid \exists \Delta, A. \Delta \vdash_\Sigma t : A\}$. The *ARS for a program* $\mathcal{P}$ is $(\mathsf{Term}_\Sigma, \longrightarrow_{\mathcal{P}})$. Let $\mathcal{P}, \mathcal{P}'$ be programs for typed languages Σ, Σ', respectively. $\mathcal{P}'$ is an extension of $\mathcal{P}$ iff $\Sigma \subseteq_{\mathsf{F}} \Sigma'$. If $\mathcal{P}'$ is an extension of $\mathcal{P}$, then $\mathcal{P}'$ is a *conservative*, *SN-preserving*, or *WN-preserving* extension of $\mathcal{P}$ if the corresponding condition holds for the ARSs $(\mathsf{Term}_\Sigma, \longrightarrow_{\mathcal{P}})$ and $(\mathsf{Term}_{\Sigma'}, \longrightarrow_{\mathcal{P}'})$.

We will define a back-interpretations by replacing in terms $g\, t_1 \ldots t_n$ the new constants g by a term of the original language. Due to lack of λ-abstraction, we only get a term of the original language if g is applied to n arguments. So, for our back translation, we need an $\mathsf{arity}(g) = n$ of new constants, and an interpretation $\mathsf{Int}(g)$ of those terms:

Assume $\Sigma \subseteq_{\mathsf{F}} \Sigma'$. A *concrete back-interpretation* $(\mathsf{arity}, \mathsf{Int})$ *of* Σ' *into* Σ is given by the following:

- An arity $\mathsf{arity}(g) = n$ assigned to each new constant g of Σ' such that $\Sigma'(g) = A_1 \to \cdots \to A_n \to A$ for some types $A_1, \ldots, A_n, A$. Here, A (as well as any A_i) might be a function type.

- For every new constant g of Σ' with $\mathsf{arity}(g) = n$ and $\Sigma'(g) = A_1 \to \cdots \to A_n \to A$ a term $\mathsf{Int}(g) = t$ of Σ such that $x_1 : A_1, \ldots, x_n : A_n \vdash t : A$. In this case, we write $\mathsf{Int}(g)[\vec{t}]$ for $t[\vec{x} := \vec{t}]$.

Assume that (arity, Int) is a concrete back-interpretation of Σ' into Σ.

- The set $\mathsf{Good}_{\mathsf{arity},\mathsf{Int}}$ of good terms is given by the set of $t \in \mathsf{Term}_\Sigma$ such that each occurrence of a new constant g of arity n in t is applied to at least n arguments.
- If $t \in \mathsf{Good}_{\mathsf{arity},\mathsf{Int}}$, then $\mathsf{int}_{\mathsf{arity},\mathsf{Int}}(t)$, in short $\mathsf{int}(t)$, is obtained by inductively replacing all occurrences of $g\,\vec{t}$ for new constants g by $\mathsf{Int}(g)[\mathsf{int}(\vec{t})]$.

Trivially, concrete back-interpretations are back-interpretations. We now have the definitions in place to prove SN+WN-conservativity of our translation.

Lemma 5 (Some simple facts)

(a) If $f : A \lhd| \; Q \; (\Delta \vdash q : A)$ then each variable in Δ occurs exactly once in q.
(b) If x is a variable occurring in pattern q, then t is a subterm of $q[x := t]$.
(c) Assume s is a maximal subterm of t, i.e. s is a subterm such that there is no term s' such that $s\ s'$ is a subterm starting at the same occurrence as s in t. If t is good, then s is good as well.

Theorem 6 (Correctness of Translation). Let $\mathcal{P}$ be a program for Σ. Then there exists a typed language $\Sigma' \supseteq_{\mathsf{F}} \Sigma$ and a simple program $\mathcal{P}'$ for Σ', which is a conservative extension of $\mathcal{P}$ preserving SN and WN.

Proof: Define for a program $\mathcal{P}$ the height of its derivation $\mathsf{height}(\mathcal{P})$ as the sum of the heights of the derivations of those covering patterns in $\mathcal{P}$, which are not simple covering patterns. The proof is by induction on $\mathsf{height}(\mathcal{P})$.

The case $\mathsf{height}(\mathcal{P}) = 0$ is trivial, since $\mathcal{P}$ is simple. Assume $\mathsf{height}(\mathcal{P}) > 0$. We obtain a $\Sigma' \supseteq_{\mathsf{F}} \Sigma$ and corresponding program $\mathcal{P}'$ for Σ' by applying one step of Algorithm 3.3 to $\mathcal{P}$. We show below that $\mathcal{P}'$ is a conservative extension of $\mathcal{P}$ preserving SN and WN. Since the derivations for the coverage complete pattern sets in $\mathcal{P}'$ are the same as for $\mathcal{P}$, except for the one for $\mathcal{P}'_f$, which is reduced in height by one as the algorithm takes out the last derivation of the coverage derivation of $\mathcal{P}_f$, and that for $\mathcal{P}'_g$, which is simple, we have $\mathsf{height}(\mathcal{P}') = \mathsf{height}(\mathcal{P}) - 1$. By induction hypothesis there exists a conservative extension $\mathcal{P}''$ of $\mathcal{P}'$ preserving SN and WN, which is simple, which is as well a conservative extension of $\mathcal{P}$ preserving SN and WN. This extension is obtained by the recursive call made by the algorithm.

So we need to show that $\mathcal{P}'$ is a conservative extension of $\mathcal{P}$ preserving SN and WN. Let $f, g, \Delta', \vec{y}, q, A, I, \Delta_i, q_i, t_i, C_i, q'_i$ be as stated in Algorithm 3.3, $\Delta_i = \vec{y}_i : \vec{A}_i$, and n be the length of Δ'.

We introduce a concrete back-interpretation of $\mathcal{P}'$ into $\mathcal{P}$ by $\mathsf{arity}(g) := n$ and $\mathsf{Int}(g)[\vec{y}] := q$. Let $\mathsf{m}(t)$ be the number of occurrences of f in t. Let (Good, int) be the corresponding back interpretation.

Assume $\mathcal{P}'$ fulfils with the given q'_i the following conditions:

(1) $\mathsf{int}(q'_i) = q_i \longrightarrow_{\mathcal{P}'} q'_i$
(2) If $q[\vec{x} := \vec{s}]\ \vec{t} = q_i$, then $g\ \vec{s}\ \vec{t} = q'_i$, where t_i are terms or of the form $.d$.

Then $(\mathsf{Good}, \mathsf{int})$ fulfils (SN 1), (SN 2), and (WN), and therefore $\mathcal{P}'$ is a conservative extension of $\mathcal{P}$ preserving SN and WN:
(SN 1) holds since the only changed derivation is based on the original redex $q_i[\vec{y}_i := \vec{t}] \longrightarrow_{\mathcal{P}} t_i[\vec{y}_i := \vec{t}]$ and by (1) $q_i[\vec{y}_i := \vec{t}] \longrightarrow_{\mathcal{P}'} q_i'[\vec{y}_i := \vec{t}] \longrightarrow_{\mathcal{P}'} t_i[\vec{y}_i := \vec{t}]$.
(SN 2) holds since the new redexes are the following:

- $q[\vec{y} := \vec{t}] \longrightarrow_{\mathcal{P}'} g\ \vec{t}$, where $q[\vec{y} := \vec{t}]$ is good. Since it is good and variables in a pattern are not applied to other terms, by Lem. 5 $\vec{t}$ is good as well, and therefore as well $g\ \vec{t}$. We have $\mathsf{int}(q[\vec{y} := \vec{t}]) = q[\vec{y} := \mathsf{int}(\vec{t})] = \mathsf{int}(g\ \vec{t})$. Furthermore, $\mathsf{m}(q[\vec{y} := \vec{t}]) = \mathsf{m}(g\ \vec{t}) + 1 > \mathsf{m}(g\ \vec{t})$, since pattern q starts with f, and each variable in $\vec{y}$ occurs by Lem. 5 exactly once in q.
- $q_i'[\vec{y}_i := \vec{t}] \longrightarrow_{\mathcal{P}'} t_i[\vec{y}_i := \vec{t}]$. Since $q_i'[\vec{y}_i := \vec{t}]$ is good, as in (a) $\vec{t}$ are good and therefore $t_i[\vec{y}_i := \vec{t}]$ is good. Furthermore, by (1) $\mathsf{int}(q_i'[\vec{y}_i := \vec{t}]) = \mathsf{int}(q_i')[\vec{y}_i := \mathsf{int}(\vec{t})]) = q_i[\vec{y}_i := \mathsf{int}(\vec{t})] \longrightarrow_{\mathcal{P}} t_i[\vec{y}_i := \mathsf{int}(\vec{t})] = \mathsf{int}(t_i[\vec{y}_i := \vec{t}])$.

Proof of (WN): We first show that (2) implies
(3) If $s \in \mathsf{Good}$, $\mathsf{int}(s) = q_i$ then $s = q_i \lor s = q_i'$
Since q_i starts with f, s must start with f or g. The only occurrence of a constant in q_i is at the beginning, therefore $s = f\ \vec{r}$ or $s = g\ \vec{r}$ where $\mathsf{int}(\vec{r}) = \vec{r}$. If $s = f\ \vec{r}$ then $s = \mathsf{int}(s) = q_i$. If $s = g\ \vec{r} = g\ \vec{s}\ \vec{t}$, $q[\vec{x} := \vec{s}]\ \vec{t} = \mathsf{int}(s) = q_i$, therefore by (2) $s = g\ \vec{s}\ \vec{t} = q_i'$.
Using (3), assume $s \in \mathsf{Good}$, $s \in \mathsf{NF}'$, and show $\mathsf{int}(s) \in \mathsf{NF}$. Assume $\mathsf{int}(s) \notin \mathsf{NF}$, $\mathsf{int}(s)$ has redex $\widetilde{q}[\vec{x} := \vec{r}]$ for a pattern $\widetilde{q}$ of $\mathcal{P}$. If $\widetilde{q} \neq q_i$, $\widetilde{q}$ starts with some $h \neq f, g$, and has no occurrences of f, g. Then s contains $\widetilde{q}[\vec{x} := \vec{r'}]$ where $\mathsf{int}(\vec{r'}) = \vec{r}$, and has therefore a redex, contradicting $s \in \mathsf{NF}'$. Therefore $\widetilde{q} = q_i$ for some i. Therefore s contains a subterm $s'[\vec{x} := \vec{r'}]$ such that $\mathsf{int}(s') = q_i$, $\mathsf{int}(\vec{r'}) = \vec{r}$. But then by (1), (3) $s'[\vec{x} := \vec{r'}]$ has a reduction, again a contradiction.

So the proof is complete provided conditions (1), (2) are fulfilled. We verify the case when the last rule is $(\mathrm{C_{Dest}})$, the other cases follow similarly:
(1) $\mathsf{int}(q_d') = \mathsf{int}(g\ \vec{x}\ .d) = \mathsf{Int}(g)[\vec{x}]\ .d = q\ .d = q_d \longrightarrow_{\mathcal{P}'} g\ \vec{x}\ .d = q_d'$.
(2) If $q[\vec{x} := \vec{s}]\ \vec{t} = q_d = q\ .d$, $\vec{s} = \vec{x}$, $\vec{t} = .d$, $g\ \vec{s}\ \vec{t} = g\ \vec{x}\ .d = q_d'$.

6 Conclusion

We have described a reduction of deep copattern matching to shallow copattern matching. The translation preserves weak and strong normalization. It is conservative, thus establishing a weak bisimulation between the original and the translated program. The translated programs can be used for more efficient evaluation in a checker for dependent types or can serve as intermediate code for translation into a more low-level language that has no concept of pattern at all.

There are two more translations of interest. The first one, which we have mostly worked out, is a translation into a variable-free language of combinators, including a proof of conservativity and preservation of normalization. Our techniques were developed more generally in order to prove correctness for this translation as well. A second translation would be to a call-by-need lambda-calculus with lazy record constructors. This would allow us to map definitions of infinite structures by copatterns back to Haskell style definitions by lazy evaluation. While there seems to be no (weak) bisimulation in this case, one still can hope for preservation of normalization, maybe established by logical relations.

Acknowledgements. The authors want to thank the referees for many detailed and valuable comments, especially regarding preservation of weak normalisation and generalisation to ARS. Anton Setzer acknowledges support by EPSRC (Engineering and Physical Science Research Council, UK) grant EP/C0608917/1. Andreas Abel acknowledges support by a Vetenskapsrådet framework grant 254820104 (Thierry Coquand) to the Department of Computer Science and Engineering at Gothenburg University. Brigitte Pientka acknowledges support by NSERC (National Science and Engineering Research Council Canada). David Thibodeau acknowledges support by a graduate scholarship of Les Fonds Québécois de Recherche Nature et Technologies (FQRNT).

References

1. Abel, A.: A Polymorphic Lambda-Calculus with Sized Higher-Order Types. PhD thesis, Ludwig-Maximilians-Universität München (2006)
2. Abel, A., Pientka, B., Thibodeau, D., Setzer, A.: Copatterns: Programming infinite structures by observations. In: Proc. of the 40th ACM Symp. on Principles of Programming Languages, POPL 2013, pp. 27–38. ACM Press (2013)
3. Augustsson, L.: Compiling pattern matching. In: Jouannaud, J.-P. (ed.) FPCA 1985. LNCS, vol. 201, pp. 368–381. Springer, Heidelberg (1985)
4. Barthe, G., Frade, M.J., Giménez, E., Pinto, L., Uustalu, T.: Type-based termination of recursive definitions. Math. Struct. in Comput. Sci. 14(1), 97–141 (2004)
5. Brady, E.: Idris, a general purpose dependently typed programming language: Design and implementation (2013), `http://www.cs.st-andrews.ac.uk/~eb/drafts/impldtp.pdf`
6. Cockett, R., Fukushima, T.: About Charity. Technical report, Department of Computer Science, The University of Calgary, Yellow Series Report No. 92/480/18 (June 1992)
7. Hagino, T.: A typed lambda calculus with categorical type constructors. In: Pitt, D.H., Rydeheard, D.E., Poigné, A. (eds.) Category Theory and Computer Science. LNCS, vol. 283, pp. 140–157. Springer, Heidelberg (1987)
8. Hagino, T.: Codatatypes in ML. J. Symb. Logic 8(6), 629–650 (1989)
9. INRIA. The Coq Proof Assistant Reference Manual. INRIA, version 8.4 edition (2012)
10. Norell, U.: Towards a Practical Programming Language Based on Dependent Type Theory. PhD thesis, Dept. of Computer Science and Engineering, Chalmers, Göteborg, Sweden (2007)
11. Severi, P.G.: Normalisation in lambda calculus and its relation to type inference. PhD thesis, Technische Universiteit Eindhoven, Eindhoven, The Netherlands (1996)
12. Terese. Term Rewriting Systems. Cambridge University Press (2003)
13. van Raamsdonk, F.: Concluence and Normalisation for Higher-Order Rewriting. PhD thesis, Vrije Universiteit, Amsterdam, The Netherlands (1996)

Proving Confluence of Term Rewriting Systems via Persistency and Decreasing Diagrams

Takahito Aoto, Yoshihito Toyama, and Kazumasa Uchida

RIEC, Tohoku University,
2-1-1 Katahira, Aoba-ku, Sendai 980-8577, Japan
{aoto,toyama,uchida}@nue.riec.tohoku.ac.jp

Abstract. The decreasing diagrams technique (van Oostrom, 1994) has been successfully used to prove confluence of rewrite systems in various ways; using rule-labelling (van Oostrom, 2008), it can also be applied directly to prove confluence of some linear term rewriting systems (TRSs) automatically. Some efforts for extending the rule-labelling are known, but non-left-linear TRSs are left beyond the scope. Two methods for automatically proving confluence of non-(left-)linear TRSs with the rule-labelling are given. The key idea of our methods is to combine the decreasing diagrams technique with persistency of confluence (Aoto & Toyama, 1997).

Keywords: Confluence, Persistency, Decreasing Diagrams, Rule-Labelling, Non-Linear, Term Rewriting Systems.

1 Introduction

Decreasing diagrams [11] give a characterization of confluence of abstract rewrite systems; the criterion based on decreasing diagrams can be adapted to prove confluence of rewrite systems in various ways. In particular, *rule-labelling* [12] has been adapted to prove confluence of *left-linear* TRSs [1,8,19] automatically. A property of TRSs is said to be *persistent* if the property is preserved under elimination of sorts [20]. It is shown in [2] that confluence is persistent, that is, if a many-sorted TRS is confluent on (many-sorted) terms then so is the underlying unsorted TRS on all (i.e. including ill-sorted) terms.

In this paper, the decreasing diagrams technique and persistency of confluence are combined to give methods for proving confluence of *non-linear* TRSs automatically. For proving confluence of TRSs $\mathcal{R}$, we consider a subsystem $\mathcal{R}^{\tau}_{nl}$ which is obtained from some many-sorted version $\mathcal{R}^{\tau}$ of $\mathcal{R}$. Based on assumptions on the subsystem $\mathcal{R}^{\tau}_{nl}$, we develop two confluence criteria based on decreasing diagrams with rule-labelling—one of the criteria is based on the assumption that $\mathcal{R}^{\tau}_{nl}$ is terminating, and the other is based on the assumption that $\mathcal{R}^{\tau}_{nl}$ is innermost normalizing. These two criteria are incomparable, and the proofs of the correctness are given independently. Both of the criteria, however, can be applied to prove confluence of non-left-linear non-terminating TRSs, for which no

G. Dowek (ed.): RTA-TLCA 2014, LNCS 8560, pp. 46–60, 2014.

decreasing diagrams technique with rule-labelling has been known and only few techniques for proving confluence have been known.

The rest of the paper is organized as follows. Section 2 covers preliminaries; some common notions and notations to be used in Sections 3 and 4 are also presented. In Section 3, we introduce the class of strongly quasi-linear TRSs and show a confluence criterion for TRSs in this class. In Section 4, we introduce the class of quasi-linear TRSs and show a confluence criterion for TRSs in this class. We also show that these two criteria are incomparable. In Section 5, we report on an implementation of these criteria in our confluence prover ACP [3] and on experiments. Related work is also explained in Section 5. Section 6 concludes.

2 Preliminaries

We fix notations assuming basic familiarity with term rewriting [4].

The transitive (reflexive, transitive and reflexive, equivalence) closure of a relation $\rightarrow$ (on a set A) is denoted by $\stackrel{+}{\rightarrow}$ ($\stackrel{=}{\rightarrow}$, $\stackrel{*}{\rightarrow}$, $\stackrel{*}{\leftrightarrow}$, respectively). An element a is a *normal form* if $a \rightarrow b$ for no b; *normalizing* if $a \stackrel{*}{\rightarrow} b$ for some normal form b; *terminating* if there exists no infinite sequence $a = a_0 \rightarrow a_1 \rightarrow \cdots$. The relation $\rightarrow$ is *normalizing* (*terminating*) if so are all $a \in A$; *confluent* if $\stackrel{*}{\leftarrow} \circ \stackrel{*}{\rightarrow} \subseteq \stackrel{*}{\rightarrow} \circ \stackrel{*}{\leftarrow}$.

We denote a set of (arity-fixed) function symbols by $\mathcal{F}$, an enumerable set of variables by $\mathcal{V}$, and the set of terms by $\mathrm{T}(\mathcal{F},\mathcal{V})$. A variable in a term t is *linear* if it occurs only once in t, otherwise *non-linear*. The set of variables (linear variables, non-linear variables) in t is denoted by $\mathcal{V}(t)$ ($\mathcal{V}_l(t)$, $\mathcal{V}_{nl}(t)$, respectively). A term t is *ground* if $\mathcal{V}(t) = \emptyset$. A *position* is a sequence of positive integers, where ϵ stands for the empty sequence. The set of positions (function positions, variable positions) of a term t is denoted by $\mathrm{Pos}(t)$ ($\mathrm{Pos}_{\mathcal{F}}(t)$, $\mathrm{Pos}_{\mathcal{V}}(t)$, respectively). We use $\leqslant$ for the *prefix order* on positions. Positions p and q are *disjoint* ($p \parallel q$) if $p \nleqslant q$ and $q \nleqslant p$. The symbol (subterm) of a term t at the position p is denoted by $t(p)$ ($t|_p$, respectively). The *subterm relation* is denoted by $\trianglelefteq$; its strict part is by $\triangleleft$. We write $\theta : X \rightarrow T$ to if the substitution θ satisfies $\theta(x) = x$ for all $x \in \mathcal{V} \setminus X$ and $\theta(x) \in T$ for any $x \in X$. The *most general unifier* of s and t is denoted by $\mathrm{mgu}(s,t)$. A *rewrite rule* $l \rightarrow r$ satisfies $l \notin \mathcal{V}$ and $\mathcal{V}(r) \subseteq \mathcal{V}(l)$. Rewrite rules are identified modulo renaming of variables. A rewrite rule $l \rightarrow r$ is *linear* if l and r are linear. The set of non-linear variables of a rewrite rule $l \rightarrow r$ is given by $\mathcal{V}_{nl}(l \rightarrow r) = \mathcal{V}_{nl}(l) \cup \mathcal{V}_{nl}(r)$; that of linear variables is by $\mathcal{V}_l(l \rightarrow r) = \mathcal{V}(l) \setminus \mathcal{V}_{nl}(l \rightarrow r)$. A *term rewriting system* (*TRS*) is a set $\mathcal{R}$ of rewrite rules; $\mathcal{R}$ is linear if so are all its rewrite rules. A *rewrite step* $s \rightarrow_{\mathcal{R}} t$ is written as $s \rightarrow_{p,l\rightarrow r,\theta} t$ to specify the position p, the rewrite rule $l \rightarrow r \in \mathcal{R}$ and the substitution θ employed. If $s \rightarrow_{p,l\rightarrow r,\theta} t$ or $s \leftarrow_{p,l\rightarrow r,\theta} t$, we (ambiguously) write $s \leftrightarrow_{p,l\rightarrow r,\theta} t$. If not necessary, subscripts $p, l \rightarrow r, \theta, \mathcal{R}$ will be dropped. The set of normal forms (w.r.t. the *rewrite relation* $\rightarrow_{\mathcal{R}}$) is denoted by $\mathrm{NF}_{\mathcal{R}}(\mathcal{F},\mathcal{V})$, or just $\mathrm{NF}(\mathcal{F},\mathcal{V})$. A TRS $\mathcal{R}$ is normalizing (terminating, confluent) if so is $\rightarrow_{\mathcal{R}}$. We write $s \rightarrow^{im} t$ if $s \rightarrow_p t$ is *innermost*, i.e. any proper subterm of $s|_p$ is a normal form. A term or a TRS is *innermost normalizing* (*innermost terminating*) if it is normalizing (terminating, respectively) w.r.t. $\rightarrow^{im}$.

A *conversion* $\gamma : s_1 \leftrightarrow_{l_1 \to r_1} s_2 \leftrightarrow_{l_2 \to r_2} \cdots \leftrightarrow_{l_{n-1} \to r_{n-1}} s_n$ is specified as $\gamma : s_1 \overset{*}{\leftrightarrow} s_n$ if the detail is not necessary. We put $\mathit{Rules}(\gamma) = \{l_i \to r_i \mid 1 \leq i < n\}$, or $\mathit{Rules}(s_1 \overset{*}{\leftrightarrow} s_n) = \{l_i \to r_i \mid 1 \leq i < n\}$ for brevity. A conversion $t_1 \leftarrow_{p_1, l_1 \to r_1, \theta_1} s \to_{p_2, l_2 \to r_2, \theta_2} t_2$ is called a *peak*; it is a *disjoint* peak if $p_1 \parallel p_2$, it is a *variable* peak if $p_1 = p_2.o.q$ for some $o \in \mathrm{Pos}_{\mathcal{V}}(l_2)$ and q or the other way round, it is a *overlap* peak if $p_1 = p_2.o$ for some $o \in \mathrm{Pos}_{\mathcal{F}}(l_2)$ or the other way round; furthermore, an overlap peak is *trivial* if $p_1 = p_2$ and $l_1 \to r_1 = l_2 \to r_2$. For rewrite rules $l_1 \to r_1, l_2 \to r_2 \in \mathcal{R}$ (w.l.o.g. $\mathcal{V}(l_1) \cap \mathcal{V}(l_2) = \emptyset$), any non-trivial overlap peak of the form $l_2[r_1]_q\theta \leftarrow_{q, l_1 \to r_1, \theta} l_2\theta \to_{\epsilon, l_2 \to r_2, \theta} r_2\theta$ is called a *critical* peak, if $q \in \mathrm{Pos}_{\mathcal{F}}(l_2)$ and $\theta = \mathrm{mgu}(l_1, l_2|_q)$. The set of critical peaks of rules from $\mathcal{R}$ is denoted by $\mathrm{CP}(\mathcal{R})$.

Decreasing Diagrams. Let $\succ$ be a partial order on a set $\mathcal{L}$ of labels. For $\alpha, \beta \in \mathcal{L}$, subsets $\curlyvee\alpha, \curlyvee\alpha\vee\beta \subseteq \mathcal{L}$ are given by $\curlyvee\alpha = \{\gamma \in \mathcal{L} \mid \gamma \prec \alpha\}$ and $\curlyvee\alpha\vee\beta = \{\gamma \in \mathcal{L} \mid \gamma \prec \alpha \vee \gamma \prec \beta\}$. Let A be a set and $\to_\alpha$ be a relation on A for each $\alpha \in \mathcal{L}$. We let $\to_\ell = \bigcup_{\alpha \in \ell} \to_\alpha$ for $\ell \subseteq \mathcal{L}$. Then the relation $\to_{\mathcal{L}}$ is said to be *locally decreasing w.r.t.* $\succ$ if, for any $\alpha, \beta \in \mathcal{L}$, $\leftarrow_\alpha \circ \to_\beta \subseteq \overset{*}{\leftrightarrow}_{\curlyvee\alpha} \circ \overset{=}{\to}_\beta \circ \overset{*}{\leftrightarrow}_{\curlyvee\alpha\vee\beta} \circ \overset{=}{\leftarrow}_\alpha \circ \overset{*}{\leftrightarrow}_{\curlyvee\beta}$.

Proposition 2.1 (Confluence by decreasing diagrams [12]). *A relation* $\to_{\mathcal{L}}$ *is confluent if it is locally decreasing w.r.t. some well-founded partial order* $\succ$ *on* $\mathcal{L}$.

In order to apply this proposition for proving the confluence of a TRS $\mathcal{R}$, we need to set relations $\to_\alpha$ $(\alpha \in \mathcal{L})$ on $\mathrm{T}(\mathcal{F}, \mathcal{V})$ such that $\bigcup_{\alpha \in \mathcal{L}} \to_\alpha = \to_{\mathcal{R}}$. For this, we consider a *labelling function*, say *lab*, that assigns a label to each rewrite step, and put $s \to_\alpha t$ if $\alpha = \mathit{lab}(s \to t)$. We say a peak $t_1 \leftarrow_\alpha s \to_\beta t_2$ is *decreasing w.r.t. lab (and* $\succ$*)* if there exists a conversion $t_1 \overset{*}{\leftrightarrow}_{\curlyvee\alpha} \circ \overset{=}{\to}_\beta \circ \overset{*}{\leftrightarrow}_{\curlyvee\alpha\vee\beta} \circ \overset{=}{\leftarrow}_\alpha \circ \overset{*}{\leftrightarrow}_{\curlyvee\beta} t_n$ (Figure 1). Then, by the proposition, $\mathcal{R}$ is confluent if there exist a labelling function *lab* such that any peak is decreasing w.r.t. *lab*.

Persistency. Let $\mathcal{S}$ be a set of *sorts*. A *sort assignment* τ assigns $\tau(x) \in \mathcal{S}$ to each variable $x \in \mathcal{V}$ and $\tau(f) \in \mathcal{S}^{n+1}$ to each function symbol $f \in \mathcal{F}$ of arity n, in such a way that $\{x \in \mathcal{V} \mid \tau(x) = \sigma\}$ is infinite for any $\sigma \in \mathcal{S}$. Sort assignment τ induces a *many-sorted signature*—the set of well-sorted terms is denoted by $\mathrm{T}(\mathcal{F}, \mathcal{V})^\tau$. We write t^τ to denote $t \in \mathrm{T}(\mathcal{F}, \mathcal{V})^\tau$; $\tau(t) = \sigma$ if the sort of $t \in \mathrm{T}(\mathcal{F}, \mathcal{V})^\tau$ is σ. A quasi-order $\gtrsim$ on $\mathcal{S}$ is given like this: $\sigma \gtrsim \rho$ if there exists a well-sorted term of sort σ having a subterm of sort ρ.

A sort assignment τ is *consistent* with a TRS $\mathcal{R}$ if (l and r are well-sorted and) $\tau(l) = \tau(r)$ for all $l \to r \in \mathcal{R}$ where w.l.o.g. the sets of variables in rewrite rules are supposed to be mutually disjoint. A sort assignment τ consistent with a TRS $\mathcal{R}$ induces a *many-sorted TRS* $\mathcal{R}^\tau$; the rewrite relation of $\mathcal{R}^\tau$ (and hence the notions of confluence, etc.) is defined on $\mathrm{T}(\mathcal{F}, \mathcal{V})^\tau$. If no confusion arises, many-sorted TRSs are called TRSs for simplicity.

Proposition 2.2 (Persistency of confluence [2]). *For any sort assignment* τ *consistent with* $\mathcal{R}$, $\mathcal{R}^\tau$ *is confluent iff* $\mathcal{R}$ *is confluent.*

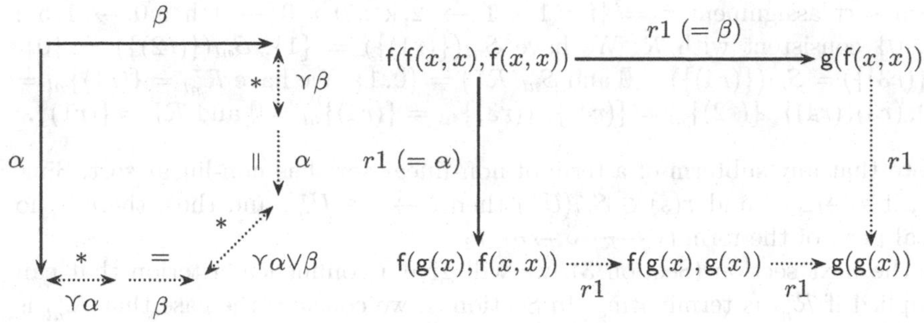

Fig. 1. Decreasing peak and an non-decreasing peak by $r1 : \mathsf{f}(x, x) \to \mathsf{g}(x)$

2.1 Non-Linear Sorts

We will give confluence criteria based on decreasing diagrams in the following two sections, one in each of sections; these two criteria are incomparable and their correctness are proven independently. Some notions, however, are shared—we introduce these notions in this subsection.

Our confluence criteria based on decreasing diagrams aim at dealing with non-linear rewrite rules. For this, we extend the *rule-labelling* [12], which considers a function $\delta : \mathcal{R} \to \mathcal{L}$ and labels each rewrite step by $lab(s \to_{l\to r} t) = \delta(l \to r)$. To show the confluence of a TRS $\mathcal{R}$ by decreasing diagrams with only rule-labelling, $\mathcal{R}$ needs to be linear [12,1]—if a non-linear rewrite rule is contained then one always obtains non-decreasing peaks (Figure 1). Our idea to deal with such cases is to restrict rewrite rules usable in instantiations of non-linear variables (of other rewrite rules) by considering well-sorted terms. The set of rewrite rules usable in instantiations of non-linear variables of rewrite rules in U^τ is denoted by U^τ_{nl} and is formally given as below.

Definition 2.3 (non-linear sort, many-sorted TRS $\mathcal{R}^\tau_{nl}$). *Let $\mathcal{R}$ be a TRS, τ a sort assignment consistent with $\mathcal{R}$ and $U^\tau \subseteq \mathcal{R}^\tau$. A sort $\sigma \in \mathcal{S}$ is said to be a* non-linear sort *of U^τ if there exist $l \to r \in U^\tau$ and $x \in \mathcal{V}_{nl}(l \to r)$ such that $\tau(x) = \sigma$. The set of non-linear sorts of U^τ is denoted by $\mathcal{S}_{nl}(U^\tau)$. By non-linear sorts, we mean non-linear sorts of $\mathcal{R}^\tau$. We define the set $U^\tau_{nl} \subseteq \mathcal{R}^\tau$ as*

$$U^\tau_{nl} = \{l \to r \in \mathcal{R}^\tau \mid \exists \sigma \in \mathcal{S}_{nl}(U^\tau).\ \tau(l) \lesssim \sigma\}.$$

(U^τ_{nl} is written as $\mathcal{R}^\tau_{nl}$ if we take $U^\tau = \mathcal{R}^\tau$.) We also put $U^\tau_l = \mathcal{R}^\tau \setminus U^\tau_{nl}$.

Clearly, $\mathcal{S}_{nl}$ and $()_{nl}$ are monotone, i.e. $U^\tau \subseteq T^\tau$ implies $\mathcal{S}_{nl}(U^\tau) \subseteq \mathcal{S}_{nl}(T^\tau)$ and $U^\tau_{nl} \subseteq T^\tau_{nl}$.

Example 2.4. Let $\mathcal{S} = \{0, 1, 2\}$ and

$$\mathcal{R} = \left\{ \begin{array}{ll} (r1)\ \mathsf{f}(x, x) \to \mathsf{f}(\mathsf{h}(\mathsf{b}), \mathsf{h}(\mathsf{a})) & (r2)\ \mathsf{h}(x) \to \mathsf{k}(x, x) \\ (r3)\ \mathsf{k}(\mathsf{a}, \mathsf{b}) \to \mathsf{h}(\mathsf{a}) & (r4)\ \mathsf{a} \to \mathsf{b} \end{array} \right\}.$$

Take a sort assignment $\tau = \{\mathsf{f} : 1 \times 1 \to 2, \mathsf{k} : 0 \times 0 \to 1, \mathsf{h} : 0 \to 1, \mathsf{a} : 0, \mathsf{b} : 0\}$ consistent with $\mathcal{R}$. We have $\mathcal{S}_{nl}(\{(r1)\}) = \{1\}$, $\mathcal{S}_{nl}(\{(r2)\}) = \{0\}$, $\mathcal{S}_{nl}(\{(r3)\}) = \mathcal{S}_{nl}(\{(r4)\}) = \emptyset$ and $\mathcal{S}_{nl}(\mathcal{R}^\tau) = \{0,1\}$. We have $\mathcal{R}^\tau_{nl} = \{(r1)\}_{nl} = \{(r2),(r3),(r4)\}$, $\{(r2)\}_{nl} = \{(r4)\}$, $\{(r3)\}_{nl} = \{(r4)\}_{nl} = \emptyset$ and $\mathcal{R}^\tau_l = \{(r1)\}$.

Note that any subterm of a term of non-linear sort has non-linear sort. Similarly, if $s \to_{l \to r} t$ and $\tau(s) \in \mathcal{S}_{nl}(U^\tau)$ then $l \to r \in U^\tau_{nl}$, and thus, there is no critical peak of the form $t_1 \leftarrow_{\mathcal{R}^\tau_l} \circ \to_{\mathcal{R}^\tau_{nl}} t_2$.

In the next section (Section 3), we will give a confluence criterion that can be applied if $\mathcal{R}^\tau_{nl}$ is terminating. In Section 4, we consider the case that $\mathcal{R}^\tau_{nl}$ is (possibly not terminating but) innermost normalizing.

3 Confluence of Strongly Quasi-Linear TRSs

In this section, we give a confluence criterion based on the decreasing diagrams and strong quasi-linearity, a notion for many-sorted TRSs given as follows.

Definition 3.1 (strongly quasi-linear). *A many-sorted TRS $\mathcal{R}^\tau$ is* strongly quasi-linear *if the many-sorted TRS $\mathcal{R}^\tau_{nl}$ is terminating.*

Clearly, if $\mathcal{R}^\tau$ is strongly quasi-linear, any (well-sorted) term of non-linear sort is terminating. Note that any (well-sorted) term is terminating w.r.t. $\mathcal{R}^\tau_{nl}$. For strongly quasi-linear TRSs, the following labelling function is considered.

Definition 3.2 (labelling for strongly quasi-linear TRS). *Let $\mathcal{R}^\tau$ be a strongly quasi-linear TRS.*

1. *Let $\mathcal{L}$ be a set and $>$ a well-founded partial order on it. We consider the set $\mathcal{L} \cup \mathrm{T}(\mathcal{F} \cup \mathcal{V})^\tau$ as the set of labels.*
2. *We define a relation $\succ$ on $\mathcal{L} \cup \mathrm{T}(\mathcal{F} \cup \mathcal{V})^\tau$ as follows: $\alpha \succ \beta$ if either (i) $\alpha, \beta \in \mathcal{L}$ and $\alpha > \beta$, (ii) $\alpha \in \mathcal{L}$ and $\beta \in \mathrm{T}(\mathcal{F}, \mathcal{V})^\tau$, or (iii) $\alpha, \beta \in \mathrm{T}(\mathcal{F}, \mathcal{V})^\tau$ and $\alpha \xrightarrow{+}_{\mathcal{R}^\tau_{nl}} \beta$.*
3. *Let $\delta : \mathcal{R}^\tau_l \to \mathcal{L}$. The labelling function lab_δ from the rewrite steps of $\mathcal{R}^\tau$ to $\mathcal{L} \cup \mathrm{T}(\mathcal{F} \cup \mathcal{V})^\tau$ is given like this:*

$$lab_\delta(s \to_{l \to r} t) = \begin{cases} \delta(l \to r) & \text{if } l \to r \in \mathcal{R}^\tau_l \\ s & \text{if } l \to r \in \mathcal{R}^\tau_{nl} \end{cases}$$

The labelling given like $lab_\delta(s \to t) = s$ is called *source-labelling* [12]. Thus, our labelling is a combination of the rule-labelling and the source-labelling[1].

In the rest of this section, we assume that τ is a sort assignment consistent with $\mathcal{R}$ and that $\mathcal{R}^\tau$ is strongly quasi-linear. Furthermore, we suppose a set $\mathcal{L}$ of labels with a well-founded partial order $>$ and $\delta : \mathcal{R}^\tau_l \to \mathcal{L}$ are fixed.

The next lemma is an immediate corollary of well-foundedness of the partial order $>$ on $\mathcal{L}$ and the termination of $\mathcal{R}^\tau_{nl}$.

---
[1] A similar idea has been adapted in Theorem 5 of [12].

Lemma 3.3. *The relation $\succ$ on $\mathcal{L} \cup \mathrm{T}(\mathcal{F}, \mathcal{V})^\tau$ is a well-founded partial order.*

It is trivial to show that disjoint peaks are decreasing. The proof that variable peaks are decreasing is straightforward but interesting, as it reveals why our choice of the labelling function matters (and other variations do not work).

Lemma 3.4. *Any disjoint peak is decreasing w.r.t. lab_δ.*

Lemma 3.5. *Any variable peak is decreasing w.r.t. lab_δ.*

Thus, it remains to show that overlap peaks are decreasing, but this does not hold in general. We reduce decreasingness of overlap peaks to that of critical peaks, where decreasingness of critical peaks is guaranteed by a sufficient criterion, which we introduce below.

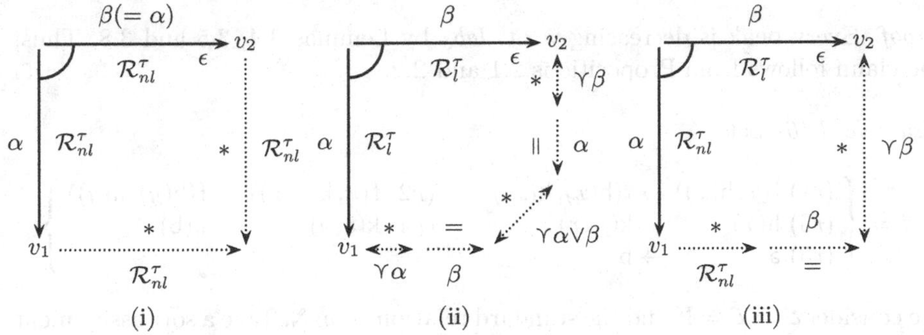

Fig. 2. Hierarchically decreasing critical peaks

Definition 3.6 (hierarchical decreasingness). *Any critical peak $v_1 \leftarrow_{l_1 \to r_1} \circ \to_{l_2 \to r_2} v_2$ is said to be* hierarchically decreasing w.r.t. δ and $>$ *if either one of the following conditions (i)–(iii) holds (Figure 2):*

(i) $l_1 \to r_1, l_2 \to r_2 \in \mathcal{R}^\tau_{nl}$ *and* $v_1 \overset{*}{\to} \circ \overset{*}{\leftarrow} v_2$ *(and hence* $v_1 \overset{*}{\to}_{\mathcal{R}^\tau_{nl}} \circ \overset{*}{\leftarrow}_{\mathcal{R}^\tau_{nl}} v_2$*).*
(ii) $l_1 \to r_1, l_2 \to r_2 \in \mathcal{R}^\tau_l$ *and* $v_1 \overset{*}{\leftrightarrow}_{\curlyvee\alpha} \circ \overset{=}{\to}_\beta \circ \overset{*}{\leftrightarrow}_{\curlyvee\alpha\vee\beta} \circ \overset{=}{\leftarrow}_\alpha \circ \overset{*}{\leftrightarrow}_{\curlyvee\beta} v_2$ *and*
(iii) $l_1 \to r_1 \in \mathcal{R}^\tau_{nl}$, $l_2 \to r_2 \in \mathcal{R}^\tau_l$ *and* $v_1 \overset{*}{\to}_{\mathcal{R}^\tau_{nl}} \circ \overset{=}{\to}_\beta \circ \overset{*}{\leftrightarrow}_{\curlyvee\beta} v_2$,

where $\alpha = \delta(l_1 \to r_1)$ and $\beta = \delta(l_2 \to r_2)$. A many-sorted TRS $\mathcal{R}^\tau$ is said to be hierarchically decreasing (w.r.t. δ and $>$) if so are all critical peaks of $\mathcal{R}^\tau$.

Note that the remaining case, i.e. the case of $l_1 \to r_1 \in \mathcal{R}^\tau_l$ and $l_2 \to r_2 \in \mathcal{R}^\tau_{nl}$ needs not be considered (see a remark below Example 2.4). It may look the conditions (i) and (iii) can be obtained by reducing the decreasingness by using the fact that any label of rewrite steps of $\mathcal{R}^\tau_{nl}$ is smaller than any label of rewrite steps of $\mathcal{R}^\tau_l$, but this is not true; in fact, these conditions are weaker than what are possible according to the definition of decreasingness.

The following properties are used to reduce the decreasingness of overlap peaks to that of critical peaks. For a rewrite step $\gamma : s \to_{l \to r} t$, a context C and a substitution θ, we put $C[\gamma\theta] : C[s\theta] \to_{l \to r} C[t\theta]$ ([19]).

Lemma 3.7. *Let γ, γ' be rewrite steps of $\mathcal{R}^\tau$, C a context, θ a substitution and $\propto \in \{=, \prec\}$. If $lab_\delta(\gamma) \propto lab_\delta(\gamma')$ then $lab_\delta(C[\gamma\theta]) \propto lab_\delta(C[\gamma'\theta])$.*

Lemma 3.8. *If $\mathcal{R}^\tau$ is hierarchically decreasing w.r.t. δ, then any overlap peak is decreasing w.r.t. lab_δ.*

Proof. It follows from the definition of hierarchical decreasingness that any critical peak is decreasing. Then the claim follows using Lemma 3.7. □

Now we arrive at the main theorem of this section.

Theorem 3.9 (confluence of strongly quasi-linear TRSs). *If $\mathcal{R}^\tau$ is strongly quasi-linear and hierarchically decreasing, then $\mathcal{R}$ is confluent.*

Proof. Every peak is decreasing w.r.t. lab_δ by Lemmas 3.4, 3.5 and 3.8. Thus, the claim follows from Propositions 2.1 and 2.2. □

Example 3.10. Let

$$\mathcal{R} = \left\{ \begin{array}{llll} (r1)\ \mathsf{f}(x, \mathsf{h}(x)) & \to \mathsf{f}(\mathsf{h}(x), \mathsf{h}(x)) & (r2)\ \mathsf{f}(x, \mathsf{k}(y, z)) & \to \mathsf{f}(\mathsf{h}(y), \mathsf{h}(y)) \\ (r3)\ \mathsf{h}(x) & \to \mathsf{k}(x, x) & (r4)\ \mathsf{k}(\mathsf{a}, \mathsf{a}) & \to \mathsf{h}(\mathsf{b}) \\ (r5)\ \mathsf{a} & \to \mathsf{b} & & \end{array} \right\}$$

We consider $\mathcal{S} = \mathcal{L} = \mathbb{N}$ and the standard relation $>$ on $\mathbb{N}$. Take a sort assignment $\tau = \{\mathsf{f} : 0 \times 0 \to 1, \mathsf{k} : 0 \times 0 \to 0, \mathsf{h} : 0 \to 0, \mathsf{a} : 0, \mathsf{b} : 0\}$ consistent with $\mathcal{R}$. Then $\mathcal{S}_{nl}(\mathcal{R}^\tau) = \{0\}$ and $\mathcal{R}^\tau_{nl} = \{(r3), (r4), (r5)\}$ is terminating. Thus $\mathcal{R}^\tau$ is strongly quasi-linear. Take $\delta = \{(r1) \mapsto 0, (r2) \mapsto 0\} : \mathcal{R}^\tau_l \to \mathbb{N}$. We have

$$\mathrm{CP}(\mathcal{R}) = \left\{ \begin{array}{ll} (cp1) & \mathsf{f}(x, \mathsf{k}(x, x)) \leftarrow_{(r3)} \mathsf{f}(x, \mathsf{h}(x)) \to_{(r1)} \mathsf{f}(\mathsf{h}(x), \mathsf{h}(x)) \\ (cp2) & \mathsf{f}(x, \mathsf{h}(\mathsf{b})) \leftarrow_{(r4)} \mathsf{f}(x, \mathsf{k}(\mathsf{a}, \mathsf{a})) \to_{(r2)} \mathsf{f}(\mathsf{h}(\mathsf{a}), \mathsf{h}(\mathsf{a})) \\ (cp3) & \mathsf{k}(\mathsf{b}, \mathsf{a}) \leftarrow_{(r5)} \mathsf{k}(\mathsf{a}, \mathsf{a}) \to_{(r4)} \mathsf{h}(\mathsf{b}) \\ (cp4) & \mathsf{k}(\mathsf{a}, \mathsf{b}) \leftarrow_{(r5)} \mathsf{k}(\mathsf{a}, \mathsf{a}) \to_{(r4)} \mathsf{h}(\mathsf{b}) \end{array} \right\}.$$

We now check that every critical peak is hierarchically decreasing.

- (cp1) We have $(r3) \in \mathcal{R}^\tau_{nl}$, $(r1) \in \mathcal{R}^\tau_l$ and $\delta((r1)) = 0$. Thus $\mathsf{f}(x, \mathsf{k}(x, x)) \leftarrow_{\mathcal{R}^\tau_{nl}} \circ \to_0 \mathsf{f}(\mathsf{h}(x), \mathsf{h}(x))$. Since $\mathsf{f}(x, \mathsf{k}(x, x)) \to_0 \mathsf{f}(\mathsf{h}(x), \mathsf{h}(x))$, the condition (iii) of hierarchical decreasingness holds.
- (cp2) We have $(r4) \in \mathcal{R}^\tau_{nl}$ $(r2) \in \mathcal{R}^\tau_l$ and $\delta((r2)) = 0$. $\mathsf{f}(x, \mathsf{h}(\mathsf{b})) \leftarrow_{\mathcal{R}^\tau_{nl}} \circ \to_0 \mathsf{f}(\mathsf{h}(\mathsf{a}), \mathsf{h}(\mathsf{a}))$. Since $\mathsf{f}(x, \mathsf{h}(\mathsf{b})) \to_{\mathcal{R}^\tau_{nl}} \mathsf{f}(x, \mathsf{k}(\mathsf{b}, \mathsf{b})) \to_0 \mathsf{f}(\mathsf{h}(\mathsf{b}), \mathsf{h}(\mathsf{b})) \leftarrow_{\mathcal{R}^\tau_{nl}} \mathsf{f}(\mathsf{h}(\mathsf{a}), \mathsf{h}(\mathsf{b})) \leftarrow_{\mathcal{R}^\tau_{nl}} \mathsf{f}(\mathsf{h}(\mathsf{a}), \mathsf{h}(\mathsf{a}))$, the condition (iii) of hierarchical decreasingness holds.
- (cp3), (cp4) We have $(r4), (r5) \in \mathcal{R}^\tau_{nl}$. It is easy to check the condition (i) of hierarchical decreasingness holds.

Thus, every critical peak is hierarchically decreasing. Hence, by Theorem 3.9, it follows that $\mathcal{R}$ is confluent.

4 Confluence of Quasi-Linear TRSs

In this section, we give a confluence criterion based on the decreasing diagrams and quasi-linearity, a notion for many-sorted TRSs obtained by replacing "termination of $\mathcal{R}^{\tau}_{nl}$" of strong quasi-linearity by "innermost normalization of $\mathcal{R}^{\tau}_{nl}$."

Definition 4.1 (quasi-linear). *A many-sorted TRS $\mathcal{R}^{\tau}$ is* quasi-linear *if the many-sorted TRS $\mathcal{R}^{\tau}_{nl}$ is innermost normalizing.*

Clearly, strongly quasi-linear (many-sorted) TRSs are quasi-linear but not vice versa.

To deal with non-linear TRSs, we here introduce a many-sorted linear TRS $\mathcal{R}^{\tau}_{nf}$, which is obtained by *instantiating* non-linear variables by ground normal forms. We will give a translation from a *quasi-linear* TRS $\mathcal{R}^{\tau}$ to a many-sorted TRS $\mathcal{R}^{\tau}_{nf}$ with *infinite* number of *linear* rewrite rules. Then we show that confluence of $\mathcal{R}^{\tau}_{nf}$ implies that of $\mathcal{R}^{\tau}$.

We will distinguish object terms to be rewritten and rewrite rules. To deal with confluence, variables in object terms can always be regarded as constants. Thus, consider constants c_x corresponding to each variable x, and let $\mathcal{C}_{\mathcal{V}} = \{c_x \mid x \in \mathcal{V}\}$. Now, we consider the set $\mathrm{T}(\mathcal{F} \cup \mathcal{C}_{\mathcal{V}})$ as the set of object terms to be rewritten. Suppose $t \in \mathrm{T}(\mathcal{F}, \mathcal{V})$ and $\mathcal{V}(t) = \{x_1, \ldots, x_n\}$. Let t^c be the term in $\mathrm{T}(\mathcal{F} \cup \mathcal{C}_{\mathcal{V}})$ obtained by replacing each x_i with c_{x_i} $(1 \leq i \leq n)$. Then $s \to_{\mathcal{R}} t$ iff $s^c \to_{\mathcal{R}} t^c$. Hence $\mathcal{R}$ is confluent on $\mathrm{T}(\mathcal{F}, \mathcal{V})$ iff $\mathcal{R}$ is confluent on $\mathrm{T}(\mathcal{F} \cup \mathcal{C}_{\mathcal{V}})$. Similarly, by extending sort assignment τ by $\tau(c_x) = \tau(x)$, it follows that $\mathcal{R}^{\tau}$ is confluent on $\mathrm{T}(\mathcal{F}, \mathcal{V})^{\tau}$ iff $\mathcal{R}^{\tau}$ is confluent on $\mathrm{T}(\mathcal{F} \cup \mathcal{C}_{\mathcal{V}})^{\tau}$. Henceforth, let $\mathrm{NF}(\mathcal{F} \cup \mathcal{C}_{\mathcal{V}})^{\tau}$ be the set of normal forms from $\mathrm{T}(\mathcal{F} \cup \mathcal{C}_{\mathcal{V}})^{\tau}$ (w.r.t. $\to_{\mathcal{R}^{\tau}}$).

Definition 4.2 (linearization of quasi-linear TRSs). *Let $\mathcal{R}^{\tau}$ be a quasi-linear TRS. For $U^{\tau} \subseteq \mathcal{R}^{\tau}$, we define a many-sorted TRS U^{τ}_{nf} by*

$$U^{\tau}_{nf} = \bigcup_{l \to r \in U^{\tau}} \{l\hat{\theta} \to r\hat{\theta} \mid \hat{\theta} : \mathcal{V}_{nl}(l \to r) \to \mathrm{NF}(\mathcal{F} \cup \mathcal{C}_{\mathcal{V}})^{\tau}\}.$$

(U^{τ}_{nf} is written as $\mathcal{R}^{\tau}_{nf}$ if we take $U^{\tau} = \mathcal{R}^{\tau}$.) We write a rewrite rule of U^{τ}_{nf} as $l\hat{\theta} \to r\hat{\theta}$, for brevity, to denote $l \to r \in U^{\tau}$ and $\hat{\theta} : \mathcal{V}_{nl}(l \to r) \to \mathrm{NF}(\mathcal{F} \cup \mathcal{C}_{\mathcal{V}})^{\tau}$.

Example 4.3. Let $\mathcal{S} = \{0, 1, 2\}$ and

$$\mathcal{R} = \big\{\, (r1)\ \mathsf{f}(x, x, y) \to \mathsf{f}(x, \mathsf{g}(x), y) \quad (r2)\ \mathsf{f}(x, y, z) \to \mathsf{h}(\mathsf{a}) \,\big\}.$$

Take a sort assignment $\tau = \{\mathsf{f} : 0 \times 0 \times 1 \to 2, \mathsf{g} : 0 \to 0, \mathsf{h} : 0 \to 2, \mathsf{a} : 0\}$ consistent with $\mathcal{R}$. Since $\mathcal{V}_{nl}(r1) = \{x\}$ and $\mathcal{V}_{nl}(r2) = \emptyset$, we obtain $\mathcal{R}^{\tau}_{nf} = \{\mathsf{f}(s, s, y) \to \mathsf{f}(s, \mathsf{g}(s), y) \mid s \in \mathrm{NF}(\mathcal{F} \cup \mathcal{C}_{\mathcal{V}})^{\tau}, \tau(s) = 0\} \cup \{\mathsf{f}(x, y, z) \to \mathsf{h}(\mathsf{a})\}$.

It is clear that $\mathcal{R}^{\tau}_{nf}$ is a linear TRS, as all non-linear variables of rewrite rules are instantiated by ground terms. Since there are infinitely many instantiations of each rewrite rule, $\mathcal{R}^{\tau}_{nf}$ has infinitely many numbers of rewrite rules.

In the rest of this section, we assume that τ is a sort assignment consistent with $\mathcal{R}$ and that $\mathcal{R}^{\tau}$ is quasi-linear. We also abbreviate $\to_{\mathcal{R}^{\tau}}$ and $\to_{\mathcal{R}^{\tau}_{nf}}$ by $\to$ and $\underset{nf}{\to}$, respectively. The next lemma is used to show Lemma 4.5.

Lemma 4.4. *Suppose* $s \xrightarrow{*}^{im} t$ *and let* $U^\tau = Rules(s \xrightarrow{*}^{im} t)$. *Then* $s \xrightarrow{*}_{U^\tau_{nf}} t$.

Lemma 4.5. *Let* $l \to r \in \mathcal{R}^\tau$ *and* $U^\tau = \{l \to r\}_{nl}$. *Then,* $\to_{l \to r} \subseteq \xrightarrow{*}_{U^\tau_{nf}} \circ \to_{\{l \to r\}_{nf}} \circ \xleftarrow{*}_{U^\tau_{nf}}$ *on* $\mathrm{T}(\mathcal{F} \cup \mathcal{C}_\mathcal{V})^\tau$.

A corollary of the previous lemma is the following sufficient criterion of confluence of $\mathcal{R}^\tau$ in terms of $\mathcal{R}^\tau_{nf}$, on which our analysis will be based.

Lemma 4.6. *A quasi-linear TRS* $\mathcal{R}^\tau$ *is confluent on* $\mathrm{T}(\mathcal{F} \cup \mathcal{C}_\mathcal{V})^\tau$ *if so is its linearization* $\mathcal{R}^\tau_{nf}$.

Proof. The claim easily follows from $\to \subseteq \underset{nf}{\to} \subseteq \underset{nf}{\overset{*}{\leftrightarrow}}$, which holds by Lemma 4.5 and the definition of $\mathcal{R}^\tau_{nf}$. □

The next lemma is used to analyze overlap peaks of $\mathcal{R}^\tau_{nf}$ by those of $\mathcal{R}^\tau$.

Lemma 4.7. *Let* $v_1 \leftarrow_{l_1\hat{\theta}_1 \to r_1\hat{\theta}_1} \circ \to_{l_2\hat{\theta}_2 \to r_2\hat{\theta}_2} v_2$ *be a critical peak of* $\mathcal{R}^\tau_{nf}$. *Then there exist a critical peak* $u_1 \leftarrow_{l_1 \to r_1} \circ \to_{l_2 \to r_2} u_2$ *of* $\mathcal{R}^\tau$ *and a substitution* θ *such that* $u_i\theta = v_i$ $(i = 1, 2)$.

We note the converse of Lemma 4.7 does not hold in general.

Let $\mathcal{L}$ stand for the set of labels and $\succ$ be a well-founded partial order on $\mathcal{L}$.

Definition 4.8 (labelling on $\mathcal{R}^\tau_{nf}$). *Let* $lab : \mathcal{R}^\tau \to \mathcal{L}$. *We extend*[2] *lab to a function* $\mathcal{R}^\tau_{nf} \to \mathcal{L}$ *by* $lab(l\hat{\theta} \to r\hat{\theta}) = lab(l \to r)$. *Furthermore, we label the rewrite steps* $s \to_{l\hat{\theta} \to r\hat{\theta}} t$ *of* $\mathcal{R}^\tau_{nf}$ *by* $lab(l\hat{\theta} \to r\hat{\theta})$.

In the following, we assume some $lab : \mathcal{R}^\tau \to \mathcal{L}$ is fixed. Let ℓ (literally) stands for α or $\alpha \vee \beta$ $(\alpha, \beta \in \mathcal{L})$. For any $U^\tau \subseteq \mathcal{R}^\tau$, we write $U^\tau \prec \ell$ ($U^\tau_{nf} \prec \ell$) if $lab(l \to r) \prec \ell$ for all $l \to r \in U^\tau$ ($l \to r \in U^\tau_{nf}$, respectively). Note $U^\tau \prec \ell$ iff $U^\tau_{nf} \prec \ell$ for any $U^\tau \subseteq \mathcal{R}^\tau$. Let $Rules_{nl}(\gamma) = (Rules(\gamma))_{nl}$ for any conversion γ.

The next technical lemma, to be used in our key lemma (Lemma 4.11), is an immediate consequence of Lemma 4.5.

Lemma 4.9. *Let* $\gamma : s \overset{*}{\leftrightarrow}_{\curlyvee\ell} s' \overset{=}{\to}_\beta t$ *be a conversion on* $\mathrm{T}(\mathcal{F} \cup \mathcal{C}_\mathcal{V})^\tau$. *If* $Rules_{nl}(\gamma) \prec \ell$ *then* $s \underset{nf}{\overset{*}{\leftrightarrow}}_{\curlyvee\ell} \hat{s} \underset{nf}{\overset{=}{\to}}_\beta \hat{t} \underset{nf}{\overset{*}{\leftarrow}}_{\curlyvee\ell} t$ *on* $\mathrm{T}(\mathcal{F} \cup \mathcal{C}_\mathcal{V})^\tau$. *Furthermore,* $s' = t$ *implies* $\hat{s} = \hat{t}$.

Definition 4.10 (linearized-decreasingness). *Any critical peak* $v_1 \leftarrow_\alpha \circ \to_\beta v_2$ *of* $\mathcal{R}^\tau$ *is said to be* linearized-decreasing *w.r.t. lab* $: \mathcal{R}^\tau \to \mathcal{L}$ *and* $\succ$ *if there exists a conversion*

$$v_1 \overset{*}{\leftrightarrow}_{\curlyvee\alpha} \circ \overset{=}{\to}_\beta u_1 \overset{*}{\leftrightarrow}_{\curlyvee\alpha\vee\beta} u_2 \overset{=}{\leftarrow}_\alpha \circ \overset{*}{\leftrightarrow}_{\curlyvee\beta} v_2$$

on $\mathrm{T}(\mathcal{F}, \mathcal{V})^\tau$ *such that the following conditions (i)–(iii) are satisfied (Figure 3):*

[2] Thus, strictly speaking, $l\hat{\theta} \to r\hat{\theta}$ should be considered as $\langle l \to r, \hat{\theta} \rangle$, to distinguish common instances of different rewrite rules.

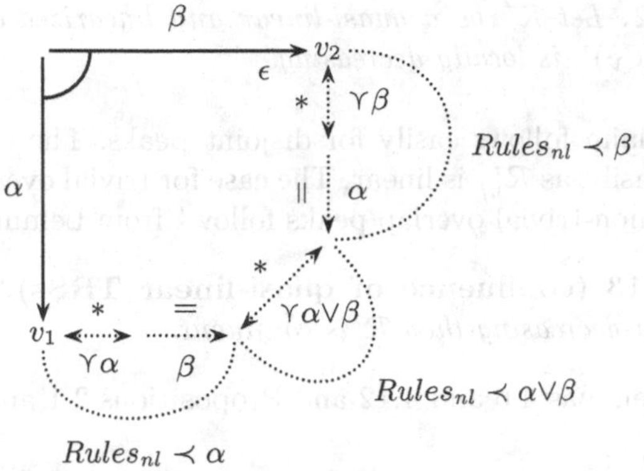

Fig. 3. Linearized-decreasing critical peak

(i) $Rules_{nl}(v_1 \overset{*}{\leftrightarrow}_{\curlyvee\alpha} \circ \overset{=}{\rightarrow}_\beta u_1)) \prec \alpha$,
(ii) $Rules_{nl}(u_1 \overset{*}{\leftrightarrow}_{\curlyvee\alpha\vee\beta} u_2) \prec \alpha \vee \beta$, *and*
(iii) $Rules_{nl}(u_2 \overset{=}{\leftarrow}_\alpha \circ \overset{*}{\leftrightarrow}_{\curlyvee\beta} v_2) \prec \beta$.

A many-sorted TRS $\mathcal{R}^\tau$ is said to be linearized-decreasing (w.r.t. lab and $\succ$) if so are all critical peaks of $\mathcal{R}^\tau$.

Lemma 4.11. *Let $\mathcal{R}^\tau$ be a quasi-linear and linearized-decreasing TRS. Let $w_1 \leftarrow_\alpha \circ \rightarrow_\beta w_2$ be a critical peak of $\mathcal{R}^\tau_{nf}$ and $w_1\theta, w_2\theta \in \mathrm{T}(\mathcal{F} \cup \mathcal{C}_\mathcal{V})^\tau$. Then, $w_1\theta \overset{*}{\underset{nf}{\leftrightarrow}}_{\curlyvee\alpha} \circ \overset{=}{\underset{nf}{\rightarrow}}_\beta \circ \overset{*}{\underset{nf}{\leftrightarrow}}_{\curlyvee\alpha\vee\beta} \circ \overset{=}{\underset{nf}{\leftarrow}}_\alpha \circ \overset{*}{\underset{nf}{\leftrightarrow}}_{\curlyvee\beta} w_2\theta$ on $\mathrm{T}(\mathcal{F} \cup \mathcal{C}_\mathcal{V})^\tau$.*

Proof. Let $w_1 \leftarrow_{l_1\hat{\theta}_1 \to r_1\hat{\theta}_1} \circ \rightarrow_{l_2\hat{\theta}_2 \to r_2\hat{\theta}_2} w_2$. Then, by Lemma 4.7, there exists critical peak $v_1 \leftarrow_{l_1 \to r_1} \circ \rightarrow_{l_2 \to r_2} v_2$ such that $w_1 = v_1\theta'$ and $w_2 = v_2\theta'$ for some θ'. Then, by the definition of labelling of rewrite steps of $\mathcal{R}^\tau_{nf}$, we have $v_1 \leftarrow_\alpha \circ \rightarrow_\beta v_2$. Thus, by assumption, there exists a conversion $v_1 \overset{*}{\leftrightarrow}_{\curlyvee\alpha} \circ \overset{=}{\rightarrow}_\beta u_1 \overset{*}{\leftrightarrow}_{\curlyvee\alpha\vee\beta} u_2 \overset{=}{\leftarrow}_\alpha \circ \overset{*}{\leftrightarrow}_{\curlyvee\beta} v_2$ on $\mathrm{T}(\mathcal{F}, \mathcal{V})^\tau$ satisfying conditions (i)–(iii) of Definition 4.10. Now, apply the substitution $\theta \circ \theta'$ to this conversion to obtain

$$w_1\theta = v_1\theta'\theta \overset{*}{\leftrightarrow}_{\curlyvee\alpha} \circ \overset{=}{\rightarrow}_\beta u_1\theta'\theta \overset{*}{\leftrightarrow}_{\curlyvee\alpha\vee\beta} u_2\theta'\theta \overset{=}{\leftarrow}_\alpha \circ \overset{*}{\leftrightarrow}_{\curlyvee\beta} v_2\theta'\theta = w_2\theta.$$

Here, w.l.o.g. one can extend θ with $x \mapsto c_x$ so that this conversion is on $\mathrm{T}(\mathcal{F} \cup \mathcal{C}_\mathcal{V})^\tau$. Furthermore, conditions (i)–(iii) of Definition 4.10 imply

(i') $Rules_{nl}(v_1\theta'\theta \overset{*}{\leftrightarrow}_{\curlyvee\alpha} \circ \overset{=}{\rightarrow}_\beta u_1\theta'\theta) \prec \alpha$,
(ii') $Rules_{nl}(u_1\theta'\theta \overset{*}{\leftrightarrow}_{\curlyvee\alpha\vee\beta} u_2\theta'\theta) \prec \alpha \vee \beta$, and
(iii') $Rules_{nl}(u_2\theta'\theta \overset{=}{\leftarrow}_\alpha \circ \overset{*}{\leftrightarrow}_{\curlyvee\beta} v_2\theta'\theta) \prec \beta$.

Then, by Lemma 4.9, $v_1\theta'\theta \overset{*}{\underset{nf}{\leftrightarrow}}_{\curlyvee\alpha} \circ \overset{=}{\underset{nf}{\rightarrow}}_\beta \circ \overset{*}{\underset{nf}{\leftarrow}}_{\curlyvee\alpha} u_1\theta'\theta \overset{*}{\underset{nf}{\leftrightarrow}}_{\curlyvee\alpha\vee\beta} u_2\theta'\theta \overset{*}{\underset{nf}{\rightarrow}}_{\curlyvee\beta} \circ \overset{=}{\underset{nf}{\leftarrow}}_\alpha \circ \overset{*}{\underset{nf}{\leftrightarrow}}_{\curlyvee\beta} v_2\theta'\theta$ on $\mathrm{T}(\mathcal{F} \cup \mathcal{C}_\mathcal{V})^\tau$. As $\underset{nf}{\leftarrow}_{\curlyvee\alpha}, \underset{nf}{\rightarrow}_{\curlyvee\beta} \subseteq \underset{nf}{\leftrightarrow}_{\curlyvee\alpha\vee\beta}$, the claim follows. □

Lemma 4.12. *Let $\mathcal{R}^\tau$ be a quasi-linear and linearized-decreasing TRS. Then $\underset{nf}{\rightarrow}$ on $\mathrm{T}(\mathcal{F} \cup \mathcal{C}_\mathcal{V})^\tau$ is locally decreasing.*

Proof. The claim follows easily for disjoint peaks. The case for variable peaks follows also easily, as $\mathcal{R}^\tau_{nf}$ is linear. The case for trivial overlap peaks are obvious. The case for non-trivial overlap peaks follows from Lemma 4.11. □

Theorem 4.13 (confluence of quasi-linear TRSs). *If $\mathcal{R}^\tau$ is quasi-linear and linearized-decreasing then $\mathcal{R}$ is confluent.*

Proof. Use Lemmas 4.6 and 4.12 and Propositions 2.1 and 2.2. □

Example 4.14. Let

$$\mathcal{R} = \left\{ \begin{array}{llll} (r1)\ \mathsf{f}(x,y) & \rightarrow \mathsf{f}(\mathsf{g}(x),\mathsf{g}(x)) & (r2)\ \mathsf{f}(\mathsf{g}(x),x) & \rightarrow \mathsf{f}(x,\mathsf{g}(x)) \\ (r3)\ \mathsf{g}(x) & \rightarrow \mathsf{h}(x) & (r4)\ \mathsf{h}(\mathsf{g}(x)) & \rightarrow \mathsf{g}(\mathsf{g}(x)) \end{array} \right\}$$

We consider $\mathcal{S} = \mathcal{L} = \mathbb{N}$ and the standard relation $>$ on $\mathbb{N}$. Take a sort assignment $\tau = \{\mathsf{f} : 0 \times 0 \rightarrow 1, \mathsf{g} : 0 \rightarrow 0, \mathsf{h} : 0 \rightarrow 0\}$ consistent with $\mathcal{R}$. We have $\mathcal{S}_{nl}(\mathcal{R}^\tau) = \{0\}$ and $\mathcal{R}^\tau_{nl} = \{(r3),(r4)\}$ is innermost normalizing. Thus $\mathcal{R}^\tau$ is quasi-linear. We have

$$\mathrm{CP}(\mathcal{R}) = \left\{ \begin{array}{ll} (cp1) & \mathsf{f}(\mathsf{h}(x),x) \leftarrow_{(r3)} \mathsf{f}(\mathsf{g}(x),x) \rightarrow_{(r2)} \mathsf{f}(x,\mathsf{g}(x)) \\ (cp2) & \mathsf{h}(\mathsf{h}(x)) \leftarrow_{(r3)} \mathsf{h}(\mathsf{g}(x)) \rightarrow_{(r4)} \mathsf{g}(\mathsf{g}(x)) \\ (cp3) & \mathsf{f}(x,\mathsf{g}(x)) \leftarrow_{(r2)} \mathsf{f}(\mathsf{g}(x),x) \rightarrow_{(r1)} \mathsf{f}(\mathsf{g}(\mathsf{g}(x)),\mathsf{g}(\mathsf{g}(x))) \\ (cp3') & \mathsf{f}(\mathsf{g}(\mathsf{g}(x)),\mathsf{g}(\mathsf{g}(x))) \leftarrow_{(r1)} \mathsf{f}(\mathsf{g}(x),x) \rightarrow_{(r2)} \mathsf{f}(x,\mathsf{g}(x)) \end{array} \right\}.$$

Take $lab = \{(r1) \mapsto 2, (r2) \mapsto 3, (r3) \mapsto 0, (r4) \mapsto 1\} : \mathcal{R} \rightarrow \mathbb{N}$.

- (cp1) We have $\gamma_1 : \mathsf{f}(\mathsf{h}(x),x) \rightarrow_{(r1)} \mathsf{f}(\mathsf{g}(\mathsf{h}(x)),\mathsf{g}(\mathsf{h}(x))) \leftarrow_{(r1)} \mathsf{f}(\mathsf{h}(x),\mathsf{g}(x)) = \gamma_2 : \mathsf{f}(\mathsf{h}(x),\mathsf{g}(x)) \leftarrow_{(r3)} \mathsf{f}(\mathsf{g}(x),\mathsf{g}(x)) \leftarrow_{(r1)} \mathsf{f}(x,\mathsf{g}(x))$, and $Rules_{nl}(\gamma_1) = \{(r3),(r4)\} \prec 0 \vee 3$ and $Rules_{nl}(\gamma_2) = \{(r3),(r4)\} \prec 3$. Thus the critical peak is linearized-decreasing.
- (cp2) We have $\gamma : \mathsf{h}(\mathsf{h}(x)) \leftarrow_{(r3)} \mathsf{g}(\mathsf{h}(x)) \leftarrow_{(r3)} \mathsf{g}(\mathsf{g}(x))$, and $Rules_{nl}(\gamma) = \emptyset$. Thus the critical peak is linearized-decreasing.
- (cp3) We have $\gamma : \mathsf{f}(x,\mathsf{g}(x)) \rightarrow_{(r1)} \mathsf{f}(\mathsf{g}(x),\mathsf{g}(x)) \rightarrow_{(r1)} \mathsf{f}(\mathsf{g}(\mathsf{g}(x)),\mathsf{g}(\mathsf{g}(x)))$ Since $Rules_{nl}(\gamma) = \{(r3),(r4)\} \prec 2$, the critical peak is linearized-decreasing. The case (cp3′) follows similarly.

Hence, by Theorem 4.13, it follows that $\mathcal{R}$ is confluent.

We now remark that Theorems 3.9 and 4.13 are incomparable. First, $\mathcal{R}$ in Example 4.14 is not strongly quasi-linear, as $\mathcal{R}^\tau_{nl}$ is not terminating. Thus, Theorem 3.9 is not subsumed by Theorem 4.13. In the next example, we show that Theorem 4.13 is not subsumed by Theorem 3.9.

Example 4.15. Let us consider $\mathcal{R}$ in Example 3.10. First note that $\{(r1)\}_{nl} = \{(r2)\}_{nl} = \{(r3)\}_{nl} = \{(r3),(r4),(r5)\}$. We consider conversions for the critical peak (cp2): $v_1 = \mathsf{f}(x,\mathsf{h}(\mathsf{b})) \leftarrow_{(r4)} \mathsf{f}(x,\mathsf{k}(\mathsf{a},\mathsf{a})) \rightarrow_{(r2)} \mathsf{f}(\mathsf{h}(\mathsf{a}),\mathsf{h}(\mathsf{a})) = v_2$. It is easy to see $x \overset{*}{\leftrightarrow} t$ implies $x = t$ for any t. From this, it follows that the conversion has the form $v_1 \overset{*}{\leftrightarrow} w_1 \rightarrow_{(r2)} w_2 \overset{*}{\leftrightarrow} v_2$. We now consider decreasing conversion $v_1 \overset{*}{\leftrightarrow}_{\curlyvee lab(r4)} \circ \overset{=}{\rightarrow}_{lab(r2)} u_1 \overset{*}{\leftrightarrow}_{\curlyvee lab(r4) \vee lab(r2)} u_2 \overset{=}{\leftarrow}_{lab(r4)} u_3 \overset{*}{\leftrightarrow}_{\curlyvee lab(r2)} v_2$ and distinguish cases by in which part of this conversion the rewrite step $w_1 \rightarrow_{(r2)} w_2$ is involved.

- Case $w_1 \rightarrow w_2$ is in $v_1 \overset{*}{\leftrightarrow}_{\curlyvee lab(r4)} \circ \overset{=}{\rightarrow}_{lab(r2)} u_1$. Then by $(r4) \in \{(r2)\}_{nl}$, one requires $lab(r4) \prec lab(r4)$, which is impossible.
- Case $w_1 \rightarrow w_2$ is in $u_1 \overset{*}{\leftrightarrow}_{\curlyvee lab(r4) \vee lab(r2)} u_2$. Then one needs $lab(r2) \prec lab(r4) \vee lab(r2)$ and $lab(r4) \prec lab(r4) \vee lab(r2)$. This is again impossible.
- Case $w_1 \rightarrow w_2$ is in $u_2 \overset{=}{\leftarrow}_{lab(r4)} u_3$. This is impossible because of the direction of the rewrite steps does not coincide.
- Case $w_1 \rightarrow w_2$ is in $u_3 \overset{*}{\leftrightarrow}_{\curlyvee lab(r2)} v_2$. Then one needs $lab(r2) \prec lab(r2)$, which is impossible.

Thus, the critical peak (cp2) is not linearized-decreasing.

Relations between Theorem 4.13 and Theorem 3.9 for some particular classes of TRSs follow. For non-overlapping TRSs, Theorem 4.13 is strictly subsumed by Theorem 3.9. For linear TRSs, Theorem 4.13 and Theorem 3.9 (and the original rule-labelling) are equivalent.

Finally, we note that one can generally include linear rules to $\mathcal{R}^{\tau}_{nl}$ in Theorem 3.9, and then Knuth-Bendix's criterion is obtained from Theorem 3.9. This is, however, not surprising as it is known that Knuth-Bendix's criterion can be given by decreasing diagrams with the source-labelling (Example 12 of [12]).

5 Implementation, Experiments and Related Work

The confluence criteria of the paper have been implemented in the confluence prover ACP [3]. We straightforwardly adapt techniques for automating decreasing diagrams based on rule-labelling [1,8]. We use SML/NJ [13] for the implementation language and the constraint solver Yices [5] to check the satisfiability of constraints encoding existence of a labelling function satisfying our criteria.

Some heuristics and approximation employed in our implementation follow. To construct many-sorted TRS $\mathcal{R}^{\tau}$ from an unsorted TRS $\mathcal{R}$, it suffices to compute sort assignment τ consistent with $\mathcal{R}$. In practice, its enough to choose such a sort assignment that maximally distinct sorts, in order to maximize the applicability of the criteria. This can be done by first assigning fresh sorts for each sort declarations of function symbols and for variables, and then solving the constraint on these sorts that arises from the requirement that lhs and rhs of each rewrite rule are well-sorted terms having the same sort. To check the quasi-linearity of TRSs, one has to check innermost normalization of TRSs. To the best of our knowledge, no works concentrated on proving innermost normalization are known; thus, the check is approximated by checking innermost termination. To check decreasing diagram criteria, one has to find, for each critical peak $v_1 \leftarrow u \rightarrow v_2$, some conversions $v_1 \overset{*}{\leftrightarrow} v_2$ that are used as the candidates

Table 1. Experiments with the state-of-art confluence provers

	ACP	CSI	Saigawa	Thm. 3.9	Thm. 4.13	Thm. 3.9&4.13
Example 3.10	×	×	×	✓	×	✓
Example 4.14	×	×	×	×	✓	✓
9 examples from [14]	8	1	1	9	9	9
11 new examples	0	0	0	9	10	11

for $v_1 \overset{*}{\leftrightarrow}_{\curlyvee\alpha} \circ \overset{=}{\rightarrow}_{\beta} u_1 \overset{*}{\leftrightarrow}_{\curlyvee\alpha\vee\beta} u_2 \overset{=}{\leftarrow}_{\alpha} \circ \overset{*}{\leftrightarrow}_{\curlyvee\beta} v_2$. For this, our implementation uses sets of conversions $v_1 \overset{\leq 4}{\rightarrow}_{\mathcal{R}^{\leftrightarrow}} \circ \overset{\leq 4}{\leftarrow}_{\mathcal{R}^{\leftrightarrow}} v_2$ as the sets of candidates, where $\mathcal{R}^{\leftrightarrow} = \mathcal{R} \cup \{r \to l \mid l \to r \in \mathcal{R}, r \notin \mathcal{V}, \mathcal{V}(l) \subseteq \mathcal{V}(r)\}$ [8] and $s \overset{\leq 4}{\to} t$ means $s \overset{*}{\to} t$ in less than four rewrite steps. Then applicability of the criterion for all possible choice of u_1, u_2 in these sequences is encoded in a constraint [1].

In [14], a critical pair criterion for quasi-left-linear TRSs have been given. The main differences between quasi-left-linear TRSs and quasi-linear TRSs are that (1) the former considers only non-linear variables on lhs of the rewrite rules while the latter considers non-linear variables on lhs or rhs of the rewrite rules, and (2) in the the former, $\mathcal{R}^{\tau}_{nf}$ is obtained by instantiating all variables of non-left-linear sorts, while in the latter $\mathcal{R}^{\tau}_{nf}$ is obtained by instantiating non-linear variables. We have adapted decreasing diagrams with rule-labelling for proving confluence of $\mathcal{R}^{\tau}_{nf}$ but in [14] critical pair criteria for left-linear TRSs (e.g. [9,15]) are applied.

We now report on experiments for the collection of 9 examples from [14], and 11 new examples constructed in the course of experiments, including Examples 3.10 and 4.14. These are all non-left-linear and non-terminating TRSs. Tests are performed on a PC with one 2.50GHz CPU and 4G memory; the timeout is set to 60 seconds. For comparison with the state-of-art confluence provers, ACP (ver. 0.41) [3], CSI (ver. 0.4.1) [18] and Saigawa (ver. 1.5) [7] are used. The summary of experiments is shown in Table 1. For examples, ✓ denotes success and × denotes failure. Examples 3.10 and 4.14 are solved by none of the state-of-art confluence provers. For the collections, the number of successes is shown. ACP implements the technique given in [14], and thus can solve all examples but the last one which has been left open for automated confluence proving in [14]. Both of our new criteria prove this last example from [14]. All provers fail at solving all our new examples. Hence, in particular, the technique of [14] is not effective for all of our new examples. The difference of Thm. 3.9 and Thm. 4.13 appears on only few examples. In particular, Example 3.10 is only solved by the criterion of Thm. 3.9, and Example 4.14 is only solved by the criterion of Thm. 4.13.

Next we discuss other related work. In [1,12,19], the rule-labelling is extended to (non-linear) left-linear TRSs, where the one in latest [19] subsumes those in the others. This technique essentially depends on the path information to the duplicating variables in the rewrite rules to ensure the decreasingness of variable peaks. In [10], a criterion for proving confluence of non-left-linear TRSs using relative termination has been developed, whose correctness is proved based on the decreasing diagrams with source-labelling. This criterion essentially requires termination of $\mathcal{R}_1$ relative to $\mathcal{R}_2$ to show the confluence of $\mathcal{R} = \mathcal{R}_1 \cup \mathcal{R}_2$. Several other confluence criteria applicable for non-left-linear TRSs have been developed

Table 2. Experiments of confluence criteria for non-left-linear non-terminating TRSs

Criterion (or tool)	42 Cops	10 Cops
Thm. 3.9	13	1
Thm. 4.13	12	1
Criterion for quasi-left-linear TRSs [14] (in ACP)	9	0
Criterion for weight-decreasing TRSs [6] (in ACP)	13	0
Criterion for simple-right-linear TRSs [17] (in ACP)	1	0
Saigawa (including [10])	12	0

in [6,17]; these criteria require some restrictions on the form of rewrite rules. All of these criteria are incomparable with the techniques developed in the present paper is witnessed in Table 1, as (the automatizable parts of) these techniques have been involved in some of the confluence provers ACP, CSI and Saigawa.

Next we compare strength of confluence criteria for non-terminating non-left-linear TRSs experimentally. For this, we use two collections of non-terminating non-left-linear problems from Cops (Confluence Problems) database: (i) the collection of 42 problems from CoCo 2013 that are not solved as '`non_confluent`' by any tool, and (ii) the collection of 10 problems from CoCo 2013 that are not solved by any tool. The criterion of [10] is approximated by Saigawa (Saigawa does not facilitate to choose a single technique employed). For other criteria, ACP is adapted to single out each criterion. In Table 2, the numbers of successes for each criterion are shown. We observe that the problems in CoCo 2013 do not differentiate strength of most of techniques very much. We note the current implementation of the technique of [17] in ACP is not very elaborated. The problem that is not solved in any provers in CoCo 2013 but solved by our new criteria is the last example from [14] mentioned before.

The collection of new examples and details of the experiments are available on the webpage `http://www.nue.riec.tohoku.ac.jp/tools/acp/experiments/rtatlca14/all.html`.

In this paper, sort constraint is used to limit instantiations of non-linear variables of rewrite rules. Imposing such limitation more abstractly leads to the framework of membership conditional rewriting systems and a confluence criterion for such systems [16].

6 Conclusion

We have presented two criteria for confluence of TRSs $\mathcal{R}$ based on decreasing diagrams with rule-labelling and persistency: (1) $\mathcal{R}^\tau$ is strongly quasi-linear and hierarchically decreasing, and (2) $\mathcal{R}^\tau$ is quasi-linear and linearized-decreasing. We have also shown that these criteria are incomparable. These criteria are particularly useful for proving non-linear TRSs confluent, including non-terminating non-left-linear TRSs for which only few confluence criteria have been known. Our criteria have been implemented in the confluence prover ACP. We have shown that our criteria are successfully used in confluence provers for proving confluence of TRSs for which none of the state-of-art confluence provers succeed.

Acknowledgements. Thanks are due to anonymous referees for helpful and insightful comments. This work was partially supported by grants from JSPS, Nos. 25330004 and 25280025.

References

1. Aoto, T.: Automated confluence proof by decreasing diagrams based on rule-labelling. In: Proc. of 21st RTA. LIPIcs, vol. 6, pp. 7–16. Schloss Dagstuhl (2010)
2. Aoto, T., Toyama, Y.: Persistency of confluence. Journal of Universal Computer Science 3(11), 1134–1147 (1997)
3. Aoto, T., Yoshida, J., Toyama, Y.: Proving confluence of term rewriting systems automatically. In: Treinen, R. (ed.) RTA 2009. LNCS, vol. 5595, pp. 93–102. Springer, Heidelberg (2009)
4. Baader, F., Nipkow, T.: Term Rewriting and All That. Cambridge University Press (1998)
5. Dutertre, B., de Moura, L.: The YICES SMT solver, `http://yices.csl.sri.com/`
6. Gomi, H., Oyamaguchi, M., Ohta, Y.: On the Church-Rosser property of root-E-overlapping and strongly depth-preserving term rewriting systems. Transactions of IPSJ 39(4), 992–1005 (1998)
7. Hirokawa, N., Klein, D.: Saigawa: A confluence tool. In: Proc. of 1st IWC, p. 49 (2012)
8. Hirokawa, N., Middeldorp, A.: Decreasing diagrams and relative termination. Journal of Automated Reasoning 47(4), 481–501 (2011)
9. Huet, G.: Confluent reductions: Abstract properties and applications to term rewriting systems. Journal of the ACM 27(4), 797–821 (1980)
10. Klein, D., Hirokawa, N.: Confluence of non-left-linear TRSs via relative termination. In: Bjørner, N., Voronkov, A. (eds.) LPAR-18 2012. LNCS, vol. 7180, pp. 258–273. Springer, Heidelberg (2012)
11. van Oostrom, V.: Confluence by decreasing diagrams. Theoretical Computer Science 126(2), 259–280 (1994)
12. van Oostrom, V.: Confluence by decreasing diagrams. In: Voronkov, A. (ed.) RTA 2008. LNCS, vol. 5117, pp. 306–320. Springer, Heidelberg (2008)
13. Standard ML of New Jersey, `http://www.sml.org/`
14. Suzuki, T., Aoto, T., Toyama, Y.: Confluence proofs of term rewriting systems based on persistency. Computer Software 30(3), 148–162 (2013) (in Japanese)
15. Toyama, Y.: Commutativity of term rewriting systems. In: Fuchi, K., Kott, L. (eds.) Programming of Future Generation Computers II, pp. 393–407. North-Holland, Amsterdam (1988)
16. Toyama, Y.: Membership conditional term rewriting systems. IEICE Transactions E72(11), 1224–1229 (1989)
17. Toyama, Y., Oyamaguchi, M.: Conditional linearization of non-duplicating term rewriting systems. IEICE Trans. Information and Systems E84-D(5), 439–447 (2001)
18. Zankl, H., Felgenhauer, B., Middeldorp, A.: CSI – A confluence tool. In: Bjørner, N., Sofronie-Stokkermans, V. (eds.) CADE 2011. LNCS, vol. 6803, pp. 499–505. Springer, Heidelberg (2011)
19. Zankl, H., Felgenhauer, B., Middeldorp, A.: Labelings for decreasing diagrams. In: Proc. of 22nd RTA. LIPIcs, vol. 10, pp. 377–392. Schloss Dagstuhl (2011)
20. Zantema, H.: Termination of term rewriting: interpretation and type elimination. Journal of Symbolic Computation 17, 23–50 (1994)

Predicate Abstraction of Rewrite Theories

Kyungmin Bae and José Meseguer

Department of Computer Science,
University of Illinois at Urbana-Champaign, Urbana IL 61801
{kbae4,meseguer}@cs.uiuc.edu

Abstract. For an infinite-state concurrent system $\mathcal{S}$ with a set *AP* of state predicates, its *predicate abstraction* defines a finite-state system whose states are subsets of *AP*, and its transitions $s \to s'$ are witnessed by concrete transitions between states in $\mathcal{S}$ satisfying the respective sets of predicates s and s'. Since it is not always possible to find such witnesses, an over-approximation adding extra transitions is often used. For systems $\mathcal{S}$ described by formal specifications, predicate abstractions are typically built using various automated deduction techniques. This paper presents a new method—based on rewriting, semantic unification, and variant narrowing—to automatically generate a predicate abstraction when the formal specification of $\mathcal{S}$ is given by a conditional rewrite theory. The method is illustrated with concrete examples showing that it naturally supports abstraction refinement and is quite accurate, i.e., it can produce abstractions not needing over-approximations.

1 Introduction

To automatically verify a temporal logic property φ of an infinite-state system $\mathcal{S}$ by model checking, two methods can be used: (i) we can try to use an *infinite-state* model checking procedure to *directly* verify that $\mathcal{S} \models \varphi$, or (ii) we can *abstract* $\mathcal{S}$ into a *finite-state* system $\mathcal{A}$, verify that $\mathcal{A} \models \varphi$ holds using a standard model checking method, and use general abstraction results to *prove* that this ensures $\mathcal{S} \models \varphi$ for the original system $\mathcal{S}$. These two methods have complementary strengths. On the one hand, a direct, infinite-state model checking method can settle the question once and for all, but may not always terminate, and may only be applicable under some restrictions on $\mathcal{S}$ and φ. On the other hand, abstractions allow the use of efficient finite-state model checking algorithms and tools, but may exhibit *spurious counterexamples*; that is, counterexamples showing that $\mathcal{A} \not\models \varphi$, while in fact $\mathcal{S} \models \varphi$ holds. In such cases, *abstraction refinement* [6] can be used to find a less abstract, yet still finite, abstraction $\mathcal{A}'$ where we may show $\mathcal{A}' \models \varphi$ if indeed φ holds for the system $\mathcal{S}$.

Predicate abstraction is one of the simplest and most widely used abstraction methods. We assume that several *state predicates* $AP = \{p_1, \ldots, p_n\}$ are defined on the states of $\mathcal{S}$. Then, the abstract system $\mathcal{S}/AP$ has set of states $\mathcal{P}(AP)$, and a transition $s \to s'$ exists for $s, s' \in \mathcal{P}(AP)$ iff there exists a concrete transition $u \to v$ in $\mathcal{S}$ such that u (resp. v) satisfies exactly the predicates in s (resp. in s').

G. Dowek (ed.): RTA-TLCA 2014, LNCS 8560, pp. 61–76, 2014.

Since it may not be always possible to prove the existence of such a concrete transition $u \to v$, an *over-approximation* $\alpha(\mathcal{S}/AP)$, which adds extra transitions when in doubt, may instead be used. Predicate abstraction can be used either for: (i) systems $\mathcal{S}$ defined by *programs* in a conventional language like C, or (ii) systems $\mathcal{S}$ described by *formal specifications*, in which case some theorem proving techniques are used to build a predicate abstraction $\alpha(\mathcal{S}/AP)$.

This paper presents an automatic method to build a predicate abstraction $\alpha(\mathcal{R}/AP)$ for a concurrent system specified as a *rewrite theory* $\mathcal{R} = (\Sigma, E, R)$ with equations E and rewrite rules R. We systematically exploit rewriting and unification techniques to *automate* the predicate abstraction process by a method that, as we illustrate with examples, has a good chance of building abstractions that are as *accurate* as possible. What is exploited is *both* the simplicity of the specification as a rewrite theory $\mathcal{R}$, and the power and automatic nature of rewriting and unification techniques, including the recently developed *generic* method to automatically derive unification algorithms by *variant narrowing* [14], already available in Maude. Although several abstraction methods have been studied before for rewrite theories (for a detailed survey see § 3.12 in [20]), except for the technical report [23] based on theorem proving, predicate abstraction of rewrite theories has remained undeveloped until now.

Our Contributions. Our main contributions are as follows:

- *Automatic method*: unlike the quite different semi-automatic method in [23], which generated proof obligations for an interactive theorem prover, our method is fully automatic, although it may produce an over-approximation.
- *Wide applicability beyond narrowing-based methods*: narrowing-based model checking techniques (e.g., in [4,13]) require equations E to have a finitary E-unification algorithm and rules R to be unconditional; no such restrictions apply to $\mathcal{R} = (\Sigma, E, R)$ in our abstraction method; and
- *Effective unsatisfiability checking procedure*, which, although incomplete, is automatic and can be used in practice to prove a conjunction of equations E-unsatisfiable. We believe this method can be useful not just for building predicate abstractions, but for other formal verification tasks.

Related Work. Predicate abstraction was first introduced in [16], and has been widely applied to both conventional programming languages, e.g., [5,25], and formal specifications, e.g., [10,17]. However, predicate abstraction for rewrite theories has *not* been much developed, except for the semi-automatic method in [23] using theorem proving. This work is also related to existing rewriting-based methods for model checking infinite-state systems, for example, using equational abstractions [21], narrowing [4,13], or tree automata [15,22]. The present work differs from and complements those approaches in that predicate abstraction always reduces such an infinite-state system to a finite-state one. Finally, our method to check unsatisfiability of equational constraints is related to the techniques in [2,12] to check unfeasiblity of conditional critical pairs in the context of proving confluence of conditional rewrite rules.

2 Preliminaries

Equational Theories. An order-sorted signature is a triple $\Sigma = (S, \leq, \Sigma)$ with poset of sorts $(S, \leq)$ and operators $\Sigma = \{\Sigma_{w,s}\}_{(w,s)\in S^*\times S}$ typed in $(S, \leq)$. A *Σ-algebra* A consists of a set A_{s} for each sort s, a subset inclusion $A_{\mathsf{s}} \subseteq A_{\mathsf{s}'}$ for each subsort relation $\mathsf{s} \leq \mathsf{s}'$, and a function $A_f : A_{\mathsf{s}_1} \times \cdots \times A_{\mathsf{s}_n} \to A_{\mathsf{s}}$ for each operator $f \in \Sigma_{\mathsf{s}_1\cdots\mathsf{s}_n,\mathsf{s}}$. The set $\mathcal{T}_\Sigma(\mathcal{X})_{\mathsf{s}}$ denotes the set of Σ-terms of sort s over $\mathcal{X}$ an infinite set of S-sorted variables, and $\mathcal{T}_{\Sigma,\mathsf{s}}$ denotes the set of ground Σ-terms of sort s. We assume that $\mathcal{T}_{\Sigma,\mathsf{s}} \neq \emptyset$ for each sort s in Σ. A *substitution* $\sigma : Y \to \mathcal{T}_\Sigma(\mathcal{X})$ is a function that maps variables $Y \subseteq \mathcal{X}$ to terms of the same sort, and is homomorphically extended to $\mathcal{T}_\Sigma(\mathcal{X})$ in a natural way.

A *Σ-equation* is an unoriented pair $t = t'$ of terms $t, t' \in \mathcal{T}_\Sigma(\mathcal{X})_{\mathsf{s}}$ for some sort $s \in \Sigma$. For a Σ-algebra A and a valuation $a : \mathcal{X} \to A$ assigning a value in A_{s} to each variable $x \in \mathcal{X}$ of sort s, if $\bar{a} : \mathcal{T}_\Sigma(\mathcal{X}) \to A$ is the homomorphic extension of a to terms, by definition, $(A, a) \models t = t'$ iff $\bar{a}(t) = \bar{a}(t')$. An *equational theory* (Σ, E) with a set of Σ-equations E induces a congruence relation $=_E$ such that $u =_E u'$ iff $(\Sigma, E) \vdash (\forall \mathcal{X})\ u = u'$ iff $(A, a) \models u = u'$ for any model A of E (i.e., for each equation $t = t' \in E$ and valuation a, $(A, a) \models t = t'$ holds) [18].

An *E-unifier* of an equation $t = t'$ is a substitution σ such that $\sigma t =_E \sigma t'$. A set $CSU_E(t = t')$ is a *complete set of E-unifiers* in which any E-unifier ρ of $t = t'$ has a more general E-unifier $\sigma \in CSU_E(t = t')$, i.e., $(\exists \eta)\ \rho =_E \eta \circ \sigma$. For a set of equations E and a set of equational axioms B (such as associativity, commutativity, and identity), if $E \cup B$ has the *finite variant property*, there is a finitary $E \cup B$-unification algorithm to find *finite* $CSU_{E\cup B}(t = t')$ [9,14]. As explained in [9], $E \cup B$ has the finite variant property iff for every term t there is a bound n such that the normal form of θt for a normalized substitution θ is reachable from t by applying E modulo B less than n times.

Rewrite Theories. A rewrite theory is a formal specification of a concurrent system [19]. We consider *order-sorted rewrite theories* $\mathcal{R} = (\Sigma, E, R)$ with (Σ, E) an equational theory and R a set of *rewrite rules* of the form $l \longrightarrow r$ **if** C, where $l, r \in \mathcal{T}_\Sigma(\mathcal{X})_{\mathsf{s}}$ for some sort s and C is a conjunction of Σ-equations. The system's states are modeled as the initial algebra $\mathcal{T}_{\Sigma/E}$ (i.e., each state is an E-equivalence class $[t]_E \in \mathcal{T}_{\Sigma/E}$ of ground terms). Each rule $(l \longrightarrow r$ **if** $C) \in R$ defines a *one-step rewrite* $t \longrightarrow_{R,E} t'$, specifying the system's transitions, iff there exist a subterm u of t and a substitution σ such that $u =_E \sigma l$ and t' is the term obtained from t by replacing u by σr, provided that $\sigma u =_E \sigma v$ for each condition $u = v$ in C.

A rewrite theory $\mathcal{R}$ is called *topmost* iff there exists a sort State at the top of one of the connected component of $(S, \leq)$ such that: (i) for each rewrite rule $l \longrightarrow r$ **if** C, both l and r have the top sort State; and (ii) no operator in Σ has State or any of its subsorts as an argument sort. This ensures that all rewrites with rules in R must take place at the top of the term. Throughout this paper we assume that E decomposes as a disjoint union $E = E_o \cup B$ with B a set of equational axioms (such as associativity, commutativity, and identity) and E_o *convergent* (i.e., sort-decreasing, terminating, confluent, and coherent modulo B) [11], and that R is coherent with E_o module B [24].

LTL Model Checking. A linear temporal logic (LTL) formula is constructed by state propositions $p \in AP$ and temporal logic operators such as $\neg$ (negation), $\wedge$, $\vee$, $\rightarrow$ (implication), $\Box$ (always), $\Diamond$ (eventually), **U** (until), and $\bigcirc$ (next). The semantics of LTL is defined on a *Kripke structure* $\mathcal{K} = (S, AP, \mathcal{L}, \longrightarrow_{\mathcal{K}})$ [7], where S is a set of *states*, AP is a set of *state propositions*, $\mathcal{L} : S \rightarrow \mathcal{P}(AP)$ is a *state-labeling function*, and $\longrightarrow_{\mathcal{K}} \subseteq S \times S$ is a *transition relation.*

Given two Kripke structures $\mathcal{K}_i = (S_i, AP, \mathcal{L}_i, \longrightarrow_{\mathcal{K}_i})$, $i = 1, 2$, a relation $H \subseteq S_1 \times S_2$ is a *simulation* iff (i) if $s_1 \, H \, s_2$, then $\mathcal{L}_1(s_1) = \mathcal{L}_2(s_2)$, and (ii) if $s_1 \, H \, s_2$ and $s_1 \longrightarrow_{\mathcal{K}_1} s_1'$, there exists $s_2' \in S_2$ such that $s_1' \, H \, s_2'$ and $s_2 \longrightarrow_{\mathcal{K}_2} s_2'$. A simulation $H \subseteq S_1 \times S_2$ is *total* iff for each $s_1 \in S_1$ there exists $s_2 \in S_2$ such that $s_1 H s_2$, and H is a *bisimulation* iff both H and H^{-1} are simulations.

Lemma 1. *[7] Given a simulation H between two Kripke structures $\mathcal{K}_1$ and $\mathcal{K}_2$, if $s_0^1 \, H \, s_0^2$, then for any LTL formula φ, $\mathcal{K}_2, s_0^2 \models \varphi$ implies $\mathcal{K}_1, s_0^1 \models \varphi$.*

We can associate to a rewrite theory $\mathcal{R} = (\Sigma, E, R)$ a corresponding Kripke structure. A state proposition p is defined as a term of sort Prop, whose meaning is defined by equations using the auxiliary operator $\_\models\_$: State Prop $\rightarrow$ Bool. By definition, $p \in \mathcal{T}_{\Sigma/E,\mathsf{Prop}}$ is satisfied on $[t]_E$ iff $(t \models p) =_E$ *true*. We assume that: (i) sort Bool has two constants *true* and *false* with *true* $\neq_E$ *false*; (ii) any $t \in \mathcal{T}_{\Sigma,\mathsf{Bool}}$ is provably equal to either *true* or *false*; and (iii) $\mathcal{R}$ is deadlock-free, since $\mathcal{R}$ can be easily transformed into an equivalent deadlock-free theory [21].

Definition 1. *Given $\mathcal{R} = (\Sigma, E, R)$ and a set $AP \subseteq \mathcal{T}_{\Sigma/E,\mathsf{Prop}}$ defined by E, the corresponding Kripke structure is $\mathcal{K}(\mathcal{R})_{AP} = (\mathcal{T}_{\Sigma/E,\mathsf{State}}, AP, \mathcal{L}_{AP}, \longrightarrow_{R,E})$, where $\mathcal{L}_{AP}([t]_E) = \{p \in AP \mid (t \models p) =_E \mathit{true}\}$.*

3 *AP*-Abstractions of Rewrite Theories

This section shows how a predicate abstraction of a topmost rewrite theory $\mathcal{R} = (\Sigma, E, R)$ can be constructed by using E-unification. Following the usual predicate abstraction methods, we consider abstract states as subsets of a set of state propositions $AP = \{p_1, \ldots, p_n\}$, where a transition $s \longrightarrow s' \in \mathcal{P}(AP)^2$ is defined if there exists a *concrete* one-step rewrite $t \longrightarrow_{R,E} t'$ such that:

$$s = \{p \in AP \mid (t \models p) =_E \mathit{true}\} \ \wedge \ s' = \{p \in AP \mid (t' \models p) =_E \mathit{true}\} \qquad (1)$$

Our approach is motivated by the following observation. For a topmost rewrite theory $\mathcal{R} = (\Sigma, E, R)$, a *concrete* one-step rewrite $t \longrightarrow_{R,E} t'$ exists iff for a rule $(l \longrightarrow r \textbf{ if } C) \in R$ and a *ground* substitution σ:

$$t =_E \sigma l \quad \wedge \quad t' =_E \sigma r \quad \wedge \quad (\forall u = v \in C) \ \sigma u =_E \sigma v \qquad (2)$$

The abstraction $s \longrightarrow s'$ of the transition $t \longrightarrow_{R,E} t'$ is given by Condition (1). Since $t =_E \sigma l$ and $t' =_E \sigma r$ by Condition (2), we can replace t and t' in (1) by σl and σr, respectively, and we obtain: $s = \{p \in AP \mid (\sigma l \models p) =_E \mathit{true}\}$, $s' = \{p \in AP \mid (\sigma r \models p) =_E \mathit{true}\}$, and $s \longrightarrow s'$ **if** $(\forall u = v \in C) \ \sigma u =_E \sigma v$. That is, $s \longrightarrow s'$ holds if there exist a rewrite rule $(l \longrightarrow r \textbf{ if } C) \in R$ and a ground substitution σ that satisfy these E-equality constraints.

Abstract Kripke Structures. We construct an *AP*-abstract Kripke structure of $\mathcal{R}$, whose states are elements of $\mathcal{P}(AP)$, and whose transitions are decided by solving E-equality constraints. Let $p \in \_ : \mathcal{P}(AP) \to \{true, false\}$ be the truth function such that $(p \in s) = true$ iff $p \in s$, and $(p \in s) = false$ iff $p \notin s$.

Definition 2. *Given a topmost rewrite theory* $\mathcal{R} = (\Sigma, E, R)$ *and a finite set of state propositions AP defined by E, the AP-*abstract Kripke structure *of* $\mathcal{R}$ *is a Kripke structure* $\mathcal{R}/AP = (\mathcal{P}(AP), AP, id_{\mathcal{P}(AP)}, \longrightarrow_{\mathcal{R}/AP})$*, where:*

- $id_{\mathcal{P}(AP)} : \mathcal{P}(AP) \to \mathcal{P}(AP)$ *is the identity labeling function, and*
- $s \longrightarrow_{\mathcal{R}/AP} s'$ *iff there exists a rewrite rule* $(l \longrightarrow r \textbf{ if } C) \in R$ *such that the following constraints are E-satisfiable:*

$$\bigwedge_{p \in AP}(l \models p) = (p \in s) \;\wedge\; \bigwedge_{p \in AP}(r \models p) = (p \in s') \;\wedge\; \bigwedge_{u=v \in C} u = v \qquad (\dagger)$$

If there exists a finitary E-unification algorithm (e.g., E has the finite variant property), the satisfiability of (†) can be decided by checking for the emptiness of the *finite* complete set of the E-unifiers:

$$CSU_E(\textstyle\bigwedge_{p \in AP}(l \models p) = (p \in s) \wedge \bigwedge_{p \in AP}(r \models p) = (p \in s') \wedge \bigwedge_{u=v \in C} u = v).$$

However, checking satisfiability of the constraints (†) by E-unification is in general undecidable [3]. Therefore, $\mathcal{R}/AP$ may *not* have an effective procedure to *precisely* decide its transitions. In practice, there are three cases:

1. For *some* rule in R, we can prove the *satisfiability* of (†) for states s and s', in which case we know that $s \longrightarrow_{\mathcal{R}/AP} s'$ holds.
2. For *every* rule in R, we can prove the *unsatisfiability* of (†) for states s and s', in which case we know that $s \longrightarrow_{\mathcal{R}/AP} s'$ does *not* hold.
3. Otherwise, we cannot decide whether $s \longrightarrow_{\mathcal{R}/AP} s'$ holds or not. In this case we can *add* $s \longrightarrow_{\alpha(\mathcal{R}/AP)} s'$ to a Kripke structure $\alpha(\mathcal{R}/AP)$ *approximating* (and therefore simulating) $\mathcal{R}/AP$ with possibly more transitions.

By definition, a Kripke structure $\alpha(\mathcal{K}) = (S, AP, \mathcal{L}, \longrightarrow_{\alpha(\mathcal{K})})$ is an *approximation* of $\mathcal{K} = (S, AP, \mathcal{L}, \longrightarrow_{\mathcal{K}})$ iff $\longrightarrow_{\mathcal{K}} \subseteq \longrightarrow_{\alpha(\mathcal{K})}$. In Section 4 we propose some procedures for checking satisfiability of the constraints (†).

Theorem 1. *For an LTL formula* φ *and a state* $t \in \mathcal{T}_{\Sigma,\mathsf{State}}$*, if* $s = \mathcal{L}_{AP}([t]_E)$*, then* $\alpha(\mathcal{R}/AP), s \models \varphi$ *implies* $\mathcal{K}(\mathcal{R})_{AP}, [t]_E \models \varphi$.

Proof. It suffices to show that the state-labeling function $\mathcal{L}_{AP}$ of $\mathcal{K}(\mathcal{R})_{AP}$, where $\mathcal{L}_{AP}([u]_E) = \{p \in AP \mid (u \models p) =_E true\}$, is a total simulation from $\mathcal{K}(\mathcal{R})_{AP}$ to $\mathcal{R}/AP$. Then, $(id_{\mathcal{P}(AP)} \circ \mathcal{L}_{AP}) = \mathcal{L}_{AP}$ becomes a simulation from $\mathcal{K}(\mathcal{R})_{AP}$ to $\alpha(\mathcal{R}/AP)$, since $id_{\mathcal{P}(AP)}$ is a simulation from $\mathcal{R}/AP$ to $\alpha(\mathcal{R}/AP)$.

Suppose that $t \longrightarrow_{R,E} t'$ exists in $\mathcal{K}(\mathcal{R})_{AP}$. By definition, for a rewrite rule $(l \longrightarrow r \textbf{ if } C) \in R$, there is a ground substitution σ such that $t =_E \sigma l$, $t' =_E \sigma r$, and for each $u = v \in C$, $\sigma u =_E \sigma v$. Let $s = \mathcal{L}_{AP}([t]_E)$ and $s' = \mathcal{L}_{AP}([t']_E)$. Since $t =_E \sigma l$ and $t' =_E \sigma r$, we have $s = \{p \in AP \mid (\sigma l \models p) =_E true\}$ and $s' = \{p \in AP \mid (\sigma r \models p) =_E true\}$. Hence, $\bigwedge_{p \in AP}(\sigma l \models p) =_E (p \in s)$ and $\bigwedge_{p \in AP}(\sigma r \models p) =_E (p \in s')$ hold. That is, σ is a solution of (†), so that $s \longrightarrow_{\mathcal{R}/AP} s'$. Therefore, $\mathcal{L}_{AP}$ is a total simulation from $\mathcal{K}(\mathcal{R})_{AP}$ to $\mathcal{R}/AP$. □

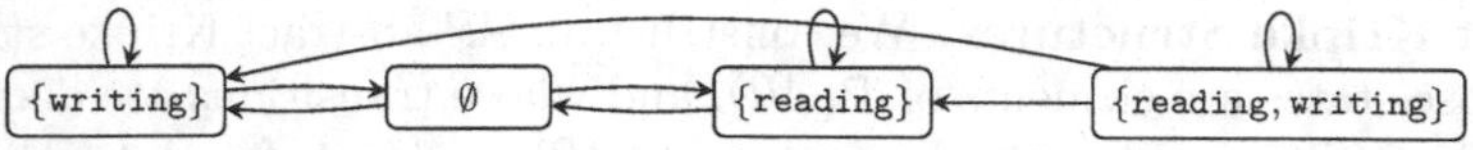

Fig. 1. The *AP*-abstract Kripke structure with the initial state ∅

Rule	$\langle 0,0\rangle \to \langle 0,s(0)\rangle$	$\langle r,0\rangle \to \langle s(r),0\rangle$	$\langle r,s(w)\rangle \to \langle r,w\rangle$	$\langle s(r),w\rangle \to \langle r,w\rangle$
$\emptyset \to$ {w}	$\langle 0,0\rangle \to \langle 0,s(0)\rangle$			
$\emptyset \to$ {r}		$\langle 0,0\rangle \to \langle s(0),0\rangle$		
{w} $\to \emptyset$			$\langle 0,s(0)\rangle \to \langle 0,0\rangle$	
{w} $\to$ {w}			$\langle 0,s(s(w))\rangle \to \langle 0,s(w)\rangle$	
{r} $\to \emptyset$				$\langle s(0),0\rangle \to \langle 0,0\rangle$
{r} $\to$ {r}		$\langle s(r),0\rangle \to \langle s(s(r)),0\rangle$		$\langle s(s(r)),0\rangle \to \langle s(r),0\rangle$
{r,w} $\to$ {w}				$\langle s(0),s(w)\rangle \to \langle 0,s(w)\rangle$
{r,w} $\to$ {r}			$\langle s(r),s(0)\rangle \to \langle s(r),0\rangle$	
{r,w} $\to$ {r,w}			$\langle s(r),s(s(w))\rangle \to \langle s(r),s(w)\rangle$	$\langle s(s(r)),s(w)\rangle \to \langle s(r),s(w)\rangle$

Fig. 2. The rule instances for each abstract transition $s \to s' \in \mathcal{P}(AP)^2$

Example. We illustrate our ideas with a simple model of the readers-writers problem (adapted from [8]). Each state is modeled as a pair $\langle r,w\rangle \in \mathbb{N}^2$ in which r is the number of readers and w is the number of writers, given by the operator $\langle \_,\_\rangle$: Nat Nat $\to$ State. Natural numbers are expressed in Peano notation using the successor function s : Nat $\to$ Nat and the zero constant 0 of sort Nat. The behavior is defined by the following four unconditional rewrite rules: $\langle 0,0\rangle \to \langle 0,s(0)\rangle$, $\langle r,0\rangle \to \langle s(r),0\rangle$, $\langle r,s(w)\rangle \to \langle r,w\rangle$, and $\langle s(r),w\rangle \to \langle r,w\rangle$. Mutual exclusion is expressed by the LTL formula $\Box\neg(\mathit{reading} \wedge \mathit{writing})$, where the state propositions are defined by the equations: $\langle s(r),w\rangle \models \mathit{reading} = \mathit{true}$, $\langle 0,w\rangle \models \mathit{reading} = \mathit{false}$, $\langle r,s(w)\rangle \models \mathit{writing} = \mathit{true}$, and $\langle r,0\rangle \models \mathit{writing} = \mathit{false}$, satisfying the finite variant property since the right-hand sides are constants.

This system is infinite-state, since the number of readers r is unbounded. However, we obtain the *finite* *AP*-abstract Kripke structure $\mathcal{R}/AP$ in Figure 1, whose transitions are decided by using E-unification. Figure 2 shows the rule instances $\zeta l \to \zeta r$ for each rule $l \longrightarrow r \in R$, transition $s \longrightarrow s' \in \mathcal{P}(AP)^2$, and E-unifier $\zeta \in CSU_E(\bigwedge_{p\in AP}(l \models p) = (p \in s) \wedge \bigwedge_{p\in AP}(r \models p) = (p \in s'))$ that represent the transitions of $\mathcal{R}/AP$. The property $\Box\neg(\mathit{reading} \wedge \mathit{writing})$ holds from the initial state $\emptyset$ in $\mathcal{R}/AP$, since $\{\mathit{reading}, \mathit{writing}\}$ is *not* reachable. Therefore, $\Box\neg(\mathit{reading} \wedge \mathit{writing})$ also holds from $\langle 0,0\rangle$ in $\mathcal{K}(\mathcal{R})_{AP}$, thanks to Theorem 1.

If we consider the LTL formula $\Box\Diamond\neg\mathit{writing}$ (i.e., infinitely often not writing), there exists the spurious counterexample $\emptyset \to \{\mathit{writing}\} \to \{\mathit{writing}\} \to \cdots$. As usual for predicate abstraction methods, we can *refine* the *AP*-abstraction by adding the state proposition *1w* to further specialize the abstract state space, meaning that there exists only one writer, defined by the three equations: $\langle R,s(0)\rangle \models \mathit{1w} = \mathit{true}$, $\langle R,0\rangle \models \mathit{1w} = \mathit{false}$, and $\langle R,s(s(w))\rangle \models \mathit{1w} = \mathit{false}$. We have then the refined *AP*-abstract Kripke structure in Figure 3 in which the formula $\Box\Diamond\neg\mathit{writing}$ holds from the initial state $\emptyset$. Again, by Theorem 1, $\Box\Diamond\neg\mathit{writing}$ also holds from $\langle 0,0\rangle$ in the concrete Kripke structure $\mathcal{K}(\mathcal{R})_{AP}$.

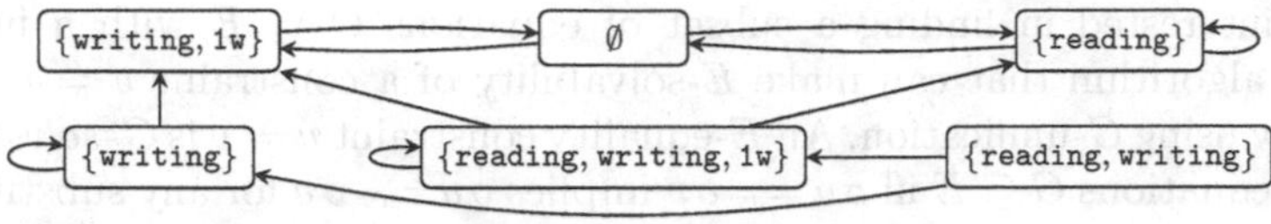

Fig. 3. Refined *AP*-abstract Kripke structure for the readers-writers problem

4 Effective Procedures for Equality Constraints

In Section 3 we defined a predicate abstraction $\mathcal{R}/AP$ of a topmost rewrite theory $\mathcal{R} = (\Sigma, E, R)$ using E-equality constraints when the set of state propositions AP is specified by E. If there exists a finitary E-unification algorithm, such as E satisfying the finite variant property, then we can automatically construct $\mathcal{R}/AP$ by using E-unification. This section shows how to check the *unsatisfiability* of E-equality constraints. Since the problem is in general semi-decidable [3], we are interested in a *sound but incomplete* terminating procedures that can be automated, so that an over-approximation $\alpha(\mathcal{R}/AP)$ can be built when the procedure fails to give an answer. A semi-decidable E-unification algorithm does not give such a procedure because it may *not* terminate when no solution exists.

Our method relies on the fact that a state proposition p is typically defined using only a subset of equations $E_p \subseteq E$, so that solving constraints for p may only involve E_p, *not* all of E. If a finitary E_p-unification algorithm is available, then we can discharge the constraints for p using E_p-unification. After resolving all such *solvable* constrains, we apply a sound procedure based on E-reduction to test if the remaining constraint are inconsistent or not.

Decomposition of Constraints. In practice, the equational semantics of a state proposition p can be restricted to a subset of the equations $E_p \subseteq E$. As assumed throughout this paper, E decomposes as a disjoint union $E = E_o \cup B$ with B a set of equational axioms and E_o convergent modulo B.

Definition 3. *Let $\Omega \subseteq \Sigma$ be a set of free constructors for E_o modulo B. Given a set of patterns $U = \{u_1, \ldots, u_n\} \subseteq \mathcal{T}_\Sigma(\mathcal{X})$, we define:*

$$[\![U]\!] = \{t \in \mathcal{T}_\Sigma(\mathcal{X}) \mid (\exists \sigma : \mathcal{X} \to \mathcal{T}_\Omega(\mathcal{X}), \exists u \in U)\ t =_B \sigma u\},$$
$$E_U = \{u = v \in E_o \mid u, v \in [\![U]\!]\}.$$

We call E_o syntactically independent *with respect to U iff E_U is convergent modulo B and satisfies: (i) any proper subterm v of $u \in U$ is strongly E_o, B-irreducible (i.e., γv is a normal form for any normalized substitution γ), and (ii) for every $t = t' \in E - E_U$ and $u \in U$, $CSU_B(t = u) = \emptyset$.*

For the readers-writers problem in Section 3, every state proposition p has its syntactically independent patterns U_p; e.g., for the state proposition *reading*, for $U_{reading} = \{(S \models reading), true, false\}$ for a variable S of sort State, where $E_{U_{reading}} = \{\langle s(r), w\rangle \models reading = true,\ \langle 0, w\rangle \models reading = false\}$.

We are interested in finding a subset of equations $G \subseteq E$ with a finitary unification algorithm that can make E-solvability of a constraint $u = v$ in (†) decidable by using G-unification. An E-equality constraint $u = v$ is *G-solvable* for a subset of equations $G \subseteq E$ iff $\sigma u =_E \sigma v$ implies $\sigma u =_G \sigma v$ for any substitution σ. Since $G \subseteq E$, if $u = v$ is G-solvable, then $\sigma u =_E \sigma v \iff \sigma u =_G \sigma v$.

Proposition 1. *If E_o is syntactically independent with respect to U, then for any $u, v \in [\![U]\!]$, an E-equality constraint $u = v$ is $E_U \cup B$-solvable.*

Proof. Since we assume that $\mathcal{T}_\Sigma(\mathcal{X})_{\mathsf{s}} \neq \emptyset$ for each sort s, a constraint $u = v$ is $E_o \cup B$-solvable iff for some normalized ground substitution $\sigma : \mathcal{X} \to \mathcal{T}_\Omega$, $can_{E_o/B}(\sigma u) =_B can_{E_o/B}(\sigma v)$, where $can_{E_o/B}(t)$ denotes an E_o/B-canonical form of t. Since $\sigma u, \sigma v \in [\![U]\!]$, by Ω-terms being free modulo B, convergence of E_U modulo B, and (i)–(ii) in Definition 3, $can_{E_o/B}(\sigma u) =_B can_{E_U/B}(\sigma u)$ and $can_{E_o/B}(\sigma v) =_B can_{E_U/B}(\sigma v)$ hold. The lifting lemma for narrowing modulo B then forces $u = v$ to be $E_U \cup B$-solvable by narrowing with E_U modulo B. □

Even when $E_U \cup B$ does *not* have the finite variant property, there may still exist a subset $\tilde{E}_U \subseteq E_U$ where $\tilde{E}_U \cup B$ has the finite variant property and a constraint $u = v$ is $\tilde{E}_U \cup B$-solvable (see Section 5 for an example).

For a set of E-equality constraints D, if $G \subseteq E$ has a finitary G-unification algorithm and a constraint $u = v \in D$ is G-solvable, then $CSU_G(u = v)$ is finite, and $u = v$ is E-satisfiable iff $CSU_G(u = v) \neq \emptyset$. Therefore, we can decompose the problem D into finding $\zeta \in CSU_G(u = v)$ and solving the remaining constraints.

Lemma 2. *Given a set of E-equality constraints D, if $G \subseteq E$ has a finitary unification algorithm and $u = v \in D$ is G-solvable, then D is E-satisfiable iff for some $\zeta \in CSU_G(u = v)$, $\{\zeta u = \zeta v \mid u = v \in D - \{u = v\}\}$ is E-satisfiable*

Proof. Suppose that there exists a substitution ρ such that $\bigwedge_{u=v \in D} \rho u =_E \rho v$. Since $u = v$ is G-solvable, $\rho u =_{E_{FV}} \rho v$ also holds. Therefore, there exists a substitution $\zeta \in CSU_{E_{FV}}(u = v)$ such that $(\exists \eta)\ \rho =_{E_{FV}} \eta \circ \zeta$. Notice that $\rho =_E \eta \circ \zeta$, since $G \subseteq E$. Therefore, $\bigwedge_{u=v \in D-\{u=v\}} \eta(\zeta u) =_E \eta(\zeta v)$ holds. □

We can repeatedly apply this procedure to solve each G-solvable constraint in (†) to determine $s \longrightarrow_{\mathcal{R}/AP} s'$, provided that G has a finitary unification algorithm. If there exists *no* common solution, then $s \longrightarrow_{\mathcal{R}/AP} s'$ does *not* hold.

Unfeasibility of Constraints. Applying Lemma 2, we can transform a set D of E-equality constraints into an equivalent set F having *no* G-solvable constraints. We now present a sound but incomplete procedure to test for *E-unsatisfiability* of F. We have assumed that E decomposes as a disjoint union $E = E_o \cup B$, where E_o is convergent and a finitary B-unification algorithm exists, and $\mathcal{T}_{\Sigma,\mathsf{s}} \neq \emptyset$ for each sort $\mathsf{s} \in \Sigma$. Therefore, we can use the *canonical term algebra* $\mathcal{C}_{\Sigma/E_o,B}$, whose elements are B-equivalence classes of Σ-terms in E_o/B-canonical form, and which is an initial $E_o \cup B$-algebra. That is, F is E-satisfiable iff there exists a normalized ground substitution θ such that $(\mathcal{C}_{\Sigma/E_o,B}, q_B \circ \theta) \models F$, where $q_B : \mathcal{T}_\Sigma \to \mathcal{T}_{\Sigma/B}$ is the quotient map $t \mapsto [t]_B$ for B.

Given the set X of the variables in F, let $\bar{X} = \{\bar{x} \mid x \in X\}$ be the set in which each variable $x \in X$ of sort s is turned into the constant $\bar{x}$ of the same sort s, where $\bar{X} \cap \Sigma = \emptyset$, and let $\bar{F} = \{\bar{u} = \bar{v} \mid u = v \in F\}$ be the set of the *ground* constraints obtained from F by replacing each $x \in X$ by $\bar{x} \in \bar{X}$. Recall that a $\Sigma \cup \bar{X}$-algebra is exactly a pair (A, a) with a valuation $a : \bar{X} \to A$. Therefore, if $(\mathcal{C}_{\Sigma/E_o,B}, q_B \circ \theta) \models F$, then the valuation $q_B \circ \hat{\theta}$, with $\hat{\theta}(\bar{x}) = \theta(x)$ for $x \in X$, gives us a $\Sigma \cup \bar{X}$-algebra $(\mathcal{C}_{\Sigma/E_o,B}, q_B \circ \hat{\theta})$ satisfying *both* E and $\bar{F}$. Soundness of equational logic then ensures that whenever $(\Sigma \cup \bar{X}, E \cup \bar{F}) \vdash (\forall \emptyset)\ \bar{u} = \bar{v}$, where $vars(u) \cup vars(v) \subseteq X$, we must have $(\mathcal{C}_{\Sigma/E_o,B}, q_B \circ \hat{\theta}) \models (\forall \emptyset)\ \bar{u} = \bar{v}$.

Our sound procedure for testing unsatisfiability of F is based on the idea of obtaining a proof of the form $(\Sigma \cup \bar{X}, E \cup \bar{F}) \vdash (\forall \emptyset)\ \bar{u} = \bar{v}$, where u and v are *strongly* E_o, B*-irreducible* Σ-terms (i.e., each normalized substitution γ makes both γu and γv normalized) and $CSU_B(u = v) = \emptyset$. This gives a contradiction to the assumption that F is satisfiable; if F is satisfied by a normalized substitution θ, then $(\mathcal{C}_{\Sigma/E_o,B}, q_B \circ \hat{\theta}) \models \bar{u} = \bar{v}$, i.e., $[\theta u]_B = [\theta v]_B$, since u and v are strongly E_o, B-irreducible, but $CSU_B(u = v) = \emptyset$ implies $[\theta u]_B \neq [\theta v]_B$.

A practical way of obtaining such a proof $(\Sigma \cup \bar{X}, E \cup \bar{F}) \vdash (\forall \emptyset)\ \bar{u} = \bar{v}$ is by rewriting modulo B. That is, we can use the set of rewrite rules $E_o^{\rightarrow} \cup \bar{F}^{\rightarrow}$ with $E_o^{\rightarrow}$ the oriented equations and for a B-compatible order $\succ$ on ground terms:

$$\bar{F}^{\rightarrow} = \{can_{E_o/B}(\bar{w}) \to can_{E_o/B}(\bar{w}') \mid w = w' \in F \text{ or } w' = w \in F, \\ can_{E_o/B}(\bar{w}) \succ can_{E_o/B}(\bar{w}')\},$$

where $can_{E_o/B}(w)$ denotes an E_o/B-canonical form of w. Then, if we obtain $\bar{u} \longleftrightarrow^{*}_{E_o^{\rightarrow} \cup \bar{F}^{\rightarrow}, B} \bar{v}$ by rewriting modulo B, then by equational reasoning, we have a *fortiori* derived $(\Sigma \cup \bar{X}, E \cup \bar{F}) \vdash (\forall \emptyset)\ \bar{u} = \bar{v}$. In summary:

Theorem 2. *For a set of E-equality constraints F with $X = vars(F)$, if there exist strongly E_o, B-irreducible Σ-terms u and v such that $vars(u) \cup vars(v) \subseteq X$, $CSU_B(u = v) = \emptyset$, and $\bar{u} \longleftrightarrow^{*}_{E_o^{\rightarrow} \cup \bar{F}^{\rightarrow}, B} \bar{v}$, then F is E-unsatisfiable.*

Example. We consider a model of the readers-writers problem with explicit shared variables and processes, adapted from [1] (which is a more sophisticated version of the model in Section 3). Each state has the form $\langle n, b \mid p_1; \cdots; p_n \rangle$, given by the operator $\langle \_, \_ | \_ \rangle$: Nat Bool ProcSet $\to$ State, where n denotes the number of readers, b is a Boolean flag to denote *no* readers and *no* writers, and $p_1; \cdots; p_n$ is a multiset of processes, each in a status $p_i \in \{idle, read, write\}$. The behavior of the system is specified by the following (conditional) rewrite rules:

$$\begin{aligned}
\langle n, b \mid idle\ ; PS \rangle &\longrightarrow \langle s(n), b' \mid read\ ; PS \rangle \quad \textbf{if } c(n, b, b') = true \\
\langle s(n), b \mid read\ ; PS \rangle &\longrightarrow \langle n, b' \mid idle\ ; PS \rangle \quad \textbf{if } c(n, b', b) = true \\
\langle n, true \mid idle\ ; PS \rangle &\longrightarrow \langle n, false \mid write\ ; PS \rangle \\
\langle n, false \mid write\ ; PS \rangle &\longrightarrow \langle n, true \mid idle\ ; PS \rangle
\end{aligned}$$

where the function $c(n, b, b')$ returns *true* iff $b = true$ and $b' = false$ when $n = 0$, or $b = b'$ when $n > 0$, defined by the equations:

$$c(0, true, false) = true \qquad c(0, false, b) = false \qquad c(0, b, true) = false$$
$$c(s(n), b, b) = true \quad c(s(n), true, false) = false \quad c(s(n), false, true) = false$$

Mutual exclusion is expressed by the formula $\Box\neg(reading \wedge writing)$. We can define sorts ProcIdleSet (denoting processes in *idle*), ProcReadSet (for processes in *idle* or *read*), and ProcWriteSet (denoting processes in *idle* or *write*), having subsort relations ProcIdleSet $\leq$ ProcReadSet ProcWriteSet $\leq$ ProcSet. The state propositions are then defined by the following equations, where RS is a variable of sort ProcReadSet and WS is a variable of sort ProcWriteSet:

$$\langle n, b \mid read \;;\, PS\rangle \models reading = true \quad \langle n, b \mid WS\rangle \models reading = false$$
$$\langle n, b \mid write \;;\, PS\rangle \models writing = true \quad \langle n, b \mid RS\rangle \models writing = false$$

This system is finite-state for a *fixed* set of processes; but the system is actually infinite-state, since the number of processes is unbounded. We are interested in verifying $\Box\neg(reading \wedge writing)$ for an unbounded number of processes.

In order to verify $\Box\neg(reading \wedge writing)$, we need to have two additional state propositions *free* and *good*, defined by the following equations, where $good(n, PS)$ returns *true* iff n is equal to the number of readers in PS:

$$\langle n, true \mid PS\rangle \models free = true \qquad \langle n, false \mid PS\rangle \models free = false$$
$$\langle n, b \mid PS\rangle \models good = good(n, PS) \quad good(s(n), read \;;\, PS) = good(n, PS)$$
$$good(n, write \;;\, PS) = good(n, PS) \qquad good(n, idle \;;\, PS) = good(n, PS)$$
$$good(0, WS) = true \quad good(s(n), WS) = false \quad good(0, read \;;\, PS) = false$$

Notice that every state proposition has syntactically independent equations. Hence, we can easily see that every constraint for c, *reading*, *writing*, and *free* is solvable by a set of equations satisfying the finite variant property. However, the equations defining *good* do *not* have the finite variant property.

We can obtain a finite $\mathcal{R}/AP$ in which only two states $\{reading, good\}$ and $\{writing, good\}$ are reachable from the initial state $\{free, good\}$. For example, from $\{free, good\}$, after resolving each $E_p \cup B$-solvable constraint, we have the following sets of the remaining constraints:

$$\langle 0, true \mid idle \;;\, IS\rangle \models good = true \;\wedge\; \langle s(0), false \mid read \;;\, IS\rangle \models good = b_1,$$
$$\langle s(n), true \mid idle \;;\, IS\rangle \models good = true \;\wedge\; \langle s(s(n)), true \mid read \;;\, IS\rangle \models good = b_2,$$
$$\langle n, true \mid idle \;;\, IS\rangle \models good = true \;\wedge\; \langle n, false \mid write \;;\, IS\rangle \models good = b_3,$$

where IS has sort ProcIdleSet and $b_1, b_2, b_3 \in \{true, false\}$. By normalizing each constraint after replacing the variables into the constants, we have the oriented constraint sets $\{true \to true, true \to b_1\}$, $\{true \to false, b_2 \to false\}$, and $\{good(\overline{n}, \overline{IS}) \to true, good(\overline{n}, \overline{IS}) \to b_3\}$. Therefore, the cases of $b_1 = false$, $b_2 \in \{true, false\}$, and $b_3 = false$ are unsatisfiable. That is, $\{free, good\}$ has only two next states $\{reading, good\}$ and $\{writing, good\}$. Similarly, $\{reading, good\}$ has the next states $\{free, good\}$ and $\{reading, good\}$, and $\{writing, good\}$ has the next states $\{free, good\}$ and $\{writing, good\}$. Therefore, $\Box\neg(reading \wedge writing)$ holds in $\mathcal{R}/AP$ from $\{free, good\}$. Thanks to Theorem 1, the formula also holds in $\mathcal{K}(\mathcal{R})_{AP}$ for an unbounded number of processes.

```
findNextStates(s ∈ 𝒫(AP)):
  NextStates := ∅;
  foreach s' ∈ 𝒫(AP) do
    foreach (l ⟶ r if C) ∈ R do
      CTR := getConstraints(s, s', l, r, C); // CTR = the set of constraints (†)
      (D,F,G) := findSolvable(CTR); // D = a set of G-solvable equations
      if isSatisfiable(D,F,G) then add s' to NextStates;
  return NextStates;

isSatisfiable(D, F, G):
  if D ≠ ∅ then
    choose u = v from D;
    foreach ζ ∈ CSU_G(u = v) do
      if isSatisfiable(ζ(D − {u = v}), ζF, G) then return true;
    return false;
  else
    return (if F = ∅ then true else ¬ testUnsatisfiable(F));
```

Fig. 4. Predicate Abstraction Algorithm for $\mathcal{R} = (\Sigma, E, R)$

Predicate Abstraction Algorithm. Figure 4 shows an algorithm to generate a predicate abstraction $\alpha(\mathcal{R}/AP)$ of a rewrite theory $\mathcal{R} = (\Sigma, E, R)$. There exists a transition $s \longrightarrow s'$ in $\alpha(\mathcal{R}/AP)$ iff $s' \in$ findNextStates(s), where the function findNextStates(s) returns a set of next *abstract* states from $s \in \mathcal{P}(AP)$.

The function findNextStates uses a number of subroutines that correspond to methods in Sections 3 and 4. Given a set of E-equality constraints CTR, using Proposition 1, findSolvable(CTR) returns a triple (D, F, G) such that D is a set of G-solvable constraints, G has the finite variant property, and $CTR = D \cup F$. Then, using Lemma 2, isSatisfiable(D, F, G) returns *false* if the set of constraints $D \cup F$ is unsatisfiable, where testUnsatisfiable(F) returns *true* if F is shown to be unsatisfiable using Theorem 2. More details on each subroutine, as well as some demos based on several components of this algorithm, can be found at `http://formal.cs.illinois.edu/kbae/pred`.

5 Case Study

We illustrate our predicate abstraction method with another nontrivial infinite-state system, namely, Lamport's bakery protocol for mutual exclusion (adapted from [4,13]). Each state has the form $\langle n, m, A, [i_1, d_1] \ldots [i_n, d_n] \rangle$, given by the operator $\langle \_, \_, \_, \_ \rangle$: $\mathsf{Nat\ Nat\ Action\ ProcSet} \to \mathsf{State}$, where n is the current number in the bakery's number dispenser, m is the number currently being served, $A \in \{\bot, \mathit{wake}, \mathit{crit}, \mathit{exit}\}$ is the action taken in the previous step, and $[i_1, d_1] \ldots [i_n, d_n]$ is a multiset of customer processes, each with an identity i_j and in a mode d_j. A mode can be *idle* (not yet picked a number), $\mathit{wait}(n)$ (waiting with number n), or $\mathit{crit}(n)$ (being served with number n).

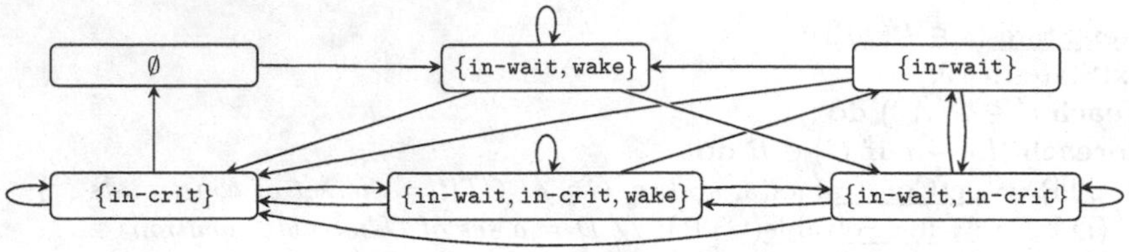

Fig. 5. *AP*-abstract Kripke structure for the livelock freedom property

The behavior is then specified by the following unconditional rewrite rules, where natural numbers are modeled as multisets of s with the multiset union operator $\_\_ : \mathsf{Nat}\ \mathsf{Nat} \to \mathsf{Nat}$ (empty syntax), satisfying laws of commutativity and associativity, and the empty multiset $\mathsf{0}$ (e.g., $0 = \mathsf{0}$, and $3 = s\,s\,s$):

$$\begin{aligned}\langle n, m, A, [i, \mathit{idle}]\, PS\rangle &\longrightarrow \langle s\,n, m, \mathit{wake}, [i, \mathit{wait}(n)]\, PS\rangle\\ \langle n, m, A, [i, \mathit{wait}(m)]\, PS\rangle &\longrightarrow \langle n, m, \mathit{crit}, [i, \mathit{crit}(m)]\, PS\rangle\\ \langle n, m, A, [i, \mathit{crit}(m)]\, PS\rangle &\longrightarrow \langle n, s\,m, \mathit{exit}, [i, \mathit{idle}]\, PS\rangle\end{aligned}$$

Notice that this system is infinite-state since: (i) the counters n and m are unbounded, and (ii) the number of customer processes is unbounded.

We can define three sorts ProcIdleSet (for processes in *idle*), ProcWaitSet (denoting processes in *idle* or $\mathit{wait}(n)$), and ProcCritSet (denoting processes in *idle* or $\mathit{crit}(n)$), yielding an order-sorted signature with subsort relations ProcIdleSet $\leq$ ProcWaitSet ProcCritSet $\leq$ ProcSet. We can define sort NWAction denoting the *non-wake* actions ($\bot$, *crit*, and *exit*), where NWAction $\leq$ Action.

Livelock Freedom. We are interested in verifying the livelock freedom "*if some process is waiting, then some (possibly different) process eventually enters the critical section*" under the fairness assumption "*wake* is *not* taken infinitely many times," expressed as the LTL formula $\Box\Diamond\neg \mathit{wake} \to \Box(\mathit{in\text{-}wait} \to \Diamond \mathit{in\text{-}crit})$. The state propositions are defined by the following equations that satisfy the finite variant property, where CS is a variable of sort ProcCritSet, WS is a variable of sort ProcWaitSet, and NA is a variable of sort NWAction:

$$\begin{aligned}\langle n, m, A, [i, \mathit{wait}(k)]\, PS\rangle \models \mathit{in\text{-}wait} = \mathit{true} &\quad \langle n, m, A, CS\rangle \models \mathit{in\text{-}wait} = \mathit{false}\\ \langle n, m, A, [i, \mathit{crit}(k)]\, PS\rangle \models \mathit{in\text{-}crit} = \mathit{true} &\quad \langle n, m, A, WS\rangle \models \mathit{in\text{-}crit} = \mathit{false}\\ \langle n, m, \mathit{wake}, PS\rangle \models \mathit{wake} = \mathit{true} &\quad \langle n, m, NA, PS\rangle \models \mathit{wake} = \mathit{false}.\end{aligned}$$

We obtain the *AP*-abstract Kripke structure in Figure 5 using E-unification, because a finitary E-unification algorithm is available. Since *in-crit* holds in every state in the second row, the formula $\Box\Diamond\neg \mathit{wake} \to \Box(\mathit{in\text{-}wait} \to \Diamond \mathit{in\text{-}crit})$ holds in the *AP*-abstract Kripke structure from the initial state $\emptyset$. Hence, by Theorem 1, the formula also holds in the concrete Kripke structure $\mathcal{K}(\mathcal{R})_{AP}$ from any state $[t]_E$ with $\mathcal{L}_{AP}([t]_E) = \emptyset$ for an unbounded number of processes.

Mutual Exclusion. We now consider the mutual exclusion "*at most one process can enter the critical section,*" expressed by the LTL formula $\Box mutex$ in which the state proposition is defined by the equations: $\langle n, m, A, WS\rangle \models mutex = true$, $\langle n, m, A, [i, crit(k)]\ WS\rangle \models mutex = true$, and

$$\langle n, m, A, [i, crit(k)]\ [i', crit(k')]\ PS\rangle \models mutex = false.$$

We need three extra state propositions to define a predicate abstraction: *mcrit*, *bound*, and *uniq*. First, *mcrit* holds in a state $\langle n, m, A, PS\rangle$ if at most one process enters the critical section with number m, defined by the equations: $\langle n, m, A, WS\rangle \models mcrit = true$, $\langle n, m, A, [i, crit(m)]\ WS\rangle \models mcrit = true$, and

$$\begin{aligned}\langle n, m, A, [i, crit(s\,k\,m)]\ WS\rangle &\models mcrit = false\\ \langle n, s\,k\,m, A, [i, crit(m)]\ WS\rangle &\models mcrit = false\\ \langle n, m, A, [i, crit(k)]\ [i', crit(k')]\ PS\rangle &\models mcrit = false.\end{aligned}$$

Next, the state proposition *bound* holds in a state $\langle n, m, A, PS\rangle$ if any ticket number of a process in PS is less than n, defined by the equations:

$$\left.\begin{array}{ll} & \langle n, m, A, PS\rangle \models bound = bd(n, PS)\\ bd(n, [i, wait(n\,k)]PS) = false & bd(n, [i, crit(n\,k)]\,PS) = false\end{array}\right] (E_b)$$

$$\begin{array}{ll} bd(n, IS) = true & bd(s\,n\,k, [i, wait(k)]\ PS) = bd(s\,n\,k, PS)\\ & bd(s\,n\,k, [i, crit(k)]\ PS) = bd(s\,n\,k, PS)\end{array}$$

Finally, the state proposition *uniq* holds in a state $\langle n, m, A, PS\rangle$ if no duplicate ticket numbers of processes in PS, defined by the equations:

$$\left.\begin{array}{ll} \langle n, m, A, PS\rangle \models uniq = q(PS) & q([i, wait(k)]\ [j, wait(k)]\ PS) = false\\ q([i, wait(k)]\ [j, crit(k)]\ PS) = false & q([i, crit(k)]\ [j, crit(k)]\ PS) = false\end{array}\right] (E_q)$$

$$\begin{array}{ll} q([i, wait(k)]\ [j, wait(s\,m\,k)]\ PS) = q([i, wait(k)]\ PS) & q([i, idle]\ PS) = q(PS)\\ q([i, wait(k)]\ [j, crit(s\,m\,k)]\ PS) = q([i, wait(k)]\ PS) & q([i, wait(k)]) = true\\ q([i, crit(k)]\ [j, wait(s\,m\,k)]\ PS) = q([i, crit(k)]\ PS) & q([i, crit(k)]) = true\\ q([i, crit(k)]\ [j, crit(s\,m\,k)]\ PS) = q([i, crit(k)]\ PS) & q(none) = true\end{array}$$

Notice that every state proposition has syntactically independent equations. In particular, the equations for *mutex* and *mcrit* have the finite variant property. The equations defining *bound* and *uniq* do *not* have the finite variant property, but the equations $E_b \cup B$ and $E_q \cup B$ that define the *negative* cases of *bound* and *uniq* do have the finite variant property. Furthermore:

Lemma 3. *For a state term $u \in \mathcal{T}_\Sigma(\mathcal{X})_{\mathsf{State}}$, a constraint $(u \models bound = false)$ is $E_b \cup B$-solvable, and $(u \models uniq = false)$ is $E_q \cup B$-solvable.*

Proof. A term u has the form $\langle n, m, A, t_{procs}\rangle$. If $\sigma u \models bound =_E false$ for some substitution σ, then σt_{procs} contains a process whose ticket number is greater than or equal to n. Since $E_b \cup B$ reduces such a negative case to *false* in 2 steps, $\sigma u \models bound =_{E_b \cup B} false$. Similarly, if $\sigma u \models uniq =_E false$, then σt_{procs} contains two processes with the same ticket number, and $\sigma u \models uniq =_{E_q \cup B} false$. $\square$

We obtain $\mathcal{R}/AP$ having a *single* reachable state from the given initial state $\{mutex, mcrit, bound, uniq\}$. By resolving each constraint for *mutex* and *mcrit*, from $\{mutex, mcrit, bound, uniq\}$, the resulting sets of remaining constraints are $\{l_k \models bound = true,\ l_k \models uniq = true,\ r_k \models bound = b_k,\ r_k \models uniq = b'_k\}$ for $1 \leq j \leq 5$, where $b_k, b'_k \in \{true, false\}$ and each pair $l_k \to_k r_k$ is given by:

$$\langle n, m, A, [i, idle]\ WS\rangle \to_1 \langle s\,n, m, wake, [i, wait(n)]\ WS\rangle$$
$$\langle n, m, A, [i, idle]\ [j, crit(m)]\ WS\rangle \to_2 \langle s\,n, m, wake, [i, wait(n)]\ [j, crit(m)]\ WS\rangle$$
$$\langle n, m, A, [i, wait(m)]\ WS\rangle \to_3 \langle n, m, crit, [i, crit(m)]\ WS\rangle$$
$$\langle n, m, A, [i, crit(m)]\ WS\rangle \to_4 \langle n, s\,m, exit, [i, idle]\ WS\rangle$$
$$\langle n, m, A, [i, wait(m)]\ [j, crit(m)]\ WS\rangle \to_5 \langle n, m, crit, [i, crit(m)]\ [j, crit(m)]\ WS\rangle$$

The case of $k = 5$ is unsatisfiable for any values of $b_5, b'_5 \in \{true, false\}$, since $l_5 \models uniq =_E false$. For the cases of $1 \leq k \leq 4$, if $b_k = false$ for *bound*, then its solution $\zeta \in CSU_{E_b \cup B}(r_k \models bound = false)$ makes $\zeta l_k \models bound =_E false$.[1] That is, the cases of $b_k = false$ are unsatisfiable. Similarly, if $b'_k = false$ for *uniq*, then $\zeta \in CSU_{E_q \cup B}(r_k \models uniq = false)$ makes either $\zeta l_k \models bound =_E false$ or $\zeta l_k \models uniq =_E false$, i.e., unsatisfiable. Therefore, $\{mutex, mcrit, bound, uniq\}$ has the one next state, itself. Clearly, $\Box mutex$ holds in $\mathcal{R}/AP$, and thus $\Box mutex$ also holds in $\mathcal{K}(\mathcal{R})_{AP}$ for an unbounded number of processes.

6 Concluding Remarks

In this work we have presented a predicate abstraction method for a wide range of infinite-state concurrent systems specified as conditional rewrite theories. One of the main ideas is to reduce a predicate abstraction problem into solving a set of E-equality constraints, which can be automated if there exists a finitary E-unification algorithm. To deal with the case where no finitary E-unification algorithm exists, we have also presented sound but incomplete procedures to check unsatisfiability of E-equality constraints, where an over-approximation can be constructed if such procedures fail to give an answer. We have illustrated our method with nontrivial infinite-state systems, namely, parameterized protocols for an unbounded number of processes.

Future work includes implementing a predicate abstraction-based LTL model checker in Maude,[2] integrating it with other infinite-state model checking methods such as equational abstractions and narrowing-based methods, and applying SMT techniques, as well as E-unification, to solve E-equality constraints.

Acknowledgments. This work has been supported in part by NSF Grant CNS 13-19109 and AFOSR Grant FA8750-11-2-0084.

[1] For example, $CSU_{E_b \cup B}(\langle s\,n, m, wake, [j, wait(n)]\ WS\rangle \models bound = false)$ has the substitution $\zeta = \{WS \leftarrow [j', wait(s\,n\,m')]\ WS'\}$, and $\zeta l_1 \models bound =_E false$.

[2] The key software components used by the algorithm (already available in Maude) and detailed experiments showing how they can automate the case studies are explained in the webpage `http://formal.cs.illinois.edu/kbae/pred`

References

1. Abdulla, P.A., Chen, Y.-F., Delzanno, G., Haziza, F., Hong, C.-D., Rezine, A.: Constrained monotonic abstraction: A CEGAR for parameterized verification. In: Gastin, P., Laroussinie, F. (eds.) CONCUR 2010. LNCS, vol. 6269, pp. 86–101. Springer, Heidelberg (2010)
2. Avenhaus, J., Loría-Sáenz, C.: On conditional rewrite systems with extra variables and deterministic logic programs. In: Pfenning, F. (ed.) LPAR 1994. LNCS, vol. 822, pp. 215–229. Springer, Heidelberg (1994)
3. Baader, F., Snyder, W.: Unification theory. In: Handbook of Automated Reasoning, pp. 445–532. Elsevier and MIT Press (2001)
4. Bae, K., Escobar, S., Meseguer, J.: Abstract logical model checking of infinite-state systems using narrowing. In: RTA. LIPIcs, vol. 21, pp. 81–96 (2013)
5. Ball, T., Majumdar, R., Millstein, T., Rajamani, S.K.: Automatic predicate abstraction of C programs. ACM SIGPLAN Notices 36(5), 203–213 (2001)
6. Clarke, E., Grumberg, O., Jha, S., Lu, Y., Veith, H.: Counterexample-guided abstraction refinement. In: Emerson, E.A., Sistla, A.P. (eds.) CAV 2000. LNCS, vol. 1855, pp. 154–169. Springer, Heidelberg (2000)
7. Clarke, E.M., Grumberg, O., Peled, D.A.: Model Checking. The MIT Press (2001)
8. Clavel, M., Durán, F., Eker, S., Lincoln, P., Martí-Oliet, N., Meseguer, J., Talcott, C.: All About Maude - A High-Performance Logical Framework. LNCS, vol. 4350. Springer, Heidelberg (2007)
9. Comon-Lundh, H., Delaune, S.: The finite variant property: How to get rid of some algebraic properties. In: Giesl, J. (ed.) RTA 2005. LNCS, vol. 3467, pp. 294–307. Springer, Heidelberg (2005)
10. Das, S., Dill, D.L., Park, S.: Experience with predicate abstraction. In: Halbwachs, N., Peled, D.A. (eds.) CAV 1999. LNCS, vol. 1633, pp. 160–171. Springer, Heidelberg (1999)
11. Dershowitz, N., Jouannaud, J.P.: Rewrite systems. In: Handbook of Theoretical Computer Science, vol. B, pp. 243–320. North-Holland (1990)
12. Durán, F., Meseguer, J.: A Church-Rosser checker tool for conditional order-sorted equational Maude specifications. In: Ölveczky, P.C. (ed.) WRLA 2010. LNCS, vol. 6381, pp. 69–85. Springer, Heidelberg (2010)
13. Escobar, S., Meseguer, J.: Symbolic model checking of infinite-state systems using narrowing. In: Baader, F. (ed.) RTA 2007. LNCS, vol. 4533, pp. 153–168. Springer, Heidelberg (2007)
14. Escobar, S., Sasse, R., Meseguer, J.: Folding variant narrowing and optimal variant termination. J. Algebraic and Logic Programming 81, 898–928 (2012)
15. Genet, T., Rusu, V.: Equational approximations for tree automata completion. Journal of Symbolic Computation 45(5), 574–597 (2010)
16. Graf, S., Saïdi, H.: Construction of abstract state graphs with PVS. In: Grumberg, O. (ed.) CAV 1997. LNCS, vol. 1254, pp. 72–83. Springer, Heidelberg (1997)
17. Lahiri, S.K., Bryant, R.E., Cook, B.: A symbolic approach to predicate abstraction. In: Hunt Jr., W.A., Somenzi, F. (eds.) CAV 2003. LNCS, vol. 2725, pp. 141–153. Springer, Heidelberg (2003)
18. Meseguer, J.: Membership algebra as a logical framework for equational specification. In: Parisi-Presicce, F. (ed.) WADT 1997. LNCS, vol. 1376, pp. 18–61. Springer, Heidelberg (1998)
19. Meseguer, J.: Conditional rewriting logic as a unified model of concurrency. Theor. Comp. Sci. 96(1), 73–155 (1992)

20. Meseguer, J.: Twenty years of rewriting logic. J. Algebraic and Logic Programming 81, 721–781 (2012)
21. Meseguer, J., Palomino, M., Martí-Oliet, N.: Equational abstractions. Theor. Comp. Sci. 403(2-3), 239–264 (2008)
22. Ohsaki, H., Seki, H., Takai, T.: Recognizing boolean closed A-tree languages with membership conditional rewriting mechanism. In: Nieuwenhuis, R. (ed.) RTA 2003. LNCS, vol. 2706, pp. 483–498. Springer, Heidelberg (2003)
23. Palomino, M.: A predicate abstraction tool for maude (2005), `http://maude.sip.ucm.es/~miguelpt/bibliography.html`
24. Viry, P.: Equational rules for rewriting logic. Theor. Comp. Sci. 285 (2002)
25. Visser, W., Havelund, K., Brat, G., Park, S., Lerda, F.: Model checking programs. Automated Software Engineering 10(2), 203–232 (2003)

Unification and Logarithmic Space*

Clément Aubert and Marc Bagnol

Aix-Marseille Université, CNRS, I2M, UMR 7373, 13453 Marseille, France

Abstract. We present an algebraic characterization of the complexity classes LOGSPACE and NLOGSPACE, using an algebra with a composition law based on unification. This new bridge between unification and complexity classes is inspired from proof theory and more specifically linear logic and Geometry of Interaction.

We show how unification can be used to build a model of computation by means of specific subalgebras associated to finite permutation groups.

We then prove that whether an observation (the algebraic counterpart of a program) accepts a word can be decided within logarithmic space. We also show that the construction can naturally encode pointer machines, an intuitive way of understanding logarithmic space computing.

Keywords: Implicit Complexity, Unification, Logarithmic Space, Proof Theory, Pointer Machines, Geometry of Interaction.

Introduction

Proof Theory and Complexity Theory. There is a longstanding tradition of relating proof theory (more specifically linear logic [1]) and implicit complexity theory that dates back to the introduction of bounded [2] and light [3] logics. Control over the modalities [4,5], type assignment [6] and stratification of exponential boxes [7], to name a few, led to a clearer understanding of the complexity bounds linear logic could entail on the cut-elimination procedure.

We propose to push further this approach by adopting a more semantical and algebraic point of view that will allow us to capture non-deterministic logarithmic space computation.

Geometry of Interaction. As the study of cut-elimination has grown as a central topic in proof theory, its mathematical modeling became of great interest. The Geometry of Interaction [8] research program led to mathematical models of cut-elimination in terms of paths in proofnets [9], token machines [10] and operator algebras [11]. Related complexity concerns have already been explored [12,13].

Recent works [13,14,15] studied the link between Geometry of Interaction and logarithmic space, relying on the theory of von Neumann algebras. Those three articles are indubitably sources of inspiration of this work, but the whole construction is made anew, in a simpler framework.

* This work was partly supported by the ANR-10-BLAN-0213 Logoi and the ANR-11-BS02-0010 Récré.

G. Dowek (ed.): RTA-TLCA 2014, LNCS 8560, pp. 77–92, 2014.

Unification. Unification is one of the key-concepts of theoretical computer science, for it is used in logic programming and is a classical subject of study for complexity theory. It was shown [16,17] that one can model cut-elimination with unification techniques.

Execution will be expressed in terms of matching in a *unification algebra*. This is a simple framework, yet expressive enough to encode the action of finite permutation groups on an unbounded tensor product, which is a crucial ingredient of our construction.

Contribution. We carry on the methodology of bridging Geometry of Interaction and complexity theory with a renewed approach. It relies on an simpler representation of execution in a unification-based algebra, proved to capture exactly logarithmic space complexity.

While the representation of inputs (words over a finite alphabet) comes from the classical Church representation, observations (the algebraic counterpart of programs) are shown to correspond very naturally to a notion of pointer machines. This correspondence allows us to prove that reversibility (of machines) is related to the algebraic notion of isometricity (of observations).

Organization of This Article. In Sect.1 we review some classical results on unification of first-order terms and use them to build the algebra that will constitute our computational setting.

We explain in Sect.2 how words and computing devices (observations) can be modeled by particular elements of this algebra. The way they interact to yield a notion of language recognized by an observation is described in Sect.3.

Finally, we show in Sect.4 that our construction captures exactly logarithmic space computation, both deterministic and non-deterministic.

1 The Unification Algebra

1.1 Unification

Unification can be generally thought of as the study of formal solving of equations between terms.

This topic was introduced by Herbrand, but became really widespread after the work of J. A. Robinson on automated theorem proving. The unification technique is also at the core of the logic programming language PROLOG and type inference for functional programming languages such as CAML and HASKELL.

Specifically, we will be interested in the following problem:

Given two terms, can they be "made equal" by replacing their variables?

Definition 1 *(terms)*
We consider the following set of first-order terms

$$\mathtt{T} ::= \; x, y, z, \ldots \mid \mathtt{a}, \mathtt{b}, \mathtt{c}, \ldots \mid \mathtt{T} \bullet \mathtt{T}$$

where $x, y, z, \cdots \in \mathtt{V}$ *are variables,* $\mathtt{a}, \mathtt{b}, \mathtt{c}, \ldots$ *are constants and* $\bullet$ *is a binary function symbol.*

For any $t \in \mathtt{T}$ *, we will write* $\mathtt{Var}(t)$ *the set of variables occurring in* t *. We say that a term is closed when* $\mathtt{Var}(t) = \varnothing$ *, and denote* $\mathtt{T_c}$ *the set of closed terms.*

Notation. *The binary function symbol* $\bullet$ *is not associative, but we will write it by convention as* right associating *to lighten notations:* $t \bullet u \bullet v \ := \ t \bullet (u \bullet v)$

Definition 2 ***(substitution)***
A substitution *is a map* $\theta : \mathtt{V} \to \mathtt{T}$ *such that the set* $\mathtt{Dom}(\theta) := \{v \in \mathtt{V} | \theta(v) \neq v\}$ *(the* domain *of* θ *) is finite. A substitution with domain* $\{ x_1, \ldots, x_n \}$ *such that* $\theta(x_1) = u_1$ *, ... ,* $\theta(x_n) = u_n$ *will be written as* $\{ x_1 \mapsto u_1 ; \ldots ; x_n \mapsto u_n \}$ *.*

If $t \in \mathtt{T}$ *is a term we write* $t.\theta$ *the term* t *where any occurrence of any variable* x *has been replaced by* $\theta(x)$ *.*
If $\theta = \{ x_i \mapsto u_i \}$ *and* $\psi = \{ y_j \mapsto v_j \}$ *, their* composition *is defined as*

$$\theta;\psi \ := \ \{ x_i \mapsto u_i.\psi \} \cup \{ y_j \mapsto v_j \mid y_j \notin \mathtt{Dom}(\theta) \}$$

Remark. *The composition of substitutions is such that* $t.(\theta;\psi) = (t.\theta).\psi$ *holds.*

Definition 3 ***(renamings and instances)***
A renaming *is a substitution* α *such that* $\alpha(\mathtt{V}) \subseteq \mathtt{V}$ *and that is bijective. A term* t' *is a* renaming *of* t *if* $t' = t.\alpha$ *for some renaming* α *.*

Two substitutions θ, ψ *are equal* up to renaming *if there is a renaming* α *such that* $\psi = \theta;\alpha$ *.*

A substitution ψ *is an* instance *of* θ *if there is a substitution* σ *such that* $\psi = \theta;\sigma$ *.*

Proposition 4
Let θ, ψ *be two substitutions. If* θ *is an instance of* ψ *and* ψ *is an instance of* θ *, then they are equal up to renaming.*

Definition 5 ***(unification)***
Two terms t, u *are* unifiable *if there is a substitution* θ *such that* $t.\theta = u.\theta$ *.*
We say that θ *is a* most general unifier (MGU) *of* t, u *if any other unifier of* t, u *is an instance of* θ *.*

Remark. *It follows from Proposition 4 that any two MGU of a pair of terms are equal up to renaming.*

We will be interested mostly in the weaker variant of unification where one can first perform renamings on terms so that their variables are distinct, we introduce therefore a specific vocabulary for it.

Definition 6 ***(disjointness and matching)***
Two terms t, u *are* matchable *if* t', u' *are unifiable, where* t', u' *are renamings (Definition 3) of* t, u *such that* $\mathtt{Var}(t') \cap \mathtt{Var}(u') = \varnothing$ *.*
If two terms are not matchable, they are said to be disjoint.

Example. *x and $\mathtt{f} \bullet x$ are not unifiable.*
But they are matchable, as $x.\{ x \mapsto y ; y \mapsto x \} = y$ which is unifiable with $\mathtt{f} \bullet x$.
More generally, disjointness is stronger than non-unifiability.

The crucial feature of first-order unification is the (decidable) existence of most general unifiers for unification problems that have a solution.

Proposition 7 *(MGU)*
If two terms are unifiable, then they have a MGU.
Whether two terms are unifiable and, in case they are, finding a MGU is decidable.

As unification grew in importance, the study of its complexity gained in attention. A complete survey [18] tells the story of the bounds getting sharpened: general first-order unification was finally proved [19] to be a PTIME-complete problem.

In this article, we are concerned with a very much simpler case of the problem: the matching (Definition 6) of linear terms (*ie.* where variables occur at most once). This case can be solved in a space-efficient way.

Proposition 8 ***(matching in logarithmic space [20, Lemma 20])***
Whether two linear terms t, u with disjoint sets of variables are unifiable, and if so finding a MGU, can be computed in logarithmic space in the size[1] *of t, u on a deterministic Turing machine*

The lemma in [20] actually states that the problem is in NC^1, a complexity class of parallel computation known to be included in LOGSPACE.
We will use only a special case of the result, matching a linear term against a closed term.

1.2 Flows and Wirings

We now use the notions we just saw to build an algebra with a product based on unification. Let us start with a monoid with a partially defined product, which will be the basis of the construction.

Definition 9 ***(flows)***
A flow *is an oriented pair written $t \leftharpoonup u$ with $t, u \in \mathtt{T}$ such that $\mathtt{Var}(t) = \mathtt{Var}(u)$.*
Flows are considered up to renaming: for any renaming α, $t \leftharpoonup u = t.\alpha \leftharpoonup u.\alpha$.
We will write $\mathcal{F}$ the set of (equivalence classes of) flows.
We set $I := x \leftharpoonup x$ and $(t \leftharpoonup u)^\dagger := u \leftharpoonup t$ so that $(.)^\dagger$ is an involution of $\mathcal{F}$.

A flow $t \leftharpoonup u$ can be thought of as a `'match ... with u -> t'` in a ML-style language. The composition of flows follows this intuition.

[1] The *size* of a term is the total number of occurrences of symbols in it.

Definition 10 ***(product of flows)***
Let $u \leftharpoondown v \in \mathcal{F}$ *and* $t \leftharpoondown w \in \mathcal{F}$. *Suppose we have chosen two representatives of the renaming classes such that their sets of variables are disjoint.*
The product *of* $u \leftharpoondown v$ *and* $t \leftharpoondown w$ *is defined if* v, t *are unifiable with MGU* θ *(the choice of a MGU does not matter because of the remark following Definition 5) and in that case:* $(u \leftharpoondown v)(t \leftharpoondown w) := u.\theta \leftharpoondown w.\theta$.

Definition 11 ***(action on closed terms)***
If $t \in \mathtt{T_c}$ *is a closed term,* $(u \leftharpoondown v)(t)$ *is defined whenever* t *and* v *are unifiable, with MGU* θ, *in that case* $(u \leftharpoondown v)(t) := u.\theta$

Examples. *Composition of flows:* $(x \bullet \mathtt{c} \leftharpoondown (\mathtt{c} \bullet \mathtt{c}) \bullet x)(y \bullet z \leftharpoondown z \bullet y) = x \bullet \mathtt{c} \leftharpoondown x \bullet \mathtt{c} \bullet \mathtt{c}$.
Action on a closed term: $(x \bullet \mathtt{c} \leftharpoondown x \bullet \mathtt{c} \bullet \mathtt{c})(\mathtt{d} \bullet \mathtt{c} \bullet \mathtt{c}) = \mathtt{d} \bullet \mathtt{c}$.

Remark. *The condition on variables ensures that the result is a closed term (because* $\mathtt{Var}(u) \subseteq \mathtt{Var}(v)$ *) and that the action is injective on its domain of definition (because* $\mathtt{Var}(v) \subseteq \mathtt{Var}(u)$ *). Moreover, the action is compatible with the product of flows:* $l(k(t)) = (l\,k)(t)$ *and both are defined at the same time.*

By adding a formal element $\perp$ (representing the failure of unification) to the set of flows, one could turn the product into a completely defined operation, making $\mathcal{F}$ an *inverse monoid*. However, we will need to consider the wider algebra of *sums* of flows that is easily defined directly from the partially defined product.

Definition 12 ***(wirings)***
Wirings *are* $\mathbb{C}$*-linear combinations of flows (formally: almost-everywhere null functions from the set of flows* $\mathcal{F}$ *to* $\mathbb{C}$ *), endowed with the following operations:*

$$\Big(\sum_i \lambda_i\, l_i\Big)\Big(\sum_j \mu_j\, k_j\Big) := \sum_{\substack{i,j \text{ such that} \\ (l_i k_j) \text{ is defined}}} \lambda_i \mu_j (l_i\, k_j) \qquad \textit{(with } \lambda_i,\mu_j \in \mathbb{C} \textit{ and } l_i, k_j \in \mathcal{F}\textit{)}$$

$$\textit{and} \qquad \Big(\sum_i \lambda_i\, l_i\Big)^\dagger := \sum_i \overline{\lambda}_i\, l_i^\dagger \qquad \textit{(where } \overline{\lambda} \textit{ is the complex conjugate of } \lambda\textit{)}$$

We write $\mathcal{U}$ *the set of wirings and refer to it as the* unification algebra.

Remark. *Indeed,* $\mathcal{U}$ *is a unital* $*$*-algebra: it is a* $\mathbb{C}$*-algebra (considering the product defined above) with an involution* $(.)^\dagger$ *and a unit* I .

Definition 13 ***(partial isometries)***
A partial isometry *is a wiring* $U \in \mathcal{U}$ *satisfying* $UU^\dagger U = U$.

Example. $(\mathtt{c} \bullet x \leftharpoondown x \bullet \mathtt{d}) + (\mathtt{d} \bullet \mathtt{c} \leftharpoondown \mathtt{c} \bullet \mathtt{c})$ *is a partial isometry.*

While $\mathcal{U}$ offers the general algebraic background to work in, we will need to consider a particular kind of wirings to study computation.

Definition 14 ***(concrete and isometric wirings)***
A wiring is concrete *whenever it is a sum of flows with all coefficients equal to* 1 .

An isometric wiring *is a concrete wiring that is also a partial isometry.*
Given a set of wirings E we write E^+ for the set of all concrete wirings of E.

Isometric wirings enjoy a direct characterization.

Proposition 15 ***(isometric wirings)***
The isometric wirings are exactly the wirings of the form $\sum_i u_i \leftharpoonup t_i$ with the u_i pairwise disjoint (Definition 6) and t_i pairwise disjoint.

It will be useful to consider the action of wirings on closed terms. For this purpose we extend Definition 11 to wirings.

Definition 16 ***(action on closed terms)***
Let $\mathbb{V}_\mathsf{c}$ be the free $\mathbb{C}$-vector space over $\mathtt{T}_\mathsf{c}$.
Wirings act on base vectors of $\mathbb{V}_\mathsf{c}$ in the following way

$$\Big(\sum_i \lambda_i\, l_i\Big)(t) := \sum_{\substack{i \text{ such that} \\ l_i(t) \text{ is defined}}} \lambda_i\big(l_i(t)\big) \quad \in \mathbb{V}_\mathsf{c}$$

which extends by linearity into an action on the whole $\mathbb{V}_\mathsf{c}$.
Isometric wirings have a particular behavior in terms of this action.

Lemma 17 ***(isometric action)***
Let F be an isometric wiring and t a closed term. We have that $F(t)$ and $F^\dagger(t)$ are either 0 or another closed term t' (seen as an element of $\mathbb{V}_\mathsf{c}$).

1.3 Tensor Product and Permutations

We define now the representation in $\mathcal{U}$ of structures that provide enough expressivity to model computation.
Unbounded tensor products will allow to represent data of arbitrary size, and finite permutations will be used to manipulate these data.

Notations. *Given any set of wirings or closed terms E, we write $\mathtt{Vect}(E)$ the vector space generated by E,* ie. *the set of* finite *linear combinations of elements of E (for instance $\mathtt{Vect}(\mathtt{T}_\mathsf{c}) = \mathbb{V}_\mathsf{c}$).*
Moreover, we set $\mathcal{I} := \{\, \lambda I \mid \lambda \in \mathbb{C}\,\}$ (with $I = x \leftharpoonup x$ as in Definition 9) which is the $*$-algebra of multiples of the identity, and $u \leftrightharpoons v := u \leftharpoonup v + v \leftharpoonup u$.

*For brevity we write "$*$-algebra" instead of the more correct "$*$-subalgebra of $\mathcal{U}$" (*ie. *a subset of $\mathcal{U}$ that is stable by linear combinations, product and $(.)^\dagger$).*

Definition 18 ***(tensor product)***
Let $u \leftharpoonup v$ and $t \leftharpoonup w$ be two flows. Suppose we have chosen representatives of these renaming classes that have their sets of variables disjoint. We define their tensor product *as $(u \leftharpoonup v) \dot{\otimes} (t \leftharpoonup w) := u {\bullet} t \leftharpoonup v {\bullet} w$. The operation is extended to wirings by bilinearity.*

Given two $*$-algebras $\mathcal{A}, \mathcal{B}$, we define their tensor product as the $*$-algebra

$$\mathcal{A} \dot{\otimes} \mathcal{B} := \mathtt{Vect}\big\{\, F \dot{\otimes} G \;\big|\; F \in \mathcal{A},\, G \in \mathcal{B} \,\big\}$$

This actually defines an embedding of the algebraic tensor product of $*$-algebras into $\mathcal{U}$, which means in particular that $(F \dot{\otimes} G)(P \dot{\otimes} Q) = (FP) \dot{\otimes} (GQ)$. It ensures also that the $\dot{\otimes}$ operation indeed yields $*$-algebras.

Notation. *As $\bullet$, the $\dot{\otimes}$ operation is not associative. We carry on our convention and write it as* right associating*:* $\mathcal{A} \dot{\otimes} \mathcal{B} \dot{\otimes} \mathcal{C} := \mathcal{A} \dot{\otimes} (\mathcal{B} \dot{\otimes} \mathcal{C})$.

Definition 19 ***(unbounded tensor)***
Let $\mathcal{A}$ be a $*$-algebra. We define the $*$-algebras $\mathcal{A}^{\otimes n}$ for all $n \in \mathbb{N}$ as

$$\mathcal{A}^{\otimes 0} := \mathcal{I} \quad \textit{and} \quad \mathcal{A}^{\otimes n+1} := \mathcal{A} \dot{\otimes} \mathcal{A}^{\otimes n}$$

and the $*$-algebra $\mathcal{A}^{\otimes \infty} := \mathtt{Vect}\left(\bigcup_{n \in \mathbb{N}} \mathcal{A}^{\otimes n} \right)$.

We will consider finite permutations, but allow them to be composed even when their domain of definition do not match.

Notations. *Let $\mathfrak{S}_n$ be the set of finite permutations over $\{1, \ldots, n\}$, if $\sigma \in \mathfrak{S}_n$, we define $\sigma_{+k} \in \mathfrak{S}_{n+k}$ as the permutation σ extended to $\{1, \ldots, n, \ldots, n+k\}$ letting $\sigma_{+k}(n+i) := n+i$ for $i \in \{1, \ldots, k\}$.*
We also write $I_k := Id_{\{1,\ldots,k\}} \in \mathfrak{S}_k$.

Definition 20 ***(representation)***
To a permutation $\sigma \in \mathfrak{S}_n$ we associate the flow

$$[\sigma] := x_1 \bullet x_2 \bullet \cdots \bullet x_n \bullet y \leftharpoonup x_{\sigma(1)} \bullet x_{\sigma(2)} \bullet \cdots \bullet x_{\sigma(n)} \bullet y$$

A permutation $\sigma \in \mathfrak{S}_n$ will act on the first n components of the unbounded tensor product (Definition 19) by swapping them and leaving the rest unchanged. The wirings $[\sigma]$ internalize this action: in the above definition, the variable y at the end stands for the components that are not affected.

Example. *Let $\tau \in \mathfrak{S}_2$ be the permutation swapping the two elements of $\{1, 2\}$ and $U_1 \dot{\otimes} U_2 \dot{\otimes} U_3 \dot{\otimes} I \in \mathcal{U}^{\otimes 3} \subseteq \mathcal{U}^{\otimes \infty}$. We have $[\tau] = x_1 \bullet x_2 \bullet y \leftharpoonup x_2 \bullet x_1 \bullet y$ and $[\tau](U_1 \dot{\otimes} U_2 \dot{\otimes} U_3 \dot{\otimes} I)[\tau]^\dagger = U_2 \dot{\otimes} U_1 \dot{\otimes} U_3 \dot{\otimes} I$.*

Proposition 21 ***(representation)***
For $\sigma \in \mathfrak{S}_n$ and $\tau \in \mathfrak{S}_{n+k}$ we have

$$[\sigma_{+k}] = [\sigma][I_{n+k}] = [I_{n+k}][\sigma] \qquad [\sigma_{+k} \circ \tau] = [\sigma][\tau] \qquad \textit{and} \qquad [\sigma^{-1}] = [\sigma]^\dagger$$

Definition 22 ***(permutation algebra)***
For $n \in \mathbb{N}$ we set $[\mathfrak{S}_n] := \{ [\sigma] \mid \sigma \in \mathfrak{S}_n \}$ and $\mathcal{S}_n := \mathtt{Vect}[\mathfrak{S}_n]$.

We define then $\mathcal{S} := \mathtt{Vect}\left(\bigcup_{n \in \mathbb{N}} \mathcal{S}_n \right)$*, which we call the* permutation algebra.

Proposition 21 ensures that the $\mathcal{S}_n$ and $\mathcal{S}$ are $*$-algebras.

2 Words and Observations

The representation of words over an alphabet in the unification algebra directly comes from the translation of Church lists in linear logic and their interpretation in Geometry of Interaction models [11,16].

This proof-theoretic origin is an useful guide for intuition, even if we give here a more straightforward definition of the notion.

From now on, we fix a set of two distinguished constant symbols $\mathtt{LR} := \{\mathtt{L}, \mathtt{R}\}$.

Definition 23 ***(word algebra)***
To a set S of closed terms, we associate the $*$-algebra

$$S^* := \mathtt{Vect}\{\, t \leftharpoonup u \mid t, u \in S \,\}$$

(which is indeed an algebra because unification of closed terms is simply equality)

The word algebra *associated to a finite set of constant symbols Σ is the $*$-algebra defined as*

$$\mathcal{W}_\Sigma := (\mathcal{I} \dot{\otimes} \Sigma^* \dot{\otimes} \mathtt{LR}^*) \otimes (\mathtt{T}_\mathtt{c}^*)^{\otimes 1}$$

($\mathtt{T}_\mathtt{c}$ is the set of all closed terms, $\mathcal{I}$ is defined at the beginning of Sect.1.3
$\dot{\otimes}$ is as in Definition 18 and $(.)^{\otimes 1}$ is the case $n = 1$ of Definition 19)

The words we consider are cyclic, with a begin/end marker $\star$, a reserved constant symbol. For example the word $\mathtt{0010}$ is to be thought of as $\star\mathtt{0010} = \mathtt{10}\star\mathtt{00} = \mathtt{0}\star\mathtt{001} = \cdots$.
We consider therefore that the alphabet Σ always contains the symbol $\star$.

Definition 24 ***(word representation)***
Let $W = \star\mathtt{c}_1 \dots \mathtt{c}_n$ be a word over Σ and $t_0, t_1, \dots, t_n$ be distinct closed terms.
The representation *$W(t_0, t_1, \dots, t_n) \in \mathcal{W}_\Sigma^+$ with respect to $t_0, t_1, \dots, t_n$ of W is an isometric wiring (Definition 14), defined as*

$$\begin{aligned} W(t_0, t_1, \dots, t_n) := \quad & x \bullet \star \bullet \mathtt{R} \bullet (t_0 \bullet y) \leftrightharpoons x \bullet \mathtt{c}_1 \bullet \mathtt{L} \bullet (t_1 \bullet y) \\ + \; & x \bullet \mathtt{c}_1 \bullet \mathtt{R} \bullet (t_1 \bullet y) \leftrightharpoons x \bullet \mathtt{c}_2 \bullet \mathtt{L} \bullet (t_2 \bullet y) \\ + \; & \cdots \\ + \; & x \bullet \mathtt{c}_n \bullet \mathtt{R} \bullet (t_n \bullet y) \leftrightharpoons x \bullet \star \bullet \mathtt{L} \bullet (t_0 \bullet y) \end{aligned}$$

We now define *observations*, programs computing on representations of words. They lie in a particular $*$-algebra based on the representation of permutations presented in Sect.1.3.

Definition 25 ***(observation algebra)***
An observation *over a finite set of symbols Σ is any element of $\mathcal{O}_\Sigma^+$ where $\mathcal{O}_\Sigma := (\mathtt{T}_\mathtt{c}^* \dot{\otimes} \Sigma^* \dot{\otimes} \mathtt{LR}^*) \dot{\otimes} \mathcal{S}$, i.e. a* finite *sum of flows of the form*

$$(s' \bullet \mathtt{c}' \bullet \mathtt{d}' \leftharpoonup s \bullet \mathtt{c} \bullet \mathtt{d}) \otimes [\sigma]$$

with s, s' closed terms, $\mathtt{c}, \mathtt{c}' \in \Sigma$, $\mathtt{d}, \mathtt{d}' \in \mathtt{LR}$ and σ is a permutation.
Moreover when an observation happens to be an isometric wiring, we will call it an isometric observation.

3 Normativity: Independence from Representations

We are going to define how observations accept and reject words. This needs to be discussed, because there is a potential issue with word representations: an observation is an element of $\mathcal{U}$ and can therefore only interact with *representations* of a word, and there are many possible representation of the same word (in Definition 24, different choices of closed terms lead to different representations). Therefore one has to ensure that acceptance or rejection is independent of the representation, so that the notion makes the intended sense. The termination of computations will correspond to the algebraic notion of *nilpotency*, which we recall here.

Definition 26 ***(nilpotency)***
A wiring F is nilpotent *if $F^n = 0$ for some n.*

Definition 27 ***(automorphism)***
An automorphism *of a $*$-algebra $\mathcal{A}$ is a linear application $\varphi : \mathcal{A} \to \mathcal{A}$ such that for all $F, G \in \mathcal{A}$: $\varphi(FG) = \varphi(F)\varphi(G)$, $\varphi(F^\dagger) = \varphi(F)^\dagger$ and φ is injective.*

Example. *$\varphi(U_1 \dot{\otimes} U_2) := U_2 \dot{\otimes} U_1$ induces an automorphism of $\mathcal{U} \dot{\otimes} \mathcal{U}$.*

Notation. *If φ is an automorphism of $\mathcal{A}$ and ψ is an automorphism of $\mathcal{B}$, we write $\varphi \dot{\otimes} \psi$ the automorphism of $\mathcal{A} \dot{\otimes} \mathcal{B}$ defined for all $A \in \mathcal{A}, B \in \mathcal{B}$ as $(\varphi \dot{\otimes} \psi)(A \dot{\otimes} B) := \varphi(A) \dot{\otimes} \psi(B)$ and extended to all $\mathcal{A} \dot{\otimes} \mathcal{B}$ by linearity.*

Definition 28 ***(normative pair)***
A pair $(\mathcal{A}, \mathcal{B})$ of $*$-algebras is a normative pair *whenever any automorphism φ of $\mathcal{A}$ can be extended into an automorphism $\overline{\varphi}$ of the $*$-algebra $\mathcal{E}$ generated by $\mathcal{A} \cup \mathcal{B}$, such that $\overline{\varphi}(F) = F$ for any $F \in \mathcal{B} \subseteq \mathcal{E}$.*

The two following propositions set the basis for a notion of acceptance/rejection independent of the representation of a word.

Proposition 29 ***(automorphic representations)***
Any two representations $W(t_0, \dots, t_n), W(u_0, \dots, u_n)$ of a word W over Σ are automorphic: there exists an automorphism φ of $(\mathtt{T}_\mathtt{c}^*)^{\otimes 1}$ such that

$$(Id_{\mathcal{U}} \dot{\otimes} \varphi)\big(W(t_0, \dots, t_n)\big) = W(u_0, \dots, u_n)$$

Proof. Consider a bijection $f : \mathtt{T}_\mathtt{c} \to \mathtt{T}_\mathtt{c}$ such that $f(t_i) = u_i$ for all i. Then set $\varphi(v \bullet x \leftharpoonup w \bullet x) := f(v) \bullet x \leftharpoonup f(w) \bullet x$, extended by linearity. □

Proposition 30 ***(nilpotency and normative pairs)***
Let $(\mathcal{A}, \mathcal{B})$ be a normative pair and φ an automorphism of $\mathcal{A}$. Let $F \in \mathcal{U} \dot{\otimes} \mathcal{A}$, $G \in \mathcal{U} \dot{\otimes} \mathcal{B}$ and let $\psi := Id_{\mathcal{U}} \dot{\otimes} \varphi$. Then GF is nilpotent if and only if $G\psi(F)$ is nilpotent.

Proof. Let $\overline{\varphi}$ be the extension of φ as in Definition 28 and $\overline{\psi} := \mathrm{Id}_{\mathcal{U}} \dot{\otimes} \overline{\varphi}$.
We have for all $n \neq 0$ that $(G\psi(F))^n = (\overline{\psi}(G)\overline{\psi}(F))^n = (\overline{\psi}(GF))^n = \overline{\psi}((GF)^n)$.
By injectivity of $\overline{\psi}$, $(G\psi(F))^n = 0$ if and only if $(GF)^n = 0$. □

Corollary 31 ***(independence)***
If $((\mathtt{T}^*_{\mathtt{c}})^{\otimes 1}, \mathcal{B})$ is a normative pair, W a word over Σ and $F \in \mathcal{U} \dot{\otimes} \mathcal{B}$. The product of F with the representation of the word, $FW(t_0, \dots, t_n)$, is nilpotent for one choice of $(t_0, \dots, t_n)$ if and only if it is nilpotent for all choices of $(t_0, \dots, t_n)$.

The basic components of the word and observation algebras we introduced earlier can be shown to form a normative pair.

Theorem 32
The pair $((\mathtt{T}^*_{\mathtt{c}})^{\otimes 1}, \mathcal{S})$ is normative.

Proof (sketch). By simple computations, the set

$$\mathcal{A} := \mathtt{Vect}\left\{ \sigma F \;\middle|\; \sigma \in \mathcal{S} \text{ and } F \in (\mathtt{T}^*_{\mathtt{c}})^{\otimes\infty} \right\}$$

can be shown to be a $*$-algebra $\mathcal{E}$, the $*$-algebra generated by $\mathcal{S} \cup (\mathtt{T}^*_{\mathtt{c}})^{\otimes 1}$.
As φ is an automorphism of $(\mathtt{T}^*_{\mathtt{c}})^{\otimes 1}$, it can be written as $\varphi(G \dot{\otimes} I) = \psi(G) \dot{\otimes} I$ for all G, with ψ an automorphism of $\mathtt{T}^*_{\mathtt{c}}$.

We set for $F = F_1 \dot{\otimes} \cdots \dot{\otimes} F_n \dot{\otimes} I \in (\mathtt{T}^*_{\mathtt{c}})^{\otimes n}$, $\tilde{\varphi}(F) := \psi(F_1) \dot{\otimes} \cdots \dot{\otimes} \psi(F_n) \dot{\otimes} I$ which extends into an automorphism of $(\mathtt{T}^*_{\mathtt{c}})^{\otimes\infty}$ by linearity. Finally, we extend $\tilde{\varphi}$ to $\mathcal{A}$ by $\overline{\varphi}(\sigma F) := \sigma \tilde{\varphi}(F)$. It is then easy to check that $\overline{\varphi}$ has the required properties. □

Remark. *Here we sketched a direct proof for brevity, but this can also be shown by involving a little more mathematical structure (actions of permutations on the unbounded tensor and crossed products) which would give a more synthetic proof.*

We can then define the notion of the language recognized by an observation, *via* Corollary 31.

Definition 33 ***(language of an observation)***
Let $\phi \in \mathcal{O}^+_\Sigma$ be an observation over Σ. The language recognized by *ϕ is the following set of words over Σ:*

$$\mathcal{L}(\phi) := \{\, W \textit{ word over } \Sigma \mid \phi W(t_0, \dots, t_n) \textit{ nilpotent for any } (t_0, \dots, t_n) \,\}$$

4 Wirings and Logarithmic Space

Now that we have defined our framework and showed how observations can compute, we study the complexity of deciding whenever an observation accepts a word (4.1), and how wirings can recognize any language in (N)LOGSPACE (4.2).

4.1 Soundness of Observations

The aim of this subsection is to prove the following theorem:

Theorem 34 ***(space soundness)***
Let $\phi \in \mathcal{O}^+_\Sigma$ be an observation over Σ.

- *$\mathcal{L}(\phi)$ is decidable in non-deterministic logarithmic space.*
- *If ϕ is isometric, then $\mathcal{L}(\phi)$ is decidable in deterministic logarithmic space.*

Actually, the result stands for the complements of these languages, but as co-NLogspace = NLogspace by the Immerman-Szelepcsényi theorem, this makes no difference.

The main tool for this purpose is the notion of *computation space*: finite dimensional subspaces of $\mathbb{V}_c$ (Definition 16) on which we will be able to observe the behavior of certain wirings. It can be understood as the place where all the relevant interactions between an observation and a representation of a word take place.

Definition 35 ***(separating space)***
A subspace E of $\mathbb{V}_c$ is separating *for a wiring F whenever $F(E) \subseteq E$ and $F^n(E) = 0$ implies $F^n = 0$.*

Observations are *finite* sums of wirings. We can naturally associate a finite-dimensional vector space to an observation and a finite set of closed terms.

Definition 36 ***(computation space)***
Let $\{t_0, \ldots, t_n\}$ be a set of distinct closed terms and $\phi \in \mathcal{O}^+_\Sigma$ an observation. Let $N(\phi)$ be the smallest integer and $\mathtt{S}(\phi)$ the smallest (finite) set of closed terms such that $\phi \in (\mathtt{S}(\phi)^* \dot{\otimes} \Sigma^* \dot{\otimes} \mathtt{LR}^*) \dot{\otimes} \mathcal{S}_{N(\phi)}$.
The computation space $\mathtt{Comp}_\phi(t_0, \ldots, t_n)$ *is the subspace of $\mathbb{V}_c$ generated by the terms*

$$s \bullet \mathtt{c} \bullet \mathtt{d} \bullet (a_1 \bullet \cdots \bullet a_{N(\phi)} \bullet \star)$$

where $s \in \mathtt{S}(\phi)$, $\mathtt{c} \in \Sigma$, $\mathtt{d} \in \mathtt{LR}$ and the $a_i \in \{t_0, \ldots, t_n\}$.
The dimension of $\mathtt{Comp}_\phi(t_0, \ldots, t_n)$ is $|\mathtt{S}(\phi)|\,|\Sigma| 2(n+1)^{N(\phi)}$ (where $|A|$ is the cardinal of A), which is polynomial in n.

Lemma 37 ***(separation)***
For any observation ϕ and any word W, the space $\mathtt{Comp}_\phi(t_0, \ldots, t_n)$ is separating for the wiring $\phi W(t_0, \ldots, t_n)$.

Proof (of Theorem 34). With these lemmas at hand, we can define the non-deterministic algorithm below. It takes as an input the representation $W(t_0, \ldots, t_n)$ of a word W of length n.

ϕ being a constant, one can compute once and for all $N(\phi)$ and $\mathtt{S}(\phi)$.

```
1: D ← |S(φ)| |Σ|2(n+1)^N(φ)
2: C ← 0
3: pick a term v ∈ Comp_φ(t_0,...,t_n)
4: while C ≤ D do
5:   if (φW(t_0,...,t_n))(v) = 0 then
6:     return ACCEPT
7:   end if
8:   pick a term v'
     in (φW(t_0,...,t_n))(v)
9:   v ← v'
10:  C ← C+1
11: end while
12: return REJECT
```

All computation paths (the "pick" at lines 3 and 8 being non-deterministic choices) accept if and only if $(\phi W(t_0,\dots,t_n))^n(\mathtt{Comp}_\phi(t_0,\dots,t_n)) = 0$ for some n lesser or equal to the dimension D of the computation space $\mathtt{Comp}_\phi(t_0,\dots,t_n)$. By Lemma 37, this is equivalent to $\phi W(t_0,\dots,t_n)$ being nilpotent.

The term chosen at lines 3 is representable by an integer of size at most D and is erased by the one chosen at line 8 every time we go through the **while**-loop. C and D are integers proportional to the dimension of the computation space, which is polynomial in n (Definition 36), thus representable in logarithmic space in the size of the input.

The computation of $(\phi W(t_0,\dots,t_n))(v)$ at line 5 and 8 and can be performed in logarithmic space by Proposition 8, as we are unifying closed terms with linear terms.

Moreover, if ϕ is an isometric wiring, $(\phi W(t_0,\dots,t_n))(v)$ consists of a single term instead of a sum by Lemma 17, and there is therefore no non-deterministic choice to be made at line 8. It is then enough to run the algorithm enumerating all possible terms of $\mathtt{Comp}_\phi(t_0,\dots,t_n)$ at line 3 to determine the nilpotency of $\phi W(t_0,\dots,t_n)$. □

4.2 Completeness: Representing Pointer Machines as Wirings

To prove the converse of Theorem 34, we prove that wirings can encode a special kind of read-only multi-head Turing Machine: pointers machines. The definition of this model will be guided by our understanding of the computation of wirings: they won't have the ability to write and acceptance will be defined as termination of all paths of computation. For a survey of this topic, one may consult the first author's thesis [21, Chap.4], the main novelty of this part of our work is to notice that reversible computation is represented by isometric operators.

Definition 38 ***(pointer machine)***
A pointer machine *over an alphabet Σ is a tuple $(N, \mathtt{S}, \Delta)$ where*

- $N \neq 0$ *is an integer, the* number of pointers,
- $\mathtt{S}$ *is a finite set, the* states *of the machine,*
- $\Delta \subseteq (\mathtt{S} \times \Sigma \times \mathtt{LR}) \times (\mathtt{S} \times \Sigma \times \mathtt{LR}) \times \mathfrak{S}_N$, *the* transitions *of the machine (we will write $(s, \mathtt{c}, \mathtt{d}) \to (s', \mathtt{c}', \mathtt{d}') \times \sigma$ the transitions, for readability).*

A pointer machine will be called deterministic *if for any $A \in \mathtt{S} \times \Sigma \times \mathtt{LR}$, there is at most one $B \in \mathtt{S} \times \Sigma \times \mathtt{LR}$ and one $\sigma \in \mathfrak{S}_N$ such that $A \to B \times \sigma \in \Delta$. In that case we can see Δ as a partial function, and we say that M is* reversible *if Δ is a partial injection.*

We call the first of the N pointers the *main* pointer, it is the only one that can move. The other pointers are referred to as the *auxiliary* pointers. An auxiliary pointer will be able to become the main pointer during the computation thanks to permutations.

Definition 39 ***(configuration)***
Given the length n *of a word* $W = \star \mathtt{c}_1 \dots \mathtt{c}_n$ *over* Σ *and a pointer machine* $M = (N, \mathtt{S}, \Delta)$, *a* configuration *of* (M, n) *is an element of*

$$\mathtt{S} \times \Sigma \times \mathtt{LR} \times \{0, 1, \dots, n\}^N$$

The element of $\mathtt{S}$ is the state of the machine and the element of Σ is the letter the main pointer points at. The element of $\mathtt{LR}$ is the direction of the next move of the main pointer, and the elements of $\{0, 1, \dots, n\}^N$ correspond to the positions of the (main and auxiliary) pointers on the input.

As the input tape is considered cyclic with a special symbol marking the beginning of the word (recall Definition 24), the pointer positions are integers *modulo* $n + 1$ for an input word of length n.

Definition 40 ***(transition)***
Let W *be a word and* $M = (N, \mathtt{S}, \Delta)$ *be a pointer machine. A* transition *of* M *on input* W *is a triple of configurations*

$$s, \mathtt{c}, \mathtt{d}, (p_1, \dots, p_N) \xrightarrow{\mathtt{MOVE}} s, \mathtt{c}', \overline{\mathtt{d}}, (p'_1, \dots, p'_N) \xrightarrow{\mathtt{SWAP}} s', \mathtt{c}'', \mathtt{d}', (p'_{\sigma(1)}, \dots, p'_{\sigma(N)})$$

such that

1. *if* $\mathtt{d} \in \mathtt{LR}$, $\overline{\mathtt{d}}$ *is the other element of* $\mathtt{LR}$,
2. $p'_1 = p_1 + 1$ *if* $\mathtt{d} = \mathtt{R}$ *and* $p'_1 = p_1 - 1$ *if* $\mathtt{d} = \mathtt{L}$,
3. $p'_i = p_i$ *for* $i \neq 1$,
4. $\mathtt{c}$ *is the letter at position* p_1 *and* $\mathtt{c}'$ *is the letter at position* p'_1,
5. *and* $(s, \mathtt{c}', \overline{\mathtt{d}}) \to (s', \mathtt{c}'', \mathtt{d}') \times \sigma$ *belongs to* Δ.

There is no constraint on c'', but every time this value differs from the letter pointed by $p'_{\sigma(1)}$, the computation will halt on the next MOVE phase, because there is a mismatch between the value that is supposed to have been read and the actual bit of W stored at this position, and that would contradict the first part of item 4. In terms of wirings, the MOVE phase corresponds to the application of the representation of the word, whereas the SWAP phase corresponds to the application of the observation.

Definition 41 ***(acceptance)***
We say that M *accepts* W *if any sequence of transitions* $(C_i \xrightarrow{\mathtt{MOVE}} C'_i \xrightarrow{\mathtt{SWAP}} C''_i)$ *such that* $C''_i = C_{i+1}$ *for all* i *is necessarily finite.*
We write $\mathcal{L}(M)$ *the set of words accepted by* M.

This means informally: we consider that a pointer machine accepts a word if it cannot ever loop, from whatever configuration it starts from. That a lot of paths of computation accepts "wrongly" is no worry, since only rejection is meaningful: our pointer machines compute in a "universally non-deterministic" way, to stick to the acceptance condition of wirings, nilpotency.

Proposition 42 ***(space and pointer machines)***
If $L \in$ NLOGSPACE, *then there exist a pointer machine* M *such that* $\mathcal{L}(M) = L$. *Moreover, if* $L \in$ LOGSPACE *then* M *can be chosen to be reversible.*

Proof (sketch). It is well-known [22] that read-only Turing Machines – or equivalently (non-)Deterministic Multi-Head Finite Automata – characterize (N)LOGSPACE. It takes little effort to see that our pointer machines are just a reasonable rearrangement of this model, since it is always possible to encode the missing information in the states of the machine.
That acceptance and rejection are "reversed" is harmless in the deterministic (or equivalently reversible [23]) case, and uses that CO-NLOGSPACE = NLOGSPACE to get the expected result in the non-deterministic case. □

As we said, our pointer machines are designed to be easily simulated by wirings, so that we get the expected result almost for free.

Theorem 43 ***(space completeness)***
If $L \in$ NLOGSPACE, *then there exist an observation* $\phi \in \mathcal{O}^+_\Sigma$ *such that* $\mathcal{L}(\phi) = L$. *Moreover, if* $L \in$ LOGSPACE *then* ϕ *is an isometric wiring.*

Proof. By Proposition 42, there exists a pointer machine $M = (N, \mathtt{S}, \Delta)$ such that $\mathcal{L}(M) = L$. We associate to the set $\mathtt{S}$ a set of distinct closed terms $[\mathtt{S}]$ and write $[s]$ the term associated to s. To any element $D = (s, \mathtt{c}, \mathtt{d}) \to (s', \mathtt{c}', \mathtt{d}') \times \sigma$ of Δ we associate the flow

$$[D] := ([s'] \bullet \mathtt{c}' \bullet \mathtt{d}' \leftharpoonup [s] \bullet \mathtt{c} \bullet \mathtt{d}) \dot{\otimes} [\sigma] \in ([\mathtt{S}]^* \dot{\otimes} \Sigma^* \dot{\otimes} \mathtt{LR}^*) \dot{\otimes} \mathcal{S}_n \subseteq \mathcal{O}^+_\Sigma$$

and we define the observation $[M] \in \mathcal{O}^+_\Sigma$ as $\sum_{D \in \Delta} [D]$.

One can easily check that this translation preserves the language recognized (there is even a step by step simulation of the computation on the word W by the wiring $[M]W(t_0, \dots, t_n)$) and relates reversibility with isometricity: in fact, M is reversible if and only if $[M]$ is an isometric wiring. Then, if $L \in$ LOGSPACE, M is deterministic and can always be chosen to be reversible [23]. □

Discussion

The language of the unification algebra gives us a twofold point of view on computation, either through algebraic structures (that are described finitely by wirings) or pointer machines. We may therefore start exploring possible variations of the construction, combining intuitions from both worlds.

For instance, the choice of a normative pair can affect the expressivity of the construction: the more restrictive the notion of representation of a word is, the more liberal that of an observation can become, as suggested by T. Seiller. Whether and how this can affect the corresponding complexity class is definitely a direction for future work.

Another pending question about this approach to complexity classes is to delimit the minimal prerequisites of the construction, its core.
Earlier works [13,14,15] made use of von Neumann algebras to get a setting that is expressive enough, we lighten the construction by using simpler objects. Yet, the possibility of representing the action of permutations on a unbounded tensor product is a common denominator that seems deeply related to logarithmic space and pointer machines.

The logical counterpart of this work also needs clarifying. Indeed, the idea of representation of words comes directly from proof-theory, while the notion of observation does not seem to correspond to any known logical construction.

Finally, execution in our setting being based on iteration of matching, which is computable efficiently by a parallel machine, it seems possible to relate our modelisation with parallel computation.

References

1. Girard, J.Y.: Linear logic. Theoret. Comput. Sci. 50(1), 1–101 (1987)
2. Girard, J.Y., Scedrov, A., Scott, P.J.: Bounded linear logic: a modular approach to polynomial-time computability. Theoret. Comput. Sci. 97(1), 1–66 (1992)
3. Girard, J.Y.: Light linear logic. In: Leivant, D. (ed.) LCC 1994. LNCS, vol. 960, pp. 145–176. Springer, Heidelberg (1995)
4. Schöpp, U.: Stratified bounded affine logic for logarithmic space. In: LICS, pp. 411–420. IEEE Computer Society (2007)
5. Dal Lago, U., Hofmann, M.: Bounded linear logic, revisited. LMCS 6(4) (2010)
6. Gaboardi, M., Marion, J.Y., Ronchi Della Rocca, S.: An implicit characterization of PSPACE. ACM Trans. Comput. 13(2), 18:1–18:36 (2012)
7. Baillot, P., Mazza, D.: Linear logic by levels and bounded time complexity. Theoret. Comput. Sci. 411(2), 470–503 (2010)
8. Girard, J.Y.: Towards a geometry of interaction. In: Gray, J.W., Ščedrov, A. (eds.) Proceedings of the AMS Conference on Categories, Logic and Computer Science. Categories in Computer Science and Logic, vol. 92, pp. 69–108. AMS (1989)
9. Asperti, A., Danos, V., Laneve, C., Regnier, L.: Paths in the lambda-calculus. In: LICS, pp. 426–436. IEEE Computer Society (1994)
10. Laurent, O.: A token machine for full geometry of interaction (extended abstract). In: Abramsky, S. (ed.) TLCA 2001. LNCS, vol. 2044, pp. 283–297. Springer, Heidelberg (2001)
11. Girard, J.Y.: Geometry of interaction 1: Interpretation of system F. Studies in Logic and the Foundations of Mathematics 127, 221–260 (1989)
12. Baillot, P., Pedicini, M.: Elementary complexity and geometry of interaction. Fund. Inform. 45(1-2), 1–31 (2001)
13. Girard, J.Y.: Normativity in logic. In: Dybjer, P., Lindström, S., Palmgren, E., Sundholm, G. (eds.) Epistemology Versus Ontology. Logic, Epistemology, and the Unity of Science, vol. 27, pp. 243–263. Springer (2012)

14. Aubert, C., Seiller, T.: Characterizing co-NL by a group action. Arxiv preprint abs/1209.3422 (2012)
15. Aubert, C., Seiller, T.: Logarithmic space and permutations. Arxiv preprint abs/1301.3189 (2013)
16. Girard, J.Y.: Geometry of interaction III: accommodating the additives. In: Girard, J.Y., Lafont, Y., Regnier, L. (eds.) Advances in Linear Logic. London Mathematical Society Lecture Note Series, vol. 222, pp. 329–389. CUP (1995)
17. Girard, J.Y.: Three lightings of logic. In: Ronchi Della Rocca, S. (ed.) CSL. LIPIcs, vol. 23, pp. 11–23. Schloss Dagstuhl - Leibniz-Zentrum für Informatik (2013)
18. Knight, K.: Unification: A multidisciplinary survey. ACM Comput. Surv. 21(1), 93–124 (1989)
19. Dwork, C., Kanellakis, P.C., Mitchell, J.C.: On the sequential nature of unification. J. Log. Program. 1(1), 35–50 (1984)
20. Dwork, C., Kanellakis, P.C., Stockmeyer, L.J.: Parallel algorithms for term matching. SIAM J. Comput. 17(4), 711–731 (1988)
21. Aubert, C.: Linear Logic and Sub-polynomial Classes of Complexity. PhD thesis, Université Paris 13–Sorbonne Paris Cité (2013)
22. Hartmanis, J.: On non-determinancy in simple computing devices. Acta Inform. 1(4), 336–344 (1972)
23. Lange, K.J., McKenzie, P., Tapp, A.: Reversible space equals deterministic space. J. Comput. System Sci. 60(2), 354–367 (2000)

Ramsey Theorem as an Intuitionistic Property of Well Founded Relations

Stefano Berardi and Silvia Steila

Dipartimento di Informatica, Università di Torino, Torino, Italy
{stefano,steila}@di.unito.it

Abstract. Ramsey Theorem for pairs is a combinatorial result that cannot be intuitionistically proved. In this paper we present a new form of Ramsey Theorem for pairs we call H-closure Theorem. H-closure is a property of well-founded relations, intuitionistically provable, informative, and simple to use in intuitionistic proofs. Using our intuitionistic version of Ramsey Theorem we intuitionistically prove the Termination Theorem by Poldenski and Rybalchenko. This theorem concerns an algorithm inferring termination for while-programs, and was originally proved from the classical Ramsey Theorem, then intuitionistically, but using an intuitionistic version of Ramsey Theorem different from our one. Our long-term goal is to extract effective bounds for the while-programs from the proof of Termination Theorem, and our new intuitionistic version of Ramsey Theorem is designed for this goal.

Keywords: Intuitionism, Ramsey Theorem, inductive definitions, termination of while-programs.

1 Introduction

Podelski and Rybalchenko [1] defined an algorithm taking in input an imperative program made with the instructions **while**, **if** and assignment, and able to decide in some case whether the program is terminating or not, and in some other cases leaving the question open. The authors prove a result they call the Termination Theorem, stating the correctness of their algorithm. The authors use in their proof Ramsey Theorem for pairs [2], from now on called just "Ramsey" for short. Ramsey is a classical result that cannot be intuitionistically proved: we refer to [3] for a detailed analysis of the minimal classical principle required to prove Ramsey. According to the Π^0_2-conservativity of Classical Analysis w.r.t. Intuitionistic Analysis [4], the proof of Termination Theorem hides some effective bounds for the while program which the theorem shows to terminate. Our long-term goal is to find them, by first turning the proof of Termination Theorem into an intuitionistic proof.

Our first step is to formulate a version of Ramsey which has a purely intuitionistic proof, that is, a proof which does not use Excluded Middle, nor Brouwer Thesis nor Choice. Our version of Ramsey is informative, in the sense that it has no negation, while it has a disjunction. We say that a relation R

G. Dowek (ed.): RTA-TLCA 2014, LNCS 8560, pp. 93–107, 2014.

is H-well-founded if the tree of all R-decreasing transitive sequences is well-founded. We express Ramsey as a property of well-founded relations, saying that H-well-founded relations are closed under finite unions. For short we will call this statement the *H-closure Theorem*. Thus, we are able to split the proof of Ramsey into two parts: the intuitionistic proof of the H-closure Theorem, followed by an easy classical proof of the equivalence between Ramsey and the H-closure Theorem.

The result closest to H-closure we could find is by Coquand [5]. Coquand, as Veldman and Bezem did before him [6], considers *almost full relations* and proves that they are closed under finite *intersections*. Veldman and Bezem use Choice Axiom of type 0 (if $\forall x \in \mathbb{N}.\exists y \in \mathbb{N}.C(x, y)$, then $\exists f : \mathbb{N} \to \mathbb{N}.\forall x \in \mathbb{N}.C(x, f(x))$) and Brouwer's thesis. Coquand's proof, instead, is purely intuitionistic, and it may be used to give a purely intuitionistic proof of the Termination Theorem [7]. However, it is not evident what are the effective bounds hidden in Coquand's proof of Termination Theorem. If we compare H-closure with the Almost Full Theorem, in the most recent version by Coquand [5], we find no easy way to intuitionistically deduce one from the other, due to the use of de' Morgan laws to move from the definition of almost full to the definition of H-closure. H-closure is in a sense more similar to the original Ramsey Theorem, because it was obtained from it with just one classical step, a contrapositive (see §2), while almost fullness requires one application of de' Morgan Law, followed by a contrapositive. We expect that H-closure, hiding one application less of de' Morgan laws, should be a version of Ramsey simpler to use in intuitionistic proofs and for extracting bounds.

Our motivation for producing a new intuitionistic version of Ramsey is to provide a new intuitionistic proof of the Termination Theorem. We expect that, by analysing this new proof, we will be able to extract effective bounds from the Termination Theorem, and possibly, from other concrete applications of Ramsey.

This is the plan of the paper. In section 2 we present Ramsey Theorem for pairs and we informally introduce H-closure. In section 3 we formally define inductive well-foundedness and H-well-foundedness, whose main properties are stated in section 4. The goal of section 5 is to present what we call Nested Fan Theorem, which is a part of the proof of the H-closure Theorem, as shown in section 6. In section 7 we intuitionistically prove the Termination Theorem. In section 8 we compare our result with the previous works along the same line and we draw some conclusions. Unless explicitly stated, our proofs use intuitionistic second order arithmetic, without Choice Axiom, Brouwer Thesis, Bar-Induction.

2 Ramsey Theorem and a Variant of It, H-Closure

We first recall the statement of Ramsey Theorem for pairs, just Ramsey for short. Assume G is a countable non-oriented graph which is complete, i.e., between any two different elements of G there is exactly one edge in G. Assume we "colored" the edges of G with $n > 0$ different colors, that is, we partioned the edges of G into n sets. Then there is an infinite set $X \subseteq G$ such that all the edges between any two different $x, y \in X$ have the same color: for some $k = 1, \ldots, n$, all the edges of X fall in the k-th subset of the partition. We call X an *homogenous* set of color k.

Assume $\sigma = x_0, x_1, \ldots, x_n, \ldots$ is an injective enumeration of the elements of G, that is: $G = \mathbf{range}(\sigma)$. We represent a non-oriented edge, between two points x_i, x_j in G with $j < i$, by the pair (i, j), arbitrarily oriented from i to j. The opposite edge from x_j to x_i is the same edge of G, and it is again represented with (i, j). Thus, a partition of edges in n sets $S_1, \ldots, S_n$ may be represented by a partition of the set $\{(x_i, x_j) : j < i\}$ into n binary relations $S_1, \ldots, S_n$. Therefore one possible formalization of Ramsey is the following.

Theorem 1 (Ramsey for pairs [2]). *Assume I is a set having some injective enumeration $\sigma = x_0, x_1, \ldots, x_i, \ldots$. Assume $S_1, \ldots, S_n$ are binary relations on I which are a partition of $\{(x_i, x_j) \in I \times I : j < i\}$, that is:*

1. $S_1 \cup \cdots \cup S_n = \{(x_i, x_j) \in I \times I : j < i\}$
2. *for all* $1 \leq k < h \leq n$: $S_k \cap S_h = \emptyset$.

Then for some $k = 1, \ldots, n$ there exists some infinite $X \subseteq \mathbb{N}$ such that: $\forall i, j \in X.(j < i \implies x_i S_k x_j)$.

In the statement above three assumptions may be dropped.

1. First of all, we may drop the assumption that $S_1, \ldots, S_n$ are pairwise disjoint. Suppose we do. Then, if we set $S'_1 = S_1$, $S'_2 = S_2 \setminus S'_1$, $S'_3 = S_3 \setminus (S'_1 \cup S'_2)$, $\ldots$, we obtain a partition $S'_1, \ldots, S'_n$ of $\{(x_i, x_j) : j < i\}$. Therefore there exists a $k = 1, \ldots, n$ and some infinite $X \subseteq \mathbb{N}$, such that $\forall i, j \in X.(j < i \implies x_i S'_k x_j)$, and with more reason, $\forall i, j \in X.(j < i \implies x_i S_k x_j)$.
2. Second, we may drop the assumption "σ is injective" (in this case, $\mathbf{range}(\sigma)$ may be a finite set). Assume we do. Then, if we set $S'_k = \{(i, j) : x_i S_k x_j\}$ for all $k = 1, \ldots, n$, we obtain n relations $S'_1, \ldots, S'_k$ on $\mathbb{N}$, whose union is the set $\{(i, j) \in \mathbb{N} \times \mathbb{N} : j < i\}$. Therefore there exists a $k = 1, \ldots, n$ and some infinite $X \subseteq \mathbb{N}$, such that $\forall i, j \in X.(j < i \implies i S'_k j)$, and with more reason, $\forall i, j \in X.(j < i \implies x_i S_k x_j)$.
3. Third, we may drop the assumption that σ is an enumeration of I. Suppose we do. Then, if we restrict $S_1, \ldots, S_n$ to $I_0 = \mathbf{range}(\sigma)$, we obtain some binary relations $S'_1, \ldots, S'_n$ on I_0 such that $S'_1 \cup \ldots \cup S'_n = \{(x_i, x_j) \in I_0 \times I_0 : j < i\}$. Again, we conclude that there exists some $k = 1, \ldots, n$ and some infinite $X \subseteq \mathbb{N}$, such that $\forall x_i, x_j \in X.(j < i \implies x_i S'_k x_j)$, and with more reason, $\forall i, j \in X.(j < i \implies x_i S_k x_j)$.

Summing up, we showed that, classically, we may restate Ramsey Theorem as follows:

For all sequences $\sigma = x_0, x_1, x_2, \ldots$ on I, if $\forall i, j \in \mathbb{N}.(j < i \implies x_i (S_1 \cup \ldots \cup S_n) x_j)$, then for some k there is some infinite $X \subseteq \mathbb{N}$, such that $\forall i, j \in X.(j < i \implies x_i S_k x_j)$.

It is likely that even this statement cannot be intuitionistically proved, because the sequence τ is akin to an homogeneous set, and there is no effective way to produce homogeneous sets (see for instance [3]). By taking the contrapositive, we obtain the following corollary:

If for all $k = 1, \ldots, n$, *all sequences* $\tau = y_0, \ldots, y_n, \ldots$ *such that* $\forall i, j \in \mathbb{N}.(j < i \implies y_i S_k y_j)$ *are finite, then all sequences* $\sigma = x_0, \ldots, x_n, \ldots$ *such that* $\forall i, j \in \mathbb{N}.(j < i \implies x_i(S_1 \cup \ldots \cup S_n)x_j)$ *are finite.*

It is immediate to check that, classically, this is yet another version of Ramsey. We call this property *classical H-closure.*

Let us call $H(S)$ the set of all lists such that $1 \leq j < i \leq n$ implies $x_i S x_j$. Then classical H-closure may be restated as follows: if $S_1, \ldots, S_n$ are binary relations over some set I, and $H(S_1), \ldots, H(S_k)$ are sets of lists well-founded by extension, then $H(S_1 \cup \ldots \cup S_k)$ is a set of list well-founded by extension. Thus, classical H-closure is a property classically equivalent to Ramsey Theorem, but which is about well-founded relations. In Proof Theory, there is plenty of examples of classical proofs of well-foundedness which are turned into intuitionistic proofs, and indeed from H-closure we will obtain an intuitionistic version of Ramsey.

There is a last step to be done. We call *intuitionistic H-closure*, or just H-closure for short, the statement obtained by replacing, in classical H-closure, the classical definition of well-foundedness (all decreasing sequences are finite) with the inductive definition of well-foundedness, which is customary in intuitionistic logic. We will recall the inductive definition of well-foundedness in §3.1: thus, for the formal definition of H-well-foundedness we have to wait until §3.2.

3 Well-Founded Relations

In this section we introduce the main objects we will deal with in this paper: well-founded relations.

We will use I, J, ... to denote sets, R, S, T, U will denote binary relations, X, Y, Z will be subsets, and x, y, z, t, ... elements. We identify the properties $P(\cdot)$ of elements of I with their extensions $X = \{x \in I : P(x)\} \subseteq I$.

Let R be a binary relation on I. Classically $x \in I$ is R-well-founded if there is no infinite decreasing R-chain $\ldots x_n R x_{n-1} R \ldots x_1 R x_0 = x$ from x in I. Classically R is well-founded if and only if every $x \in I$ is R-well-founded.

The inductive definition of well-founded relations is more suitable than the classical one in the intuitionistic proofs. In the first subsection we introduce this definition; in the last subsection we present the definition of H-well-foundedness, which is fundamental to state the new intuitionistic form of Ramsey Theorem.

3.1 Intuitionistic Well-Founded Relations

The intuitionistic definition of well-founded relation uses the definition of inductive property. For short we will say that a relation is "well-founded" to say that it is intuitionistically well-founded.

Let R be a binary relation on I. A property is R-inductive if whenever it is true for all R-predecessors of a point it is true for the point. $x \in I$ is

R-well-founded if and only if it belongs to every R-inductive property; R is well-founded if every x in I is R-well-founded. Formally:

Definition 1. *Let R be a binary relation on I.*

- *A property $X \subseteq I$ is* R-inductive *if and only if* IND_X^R*; where* $\mathrm{IND}_X^R := \forall y. (\forall z. (zRy \implies z \in X) \implies y \in X)$.
- *An element $x \in I$* is R-well-founded *if and only if* $\mathrm{WF}^R(x)$*; where* $\mathrm{WF}^R(x) := \forall X. \left(\mathrm{IND}_X^R \implies x \in X\right)$.
- R is well-founded *if and only if* $\mathrm{WF}(R)$*; where* $\mathrm{WF}(R) := \forall x. \mathrm{WF}^R(x)$.

A binary structure, just a *structure* for short, is a pair (I, R), where R is a binary relation on I. We say that (I, R) is well-founded if R is well-founded.

We need to introduce also the notion of co-inductivity. A property X is R-co-inductive in $y \in I$ if it satisfies the inverse property of R-inductive: if the property X holds for a point, then it holds also for all its R-predecessors. Formally:

Definition 2. *Let R be a binary relation on I.*

- *A property X is* R-co-inductive in $y \in I$ *if and only if* $\mathrm{CoIND}_X^R(y)$*; where*
$$\mathrm{CoIND}_X^R(y) := \forall z. (zRy \implies z \in X).$$
- *A property X is* R-co-inductive *if and only if* CoIND_X^R*; where*
$$\mathrm{CoIND}_X^R := \forall y. (y \in X \implies \forall z. (zRy \implies z \in X)).$$

In general we will intuitionistically prove that if there exists an infinite decreasing R-chain from x then x is not R-well-founded. Classically, and by using the Axiom of Choice, x is R-well-founded if and only if there are no infinite decreasing R-chains from x, and R is well-founded if and only if there are no infinite decreasing R-chains in I.

3.2 H-Well-Founded Relations

In order to define H-well-foundedness we need to introduce some notations. We denote a list on I with $\langle x_1, \ldots, x_n \rangle$; $\langle\rangle$ is the empty list. We define the operation of concatenation of two lists on I in the natural way as follows: $\langle x_1, \ldots, x_n \rangle * \langle y_1, \ldots, y_m \rangle = \langle x_1, \ldots, x_n, y_1, \ldots, y_m \rangle$. We define the relation of one-step expansion $\succ$ between two lists (L, M) on the same I, as $L \succ M \iff L = M * \langle y \rangle$, for some y.

Definition 3. *Let R be a binary relation on I.*

- *$H(R)$ is the set of the R-decreasing transitive finite sequences on I:*
$$\langle x_1, \ldots, x_n \rangle \in H(R) \iff \forall i, j \in [1, n]. i < j \implies x_j R x_i.$$
- *R is H-well-founded if $H(R)$ is $\succ$-well-founded.*

H-well-founded relations are more common than well-founded relations.

Proposition 1. *1. R well-founded implies that R H-well-founded.*
2. R H-well-founded and R transitive imply that R well-founded

4 Basic Properties of Well-Founded Relations

There are several methods to intuitionistically prove that a binary relation R is well-founded by using the well-foundedness of another binary relation S.

The goal of this section is to prove these results. In §4.1 we are going to define simulation relations, in §4.2 we introduce some operations which preserve well-foundedness, while in §4.3 we will show the main properties of well-foundedness.

4.1 Simulation Relations

A simulation relation is a binary relation which correlates two other binary relations.

Definition 4. *Let R be a binary relation on I and S be a binary relation on J. Let T be a binary relation on $I \times J$.*

- *Domain of T.* $\mathrm{dom}(T) = \{x \in I : \exists y \in J.xTy\}$.
- *Morphism. $f : (I, R) \to (I, S)$ is a morphism if f is a function such that $\forall x, y \in I.xRy \implies f(x)Sf(y)$.*
- *Simulation. T is a simulation of R in S if and only if it is a relation and*
$$\forall x, z \in I.\forall y \in J.\,((xTy \wedge zRx) \implies \exists t \in J.\,(tSy \wedge zTt))$$
- *Total simulation. A simulation relation T of R in S is total if* $\mathrm{dom}(T) = I$.
- *Simulable. R is simulable in S if there exists a* total *simulation relation T of R in S.*

If we have a simulation T of R in S and xTy holds, we can transform each finite decreasing R-chain in I from x in a finite decreasing S-chain in J from y. By using the Axiom of Choice this result holds also for infinite decreasing R-chains from a point in $\mathrm{dom}(T)$. Then if there are no infinite decreasing S-chains in J there are no infinite decreasing R-chains in $\mathrm{dom}(T)$. If, furthermore, the simulation is total there are no infinite decreasing R-chains in I. By using classical logic and the Axiom of Choice we may conclude that if S is well-founded and T is a total simulation relation of R in S then R is well-founded. In the last subsection of this section we will present an intuitionistic proof of this result that does not use the Axiom of Choice.

We may see binary relations as abstract reduction relations. Recall that an abstract reduction relation is a simply binary relation (for example rewriting relations). A reduction relation is said to be terminating or strongly normalizing if and only if there are no infinite chains [8]. Observe that we use simulation to prove well foundedness and this is the same method used for labelled state transition systems [9]; for us every set of labels is a singleton.

4.2 Some Operations on Binary Structures

In this subsection we introduce some operations mapping binary structures into binary structures. In §4.3 we prove that these operations preserves well-foundedness.

The first operation is the successor operation (adding a top element).

Definition 5. *Let R be a relation on I and let $\top$ be an element not in I. We define the relation $R+1 = R \cup \{(x,\top) : x \in I\}$ on $I+1 = I \cup \{\top\}$. We define the successor structure of (I,R) as $(I,R)+1 = (I+1, R+1)$.*

Another operation on binary structures is the relation defined by components, inspired by the order by components.

Definition 6. *Let R be a binary relation on I, and let S be a binary relation on J. The relation $R \otimes S$ of components R, S is defined as below:*

$$R \otimes S := (R \times \mathrm{Diag}(J)) \cup (\mathrm{Diag}(I) \times S) \cup (R \times S),$$

where $\mathrm{Diag}(X) = \{(x,x) : x \in X\}$.

Equivalently $R \otimes S$ is defined for all $x, x' \in I$ and for all $y, y' \in J$ by:

$$(x,y) R \otimes S (x',y') \iff$$
$$((xRx') \wedge (y = y')) \vee ((x = x') \wedge (ySy')) \vee ((xRx') \wedge (ySy')) .$$

If R, S are orderings then $R \otimes S$ is the componentwise ordering, also called the product ordering. In this case $R \otimes S = R \times S$, while in general $R \otimes S \supseteq R \times S$.

4.3 Properties of Well-Foundedness

Now we may list the main intuitionistic properties of well-founded relations.

Proposition 2. *Let R be a binary relation on I, and let S be a binary relation on J.*

1. *Well-foundedness is both an inductive and a co-inductive property:*

   $$x \text{ is } R\text{-well-founded} \iff \forall y.(yRx \implies y \text{ is } R\text{-well-founded}).$$

2. *If R, S are well-founded, then $R \otimes S$ is well-founded.*
3. *If T is a simulation of R in S and if xTy and y is S-well-founded, then x is R-well-founded.*
4. *If T is a simulation of R in S and S is well-founded, then $\mathrm{dom}(T)$ is R-well-founded.*
5. *If R is simulable in S and S is well-founded, then R is well-founded.*
6. *If $f : (I,R) \to (J,S)$ is a morphism and if S is well-founded, then R is well-founded.*
7. *If R is included in S and S is well-founded then R is well-founded.*

Corollary 1. *Let R be a binary relation on I. (I,R) well-founded implies that $(I,R)+1$ well-founded.*

Corollary 2. *Let R be a binary relation on I and $x \in I$. If there exists an infinite decreasing R-chain from x, then x is not R-well-founded.*

So the intuitionistic definition of well-founded intuitionistically implies the classical definition; while the other implication is purely classical.

When I and R are finite, we may characterize the well-foundedness and the H-well-foundedness in an elementary way.

Definition 7. *Let R be a binary relation on I and $x \in I$. A finite sequence $\langle x_0, \ldots, x_n \rangle$ is an R-cycle from x if $n > 0$ and*

$$x = x_n R x_{n-1} R x_{n-2} R \ldots R x_0 = x.$$

If $n = 1$ (that is, if xRx), we call the R-cycle an R-loop.

Proposition 3. *Assume $I = \{x_1, \ldots, x_k\}$ for some $k \in \mathbb{N}$. Let R be any binary relation on I.*

1. *R is well-founded if and only if there are no R-cycles.*
2. *R is H-well-founded if and only if there are no R-loops.*

Thanks to Proposition 3 we may prove H-closure Theorem if $R_1, \ldots, R_n$ are relations over a finite set I. In fact $R = (R_1 \cup R_2 \cup \cdots \cup R_n)$ is H-well-founded if and only if there are no R-loops. This is equivalent to: there are no R_i-loops for any $i \in [1, n]$. Hence R is H-well-founded if and only if for each $i \in [1, n]$, R_i is H-well-founded.

Now we want to prove H-closure Theorem for any set I.

5 An Intuitionistic Version of König's Lemma

In this section we deal with binary trees. In the first part we introduce binary trees with some equivalent definitions, while in the second part we use binary trees to prove an intuitionistic version of König Lemma for nested binary trees (binary trees whose nodes are themselves binary trees), which we call Nested Fan Theorem. As in the classical case [3], there is a strong link between intuitionistic Ramsey Theorem and Nested Fan Theorem.

5.1 Binary Trees

Let R be a binary relation. Then we can define the set of all binary trees where each child node is in relation R with its father node. If R is well-founded, this set will be well-founded with respect to the relation "one-step extension" between trees.

A finite binary tree may be defined in many ways, the most common runs as follows.

Definition 8. *A finite binary tree on I is defined inductively as an empty tree, called* Nil, *or a triple composed by one element of I and two trees, called immediate subtrees: so we have* $\mathrm{Tr} = \mathrm{Nil}$ *or* $\mathrm{Tr} = \langle x, \mathrm{Tr}_1, \mathrm{Tr}_2 \rangle$.

$$\mathrm{BinTr} = \{\mathrm{Tr} : \mathrm{Tr} \textit{ is a binary tree } \}$$

Let $\mathrm{Tr} = \langle x, \mathrm{Tr}_1, \mathrm{Tr}_2 \rangle$, then

- Tr is a tree with root x;
- if $\mathrm{Tr}_1 = \mathrm{Tr}_2 = \mathrm{Nil}$, we will say that Tr is a leaf-tree;
- if $\mathrm{Tr}_1 \neq \mathrm{Nil}$ and $\mathrm{Tr}_2 = \mathrm{Nil}$, we will say that Tr has exactly one left child;
- if $\mathrm{Tr}_1 = \mathrm{Nil}$ and $\mathrm{Tr}_2 \neq \mathrm{Nil}$, we will say that Tr has exactly one right child;
- if $\mathrm{Tr}_1 \neq \mathrm{Nil}$ and $\mathrm{Tr}_2 \neq \mathrm{Nil}$, we will say that Tr has two children: one right child and one left child.

A binary tree may be also define as a labelled oriented graph on I, empty (if $\mathrm{Tr} = \mathrm{Nil}$) or with a special element, called root, which has exactly one path from the root to any node. Each edge is labelled with a color $c \in C = \{1, 2\}$ in such a way that from each node there is at most one edge in each color.

Equivalently we may define firstly colored lists and then the binary tree as a set of some colored lists.

Definition 9. *A colored list* (L, f) *is a pair, where* $L = \langle x_1, \ldots, x_n \rangle$ *is a list on* I *equipped with a list* $f = \langle c_1, \ldots, c_{n-1} \rangle$ *on* $C = \{1, 2\}$. nil $= (\langle\rangle, \langle\rangle)$ *is the empty colored list and* $\mathrm{ColList}(C)$ *is the set of the colored lists with colors in* C.

We should imagine that the list L is drawn as a sequence of its elements and that for each $i \in [1, n-1]$ the segment (x_i, x_{i+1}) has color c_i. Observe that if $L = \langle\rangle$ or if $L = \langle x \rangle$, then $f = \langle\rangle$: if there are no edges in L, then there are no colors (L, f).

We use $\lambda, \mu, \ldots$ to denote colored lists in $\mathrm{ColList}(C)$. Let $c \in C$. We define the composition of color c of two colored lists by connecting the last element of the first list (if any) with the first of the second list (if any) with an edge of color c. Formally we set $\mathrm{nil} *_c \lambda = \lambda *_c \mathrm{nil} = \lambda$, and $(L, f) *_c (M, g) = (L*M, f*\langle c \rangle *g)$ whenever $L, M \neq \mathrm{nil}$.

We can define the relation one-step extension on colored lists: $\succ_c$ is the one-step extension of color c and $\succ_{\mathrm{col}}$ is the one-step extension of any color. Assume $C = \{1, 2\}$ and $x \in I$ and $\lambda, \mu \in \mathrm{ColList}(C)$. Then we set:

- $\lambda *_c (\langle x \rangle, \langle\rangle) \succ_c \lambda$.
- $\lambda \succ_{\mathrm{col}} \mu$ if $\lambda \succ_c \mu$ for some $c \in C$.

Now we can equivalently define a binary tree on I as a particular set of some colored lists.

Definition 10. *A binary tree* Tr *is a set of colored lists on* I, *such that:*

1. nil *is in* Tr;
2. *If* $\lambda \in \mathrm{Tr}$ *and* $\lambda \succ_{\mathrm{col}} \mu$, *then* $\mu \in \mathrm{Tr}$;
3. *Each list in* Tr *has at most one one-step extension for each color* $c \in C$: *if* $\lambda_1, \lambda_2, \lambda \in \mathrm{Tr}$ *and* $\lambda_1, \lambda_2 \succ_c \lambda$, *then* $\lambda_1 = \lambda_2$.

For all sets $\mathcal{L} \subseteq \mathrm{ColList}(C)$ *of colored lists,* $\mathrm{BinTr}(\mathcal{L})$ *is the set of binary trees whose branches are all in* $\mathcal{L}$.

For instance the empty tree is the set Nil $=\{\text{nil}\}$. From $(\langle x\rangle,\langle\rangle)\succ_c \text{nil}$ we deduce that there is at most one $(\langle x\rangle,\langle\rangle)\in \text{Tr}$: x is root of Tr. The leaf-tree of root x is equal to $\{(\langle x\rangle,\langle\rangle),\text{nil}\}$. The tree with only one root x and two children y, z is equal to

$$\{(\langle x,y\rangle,\langle 1\rangle),(\langle x,z\rangle,\langle 2\rangle),(\langle x\rangle,\langle\rangle),\text{nil}\}.$$

The last definition we need is the one-step extension $\succ_T$ between binary trees; $\text{Tr}'\succ_T \text{Tr}$ if Tr' has one leaf more than Tr.

Definition 11 (One-step extension for binary trees). *If* Tr *is a binary tree and* $\lambda\in\text{Tr}$ *and* $\mu\succ_c\lambda$ *and* $\lambda'\succ_c\lambda$ *for no* $\lambda'\in\text{Tr}$, *then*

$$\text{Tr}\cup\{\mu\}\succ_T \text{Tr}$$

5.2 Nested Fan Theorem

König Lemma is a result of classical logic which guarantees that if every branch of a binary tree is finite then the tree is finite.

There exists a corresponding intuitionistic result, intuitionistically weaker than the original one that we may state as follows.

Lemma 1 (Fan Theorem). *Each inductively well-founded binary tree is finite.*

Here we are interested to an intuitionistic version of Fan Theorem for nested trees (trees whose nodes are trees), that we will call Nested Fan Theorem.

Let consider a tree Tr whose nodes are finite binary trees, and whose father/child relation between nodes is the one-step extension $\succ_T$. Classically we may say: if for each branch of Tr the union of the nodes in this branch is a binary tree with only finite branches, then each branch of Tr is finite.

In the intuitionistic proof of the intuitionistic Ramsey Theorem we will use an intuitionistic version of this statement, in which the finitess of the branches is replaced by inductive well-foundedness of branches. Intuitionistic Nested Fan Theorem states that if a set of colored lists $\mathcal{L}$ is well-founded then the set $\text{BinTr}(\mathcal{L})$, of all binary trees whose branches are all in $\mathcal{L}$, is well-founded.

Lemma 2 (Intuitionistic Nested Fan Theorem). *Let* $C=\{1,2\}$ *be a set of colors and let* $\mathcal{L}\subseteq\text{ColList}(C)$ *be any set of colored lists with all colors in* C. *Then*

$$(\mathcal{L},\succ_{col}) \textit{ is well-founded} \implies (\text{BinTr}(\mathcal{L}),\succ_T) \textit{ is well-founded.}$$

Proof (sketch). Let $c\in C$, $\lambda\in\text{ColList}(C)$. We define $\text{BinTr}(\mathcal{L},\lambda,c)$ as the set of binary trees $\{\text{Tr}\in\text{BinTr}(\mathcal{L}):\lambda *_c \text{Tr}\subseteq\mathcal{L}\}$. $\text{BinTr}(\mathcal{L},\lambda,c)$ is the set of trees occurring in some tree of $\text{BinTr}(\mathcal{L})$, as immediate subtree number c of the last node of the branch λ. For instance, $\text{BinTr}(\mathcal{L},\text{nil},c)=\text{BinTr}(\mathcal{L})$.

Since $\mathcal{L}$ is well-founded, it can be proved that $(\text{BinTr}(\mathcal{L},\lambda,c),\succ_T)$ is well-founded for all $\lambda\in\mathcal{L}$ by induction over λ. The thesis will follow if we set $\lambda=\text{nil}$, $c=1$ (a dummy value).

6 An Intuitionistic Form of Ramsey Theorem

In this section we present a new intuitionistic version of Ramsey Theorem, the H-closure Theorem. In the first part of the section we state it and we prove the easy classical equivalence between it and Ramsey Theorem, in the second part we prove the H-closure Theorem.

6.1 Stating an Intuitionistic Form of Ramsey Theorem

In [3] we proved that the first order fragment of Ramsey Theorem is equivalent to the purely classical principle Σ^0_3- LLPO [10], so it is not an intuitionistic result. The H-closure Theorem is a version of Ramsey Theorem intuitionistically valid.

Theorem 2. *[H-closure Theorem] The H-well-founded relations are closed under finite unions:*

$$(R_1, \ldots, R_n \text{ } H\text{-well-founded}) \implies ((R_1 \cup \cdots \cup R_n) \text{ } H\text{-well-founded}).$$

H-closure Theorem is classically true, because there exists a simple classical proof of the equivalence between Ramsey Theorem and H-closure Theorem. This is one reason for finding an intuitionistic proof of H-closure Theorem: it splits the proof of Ramsey Theorem into two parts, one intuitionistic and the other classical but simple (it could be proved using the sub-classical principle LLPO-3 [3]). We claim we may derive Ramsey for recursive colorings in Heyting Arithmetic plus the following sub-classical schema:

Assume T is an infinite r.e. k-branching tree. There is an arithmetical formula defining a branch r of T and some $i \leq k$ such that r includes infinitely many "i-children".

6.2 Proving the Intuitionistic Form Ramsey Theorem

We introduce a particular set of colored lists: the (R_1, R_2)-colored lists. This set will be well-founded if R_1, R_2 are H-well-founded. Let (L, f) be a colored list. We say that (L, f) is a (R_1, R_2)-colored list if for every segment (x_i, x_{i+1}) of (L, f), if it has color $k \in \{1, 2\}$ then x_i is R_k-greater than all the elements of L that follows it. Informally, a sequence is a (R_1, R_2)-colored list if whenever the sequence decreases w.r.t. R_i, then it remains smaller w.r.t. to R_i. Formally:

Definition 12. $(L, f) \in \text{ColList}(C)$ *is a R_1, R_2-colored list if either $L = \langle\rangle$ and $f = \langle\rangle$ or $L = \langle x_1, \ldots, x_n\rangle$, $f = \langle c_1, \ldots, c_{n-1}\rangle$, and*

$$\forall i \in [1, n-1].(c_i = k \implies (\forall j \in [1, n].i < j \implies (x_j R_k x_i))).$$

$\text{ColList}(R_1, R_2) \subseteq \text{ColList}(C)$ *is the set of (R_1, R_2)-colored lists.*

We may think of a (R_1, R_2)-colored list as a simultaneous construction of one R_1-decreasing transitive list and one R_2-decreasing transitive list. We call an *Erdős-tree over R_1, R_2*, a (R_1, R_2)-tree for short, any binary tree whose branches are all in $\mathrm{ColList}(R_1, R_2)$. Erdős-trees are inspired by the trees used first by Erdős then by Jockusch in their proofs of Ramsey [11], hence the name. We may think of a (R_1, R_2)-tree as a simultaneous construction of many R_1-decreasing transitive lists and many R_2-decreasing transitive lists.

$\mathrm{BinTr}(\mathrm{ColList}(R_1, R_2))$ is the set of all (R_1, R_2)-trees. We will considering the one-step extension $\succ_{\mathrm{col}}$ on colored lists in $\mathrm{ColList}(R_1, R_2)$, and the one-step extension $\succ_T$ on binary trees in $\mathrm{BinTr}(\mathrm{ColList}(R_1, R_2))$.

Now we note that each one-step step extension in a $R_1 \cup R_2$-decreasing transitive list may be simulated as an one-step step extension of some Erdős-tree on (R_1, R_2), that is, as an one-step extension either of one R_1-decreasing transitive list or of one R_2-decreasing transitive list, among those associated to the branches of the (R_1, R_2)-tree. From the well-foundedness of the set $\mathrm{BinTr}(\mathrm{ColList}(R_1, R_2))$ of Erdős-trees we will derive our intuitionistic version of Ramsey Theorem.

Lemma 3. *(Simulation) Let R_1, R_2 be binary relations on a set I.*

1. $(\mathrm{ColList}(R_1, R_2), \succ_{\mathrm{col}})$ *is simulable in* $(H(R_1) \times H(R_2), \succ \otimes \succ) + 1$.
2. $H(R_1 \cup R_2, \succ)$ *is simulable in* $(\mathrm{BinTr}(\mathrm{ColList}(R_1, R_2)), \succ_T)$.

Corollary 3. *Let R_1, R_2 be binary relations H-well-founded on a set I.*

1. *The set* $(\mathrm{ColList}(R_1, R_2), \succ_{\mathrm{col}})$ *of R_1, R_2-colored lists is well-founded.*
2. *The set* $(\mathrm{BinTr}(\mathrm{ColList}(R_1, R_2)), \succ_T)$ *is well-founded.*

Proof. 1. $(H(R_1) \times H(R_2), \succ \otimes \succ)$ is well-founded by Proposition 2.2, since its components are. By Corollary 1, $(H(R_1) \times H(R_2), \succ \otimes \succ) + 1$ is well-founded. Since $(\mathrm{ColList}(R_1, R_2), \succ_{\mathrm{col}})$ is simulable in $(H(R_1) \times H(R_2), \succ \otimes \succ) + 1$ by Lemma 3, then it is well-founded by Proposition 2.5.
2. Since $(\mathrm{ColList}(R_1, R_2), \succ_{\mathrm{col}})$ is well-founded thanks to the previous point, $(\mathrm{BinTr}(\mathrm{ColList}(R_1, R_2)), \succ_T)$ is well-founded by Lemma 2.

Let $\emptyset$ be the empty binary relation on I. Then $H(\emptyset)$ does not contain lists of length greater or equal than 2. Hence $H(\emptyset) = \{\langle x \rangle : x \in I\} \cup \{\langle\rangle\}$. $H(V)$ is $\succ$-well-founded since each $\langle x \rangle$ is $\succ$-minimal, and $\langle\rangle$ has height less or equal than 1. Thus, the empty relation is H-well-founded.

Theorem 3. *Let $n \in \mathbb{N}$. If $R_1, \ldots, R_n$ H-well-founded then $(R_1 \cup \cdots \cup R_n)$ is H-well-founded.*

Proof. We may prove it by induction on $n \in \omega$. If $n = 0$ then $(R_1 \cup \cdots \cup R_n) = \emptyset$: we already considered this case. Assume that $n > 0$, and that the thesis holds for any $m < n$. Then $R_1 \cup \cdots \cup R_{n-1}$ is H-well-founded. Thus, in order to prove that $(R_1 \cup \cdots \cup R_n)$ is H-well-founded, it is enough to consider the case $n = 2$.

Assume R_1, R_2 are H-well-founded relations: then by applying Corollary 3.2, $(\mathrm{BinTr}(\mathrm{ColList}(R_1, R_2)), \succ_T)$ is well-founded. By Lemma 3, $(H(R_1 \cup R_2), \succ)$ is simulable in $(\mathrm{BinTr}(\mathrm{ColList}(R_1, R_2)), \succ_T)$, well-founded, therefore it is itself well-founded by Proposition 2.5.

Corollary 4. *Let $n \in \mathbb{N}$. $R_1, \ldots, R_n$ are H-well-founded if and only if $(R_1 \cup \cdots \cup R_n)$ is H-well-founded.*

Proof. $\Rightarrow$ Theorem 3.

$\Leftarrow$ If R and S are binary relations such that $R \subseteq S$, then S is H-well-founded implies that R is H-well-founded. In fact we have $H(R) \subseteq H(S)$; so by Proposition 2.7, if $(H(S), \succ)$ is well-founded then $(H(R), \succ)$ is well-founded. Since $\forall i \in [1, n].R_i \subseteq R_1 \cup \cdots \cup R_n$, then R_i is H-well-founded.

7 Podelski and Rybalchenko's Termination Theorem

In this last section we prove that the Termination Theorem [1, Theorem 1] is intuitionistically valid. For all details we refer to this paper: here we only include the definitions of program, computation, transition invariant and disjunctively well-founded relations that Podelski and Rybalchenko used.

Definition 13 (Transition Invariants). *As in [1]:*

- *A program $P = (W, I, R)$ consists of:*
  - *W: a set of states,*
  - *I: a set of starting states, such that $I \subseteq W$,*
  - *R: a transition relation, such that $R \subseteq W \times W$.*
- *A computation is a maximal sequence of states $s_1, s_2, \ldots$ such that*
  - *$s_1 \in I$,*
  - *$(s_i, s_{i+1}) \in R$ for all $i \geq 1$.*
- *The set* Acc *of accessible states consists of all states that appear in some computation.*
- *A transition invariant T is a superset of the transitive closure of the transition relation R restricted to the accessible states* Acc. *Formally,*

$$R^+ \cap (\mathrm{Acc} \times \mathrm{Acc}) \subseteq T.$$

- *The program P is terminating if and only if $R \cap (Acc \times Acc)$ is well-founded.*
- *A relation T is disjunctively well-founded if it is a finite union $T = T_1 \cup \cdots \cup T_n$ of well-founded relations.*

Lemma 4. *If $T = R \cap (\mathrm{Acc} \times \mathrm{Acc})$ is well-founded then $U = R^+ \cap (\mathrm{Acc} \times \mathrm{Acc})$ is well-founded.*

Theorem 4 (Termination). *The program P is terminating if and only if there exists a disjunctively well-founded transition invariant for P.*

Proof. $\Leftarrow$ Let $T = T_1 \cup \cdots \cup T_n$ with $T_1, \ldots, T_n$ well-founded and T transitive, then by H-closure Theorem 3 and thanks to the Proposition 1 we obtain T is well-founded, so P is terminating.

$\Rightarrow$ Let P be terminating then $R \cap (\mathrm{Acc} \times \mathrm{Acc})$ is well-founded. By Lemma 4 $R^+ \cap (\mathrm{Acc} \times \mathrm{Acc})$ is well-founded. Then we are done.

8 Related Works and Conclusions

In [3] we studied how much Excluded Middle is needed to prove Ramsey Theorem. The answer was that the first order fragment of Ramsey Theorem is equivalent in HA to Σ^0_3-LLPO, a classical principle strictly between Excluded Middle for 3-quantifiers arithmetical formulas and Excluded Middle for 2-quantifiers arithmetical formulas [10]. Σ^0_3-LLPO may be interpreted as König's Lemma restricted to trees definable by some Δ^0_3-predicate (see again [10]).

However, Ramsey Theorem in the proof of the Termination Theorem [1] may be replaced by H-closure, obtaining a fully intuitionistic proof. It is worth noticing that we obtained the result of H-closure by analyzing the proof of Termination Theorem, not by building over any existing intuitionistic interpretation.

We could not find any evident connection with the intuitionistic interpretations by Bellin, Oliva and Powell. Bellin [12] applied the no-counterexample interpretation to Ramsey theorem, while Oliva and Powell [13] used the dialectica interpretation. They approximated the homogeneous set by a set which may stand any test for some initial segment (a segment dependent by the try itself). Instead we proved a well-foundedness result.

Instead, we found interesting connections with the intuitionistic interpretations expressing Ramsey Theorem as a property of well-founded relations. This research line started in 1974: the very first intuitionistic proof used Bar Induction. We refer to §10 of [6] for an account of this earlier stage of the research. Until 1990, all intuitionistic versions of Ramsey were negated formulas, hence non-informative. In 1990 [6] Veldman and Bezem proved, using Choice Axiom and Brouwer thesis, the first intuitionistic negation-free version of Ramsey: *almost full relations are closed under finite intersections*, from now on the *Almost-Full Theorem.*

We explain the Almost-full theorem. *Brouwer thesis* says: a relation R is inductively well-founded if and only if all R-decreasing sequences are finite. Brouwer thesis is classically true, yet it is not provable using the rules of intuitionistic natural deduction. In [5] (first published in 1994, updated in 2011) Coquand showed that we may bypass the need of Choice Axiom and Brouwer thesis in the Almost Full Theorem, provided we take as definition of well-founded directly the inductive definition of well-founded (as we do in this paper).

In [6], a binary relation R over a set is *almost full* if for all infinite sequences $x_0, x_1, x_2, \ldots, x_n, \ldots$ on I there are *some* $i < j$ such that $x_i R x_j$. We claim that, classically, the set of almost full relations R is the set of relations such that *the complement of the inverse of* R is H-well-founded. Indeed, let $\neg R^{-1}$ be the complement of the inverse of R: then, classically, $\neg R^{-1}$ almost full means that in all infinite sequences we have $x_i \neg R^{-1} x_j$ for some $i < j$, that is, $x_j \neg R x_i$ for some $i < j$, that is, all sequences such that $x_j R x_i$ for all $i < j$ are finite. Classically, this is equivalent to H-well-foundedness of R. The fact that the relationship between H-well-founded and almost full requires a complement explains why we prove closure under finite unions, while Veldman, Bezem and Coquand proved the closure under finite *intersections*.

For the future, we plan to use our proof to extract some effective bounds for the Termination Theorem. Another challenge is to extract the bounds implicit in the intuitionistic proof [7], which, as we said, uses Ramsey Theorem in the form: "almost full relations are closed under intersection", and to compare the two bounds.

References

1. Podelski, A., Rybalchenko, A.: Transition invariants. In: LICS, pp. 32–41 (2004)
2. Ramsey, F.P.: On a problem in formal logic. Proc. London Math. Soc. 30, 264–286 (1930)
3. Berardi, S., Steila, S.: Ramsey Theorem for pairs as a classical principle in Intuitionistic Arithmetic. Accepted in Types 2013 Postproceedings (2013)
4. Friedman, H.: Classically and intuitionistically provably recursive functions, vol. 699 (1978)
5. Coquand, T.: A direct proof of Ramsey's Theorem. Author's website (1994) (revised in 2011)
6. Veldman, W., Bezem, M.: Ramsey's theorem and the pigeonhole principle in intuitionistic mathematics. Journal of the London Mathematical Society s2-47(2), 193–211 (1993)
7. Vytiniotis, D., Coquand, T., Wahlstedt, D.: Stop when you are almost-full - adventures in constructive termination. In: Beringer, L., Felty, A. (eds.) ITP 2012. LNCS, vol. 7406, pp. 250–265. Springer, Heidelberg (2012)
8. Harrison, J.: Handbook of Practical Logic and Automated Reasoning, 1st edn. Cambridge University Press (2009)
9. Park, D.: Concurrency and automata on infinite sequences. In: Deussen, P. (ed.) Proceedings of the 5th GI-Conference on Theoretical Computer Science. LNCS, vol. 104, pp. 167–183. Springer, Heidelberg (1981)
10. Akama, Y., Berardi, S., Hayashi, S., Kohlenbach, U.: An Arithmetical Hierarchy of the Law of Excluded Middle and Related Principles. In: LICS, pp. 192–201. IEEE Computer Society (2004)
11. Jockusch, C.: Ramsey's Theorem and Recursion Theory. J. Symb. Log. 37(2), 268–280 (1972)
12. Bellin, G.: Ramsey interpreted: a parametric version of Ramsey's Theorem. In: AMS (ed.) Logic and Computation: Proceedings of a Symposium held at Carnegie-Mellon University, vol. 106, pp. 17–37 (1990)
13. Oliva, P., Powell, T.: A Constructive Interpretation of Ramsey's Theorem via the Product of Selection Functions. CoRR abs/1204.5631 (2012)

A Model of Countable Nondeterminism in Guarded Type Theory

Aleš Bizjak[1], Lars Birkedal[1], and Marino Miculan[2]

[1] Aarhus University, Aarhus, Denmark
{abizjak,birkedal}@cs.au.dk
[2] University of Udine, Udine, Italy
marino.miculan@uniud.it

Abstract. We show how to construct a logical relation for countable nondeterminism in a guarded type theory, corresponding to the internal logic of the topos $\mathbf{Sh}(\omega_1)$ of sheaves over ω_1. In contrast to earlier work on abstract step-indexed models, we not only construct the logical relations in the guarded type theory, but also give an internal proof of the adequacy of the model with respect to standard contextual equivalence. To state and prove adequacy of the logical relation, we introduce a new propositional modality. In connection with this modality we show why it is necessary to work in the logic of $\mathbf{Sh}(\omega_1)$.

1 Introduction

Countable nondeterminism arises naturally when modeling properties of concurrent systems or systems with user input, etc. Still, semantic models for reasoning about *must-contextual equivalence* of higher-order programming languages with countable nondeterminism are challenging to construct [3,7,1,10,11,12,13,17]. Recently, it was shown how step-indexed logical relations, indexed over the first uncountable ordinal ω_1, can be used to give a simple model of a higher-order programming language $\mathbf{F}^{\mu,?}$ with recursive types and countable nondeterminism [4], allowing one to reason about must-contextual equivalence. Using step-indexed logical relations is arguably substantially simpler than using other models, but still involves some tedious reasoning about indices, as is characteristic of any concrete step-indexed model.

In previous work [8,5], the guarded type theory corresponding to the internal logic of the topos $\mathbf{Sh}(\omega)$ of sheaves[1] on ω has been proved very useful for developing abstract accounts of step-indexed models indexed over ω. Such abstract accounts eliminate much of the explicit tedious reasoning about indices. We recall that the internal logic of $\mathbf{Sh}(\omega)$ can be thought of as a logic of discrete time, with time corresponding to ordinals and smaller ordinals being the future. In the application to step-indexed logical relations, the link between steps in the

[1] Considered as sheaves on the topological space ω equipped with the Alexandrov topology.

G. Dowek (ed.): RTA-TLCA 2014, LNCS 8560, pp. 108–123, 2014.

operational semantics and the notion of time provided by the internal logic of $\mathbf{Sh}(\omega)$ is made by defining the operational semantics using guarded recursion [5].

In this paper we show how to construct a logical relation for countable nondeterminism in a guarded type theory GTT corresponding to the internal logic of the topos $\mathbf{Sh}(\omega_1)$ of sheaves over ω_1. For space reasons we only consider the case of must-equivalence; the case for may-equivalence is similar. In contrast to earlier work on abstract step-indexed models [8,5], we not only construct the logical relation in the guarded type theory, but also give an internal proof of the adequacy of the model with respect to must-contextual equivalence. To state and prove adequacy of the logical relation we introduce a new propositional modality $\Box$: intuitively, $\Box\varphi$ holds if φ holds at all times. Using this modality we give a logical explanation for why it is necessary to work in the logic of $\mathbf{Sh}(\omega_1)$: a certain logical equivalence involving $\Box$ holds in the internal logic of $\mathbf{Sh}(\omega_1)$ but not in the internal logic of $\mathbf{Sh}(\omega)$ (see Lemma 4).

To model *must*-equivalence, we follow [4] and define the logical relation using *biorthogonality*. Typically, biorthogonality relies on a definition of convergence; in our case, it would be must-convergence. In an abstract account of step-indexed models, convergence would need to be defined by guarded recursion (to show the fundamental lemma). However, that is not possible in the logic of $\mathbf{Sh}(\omega_1)$. There are two ways to understand that. If one considers the natural guarded-recursive definition of convergence,[2] using Löb induction one could show that a non-terminating computation would converge! Another way to understand this issue is in terms of the model. The stratified convergence predicate $\Downarrow_\beta$ from [4] is not a well-defined subobject in $\mathbf{Sh}(\omega_1)$. Intuitively, the reason is that all predicates in GTT are closed wrt. the future (smaller ordinals), but if an expression converges to a value in, say, 15 computation steps, then it does not necessarily converge to a value in 14 steps. Instead we observe that the dual of stratified must-convergence, the stratified *may-divergence*, is a subobject of $\mathbf{Sh}(\omega_1)$ and can easily be defined as a predicate in GTT using guarded recursion. Thus we use the stratified may-divergence predicate to define biorthogonality, modifying the definition accordingly.

The remainder of the paper is organized as follows. In Section 2 we explain the guarded type theory GTT, which we use to define the operational semantics of the higher-order programming language $\mathbf{F}^{\mu,?}$ with countable nondeterminism (Section 3) and to define the adequate logical relation for reasoning about contextual equivalence (Section 4). We include an example to demonstrate how reasoning in the resulting model avoids tedious step-indexing. Finally, in Section 5 we show that the guarded type theory GTT is consistent by providing a model thereof in $\mathbf{Sh}(\omega_1)$. Thus, most of the paper can be read without understanding the details of the model $\mathbf{Sh}(\omega_1)$. For reasons of space, most proofs have been omitted; they can be found in the accompanying technical report [6].

[2] $\text{must-converge}(e) \leftrightarrow \forall e', e \rightsquigarrow e' \rightarrow \triangleright(\text{must-converge}(e'))$.

2 The logic GTT

The logic GTT is the internal logic of $\mathbf{Sh}(\omega_1)$. In this section we explain some of the key features of the logic; in the subsequent development we will also use a couple of additional facts, which will be introduced as needed.

The logic is an extension of a multisorted intuitionistic higher-order logic with two modalities $\triangleright$ and $\square$, pronounced "later" and "always" respectively. Types (aka sorts) are ranged over by X, Y; we denote the type of propositions by Ω and the function space from X to Y as Y^X. We write $\mathcal{P}(X) = \Omega^X$ for the type of the power set of X. We think of types as *variable* sets (although in the logic we will not deal with indices explicitly). There is a subset of types which we call *constant sets*; given a set a, we denote by $\Delta(a)$ the type which is constantly equal to a. Constant sets are closed under product and function space. For each type X there is a type $\blacktriangleright X$ and a function symbol $\mathrm{next}^X : X \to \blacktriangleright X$. Intuitively $\blacktriangleright X$ is "one time step later" than the type X, so we can only use it *later*, i.e. after one time step and $\mathrm{next}^X(x)$ freezes x for a time step so it is only available later.

We also single out the space of *total* types. Intuitively, these are the types whose elements at each stage have evolved from some elements from previous stages, i.e. they do not appear out of nowhere.

Definition 1. *For a type X we define* Total(X) *to mean that* next^X *is surjective*

$$\mathrm{Total}(X) \overset{\triangle}{\leftrightarrow} \forall x : \blacktriangleright X, \exists x' : X, \mathrm{next}^X(x') = x$$

and say that X *is* total *when* Total(X) *holds.*

Note that for each X, Total(X) is a formula of the logic, but Total itself is not a predicate of the logic. Constant sets $\Delta(a)$ for an inhabited a are total. For simplicity, we do not formalize how to construct constant sets. In the following, we shall instead just state for some of the types that we use that they are constant; these facts can be shown using the model in Section 5.

We will adopt the usual "sequent-in-context" judgment of the form $\Gamma \mid \Xi \vdash \varphi$ for saying that the formula φ is a consequence of formulas in Ξ, under the typing context Γ.

The $\triangleright$ modality on *formulas* is used to express that a formula holds only "later", that is, after a time step. More precisely, there is a function symbol $\triangleright : \Omega \to \Omega$ which we extend to formulas by composition. We require $\triangleright$ to satisfy the following properties (Γ is an arbitrary context).

1. (Monotonicity) $\Gamma \mid \varphi \vdash \triangleright\varphi$
2. (Löb induction rule) $\Gamma \mid (\triangleright\varphi \to \varphi) \vdash \varphi$
3. $\triangleright$ commutes over $\top$, $\wedge$, $\to$ and $\vee$ (but does not preserve $\bot$).
4. For all X, Y and φ we have $\Gamma, x : X \mid \exists y : Y, \triangleright\varphi(x, y) \vdash \triangleright(\exists y : Y, \varphi(x, y))$.
5. For all X, Y and φ we have $\Gamma, x : X \mid \triangleright(\forall y : Y, \varphi(x, y)) \vdash \forall y : Y, \triangleright\varphi(x, y)$.
   The converse entailment in the last rule holds if Y is total.

Following [5, Definition 2.8] we define a notion of contractiveness which will be used to construct unique fixed points of morphisms on total types.

Definition 2. *We define the predicate* Contr *on* Y^X *as*

$$\mathrm{Contr}(f) \stackrel{\triangle}{\leftrightarrow} \forall x, x' : X, \triangleright(x = x') \to f(x) = f(x')$$

and we say that f *is* (internally) contractive *if* $\mathrm{Contr}(f)$ *holds.*

Intuitively, a function f is contractive if $f(x)$ *now* depends only on the value of x later, in the future. The following theorem holds in the logic.

Theorem 1 (Internal Banach's fixed point theorem). *Internally, any contractive function* f *on a total object* X *has a unique fixed point. More precisely, the following formula is valid in the logic of* $\mathbf{Sh}(\omega_1)$*:*

$$\mathrm{Total}(X) \to \forall f : X^X, \mathrm{Contr}(f) \to \exists! x : X, f(x) = x.$$

We will use Theorem 1 in Section 4 on a function of type $\mathcal{P}(X) \to \mathcal{P}(X)$ for a constant set X. We thus additionally assume that $\mathrm{Total}(\mathcal{P}(X))$ holds for any constant set X.

The $\Box$ modality is used to express that a formula holds for all time steps. It is thus analogous to the $\Box$ modality in temporal logic. It is defined as the *right* adjoint to the $\neg\neg$-closure operation on formulas and behaves as an interior operator. More precisely, for a formula φ in context Γ, $\Box\varphi$ is another formula in context Γ. In contrast to the $\triangleright$ modality, $\Box$ on formulas does not arise from a function on Ω and consequently does not commute with substitution, i.e., in general $(\Box\varphi)[t/x]$ is not equivalent to $\Box(\varphi[t/x])$, although $(\Box\varphi)[t/x]$ always implies $\Box(\varphi[t/x])$ which is useful for instantiating universally quantified assumptions. Thus, to be precise, we would have to annotate the $\Box$ with the context in which it is used. However, restricting to contexts consisting of constant types, $\Box$ does commute with substitution and since we will only use it in such contexts we will omit explicit contexts.

The basic rules for the $\Box$ modality are the following. In particular, note the first rule which characterizes $\Box$ as the right adjoint to the $\neg\neg$-closure.

$$\frac{\Gamma \mid \neg\neg\varphi \vdash \psi}{\Gamma \mid \varphi \vdash \Box\psi} \qquad \frac{\Gamma \mid \varphi \vdash \psi}{\Gamma \mid \Box\varphi \vdash \Box\psi} \qquad \frac{}{\Gamma \mid \Box\varphi \vdash \varphi}$$

$$\frac{}{\Gamma \mid \Box\varphi \vdash \Box\Box\varphi} \qquad \frac{}{\Gamma \mid \neg\neg(\Box\varphi) \vdash \Box\varphi} \qquad \frac{}{\Gamma \mid \neg\neg\varphi \vdash \Box(\neg\neg\varphi)}$$

Note that some of the rules can be derived from others. A simple consequence of the rules is that $\neg\neg\varphi \leftrightarrow \Box(\neg\neg\varphi)$ and $\neg\neg(\Box\varphi) \leftrightarrow \Box\varphi$. Thus one way to understand $\Box\varphi$ is as the largest predicate that implies φ and is $\neg\neg$-closed.

Proposition 1. *Using the rules for* $\Box$ *stated above we can prove the following in the logic.*

$$\Box\top \leftrightarrow \top \text{ and } \Box\bot \leftrightarrow \bot \qquad \Gamma \mid \emptyset \vdash \Box(\varphi \wedge \psi) \leftrightarrow \Box\varphi \wedge \Box\psi$$

$$\Gamma \mid \emptyset \vdash \Box(\forall x : X, \varphi) \leftrightarrow \forall x : X, \Box\varphi \qquad \Gamma \mid \emptyset \vdash \Box(\varphi \to \psi) \to \Box\varphi \to \Box\psi$$

$$
\begin{aligned}
\tau &::= \alpha \mid \mathbf{1} \mid \tau_1 \times \tau_2 \mid \tau_1 + \tau_2 \mid \tau_1 \to \tau_2 \mid \mu\alpha.\tau \mid \forall\alpha.\tau \mid \exists\alpha.\tau \\
e &::= x \mid \langle\rangle \mid \langle e_1, e_2\rangle \mid \texttt{inl}\ e \mid \texttt{inr}\ e \mid \lambda x.e \mid \Lambda.e \mid \texttt{pack}\, e \mid \texttt{unfold}\, e \mid \texttt{fold}\, e \\
&\quad \mid \texttt{?} \mid \texttt{proj}_i\ e \mid e_1\, e_2 \mid \texttt{case}\,(e, x_1.e_1, x_2.e_2) \mid e[] \mid \texttt{unpack}\ e_1\ \texttt{as}\ x\ \texttt{in}\ e_2 \\
E &::= - \mid \langle E, e\rangle \mid \langle v, E\rangle \mid \texttt{inl}\ E \mid \texttt{inr}\ E \mid \texttt{pack}\, E \mid \texttt{proj}_i\, E \mid E\, e \mid v\, E \mid E[] \\
&\quad \mid \texttt{case}\,(E, x_1.e_1, x_2.e_2) \mid \texttt{unpack}\ E\ \texttt{as}\ x\ \texttt{in}\ e \mid \texttt{unfold}\, E \mid \texttt{fold}\, E
\end{aligned}
$$

Fig. 1. Syntax of $\mathbf{F}^{\mu,?}$: types τ, terms e and evaluation contexts E. $\texttt{inl}\ e$ and $\texttt{inr}\ e$ introduce terms of sum type. $\texttt{case}\,(e, x_1.e_1, x_2.e_2)$ is the pattern matching construct that eliminates a term e of the sum type with the left branch being e_1 and right branch e_2. $\texttt{pack}\, e$ and $\texttt{unpack}\ e_1\ \texttt{as}\ x\ \texttt{in}\ e_2$ introduce and eliminate terms of existential types and $\Lambda.e$ and $e[]$ introduce and eliminate terms of universal types.

A useful derived introduction rule for the $\Box$ modality is the well-known $\Box$-introduction rule for S4. It states that if we can prove φ using only $\Box$'ed facts, then we can also conclude $\Box\varphi$. Formally:

$$\frac{\Gamma \mid \Xi \vdash \varphi}{\Gamma \mid \Xi \vdash \Box\varphi}\, \Xi = \Box\varphi_1, \Box\varphi_2, \ldots, \Box\varphi_n$$

3 The Language $\mathbf{F}^{\mu,?}$

In this section we introduce $\mathbf{F}^{\mu,?}$, a call-by-value functional language akin to System F, i.e., with impredicative polymorphism, existential and general recursive types, extended with a countable choice expression ?. We work informally in the logic outlined above except where explicitly stated.

Syntax We assume disjoint, countably infinite sets of *type variables*, ranged over by α, and *term variables*, ranged over by x. The syntax of types, terms and evaluation contexts is defined in Figure 1. Values v and contexts (terms with a hole) C can be defined in the usual way. The free type variables in a type $ftv(\tau)$ and free term variables in a term $fv(e)$, are defined in the usual way. The notation $\sigma[\boldsymbol{\tau}/\boldsymbol{\alpha}]$ denotes the simultaneous capture-avoiding substitution of types $\boldsymbol{\tau}$ for the free type variables $\boldsymbol{\alpha}$ in the type σ; similarly, $e[\boldsymbol{v}/\boldsymbol{x}]$ denotes simultaneous capture-avoiding substitution of values $\boldsymbol{v}$ for the free term variables $\boldsymbol{x}$ in e. We define the type of natural numbers as $\texttt{nat} = \mu\alpha.1 + \alpha$ and the corresponding numerals as $\underline{0} = \texttt{fold}\,(\texttt{inl}\ \langle\rangle)$ and $\underline{n+1} = \texttt{fold}\,(\texttt{inr}\ \underline{n})$ by induction on n.

The judgment $\Delta \vdash \tau$ expresses $ftv(\tau) \subseteq \Delta$. The typing judgment $\Delta \mid \Gamma \vdash e : \tau$ expresses that e has type τ in type variable context Δ and term variable context Γ. Typing rules are the same as for system F with recursive types, apart from the typing of the ?, which has type $\texttt{nat}$ in any well-formed context.

We write **Type** for the set of closed types τ, i.e. types τ satisfying $ftv(\tau) = \emptyset$. We write $\mathbf{Val}\,(\tau)$ and $\mathbf{Tm}\,(\tau)$ for the sets of closed values and terms of type τ, respectively. $\mathbf{Stk}\,(\tau)$ denotes the set of evaluation contexts E with the hole of type τ. The typing of evaluation contexts can be defined as in [4] by an inductive relation. We write **Val** and **Tm** for the set of all closed values and closed terms, respectively, and **Stk** for the set of all evaluation contexts.

$$\begin{array}{llll}
\mathtt{proj}_i\,\langle v_1, v_2\rangle \longmapsto v_i & & \mathtt{unfold}\,(\mathtt{fold}\,v) \longmapsto v \\
(\lambda x.e)\,v \longmapsto e[v/x] & & \mathtt{unpack}\,(\mathtt{pack}\,v)\ \mathtt{as}\ x\ \mathtt{in}\ e \longmapsto e[v/x] \\
(\Lambda.e)[\,] \longmapsto e & & \mathtt{case}\,(\mathtt{inl}\,v, x_1.e_1, x_2.e_2) \longmapsto e_1[v/x_1] \\
? \longmapsto \underline{n} & (n \in \mathbb{N}) & \mathtt{case}\,(\mathtt{inr}\,v, x_1.e_1, x_2.e_2) \longmapsto e_2[v/x_2]
\end{array}$$

$$E[e] \rightsquigarrow E[e'] \qquad \text{if } e \longmapsto e'$$

Fig. 2. Operational semantics of $\mathbf{F}^{\mu,?}$: basic reductions $\longmapsto$ and one step reduction $\rightsquigarrow$

Using the model in Section 5, we can show that the types of terms, values, evaluation contexts and contexts are constant sets. We use this fact in the proof of adequacy in Section 4.

Operational Semantics. The operational semantics of $\mathbf{F}^{\mu,?}$ is given in Figure 2 by a one-step reduction relation $e \rightsquigarrow e'$. The rules are standard apart from the rule for ? which states that the countable choice expression ? evaluates nondeterministically to any numeral $\underline{n}$ $(n \in \mathbb{N})$. We extend basic reduction $\longmapsto$ to the single step reduction relation $\rightsquigarrow$ using evaluation contexts E.

To define the logical relation we need further restricted reduction relations. These will allow us to ignore most reductions in the definition of the logical relation, except the ones needed to prove the fundamental property (Corollary 1).

Let $\rightsquigarrow^*$ be the reflexive transitive closure of $\rightsquigarrow$. Following [4] we call *unfold-fold* reductions those of the form $\mathtt{unfold}\,(\mathtt{fold}\,v) \longmapsto v$, and *choice* reductions those of the form $? \longmapsto \underline{n}$ $\;(n \in \mathbb{N})$. Choice reductions are important because these are the only ones that do not preserve equivalence. We define

- $e \overset{p}{\rightsquigarrow} e'$ if $e \rightsquigarrow^* e'$ and *none* of the reductions is a choice reduction;
- $e \overset{0}{\rightsquigarrow} e'$ if $e \rightsquigarrow^* e'$ and *none* of the reductions is an unfold-fold reduction;
- $e \overset{1}{\rightsquigarrow} e'$ if $e \rightsquigarrow^* e'$ and *exactly one* of the reductions is an unfold-fold reduction;
- $e \overset{p,0}{\rightsquigarrow} e'$ if $e \overset{p}{\rightsquigarrow} e'$ and $e \overset{0}{\rightsquigarrow} e'$;
- $e \overset{p,1}{\rightsquigarrow} e'$ if $e \overset{p}{\rightsquigarrow} e'$ and $e \overset{1}{\rightsquigarrow} e'$.

The $\overset{1}{\rightsquigarrow}$ reduction relation will be used in the stratified definition of divergence and the other reduction relations will be used to state additional properties of the logical relation in Lemma 1. Note that although some of the relations are described informally using negation they can be described constructively in a positive way. For instance, $\overset{p}{\rightsquigarrow}$ can be defined in the same way as the $\rightsquigarrow^*$ but using a subset of the one step relation $\rightsquigarrow$.

Divergence Relations. We define the logical relation using biorthogonality. As we explained in the introduction we use two *may-divergence* predicates, which are, informally, the negations of the two *must-convergence* relations from [4]. Thus we define, in the logic, the *stratified may-divergence predicate* $\uparrow$ as the unique fixed point of $\Psi : \mathcal{P}(\mathbf{Tm}) \to \mathcal{P}(\mathbf{Tm})$ given as

$$\Psi(A) = \left\{ e : \mathbf{Tm} \;\middle|\; \exists e' : \mathbf{Tm}, e \overset{1}{\rightsquigarrow} e' \wedge \triangleright (e' \in A) \right\}.$$

Ψ is internally contractive and since **Tm** is a constant set $\mathcal{P}(\mathbf{Tm})$ is total. By Theorem 1, Ψ has a unique fixed point.

We also define the non-stratified may-divergence predicate $\uparrow$ as the *greatest* fixed-point of $\Phi : \mathcal{P}(\mathbf{Tm}) \to \mathcal{P}(\mathbf{Tm})$ given as

$$\Phi(A) = \left\{ e : \mathbf{Tm} \mid \exists e' : \mathbf{Tm}, e \rightsquigarrow e' \wedge e' \in A \right\}.$$

Since Φ is monotone and $\mathcal{P}(\mathbf{Tm})$ is a complete lattice, the greatest fixed point exists by Knaster-Tarski's fixed-point theorem, which holds in our logic.[3] Observe that Ψ is almost the same as $\Phi \circ \triangleright$, apart from using a different reduction relation. We write $e\uparrow$ and $e\underline{\uparrow}$ for $e \in \uparrow$ and $e \in \underline{\uparrow}$, respectively.

The predicates $\underline{\uparrow}$ and $\uparrow$ are closed under some, but not all, reductions.

Lemma 1. *Let* $e, e' : \mathbf{Tm}$. *The following properties hold in the logic GTT.*

$$\textit{if } e \overset{p}{\rightsquigarrow} e' \textit{ then } e\uparrow \leftrightarrow e'\uparrow \qquad \textit{if } e \overset{p,0}{\rightsquigarrow} e' \textit{ then } e\underline{\uparrow} \leftrightarrow e'\underline{\uparrow}$$
$$\textit{if } e \overset{0}{\rightsquigarrow} e' \textit{ then } e'\underline{\uparrow} \to e\underline{\uparrow} \qquad \textit{if } e \overset{1}{\rightsquigarrow} e' \textit{ then } \triangleright(e'\underline{\uparrow}) \to e\underline{\uparrow}$$

Must-contextual approximation Contexts can be typed as second-order terms, by means of a typing judgment of the form $C : (\Delta \mid \Gamma \Rrightarrow \tau) \looparrowright (\Delta' \mid \Gamma' \Rrightarrow \sigma)$, stating that whenever $\Delta \mid \Gamma \vdash e : \tau$ holds, $\Delta' \mid \Gamma' \vdash C[e] : \sigma$ also holds. The typing of contexts can be defined as an inductive relation defined by suitable typing rules, which we omit here due to lack of space; see [2]. We write $C : (\Delta \mid \Gamma \Rrightarrow \tau)$ to mean there exists a type σ, such that $C : (\Delta \mid \Gamma \Rrightarrow \tau) \looparrowright (\varnothing \mid \varnothing \Rrightarrow \sigma)$ holds.

We define *contextual must-approximation* using the may-divergence predicate. This is in contrast with the definition in [4] which uses the must-convergence predicate. However externally, in the model, the two definitions coincide.

Definition 3 (Must-contextual approximation). *In GTT, we define must-contextual approximation* $\Delta \mid \Gamma \vdash e_1 \lesssim^{ctx}_{\Downarrow} e_2 : \tau$ *as*

$$\Delta \mid \Gamma \vdash e_1 : \tau \wedge \Delta \mid \Gamma \vdash e_2 : \tau \ \wedge \forall C, (C : (\Delta \mid \Gamma \Rrightarrow \tau)) \wedge C[e_2]\uparrow \to C[e_1]\uparrow.$$

Note the order in the implication: if $C[e_2]$ may-diverges then $C[e_1]$ may-diverges. This is the contrapositive of the definition in [4] which states that if $C[e_1]$ must-converges then $C[e_2]$ must-converges. Must-contextual approximation defined explicitly using contexts can be shown to be the largest compatible adequate and transitive relation, so it coincides with contextual approximation in [4].

4 Logical Relation

In this section we give an abstract account of the concrete step-indexed model from [4] by defining a logical relation interpretation of types in GTT. The result is a simpler model without a proliferation of step-indices, as we will demonstrate in the example at the end of the section.

[3] Knaster-Tarski's fixed point theorem holds in the internal language of any topos.

$$\begin{aligned}
\llbracket \Delta \vdash \alpha \rrbracket (\varphi) &= \varphi_r(\alpha) \\
\llbracket \Delta \vdash \mathbf{1} \rrbracket (\varphi) &= \mathbf{Id_1} \\
\llbracket \Delta \vdash \tau_1 \times \tau_2 \rrbracket (\varphi) &= \{(\langle v, u\rangle, \langle v', u'\rangle) \mid (v, v') \in \llbracket \Delta \vdash \tau_1 \rrbracket (\varphi), (u, u') \in \llbracket \Delta \vdash \tau_2 \rrbracket (\varphi)\} \\
\llbracket \Delta \vdash \tau_1 + \tau_2 \rrbracket (\varphi) &= \{(\mathtt{inl}\, v, \mathtt{inl}\, v') \mid (v, v') \in \llbracket \Delta \vdash \tau_1 \rrbracket (\varphi)\} \cup \\
&\quad\ \{(\mathtt{inr}\, u, \mathtt{inr}\, u') \mid (u, u') \in \llbracket \Delta \vdash \tau_2 \rrbracket (\varphi)\} \\
\llbracket \Delta \vdash \tau_1 \to \tau_2 \rrbracket (\varphi) &= \{(\lambda x.e, \lambda y.e') \mid \forall (v, v') \in \llbracket \Delta \vdash \tau_1 \rrbracket (\varphi), \\
&\quad\ (e[v/x], e'[v'/y]) \in \llbracket \Delta \vdash \tau_2 \rrbracket (\varphi)^{\top\top}\} \\
\llbracket \Delta \vdash \forall \alpha.\tau \rrbracket (\varphi) &= \{(\Lambda.e, \Lambda.e') \mid \forall \sigma, \sigma' \in \mathbf{Type}, \forall s \in \mathbf{VRel}(\sigma, \sigma'), \\
&\quad\ (e, e') \in \llbracket \Delta, \alpha \vdash \tau \rrbracket (\varphi[\alpha \mapsto (\sigma, \sigma', s)])^{\top\top}\} \\
\llbracket \Delta \vdash \exists \alpha.\tau \rrbracket (\varphi) &= \{(\mathtt{pack}\, v, \mathtt{pack}\, v') \mid \exists \sigma, \sigma' \in \mathbf{Type}, \exists s \in \mathbf{VRel}(\sigma, \sigma'), \\
&\quad\ (v, v') \in \llbracket \Delta, \alpha \vdash \tau \rrbracket (\varphi[\alpha \mapsto (\sigma, \sigma', s)])\} \\
\llbracket \Delta \vdash \mu\alpha.\tau \rrbracket (\varphi) &= \mathtt{fix}\,(\lambda s.\{(\mathtt{fold}\, v, \mathtt{fold}\, v') \mid \triangleright((v, v') \in \llbracket \Delta, \alpha \vdash \tau \rrbracket (\varphi[\alpha \mapsto s]))\})
\end{aligned}$$

where the $\cdot^{\top\top} : \mathbf{VRel}(\tau, \tau') \to \mathbf{TRel}(\tau, \tau')$ is defined with the help of $\cdot^{\top} : \mathbf{VRel}(\tau, \tau') \to \mathbf{SRel}(\tau, \tau')$ as follows

$$\begin{aligned}
r^{\top} &= \{(E, E') \mid \forall (v, v') \in r, E'[v']\uparrow \to E[v]\uparrow\} \\
r^{\top\top} &= \{(e, e') \mid \forall (E, E') \in r^{\top}, E'[e']\uparrow \to E[e]\uparrow\}.
\end{aligned}$$

Fig. 3. Interpretation of types. All the relations are on *typeable* terms and contexts

Relational Interpretation of Types. Let $\mathbf{Type}(\Delta) = \{\tau \mid \Delta \vdash \tau\}$ be the set of types well-formed in context Δ. Given $\tau, \tau' \in \mathbf{Type}$ let $\mathbf{VRel}(\tau, \tau') = \mathcal{P}(\mathbf{Val}(\tau) \times \mathbf{Val}(\tau'))$, $\mathbf{TRel}(\tau, \tau') = \mathcal{P}(\mathbf{Tm}(\tau) \times \mathbf{Tm}(\tau'))$ and $\mathbf{SRel}(\tau, \tau') = \mathcal{P}(\mathbf{Stk}(\tau) \times \mathbf{Stk}(\tau'))$. We implicitly use the inclusion $\mathbf{VRel}(\tau, \tau') \subseteq \mathbf{TRel}(\tau, \tau')$. For a type variable context Δ, we define $\mathbf{VRel}(\Delta)$ to be

$$\left\{(\varphi_1, \varphi_2, \varphi_r) \;\middle|\; \varphi_1, \varphi_2 : \Delta \to \mathbf{Type}, \forall \alpha \in \Delta, \varphi_r(\alpha) \in \mathbf{VRel}(\varphi_1(\alpha), \varphi_2(\alpha))\right\}$$

where the first two components give syntactic types for the left and right hand sides of the relation and the third component is a relation between those types. The interpretation of types, $\llbracket \cdot \vdash \cdot \rrbracket$, is shown in Figure 3. The definition is by induction on the judgement $\Delta \vdash \tau$. Given a judgment $\Delta \vdash \tau$, and $\varphi \in \mathbf{VRel}(\Delta)$, we have $\llbracket \Delta \vdash \tau \rrbracket (\varphi) \in \mathbf{VRel}(\varphi_1(\tau), \varphi_2(\tau))$ where φ_1 and φ_2 are the first two components of φ, and $\varphi_i(\tau)$ denotes substitution of types in φ_i for free type variables in τ. Since we are working in the logic GTT, the interpretations of all type constructions are simple and intuitive. For instance, functions are related when they map related values to related results, two values of universal type are related if they respect all value relations. In particular, there are no admissibility requirements on the relations, nor any step-indexing — but just a use of $\triangleright$ in the interpretation of recursive types, to make it well-defined as a consequence of Theorem 1, using that the type $\mathcal{P}(\mathbf{Tm} \times \mathbf{Tm})$ is total.

The definition of $\top\top$-closure is where we connect operational semantics and the $\triangleright$ modality, using the stratified may-divergence predicate $\underline{\uparrow}$. $\top\top$-closed relations are closed under some reductions. More precisely, the following holds.

Lemma 2. *Let τ, τ' :* **Type** *and $r \in$* **VRel** (τ, τ').

- *If $e \overset{p,0}{\rightsquigarrow} e_1$ and $e' \overset{p}{\rightsquigarrow} e_1'$ then $(e, e') \in r^{\top\top} \leftrightarrow (e_1, e_1') \in r^{\top\top}$.*
- *If $e \overset{1}{\rightsquigarrow} e_1$ then for all e' :* **Tm**, *if $\triangleright((e_1, e') \in r^{\top\top})$ then $(e, e') \in r^{\top\top}$.*

We use this fact extensively in the proofs of the fundamental property and example equivalences.

In order to define logical relations, we need first to extend the interpretation of types to the interpretation of contexts (note that in particular, related substitutions map into well-typed values):

$$\llbracket \Delta \vdash \Gamma \rrbracket (\varphi) = \{(\gamma, \gamma') \mid \gamma, \gamma' : \mathbf{Val}^{\mathrm{dom}(\Gamma)}, \\ \forall x \in \mathrm{dom}(\Gamma), (\gamma(x), \gamma'(x)) \in \llbracket \Delta \vdash \Gamma(x) \rrbracket (\varphi)\}$$

The logical relation and its fundamental property We define the logical relation on open terms by reducing it to relations on closed terms by substitution.

Definition 4 (Logical relation). *$\Delta \mid \Gamma \vdash e_1 \lesssim_{\Downarrow}^{log} e_2 : \tau$ if*

$$\forall \varphi \in \mathbf{VRel}(\Delta), \forall (\gamma, \gamma') \in \llbracket \Delta \vdash \Gamma \rrbracket (\varphi), (e_1\gamma, e_2\gamma') \in \llbracket \Delta \vdash \tau \rrbracket (\varphi)^{\top\top}.$$

To prove the fundamental property of logical relations and connect the logical relation to contextual-must approximation we start with some simple properties relating evaluation contexts and relations. All the lemmata are essentially of the same form: given two related evaluation contexts at a suitable type, the contexts extended with an elimination form are also related at a suitable type. We only state the case for `unfold`, since it shows the interplay between unfold-fold reductions and the stratified may divergence predicate.

Lemma 3. *If $(E, E') \in \llbracket \Delta \vdash \tau[\mu\alpha.\tau/\alpha] \rrbracket (\varphi)^{\top}$ then*

$$(E \circ (\mathtt{unfold}\,[]), E' \circ (\mathtt{unfold}\,[])) \in \llbracket \Delta \vdash \mu\alpha.\tau \rrbracket (\varphi)^{\top}.$$

Proof. Given $(\mathtt{fold}\,v, \mathtt{fold}\,v') \in \llbracket \Delta \vdash \mu\alpha.\tau \rrbracket (\varphi)$ suppose $E'\,[\mathtt{unfold}\,(\mathtt{fold}\,v')] \uparrow$. By Lemma 1 we have $E'\,[v'] \uparrow$ and so $\triangleright(E'[v'] \uparrow)$. By definition of interpretation of recursive types we have $\triangleright((v, v') \in \llbracket \Delta \vdash \tau[\mu\alpha.\tau/\alpha] \rrbracket (\varphi))$. Thus $\triangleright(E[v] \underline{\uparrow})$ and so by Lemma 1 we have $E[\mathtt{unfold}\,(\mathtt{fold}\,v)] \underline{\uparrow}$. □

Note that the proof would not work, were we to use the $\uparrow$ relation in place of $\underline{\uparrow}$ in the definition of the $\top\top$ closure since the last implication would not hold.

Proposition 2. *The logical approximation relation is compatible with the typing rules (see also [6, Prop. 2.4.17]).*

Proof. We only give two cases, to show how to use the context extension lemmata. *Elimination of recursive types:* we need to show

$$\frac{\Delta \mid \Gamma \vdash e \lesssim_{\Downarrow}^{log} e' : \mu\alpha.\tau}{\Delta \mid \Gamma \vdash \texttt{unfold}\, e \lesssim_{\Downarrow}^{log} \texttt{unfold}\, e' : \tau[\mu\alpha.\tau/\alpha]}.$$

So take $\varphi \in \mathbf{VRel}(\Delta)$ and $(\gamma, \gamma') \in [\![\Delta \vdash \Gamma]\!](\varphi)$. Let $f = e\gamma$ and $f' = e'\gamma'$. We have to show $(\texttt{unfold}\, f, \texttt{unfold}\, f') \in [\![\Delta \vdash \tau[\mu\alpha.\tau/\alpha]]\!](\varphi)^{\top\top}$. So take $(E, E') \in [\![\Delta \vdash \tau[\mu\alpha.\tau/\alpha]]\!](\varphi)^{\top}$. By assumption $(f, f') \in [\![\Delta \vdash \mu\alpha.\tau]\!](\varphi)^{\top\top}$ so it suffices to show $(E \circ (\texttt{unfold}\, []), E' \circ (\texttt{unfold}\, [])) \in [\![\Delta \vdash \mu\alpha.\tau]\!](\varphi)^{\top}$ and this is exactly the content of Lemma 3.

The ? expression: we need to show $\Delta \mid \Gamma \vdash ? \lesssim_{\Downarrow}^{log} ? : \texttt{nat}$. It is easy to see by induction that for all $n \in \mathbb{N}$, $(\underline{n}, \underline{n}) \in [\![\vdash \texttt{nat}]\!]$. So take $(E, E') \in [\![\vdash \texttt{nat}]\!]^{\top}$ and assume $E'[?]\uparrow$. By definition of the $\uparrow$ relation there exists an e', such that $? \rightsquigarrow e'$ and $E'[e']\uparrow$. Inspecting the operational semantics we see that $e' = \underline{n}$ for some $n \in \mathbb{N}$. This implies $E[\underline{n}]\uparrow$ and so by Lemma 1 we have $E[?]\uparrow$. □

Corollary 1 (Fundamental property of logical relations). *If $\Delta \mid \Gamma \vdash e : \tau$ then $\Delta \mid \Gamma \vdash e \lesssim_{\Downarrow}^{log} e : \tau$*

Proof. By induction on the typing derivation $\Delta \mid \Gamma \vdash e : \tau$, using Prop. 2. □

We need the next corollary to relate the logical approximation relation to must-contextual approximation.

Corollary 2. *For any expressions e, e' and context C, if $\Delta \mid \Gamma \vdash e \lesssim_{\Downarrow}^{log} e' : \tau$ and $C : (\Delta \mid \Gamma \Rrightarrow \tau) \looparrowright (\Delta' \mid \Gamma' \Rrightarrow \sigma)$ then $\Delta' \mid \Gamma' \vdash C[e] \lesssim_{\Downarrow}^{log} C[e'] : \tau'$.*

Proof. By induction on the judgment $C : (\Delta \mid \Gamma \Rrightarrow \tau) \looparrowright (\Delta' \mid \Gamma' \Rrightarrow \sigma)$, using Proposition 2. □

Adequacy We now wish to show soundness of the logical relation with respect to must-contextual approximation. However, the implication

$$\Delta \mid \Gamma \vdash e \lesssim_{\Downarrow}^{log} e' : \tau \rightarrow \Delta \mid \Gamma \vdash e \lesssim_{\Downarrow}^{ctx} e' : \tau$$

does not hold, due to the different divergence relations used in the definition of the logical relation. To see precisely where the proof fails, let us attempt it. Let $\Delta \mid \Gamma \vdash e \lesssim_{\Downarrow}^{log} e' : \tau$ and take a well-typed closing context C with result type σ. Then by Corollary 2, $\varnothing \mid \varnothing \vdash C[e] \lesssim_{\Downarrow}^{log} C[e'] : \sigma$. Unfolding the definition of the logical relation we get $(C[e], C[e']) \in [\![\varnothing \vdash \sigma]\!]^{\top\top}$. It is easy to see that $(-,-) \in [\![\varnothing \vdash \sigma]\!]^{\top}$ and so we get by definition of $\top\top$ that $C[e']\uparrow \rightarrow C[e]\underset{\sim}{\uparrow}$. However the definition of contextual equivalence requires the implication $C[e']\uparrow \rightarrow C[e]\uparrow$, which is not a consequence of the previous one.

Intuitively, the gist of the problem is that $\uparrow$ defines a time-independent predicate, whereas $\underset{\sim}{\uparrow}$ is time-dependent, since it is defined by guarded recursion. However, in the model in Section 5, we can show the validity of a formula expressing a connection between $\uparrow$ and $\underset{\sim}{\uparrow}$:

Lemma 4. $e : \mathbf{Tm} \mid \varnothing \vdash \Box(e\underline{\uparrow}) \to e\uparrow$ *holds in the logic GTT.*

Thus we additionally assume this principle in our logic. Note that this lemma is *not* valid in the logic of the topos of trees [5] and this is the reason we must work in the logic of $\mathbf{Sh}(\omega_1)$. We sketch a proof of the lemma at the end of Section 5 which shows the role of $\Box$ and why the lemma does not hold in the topos of trees. Using Lemma 4 we are led to the following corrected statement of adequacy using the $\Box$ modality.[4]

Theorem 2 (Adequacy). *If e and e' are of type τ in context $\Delta \mid \Gamma$ then $\Box(\Delta \mid \Gamma \vdash e \lesssim_{\Downarrow}^{log} e' : \tau)$ implies $\Delta \mid \Gamma \vdash e \lesssim_{\Downarrow}^{ctx} e' : \tau$.*

To prove this theorem we first observe that all the lemmata used in the proof of Corollary 2 are proved in constant contexts, using only other constant facts. Hence, Corollary 2 can be strengthened, yielding the following restatement.

Proposition 3. $\Box[\forall \Delta, \Delta', \Gamma, \Gamma', \tau, \sigma, C, e, e', C : (\Delta \mid \Gamma \Rrightarrow \tau) \looparrowright (\Delta' \mid \Gamma' \Rrightarrow \sigma)$
$\to \Delta \mid \Gamma \vdash e \lesssim_{\Downarrow}^{log} e' : \tau \to \Delta' \mid \Gamma' \vdash C[e] \lesssim_{\Downarrow}^{log} C[e'] : \tau']$.

Note that all the explicit universal quantification in the proposition is over constant types. One additional ingredient we need to complete the proof is the fact that $\uparrow$ is $\neg\neg$-closed, i.e. $e\uparrow \leftrightarrow \neg\neg(e\uparrow)$. We can show this in the logic using the fact that $\uparrow$ is the greatest post-fixed point by showing that $\neg\neg\uparrow$ is another one. This fact further means that $\Box(e\uparrow) \leftrightarrow (e\uparrow)$ (using the adjoint rule relating $\neg\neg$ and $\Box$ in Section 2). We are now ready to proceed with the proof of Theorem 2.

Proof (Theorem 2). Continuing the proof we started above we get, using Proposition 1, that $\Box(C[e']\uparrow \to C[e]\underline{\uparrow})$ and thus also $\Box(C[e']\uparrow) \to \Box(C[e]\underline{\uparrow})$. Moreover, $\Box(C[e']\uparrow) \leftrightarrow C[e']\uparrow$ and, by Lemma 4, $\Box(C[e]\underline{\uparrow}) \to C[e]\uparrow$. We thus conclude $C[e']\uparrow \to C[e]\uparrow$, as required. □

Thus, if we can prove that e and e' are logically related relying only on constant facts we can use this theorem to conclude that e must-contextually approximates e'. In particular, the fundamental property (Corollary 1) can be strengthened to a "boxed" statement.

Completeness. As in [4] we also get completeness with respect to contextual approximation. The proof proceeds as in [4] via the notion of CIU-approximation [15,4]. This property relies on the fact that we have built the logical relation using biorthogonality and using typeable realizers.

Theorem 3. *For any Δ, Γ, e, e' and τ,*

$$\Delta \mid \Gamma \vdash e \lesssim_{\Downarrow}^{CIU} e' : \tau \leftrightarrow \Delta \mid \Gamma \vdash e \lesssim_{\Downarrow}^{ctx} e' : \tau \leftrightarrow \Box(\Delta \mid \Gamma \vdash e \lesssim_{\Downarrow}^{log} e' : \tau)$$

4 Readers who are familiar with concrete step-indexed models will note that the $\Box$ modality captures the universal quantification over all steps used in the the definition of concrete step-indexed logical relations.

Applications. We can now use the logical relation to prove contextual equivalences. Indeed, the accompanying technical report [6] provides internal proofs of all the examples done in the concrete step-indexed model in [4]; these proofs are simpler than the ones in [4]. As an example, in this paper we include the proof of syntactic minimal invariance for *must*-equivalence. Remarkably, the proof below is just as simple as the proof of the minimal invariance property in the abstract account of a step-indexed model for the *deterministic* language $\mathbf{F}^{\mu}$ [8].

Let $\texttt{fix} : \forall\alpha,\beta.((\alpha\to\beta)\to(\alpha\to\beta)) \to (\alpha\to\beta)$ be the term $\Lambda.\Lambda.\lambda f.\delta_f(\texttt{fold}\,\delta_f)$ where δ_f is the term $\lambda y.\texttt{let}\ y' = \texttt{unfold}\,y\ \texttt{in}\ f\,(\lambda x.y'\,y\,x)$.

Consider the type $\tau = \mu\alpha.\texttt{nat} + \alpha \to \alpha$. Let $id = \lambda x.x$ and consider the term

$$f \;\equiv\; \lambda h, x.\texttt{case}\,(\texttt{unfold}\,x, y.\texttt{fold}\,(\texttt{inl}\,y), g.\texttt{fold}\,(\texttt{inr}\,\lambda y.h(g(h\,y))))\,.$$

We show that $\texttt{fix}[][]\,f \lesssim_{\Downarrow}^{log} id : \tau \to \tau$. The other direction is essentially the same. Since we prove this in the context of constant facts we can use Theorem 3 to conclude that the terms are contextually equivalent.

We show by Löb induction that $(\texttt{fix}[][]\,f, id) \in [\![\tau\to\tau]\!]^{\top\top}$. It is easy to see that $\texttt{fix}[][]\,f \overset{p,1}{\rightsquigarrow} \lambda x.\texttt{case}\,(\texttt{unfold}\,x, y.\texttt{fold}\,(\texttt{inl}\,y), g.\texttt{fold}\,(\texttt{inr}\,\lambda y.h(g(h\,y))))$ where $h = \lambda x.\delta_f(\texttt{fold}\,\delta_f)\,x$. Let

$$\varphi = \lambda x.\texttt{case}\,(\texttt{unfold}\,x, y.\texttt{fold}\,(\texttt{inl}\,y), g.\texttt{fold}\,(\texttt{inr}\,\lambda y.h(g(h\,y))))\,.$$

We now show directly that $(\varphi, id) \in [\![\tau\to\tau]\!]$ which suffices by Lemma 2.

Let us take $(u, u') \in [\![\tau]\!]$. By the definition of the interpretation of recursive and sum types there are two cases:

- $u = \texttt{fold}\,(\texttt{inl}\,\underline{n})$ and $u' = \texttt{fold}\,(\texttt{inl}\,\underline{n})$ for some $n \in \mathbb{N}$: immediate.
- $u = \texttt{fold}\,(\texttt{inr}\,g)$, $u' = \texttt{fold}\,(\texttt{inr}\,g')$ for some g, g' such that $\triangleright((g,g') \in [\![\tau\to\tau]\!])$. We then have that $\varphi\,u \overset{p,1}{\rightsquigarrow} \texttt{fold}\,(\texttt{inr}\,\lambda y.h(g(h\,y)))$ and $id\,u' \overset{p}{\rightsquigarrow} u'$ and so it suffices to show $\triangleright(\lambda y.\,(h(g(h\,y)), g') \in [\![\tau\to\tau]\!])$. We again show that these are related as values so take $\triangleright((v,v') \in [\![\tau]\!])$ and we need to show $\triangleright\left((h(g(h\,v)), g'\,v') \in [\![\tau]\!]^{\top\top}\right)$. Take $\triangleright((E,E') \in [\![\tau]\!]^{\top})$. Löb induction hypothesis gives us that $\triangleright((h', id) \in [\![\tau\to\tau]\!]^{\top\top})$, where h' is the body of h, i.e $h = \lambda x.h'\,x$. It is easy to see that this implies $\triangleright((h, id) \in [\![\tau\to\tau]\!]^{\top\top})$ and so by extending the contexts three times using lemmata analogous to Lemma 3 we get $\triangleright\left((E[h\,(g\,(h\,[]))], E'[g'\,[]]) \in [\![\tau]\!]^{\top}\right)$.
  So, assuming $\triangleright(E'[g'\,v']\!\uparrow)$ we get $\triangleright(E[h\,(g\,(h\,v))]\!\uparrow)$, concluding the proof.

5 The Model for GTT

In this section, we present a model for the logic GTT, where all the properties we have used in the previous sections are justified. The model we consider is the topos of sheaves over the first uncountable ordinal ω_1 (in fact, any ordinal $\alpha \geq \omega_1$ would suffice). We assume some basic familiarity with topos theory, on the level described in [14]. We briefly recall the necessary definitions.

The objects of $\mathbf{Sh}(\omega_1)$ are sheaves over ω_1 considered as a topological space equipped with the Alexandrov topology. Concretely, this means that objects of $\mathbf{Sh}(\omega_1)$ are continuous functors from $(\omega_1+1)^{\mathrm{op}}$ to $\mathbf{Set}$. We think of ordinals as time, with smaller ordinals being the future. The restriction maps then describe the evolution of elements through time.

$\mathbf{Sh}(\omega_1)$ is a full subcategory of the category of presheaves $\mathbf{PSh}(\omega_1+1)$. The inclusion functor i has a left adjoint $\mathbf{a} : \mathbf{PSh}(\omega_1+1) \to \mathbf{Sh}(\omega_1)$ called the *associated sheaf functor*. Limits and exponentials are constructed as in presheaf categories. Colimits *are not* constructed pointwise as in presheaf categories, but they require also the application of the associated sheaf functor.

There is an essential geometric morphism $\Pi_1 \dashv \Delta \dashv \Gamma : \mathbf{Sh}(\omega_1) \to \mathbf{Set}$, with Δ the *constant sheaf* functor, Γ the global sections functor and $\Pi_1(X) = X(1)$ the evaluation at 1 (we consider 0 to be the first ordinal). Given a set a, the constant sheaf $\Delta(a)$ is not the constant presheaf: rather it is equal to the singleton set 1 at stage 0, and to a at all other stages. For a sheaf X, an element $\xi \in X(\nu)$ and $\beta \leq \nu$ we write $\xi|_\beta$ for the restriction $X(\beta \leq \nu)(\xi)$.

Analogously to the topos of trees [5], there is a "later" modality on *types*, i.e. a functor $\blacktriangleright : \mathbf{Sh}(\omega_1) \to \mathbf{Sh}(\omega_1)$ defined as (we consider 0 a limit ordinal)

$$\blacktriangleright X(\nu+1) = X(\nu), \qquad \blacktriangleright X(\alpha) = X(\alpha) \text{ for } \alpha \text{ limit ordinal.}$$

There is an obvious natural transformation $\mathrm{next}^X : X \to \blacktriangleright X$.

The subobject classifier Ω is given by $\Omega(\nu) = \{\beta \mid \beta \leq \nu\}$ and its restriction maps are given by minimum. There is a natural transformation $\triangleright : \Omega \to \Omega$ given as $\triangleright_\nu(\beta) = \min\{\beta+1, \nu\}$.

Kripke-Joyal semantics [9] is a way to translate formulas in the logic to statements about objects and morphisms of $\mathbf{Sh}(\omega_1)$; we refer to [14, Section VI.5] for a detailed introduction and further references. We now briefly explain the Kripke-Joyal semantics of GTT.

Let X be a sheaf and φ, ψ formulas in the internal language with a free variable of type X. Intuitively, for an ordinal ν and an element $\xi \in X(\nu)$, $\nu \Vdash \varphi(\xi)$ means that φ holds for ξ at stage ν. A formula φ is *valid* if it holds for all ξ and at all stages.

Let $\nu \leq \omega_1$ and $\xi \in X(\nu)$. The rules of Kripke-Joyal semantics are the usual ones (see, e.g., [14, Theorem VI.7.1]), specialized for our particular topology:

- $\nu \Vdash \bot$ iff $\nu = 0$;
- $\nu \Vdash \top$ always;
- $\nu \Vdash \varphi(t)(\xi)$ iff $[\![\varphi]\!]_\nu ([\![t]\!]_\nu(\xi)) = \nu$, for a predicate symbol φ on X;
- $\nu \Vdash \varphi(\xi) \wedge \psi(\xi)$ iff $\nu \Vdash \varphi(\xi)$ and $\nu \Vdash \psi(\xi)$;
- $\nu \Vdash \varphi(\xi) \vee \psi(\xi)$ iff $\nu \Vdash \varphi(\xi)$ or $\nu \Vdash \psi(\xi)$;
- $\nu \Vdash \varphi(\xi) \to \psi(\xi)$ iff for all $\beta \leq \nu, \beta \Vdash \varphi(\xi|_\beta)$ implies $\beta \Vdash \psi(\xi|_\beta)$;
- $\nu \Vdash \neg\varphi(\xi)$ iff for all $\beta \leq \nu, \beta \Vdash \varphi(\xi|_\beta)$ implies $\beta = 0$.

Note that $0 \Vdash \varphi$ for any φ, as is usual in Kripke-Joyal semantics for sheaves over a space: intuitively, the stage 0 represents the impossible world. Moreover, if φ is a formula with free variables $x : X$ and $y : Y$, $\nu \leq \omega_1$ and $\xi \in X(\nu)$ then:

- For ν a successor ordinal: $\nu \Vdash \exists y : Y, \varphi(\xi, y)$ iff there exists $\xi' \in Y(\nu)$ such that $\nu \Vdash \varphi(\xi, \xi')$;
- For ν a limit ordinal: $\nu \Vdash \exists y : Y, \varphi(\xi, y)$ iff for all $\beta < \nu$ there exists $\xi_\beta \in Y(\beta)$ such that $\beta \Vdash \varphi(\xi|_\beta, \xi_\beta)$;
- $\nu \Vdash \forall y : Y, \varphi(\xi, y)$ iff for all $\beta \leq \nu$ and for all $\xi_\beta \in Y(\beta)$: $\beta \Vdash \varphi(\xi|_\beta, \xi_\beta)$.

The semantics of $\triangleright$ is as follows. Let φ be a predicate on X, then

$$\nu \Vdash \triangleright\varphi(\alpha) \text{ iff for all } \beta < \nu, \beta \Vdash \varphi(\alpha|_\beta).$$

For successor ordinals $\nu = \nu' + 1$ this reduces to

$$\nu + 1 \Vdash \triangleright\varphi(\alpha) \text{ iff } \nu' \Vdash \varphi(\alpha|_{\nu'}).$$

The predicate Total (X) in Definition 1 internalizes the property that all X's restriction maps are surjections which intuitively means that elements at any stage β evolve from elements in the past. Total sheaves are also called *flabby* in homological algebra literature, but we choose to use the term total since it was used in previous work on guarded recursion to describe an analogous property.

The properties of $\triangleright$ stated in Section 2 can be proved easily using the Kripke-Joyal semantics. The rules are similar to the rules in [5, Theorem 2.7], except the case of the existential quantifier in which the converse implication *does not hold*, even if we restrict to total and inhabited types, or even to constant sets. As a consequence, we cannot prove the internal Banach's fixed point theorem in the logic in the same way as in the topos of trees, cf. [5, Lemma 2.10].

In contrast to that in the topos of trees [5, Theorem 2.9], which requires the type X only to be inhabited, the internal Banach's fixed point theorem in $\mathbf{Sh}(\omega_1)$ (Theorem 1) has stronger assumptions: we require X to be total, which implies that it is inhabited. The additional assumption seems to be necessary and is satisfied in all the instances where we use the theorem. In particular, for a constant X, $\mathcal{P}(X)$ is total.

The operator $\neg\neg : \Omega \to \Omega$ gives rise to a function $\neg\neg_X$ on the lattice of subobjects $\mathbf{Sub}(X)$. In $\mathbf{Sh}(\omega_1)$, $\neg\neg_X$ preserves suprema[5] on each $\mathbf{Sub}(X)$ and therefore has a right adjoint $\Box_X : \mathbf{Sub}(X) \to \mathbf{Sub}(X)$ defined as

$$\Box_X P = \bigvee \{Q \mid \neg\neg Q \leq P\}.$$

If $X = \Delta(a)$ then $\Box_X P$ has a simpler description:

$$\Box_{\Delta(a)}(P)(\nu) = \begin{cases} 1 & \text{if } \nu = 0 \\ \bigcap_{\beta=1}^{\omega_1} P(\beta) & \text{otherwise.} \end{cases}$$

Thus for a predicate P on a constant set $\Delta(a)$, $\Box(P)$ contains only those elements for which P holds at all stages.

[5] Recall that this is *not* the case in every topos.

However, in contrast to $\neg\neg$ which commutes with reindexing, $\Box$ does not. There is a general reason for this: in any category with pullbacks, any deflationary operation $\Box$ that preserves the top element and *is natural*, i.e. commutes with reindexing, is necessarily the identity [16, Proposition 4.2]. However $\Delta(f)^* \left(\Box_{\Delta(a)}(P)\right) = \Box_{\Delta(b)}(\Delta(f)^*(P))$ for any $f : a \to b$ in **Set** and since Δ preserves products we do get that $\Box$ in the logic commutes with substitution when restricted to constant contexts.

The external interpretation of $\uparrow$ is exactly the negation of the must-convergence predicate $\Downarrow$ from [4]. In particular, $\uparrow$ is a constant predicate. In contrast, $\Uparrow(\nu)$ is a set of expressions e such that there exists a reduction of length at least ν starting with e. This can easily be seen using the description of Kripke-Joyal semantics above. Thus, $\Uparrow$ is externally the pointwise complement of the stratified must-convergence predicate $\{\Downarrow_\beta\}_{\beta<\omega_1}$ from [4]. Then, the proof that $\Box\Uparrow \;\to \uparrow$ corresponds to the proof that $\Downarrow \subseteq \bigcup_{\beta<\omega_1} \Downarrow_\beta$ in [4]. Here we technically see the need for indexing over ω_1.

Acknowledgments. This research was supported in part by the ModuRes Sapere Aude Advanced Grant from The Danish Council for Independent Research for the Natural Sciences (FNU) and in part by Microsoft Research through its PhD Scholarship Programme.

References

1. Agha, G., Mason, I.A., Smith, S.F., Talcott, C.L.: A foundation for actor computation. Journal of Functional Programming 7(1), 1–72 (1997)
2. Ahmed, A.: Step-indexed syntactic logical relations for recursive and quantified types. Tech. rep., Harvard University (2006), `http://www.ccs.neu.edu/home/amal/papers/lr-recquant-techrpt.pdf`
3. Apt, K.R., Plotkin, G.D.: Countable nondeterminism and random assignment. Journal of the ACM 33(4), 724–767 (1986)
4. Birkedal, L., Bizjak, A., Schwinghammer, J.: Step-indexed relational reasoning for countable nondeterminism. Logical Methods in Computer Science 9(4) (2013)
5. Birkedal, L., Møgelberg, R.E., Schwinghammer, J., Støvring, K.: First steps in synthetic guarded domain theory: step-indexing in the topos of trees. Logical Methods in Computer Science 8(4) (2012)
6. Bizjak, A., Birkedal, L., Miculan, M.: A model of countable nondetermism in guarded type theory (2014), `http://cs.au.dk/~abizjak/documents/trs/cntbl-gtt-tr.pdf`
7. Di Gianantonio, P., Honsell, F., Plotkin, G.D.: Uncountable limits and the lambda calculus. Nordic Journal of Computing 2(2), 126–145 (1995)
8. Dreyer, D., Ahmed, A., Birkedal, L.: Logical step-indexed logical relations. Logical Methods in Computer Science 7(2) (2011)
9. Lambek, J., Scott, P.: Introduction to Higher-Order Categorical Logic. Cambridge Studies in Advanced Mathematics. Cambridge University Press (1988)
10. Lassen, S.B.: Relational Reasoning about Functions and Nondeterminism. Ph.D. thesis, University of Aarhus (1998)

11. Lassen, S.B., Moran, A.: Unique fixed point induction for McCarthy's amb. In: Kutyłowski, M., Wierzbicki, T., Pacholski, L. (eds.) MFCS 1999. LNCS, vol. 1672, pp. 198–208. Springer, Heidelberg (1999)
12. Lassen, S.B., Pitcher, C.: Similarity and bisimilarity for countable non-determinism and higher-order functions. Electronic Notes in Theoretical Computer Science 10 (1997)
13. Levy, P.B.: Infinitary Howe's method. In: Coalgebraic Methods in Computer Science, pp. 85–104 (2006)
14. MacLane, S., Moerdijk, I.: Sheaves in Geometry and Logic: A First Introduction to Topos Theory. Mathematical Sciences Research Institute Publications. Springer, New York (1992)
15. Mason, I.A., Talcott, C.L.: Equivalence in functional languages with effects. Journal of Functional Programming 1(3), 287–327 (1991)
16. Reyes, G., Zolfaghari, H.: Bi-heyting algebras, toposes and modalities. Journal of Philosophical Logic 25(1), 25–43 (1996)
17. Sabel, D., Schmidt-Schauß, M.: A call-by-need lambda calculus with locally bottom-avoiding choice: context lemma and correctness of transformations. Mathematical Structures in Computer Science 18(3), 501–553 (2008)

Cut Admissibility by Saturation

Guillaume Burel

ÉNSIIE/Cédric, 1 square de la résistance, 91025 Évry cedex, France
guillaume.burel@ensiie.fr
http://www.ensiie.fr/~guillaume.burel/

Abstract. Deduction modulo is a framework in which theories are integrated into proof systems such as natural deduction or sequent calculus by presenting them using rewriting rules. When only terms are rewritten, cut admissibility in those systems is equivalent to the confluence of the rewriting system, as shown by Dowek, RTA 2003, LNCS 2706. This is no longer true when considering rewriting rules involving propositions. In this paper, we show that, in the same way that it is possible to recover confluence using Knuth-Bendix completion, one can regain cut admissibility in the general case using standard saturation techniques. This work relies on a view of proposition rewriting rules as oriented clauses, like term rewriting rules can be seen as oriented equations. This also leads us to introduce an extension of deduction modulo with *conditional* term rewriting rules.

Whatever their origin, proofs rarely need to be searched for without context: Program verification requires arithmetic, theories of lists or arrays, etc. Mathematical theorems are in general not proved in pure predicate logic. Consequently, even if (automated and interactive) proof systems have achieved a high degree of maturity, they need to be able to deal with theories in an efficient way. This explains the particular interest focused on SMT (Satisfiability Modulo Theory) provers in the latter years. However, one of the drawbacks of the SMT approach is that the way theories are integrated is not completely generic, in the sense that each theory needs a special treatment.

A more generic approach to integrating theories into a proof system was proposed by Dowek, Hardin and Kirchner [14]. In Deduction Modulo[1], a theory is represented by a congruence over formulæ, and proofs are searched for modulo this congruence. In practice, this congruence is most often described as a rewriting system. However, using only term rewriting rules would not be enough to capture interesting theories. For instance, Vorobyov [21] showed that even quantifier-free Presburger arithmetic cannot be presented as a convergent term rewriting system. To overcome this, Deduction Modulo also deals with proposition rewriting rules, that rewrite atomic formulæ into formulæ. Thanks to this, it was possible to present many theories in Deduction Modulo: simple type theory (also known as higher-order logic), arithmetic, B set theory [18], any pure

[1] Although it may sound rather strange, the absence of subsequent to the term "modulo" follows the original works about this field.

G. Dowek (ed.): RTA-TLCA 2014, LNCS 8560, pp. 124–138, 2014.

type system, including the calculus of constructions which is the foundation of the proof assistant Coq [7], or, in fact, any first-order theory [5]. It is then possible to use automated theorem provers based on Deduction Modulo, such as iProver Modulo [6] or Zenon Modulo [9]. Moreover, proofs in those theories can be checked using Dedukti, a proof checker based on Deduction Modulo (`https://www.rocq.inria.fr/deducteam/Dedukti/`). Note that if one wants that the proof systems modulo a rewriting system behave well, in particular, if one wants the proof search methods to be complete, or the proof calculus to enjoy usual proof-theoretical properties such as the subformula property or the witness property, the rewriting system must have the following feature: The cut rule must be admissible in the sequent calculus modulo the rewriting system. This is true for the presentations of theories cited above.

Even if any first-order theory can be presented as a rewriting system with cut admissibility, these presentations may be quite unnatural. This is particularly the case when equality is involved. Indeed, the work [5] does not handle the equality predicate $\simeq$ in a special way, and for instance an axiom $s \simeq t$ would be presented by a proposition rewriting rule $s \simeq t \to \top$ and not by a term rewriting rule $s \to t$. There are also cases in which the most natural candidate to present an axiom as a rewriting rule would be a conditional rewriting rule, for instance in the case of an axiom of the shape $A(x) \Rightarrow s(x) \simeq t(x)$. In particular, this is the case of one of the axioms of the theory used in the provers of the HOL family (HOL4, HOL Light, or even Isabelle/HOL). In the translation of proofs in the OpenTheory format [17] into proofs that can be checked by Dedukti [1], this axiom could not be easily presented as a rewriting rule, and should therefore remain as an axiom, losing partially the benefit of working modulo the theory. As we will see in Example 8, this axiom can be naturally presented as a conditional rewriting rule. In this paper, we therefore introduce Deduction Modulo Conditional Rewriting Rules, which strictly subsumes the usual presentation of Deduction Modulo.

We therefore need a criterion that ensures that cut admissibility holds in the sequent calculus modulo the conditional rewriting system. To do so, we study links between saturation processes and cut admissibility. In [12], Dowek proved that in the case were there are only *term* rewriting rules, cut admissibility is equivalent to the confluence of the rewriting system. In the case where there are proposition rewriting rules, this is no longer true; for instance the rule $A \to A \Rightarrow B$ is confluent but does not admit cuts. Now, consider a term rewriting system that does not admit cuts. Equivalently, it is not confluent. One way to recover confluence, and thus cut admissibility, is to use the completion technique of Knuth and Bendix [19], that has been refined into Unfailing Completion [2]. Unfailing Completion is a saturation process: starting from a set of equations, new equations are generated, and older ones are simplified, until all newly generated equations are redundant. The set of equations is then called saturated, and in the case of Unfailing Completion, the corresponding rewriting system is convergent on ground terms. Consequently, cut admissibility is ensured for ground terms, which is enough for cut admissibility since we can restrict ourselves to ground sequents (if one considers Eigenvariables as constants). In other words, when it succeeds, Unfailing Completion allows to recover cut admissibility. In this paper, we investigate how a saturation technique can help at regaining cut admissibility in the more general case when there are *proposition* rewriting rules.

To better apprehend how it works, let us remark that there are usually two ways to see rewriting systems: The first one is to consider them as particular cases of abstract reduction systems whose objects are terms. The second one is to consider them as a set of equations oriented by some reduction ordering. It is this second point of view that is considered in Unfailing Completion, and more generally in the automated theorem proving community. Of course, the two views generally coincide, in particular in the case of terminating rewrite systems. Let us now look at what would correspond to a proposition rewriting rule following the second point of view. According to Dowek [13], a rewriting rule $P \to C$ would coincide to what he calls a one-way clause $\neg P \vee C$, where $\neg P$ is selected, which means it is the only literal that can be used to resolve the clause in the Resolution method. This idea of selected literal is reminiscent of Ordered Resolution with Selection [4], where literals are selected according to a well-founded ordering and a selection function choosing negative literals. Therefore, the analogue of seeing term rewriting rules as equations oriented by an ordering is to see proposition rewriting rules as clauses oriented by an ordering and a selection function. Then, Ordered Resolution with Selection can be used as a saturation process that allows to recover cut admissibility, as we prove in Theorem 7.

We can go a step further. Unfailing Completion and Ordered Resolution with Selection can be combined into Superposition, which is therefore a proof search method for first-order logic with equality. Superposition includes in particular the following inference rule:

$$\mathsf{Superposition}\ \frac{s \simeq u \vee C \qquad L[t]_{\mathfrak{p}} \vee D}{\sigma(L[u]_{\mathfrak{p}} \vee C \vee D)}\ \sigma = mgu(s,t)$$

with ordering restrictions to prevent the proliferation of such inferences. If we look at the inference rule, it behaves as if $L[t]_{\mathfrak{p}}$ was rewritten (or more precisely narrowed) into $L[u]_{\mathfrak{p}}$, provided no condition in C holds. Following our analogy between rewriting rules, equations and clauses, we can therefore see the clause $s \simeq u \vee C$ as a conditional rewriting rule $s \to u$ if $\neg C$. We then prove that when a set of clauses is saturated using Superposition, its corresponding rewriting system, consisting of both proposition rewriting rules and conditional term rewriting rules, admits cuts (Theorem 7 again, since Ordered Resolution with Selection is a special case of Superposition when equality is not present).

In the following section, we will present Deduction Modulo in more details. Then in Section 2 we say a few words about saturation processes, in particular saturation up to compositeness which is a modular form of redundancy. In Section 3 we introduce Deduction Modulo Conditional Rewriting Rules, in particular by means of a sequent calculus. In Section 4, we prove the main result of this paper, namely that when a set of clauses is saturated up to compositeness, then a corresponding conditional rewriting system admits cuts.

1 Deduction Modulo

1.1 Sequent Calculi Modulo

We use standard definitions for terms, predicates, propositions (with connectives $\neg, \Rightarrow, \wedge, \vee$ and quantifiers $\forall, \exists$), sequents, substitutions, term rewriting rules and term rewriting. The substitution of a variable x by a term t in a term or a proposition A is denoted by $\{t/x\}A$, and more generally the application of a substitution σ in a term or a proposition A by σA. A literal is an atomic proposition or the negation of an atomic proposition. The negation of a literal $L^\perp$ is defined by $P^\perp = \neg P$ and $\neg P^\perp = P$. A proposition is in clausal form if it is the universal quantification of a disjunction of literals $\forall x_1, \ldots, x_n.\ L_1 \vee \ldots \vee L_p$ where $x_1, \ldots, x_n$ are the free variables of $L_1, \ldots, L_p$. In the following, we will often omit the quantifiers, and we will identify propositions in clausal form with clauses (i.e. set of literals) as if $\vee$ were associative, commutative and idempotent. This will be justified in Section 3. The symbol $\Box$ represents the empty clause. The polarity of a position in a proposition can be defined as follows: the root is positive, and the polarity switches when going under a $\neg$ or on the left of a $\Rightarrow$.

In deduction modulo, term rewriting is extended to propositions by congruence on the proposition structure. In addition, there are also proposition rewriting rules whose left-hand side is an atomic proposition and whose right-hand side can be any proposition. Such rules can also be applied to non-atomic propositions by congruence on the proposition structure. We call a rewriting system the combination of a term rewriting system and a proposition rewriting system. Deduction modulo consists in applying the inference rules of an existing proof system modulo such a rewriting system.

In this setting, rewriting rules can be applied indifferently to the left- or the right-hand side of a sequent. Consequently, they can be considered semantically as an equivalence between their left- and right-hand sides. To be able to consider implications, a polarized version of deduction modulo was introduced [11]. Proposition rewriting rules are tagged with a polarity + or −; they are then called polarized rewriting rules. A proposition A is rewritten positively into a proposition B ($A \longrightarrow^+ B$) if it is rewritten by a positive rule at a positive position or by a negative rule at a negative position. It is rewritten negatively ($A \longrightarrow^- B$) if it is rewritten by a positive rule at a negative position or by a negative rule at a positive position. Intuitively, a positive rule $A \rightarrow^+ B$ (resp. a negative rule $B \rightarrow^- A$) corresponds to an implication $B \Rightarrow A$. Term rewriting rules (but not proposition rewriting rules) are considered as both positive and negative. $\xrightarrow{*}{}^{\pm}$ is the reflexive transitive closure of $\longrightarrow^{\pm}$. This gives the polarized sequent calculus modulo, some of whose rules are presented in Figure 1.

Example 1. Consider the polarized rewriting system

$$A \subseteq B \rightarrow^- \forall x.\ x \in A \Rightarrow x \in B \qquad A \subseteq B \rightarrow^+ \neg dw(A,B) \in A$$
$$A \subseteq B \rightarrow^+ dw(A,B) \in B$$

$$\vdash \frac{}{\Gamma, A \vdash B, \Delta} A \xrightarrow[\mathcal{R}]{*} {}^{-}C {}^{+} \xleftarrow[\mathcal{R}]{*} B \qquad \vdash \frac{\Gamma, A \vdash \Delta \quad \Gamma \vdash B, \Delta}{\Gamma \vdash \Delta} A {}^{-} \xleftarrow[\mathcal{R}]{*} C \xrightarrow[\mathcal{R}]{*} {}^{+}B$$

$$\Rightarrow\vdash \frac{\Gamma, B \vdash \Delta \qquad \Gamma \vdash A, \Delta}{\Gamma, C \vdash \Delta} C \xrightarrow[\mathcal{R}]{*} {}^{-}A \Rightarrow B \qquad \vdash\neg \frac{\Gamma, A \vdash \Delta}{\Gamma \vdash B, \Delta} B \xrightarrow[\mathcal{R}]{*} {}^{+}\neg A$$

$$\forall\vdash \frac{\Gamma, \{t/x\}A \vdash \Delta}{\Gamma, B \vdash \Delta} B \xrightarrow[\mathcal{R}]{*} {}^{-}\forall x.\ A \qquad \vdash\because \frac{\Gamma \vdash A, B, \Delta}{\Gamma \vdash C, \Delta} \begin{matrix} C \xrightarrow[\mathcal{R}]{*} {}^{+}A \\ C \xrightarrow[\mathcal{R}]{*} {}^{+}B \end{matrix}$$

Fig. 1. Some inference rules of the Polarized Sequent Calculus Modulo $\mathcal{R}$

(dw can be seen as the Skolem symbol introduced by the CNF transformation of the definition of the subset relation, $dw(A, B)$ is a witness that A is not included in B if it is the case.) We can build the following proof of the transitivity of the inclusion in the polarized sequent calculus modulo this system:

$$\cfrac{\cfrac{\cfrac{\cfrac{\cfrac{\vdash\frac{}{dw(A,C) \in C \vdash A \subseteq C} \qquad \vdash\frac{}{dw(A,C) \in B \vdash dw(A,C) \in B}}{dw(A,C) \in B \Rightarrow dw(A,C) \in C, dw(A,C) \in B \vdash A \subseteq C}\Rightarrow\vdash}{B \subseteq C, dw(A,C) \in B \vdash A \subseteq C}\forall\vdash \qquad \vdash\frac{}{dw(A,C) \in A \vdash dw(A,C) \in A}}{dw(A,C) \in A \Rightarrow dw(A,C) \in B, B \subseteq C, dw(A,C) \in A \vdash A \subseteq C}\Rightarrow\vdash}{A \subseteq B, B \subseteq C, dw(A,C) \in A \vdash A \subseteq C}\forall\vdash}{\cfrac{A \subseteq B, B \subseteq C \vdash A \subseteq C, A \subseteq C}{A \subseteq B, B \subseteq C \vdash A \subseteq C}\vdash\because}\vdash\neg$$

We denote by $\Gamma \vdash_{\mathcal{R}} \Delta$ the fact that the sequent $\Gamma \vdash \Delta$ is provable in the Polarized Sequent Calculus Modulo $\mathcal{R}$. A theory Γ and a rewriting system $\mathcal{R}$ are called *compatible* if for all formulæ A, then $\Gamma \vdash A$ (without rewriting) if and only if $\vdash_{\mathcal{R}} A$.

The cut rule is admissible in the sequent calculus modulo $\mathcal{R}$ if, whenever a sequent can be proved in it, then it can be proved without using the cut rule ($\vdash$ in Figure 1). Abusing terminology, we say that a rewriting system $\mathcal{R}$ admits cut if the cut rule is admissible in the sequent calculus modulo $\mathcal{R}$. The admissibility of the cut rule has a strong proof-theoretical as well as practical importance: it entails that normal forms exist for proofs; it implies the consistency of the theory associated to $\mathcal{R}$; it is equivalent to the completeness of the proof search procedures based on deduction modulo $\mathcal{R}$; etc.

1.2 Resolution Modulo and One-Way Clauses

An extension of resolution based on deduction modulo, called ENAR for Extended Narrowing and Resolution, was proposed by Dowek, Hardin and Kirchner [14]. It consists of adding a new inference rule to the method of Robinson [20]. This rule, called Extended Narrowing, narrows a clause using the rewrite system modulo which the proof is searched for. However, in ENAR, there is a need to

$$\text{Resolution}\ \frac{P \vee C \quad \neg Q \vee D}{\sigma(C \vee D)}\ \sigma = mgu(P,Q) \qquad \text{Factoring}\ \frac{L \vee K \vee C}{\sigma(L \vee C)}\ \sigma = mgu(L,K)$$

$$\text{Ext. Narr.}^-\ \frac{P \vee C}{\sigma(D \vee C)}\ \begin{array}{l}\sigma = mgu(P,Q)\\ Q \to^- D \in \mathcal{R}\end{array} \qquad \text{Ext. Narr.}^+\ \frac{\neg Q \vee D}{\sigma(C \vee D)}\ \begin{array}{l}\sigma = mgu(P,Q)\\ P \to^+ \neg C \in \mathcal{R}\end{array}$$

Fig. 2. Polarized Resolution Modulo

transform formulæ into clausal normal form during proof search, and not only before as it is usually the case with resolution methods. Therefore, Dowek refined ENAR into Polarized Resolution Modulo, whose rules are presented in Figure 2. In Polarized Resolution Modulo, proposition rewrite rules are assumed to be *clausal*, which means that positive rewrite rules are of the form $P \to^+ \neg C$, and negative rules are of the form $P \to^- C$, where C is in clausal form. This ensures that formulæ generated by Extended Narrowing are still in clausal form.

Applying Extended Narrowing to a clause $P \vee C$ using the rule $Q \to^- D$ produces the same clause (namely $\sigma(D \vee C)$, where $\sigma = mgu(P,Q)$) as applying Resolution to this clause $P \vee C$ and the clause $\neg Q \vee D$. Similarly, narrowing with $P \to^+ \neg C$ amounts to resolving with $P \vee C$. Therefore, the polarized rewrite rule $Q \to^- D$ (resp. $P \to^+ \neg C$) can be identified with what Dowek [13] called the one-way clause $\underline{\neg Q} \vee D$ (resp. $\underline{P} \vee C$) where

– two one-way clauses cannot be resolved together;
– only the selected (underlined) literal of a one way-clause can be used in resolution.

Conversely, given a clause C and a literal L in C, it is always possible to associate a polarized rewrite rule $polar(C,L)$: $polar(P \vee C, P)$ is $P \to^+ \neg C$ and $polar(\neg Q \vee D, \neg Q)$ is $Q \to^- D$. Therefore, the same way that it is possible to see a term rewriting rule as an equation in which one side is selected, it is possible to see clausal polarized rewriting rules as clauses in which a literal is selected.

2 Saturation

If there are only term rewriting rules in $\mathcal{R}$, and no proposition rewriting rules, Dowek [12] showed that cut admissibility in the asymmetric sequent calculus modulo $\mathcal{R}$ is equivalent to confluence of $\mathcal{R}$. If $\mathcal{R}$ is not confluent, a way to get an equivalent rewriting system which is confluent is to apply the Knuth-Bendix standard completion [19], which was extended into Unfailing Completion [2]. Unfailing completion can be seen as a saturation process: one applies all possible inferences to a starting set of formulæ (in that case, positive unit equations) until all newly inferred formulæ are redundant, that is, can be simplified. In that case, the resulting set is called saturated, and the correctness of the procedure shows that a saturated set has the required property, namely ground convergence in the case of Unfailing Completion. Of course, since the required property is in general not decidable, the saturation process may not terminate. Resolution and its refinements can also be seen as saturation processes: the set of clauses is completed until either the empty clause is generated, in which case the initial set was inconsistent, or until all newly generated clauses are redundant, in which case it is possible to construct a model of the saturated set of clauses.

It would be preferable that the saturation process were modular, in the sense that, if a set Γ of formulæ is saturated, then saturating $\Gamma \cup \Delta$ should not need to apply inferences between formulæ of Γ only. (This is in particular crucial for implementing resolution using the given clause algorithm, to ensure that clauses that were redundant remains redundant when a new given clause is chosen.) Therefore, redundancy should be modular, in the sense that if C is redundant in Γ, then it should be redundant in $\Gamma \cup \Delta$. This refinement of redundancy is called compositeness by Bachmair and Ganziger [3]. In fact, resolution-based provers saturate their input in general not up to redundancy but up to compositeness (Bachmair and Ganziger call saturation up to compositeness completeness, but we will keep writing "saturation up to compositeness" to keep things clear.) Of course, saturation up to compositeness implies saturation up to redundancy.

To deal with full first-order logic with equality, and not only unit clauses, Unfailing Completion can be extended into Superposition [3], which is consequently a complete proof-search method for first-order logic with equality. In pure Superposition, the only predicate is the equality predicate (noted $\simeq$), and clauses are therefore sets of equations and inequations. It is possible to encode other predicates using function symbols, as is done for instance in the prover E. However, to separate more clearly reasoning about equality and about propositions, we will use the inference for Superposition in addition to the rules for Ordered Resolution with Selection [4] (a refinement of resolution inspired by Superposition), as is done in the prover SPASS. As in Unfailing Completion, we consider a reduction ordering $\succ$, that is, an ordering that is stable under substitution and context. Literals are compared as the multisets of multisets $\{\{s\}, \{t\}\}$ for the positive literal $s \simeq t$ and $\{\{s, \flat\}, \{t, \flat\}\}$ for the negative literal $s \not\simeq t$, where $\flat$ is a special symbol not part as the signature, which is assumed to be smaller than any term. A clause $s \simeq t \vee C$ is *reductive* for $s \simeq t$ if $t \not\succeq s$ and $s \simeq t$ is strictly maximal in $s \simeq t \vee C$. We also consider a selection function S that, given a clause C, returns a subset of the negative literals of C. Without considering simplifications, Superposition consists of the four inference rules presented in Figure 3, in addition to which we also consider the two rules of Ordered Resolution with Selection presented in Figure 4.

As we can see, Superposition has strong restrictions on the application of inference rules, which explain in part its efficiency. In particular, ordering restrictions are performed after the application of the unifier σ. Let us note notwithstanding that, thanks to the stability of $\succ$ by substitution, the calculus remains of course complete if the restriction is applied on the premises, although it makes the proof search space bigger.

In Superposition, compositeness can be defined as follows: A ground clause C is called composite with respect to Γ if there exists ground instances $C_1, \ldots, C_n$ of clauses of Γ such that $C_1, \ldots, C_n$ entails C and $C \succ C_j$ for all $1 \leq j \leq n$. A

$$\text{Equality Resolution}\ \frac{s \not\simeq t \vee C}{\sigma(C)} \qquad \text{Negative Superposition}\ \frac{s \simeq u \vee C \quad \neg P[t]_{\mathfrak{p}} \vee D}{\sigma(\neg P[u]_{\mathfrak{p}} \vee C \vee D)}$$

$$\text{Positive Superposition}\ \frac{s \simeq u \vee C \quad P[t]_{\mathfrak{p}} \vee D}{\sigma(P[u]_{\mathfrak{p}} \vee C \vee D)} \qquad \text{Eq. Factoring}\ \frac{s \simeq u \vee t \simeq v \vee C}{\sigma(u \not\simeq v \vee t \simeq v \vee C)}$$

where

1. in all rules above, $\sigma = mgu(s,t)$;
2. in Equality Resolution, either $s \not\simeq t \in S(s \simeq u \vee C)$, or ($S(s \simeq u \vee C) = \emptyset$ and $\sigma(s \not\simeq t)$ is maximal in $\sigma(s \simeq u \vee C)$;
3. in both Superpositions, $\sigma(s \simeq u \vee C)$ is reductive for $\sigma(s \simeq u)$ and t is not a variable;
4. in Negative Superposition, either $\neg P[t]_{\mathfrak{p}} \in S(\neg P[t]_{\mathfrak{p}}, D)$ or $S(\neg P[t]_{\mathfrak{p}} \vee D) = \emptyset$ and $\sigma(\neg P[t]_{\mathfrak{p}})$ is maximal in $\sigma(\neg P[t]_{\mathfrak{p}} \vee D)$;
5. in Positive Superposition, $S(P[t]_{\mathfrak{p}} \vee D) = \emptyset$ and $\sigma(P[t]_{\mathfrak{p}} \vee D)$ is reductive for $\sigma(P[t]_{\mathfrak{p}})$;
6. in Equality Factoring, $S(s \simeq u \vee t \simeq v \vee C) = \emptyset$ and $\sigma(s \simeq u)$ is maximal in $\sigma(s \simeq u \vee t \simeq v \vee C)$;
7. in both Superpositions, if $P[t]_{\mathfrak{p}}$ is an equation $v[t]_{\mathfrak{p}'} \simeq w$, then $\sigma w \not\succeq \sigma v[t]_{\mathfrak{p}'}$.

Fig. 3. Inference Rules of Superposition

$$\text{Resolution}\ \frac{P \vee C \qquad \neg Q \vee D}{\sigma(C \vee D)} \qquad \text{Factoring}\ \frac{P \vee Q \vee C}{\sigma(P \vee C)}$$

where

1. in both cases, $\sigma = mgu(P,Q)$;
2. in Resolution, σP is strictly maximal in $\sigma(P \vee C)$ and $S(\sigma(P \vee C)) = \emptyset$;
3. in Resolution, either $\sigma(\neg Q) \in S(\sigma(\neg Q \vee D)$ or $S(\sigma(\neg Q \vee D) = \emptyset$ and $\sigma(\neg Q)$ is maximal in $\sigma(\neg Q \vee D)$;
4. in Factoring, σP is maximal in $\sigma(P \vee C)$ and $S(\sigma(P \vee C)) = \emptyset$.

Fig. 4. Inference Rules of Ordered Resolution with Selection

non-ground clause is called composite with respect to Γ if all its ground instances are. Lemma 11 of [3] tells us that if C is composite in Γ, then it is composite in $\Gamma \cup \Delta$, and that all composite clauses can be safely removed from Γ, as expected.

3 Deduction Modulo Conditional Rewriting Rules

We now present an extension of deduction modulo to the case of conditional rewriting rules.

Definition 2 (Conditional rewriting rule). *A conditional rewriting rule is given by a pair of terms t and s and a set of formalæ Γ. It is denoted by $s \to t$ if Γ.*

A term u rewrites to a term v and conditions Δ using the conditional rewriting rule $s \to t$ if Γ if there is a position $\mathfrak{p}$ and a substitution θ such that $u_{|\mathfrak{p}} = \theta s$, $v = u[\theta t]_{\mathfrak{p}}$ and $\Delta = \theta\Gamma$. This rewrite relation is denoted by $u \longrightarrow v \wr \Delta$. This is extended to propositions by congruence on the proposition structure.

Note that our definition of conditional rewriting differs from the usual definition as can be found in [10], because conditions are not checked before applying the rewriting rule, but they are delayed and, as we will see, they are checked using a proof system and not just using normalization. This allows to have arbitrary first-order formulæ as conditions.

We can combine conditional (term) rewriting rules $\mathcal{C}$ with polarized (proposition) rewriting rules $\mathcal{P}$ to get what we call a polarized and conditional rewriting system : A proposition A rewrites positively to B and Δ in $\mathcal{CP}$ ($A \xrightarrow[\mathcal{CP}]{*}{}^{+} B \wr \Delta$) if

- either $A = B$ and $\Delta = \emptyset$
- or $A \xrightarrow[\mathcal{CP}]{*}{}^{+} A' \wr \Delta_1$ and
  - either $A' \xrightarrow[\mathcal{P}]{}{}^{+} B$ and $\Delta = \Delta_1$
  - or $A' \xrightarrow[\mathcal{C}]{} B \wr \Delta_2$ and $\Delta = \theta\Delta_1 \cup \Delta_2$ where θ is a renaming of the free variables of Δ_1 to avoid clashes with those of Δ_2.

Negative rewriting can be defined similarly. Let us remark that a conditional rewriting step is therefore both a positive or a negative step.

Definition 3. *A formula* does not involve equality *if* $\simeq$ *is not present in it. A conditional rewriting rule* $s \to t$ *if* C does not involve equality *if all formulæ in* C *do not involve equality.*

Given a rewriting system that does not involve equality, we can define the Sequent Calculus Modulo Polarized and Conditional Rules. We only give the main rules in Figure 5, the others can be induced from them:

$$\overset{\frown}{\vdash}\ \frac{\Gamma \vdash \theta A_1, \Delta \quad \dots \quad \Gamma \vdash \theta A_n, \Delta}{\Gamma, A \vdash B, \Delta}\ A \xrightarrow[\mathcal{CP}]{*}{}^{-} C \wr \{A_1; \dots; A_i\} \text{ and } B \xrightarrow[\mathcal{CP}]{*}{}^{+} C \wr \{A_{i+1}; \dots; A_n\}$$

$$\underset{\smile}{\vdash}\ \frac{\Gamma, A \vdash \Delta \quad \Gamma \vdash B, \Delta \quad \Gamma \vdash \theta A_1, \Delta \quad \dots \quad \Gamma \vdash \theta A_n, \Delta}{\Gamma \vdash \Delta}\ \begin{array}{l} C \xrightarrow[\mathcal{CP}]{*}{}^{-} A \wr \{A_1; \dots; A_i\} \\ C \xrightarrow[\mathcal{CP}]{*}{}^{+} B\{A_{i+1}; \dots; A_n\} \end{array}$$

$$\Rightarrow\vdash\ \frac{\Gamma, B \vdash \Delta \quad \Gamma \vdash A, \Delta \quad \Gamma \vdash \theta A_1, \Delta \quad \dots \quad \Gamma \vdash \theta A_n, \Delta}{\Gamma, C \vdash \Delta}\ C \xrightarrow[\mathcal{CP}]{*}{}^{-} A \Rightarrow B \wr \{A_1; \dots; A_n\}$$

$$\vdash\top\ \frac{\Gamma \vdash \theta A_1, \Delta \quad \dots \quad \Gamma \vdash \theta A_n, \Delta}{\Gamma \vdash A, \Delta}\ A \xrightarrow[\mathcal{CP}]{*}{}^{+} \top \wr \{A_1; \dots; A_n\}$$

$$\forall\vdash\ \frac{\Gamma, \{t/x\}A \vdash \Delta \quad \Gamma \vdash \theta A_1, \Delta \quad \dots \quad \Gamma \vdash \theta A_n, \Delta}{\Gamma, B \vdash \Delta}\ B \xrightarrow[\mathcal{CP}]{*}{}^{-} \forall x.\ A \wr \{A_1; \dots; A_n\}$$

where θ is a substitution of the free variables of all A_i.

Fig. 5. Some Inference Rules of the Sequent Calculus Modulo Polarized and Conditional Rules

Example 4. Consider the following polarized and conditional rewriting system, inspired by Collatz conjecture:

$$\begin{array}{ll} syracuse(X) \rightarrow syracuse(half(X)) \text{ if } \{even(X)\} & half(s(o)) \rightarrow o \text{ if } \emptyset \\ syracuse(X) \rightarrow syracuse(tnpo(X)) \text{ if } \{odd(X)\} & half(o) \rightarrow o \text{ if } \emptyset \\ half(s(s(X)) \rightarrow s(half(X)) \text{ if } \emptyset & even(s(X)) \rightarrow^{+} odd(x) \\ tnpo(s(X)) \rightarrow s(s(s(tnpo(X)))) \text{ if } \emptyset & odd(s(X)) \rightarrow^{+} even(x) \\ tnpo(o) \rightarrow s(o) \text{ if } \emptyset & even(o) \rightarrow^{+} \top \end{array}$$

and let us denote by $\underline{n}$ the term $\underbrace{s(\ldots s(}_{n \text{ times}} o))$.

Then, $syracuse(\underline{5}) \xrightarrow{*} \underline{1} \wr \{odd(\underline{5}); even(\underline{16}); even(\underline{8}); even(\underline{4}); even(\underline{2})\}$, and we have the following proof of $odd(syracuse(\underline{5}))$:

$$\dfrac{\dfrac{\dfrac{}{\vdash odd(\underline{5})}\vdash\top}{\vdash even(\underline{8})}\vdash\top \quad \dfrac{}{\vdash even(\underline{4})}\vdash\top \quad \dfrac{\dfrac{}{\vdash even(\underline{16})}\vdash\top}{\vdash even(\underline{2})}\vdash\top}{\vdash odd(syracuse(\underline{5}))}\vdash\top$$

Note that Deduction Modulo Polarized and Conditional Rules strictly subsumes Polarized Deduction Modulo, which is exactly the case when there are no conditions in the term rewriting rules.

Lemma 5. *The inference rules $\forall\vdash$, $\vee\vdash$, $\neg\vdash$ and $\vdash\neg$ are invertible in the Sequent Calculus Modulo Polarized and Conditional Rules, which means that their conclusion is provable if and only if their premises are. Moreover, if the proof of the conclusion does not use $\vdash_{\rightarrow}$, neither do the proofs of the premises.*

Proof. By induction on the proof of the conclusion. □

This lemma implies that we can handle formulæ in clausal normal form as clauses, that is, set of literals. See for instance [16].

4 Ensuring Cut Admissibility Using Saturation

Given an ordering $\succ$ and a selection function S, we define the polarized and conditional rewriting system associated to a set of clauses, and state that the saturation of the set of clauses implies the cut admissibility for the corresponding rewriting system.

Definition 6. *Given a set of clauses Γ, an ordering on terms $\succ$ and a selection function S, then the rewrite system $\mathcal{CS}(\Gamma, \succ, S)$ contains all rewrites rules*

- *$polar(C, L)$ for all clauses C such that $S(C) \neq \emptyset$ and for all L in $S(C)$ that does not involve equality;*
- *$polar(C, L)$ for all clauses C such that $S(C) = \emptyset$ and for all L maximal in C w.r.t. $\succ$ that does not involve equality;*

– $s \to t$ *if* $\{L_1^\perp, \ldots, L_n^\perp\}$ *for all clauses* $C = s \simeq t \vee L_1 \vee \cdots \vee L_n$ *such that* $S(C) = \emptyset$, $s \simeq t$ *is maximal in* C *and* $t \not\succ s$.

Theorem 7. *If the set* Γ *of clauses is saturated by Superposition up to compositeness, and* $\mathcal{CS}(\Gamma, \succ, S)$ *does not involve equality, then the sequent calculus modulo* $\mathcal{CS}(\Gamma, \succ, S)$ *is compatible with* Γ *and it admits cuts.*

Note that the Sequent Calculus Modulo Polarized and Conditional Rules is only defined to prove formulæ that do not involve equality.

Before we prove Theorem 7, let us look at an example.

Example 8. In provers of the HOL family, it is possible to define a new type corresponding to the (non-empty) set of terms that satisfies a predicate p. To do so, two function symbols *abs* and *rep* are introduced that go respectively from the initial type to the new one and conversely, as is represented in the following figure:

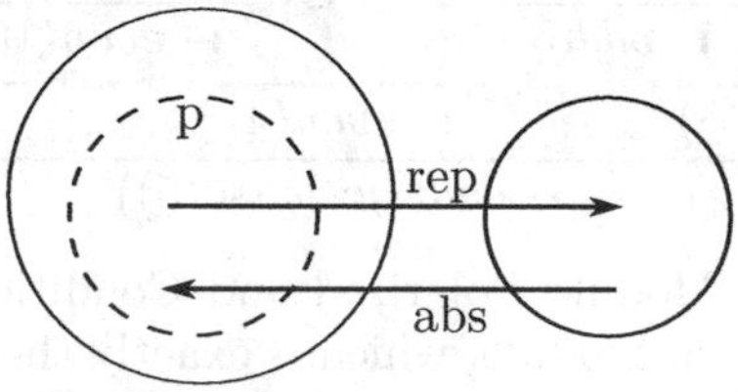

These function symbols satisfy the axioms $\forall X.\ p(X) \Leftrightarrow abs(rep(X)) \simeq X$ and $\forall Y.\ rep(abs(Y)) \simeq Y$, which correspond to the clauses

$$p(X) \vee \neg abs(rep(X)) \simeq X \quad (1)$$
$$\neg p(X) \vee abs(rep(X)) \simeq X \quad (2)$$
$$rep(abs(Y)) \simeq Y \quad (3)$$

Without a selection function, and with the lexicographic path ordering with precedence $abs \succ p$ and $rep \succ p$, the resulting conditional rewriting system is

$$abs(rep(X)) \to X \text{ if } \{p(X)\} \qquad rep(abs(Y)) \to Y \text{ if } \emptyset$$

The sequent calculus modulo this system is not compatible with the initial theory. Indeed, it is not possible to prove $\forall Y.\ p(abs(Y))$, although this is a consequence of the axioms. This comes from the fact that the set of clauses is not saturated for Superposition. To saturate the set of clause, for instance using E, we only need to add a new clause, namely $p(abs(Y))$, obtained by applying Negative Superposition on (3) and (1), and then Equality Resolution on $p(abs(X)) \vee \neg abs(X) \simeq abs(X)$, which is then composite. Note that all other generated clauses are tautologies, and therefore are composite.

Consequently, the polarized and conditional rewriting system

$$abs(rep(X)) \to X \text{ if } \{p(X)\} \quad rep(abs(Y)) \to Y \text{ if } \emptyset \quad p(abs(X)) \to^+ \neg\bot$$

admits cut for formulæ that do not involve equality.

Let us mention Holide [1], a translator of proofs in the OpenTheory [17] format into Dedukti, a proof checker based on deduction modulo. Most of the theory of HOL can be expressed as rewriting rules, except a few axioms that cannot be easily oriented. The first axiom defining *rep* and *abs* is one of these axioms, and as we have seen, it could be oriented as a conditional rewriting rule.

To prove Theorem 7, we need to prove that if A is proved in the sequent calculus modulo $\mathcal{CS}(\Gamma, \succ, S)$, then it can be proved in Γ, and that if A can be proved in Γ, then it can be proved in the sequent calculus modulo $\mathcal{CS}(\Gamma, \succ, S)$ without $\sqsubseteq$. The proof is partially based on the work of Dowek [13] who proves that a derivation in PRM can be translated into a cut-free proof in the polarized sequent calculus modulo.

Lemma 9. *If* $A \xrightarrow{*}{}^{-} B \wr \{L_1, \ldots, L_n\}$, *then* $\Gamma, A \vdash B, L_1^\perp, \ldots, L_n^\perp$.
If $A \xrightarrow{*}{}^{+} B \wr \{L_1, \ldots, L_n\}$, *then* $\Gamma, B \vdash A, L_1^\perp, \ldots, L_n^\perp$.

Proof. By induction on the length of the derivation; several steps can be combined using $\sqsubseteq$. Note the importance of renaming free variables between several steps. A single step can be proved by induction on the rewritten formula. If the rewriting occurs in a subformula, we can use the induction hypothesis to conclude. Let us therefore assume that A is atomic. We have two cases depending on whether a polarized or a conditional rule is used.

- $A \longrightarrow B \wr \{L_1, \ldots, L_n\}$. There is a rule $s \to t$ if $\{L'_1, \ldots, L'_n\}$ in $\mathcal{CS}(\Gamma, \succ, S)$ and a substitution σ such that $A_{|\mathfrak{p}} = \sigma s$, $B = A[\sigma t]_{\mathfrak{p}}$ and $L_k = \sigma L'_k$. This rule corresponds to a clause $s \simeq t \vee {L'_1}^\perp \vee \cdots \vee {L'_n}^\perp$ in Γ. From Γ, A one can therefore deduce $B \vee {L_1}^\perp \vee {L_n}^\perp$, and thus $\Gamma, A \vdash B, L_1^\perp, \ldots, L_n^\perp$.
- $A \longrightarrow^- B$. There exists a rule $P \to^- C$ in $\mathcal{CS}(\Gamma, \succ, S)$ and a substitution σ such that $A_{|\mathfrak{p}} = \sigma P$ and $B = \sigma C$. Therefore, there is a clause $\neg P \vee C$ in Γ, and from Γ, A one can therefore deduce σC, thus $\Gamma, A \vdash B$. □

Corollary 10. *If* $\Pi \vdash_{\mathcal{CS}(\Gamma,\succ,S)} \Delta$ *then* $\Gamma, \Pi \vdash \Delta$.

Proof. By induction on the proof, using Lemma 9 to convert rewriting steps. □

Lemma 11. *If the set Γ of clauses is saturated by Superposition up to compositeness, and $\mathcal{CS}(\Gamma, \succ, S)$ does not involve equality, if A does not involve equality and A is valid in Γ, then it can be proved in the sequent calculus modulo $\mathcal{CS}(\Gamma, \succ, S)$ without $\sqsubseteq$.*

Proof. Since Superposition is complete, there is a derivation of the empty clause from Γ and $\mathcal{C}\ell(\neg A)$ (the clausal normal form of $\neg A$). We are going to translate this derivation into a cut-free proof of $\mathcal{C}\ell(\neg A) \vdash$, by induction of the length of the derivation.

Let us show that all new clauses in the derivation of □ do not involve equality: Since Γ is saturated up to compositeness, all inferences using only premises in Γ are redundant and therefore can be discarded. So by induction hypothesis at least one of the premises does not involve equality. The only way to obtain an equality would be to apply Resolution or Superposition on a literal L of a clause $L \vee C$ of Γ. In the former case, the restriction on the application of Resolution implies that $polar(C, L)$ is in $\mathcal{CS}(\Gamma, \succ, S)$. Because it does not involve equality, this means that C, and thus the new clause, neither do. In the latter case, the clause involving equality is necessarily the left one in the Superposition inference rule, so that $L = s \simeq t$ for some s and t, and $C = L_1 \vee \cdots \vee L_n$. The restriction on the application of Superposition implies that $s \to t$ if $\{L_1^\perp, \ldots, L_n^\perp\}$ is in $\mathcal{CS}(\Gamma, \succ, S)$, so that $L_1, \ldots, L_n$ do not involve equality, and thus the new clause neither.

Therefore, since Γ is saturated up to compositeness, we can assume that the derivation of $\Box$ does not contain applications of Equality Resolution or Equality Factoring. Let us look at the remaining cases. To ease the proof, we can decompose the application of the inference rules into the application of an instantiation and the application of the rule without unification, as in the PEIR calculus of [13]. We have the following cases:

- Instantiation of a clause C outside Γ into σC. By induction hypothesis we have a cut-free proof of $\Delta, C, \sigma C \vdash$. We can build a cut-free proof of $\Delta, C \vdash$ by applying a contraction of C and then $\forall\vdash$ to instantiate the variables as in σ. (Remind that we omit to write quantifiers in clauses, so that C stands in fact for $\forall x_1, \ldots, x_n, C$ where $x_1, \ldots, x_n$ are the free variable of C.)
- Resolution between two clauses $P \vee C$ and $\neg P \vee D$ outside Γ. Let us suppose that we have a cut-free proof of $\Delta, P \vee C, \neg P \vee D, C \vee D \vdash$, then Proposition 7 of [13] implies that we have a proof of $\Delta, P \vee C, \neg P \vee D \vdash$.
- Resolution between a clause $P \vee C$ outside Γ and a clause obtained by instantiating a clause $\neg Q \vee D$ of Γ with substitution σ. Then $Q \to^- D$ is in $\mathcal{CS}(\Gamma, \succ, S)$, and $P \xrightarrow{*}{}^- \sigma D \wr \emptyset$. By induction hypothesis, we have a cut-free proof of $\Delta, P \vee C, C \vee \sigma D \vdash$. We can obtain a cut-free proof of $\Delta, P \vee C \vdash$ by applying a contraction of $P \vee C$ and rewriting P into σD.
- Resolution between a clause $\neg P \vee C$ outside Γ and a clause obtained by instantiating a clause $Q \vee D$ of Γ: similar to the previous case, except that we need to eliminate a double negation.
- Superposition between an clause obtained by instantiating a clause $s \simeq t \vee L_1 \vee \cdots \vee L_n$ of Γ with substitution σ and a clause $L[\sigma s]_\mathfrak{p} \vee D$ outside Γ. The restriction on the application of Superposition implies that the rule $s \to t$ if $\{L_1^\perp, \ldots, L_n^\perp\}$ is in $\mathcal{CS}(\Gamma, \succ, S)$, and consequently $L[\sigma s]_\mathfrak{p} \xrightarrow{*} L[\sigma t]_\mathfrak{p} \wr \{\sigma L_1^\perp, \ldots, \sigma L_n^\perp\}$. By induction hypothesis, we have a cut-free proof of $\Delta, L[\sigma s]_\mathfrak{p} \vee D, L[\sigma t]_\mathfrak{p} \vee D \vee L_1 \vee \cdots \vee L_n \vdash$. Since $\vee\vdash$ is cut-free invertible, we therefore have (cut-free) proofs of $\Delta, L[\sigma s]_\mathfrak{p} \vee D, L[\sigma t]_\mathfrak{p} \vee D \vdash$ and $\Delta, L[\sigma s]_\mathfrak{p} \vee D, L_i \vdash$ for all $1 \leq i \leq n$. Starting from $\Delta, L[\sigma s]_\mathfrak{p} \vee D \vdash$, we can apply contraction and rewrite $L[\sigma s]_\mathfrak{p}$, so that it remains to prove $\Delta, L[\sigma s]_\mathfrak{p} \vee D, L[\sigma t]_\mathfrak{p} \vee D \vdash$, which we have, and $\Delta, L[\sigma s]_\mathfrak{p} \vee D \vdash \sigma L_i{}^\perp$ for all $1 \leq i \leq n$, which can be obtained by application of $\vdash\neg$ or inversion of $\neg\vdash$ in the proofs above.
  Note that to make the proof more clear, we did not take the universal quantifiers into account: this is not a problem since $\forall\vdash$ is invertible, and we took care of renaming the free variables of the conditions during rewriting.

The last point to show is that from a cut-free proof of $\mathcal{C}\ell(\neg A) \vdash$ one can build a cut-free proof of $\vdash A$. This can be proved by slightly adapting the proof of Proposition 3 of [16]. □

5 Conclusion and Further Works

We have introduced an extension of Deduction Modulo that handles *conditional* rewriting rules. To get a criterion for cut admissibility in that setting, we have examined how rewriting rules can be seen as oriented equations and oriented clauses. This reflection has lead us to study how saturation techniques can help presenting a theory through a rewriting system with cut admissibility. Our main result is that whenever a set of clauses is saturated, we can build a corresponding rewriting system admitting cuts. We can therefore use state-of-the-art automated theorem provers, which are based on saturation techniques, to orient a theory so that it can be used in Deduction Modulo. These notable results could be extended in several directions.

First, the conditions in the conditional rewriting rules obtained from a saturated set of clauses are simple, since they are only a set of literals. This comes from the fact that we start from clauses, and not arbitrary formulæ. To get more interesting conditions, an idea would be to consider the work of Ganzinger and Stuber [15] that extend Superposition with formulæ that need not be in clausal normal form.

Second, our work is restricted to the case where equality appears only in the rewriting rules, not in the conditions nor in the formulæ to be proved. If we allowed equations in them, Negative Superposition could be applied to clauses of the theory in which a negative equation is selected. Therefore, these clauses could not be discarded as it is the case in Definition 6. Another issue would be the design of a sequent calculus modulo for first-order logic with equality. It could be handled by extending one of the calculi of [8].

Third, saturation implies cut admissibility, but the converse is not true in general. It would be interesting to be able to characterize cut admissibility as precisely as can be done when only terms are rewritten, where it is equivalent to the well studied notion of confluence.

Finally, it would be interesting to see how our work on conditional rewriting rules can be extended to the $\lambda\Pi$-calculus modulo, the system at the heart of the proof checker Dedukti. By doing so, we would be able to orient the theory used in the provers of the HOL family without using axioms, thus improving the performance of the translator Holide [1].

References

1. Assaf, A., Burel, G.: Translating HOL to Dedukti (2013) (submitted)
2. Bachmair, L., Dershowitz, N., Plaisted, D.: Completion without failure. In: Aït-Kaci, H., Nivat, M. (eds.) Resolution of Equations in Algebraic Structures. Rewriting Techniques, vol. 2, pp. 1–30. Academic Press inc. (1989)

3. Bachmair, L., Ganzinger, H.: Rewrite-based equational theorem proving with selection and simplification. Journal of Logic and Computation 4(3), 1–31 (1994)
4. Bachmair, L., Ganzinger, H.: Resolution theorem proving. In: Robinson, J.A., Voronkov, A. (eds.) Handbook of Automated Reasoning, pp. 19–99. Elsevier and MIT Press (2001)
5. Burel, G.: From axioms to rewriting rules, available on author's web page
6. Burel, G.: Experimenting with deduction modulo. In: Bjørner, N., Sofronie-Stokkermans, V. (eds.) CADE 2011. LNCS, vol. 6803, pp. 162–176. Springer, Heidelberg (2011)
7. Cousineau, D., Dowek, G.: Embedding pure type systems in the lambda-Pi-calculus modulo. In: Della Rocca, S.R. (ed.) TLCA 2007. LNCS, vol. 4583, pp. 102–117. Springer, Heidelberg (2007)
8. Degtyarev, A., Voronkov, A.: Equality reasoning in sequent-based calculi. In: Robinson, J.A., Voronkov, A. (eds.) Handbook of Automated Reasoning, pp. 611–706. Elsevier and MIT Press (2001)
9. Delahaye, D., Doligez, D., Gilbert, F., Halmagrand, P., Hermant, O.: Zenon modulo: When Achilles outruns the tortoise using deduction modulo. In: McMillan, K., Middeldorp, A., Voronkov, A. (eds.) LPAR-19 2013. LNCS, vol. 8312, pp. 274–290. Springer, Heidelberg (2013)
10. Dershowitz, N., Okada, M., Sivakumar, G.: Confluence of conditional rewrite systems. In: Kaplan, S., Jouannaud, J.-P. (eds.) CTRS 1987. LNCS, vol. 308, pp. 31–44. Springer, Heidelberg (1988)
11. Dowek, G.: What is a theory? In: Alt, H., Ferreira, A. (eds.) STACS 2002. LNCS, vol. 2285, pp. 50–64. Springer, Heidelberg (2002)
12. Dowek, G.: Confluence as a cut elimination property. In: Nieuwenhuis, R. (ed.) RTA 2003. LNCS, vol. 2706, pp. 2–13. Springer, Heidelberg (2003)
13. Dowek, G.: Polarized resolution modulo. In: Calude, C.S., Sassone, V. (eds.) TCS 2010. IFIP AICT, vol. 323, pp. 182–196. Springer, Heidelberg (2010)
14. Dowek, G., Hardin, T., Kirchner, C.: Theorem proving modulo. Journal of Automated Reasoning 31(1), 33–72 (2003)
15. Ganzinger, H., Stuber, J.: Superposition with equivalence reasoning and delayed clause normal form transformation. Inf. Comput. 199(1-2), 3–23 (2005)
16. Hermant, O.: Resolution is cut-free. Journal of Automated Reasoning 44(3), 245–276 (2009)
17. Hurd, J.: The OpenTheory standard theory library. In: Bobaru, M., Havelund, K., Holzmann, G.J., Joshi, R. (eds.) NFM 2011. LNCS, vol. 6617, pp. 177–191. Springer, Heidelberg (2011)
18. Jacquel, M., Berkani, K., Delahaye, D., Dubois, C.: Tableaux modulo theories using superdeduction – an application to the verification of B proof rules with the Zenon automated theorem prover. In: Gramlich, B., Miller, D., Sattler, U. (eds.) IJCAR 2012. LNCS, vol. 7364, pp. 332–338. Springer, Heidelberg (2012)
19. Knuth, D.E., Bendix, P.B.: Simple word problems in universal algebras. In: Leech, J. (ed.) Computational Problems in Abstract Algebra, pp. 263–297. Pergamon Press, Oxford (1970)
20. Robinson, J.A.: A machine-oriented logic based on the resolution principle. Journal of the ACM 12, 23–41 (1965)
21. Vorobyov, S.G.: On the arithmetic inexpressiveness of term rewriting systems. In: LICS, pp. 212–217 (1988)

Automatic Evaluation of Context-Free Grammars (System Description)*

Carles Creus and Guillem Godoy

Universitat Politècnica de Catalunya, Department of Software,
Barcelona, Spain
ccreuslopez@gmail.com, ggodoy@lsi.upc.edu

Abstract. We implement an online judge for context-free grammars. Our system contains a list of problems describing formal languages, and asking for grammars generating them. A submitted proposal grammar receives a verdict of acceptance or rejection depending on whether the judge determines that it is equivalent to the reference solution grammar provided by the problem setter. Since equivalence of context-free grammars is an undecidable problem, we consider a maximum length ℓ and only test equivalence of the generated languages up to words of length ℓ. This length restriction is very often sufficient for the well-meant submissions. Since this restricted problem is still NP-complete, we design and implement methods based on hashing, SAT, and automata that perform well in practice.

Keywords: grammars, equivalence, hashing, SAT, automata.

1 Introduction

Nowadays, there is an increasing interest in offering college-level courses online. Websites like Khan Academy [10], Coursera [13], Udacity [15] and edX [1], provide online courses on numerous topics. The users/students have access to videos and texts explaining several subjects, as well as tools for automated evaluation by means of exercises. In the specific context of computer science, the use of online judges for testing correctness of programs is used in several academic domains as a self-learning tool for students, as well as a precise method in exams for scoring their programming skills (see, e.g., [14,7,2]).

For the last two years we have developed a specific online judge for the subject of Theory of Computation [9], located at `http://racso.lsi.upc.edu/juez`. The site offers exercises about deterministic finite automata, context-free grammars, push-down automata, reductions between undecidable problems, and reductions between NP-complete problems. Users can submit their solutions, the judge evaluates them, and offers a counterexample when the submission is rejected. This is very useful to make students understand why their solutions are wrong, and to keep them motivated during the learning process. We have used

* The authors were supported by an FPU grant (first author) and the FORMALISM project (TIN2007-66523) from the Spanish Ministry of Education and Science.

G. Dowek (ed.): RTA-TLCA 2014, LNCS 8560, pp. 139–148, 2014.

the judge in the classroom, not only as a support tool for the students, but also as an evaluation method on exams. This has had a marked effect on the motivation and involvement of the students: during a fifteen-week course, each student has solved more than 150 problems in average, with more than 680 submissions. This means that each problem needed over 4 submissions to get acceptance from the judge, and the students were motivated enough to perform new attempts to reach an acceptance verdict.

In this paper we explain the techniques used to automatically evaluate the problems on context-free grammars. Each of such problems describes a language L and asks the student to submit a grammar $G_{\texttt{sub}}$ generating L. In some cases, it asks specifically for an unambiguous grammar. The judge checks the correctness of $G_{\texttt{sub}}$ by testing that it generates the same language as a reference grammar $G_{\texttt{sol}}$ provided by the problem setter. Since it is well-known that grammar equivalence is an undecidable problem [9], we cannot expect the judge to behave correctly for every input. Therefore, we focus on performing well with the well-meant grammars submitted by students to academic problems. These grammars are very simple, and when they are wrong, there is usually a small counterexample, i.e., a small word in $\mathcal{L}(G_{\texttt{sub}})\triangle\mathcal{L}(G_{\texttt{sol}})$. For this reason, we tackle the problem by fixing a length ℓ and looking for a word $w \in \mathcal{L}(G_{\texttt{sub}})\triangle\mathcal{L}(G_{\texttt{sol}})$ with length bounded by ℓ. Since this is still NP-complete, we develop methods that in practice behave well enough for small ℓ, $G_{\texttt{sub}}$, and $G_{\texttt{sol}}$. In particular, our judges[1] are based on automata and hashing techniques, and we also study possible optimizations for the reduction in [3] of the grammar equivalence to the SAT problem.

The paper is organized as follows. In Section 2 we summarize notations and basic concepts. In Section 3 we explain the different developed methods, and in Section 4 compare them with others from the literature. In Section 5 we describe our online system and our experience using it. We conclude in Section 6.

2 Preliminaries

Words are finite-length lists of symbols chosen over an underlying alphabet Σ. The length of a word w is denoted by $|w|$, and its i'th symbol, for $1 \leq i \leq |w|$, is denoted by $w[i]$. Similarly, its subword between i and j, inclusive, is denoted by $w[i..j]$. The empty word is denoted by ε. We assume that the reader is familiar with the concept of context-free grammar (CFG) as a structure $G = \langle \mathcal{V}, \Sigma, R, S \rangle$, where $\mathcal{V}$ is the set of non-terminal symbols, Σ is the alphabet of terminal symbols, $R \subset \mathcal{V} \times (\mathcal{V} \cup \Sigma)^*$ is the set of rules, and $S \in \mathcal{V}$ is the initial symbol. We denote non-terminals with uppercase letters $X, Y, Z, \ldots$ and terminals with lowercase letters $a, b, c, \ldots$, with possible subscripts. Often, grammars are just represented by a list of rules, where the non-terminal at the left-hand side of the first rule is considered the initial symbol. Also, rules with common left-hand side are usually described in compact form, e.g., two rules $X \to u$, $X \to v$ are represented by $X \to u \mid v$. In order to simplify definitions and arguments, we

[1] Source code available at `http://www.lsi.upc.edu/~ggodoy/publications.html`

assume without loss of generality that the sets of non-terminals of any two grammars are disjoint. We assume that our grammars are reduced and in CNF [9], and hence we only deal with rules of the form $X \to YZ$ and $X \to a$. Recall that the standard transformation to CNF produces a quadratic increase in size, and can be adapted to detect when ambiguity is lost due to the transformation. For a detailed definition of the language $\mathcal{L}(G)$ generated by the CFG G, and the concept of ambiguity see, e.g., [9]. We assume that the reader is familiar with the concept of deterministic finite automata (DFA) and their properties [9].

3 Judging Methods

3.1 Exhaustive

The JUDGEEXHAUSTIVE approach consists in enumerating all the words up to length ℓ that can be generated by each of the grammars and checking whether there exists some word w that is generated by just one of them. This brute force solution has some benefits in our setting. First, it is trivial to give the minimal counterexample in size, whenever one exists in our search space. Second, besides enumerating the words, it is easy to count the amount of different derivations that generate each of them. This additional information allows to check whether the grammar is ambiguous in the subset of words with length bounded by ℓ.

3.2 Hash

The JUDGEHASH approach is based on a hash function $\mathcal{H}$ that maps languages to natural numbers. We focus on the subsets $L_{G_{\mathtt{sub}},\ell} \subseteq \mathcal{L}(G_{\mathtt{sub}})$ and $L_{G_{\mathtt{sol}},\ell} \subseteq \mathcal{L}(G_{\mathtt{sol}})$ of words of length ℓ of $\mathcal{L}(G_{\mathtt{sub}})$ and $\mathcal{L}(G_{\mathtt{sol}})$, respectively, and check $L_{G_{\mathtt{sub}},\ell} = L_{G_{\mathtt{sol}},\ell}$ indirectly with $\mathcal{H}(L_{G_{\mathtt{sub}},\ell}) = \mathcal{H}(L_{G_{\mathtt{sol}},\ell})$. Note that by using hash functions we may obtain false positives due to collisions, but never false negatives. We use a typical definition [11] for a hash function for words:

$$\mathcal{H}(w) = \left(\sum_{i=1}^{|w|} w[i] \cdot \mathfrak{b}^{i-1}\right) \bmod \mathfrak{m}$$

where $\mathfrak{m}$ is a "big" prime and $\mathfrak{b}$ is a "small" prime satisfying $\mathfrak{b} > |\Sigma|$. Note that we interpret the terminal symbols in Σ as numbers, assuming that they are in $\{1, \ldots, \mathfrak{b} - 1\}$ and are pairwise different. An extension of $\mathcal{H}$ to languages like

$$\mathcal{H}(L) = \left(\sum_{w \in L} \mathcal{H}(w)\right) \bmod \mathfrak{m}$$

suffices to detect when $L_{G_{\mathtt{sub}},\ell}$ and $L_{G_{\mathtt{sol}},\ell}$ differ, and in such case a counterexample $w \in L_{G_{\mathtt{sub}},\ell} \triangle L_{G_{\mathtt{sol}},\ell}$ can easily be constructed one symbol at a time: it suffices to check that the symbol appended to w is valid for w to become the counterexample, i.e., to check that $\mathcal{H}(\{wu \in L_{G_{\mathtt{sub}},\ell}\}) \neq \mathcal{H}(\{wv \in L_{G_{\mathtt{sol}},\ell}\})$.

Due to the lack of space and for explanation purposes, instead of giving the formal definition of the efficient computation of $\mathcal{H}(L_{G,\ell})$ making use of the structure of G, we just give an example. Consider the language $L = \{a^n b^n \mid n > 0\}$

and the following CFG generating L (already in CNF):

$$\begin{aligned} S &\to AX \mid AB \\ X &\to SB \\ A &\to a \\ B &\to b \end{aligned}$$

By $\mathcal{H}(W,\ell)$ we denote $\mathcal{H}(\{w \in \Sigma^* \mid W \to_G^* w \wedge |w| = \ell\})$ and by $\mathcal{C}(W,\ell)$ we denote $|\{w \in \Sigma^* \mid W \to_G^* w \wedge |w| = \ell\}|$. Such values can be recursively obtained using the structure of G. For the direct cases we have $\mathcal{H}(A,1) = a$, $\mathcal{H}(B,1) = b$, $\mathcal{C}(A,1) = \mathcal{C}(B,1) = 1$, and $\mathcal{H}(A,n) = \mathcal{H}(B,n) = \mathcal{C}(A,n) = \mathcal{C}(B,n) = 0$ for $n > 1$. Since the right-hand sides of rules of S and X have size 2, $\mathcal{H}(S,1) = \mathcal{H}(X,1) = \mathcal{C}(S,1) = \mathcal{C}(X,1) = 0$. Since X only has the rule $X \to SB$, $\mathcal{C}(X,2) = \mathcal{C}(S,1) \cdot \mathcal{C}(B,1) = 0$, and $\mathcal{H}(X,2) = \mathcal{H}(S,1) \cdot \mathcal{C}(B,1) + \mathcal{C}(S,1) \cdot \mathcal{H}(B,1) \cdot \mathfrak{b} = 0$. Proceeding analogously, we obtain $\mathcal{C}(S,2) = 1$, $\mathcal{H}(S,2) = a + b\mathfrak{b}$. Since S has the rules $S \to AX$, $S \to AB$,

$$\begin{aligned} \mathcal{C}(S,3) &= \mathcal{C}(A,1) \cdot \mathcal{C}(X,2) + \mathcal{C}(A,2) \cdot \mathcal{C}(X,1) + \\ &\quad \mathcal{C}(A,1) \cdot \mathcal{C}(B,2) + \mathcal{C}(A,2) \cdot \mathcal{C}(B,1) = 0 \\ \mathcal{H}(S,3) &= \mathcal{H}(A,1) \cdot \mathcal{C}(X,2) + \mathcal{C}(A,1) \cdot \mathcal{H}(X,2) \cdot \mathfrak{b} + \\ &\quad \mathcal{H}(A,2) \cdot \mathcal{C}(X,1) + \mathcal{C}(A,2) \cdot \mathcal{H}(X,1) \cdot \mathfrak{b}^2 + \\ &\quad \mathcal{H}(A,1) \cdot \mathcal{C}(B,2) + \mathcal{C}(A,1) \cdot \mathcal{H}(B,2) \cdot \mathfrak{b} + \\ &\quad \mathcal{H}(A,2) \cdot \mathcal{C}(B,1) + \mathcal{C}(A,2) \cdot \mathcal{H}(B,1) \cdot \mathfrak{b}^2 = 0 \end{aligned}$$

Proceeding analogously, we obtain $\mathcal{C}(X,3) = 1$, $\mathcal{H}(X,3) = a + b\mathfrak{b} + b\mathfrak{b}^2$.

JUDGEHASH only works correctly when $G_{\mathtt{sol}}$ is unambiguous because derivations generating the same word are counted independently. The generated counterexample w to the correctness of $G_{\mathtt{sub}}$ will be either a word in $\mathcal{L}(G_{\mathtt{sub}}) \triangle \mathcal{L}(G_{\mathtt{sol}})$ or a word ambiguously generated by $G_{\mathtt{sub}}$. A membership test can determine which one of these cases takes place.

3.3 SAT

The JUDGESAT is based on the work of [3] and consists in testing equivalence of $G_{\mathtt{sub}}$ and $G_{\mathtt{sol}}$ by reducing the problem to the satisfiability of boolean propositional formulas. More specifically, the idea is to first construct a formula $\mathcal{F}_{\ell,G_{\mathtt{sub}},G_{\mathtt{sol}}}$ such that it is satisfiable if and only if there exists a counterexample word of length at most ℓ, and then to solve the formula with a state-of-the-art SAT solver. We have reimplemented this method with the idea of trying some possible optimizations. One of them consists in splitting $\mathcal{F}_{\ell,G_{\mathtt{sub}},G_{\mathtt{sol}}}$ in two independent formulas $\mathcal{F}_{\ell,G_{\mathtt{sub}} \setminus G_{\mathtt{sol}}}$ and $\mathcal{F}_{\ell,G_{\mathtt{sol}} \setminus G_{\mathtt{sub}}}$, where $\mathcal{F}_{\ell,G_i \setminus G_j}$ is satisfiable if and only if there exists a word of length ℓ in $\mathcal{L}(G_i) \setminus \mathcal{L}(G_j)$. We recall the reduction process of [3] just for $\mathcal{F}_{\ell,G_{\mathtt{sub}} \setminus G_{\mathtt{sol}}}$.

The formula $\mathcal{F}_{\ell,G_{\mathtt{sub}} \setminus G_{\mathtt{sol}}}$ is defined by means of two kinds of propositional variables: $\mathcal{X}_i^a$ and $\mathcal{X}_{i,j}^X$, where $1 \leq i \leq j \leq \ell$, $a \in \Sigma$ and $X \in \mathcal{V}$. The first kind of variable, $\mathcal{X}_i^a$, represents the fact that the counterexample w has the terminal a at position i. The second kind of variable, $\mathcal{X}_{i,j}^X$, represents the fact that the subword $w[i..j]$ can be generated from the non-terminal X. The formula $\mathcal{F}_{\ell,G_{\mathtt{sub}}\setminus G_{\mathtt{sol}}}$ can be decomposed into four different parts. First, it guarantees that the counterexample $w[1..\ell]$ is a valid word in Σ^* by forcing each position of the word to contain exactly one terminal symbol of Σ:

$$\bigwedge_{i=1}^{\ell}\big(\bigvee_{a\in\Sigma}\mathcal{X}_i^a\big)$$
$$\bigwedge_{i=1}^{\ell}\bigwedge_{a\in\Sigma}\Big(\mathcal{X}_i^a\rightarrow\bigwedge_{b\in\Sigma\setminus\{a\}}\neg\mathcal{X}_i^b\Big)$$

Second, $\mathcal{F}_{\ell,G_{\mathtt{sub}}\setminus G_{\mathtt{sol}}}$ states that $w[1..\ell]$ is generated by $G_{\mathtt{sub}}$ but not by $G_{\mathtt{sol}}$ with the two unit clauses $(\mathcal{X}_{1,\ell}^{S_{\mathtt{sub}}})$ and $(\neg\mathcal{X}_{1,\ell}^{S_{\mathtt{sol}}})$, where $S_{\mathtt{sub}}$ and $S_{\mathtt{sol}}$ are the starting symbols of $G_{\mathtt{sub}}$ and $G_{\mathtt{sol}}$, respectively. Third, it formalizes the fact that $G_{\mathtt{sub}}$ generates $w[1..\ell]$:

$$\bigwedge_{i=1}^{\ell-1}\bigwedge_{j=i+1}^{\ell}\bigwedge_{X\in\mathcal{V}_{\mathtt{sub}}}\Big(\mathcal{X}_{i,j}^X\rightarrow\bigvee_{(X\rightarrow YZ)\in R_{\mathtt{sub}}}\bigvee_{s=i}^{j-1}(\mathcal{X}_{i,s}^Y\wedge\mathcal{X}_{s+1,j}^Z)\Big)$$
$$\bigwedge_{i=1}^{\ell}\bigwedge_{X\in\mathcal{V}_{\mathtt{sub}}}\Big(\mathcal{X}_{i,i}^X\rightarrow\bigvee_{(X\rightarrow a)\in R_{\mathtt{sub}}}\mathcal{X}_i^a\Big)$$

where $\mathcal{V}_{\mathtt{sub}}$ and $R_{\mathtt{sub}}$ are the sets of non-terminals and rules of $G_{\mathtt{sub}}$. And fourth, it formalizes the fact that $G_{\mathtt{sol}}$ does not generate $w[1..\ell]$ with a formula analogous to the previous one, but with the direction of the implications reversed.

It is clear that a counterexample w of length ℓ exists if and only if either $\mathcal{F}_{\ell,G_{\mathtt{sub}}\setminus G_{\mathtt{sol}}}$ or $\mathcal{F}_{\ell,G_{\mathtt{sol}}\setminus G_{\mathtt{sub}}}$ is satisfiable. Moreover, w can be derived from any assignment η that satisfies a formula by analysing all the values $\eta(\mathcal{X}_i^a)$.

One additional optimization with respect to [3] that we have considered is to simplify the formulas as follows: whenever we detect that a non-terminal X cannot generate any word of length k, we simply create a unit clause $(\neg\mathcal{X}_{i,i+k-1}^X)$ for any relevant i. This allows us to ignore for length k all the rules with X as left-hand side and all the possible split indexes $s\in\{i,\ldots,i+k-1\}$.

3.4 DFA

The approach of JudgeDFA is based on automata techniques. The idea is to construct the minimum DFA $A_{\mathtt{sub},\ell}$ and $A_{\mathtt{sol},\ell}$ recognizing the words of length ℓ generated by $G_{\mathtt{sub}}$ and $G_{\mathtt{sol}}$, and testing whether $A_{\mathtt{sub},\ell}$ and $A_{\mathtt{sol},\ell}$ are identical. In order to be able to compute the automata directly on the grammars, we use the following function $\mathcal{A}:\mathcal{V}\times\mathbb{N}\rightarrow$ DFA mapping a non-terminal X and a length ℓ to an automaton recognizing the words of length ℓ generated from X:

$$\mathcal{A}(X,\ell)=\begin{cases}\bigcup_{(X\rightarrow a)\in R}A_a & \text{if }\ell\text{=1}\\ \bigcup_{(X\rightarrow YZ)\in R}\bigcup_{i=1}^{\ell-1}\mathcal{A}(Y,i)\cdot\mathcal{A}(Z,\ell-i) & \text{if }\ell\text{>1}\end{cases}$$

where concatenation and union are not set operations, but automata operations producing automata recognizing the concatenation and union of the languages of the given automata, and A_a denotes the automaton recognizing the word $a\in\Sigma$. Moreover, we assume that the DFA recognizing the empty language is the neutral element of the automata union. Note that we do not explicitly detect whether the grammars are ambiguous, but an approximation could be checked while evaluating $\mathcal{A}$ by testing whether the unions performed are disjoint.

4 Performance

4.1 Complexity Analysis

For JUDGEEXHAUSTIVE, when we restrict to small alphabets and small ℓ, words can be encoded as natural numbers using the native representation of the computer, and then the running time is in $\mathcal{O}(|R|\cdot\ell\cdot|\Sigma|^\ell)$ and its space in $\mathcal{O}(\mathcal{V}\cdot|\Sigma|^\ell)$.

In the case of JUDGEHASH, using a dynamic programming scheme, the space requirements to compute $\mathcal{C}$ and $\mathcal{H}$ up to the current counterexample w are in $\mathcal{O}(|\mathcal{V}|\cdot\ell^2)$ and it takes time in $\mathcal{O}(|R|\cdot\ell^3)$. The construction of the counterexample w requires the recomputation of $\mathcal{C}_w$ and $\mathcal{H}_w$ at most $\ell\cdot|\Sigma|$ times, giving a global running time in $\mathcal{O}(|R|\cdot\ell^4\cdot|\Sigma|)$.

The number of variables of the form $\mathcal{X}_i^a$ and $\mathcal{X}_{i,j}^X$ of JUDGESAT is in $\mathcal{O}(\ell\cdot|\Sigma|+\ell^2\cdot|\mathcal{V}|)$, and hence, the cost of solving $\mathcal{F}_{\ell,G_i\setminus G_j}$ is in $2^{\mathcal{O}(\ell\cdot|\Sigma|+\ell^2\cdot|\mathcal{V}|)}$. Fortunately, state-of-the-art SAT solvers perform much better than this in practice, and incremental SAT-solver techniques [8,16] lead to noticeable speed-ups since the solver can reuse for $\mathcal{F}_{\ell+1,G_i\setminus G_j}$ the knowledge obtained when solving $\mathcal{F}_{\ell,G_i\setminus G_j}$.

Finally, for JUDGEDFA, first note that the size of a minimum DFA recognizing a set of words of length ℓ is in $2^{\mathcal{O}(\log(|\Sigma|)\cdot\ell)}$. Second, recall that the concatenation operation on automata might lead to exponential blowup and the union operation multiplies the sizes of the automata being considered [9]. This implies that the computations of JUDGEDFA have space requirements in $2^{2^{\mathcal{O}(\log(|\Sigma|)\cdot\ell)}}$. However, this extreme bound is only reached when the languages generated by the grammars require excessive memorization, e.g., the language of palindromes and the language of well-parenthesized words. In other cases, JUDGEDFA can "compress" the language by sharing states and obtains much better performance.

4.2 Benchmarks

We have compared the performance of our judges using the set of benchmarks from [3], comprising 35910 different pairs of grammars. As a reference, we have also tested the prototype implementation of [3], `cfganalyzer`[2]. Additionally, we have also compared our proposals with HAMPI[3], a state-of-the-art string-constraint solver developed in [6] which is expressive enough to approximate grammar equivalence. HAMPI works by internally transforming the grammars to fixed-length regular expressions and then solving the constraints for them.

Figure 1 shows the results obtained[4] for each judge when testing grammar equivalence up to length $\ell = 15$. The plot can be interpreted as follows: all the 35910 tests are run in parallel in independent machines, the abscissa is the elapsed time, and the ordinate is the number of tests that have not reached a verdict. It is easy to see in the chart that most of the tests can be solved in less

[2] Version 2012-12-26, `http://www2.tcs.ifi.lmu.de/~mlange/cfganalyzer/`

[3] Version 2012-2-13, `http://people.csail.mit.edu/akiezun/hampi/`

[4] Measurements taken on a 64-bit Intel® Pentium® T4200, at 2GHz and with 4GB of RAM, timings available at `http://www.lsi.upc.edu/~ggodoy/publications.html`

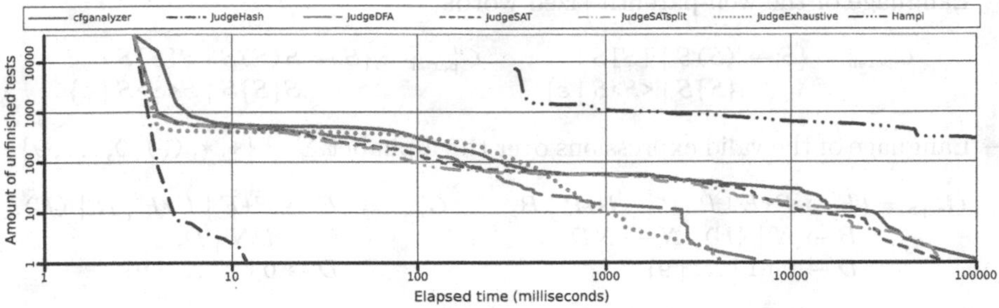

Fig. 1. Results using the benchmarks from [3], comprising 35910 grammar pairs, testing equivalence up to length $\ell = 15$

than 10 milliseconds by all the judges. This is because in such cases either the counterexamples were small, i.e., the empty word, or the grammars were deemed wrongly formatted. Timings for `cfganalyzer` and JUDGESAT correspond to the builds using the latest version of MINISAT[5] [5] (the solvers ZCHAFF [12] and PICOSAT [4] were also considered, but finally discarded due to lower performance). Clearly, the plots for JUDGESAT closely follow `cfganalyzer`, with our optimizations giving only a small benefit. The results for JUDGEEXHAUSTIVE and JUDGEDFA are comparable to `cfganalyzer`, and even significantly better when considering the most expensive tests. JUDGEEXHAUSTIVE is competitive with the rest because the alphabets are small and the languages are sparse. The timings obtained for JUDGEHASH are quite remarkable: the worst running time was just 12 milliseconds. Finally, in the case of HAMPI, we had to limit the execution time to 2 minutes and the memory to 512 MB, since in some tests the program hung consuming all the available memory. Due to these problems, we only used a subset of the tests. Overall, results for HAMPI are rather poor when compared to the rest of judges.

4.3 Stressing the Judges

To highlight the differences between our judges, we have devised an additional set of grammar pairs that focus on particular bottlenecks of each judge:

- Language of words with same number of a's and b's:

$$G_{\text{numab}} = \{S \to \mathtt{a}X\mathtt{b}S \mid \mathtt{b}Y\mathtt{a}S \mid \varepsilon,\quad X \to \mathtt{a}X\mathtt{b}X \mid \varepsilon,\quad Y \to \mathtt{b}Y\mathtt{a}Y \mid \varepsilon\} \qquad G'_{\text{numab}} = \{S \to \mathtt{a}S\mathtt{b}S \mid \mathtt{b}S\mathtt{a}S \mid \varepsilon\}$$

- Palindromes over an alphabet Σ_i with i different terminal symbols:

$$G_{\text{pal}_i} = \textstyle\bigcup_{\mathtt{a}\in\Sigma_i}\{S \to \mathtt{a}S\mathtt{a} \mid \mathtt{a} \mid \varepsilon\} \qquad G'_{\text{pal}_i} = \textstyle\bigcup_{\mathtt{a}\in\Sigma_i}\{S \to A_\mathtt{a} \mid \mathtt{a} \mid \varepsilon,\ A_\mathtt{a} \to \mathtt{a}S\mathtt{a}\}$$

[5] Version 2.2.0, `http://www.minisat.se`

– Language of the well-parenthesized words:

$$G_{\text{paren}} = \{S \to \mathtt{(}S\mathtt{)}S \mid \mathtt{[}S\mathtt{]}S \mid \mathtt{\{}S\mathtt{\}}S \mid \mathtt{<}S\mathtt{>}S \mid \varepsilon\} \qquad G'_{\text{paren}} = \{S \to S\mathtt{(}S\mathtt{)}S \mid S\mathtt{[}S\mathtt{]}S \mid S\mathtt{\{}S\mathtt{\}}S \mid S\mathtt{<}S\mathtt{>}S \mid \varepsilon\}$$

– Language of the valid expressions over the alphabet $\Sigma = \{\mathtt{+}, \mathtt{*}, \mathtt{(}, \mathtt{)}, \mathtt{0}, \ldots, \mathtt{9}\}$:

$$G_{\text{expr}} = \{E \to P\mathtt{+}E \mid P,\ P \to B\mathtt{*}P \mid B,\ B \to N \mid \mathtt{(}E\mathtt{)},\ N \to ND \mid D,\ D \to \mathtt{0} \mid \mathtt{1} \mid \ldots \mid \mathtt{9}\}$$

$$G'_{\text{expr}} = \{E \to E\mathtt{+}E \mid E\mathtt{*}E \mid N \mid \mathtt{(}E\mathtt{)},\ N \to DN \mid D,\ D \to \mathtt{0} \mid \mathtt{1} \mid \ldots \mid \mathtt{9}\}$$

Figure 2 shows the obtained running times. As expected, JUDGEEXHAUSTIVE achieves acceptable performance only when the alphabet is small and the languages sparse. The cases of JUDGEDFA and JUDGESAT are related: the latter improves the times of the former when the languages require excessive memorization. For instance, G_{pal_i}, G'_{pal_i}, G_{paren} and G'_{paren} force the automata to memorize almost all the read word. When no such excessive memorization is required, like in G_{numab}, G'_{numab}, G_{expr} and G'_{expr}, the automata of JUDGEDFA are small and it obtains better performance than JUDGESAT.

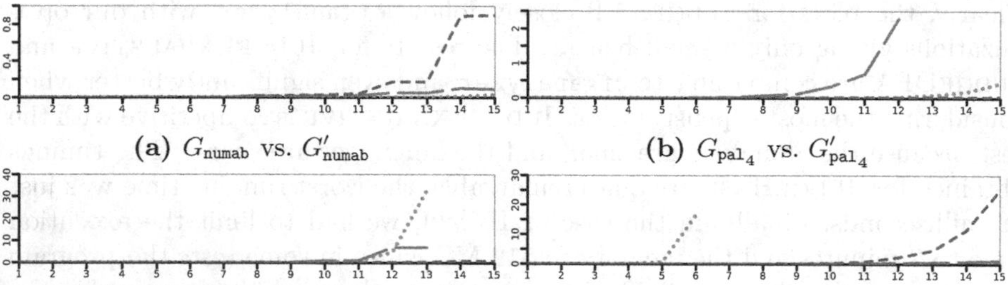

(a) G_{numab} vs. G'_{numab} (b) G_{pal_4} vs. G'_{pal_4}

(c) G_{paren} vs. G'_{paren} (d) G_{expr} vs. G'_{expr}

Fig. 2. Running times in seconds (ordinate) in terms of the maximum length ℓ tested (abscissa) for the different judges (dotted lines correspond to JUDGEEXHAUSTIVE, long-dashed to JUDGEDFA, short-dashed to JUDGESAT)

We have an extra test to stress JUDGEHASH. This is necessary since its worst case scenario is not with equivalent grammars, but when there is a big counterexample. We test the language of words of length i over $\Sigma_j = \{\mathtt{a}_1, \ldots, \mathtt{a}_j\}$:

$$G_{\text{art}_{i,j}} = \{S \to \overbrace{T \ldots T}^{i},\ T \to \mathtt{a}_1 \mid \ldots \mid \mathtt{a}_j\} \qquad G'_{\text{art}_{i,j}} = G_{\text{art}_{i,j}} \cup \{S \to \overbrace{\mathtt{a}_j \ldots \mathtt{a}_j}^{i}\}$$

Note that the lexicographically last word is ambiguously generated by $G'_{\text{art}_{i,j}}$, thus being the counterexample. Figure 3 depicts the obtained times, where it is clear that competitive performance is achieved even with rather big i and j.

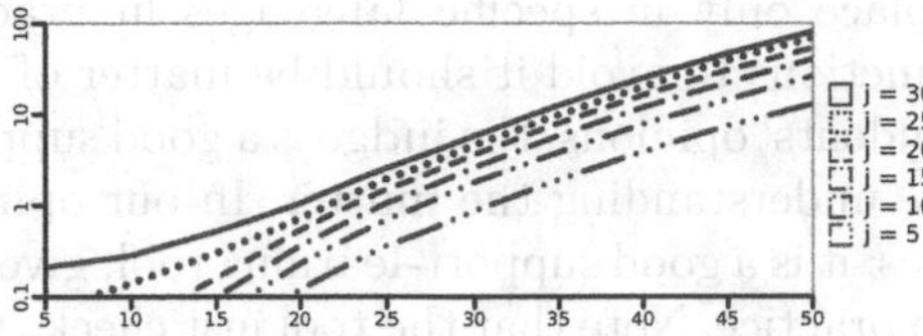

Fig. 3. JUDGEHASH's running times in seconds (ordinate) in terms of the word length i (abscissa) for $G_{\text{art}_{i,j}}$ vs. $G'_{\text{art}_{i,j}}$

5 Online Judging System

Our website currently offers 46 problems on CFG's and 21 on push-down automata, which are checked by the same judges by a prior standard transformation into CFG's. Our system is configured as follows. For problems with languages over alphabets with 2 or 3 symbols we use JUDGEEXHAUSTIVE with $\ell = 10$, because it is fast enough, has a rather uniform running time, and the answer is guaranteed to be correct up to the chosen ℓ. For problems over alphabets larger than 3 and asking for an unambiguous grammar we use JUDGEHASH and $\ell = 15$, which is combined with JUDGEEXHAUSTIVE with $\ell = 3$ in order to reduce the chances of hash collisions. For the rest of problems we use either JUDGESAT or JUDGEDFA, depending on the language, and with $\ell = 10$. Essentially, JUDGEDFA is used for those languages for which the expected natural grammar solutions produce small automata, according to the problem setter criterion.

We have been using the online judge since September 2012 with the students of the Theory of Computation subject at the Computer Science course of the Universitat Politècnica de Catalunya. For the first two semesters it was offered to the students as an optional support tool to do exercises. During the fall semester of 2013 we also used our system to hold online exams, and the students made over 12000 submissions in total, for an average of 250 submissions per student.

6 Conclusions

We have developed several techniques for determining if two given context-free grammars generate the same language. The methods we have implemented work sufficiently well in practice. In the case of the SAT-based judge, the performance of our implementation is similar to the state of the art. The hash-based method has much better performance than the others. Nevertheless, besides the fact that this method cannot be used with ambiguous solution grammars, the extension of the hash function from words to languages degrades some of its properties, and it may happen that some collisions take place independently of the chosen primes $\mathfrak{m}, \mathfrak{b}$. For instance, the following languages L_1, L_2 give rise to the same value through the hash:

$$L_1 = \{a^n b^n \mid n \geq 0\} \cup \{c^n d^n \mid n \geq 0\}$$
$$L_2 = \{a^n d^n \mid n \geq 0\} \cup \{c^n b^n \mid n \geq 0\}$$

This problem takes place only in specific languages in practice, but defining alternative hashing functions to avoid it should be matter of further work.

According to the students' opinions, the judge is a good support tool that helps them know if they are understanding the matter. In our opinion, it has all the benefits of online judges: it is a good support-learning tool, gives instant feedback, and motivates users to practice. Note that the tool just checks that the submitted solutions are correct, but neither the quality of such solutions nor that the students have understood them well enough to justify their correctness. In this sense, the professor is essential to adquire a good comprehension of the matter.

Although many problems in the list are artificial, they help students to understand the limits of expressivity of CFG's, and how context-free conditions can be combined with regular conditions. We are interested in studying whether similar techniques perform well for evaluating grammars designed for descending or ascending parsing, and for the construction of abstract syntax trees. This can be useful to develop support-learning tools for the Compilers subject.

References

1. Agarwal, A.: edX (2012), `https://www.edx.org`
2. Alexandrova, S., Balandin, A., Compeau, P., Kladov, A., Rayko, M., Sosa, E., Vyahhi, N., Dvorkin, M.: Rosalind (2012), `http://www.rosalind.info`
3. Axelsson, R., Heljanko, K., Lange, M.: Analyzing context-free grammars using an incremental SAT solver. In: Aceto, L., Damgård, I., Goldberg, L.A., Halldórsson, M.M., Ingólfsdóttir, A., Walukiewicz, I. (eds.) ICALP 2008, Part II. LNCS, vol. 5126, pp. 410–422. Springer, Heidelberg (2008)
4. Biere, A.: PicoSAT essentials. Journal on Satisfiability, Boolean Modeling and Computation 4(2-4), 75–97 (2008)
5. Eén, N., Sörensson, N.: An extensible SAT-solver. In: Giunchiglia, E., Tacchella, A. (eds.) SAT 2003. LNCS, vol. 2919, pp. 502–518. Springer, Heidelberg (2004)
6. Ganesh, V., Kieżun, A., Artzi, S., Guo, P.J., Hooimeijer, P., Ernst, M.: HAMPI: A string solver for testing, analysis and vulnerability detection. In: Gopalakrishnan, G., Qadeer, S. (eds.) CAV 2011. LNCS, vol. 6806, pp. 1–19. Springer, Heidelberg (2011)
7. García, C., Revilla, M.A.: UVa online judge (1997), `http://uva.onlinejudge.org`
8. Hooker, J.N.: Solving the incremental satisfiability problem. Journal of Logic Programming 15(1&2), 177–186 (1993)
9. Hopcroft, J.E., Motwani, R., Ullman, J.D.: Introduction to Automata Theory, Languages, and Computation, 3rd edn. Addison-Wesley (2006)
10. Khan, S.: Khan Academy (2006), `https://www.khanacademy.org`
11. Knuth, D.E.: The Art of Computer Programming, vol. III: Sorting and Searching. Addison-Wesley (1973)
12. Moskewicz, M.W., Madigan, C.F., Zhao, Y., Zhang, L., Malik, S.: Chaff: Engineering an efficient SAT solver. In: Annual ACM IEEE Design Automation Conference, pp. 530–535. ACM (2001)
13. Ng, A., Koller, D.: Coursera (2012), `https://www.coursera.org`
14. Petit, J., Giménez, O., Roura, S.: Jutge.org: an educational programming judge. In: ACM Special Interest Group on Computer Science Education, pp. 445–450 (2012)
15. Thrun, S., Stavens, D., Sokolsky, M.: Udacity (2012), `https://www.udacity.com`
16. Whittemore, J., Kim, J., Sakallah, K.A.: SATIRE: A new incremental satisfiability engine. In: Design Automation Conference, pp. 542–545 (2001)

Tree Automata with Height Constraints between Brothers*

Carles Creus and Guillem Godoy

Universitat Politècnica de Catalunya,
Departament de Llenguatges i Sistemes Informàtics, Barcelona, Spain
ccreuslopez@gmail.com, ggodoy@lsi.upc.edu

Abstract. We define the tree automata with height constraints between brothers ($\mathtt{TACBB_H}$). Constraints of equalities and inequalities between heights of siblings that restrict the applicability of the rules are allowed in $\mathtt{TACBB_H}$. These constraints allow to express natural tree languages like complete or balanced (like AVL) trees. We prove decidability of emptiness and finiteness for $\mathtt{TACBB_H}$, and also for a more general class that additionally allows to combine equality and disequality constraints between brothers.

Keywords: Tree-Automata, Constraints, Emptiness, Finiteness.

1 Introduction

For a long time, tree automata have been regarded as a fundamental tool in many theoretical areas of computer science. These automata are a generalization of word automata that work on structured input, namely trees. Due to their good computational properties, tree automata have been used as the formalism for, e.g., describing parse trees of a context-free grammar or the well-formed terms over a sorted signature [17], to characterize the solutions of formulas in monadic second-order logic [11], and to naturally capture type formalisms for XML data [19].

Unfortunately, expressiveness of plain tree automata is limited since they only have finite memory. For instance, a set like $L = \{f(t_1, t_2) \mid t_1 \neq t_2\}$ is a typical example of language that cannot be recognized by any tree automaton. In this case, the reason is that it is not possible to check that arbitrary t_1 and t_2 are different using only a finite number of states. In order to overcome such limitation, many extensions to tree automata have been considered in recent years. One of the most studied cases consists in enhancing the automaton with the possibility to test equality and disequality of certain subterms of the input tree. This kind of automata are said to have (dis)equality constraints and, depending on how the tests are defined, they can be roughly classified as either local [18] or global [12,2]. We focus on the local case. In this model, each transition rule

* The authors were supported by an FPU grant (first author) and the FORMALISM project (TIN2007-66523) from the Spanish Ministry of Education and Science.

G. Dowek (ed.): RTA-TLCA 2014, LNCS 8560, pp. 149–163, 2014.

of the automaton has an associated Boolean combination of atomic predicates of the form $p_1 = p_2$ and $p_1 \neq p_2$, for positions p_1, p_2. Such an atomic constraint holds for a given rule application if the subterms pending at p_1 and p_2, relative to the position where the rule is applied, are equal in the first case and different in the second one. The previous example language L can be recognized with such automata: it suffices to force the predicate $1 \neq 2$ to be tested at the root position. However, the increase in expressiveness obtained with constraints comes at the expense of other desirable properties. For instance, when arbitrary positions p_1 and p_2 are allowed in the atoms, emptiness and finiteness are undecidable for this kind of automata [18,7]. Nevertheless, several particular cases enjoy better properties and have been successfully used to tackle different problems: tree automata with constraints between brother positions allowed to prove decidability of fragments of quantifier-free formulas on one-step rewriting [4], as well as the recognizability problem for regular tree languages under particular cases of tree homomorphisms [3]; arbitrary local disequality constraints were used to prove EXPTIME-completeness of ground reducibility [5]; arbitrary local disequality constraints combined with a restricted version of local equality constraints, called HOM equalities, were used to prove EXPTIME-completeness of the recognizability problem for regular tree languages under tree homomorphisms [15,9]; and automata with arbitrary local constraints but with a bound on the maximum number of equality tests that can be performed at each branch of the input tree lead to the decidability of fragments of the first-order theory of reduction [10]. Other models originated in the study of secrecy in cryptographic protocols consider extending the automata with one tree-shaped memory, allowing the transition rules to update such memory and to perform comparisons [6,8]. These models are expressive enough to generalize certain cases of local equality constraints between brothers.

We propose a new kind of automata with local constraints between brother positions that differs from the previous literature in that, instead of checking subterms for either syntactic equality or some notion of equivalence (like, e.g., in [16,2]), the restrictions are imposed on the *height* of the subterms involved in the constraints. We call them tree automata with height constraints between brothers, $\mathtt{TACBB}_\mathsf{H}$ for short. More precisely, our basic atomic predicates are of the form $h(i) = h(j)$, for positions i and j of length 1, and are satisfied when the subterms pending at i and j, relative to the application of the rule, have identical height. We also consider inequality predicates of the form $h(i) < h(j)$ and comparisons introducing an integer constant x of the form $h(i) = h(j) + x$ and $h(i) < h(j) + x$, with the straightforward interpretations. It is easy to see that our notion of constraints is incomparable with syntactic equality constraints. For instance, the language of complete trees over a signature with two constant symbols a, b and a binary symbol f can be recognized by a $\mathtt{TACBB}_\mathsf{H}$ with the transition rules

$$a \to q,\ b \to q,\ f(q,q) \xrightarrow{h(1)=h(2)} q$$

but such language cannot be recognized by the automata with (dis)equality constraints between brothers (AWCBB) from [3]. Intuitively, this is because, even if two terms t_1 and t_2 were inductively guaranteed to be complete, it is not possible to check with only (dis)equality constraints whether t_1 and t_2 have equal height, and thus, whether $f(t_1, t_2)$ is also complete. Actually, it can be proved that with AWCBB it is only possible to recognize the language of complete trees when the underlying signature has a single constant symbol and a single non-constant symbol. This limitation also holds for generalizations of AWCBB like the TAB from [2], where equalities are interpreted modulo flat theories. Another interesting language that can be recognized by $\mathtt{TACBB_H}$ using the inequality predicates on height is the language of AVL trees, i.e., the set of trees where the heights of the two direct children of any internal node differ by at most one. It suffices to replace the previous rule for f with

$$f(q,q) \xrightarrow{(h(1)<h(2)+2) \wedge (h(2)<h(1)+2)} q$$

Note that the constraint forces the height of each child to be at most one more than the height of the other child.

We prove decidability of the emptiness and finiteness problems for $\mathtt{TACBB_H}$. Our approach consists in transforming the automaton into a normalized form, and obtaining a recursive formulation to describe the set of reachable states when recognizing terms of a specific height. The decision algorithm follows directly from such result. We adapt the method to automata that combine the height constraints of $\mathtt{TACBB_H}$ with the (dis)equality constraints of AWCBB, which we call $\mathtt{TACBB_{He}}$. Note that this kind of automata strictly generalizes $\mathtt{TACBB_H}$ and AWCBB, and hence, decidability of emptiness and finiteness is EXPTIME-hard [7].

The definition of TACT in [20] is incomparable with $\mathtt{TACBB_{He}}$. Nevertheless, a given tree automaton with height constraints is transformable into a TACT by preserving emptiness (but not the language). But this does not help in our setting to decide emptiness of our model, since emptiness of TACT is undecidable. The definition of $\mathrm{VTAM}^R_{\neg R}$ in [8] captures our automata models. Nevertheless, emptiness is only decidable for some particular subclasses that are incomparable with $\mathtt{TACBB_{He}}$ since, although they can recognize the particular set of complete trees, height of subtrees cannot be compared in general.

The paper is structured as follows. In Section 2 we introduce standard notations and definitions, and in Section 3 we define our model of automata. In Section 4 we tackle the case where only simple height constraints are allowed, i.e., $\mathtt{TACBB_H}$ whose atomic predicates are restricted to $h(i) = h(j)$ and $h(i) < h(j)$. In Section 5 we generalize those results to $\mathtt{TACBB_{He}}$. We conclude in Section 6.

2 Preliminaries

The size of a finite set S is denoted by $|S|$, and the powerset of S is denoted by 2^S. Given two sets of sets S_1, S_2, by $S_1 \boxtimes S_2$ we denote the set of sets that are obtained by choosing a set $\hat{S}_1$ from S_1 and a set $\hat{S}_2$ from S_2 and making their union, i.e., $\{\hat{S}_1 \cup \hat{S}_2 \mid \hat{S}_1 \in S_1 \wedge \hat{S}_2 \in S_2\}$. For example, $\{\{a,b\},\{c\}\} \boxtimes \{\{d,e\},\{f\}\} = \{\{a,b,d,e\},\{a,b,f\},\{c,d,e\},\{c,f\}\}$. A partition P of a set S is a set of non-empty sets (the parts) that are pairwise disjoint and whose union is S. By $i \equiv_P j$ we denote that i, j belong to the same part of P.

We introduce notation for terms and positions, for a survey see [1]. A *signature* consists of an alphabet Σ, i.e., a finite set of symbols, together with a mapping that assigns to each symbol in Σ a natural number (possibly 0), its *arity*. We denote by $\Sigma^{(m)}$ the subset of symbols in Σ of arity m, and by $\mathsf{maxar}(\Sigma)$ the maximum m such that $\Sigma^{(m)}$ is not empty. We write simply maxar when Σ is known by the context. Symbols in $\Sigma^{(0)}$, called *constants*, are denoted by a, b, with possible subscripts. The set $\mathcal{T}(\Sigma)$ of *ground terms* (or just terms) over Σ is the smallest set such that $f(t_1, \ldots, t_m)$ is in $\mathcal{T}(\Sigma)$ whenever $f \in \Sigma^{(m)}$, and $t_1, \ldots, t_m \in \mathcal{T}(\Sigma)$. For a term of the form $a()$ we simply write a. A *language* over Σ is a set of ground terms. A *position* is a sequence of natural numbers. The symbol λ denotes the empty sequence, also called the root position, and $p.p'$ denotes the concatenation of the positions p and p'. The set of positions of a term t, denoted $\mathsf{Pos}(t)$, is defined recursively as $\mathsf{Pos}(f(t_1, \ldots, t_m)) = \{\lambda\} \cup \{i.p \mid i \in \{1, \ldots, m\} \wedge p \in \mathsf{Pos}(t_i)\}$. The *subterm* of a term t at a position p, denoted $t|_p$, is defined recursively as $t|_\lambda = t$ and $f(t_1, \ldots, t_m)|_{i.p} = t_i|_p$. The *height* of a term t, denoted $\mathsf{height}(t)$, is defined recursively as $\mathsf{height}(t) = 0$ if t is a constant, and as $\mathsf{height}(f(t_1, \ldots, t_m)) = 1 + \max\{\mathsf{height}(t_1), \ldots, \mathsf{height}(t_m)\}$ otherwise.

Tree automata and regular tree languages are well-known concepts of theoretical computer science [13,14,7]. Here we only recall the notion of tree automata.

Definition 1. *A* tree automaton, TA *for short, is a tuple $A = \langle Q, \Sigma, F, \Delta \rangle$, where Q is a finite set of states, Σ is a signature, $F \subseteq Q$ is the set of final states, and Δ is a set of rules of the form $f(q_1, \ldots, q_m) \to q$, where $q_1, \ldots, q_m, q$ are in Q, and f is in $\Sigma^{(m)}$.*

A run *of A on a term $t \in \mathcal{T}(\Sigma)$ is a mapping $r : \mathsf{Pos}(t) \to \Delta$ such that, for each position $p \in \mathsf{Pos}(t)$, if $t|_p$ is of the form $f(t_1, \ldots, t_m)$, then $r(p)$ is a rule of the form $f(q_1, \ldots, q_m) \to q$, and the rules $r(p.1), \ldots, r(p.m)$ have $q_1, \ldots, q_m$ as right-hand sides, respectively. We say that $r(p)$ is the rule applied at position p. The run r is called* accepting *if the right-hand side of $r(\lambda)$ is a state in F. A term t is accepted by A if there exists an accepting run of A on t. The set of accepted terms by A, also called the* language recognized *by A, is denoted by $\mathcal{L}(A)$. A language L is* regular *if there exists a* TA *A such that $\mathcal{L}(A) = L$ holds.*

3 Tree Automata with Height and Equality Constraints between Brothers

In this paper we deal with several classes of tree automata with constraints between brothers. The only difference between such classes is the form of the constraints and the way such constraints are interpreted. In this section we give a general definition of tree automata with constraints between brothers that is instantiated later. To simplify notations, we consider a fixed signature Σ.

Definition 2. *A* constraint structure *is a tuple $\langle C, \mathsf{PosVar}, \models, |.|, \mathsf{Width} \rangle$. Here, C (notion of syntax) is a set of elements called constraints.* PosVar *is a function that maps each element of C to a finite set of natural numbers. Given $c \in C$, any partial mapping $I : \mathbb{N} \to \mathcal{T}(\Sigma)$ satisfying $\mathsf{PosVar}(c) \subseteq \mathsf{Dom}(I)$ is called* *an interpretation of c.* $\models$ *(notion of satisfaction) maps each pair* $c \in C$ *and interpretation* I *of* c *to either true (denoted* $I \models c$*) or false (denoted* $I \not\models c$*).* $|.|$ *(notion of size) maps each constraint* c *to a natural number.* Width *maps each constraint* c *to a natural number.*

Given two constraints c_1, c_2 *(not necessarily from the same constraint structure) holding* $\mathsf{PosVar}(c_1) = \mathsf{PosVar}(c_2)$*, we say that* c_1, c_2 *are* compatible *if there exists* I *satisfying* $I \models c_1$ *and* $I \models c_2$ *(where* $\models$ *refers to the satisfaction notion of each corresponding structure). Otherwise, we say that* c_1, c_2 *are* incompatible.

A set of constraints $\{c_1, \ldots, c_n\}$ *of the same constraint structure and holding* $\mathsf{PosVar}(c_1) = \cdots = \mathsf{PosVar}(c_n)$ *is called* deterministic *if all* c_i, c_j *are pairwise incompatible for* $1 \leq i < j \leq n$*. It is called* complete *if, for each* $I : \mathsf{PosVar}(c_1) \rightarrow \mathcal{T}(\Sigma)$*,* $I \models c_i$ *for some* $1 \leq i \leq n$*.*

Given two constraints c_1, c_2 *(not necessarily from the same constraint structure) holding* $\mathsf{PosVar}(c_1) \supseteq \mathsf{PosVar}(c_2)$*, we say that* c_1 implies c_2*, denoted* $c_1 \models c_2$*, if for each* I *satisfying* $I \models c_1$*,* $I \models c_2$ *also holds (where* $\models$ *refers to the satisfaction notion of each corresponding structure).*

Definition 3. *Let* $\mathsf{S} = \langle C, \mathsf{PosVar}, \models, |.|, \mathsf{Width}\rangle$ *be a constraint structure. A* tree automaton with constraints between brothers *based on* S*,* $\mathtt{TACBB}_{\mathsf{S}}$ *for short, is a tuple* $A = \langle Q, \Sigma, F, \Delta\rangle$*, where* Q *is a finite set of states,* Σ *is a signature,* $F \subseteq Q$ *is the set of final states, and* Δ *is a set of rules of the form* $f(q_1, \ldots, q_m) \xrightarrow{c} q$*, where* $q_1, \ldots, q_m, q$ *are in* Q*,* f *is in* $\Sigma^{(m)}$*, and* c *is a constraint of* C *satisfying* $\mathsf{PosVar}(c) \subseteq \{1, \ldots, m\}$*. The rule is* fully constrained *if* $\mathsf{PosVar}(c) = \{1, \ldots, m\}$*. The* size *of such rule is* $m + 2 + |c|$*. The size of* A *is* $|Q|$ *plus the sum of sizes of all rules in* Δ*. The* width *of a rule* $f(q_1, \ldots, q_m) \xrightarrow{c} q$ *of* A *is* $\mathsf{Width}(c)$*. The* width *of* A *is the maximum among the widths of its rules.* A *is* fully constrained*, denoted* $\mathtt{TACBB}^F_{\mathsf{S}}$*, if each of its rules is fully constrained.*

A run *of* A *on a term* $t \in \mathcal{T}(\Sigma)$ *is a mapping* $r : \mathsf{Pos}(t) \rightarrow \Delta$ *such that, for each position* $p \in \mathsf{Pos}(t)$*, if* $t|_p$ *is of the form* $f(t_1, \ldots, t_m)$*, then* $r(p)$ *is a rule of the form* $f(q_1, \ldots, q_m) \xrightarrow{c} q$*, the rules* $r(p.1), \ldots, r(p.m)$ *have* $q_1, \ldots, q_m$ *as right-hand sides, respectively, and* $I = \{1 \mapsto t|_{p.1}, \ldots, m \mapsto t|_{p.m}\} \models c$*. We say that* $r(p)$ *is the rule applied at position* p*. The run* r *is called* accepting *if the right-hand side of* $r(\lambda)$ *is a state in* F*. A term* t *is accepted by* A *if there exists an accepting run of* A *on* t*. The set of accepted terms by* A*, also called the* language recognized *by* A*, is denoted by* $\mathcal{L}(A)$*. By* $\mathcal{L}(A, q)$ *we denote the set of terms for which there exists a run* r *of* A *on them holding that the right-hand side of* $r(\lambda)$ *is* q*.*

Definition 4. *A* height and equality *constraint* c *is a boolean combination (including negation) of atoms of the form either* $h(i) = h(j)$ *or* $h(i) < h(j)$ *or* $h(i) = h(j) + x$ *or* $h(i) < h(j) + x$ *or* $i = j$ *for natural numbers* i, j *and integer number* x*. The* width *of* c*, denoted* $\mathsf{Width}(c)$*, is* 1 *plus the maximum of the absolute values of such* x*. The* size *of* c*, denoted* $|c|$*, is the number of such atoms occurring in* c*. By* $\mathsf{PosVar}(c)$ *we denote the set of such naturals* i, j *occurring in* c*. An* interpretation *of* c *is a partial mapping* $I : \mathbb{N} \rightarrow \mathcal{T}(\Sigma)$ *such that* $\mathsf{PosVar}(c) \subseteq \mathsf{Dom}(I)$*.* I satisfies *(*is a solution of*)* c*, denoted* $I \models c$ *if, by* *replacing in c each i by I(i), the expression evaluates to true by interpreting h as the* height *function,* = *as syntactic equality when it compares terms or as equality of the evaluated expressions when it compares integer numbers,* $<, +$ *as the less and addition operators on integer numbers, and the boolean operators in the usual way.*

By C_{He} *we denote the set of all height and equality constraints. By* C_H *we denote the subset of* C_{He} *of constraints with no occurrences of atoms of the form* $i = j$ *(i.e., only height constraints). By* C_{he} *we denote the subset of* C_{He} *whose atoms are of the form* $h(i) = h(j)$ *or* $h(i) < h(j)$ *or* $i = j$ *(i.e., equality constraints and only simple height constraints). By* C_h *we denote the subset of* C_{he} *whose atoms are of the form* $h(i) = h(j)$ *or* $h(i) < h(j)$ *(i.e., only simple height constraints). The constraint structures* He, he, H, h *are defined as* $\langle C, \mathsf{PosVar}, \models, |.|, \mathsf{Width}\rangle$*, by replacing* C *by* C_{He}*,* C_{he}*,* C_H*,* C_h*, respectively, and where* $\mathsf{PosVar}, \models, |.|, \mathsf{Width}$ *are defined as above.*

We will show that emptiness and finiteness of $\mathtt{TACBB}_{\mathsf{He}}$ are decidable, obtaining different time complexities depending on the restrictions of the constraints. For explanation purposes, we deal with the most particular case $\mathtt{TACBB}_{\mathsf{h}}$ in Section 4, and the most general case $\mathtt{TACBB}_{\mathsf{He}}$ in Section 5, where the results for $\mathtt{TACBB}_{\mathsf{H}}$ and $\mathtt{TACBB}_{\mathsf{he}}$ are also stated.

4 The Case with Simple Height Constraints

In this section we prove decidability of emptiness and finiteness of $\mathtt{TACBB}_{\mathsf{h}}$. A usual way to deal with automata with constraints is to transform them into a normalized form that is easy to deal with. We proceed in this way by normalizing simple height constraints according to the following definitions and lemma.

Definition 5. *A* normalized simple height constraint *is an expression c of the form* $S_1 < S_2 < \cdots < S_n$*, where all* $S_1, \ldots, S_n$ *are non-empty finite sets of natural numbers, and each natural number occurs in at most one* S_i*. By* $\mathsf{PosVar}(c)$ *we denote the set of natural numbers occurring in* $S_1 \uplus \ldots \uplus S_n$*. An interpretation of c is a partial mapping I from natural numbers to terms in* $\mathcal{T}(\Sigma)$*, whose domain includes* $\mathsf{PosVar}(c)$*. We say that I* satisfies *(*is a solution of*) c, denoted* $I \models c$ *if:*

- *for each* $i \in \{1, \ldots, n\}$*, the natural numbers occurring in* S_i *are mapped by I to terms with the same height, and*
- *for each* $i \in \{1, \ldots, n-1\}$*, each* $i_1 \in S_i$ *and each* $i_2 \in S_{i+1}$*,* $\mathsf{height}(I(i_1)) < \mathsf{height}(I(i_2))$.

The size of c is $|c| = |\mathsf{PosVar}(c)|$*. The width of c is* $\mathsf{Width}(c) = 1$ *(in the simple case the width plays no important role). The* structure of normalized simple height constraints *is the constraint structure* $\mathsf{nh} = \langle C_{nh}, \mathsf{PosVar}, \models, |.|, \mathsf{Width}\rangle$*, where* C_{nh} *is the set of all normalized simple height constraints, and* $\mathsf{PosVar}, \models, |.|, \mathsf{Width}$ *are defined as above.*

Lemma 1. *Let* $A = \langle Q, \Sigma, F, \Delta\rangle$ *be a* $\mathtt{TACBB}_{\mathsf{h}}$*. Then, a* $\mathtt{TACBB}^F_{\mathsf{nh}}$ $A' = \langle Q, \Sigma, F, \Delta'\rangle$ *satisfying* $\mathcal{L}(A') = \mathcal{L}(A)$ *can be computed in time* $\mathcal{O}(|A| \cdot 2^{\mathsf{maxar}^2})$.

Proof. Sketch: it suffices to define Δ' as $\{f(q_1,\ldots,q_m) \xrightarrow{c'} q \mid c' \in C_{nh} \wedge \mathsf{PosVar}(c') = \{1,\ldots,m\} \wedge \exists (f(q_1,\ldots,q_m) \xrightarrow{c} q) \in \Delta : (c' \models c)\}$.

To decide emptiness of a $\mathtt{TACBB}^F_{\mathsf{nh}}$ A, we will need to compute, for each h, which states are reachable by runs of A on terms with height h. To this end, we need to determine which rules can be applied at the root of such terms, and this leads to a more general computation: we need to know which part of a constraint is satisfied by taking into account runs on terms smaller than h.

Definition 6. *Let $A = \langle Q, \Sigma, F, \Delta\rangle$ be a $\mathtt{TACBB}^F_{\mathsf{nh}}$. Let h be a natural number. Let q be a state in Q. Then, we define $\mathsf{ExistTerm}_A(h,q)$ as true if there exists a term with height h in $\mathcal{L}(A,q)$, and as false otherwise. Let c be a normalized simple height constraint. Let I be a solution of c. Let N be a partial mapping from natural numbers to Q such that $\mathsf{Dom}(N) \supseteq \mathsf{Dom}(I)$. We say that N and I are* compatible *with respect to A if, for each $i \in \mathsf{Dom}(I)$, $I(i) \in \mathcal{L}(A, N(i))$. We define $\mathsf{ExistSol}_A(h,c,N)$ as true if there exists $I : \mathsf{PosVar}(c) \to \mathcal{T}(\Sigma)$ such that I is compatible with N and a solution of c, and the highest term occurring in $I(\mathsf{PosVar}(c))$ has height h, and as false otherwise. Moreover, we define $\mathsf{AccExistSol}_A(h,c,N)$ as $\bigvee_{h' \le h} \mathsf{ExistSol}_A(h',c,N)$. When A is clear from the context, we omit the subindex of $\mathsf{ExistTerm}_A$, $\mathsf{ExistSol}_A$, $\mathsf{AccExistSol}_A$.*

We define C_A as the set of pairs $\{\langle c, N\rangle \mid \exists (f(q_1,\ldots,q_m) \xrightarrow{c'} q) \in \Delta : (c$ is a non-empty prefix of c' or a set occurring in $c') \wedge N = \{1 \mapsto q_1, \ldots, m \mapsto q_m\}\}$. Note that $|C_A|$ is bounded by $2 \cdot \mathsf{maxar} \cdot |\Delta|$. The configuration *of A for height h is the description of all values $\mathsf{ExistSol}_A(h,c,N)$, $\mathsf{AccExistSol}_A(h,c,N)$, $\mathsf{ExistTerm}_A(h,q)$ for $\langle c,N\rangle \in C_A$ and $q \in Q$.*

The following lemma gives a computable definition of ExistSol, AccExistSol and ExistTerm, and a way to argue that the configuration for $h > 0$ depends only on the configuration for $h-1$. It can be proved by induction on h and $|c|$.

Lemma 2. *Let $A = \langle Q, \Sigma, F, \Delta\rangle$ be a $\mathtt{TACBB}^F_{\mathsf{nh}}$. ExistSol, AccExistSol and ExistTerm can also be defined recursively as follows:*

- *Consider the case $h = 0$. Then, $\mathsf{ExistTerm}(h,q)$ is true if and only if there exists $a \in \Sigma^{(0)}$ satisfying $a \in \mathcal{L}(A,q)$. Moreover, if c is not just a set, then $\mathsf{ExistSol}(h,c,N) = \mathsf{AccExistSol}(h,c,N) = \mathsf{False}$, and otherwise if c is a set S, $\mathsf{ExistSol}(h,c,N)$ and $\mathsf{AccExistSol}(h,c,N)$ are defined to be true if and only if $\bigwedge_{q \in N(S)} \mathsf{ExistTerm}(h,q)$ holds. In the rest of cases assume $h > 0$.*
- $\mathsf{AccExistSol}(h,c,N) = \mathsf{AccExistSol}(h-1,c,N) \vee \mathsf{ExistSol}(h,c,N)$
- $\mathsf{ExistSol}(h,c' < S,N) = \mathsf{AccExistSol}(h-1,c',N) \wedge \mathsf{ExistSol}(h,S,N)$
- $\mathsf{ExistSol}(h,S,N) = \bigwedge_{q \in N(S)} \mathsf{ExistTerm}(h,q)$
- $\mathsf{ExistTerm}(h,q) = \bigvee_{(f(q_1,\ldots,q_m) \xrightarrow{c} q) \in \Delta, m > 0} \mathsf{ExistSol}(h-1,c,\{1 \mapsto q_1,\ldots,m \mapsto q_m\})$

Corollary 1. *Let $A = \langle Q, \Sigma, F, \Delta\rangle$ be a $\mathtt{TACBB}^F_{\mathsf{nh}}$. Let h be a natural number. The configuration of A for height $h+1$ can be obtained from the configuration of A for height h with time in $\mathcal{O}(|C_A| \cdot \mathsf{maxar} + |Q| + |\Delta|)$, i.e, in $\mathcal{O}(\mathsf{maxar}^2 \cdot |\Delta| + |Q|)$.*

The previous result leads to the fact that, when we find the same configuration for heights $h_1 < h_2$, the sequence of configurations is periodic with period $h_2 - h_1$ starting from h_1. Thus, deciding emptiness corresponds to check whether there exists an accepting state q satisfying $\mathsf{ExistTerm}(h, q)$ for $h < h_2$, and deciding finiteness corresponds to check whether there exists an accepting state q satisfying $\mathsf{ExistTerm}(h, q)$ for $h_1 \leq h < h_2$. The final time complexity depends on Corollary 1 and the number of different possible configurations.

Lemma 3. *Let $A = \langle Q, \Sigma, F, \Delta \rangle$ be a $\mathtt{TACBB}^F_{\mathsf{nh}}$. The number of different configurations of A considering all possible h is bounded by $2^{|Q|} \cdot (2 \cdot \mathsf{maxar} \cdot |\Delta| + 1)$.*

Proof. Consider any two heights $0 < h_1 < h_2$ satisfying that $\mathsf{AccExistSol}$ for $h_1 - 1$ is equal to $\mathsf{AccExistSol}$ for $h_2 - 1$, and $\mathsf{ExistTerm}$ for h_1 is equal to $\mathsf{ExistTerm}$ for h_2. Since $\mathsf{ExistSol}$ for any height $h > 0$ depends only on $\mathsf{AccExistSol}$ for $h - 1$ and $\mathsf{ExistTerm}$ for h, it follows that $\mathsf{ExistSol}$ for h_1 is equal to $\mathsf{ExistSol}$ for h_2. Moreover, it also follows that $\mathsf{AccExistSol}$ for h_1 is equal to $\mathsf{AccExistSol}$ for h_2. Thus, the configurations of A for h_1 and h_2 are equal, and hence, to prove the statement it suffices to bound the number of different combinations of $\mathsf{AccExistSol}$ and $\mathsf{ExistTerm}$. To this end, note that there are at most $2^{|Q|}$ different $\mathsf{ExistTerm}$, and since $\mathsf{AccExistSol}(h, c, N) \Rightarrow \mathsf{AccExistSol}(h + 1, c, N)$ holds for any h and $\langle c, N \rangle \in C_A$, there are at most $|C_A| + 1 = 2 \cdot \mathsf{maxar} \cdot |\Delta| + 1$ different $\mathsf{AccExistSol}$.

Corollary 2. *Emptiness and finiteness of the language recognized by a $\mathtt{TACBB}^F_{\mathsf{nh}}$ $A = \langle Q, \Sigma, F, \Delta \rangle$ are decidable with time in $2^{\mathcal{O}(|Q| + \log(\mathsf{maxar} \cdot |\Delta|))}$.*

As a consequence of Lemma 1 we also obtain decidability for $\mathtt{TACBB}_{\mathsf{h}}$.

Corollary 3. *Emptiness and finiteness of the language recognized by a $\mathtt{TACBB}_{\mathsf{h}}$ $A = \langle Q, \Sigma, F, \Delta \rangle$ are decidable with time in $2^{\mathcal{O}(|Q| + \mathsf{maxar}^2 + \log(|A|))}$.*

5 Decidability of the General Case

The global approach for proving decidability of emptiness and finiteness for the general case of height and equality constraints is analogous to the case of simple height constraints. We start with a process of normalization of the automata.

Definition 7. *A* normalized height and equality constraint *is an expression c of the form $P_1 \otimes_1 P_2 \otimes_2 \cdots \otimes_{n-1} P_n$, where $P_1, \ldots, P_n$ are partitions of non-empty finite sets of natural numbers, each natural number occurs in at most one P_i, and each $\otimes_i$ is an operator of the form either $=_h$ or $\leq_h$ for a natural number $h > 0$. By $\mathsf{PosVar}(c)$ we denote the set of natural numbers occurring in $P_1, \ldots, P_n$. An interpretation of c is a partial mapping $I : \mathbb{N} \to \mathcal{T}(\Sigma)$ whose domain includes $\mathsf{PosVar}(c)$. We say that I* satisfies *(*is a solution of*) c, denoted $I \models c$, if:*

- *for each $i \in \{1, \ldots, n\}$, the natural numbers occurring in P_i are mapped to terms with the same height, and two natural numbers are mapped to the same term if and only if they occur in the same part, and*

- *for each* $i \in \{1, \ldots, n-1\}$ *such that* $\otimes_i$ *is of the form* $=_h$, *each* $i_1 \in \mathsf{PosVar}(P_i)$ *and each* $i_2 \in \mathsf{PosVar}(P_{i+1})$, $\mathsf{height}(I(i_1)) + h = \mathsf{height}(I(i_2))$, *and*
- *for each* $i \in \{1, \ldots, n-1\}$ *such that* $\otimes_i$ *is of the form* $\leq_h$, *each* $i_1 \in \mathsf{PosVar}(P_i)$ *and each* $i_2 \in \mathsf{PosVar}(P_{i+1})$, $\mathsf{height}(I(i_1)) + h \leq \mathsf{height}(I(i_2))$.

The size *of* c *is* $|c| = |\mathsf{PosVar}(c)|$. *The* width *of* c, *denoted* $\mathsf{Width}(c)$, *is the maximum among the natural numbers* h *occurring in the subscripts of the operators in* c. *The* structure of normalized height and equality constraints *is the constraint structure* $\mathsf{nHe} = \langle C_{nHe}, \mathsf{PosVar}, \models, |.|, \mathsf{Width}\rangle$, *where* C_{nHe} *is the set of all normalized height and equality constraints, and* $\mathsf{PosVar}, \models, |.|, \mathsf{Width}$ *are defined as above.*

Since we will need to compute not only if a state q is reachable by some term with height h, but also how many of such different terms reach q, we will determinize the automaton in order to ease this task.

Definition 8. *We say that a* $\mathtt{TACBB}^F_{\mathsf{nHe}}$ $A = \langle Q, \Sigma, F, \Delta\rangle$ *is* deterministic and complete *(or a* $\mathtt{dcTACBB}^F_{\mathsf{nHe}}$*) if:*

- *for each* $f \in \Sigma$ *with arity* m *and states* $q_1, \ldots, q_m \in Q$, *and each normalized height and equality constraint* c, *there is at most one* $q \in Q$ *holding* $(f(q_1, \ldots, q_m) \xrightarrow{c} q) \in \Delta$, *and*
- *for each* $f \in \Sigma$ *with arity* m *and states* $q_1, \ldots, q_m \in Q$, *the set* $\{c \mid \exists q \in Q : (f(q_1, \ldots, q_m) \xrightarrow{c} q) \in \Delta\}$ *is non-empty, deterministic, and complete.*

Note that, for each $\mathtt{dcTACBB}^F_{\mathsf{nHe}}$ A and each $t \in \mathcal{T}(\Sigma)$, there exists one and only one run of A on t. We denote by $A(t)$ the state reached by such run at λ. In order to obtain a deterministic and complete set of constraints, we will construct them from a limited number of operators.

Definition 9. *Let* m, w *be natural numbers. A* normalized constraint with respect to m *and* w *is a normalized height and equality constraint* c *over the operators* $=_1, =_2, \ldots, =_{w-1}$ *and* $\leq_w$ *and satisfying* $\mathsf{PosVar}(c) = \{1, \ldots, m\}$.

For a set of constraints to be deterministic, it must satisfy that its constraints are pairwise incompatible.

Lemma 4. *Let* m, w *be natural numbers. Let* c, d *be two different normalized constraints with respect to* m *and* w. *Then,* c *and* d *are incompatible.*

For a set of constraints to be complete, it must satisfy that any interpretation is a solution of at least one of the constraints of the set.

Definition 10. *Let* w *be a natural number. Let* $I : \mathbb{N} \to \mathcal{T}(\Sigma)$ *be a partial function holding* $\mathsf{Dom}(I) = \{1, \ldots, m\}$ *for some* m. *Let* $h_1 < h_2 < \cdots < h_n$ *be the elements of* $\mathsf{height}(I(\{1, \ldots, m\}))$. *The constraint* induced *from* I *and* w *is the normalized constraint* $P_1 \otimes_1 P_2 \otimes_2 \cdots \otimes_{n-1} P_n$ *with respect to* m *and* w, *where:*

– *for* $i \in \{1, \ldots, n\}$, P_i *is a partition of* $\{j \in \{1, \ldots, m\} \mid \mathsf{height}(I(j)) = h_i\}$ *satisfying* $(j_1 \equiv_{P_i} j_2) \Leftrightarrow (I(j_1) = I(j_2))$, *and*
– *for* $i \in \{1, \ldots, n-1\}$, $\otimes_i$ *is* $=_{h_{i+1}-h_i}$ *if* $h_{i+1} - h_i < w$, *and* $\leq_w$ *otherwise.*

Given w and an interpretation I, the fact that the constraint c induced from I and w holds $I \models c$ follows by definition.

Lemma 5. *Let* w *be a natural number. Let* $I : \mathbb{N} \to \mathcal{T}(\Sigma)$ *be a partial function holding* $\mathsf{Dom}(I) = \{1, \ldots, m\}$ *for some* m. *Let* c *be the constraint induced from* I *and* w. *Then,* $I \models c$.

Corollary 4. *Let* m, w *be natural numbers. Let* C *be the set of normalized constraints with respect to* m *and* w. *Then,* C *is deterministic and complete.*

Proof. By Lemma 4, any two different constraints in C are incompatible, and thus C is deterministic. By Lemma 5, any interpretation $I : \{1, \ldots, m\} \to \mathcal{T}(\Sigma)$ satisfies the constraint c induced from I and w. Since such c is in C by definition, it follows that C is complete.

In order to normalize a $\mathtt{TACBB}_{\mathsf{He}}$ A, we construct a $\mathtt{dcTACBB}^F_{\mathsf{nHe}}$ A' that simulates all possible runs of A. To this end, each rule of A' needs to determine exactly which rules of A are applicable. In particular, A' needs to determine which constraints are satisfied. For this reason, we will need to use the fact that the normalized constraints are, in some sense, more precise than the original ones, that is, given a normalized constraint c_1 and an original constraint c_2, either c_1 and c_2 are incompatible, or $c_1 \models c_2$ holds.

Lemma 6. *Let* m, w *be natural numbers. Let* c_1 *be a normalized constraint with respect to* m *and* w. *Let* c_2 *be a height and equality constraint whose width is smaller than or equal to* w *and satisfying* $\mathsf{PosVar}(c_2) \subseteq \{1, \ldots, m\}$. *Then, either* c_1 *and* c_2 *are incompatible, or* $c_1 \models c_2$ *holds.*

Proof. Sketch: it suffices to note that for any two solutions I_1, I_2 of c_1 and any atom c_2' occurring in c_2, either $I_1 \models c_2'$ and $I_2 \models c_2'$, or $I_1 \not\models c_2'$ and $I_2 \not\models c_2'$.

Definition 11. *Let* $A = \langle Q, \Sigma, F, \Delta \rangle$ *be a* $\mathtt{TACBB}_{\mathsf{He}}$. *Let* w *be the width of* A. *For each natural number* m, *let* C_m *be the set of normalized constraints with respect to* m *and* w. *Then, the* normalization *of* A *is defined as the* $\mathtt{TACBB}^F_{\mathsf{nHe}}$ $\langle 2^Q, \Sigma, \{F' \in 2^Q \mid F' \cap F \neq \emptyset\}, \Delta' \rangle$, *where* Δ' *is the set of rules* $\{f(S_1, \ldots, S_m) \xrightarrow{c} S \mid f \in \Sigma^{(m)} \wedge c \in C_m \wedge S_1, \ldots, S_m, S \in 2^Q \wedge S = \{q \in Q \mid \exists q_1 \in S_1, \ldots, q_m \in S_m, c' : ((f(q_1, \ldots, q_m) \xrightarrow{c'} q) \in \Delta \wedge (c \models c'))\}\}$.

Lemma 7. *Let* $A = \langle Q, \Sigma, F, \Delta \rangle$ *be a* $\mathtt{TACBB}_{\mathsf{He}}$ *having width* w. *Then, the normalization* A' *of* A *is deterministic and complete, i.e., a* $\mathtt{dcTACBB}^F_{\mathsf{nHe}}$, *and can be computed with time in* $2^{\mathcal{O}(\log(\Sigma) + \mathsf{maxar} \cdot (|Q| + \mathsf{maxar} + \log(w)) + \log(|A|))}$.

Proof. The fact that A' is a $\mathtt{dcTACBB}^F_{\mathsf{nHe}}$ follows from Definition 8, Corollary 4 and Definition 11. The time complexity follows from these observations: each rule of A' can be computed with time in $\mathcal{O}(|A|)$, the number of rules of A' is bounded by $|\Sigma| \cdot 2^{|Q| \cdot \mathsf{maxar}} \cdot |C_{\mathsf{maxar}}|$, where C_{maxar} is the set of normalized constraints with respect to maxar and w, and $|C_{\mathsf{maxar}}|$ is in $2^{\mathcal{O}(\mathsf{maxar}^2 + \mathsf{maxar} \cdot \log(w))}$.

In order to see that A' preserves the language recognized by A, we show that, for any term t, the unique state $A'(t)$ is precisely the set of states that are reachable by runs of A on t.

Lemma 8. *Let $A = \langle Q, \Sigma, F, \Delta \rangle$ be a $\mathtt{TACBB}_{\mathsf{He}}$. Let A' be the normalization of A. Let $t \in \mathcal{T}(\Sigma)$ be a term. Then, the state reached at λ by the unique run of A' on t is $S = \{q \in Q \mid t \in \mathcal{L}(A, q)\}$.*

Proof. We prove it by induction on $\mathsf{height}(t)$. Without loss of generality, let t be of the form $f(t_1, \ldots, t_m)$. By induction hypothesis, the states reached at λ by the unique runs of A' on $t_1, \ldots, t_m$ are $S_1, \ldots, S_m$, respectively, where $S_i = \{q \in Q \mid t_i \in \mathcal{L}(A, q)\}$. Let c be the constraint induced from $I = \{1 \mapsto t_1, \ldots, m \mapsto t_m\}$ and the width of A. By Definition 11, A' has the rule $f(S_1, \ldots, S_m) \xrightarrow{c} S'$ where $S' = \{q \in Q \mid \exists q_1 \in S_1, \ldots, q_m \in S_m, c' : ((f(q_1, \ldots, q_m) \xrightarrow{c'} q) \in \Delta \wedge (c \models c'))\}$. Such rule can be applied at the root of t since $I \models c$ holds by Lemma 5, and moreover, by Corollary 4 no other rule can. It remains to prove $S' = S$:

($\subseteq$) Let $q \in S'$. By definition there are $q_1 \in S_1, \ldots, q_m \in S_m$ and c' such that $(f(q_1, \ldots, q_m) \xrightarrow{c'} q) \in \Delta$ and $c \models c'$. Thus, $I \models c'$, and since we had that $t_1 \in \mathcal{L}(A, q_1), \ldots, t_m \in \mathcal{L}(A, q_m)$, it follows $t \in \mathcal{L}(A, q)$, and hence, $q \in S$.

($\supseteq$) Let $q \in S$. By definition there is a run of A on t with a rule of the form $f(q_1, \ldots, q_m) \xrightarrow{c'} q$ applied at the root. Note that $I \models c'$, and since $I \models c$, by Lemma 6, $c \models c'$. Since we had that $q_1 \in S_1, \ldots, q_m \in S_m$, it follows $q \in S'$.

Corollary 5. *Let A be a $\mathtt{TACBB}_{\mathsf{He}}$. Let A' be the normalization of A. Then, $\mathcal{L}(A') = \mathcal{L}(A)$.*

To decide emptiness of a $\mathtt{dcTACBB}^{F}_{\mathsf{nHe}}$ A, we will need to compute, for certain h's, how many runs (up to a certain bound) of A on terms with height h reach each state. We start by a previous definition describing which terms (and not only how many of them) are reached, in order to ease later arguments.

Definition 12. *Let $A = \langle Q, \Sigma, F, \Delta \rangle$ be a $\mathtt{dcTACBB}^{F}_{\mathsf{nHe}}$. Let h be a natural number. Let q be a state in Q. Then, $\mathsf{Terms}_A(h, q)$ is defined as $\{t \in \mathcal{T}(\Sigma) \mid A(t) = q \wedge \mathsf{height}(t) = h\}$. Let c be a normalized height and equality constraint. Let I be a solution of c. Let N be a partial mapping from natural numbers to Q such that $\mathsf{Dom}(N) \supseteq \mathsf{Dom}(I)$. We say that N and I are* compatible *with respect to A if, for each $i \in \mathsf{Dom}(I)$, $A(I(i)) = N(i)$. We define $\mathsf{Sols}_A(h, c, N)$ as the set of interpretations $I : \mathsf{PosVar}(c) \to \mathcal{T}(\Sigma)$ such that I is compatible with N and a solution of c, and the highest term occurring in $I(\mathsf{PosVar}(c))$ has height h. Moreover, we define $\mathsf{AccSols}_A(h, c, N)$ as $\bigcup_{h' \leq h} \mathsf{Sols}_A(h', c, N)$. When A is clear from the context, we omit the subindex of Terms_A, Sols_A, $\mathsf{AccSols}_A$.*

We define C_A as the set of pairs $\{\langle c, N \rangle \mid \exists (f(q_1, \ldots, q_m) \xrightarrow{c'} q) \in \Delta : (c$ is a non-empty prefix of c' or a partition occurring in $c') \wedge N = \{1 \mapsto q_1, \ldots, m \mapsto q_m\}\}$. Note that $|C_A|$ is bounded by $2 \cdot \mathsf{maxar} \cdot |\Delta|$.

The following lemma provides a computable definition of the sets Sols, AccSols and Terms, as well as a way to argue that the configuration for a certain h depends only on the configurations for $h-i$, with i bounded by the width of A. It can be straightforwardly proved by induction on h and on the size of c.

Lemma 9. *Let $A = \langle Q, \Sigma, F, \Delta \rangle$ be a* $\mathsf{dcTACBB}^F_{\mathsf{nHe}}$. Sols, AccSols *and* Terms *can also be defined recursively as follows:*

- *If $h < 0$, then* $\mathsf{Terms}(h,q) = \mathsf{Sols}(h,c,N) = \mathsf{AccSols}(h,c,N) = \emptyset$ *for any q, c, N.*
- *Consider the case $h = 0$. Then,* $\mathsf{Terms}(h,q) = \{a \in \Sigma^{(0)} \mid A(a) = q\}$. *Moreover, if c is not just a partition, then* $\mathsf{Sols}(h,c,N) = \mathsf{AccSols}(h,c,N) = \emptyset$*, and otherwise if c is a partition P,* $\mathsf{Sols}(h,c,N)$ *and* $\mathsf{AccSols}(h,c,N)$ *are* $\{I : \mathsf{PosVar}(P) \to \Sigma^{(0)} \mid \forall i \in \mathsf{PosVar}(P) : I(i) \in \mathsf{Terms}(h, N(i)) \wedge \forall i,j \in \mathsf{PosVar}(P) : (I(i) = I(j) \Leftrightarrow i \equiv_P j)\}$. *In the rest of cases assume $h > 0$.*
- $\mathsf{AccSols}(h,c,N) = \mathsf{AccSols}(h-1,c,N) \cup \mathsf{Sols}(h,c,N)$
- $\mathsf{Sols}(h, c' =_{h'} P, N) = \mathsf{Sols}(h-h', c', N) \boxtimes \mathsf{Sols}(h,P,N)$
- $\mathsf{Sols}(h, c' \leq_{h'} P, N) = \mathsf{AccSols}(h-h', c', N) \boxtimes \mathsf{Sols}(h,P,N)$
- $\mathsf{Sols}(h,P,N) = \{I : \mathsf{PosVar}(P) \to \mathcal{T}(\Sigma) \mid \forall i \in \mathsf{PosVar}(P) : I(i) \in \mathsf{Terms}(h, N(i)) \wedge \forall i,j \in \mathsf{PosVar}(P) : (I(i) = I(j) \Leftrightarrow i \equiv_P j)\}$
- $\mathsf{Terms}(h,q) = \{f(I(1),\ldots,I(m)) \mid (f(q_1,\ldots,q_m) \xrightarrow{c} q) \in \Delta \wedge m > 0 \wedge I \in \mathsf{Sols}(h-1, c, \{1 \mapsto q_1, \ldots, m \mapsto q_m\})\}$

As we have mentioned above, we are not interested in computing all the reached terms at a certain height h, but only how many of them reach each state, up to a certain bound: the maximum arity of the function symbols in Σ.

Definition 13. *Let $A = \langle Q, \Sigma, F, \Delta \rangle$ be a* $\mathsf{dcTACBB}^F_{\mathsf{nHe}}$. *Let h be a natural number. Let q be a state in Q. Then,* $\#\mathsf{Terms}_A(h,q)$ *is defined as* $\min(\mathsf{maxar}, |\mathsf{Terms}_A(h,q)|)$. *Let c be a normalized height and equality constraint. Let N be a partial mapping from natural numbers to Q such that* $\mathsf{Dom}(N) \supseteq \mathsf{PosVar}(c)$. *Then,* $\#\mathsf{Sols}_A(h,c,N)$ *and* $\#\mathsf{AccSols}_A(h,c,N)$ *are respectively defined as* $\min(\mathsf{maxar}, |\mathsf{Sols}_A(h,c,N)|)$ *and* $\min(\mathsf{maxar}, |\mathsf{AccSols}_A(h,c,N)|)$. *When A is clear from the context, we omit the subindex of* $\#\mathsf{Terms}_A$, $\#\mathsf{Sols}_A$, $\#\mathsf{AccSols}_A$.

The configuration *of A for height h is the description of all values* $\#\mathsf{Sols}_A(h,c,N)$, $\#\mathsf{AccSols}_A(h,c,N)$, $\#\mathsf{Terms}_A(h,q)$ *for $\langle c, N \rangle \in C_A$ and $q \in Q$.*

The following recursive definition follows from Definition 13 and Lemma 9.

Lemma 10. *Let $A = \langle Q, \Sigma, F, \Delta \rangle$ be a* $\mathsf{dcTACBB}^F_{\mathsf{nHe}}$. #Sols, #AccSols *and* #Terms *can also be defined recursively as follows, where $B(P,N,q)$ denotes $|\{P' \in P \mid N(P') = \{q\}\}|$:*

- *If $h < 0$, then* $\#\mathsf{Terms}(h,q) = \#\mathsf{Sols}(h,c,N) = \#\mathsf{AccSols}(h,c,N) = 0$ *for any q, c, N.*
- *Consider the case $h = 0$. Then,* $\#\mathsf{Terms}(h,q)$ *is defined as* $\min(\mathsf{maxar}, |\{a \in \Sigma^{(0)} \mid A(a) = q\}|)$. *Moreover, if c is not just a partition, then* $\#\mathsf{Sols}(h,c,N) = \#\mathsf{AccSols}(h,c,N) = 0$*, and otherwise if c is a*

partition P, *then* $\#\mathsf{Sols}(h,c,N)$ *and* $\#\mathsf{AccSols}(h,c,N)$ *are defined as* $\min(\mathsf{maxar}, \prod_{q\in N(\mathsf{PosVar}(P))} \binom{\#\mathsf{Terms}(h,q)}{B(P,N,q)} \cdot B(P,N,q)!)$. *In the rest of cases assume* $h > 0$.

- $\#\mathsf{AccSols}(h,c,N) = \min(\mathsf{maxar}, \#\mathsf{AccSols}(h-1,c,N) + \#\mathsf{Sols}(h,c,N))$
- $\#\mathsf{Sols}(h, c' =_{h'} P, N) = \min(\mathsf{maxar}, \#\mathsf{Sols}(h-h', c', N) \cdot \#\mathsf{Sols}(h,P,N))$
- $\#\mathsf{Sols}(h, c' \leq_{h'} P, N) = \min(\mathsf{maxar}, \#\mathsf{AccSols}(h-h', c', N) \cdot \#\mathsf{Sols}(h,P,N))$
- $\#\mathsf{Sols}(h,P,N) = \min(\mathsf{maxar}, \prod_{q\in N(\mathsf{PosVar}(P))} \binom{\#\mathsf{Terms}(h,q)}{B(P,N,q)} \cdot B(P,N,q)!)$
- $\#\mathsf{Terms}(h,q) = \min(\mathsf{maxar}, \sum_{(f(q_1,\ldots,q_m)\xrightarrow{c} q)\in\Delta, m>0} \#\mathsf{Sols}(h-1, c, \{1 \mapsto q_1, \ldots, m \mapsto q_m\}))$

Corollary 6. *Let* $A = \langle Q, \Sigma, F, \Delta\rangle$ *be a* $\mathtt{dcTACBB}^F_{\mathsf{nHe}}$ *having width* w. *Let* h *be a natural number. The configuration of* A *for height* $h+w$ *can be obtained from the configurations of* A *for heights* $h, h+1, \ldots, h+w-1$ *with time in* $\mathcal{O}((\mathsf{maxar}^2 \cdot |\Delta| + |Q|) \cdot (\log(\mathsf{maxar}) + \log(w)))$.

The previous result leads to the fact that, when we find the same w consecutive configurations starting at two different heights $h_1 < h_2$, the sequence of configurations is periodic with period $h_2 - h_1$ starting from h_1. Thus, deciding emptiness corresponds to check whether there exists an accepting state q satisfying $\#\mathsf{Terms}(h,q) \geq 1$ for $h < h_2$, and deciding finiteness corresponds to check whether there exists an accepting state q satisfying $\#\mathsf{Terms}(h,q) \geq 1$ for $h_1 \leq h < h_2$. The time complexity depends on Corollary 6 and the number of different possible groups of w consecutive configurations.

Lemma 11. *Let* $A = \langle Q, \Sigma, F, \Delta\rangle$ *be a* $\mathtt{dcTACBB}^F_{\mathsf{nHe}}$. *The number of different configurations of* A *considering all possible* h *is bounded by* $(\mathsf{maxar}+1)^{|Q|+|C_A|} \cdot (|C_A| \cdot \mathsf{maxar} + 1)$, *i.e., by* $(\mathsf{maxar}+1)^{|Q|+2\cdot\mathsf{maxar}\cdot|\Delta|} \cdot (2 \cdot \mathsf{maxar}^2 \cdot |\Delta| + 1)$.

Proof. Similar to Lemma 3, except that instead of dealing with truth values we have numbers in $\{0, \ldots, \mathsf{maxar}\}$, and that we need to take into account the number $(\mathsf{maxar}+1)^{|C_A|}$ of different $\#\mathsf{Sols}$.

Corollary 7. *Emptiness and finiteness of the language recognized by a* $\mathtt{dcTACBB}^F_{\mathsf{nHe}}$ $A = \langle Q, \Sigma, F, \Delta\rangle$ *having width* w *are decidable with time in* $2^{\mathcal{O}(w\cdot\log(\mathsf{maxar})\cdot(|Q|+\mathsf{maxar}\cdot|\Delta|))}$.

As a consequence of Lemma 7 we also obtain decidability for $\mathtt{TACBB}_{\mathsf{He}}$.

Corollary 8. *Emptiness and finiteness of the language recognized by a* $\mathtt{TACBB}_{\mathsf{He}}$ $A = \langle Q, \Sigma, F, \Delta\rangle$ *are decidable with time in* $2^{2^{\mathcal{O}(\log(|\Sigma|)+\mathsf{maxar}\cdot(|Q|+\mathsf{maxar}+\log(w))+\log(|A|))}}$, *where* w *is the width of* A.

As a final remark, we consider the simpler cases of $\mathtt{TACBB}_{\mathsf{he}}$ and $\mathtt{TACBB}_{\mathsf{H}}$. For the normalization A' of a $\mathtt{TACBB}_{\mathsf{he}}$ A, since the normalized constraints would only have operators of the form $\leq_1$, we could refine Lemma 11 as we did in Lemma 3 and ignore the number of different $\#\mathsf{Sols}$, thus obtaining $(\mathsf{maxar}+1)^{|Q'|} \cdot (2 \cdot \mathsf{maxar}^2 \cdot |\Delta'| + 1)$ as the number of possible configurations of A'. Moreover, for the sequence of configurations to become periodic it would suffice that two configurations coincided (since we look for identical groups of $w = 1$ consecutive configurations), and hence we would get the following result.

Corollary 9. *Let $A = \langle Q, \Sigma, F, \Delta \rangle$ be a $\mathtt{TACBB_{he}}$. Emptiness and finiteness of $\mathcal{L}(A)$ are decidable with time in $2^{\mathcal{O}(\log(\mathsf{maxar}) \cdot 2^{|Q|} + \log(|\Sigma|) + \mathsf{maxar} \cdot (|Q| + \mathsf{maxar}) + \log(|A|))}$.*

For the normalization A' of a $\mathtt{TACBB_H}$ A, it is not required to obtain a deterministic and complete automaton, and instead of partitions we deal with sets, as for $\mathtt{TACBB_h}$. Thus, a refined normalization process can preserve the same state set and generate the rule set Δ' combining ideas from Lemmas 1 and 7 with time in $2^{\mathcal{O}(\mathsf{maxar}^2 + \mathsf{maxar} \cdot \log(w) + \log(|A|))}$. Also, we do not need to count the number of terms for each state, just a truth value. Hence, the cost of computing a new configuration of A' (refinement of Corollary 6) would be in $\mathcal{O}((\mathsf{maxar}^2 \cdot |\Delta'| + |Q|) \cdot \log(w))$, and the number of configurations of A' (refinement of Lemma 11) would be bounded by $2^{|Q| + 2 \cdot \mathsf{maxar} \cdot |\Delta'|} \cdot (2 \cdot \mathsf{maxar} \cdot |\Delta'| + 1)$.

Corollary 10. *Let $A = \langle Q, \Sigma, F, \Delta \rangle$ be a $\mathtt{TACBB_H}$ having width w. Emptiness and finiteness of $\mathcal{L}(A)$ are decidable with time in $2^{2^{\mathcal{O}(\mathsf{maxar}^2 + \mathsf{maxar} \cdot \log(w) + \log(|A|))}}$.*

6 Conclusion

We have defined a new kind of automata with constraints between brothers, extending the current literature by allowing constraints that test the height of subterms. The obtained time complexity for deciding emptiness and finiteness is exponential for the simplest case $\mathtt{TACBB_h}$, and double exponential for $\mathtt{TACBB_H}$, $\mathtt{TACBB_{he}}$, and $\mathtt{TACBB_{He}}$. For $\mathtt{TACBB_{he}}$ and $\mathtt{TACBB_{He}}$ both problems are at least EXPTIME-hard due to the equality tests, but the precise hardness for our automata is unknown and deserves further analysis. It would also be interesting to study other extensions of the constraints. In particular, we have focused on constraints between brother positions, but emptiness and finiteness with arbitrary positions for the height constraints might also be decidable. Moreover, several classes of (dis)equality constraints with non-brother positions are known to be decidable, such as reduction automata [10] or $\mathtt{TA_{hom,\not\approx}}$ [15]. Extending those models with height constraints might preserve the decidability, and should also be considered.

References

1. Baader, F., Nipkow, T.: Term Rewriting and All That. Cambridge University Press, New York (1998)
2. Barguñó, L., Creus, C., Godoy, G., Jacquemard, F., Vacher, C.: Decidable classes of tree automata mixing local and global constraints modulo flat theories. Logical Methods in Computer Science 9(2) (2013)
3. Bogaert, B., Tison, S.: Equality and Disequality Constraints on Direct Subterms in Tree Automata. In: Finkel, A., Jantzen, M. (eds.) STACS 1992. LNCS, vol. 577, pp. 161–171. Springer, Heidelberg (1992)
4. Caron, A.-C., Seynhaeve, F., Tison, S., Tommasi, M.: Deciding the satisfiability of quantifier free formulae on one-step rewriting. In: Narendran, P., Rusinowitch, M. (eds.) RTA 1999. LNCS, vol. 1631, pp. 103–117. Springer, Heidelberg (1999)

5. Comon, H., Jacquemard, F.: Ground reducibility is EXPTIME-complete. Information and Computation 187(1), 123–153 (2003)
6. Comon, H., Cortier, V.: Tree automata with one memory, set constraints and cryptographic protocols. Theoretical Computer Science 331(1), 143–214 (2005)
7. Comon, H., Dauchet, M., Gilleron, R., Jacquemard, F., Löding, C., Lugiez, D., Tison, S., Tommasi, M.: Tree Automata Techniques and Applications (2007), `http://tata.gforge.inria.fr`
8. Comon-Lundh, H., Jacquemard, F., Perrin, N.: Visibly tree automata with memory and constraints. Logical Methods in Computer Science 4(2:8) (2008)
9. Creus, C., Gascón, A., Godoy, G., Ramos, L.: The HOM problem is EXPTIME-complete. In: Logic in Computer Science (LICS), pp. 255–264 (2012)
10. Dauchet, M., Caron, A.C., Coquidé, J.L.: Automata for reduction properties solving. Journal of Symbolic Computation 20(2), 215–233 (1995)
11. Doner, J.: Tree acceptors and some of their applications. Journal of Computer System Sciences 4, 406–451 (1970)
12. Filiot, E., Talbot, J., Tison, S.: Tree automata with global constraints. International Journal of Foundations of Computer Science 21(4), 571–596 (2010)
13. Gécseg, F., Steinby, M.: Tree Automata. Akadémiai Kiadó (1984)
14. Gécseg, F., Steinby, M.: Tree languages. In: Rozenberg, G., Salomaa, A. (eds.) Handbook of Formal Languages, vol. 3, pp. 1–68. Springer (1997)
15. Godoy, G., Giménez, O.: The HOM problem is decidable. Journal of the ACM 60(4), 23 (2013)
16. Jacquemard, F., Rusinowitch, M., Vigneron, L.: Tree automata with equality constraints modulo equational theories. Journal of Logic and Algebraic Programming 75(2), 182–208 (2008)
17. Mezei, J., Wright, J.B.: Algebraic automata and context-free sets. Information and Control 11, 3–29 (1967)
18. Mongy, J.: Transformation de noyaux reconnaissables d'arbres. Forêts RATEG. Ph.D. thesis, Laboratoire d'Informatique Fondamentale de Lille, Université des Sciences et Technologies de Lille, Villeneuve d'Ascq, France (1981)
19. Murata, M., Lee, D., Mani, M., Kawaguchi, K.: Taxonomy of XML schema languages using formal language theory. ACM Transactions of Internet Technologies 5(4), 660–704 (2005)
20. Treinen, R.: Predicate logic and tree automata with tests. In: Tiuryn, J. (ed.) FOSSACS 2000. LNCS, vol. 1784, pp. 329–343. Springer, Heidelberg (2000)

A Coinductive Confluence Proof for Infinitary Lambda-Calculus*

Łukasz Czajka

Institute of Informatics, University of Warsaw
Banacha 2, 02-097 Warszawa, Poland
lukaszcz@mimuw.edu.pl

Abstract. We give a coinductive proof of confluence, up to equivalence of root-active subterms, of infinitary lambda-calculus. We also show confluence of Böhm reduction (with respect to root-active terms) in infinitary lambda-calculus. In contrast to previous proofs, our proof makes heavy use of coinduction and does not employ the notion of descendants.

1 Introduction

Infinitary lambda-calculus is a generalization of lambda-calculus that allows infinite lambda-terms and transfinite reductions. This enables the consideration of "limits" of terms under infinite reduction sequences. For instance, for a term $M \equiv (\lambda mx.mm)(\lambda mx.mm)$ we have

$$M \to_\beta \lambda x.M \to_\beta \lambda x.\lambda x.M \to_\beta \lambda x.\lambda x.\lambda x.M \to_\beta \dots$$

Intuitively, the "value" of M is an infinite term L satisfying $L \equiv \lambda x.L$, where by $\equiv$ we denote identity of terms. In fact, L is the normal form of M in infinitary lambda-calculus.

In [6] it is shown that infinitary reductions may be defined coinductively. The standard non-coinductive definition makes explicit mention of ordinals and limits in a certain metric space [11,14,2]. Arguably, a coinductive approach is better suited to formalization in a proof-assistant.

We prove confluence of infinitary reduction up to equivalence of root-active subterms, and confluence of infinitary Böhm reduction w.r.t. root-active terms. These results have already been obtained in [11] with proofs involving the notion of descendants. However, our proof is coinductive. We show that the theory of infinitary lambda-calculus, to the extent studied here, may be entirely based on coinductive definitions and proofs, without even mentioning ordinals or metric convergence.

1.1 Related Work

Infinitary lambda-calculus was introduced in [11,10]. All results of this paper were already obtained in [11], by a different proof method. See also [14,2,5] for an

* Partly supported by NCN grant 2012/07/N/ST6/03398.

G. Dowek (ed.): RTA-TLCA 2014, LNCS 8560, pp. 164–178, 2014.

overview of various results in infinitary lambda-calculus and infinitary rewriting. The coinductive definition of infinitary reductions was introduced in [6].

Our proof differs from the proof in [11]. Instead of using the notion of descendants, we rely on coinduction. Nonetheless, the overall structure of the whole proof and proofs of some lemmas are analogous to [11].

A method of coinductive confluence proofs somewhat similar to ours was given by Joachimski in [9]. However, Joachimski's notion of reduction does not correspond to strongly convergent reductions. Essentially, it allows for infinitely many parallel contractions in one step, but only finitely many reduction steps.

There are three well-known variants of infinitary lambda-calculus: the Λ^{111}, Λ^{001} and Λ^{101} calculi [2,5,11,10]. The superscripts 111, 001, 101 indicate the depth measure used: abc means that we shall add $a/b/c$ to the depth when going down/left/right in the tree of the lambda-term [11, Definition 6]. We are concerned only with Λ^{111}. In this calculus, after addition of appropriate $\bot$-rules, every term has its Berarducci tree [10,3] as the normal form. In Λ^{001} and Λ^{101}, the normal forms are, respectively, Böhm trees and Levy-Longo trees [11,10]. With the addition of infinite eta reductions it is possible to also capture eta-Böhm trees as normal forms [13].

2 Preliminaries

2.1 Coinduction

In this section we give an introduction to coinduction, to the extent necessary to understand the proofs in this paper. Because of space limits and for the sake of readability we only present several examples from which we hope the general method should be clear. For more background on coinduction see e.g. [8,12], or see [4] for a practical introduction to the usage of coinduction in the Coq proof assistant. Our use of coinduction does not correspond exactly to the coinduction principle of Coq [7] and some of our proofs cannot be directly formalized in Coq. They could probably be formalized in recent versions of Agda extended with copatterns and sized types [1]. Also, we do not directly employ the usual coinduction principle from [8]. The correctness criterion for our proofs is that they may all be interpreted in the way outlined below.

We do not attempt here to formulate a general coinduction principle or provide a formal system in which our proofs could be easily formalized. This is an interesting question by itself, but a seperate one from simply ensuring correctness of the proofs in each particular case, by indicating how to interpret them in ordinary set theory without using coinduction. In other words, our use of coinduction may be seen as a way of leaving implicit the tedious, annoying and purely technical details which would be necessary if the proofs were to be made inductive.

Consider the following definition by a grammar of a set $\mathbb{T}$ of terms, where V is an infinite set of variables.

$$\mathbb{T} ::= V \parallel A(\mathbb{T}) \parallel B(\mathbb{T},\mathbb{T})$$

Conventionally, this definition is interpreted inductively: the set $\mathbb{T}$ consists of all *finite* terms built up from variables and the constructors A and B. When interpreted coinductively, both *finite and infinite* terms are allowed. So then e.g. a term A^ω satisfying $A^\omega \equiv A(A^\omega)$ also belongs to $\mathbb{T}$. Formally, under the coinductive interpretation $\mathbb{T}$ is a final coalgebra of an appropriate endofunctor [8]. We will not get into details here. One may think of $\mathbb{T}$ as the set of all possibly infinite labelled trees with labels specified by the grammar.

A definition of a function f with codomain $\mathbb{T}$ is by guarded corecursion if each (co)recursive call of f occurs directly inside a constructor for $\mathbb{T}$. Such a definition determines a well-defined function and with the introduction of some technicalities it may be reformulated as an ordinary inductive definition (by induction on the length of positions in a term). An example of a function defined by guarded corecursion is substitution: a function taking two terms and a variable.

$$\begin{array}{llr} x[t/x] \equiv t & & A(s)[t/x] \equiv A(s[t/x]) \\ y[t/x] \equiv y & \text{if } y \neq x & B(s_1, s_2)[t/x] \equiv B(s_1[t/x], s_2[t/x]) \end{array}$$

Now consider a relation defined by the following derivation rules.

$$\frac{}{x \to_0 x} \qquad \frac{t \to_0 t'}{A(t) \to_0 B(t', t')} \qquad \frac{t \to_0 t'}{A(t) \to_0 A(t')} \qquad \frac{s \to_0 s' \quad t \to_0 t'}{B(s,t) \to_0 B(s', t')}$$

When interpreted coinductively, in addition to finite derivations we also allow infinite ones. More formally, one may interpret the relation $\to_0$ as the greatest fixpoint of a function $F : \mathcal{P}(\mathbb{T} \times \mathbb{T}) \to \mathcal{P}(\mathbb{T} \times \mathbb{T})$ defined as follows.

$$\begin{aligned} F(R) = \{\langle t_1, t_2 \rangle \mid\ & (t_1 \equiv t_2 \equiv x) \vee \\ & \exists t, t' \, (t_1 \equiv A(t) \wedge t_2 \equiv B(t', t') \wedge R(t, t')) \vee \\ & \exists t, t' \, (t_1 \equiv A(t) \wedge t_2 \equiv A(t') \wedge R(t, t')) \vee \\ & \exists s, t, s', t' \, (t_1 \equiv B(s,t) \wedge t_2 \equiv B(s', t') \wedge R(s, s') \wedge R(t, t'))\} \end{aligned}$$

The function F is monotone, so by the Knaster-Tarski theorem its greatest fixpoint exists and may be obtained in the following way. By transfinite induction we define: $R_0 = \mathbb{T} \times \mathbb{T}$, $R_{\alpha+1} = F(R_\alpha)$, and $R_\lambda = \bigcap_{\alpha < \lambda} R_\alpha$ for λ a limit ordinal. Then there exists an ordinal ζ such that $R_\zeta = \to_0$ is the greatest fixpoint of F.

All coinductive proofs in this paper show statements of one of two forms: $\forall \overline{x} \, (\phi(\overline{x}) \to R(f(\overline{x})))$ or $\forall \overline{x} \, (\phi(\overline{x}) \to \exists y \, (R(\overline{x}, y) \wedge S(\overline{x}, y)))$, where R and S are relations defined coinductively by some derivation rules, $\overline{x}$ ranges over tuples of terms, y ranges over terms, and f is a function from tuples to tuples of terms.

In coinductive proofs we appeal to the "coinductive hypothesis", which at first sight may seem like assuming what we are supposed to prove. The trick is that we are allowed to use the result of an application of the coinductive hypothesis only in certain ways: we *have to* use it *directly* as a premise of some derivation rule, and we *must not* manipulate the resulting derivation in any other way. In contrast, in an inductive proof, the result of an application of the inductive hypothesis may be used in an arbitrary way, but there is a restriction on the parameters of the hypothesis – they should be smaller in an appropriate sense. We shall now give an example of a coinductive proof.

Example 1. We show by coinduction: if $t \in \mathbb{T}$ then $t \to_0 t$. If $t \equiv x$ then $t \to_0 t$ holds by the first rule. If $t \equiv A(s)$ then $s \to_0 s$ by the coinductive hypothesis. So $t \to_0 t$ by the third rule. If $t \equiv B(t_1, t_2)$ then $t_1 \to_0 t_1$ and $t_2 \to_0 t_2$ by the coinductive hypothesis. Hence $t \to_0 t$ by the fourth rule.

Formally, a coinductive proof of $\forall \overline{x} \, (\phi(\overline{x}) \to R(f(\overline{x})))$ may be interpreted as a proof by transfinite induction on an ordinal α of $\forall \overline{x} \, (\phi(\overline{x}) \to R_\alpha(f(\overline{x})))$ for $\alpha \leq \zeta$. The cases $\alpha = 0$ and α a limit ordinal are trivial and left implicit. A coinductive proof may be read as a proof of the inductive step for α a successor ordinal, where "coinductive hypothesis" means "inductive hypothesis" and the ordinal indices are left implicit.

In a few proofs we actually violate the above interpretation slightly by applying derivation rules to the result of an application of the coinductive hypothesis more than once. However, this is easily seen to be correct by using $F'(R) = \bigcup_{n \in \mathbb{N}_+} F^n(R)$ instead of F. We have $F^n(R_\alpha) = R_{\alpha+n} \subseteq R_{\alpha+1} = F(R_\alpha)$ for $n \in \mathbb{N}_+$, which implies $F'(R_\alpha) = F(R_\alpha)$. Hence, F' has the same greatest fixpoint as F.

The formal interpretation of a coinductive proof of a statement of the form $\forall \overline{x} \, (\phi(\overline{x}) \to \exists y \, (R(\overline{x}, y) \wedge S(\overline{x}, y)))$ is slightly more involved. With the following example, we indicate how to reduce such a proof to coinductive proofs of statements of the form $\forall \overline{x} \, (\phi(\overline{x}) \to R(f(\overline{x})))$.

Example 2. We show: if $t \to_0 t_1$ and $t \to_0 t_2$ then there exists t_3 such that $t_1 \to_0 t_3$ and $t_2 \to_0 t_3$. Coinduction with case analysis on $t \to_0 t_1$. For instance, assume $t \equiv A(t')$ and $t_1 \equiv B(t_1', t_1')$ with $t' \to_0 t_1'$. There are two cases.

1. $t_2 \equiv B(t_2', t_2')$ with $t' \to_0 t_2'$. By the coinductive hypothesis there is t_3' such that $t_1' \to_0 t_3'$ and $t_2' \to_0 t_3'$. Thus $t_1 \equiv B(t_1', t_1') \to_0 B(t_3', t_3')$ and $t_2 \equiv B(t_2', t_2') \to_0 B(t_3', t_3')$, by the last rule. Hence, we may take $t_3 \equiv B(t_3', t_3')$.
2. $t_2 \equiv A(t_2')$ with $t' \to_0 t_2'$. By the coinductive hypothesis there is t_3' such that $t_1' \to_0 t_3'$ and $t_2' \to_0 t_3'$. Hence $t_1 \equiv B(t_1', t_1') \to_0 B(t_3', t_3')$ by the last rule, and $t_2 \equiv A(t_2') \to_0 B(t_3', t_3')$ by the second rule. Thus we may take $t_3 \equiv B(t_3', t_3')$.

Formally, the above proof may be interpreted as actually showing: if $t \to_0 t_1$ and $t \to_0 t_2$ then $t_1 \downarrow t_2$, for a relation $\downarrow$ defined coinductively as follows.

$$\frac{}{x \downarrow x} \qquad \frac{t \downarrow s}{A(t) \downarrow B(s,s)} \qquad \frac{t \downarrow s}{B(t,t) \downarrow A(s)} \qquad \frac{t \downarrow s}{A(t) \downarrow A(s)} \qquad \frac{s \downarrow s' \quad t \downarrow t'}{B(s,t) \downarrow B(s',t')}$$

The rules for $\downarrow$ come from appropriate pairs of rules for $\to_0$. By guarded corecursion we define a function h which given two terms t_1, t_2 such that $t_1 \downarrow t_2$ returns a term s such that $t_1 \to_0 s$ and $t_2 \to_0 s$.

$$\begin{aligned} h(x,x) &= x \\ h(A(t), B(s,s)) &= B(h(t,s), h(t,s)) \\ h(B(t,t), A(s)) &= B(h(t,s), h(t,s)) \\ h(A(t), A(s)) &= A(h(t,s)) \\ h(B(s,t), B(s',t')) &= B(h(s,s'), h(t,t')) \end{aligned}$$

It may be shown by coinduction that if $t_1 \downarrow t_2$ then $t_1 \to_0 h(t_1,t_2)$ and $t_2 \to_0 h(t_1,t_2)$. We give a proof for $t_1 \to_0 h(t_1,t_2)$. If $t_1 \equiv x$ then $h(t_1,t_2) \equiv x$ and we are done. If $t_1 \equiv B(s,t)$ and $t_2 \equiv B(s',t')$ with $s \downarrow s'$ and $t \downarrow t'$ then by the coinductive hypothesis $s \to_0 h(s,s')$ and $t \to_0 h(t,t')$. So $t_1 \equiv B(s,t) \to_0 B(h(s,s'),h(t,t')) \equiv h(t_1,t_2)$. The other cases are analogous.

2.2 Infinitary Lambda-Calculus

In this section we define the syntax of infinitary lambda-calculus. We also provide definitions of various notions of reduction and other relations in infinitary lambda-calculus. Our definitions are coinductive. For standard introduction to infinitary lambda-calculus see e.g. [14,11].

Definition 1. The set of Λ^∞-terms is defined coinductively:

$$\Lambda^\infty ::= V \parallel \lambda V.\Lambda^\infty \parallel \Lambda^\infty \Lambda^\infty$$

where V is an infinite set of variables.

Capture-avoiding substitution is defined by guarded corecursion.

$$x[t/x] \equiv t \qquad (s_1 s_2)[t/x] \equiv s_1[t/x] s_2[t/x]$$
$$y[t/x] \equiv y \quad \text{when } x \neq y \qquad (\lambda y.s)[t/x] \equiv \lambda y.s[t/x] \quad \text{when } y \notin FV(t)$$

The relation $\to_\beta$ of *β-contraction* is defined inductively by the following rules.

$$\frac{}{(\lambda x.s)t \to_\beta s[t/x]} \qquad \frac{s \to_\beta s'}{st \to_\beta s't} \qquad \frac{t \to_\beta t'}{st \to_\beta st'} \qquad \frac{s \to_\beta s'}{\lambda x.s \to_\beta \lambda x.s'}$$

The relation $\to_\beta^*$ of *β-reduction* is the transitive-reflexive closure of $\to_\beta$.

The relation $\to_\beta^\infty$ of *infinitary β-reduction* is defined coinductively.

$$\frac{s \to_\beta^* x}{s \to_\beta^\infty x} \qquad \frac{s \to_\beta^* t_1 t_2 \quad t_1 \to_\beta^\infty t_1' \quad t_2 \to_\beta^\infty t_2'}{s \to_\beta^\infty t_1' t_2'} \qquad \frac{s \to_\beta^* \lambda x.r \quad r \to_\beta^\infty r'}{s \to_\beta^\infty \lambda x.r'}$$

We will disregard the usual problems with α-conversion. In the infinitary setting it presents some additional, but purely technical difficulties [11,6].

The idea with the definition of $\to_\beta^\infty$ is that the depth at which a redex is contracted should tend to infinity. This is achieved by defining $\to_\beta^\infty$ in such a way that always after finitely many reduction steps the subsequent contractions may be performed only under a constructor. So the depth of the contracted redex always ultimately increases. In [6] it is shown that the above definition of $\to_\beta^\infty$ coincides with the standard definition based on strongly convergent reductions.

Definition 2. Let $\bot$ be a constant, i.e. a variable which is assumed to never occur bound. A Λ^∞-term t is *root-stable* if either $t \equiv x$ with $x \not\equiv \bot$, or $t \equiv \lambda x.t'$, or $t \equiv t_1 t_2$ and there does not exist s such that $t_1 \to_\beta^\infty \lambda x.s$. A Λ^∞-term t is *root-active* if there does not exist a root-stable s such that $t \to_\beta^\infty s$.

Given $t, s \in \Lambda^\infty$, the relation $t \sim s$ is defined by coinduction.

$$\frac{t, s \text{ are root-active}}{t \sim s} \qquad \frac{}{x \sim x} \qquad \frac{t \sim s}{\lambda x.t \sim \lambda x.s} \qquad \frac{t_1 \sim s_1 \quad t_2 \sim s_2}{t_1 t_2 \sim s_1 s_2}$$

We finish this section with several lemmas concerning the introduced notions. The first three lemmas have essentially been shown in [6, Lemma 4.3-4.5].

Lemma 1. *If* $s \to_\beta^\infty s'$ *and* $t \to_\beta^\infty t'$ *then* $s[t/x] \to_\beta^\infty s'[t'/x]$.

Lemma 2. *If* $t_1 \to_\beta^\infty t_2 \to_\beta t_3$ *then* $t_1 \to_\beta^\infty t_3$.

Lemma 3. *If* $t_1 \to_\beta^\infty t_2 \to_\beta^\infty t_3$ *then* $t_1 \to_\beta^\infty t_3$.

Lemma 4. *If* s *is root-active and* $s \to_\beta^\infty t$*, then* t *is root-active.*

Proof. If $t \to_\beta^\infty t'$ for some root-stable t', then also $s \to_\beta^\infty t'$ by Lemma 3.

The following lemma was first shown in [11, Lemma 43] by a different proof.

Lemma 5. *If* $t_1, t_2 \in \Lambda^\infty$ *and* t_1 *is root-active, then so is* $t_1[t_2/x]$.

Proof. We write $s \succ_x s'$ if x is not bound in s', i.e. s' does not contain subterms of the form $\lambda x.u$, and s' may be obtained from s by changing some arbitrary subterms in s into some terms having the form $xu_1 \dots u_n$. It is easy to show by induction that

(a) if $t \to_\beta^* s$ and $t \succ_x t'$, then there exists s' such that $t' \to_\beta^* s'$ and $s \succ_x s'$,
(b) if $t' \to_\beta^* s'$ and $t \succ_x t'$, then there exists s such that $t \to_\beta^* s$ and $s \succ_x s'$.

By coinduction we show

(c) if $t \to_\beta^\infty s$ and $t \succ_x t'$, then there exists s' such that $t' \to_\beta^\infty s'$ and $s \succ_x s'$,
(d) if $t' \to_\beta^\infty s'$ and $t \succ_x t'$, then there exists s such that $t \to_\beta^\infty s$ and $s \succ_x s'$.

We only give the proof for (c), since the proof for (d) is analogous. There are three cases.

- $t \to_\beta^* s \equiv y$. Then the claim follows directly from (a).
- $t \to_\beta^* t_1t_2$, $s \equiv s_1s_2$ and $t_i \to_\beta^\infty s_i$. By (a) there is u such that $t' \to_\beta^* u$ and $t_1t_2 \succ_x u$. If u has the form $xu_1 \dots u_n$ then we may take $s' \equiv u$. Otherwise $u \equiv u_1u_2$ with $t_i \succ_x u_i$. By the coinductive hypothesis there are s_1', s_2' such that $u_i \to_\beta^\infty s_i'$ and $s_i \succ_x s_i'$. Thus we may take $s' \equiv s_1's_2'$.
- $t \to_\beta^* \lambda y.u$, $s \equiv \lambda y.u'$ and $u \to_\beta^\infty u'$. By (a) there is w such that $t' \to_\beta^* w$ and $\lambda y.u \succ_x w$. If w has the form $xu_1 \dots u_n$ then we may take $s' \equiv w$. Otherwise $w \equiv \lambda y.w_0$ with $u \succ_x w_0$. By the coinductive hypothesis there is w_1 such that $w_0 \to_\beta^\infty w_1$ and $u' \succ_x w_1$. So we may take $s' \equiv \lambda y.w_1$.

Now we show

(e) if $s \succ_x s'$ and s is root-stable, then so is s'.

So suppose $s \succ_x s'$ and s is root-stable. If s' has the form $xu_1 \dots u_n$ then it is obviously root-stable. Otherwise, $s' \equiv y$, $s' \equiv \lambda y.s''$ or $s' \equiv s_1's_2'$ with $s \equiv s_1s_2$ and $s_i \succ_x s_i'$. In the first two cases s' is root-stable. So assume $s' \equiv s_1's_2'$, $s \equiv s_1s_2$ and $s_i \succ_x s_i'$. If s' is not root-stable, then $s_1' \to_\beta^\infty \lambda z.u'$. But then by ($d$) there is w such that $s_1 \to_\beta^\infty w$ and $w \succ_x \lambda z.u'$. So w must have the form $\lambda z.u$. Contradiction.

Finally, suppose $t_1[t_2/x]$ is not root-active. Then $t_1[t_2/x] \to_\beta^\infty s$ with s root-stable. Without loss of generality $t_1[t_2/x] \succ_x t_1$. Hence by (c) there is s' such that $t_1 \to_\beta^\infty s'$ and $s \succ_x s'$. By (e) we conclude that s' is root-stable. This means that t_1 is not root-active.

3 Confluence Up to Equivalence of Root-Active Subterms

In this section we show that the relation $\to_{\beta}^{\infty}$ is confluent up to $\sim$. More precisely, we prove the following theorem.

Theorem 1. *If $t \sim t'$, $t \to_{\beta}^{\infty} s$ and $t' \to_{\beta}^{\infty} s'$, then there exist r, r' such that $s \to_{\beta}^{\infty} r$, $s' \to_{\beta}^{\infty} r'$ and $r \sim r'$.*

The general proof strategy is similar to that in [11] and is illustrated in Fig. 1. We introduce an ϵ-calculus – a modified infinitary lambda-calculus. We show confluence of infinitary reduction in the ϵ-calculus and then translate this result into confluence of $\to_{\beta}^{\infty}$ up to $\sim$.

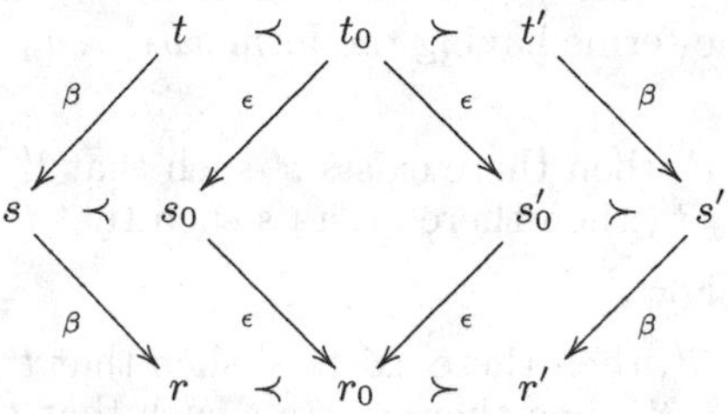

Fig. 1. Confluence proof for $\to_{\beta}^{\infty}$ up to $\sim$

Definition 3. The set of Λ^{ϵ}-terms is defined by coinduction.

$$\Lambda^{\epsilon} ::= V \parallel \lambda V.\Lambda^{\epsilon} \parallel \Lambda^{\epsilon}\Lambda^{\epsilon} \parallel \epsilon(\Lambda^{\epsilon})$$

We say that $t \in \Lambda^{\epsilon}$ *starts with* ϵ if $t \equiv \epsilon(t')$ for some t'. The relation $\to_{\epsilon}$ of ϵ-*contraction* is defined as the compatible closure of the reduction rules

$$\epsilon^{n}(\lambda x.s)t \to_{\epsilon} \epsilon(s[t/x])$$

for $n \in \mathbb{N}$. The relation $\to_{\epsilon}^{*}$ of ϵ-*reduction* is the transitive-reflexive closure of $\to_{\epsilon}$.

The relation $\to_1$ is defined coinductively.

$$\frac{}{x \to_1 x} \qquad \frac{s \to_1 s'}{\lambda x.s \to_1 \lambda x.s'} \qquad \frac{s \to_1 s' \quad t \to_1 t'}{st \to_1 s't'} \qquad \frac{t \to_1 t'}{\epsilon(t) \to_1 \epsilon(t')}$$

$$\frac{t_1[t_2/x] \to_1 t'}{\epsilon^{n}(\lambda x.t_1)t_2 \to_1 \epsilon(t')}$$

The relation $\to_{\epsilon}^{\infty}$ of *infinitary* ϵ-*reduction* is defined coinductively.

$$\frac{}{x \to_{\epsilon}^{\infty} x} \qquad \frac{s \to_{\epsilon}^{\infty} s'}{\lambda x.s \to_{\epsilon}^{\infty} \lambda x.s'} \qquad \frac{s \to_{\epsilon}^{\infty} s' \quad t \to_{\epsilon}^{\infty} t'}{st \to_{\epsilon}^{\infty} s't'} \qquad \frac{s \to_{1}^{*} \epsilon(t) \quad t \to_{\epsilon}^{\infty} t'}{s \to_{\epsilon}^{\infty} \epsilon(t')}$$

For $s \in \Lambda^\epsilon$ and $s' \in \Lambda^\infty$, the relation $s \succ s'$ is defined coinductively

$$\frac{}{\epsilon^n(x) \succ x} \qquad \frac{s' \text{ is root-active}}{\epsilon^\omega \succ s'} \qquad \frac{s \succ s'}{\epsilon^n(\lambda x.s) \succ \lambda x.s'} \qquad \frac{s \succ s' \quad t \succ t'}{\epsilon^n(st) \succ s't'}$$

where $n \in \mathbb{N}$. If $s \succ s'$ then s' is an *erasure* of s.

The purpose of our ϵ-calculus is similar to the ϵ-calculus in [11]. To make a direct coinductive confluence proof feasible, each contraction must produce at least one new constructor. In [11] the ϵ-contraction rule is $\epsilon^n(\lambda x.s)t \to \epsilon^{n+2}(s[t/x])$, so it additionally does not decrease the depth of other redexes. This is not necessary for a coinductive proof. The relation $\to_1$ is analogous to a development, but it also allows contracting some redexes which were not present in the original term. The difference is for purely technical reasons. If we used a development, then in the proof of Lemma 8 we would have to apply Lemma 7 to derivations obtained from the coinductive hypothesis. Hence, the proof would not conform to the interpretation from Sect. 2.1 which would make its correctness non-obvious.

Our first aim is to show that $\to_\epsilon^\infty$ has the Church-Rosser property. For this we need several lemmas.

Lemma 6. *Let $t_1, t_2, t_3 \in \Lambda^\epsilon$ and $y \notin FV(t_3)$. Then:*

$$t_1[t_2/y][t_3/x] \equiv t_1[t_3/x][(t_2[t_3/x])/y].$$

Proof. By coinduction with case analysis on t_1. See also [14, Chapter 12].

Lemma 7. *If $s \to_1 s'$ and $t \to_1 t'$ then $s[t/x] \to_1 s'[t'/x]$.*

Proof. Coinduction with case analysis on $s \to_1 s'$, using Lemma 6.

Lemma 8. *If $t \to_1 t_1$ and $t \to_1 t_2$ then there exists t_3 such that $t_1 \to_1 t_3$ and $t_2 \to_1 t_3$.*

Proof. By coinduction. We have the following cases.

1. $t \equiv t_1 \equiv x$. Then we must also have $t_2 \equiv x$ and we may take $t_3 \equiv x$.
2. $t \equiv \lambda x.t'$ and $t_1 \equiv \lambda x.t_1'$ with $t' \to_1 t_1'$. Then $t_2 \equiv \lambda x.t_2'$ with $t' \to_1 t_2'$. By the coinductive hypothesis, there is t_3' with $t_1' \to_1 t_3'$ and $t_2' \to_1 t_3'$. Thus take $t_3 \equiv \lambda x.t_3'$.
3. $t \equiv s_1 s_2$ and $t_1 \equiv u_1 u_2$ with $s_1 \to_1 u_1$ and $s_2 \to u_2$. Then one of the following holds.
   (a) $t_2 \equiv r_1 r_2$ with $s_1 \to_1 r_1$ and $s_2 \to_1 r_2$. By the coinductive hypothesis there are v_1, v_2 with $r_1 \to_1 v_1$, $u_1 \to_1 v_1$, $r_2 \to_1 v_2$ and $u_2 \to_1 v_2$. So $t_1 \equiv u_1 u_2 \to_1 v_1 v_2$ and $t_2 \equiv r_1 r_2 \to_1 v_1 v_2$. Thus take $t_3 \equiv v_1 v_2$.
   (b) $t_2 \equiv \epsilon(t_2')$ and $s_1 \equiv \epsilon^n(\lambda x.s_1')$ with $s_1'[s_2/x] \to_1 t_2'$. It follows directly from the definition of $\to_1$ that $u_1 \equiv \epsilon^n(\lambda x.u_1')$ with $s_1' \to_1 u_1'$. By Lemma 7, $s_1'[s_2/x] \to_1 u_1'[u_2/x]$. By the coinductive hypothesis there exists t_3' such that $u_1'[u_2/x] \to_1 t_3'$ and $t_2' \to_1 t_3'$. Hence $t_1 \equiv u_1 u_2 \equiv \epsilon^n(\lambda x.u_1')u_2 \to_1 \epsilon(t_3')$ and $t_2 \equiv \epsilon(t_2') \to_1 \epsilon(t_3')$. Thus take $t_3 \equiv \epsilon(t_3')$.

4. $t \equiv \epsilon(t')$ and $t_1 \equiv \epsilon(t_1')$ with $t' \to_1 t_1'$. Then $t_2 \equiv \epsilon(t_2')$ with $t' \to_1 t_2'$. By the coinductive hypothesis there exists t_3' such that $t_1' \to_1 t_3'$ and $t_2' \to_1 t_3'$. Hence we may take $t_3 \equiv \epsilon(t_3')$.
5. $t \equiv \epsilon^n(\lambda x.s)r$ and $t_1 \equiv \epsilon(t_1')$ with $s[r/x] \to_1 t_1'$. There are two possibilities.
   (a) $t_2 \equiv u_1 u_2$ with $\epsilon^n(\lambda x.s) \to_1 u_1$ and $r \to_1 u_2$. Then the proof is analogous to case 3(b).
   (b) $t_2 \equiv \epsilon(t_2')$ with $s[r/x] \to_1 t_2'$. By the coinductive hypothesis there exists t_3' such that $t_1' \to_1 t_3'$ and $t_2' \to_1 t_3'$. Hence we may take $t_3 \equiv \epsilon(t_3')$.

Lemma 9. *If $t_1 \to_1 t_2 \to_\epsilon^\infty t_3$ then $t_1 \to_\epsilon^\infty t_3$.*

Proof. Coinduction with case analysis on $t_2 \to_\epsilon^\infty t_3$.

Lemma 10. *If $\epsilon(t) \to_\epsilon^\infty s$ then $s \equiv \epsilon(s')$ with $t \to_\epsilon^\infty s'$.*

Proof. It follows directly from the definition of $\to_\epsilon^\infty$ that $s \equiv \epsilon(s')$ with $\epsilon(t) \to_1^* \epsilon(t')$ and $t' \to_\epsilon^\infty s'$. From the definition of $\to_1$ it follows that $t \to_1^* t'$. Thus $t \to_\epsilon^\infty s'$ by repeated application of Lemma 9.

Lemma 11. *If $s \to_\epsilon^\infty s'$ and $t \to_\epsilon^\infty t'$ then $s[t/x] \to_\epsilon^\infty s'[t'/x]$.*

Proof. Coinduction with case analysis on $s \to_\epsilon^\infty s'$, using that $s \to_1^* t$ implies $s[u/x] \to_1^* t[u/x]$, which follows from Lemma 7.

Lemma 12. *If $t \to_\epsilon^\infty t_1$ and $t \to_1 t_2$ then there exists t_3 such that $t_1 \to_1 t_3$ and $t_2 \to_\epsilon^\infty t_3$.*

Proof. Coinduction, analysing $t \to_\epsilon^\infty t_1$. There are two interesting cases.

1. $t \equiv \epsilon^n(\lambda x.s)r$, $t_1 \equiv u_1 u_2$, $t_2 \equiv \epsilon(t_2')$ with $\epsilon^n(\lambda x.s) \to_\epsilon^\infty u_1$, $r \to_\epsilon^\infty u_2$ and $s[r/x] \to_1 t_2'$. Because $\epsilon^n(\lambda x.s) \to_\epsilon^\infty u_1$, it follows from Lemma 10 and the definition of $\to_\epsilon^\infty$ that $u_1 \equiv \epsilon^n(\lambda x.u_1')$ with $s \to_\epsilon^\infty u_1'$. Hence $s[r/x] \to_\epsilon^\infty u_1'[u_2/x]$ by Lemma 11. By the coinductive hypothesis there exists t_3' such that $u_1'[u_2/x] \to_1 t_3'$ and $t_2' \to_\epsilon^\infty t_3'$. Hence $t_2 \equiv \epsilon(t_2') \to_\epsilon^\infty \epsilon(t_3')$ and $t_1 \equiv u_1 u_2 \equiv \epsilon^n(\lambda x.u_1')u_2 \to_1 \epsilon(t_3')$. Thus we may take $t_3 \equiv \epsilon(t_3')$.
2. $t_1 \equiv \epsilon(t_1')$ with $t \to_1^* \epsilon(s)$ and $s \to_\epsilon^\infty t_1'$. Since $t \to_1^* \epsilon(s)$ and $t \to_1 t_2$, it follows from Lemma 8 by an easy diagram chase that there exists s' such that $t_2 \to_1^* \epsilon(s')$ and $\epsilon(s) \to_1 \epsilon(s')$. From the definition of $\to_1$ we obtain $s \to_1 s'$. By the coinductive hypothesis there exists t' such that $s' \to_\epsilon^\infty t'$ and $t_1' \to_1 t'$. Hence $t_2 \to_\epsilon^\infty \epsilon(t')$, because $t_2 \to_1^* \epsilon(s')$ and $s' \to_\epsilon^\infty t'$. Also $t_1 \equiv \epsilon(t_1') \to_1 \epsilon(t')$. We may thus take $t_3 \equiv \epsilon(t')$.

Lemma 13. *If $t \to_\epsilon^\infty t_1$ and $t \to_\epsilon^\infty t_2$ then there exists t_3 such that $t_1 \to_\epsilon^\infty t_3$ and $t_2 \to_\epsilon^\infty t_3$.*

Proof. By coinduction. There is one non-trivial case, when e.g. $t_2 \equiv \epsilon(t_2')$ with $t \to_1^* \epsilon(s)$ and $s \to_\epsilon^\infty t_2'$. By repeated application of Lemma 12 we obtain u with $t_1 \to_1^* u$ and $\epsilon(s) \to_\epsilon^\infty u$. By Lemma 10, $u \equiv \epsilon(s')$ with $s \to_\epsilon^\infty s'$. By the coinductive hypothesis there is t_3' with $t_2' \to_\epsilon^\infty t_3'$ and $s' \to_\epsilon^\infty t_3'$. So $t_1 \to^\infty \epsilon(t_3')$, since $t_1 \to_1^* \epsilon(s')$ and $s' \to_\epsilon^\infty t_3'$. Also $t_2 \equiv \epsilon(t_2') \to_\epsilon^\infty \epsilon(t_3')$. So take $t_3 \equiv \epsilon(t_3')$.

The above lemma states the Church-Rosser property of $\to_\epsilon^\infty$. Now we need to translate this result into confluence of $\to_\beta^\infty$ up $\sim$. For this purpose we need to be able to transform infinitary ϵ-reductions on Λ^ϵ-terms into infinitary β-reductions on their erasures, and vice versa. This is achieved in Lemmas 16, 21.

Lemma 14. *If $t_1 \succ u_1$ and $t_2 \succ u_2$ then $t_1[t_2/x] \succ u_1[u_2/x]$.*

Proof. Coinduction with case analysis on $t_1 \succ u_1$, using Lemma 5.

Lemma 15. *If $t_1 \succ u_1 \to_\beta u_2$ then there exists t_2 such that $t_1 \to_\epsilon t_2 \succ u_2$. Moreover, if the contraction $u_1 \to_\beta u_2$ occurs at the root, then t_2 starts with ϵ.*

Proof. Induction on $u_1 \to_\beta u_2$. If $t_1 \equiv \epsilon^\omega$ then u_1 is root-active. By Lemma 4 so is u_2, since $u_1 \to_\beta u_2$. Thus take $t_2 \equiv \epsilon^\omega$. If $t_1 \not\equiv \epsilon^\omega$ then it suffices to consider the case when $t_1 \equiv \epsilon^m(\epsilon^n(\lambda x.s_1)s_2)$, $u_1 \equiv (\lambda x.v_1)v_2$ and $u_2 \equiv v_1[v_2/x]$ with $s_1 \succ v_1$ and $s_2 \succ v_2$. But then $t_1 \to_\epsilon \epsilon^{m+1}(s_1[s_2/x])$. Also $s_1[s_2/x] \succ v_1[v_2/x] \equiv u_2$ by Lemma 14. If $s_1[s_2/x] \equiv \epsilon^\omega$ then $\epsilon^{m+1}(s_1[s_2/x]) \equiv \epsilon^\omega$, and thus $\epsilon^{m+1}(s_1[s_2/x]) \succ u_2$. Otherwise $\epsilon^{m+1}(s_1[s_2/x]) \succ u_2$ from the definition of $\succ$. So take $t_2 \equiv \epsilon^{m+1}(s_1[s_2/x])$.

Lemma 16. *If $t_1 \succ u_1 \to_\beta^\infty u_2$ then there exists t_2 such that $t_1 \to_\epsilon^\infty t_2 \succ u_2$.*

Proof. Coinduction with case analysis on $u_1 \to_\beta^\infty u_2$. There are three cases.

1. $u_2 \equiv x$ and $u_1 \to_\beta^* x$. By repeated application of Lemma 15 we obtain t_2 such that $t_1 \to_\epsilon^* t_2 \succ x$. Then also $t_1 \to_\epsilon^\infty t_2$ and we are done.
2. $u_2 \equiv v_1'v_2'$ with $u_1 \to_\beta^* v_1v_2$ and $v_i \to_\beta^\infty v_i'$. By Lemma 15 there is s such that $t_1 \to_\epsilon^* s \succ v_1v_2$. There are two possibilities.
   (a) $s \equiv \epsilon^\omega$ and v_1v_2 is root-active. Since $v_1v_2 \to_\beta^\infty v_1'v_2' \equiv u_2$, by Lemma 4, u_2 is root-active. Thus $t_1 \to_\epsilon^\infty \epsilon^\omega \succ u_2$ and we take $t_2 \equiv \epsilon^\omega$.
   (b) $s \equiv \epsilon^n(s_1s_2)$ with $s_i \succ v_i$. If $n = 0$ then $t_1 \equiv r_1r_2$ and $u_1 \equiv w_1w_2$ with $r_i \succ w_i$. Moreover, by the second part of Lemma 15, no contraction in $u_1 \to_\beta^* v_1v_2$ occurs at the root. Thus $w_i \to_\beta^* v_i \to_\beta^\infty v_i'$. Hence $w_i \to_\beta^\infty v_i'$. By the coinductive hypothesis there are s_1', s_2' with $r_i \to_\epsilon^\infty s_i' \succ v_i'$. Hence $t_1 \equiv r_1r_2 \to_\epsilon^\infty s_1's_2' \succ v_1'v_2' \equiv u_2$ and we take $t_2 \equiv s_1's_2'$. Now assume $n > 0$. Since $s_i \succ v_i \to_\beta^\infty v_i'$, by the coinductive hypothesis there are s_1', s_2' with $s_i \to_\epsilon^\infty s_i' \succ v_i'$. Hence $\epsilon^n(s_1's_2') \succ v_1'v_2' \equiv u_2$. Also $t_1 \to_\epsilon^\infty \epsilon^n(s_1s_2)$, because $t_1 \to_1^* \epsilon(\epsilon^{n-1}(s_1s_2))$ and $s_i \to_\epsilon^\infty s_i'$ (to obtain the derivation apply once the penultimate rule and then n times the the last rule). Thus take $t_2 \equiv \epsilon^n(s_1s_2)$.
3. $u_2 \equiv \lambda x.u_2'$ with $u_1 \to_\beta^* \lambda x.u_1'$ and $u_1' \to_\beta^\infty u_2'$. By Lemma 15 there is s with $t_1 \to_\epsilon^* s \succ \lambda x.u_1'$. Obviously, $\lambda x.u_1'$ is not root-active, so the only possibility for $s \succ \lambda x.u_1'$ to hold is when $s \equiv \epsilon^n(\lambda x.s')$ with $s' \succ u_1'$. If $n = 0$ then $t_1 \equiv \lambda x.t_1'$ and $u_1 \equiv \lambda x.u_0$ with $t_1' \succ u_0 \to_\beta^* u_1' \to_\beta^\infty u_2'$. Thus $t_1' \succ u_0 \to_\beta^\infty u_2'$. By the coinductive hypothesis there is r with $t_1' \to_\epsilon^\infty r \succ u_2'$. So $t_1 \equiv \lambda x.t_1' \to_\epsilon^\infty \lambda x.r \succ \lambda x.u_2' \equiv u_2$. So take $t_2 \equiv \lambda x.r$. Now assume $n > 0$. Since $s' \succ u_1' \to_\beta^\infty u_2'$, by the coinductive hypothesis there is r with $s' \to_\epsilon^\infty r \succ u_2'$. Thus $\epsilon^n(\lambda x.r) \succ u_2'$. Because $t_1 \to_1^* s \equiv \epsilon(\epsilon^{n-1}(\lambda x.s'))$ and $s' \to_\epsilon^\infty r$, also $t_1 \to_\epsilon^\infty \epsilon^n(\lambda x.r)$. So take $t_2 \equiv \epsilon^n(\lambda x.r)$.

Lemma 17. *If $t \to_1^* \epsilon(t')$ then there exists s such that $t \to_\epsilon^* \epsilon(s)$ and $s \to_1^* t'$, where the length of $s \to_1^* t'$ is not larger than the length of $t \to_1^* \epsilon(t')$.*

Proof. Induction on the length l of the reduction $t \to_1^* \epsilon(t')$. If $t \equiv \epsilon(u)$ then $u \to_1^* t'$ follows from the definition of $\to_1$. Hence, take $s \equiv u$. Otherwise $t \equiv u_1 u_2$ and we may decompose $t \to_1^* \epsilon(t')$ into reductions: $u_1 \to_1^* \epsilon^m(\lambda x.v)$ of length l_1, $u_2 \to_1^* w$ of length l_1, $\epsilon^m(\lambda x.v)w \to_1 \epsilon(r)$ and $r \to_1^* t'$ of length l_2, where $l_1 + l_2 < l$. By applying the inductive hypothesis m times, we conclude there is u_1' with $u_1 \to_\epsilon^* \epsilon^m(u_1')$ and $u_1' \to_1^* \lambda x.v$, where the length of $u_1' \to_1^* \lambda x.v$ is at most l_1. By the definition of $\to_1$, we have $u_1' \equiv \lambda x.v_0$ with $v_0 \to_1^* v$ of length at most l_1. By repeated application of Lemma 7, there is a reduction $v_0[u_2/x] \to_1^* v[w/x]$ of length at most l_1. Since $\epsilon^m(\lambda x.v)w \to_1 \epsilon(r)$, $v[w/x] \to_1 r$. Hence, there is a reduction $v_0[u_2/x] \to_1^* v[w/x] \to_1 r \to_1^* t'$ of length at most $l_1 + l_2 + 1 \le l$. Also $t \equiv u_1 u_2 \to_\epsilon^* \epsilon^m(\lambda x.v_0)u_2 \to_\epsilon v_0[u_2/x]$. Thus take $s \equiv v_0[u_2/x]$.

Lemma 18. *If $s \to_\epsilon^\infty t$ and $t \not\equiv \epsilon^\omega$, then there exists s' such that $s \to_\epsilon^* s'$ and one of the following holds:*

- *$s' \equiv \epsilon^n(x) \equiv t$, or*
- *$s' \equiv \epsilon^n(\lambda x.r)$ and $t \equiv \epsilon^n(\lambda x.r')$ with $r \to_\epsilon^\infty r'$, or*
- *$s' \equiv \epsilon^n(r_1 r_2)$ and $t \equiv \epsilon^n(r_1' r_2')$ with $r_1 \to_\epsilon^\infty r_1'$ and $r_2 \to_\epsilon^\infty r_2'$.*

Proof. By the definition of $\to_\epsilon^\infty$ there is s' such that $s \to_1^* s'$ and one of the three conditions hold. For instance, suppose $s' \equiv \epsilon^n(r_1 r_2)$ and $t \equiv \epsilon^n(r_1' r_2')$ with $r_1 \to_\epsilon^\infty r_1'$ and $r_2 \to_\epsilon^\infty r_2'$. By applying Lemma 17 n times we obtain s_1, s_2 with $s \to_\epsilon^* \epsilon^n(s_1 s_2)$ and $s_1 s_2 \to_1^* r_1 r_2$. By the definition of $\to_1$ we have $s_1 \to_1^* r_1$ and $s_2 \to_1^* r_2$. By Lemma 9, $s_1 \to_\epsilon^\infty r_1'$ and $s_2 \to_\epsilon^\infty r_2'$. This finishes the proof.

Lemma 19. *If $t_1 \to_\epsilon t_2$ and $t_1 \succ u_1$, then there exists u_2 such that $u_1 \to_\beta u_2$ and $t_2 \succ u_2$. Moreover, if t_1 does not start with ϵ but t_2 does, then the contraction $u_1 \to_\beta u_2$ occurs at the root.*

Proof. We proceed by induction on $t_1 \to_\epsilon t_2$. All cases are trivial except when $t_1 \equiv \epsilon^m(\epsilon^n(\lambda x.s)t) \to_\epsilon \epsilon^{m+1}(s[t/x]) \equiv t_2$. Then $u_1 \equiv (\lambda x.v_1)v_2$ with $s \succ v_1$ and $t \succ v_2$. Thus $u_1 \to_\beta v_1[v_2/x]$. By Lemma 14, $s[t/x] \succ v_1[v_2/x]$. Let $u_2 \equiv v_1[v_2/x]$. If $s[t/x] \equiv \epsilon^\omega$ then $\epsilon^{m+1}(s[t/x]) \equiv \epsilon^\omega$, and thus $\epsilon^{m+1}(s[t/x]) \succ u_2$. Otherwise $\epsilon^{m+1}(s[t/x]) \succ u_2$ also holds by the definition of $\succ$.

Lemma 20. *If $t \to_\epsilon^\infty \epsilon^\omega$ and $t \succ u$, then u is root-active.*

Proof. Suppose $u \to_\beta^\infty u'$ with u' root-stable. Then by Lemma 16 there is t' with $t \to_\epsilon^\infty t' \succ u'$. Since ϵ^ω is in normal form w.r.t. infinitary ϵ-reduction, by Lemma 13, $t' \to_\epsilon^\infty \epsilon^\omega$. Since $t' \succ u'$ and u' is not root-active, $t' \equiv \epsilon^n(s)$ where s does not start with ϵ. From the definition of $\to_\epsilon^\infty$ we have $s \to_1^* \epsilon(s')$ for some s'. By Lemma 17 there is r with $s \to_\epsilon^* \epsilon(r)$. Since $t' \succ u'$ we have $s \succ u'$. By Lemma 19 there is w such that $u' \to_\beta^* w$ with at least one contraction occuring at the root. But this contradicts the fact that u' is root-stable.

Lemma 21. *If $t_1 \to_\epsilon^\infty t_2$ and $t_1 \succ u_1$, then there exists u_2 such that $t_2 \succ u_2$ and $u_1 \to_\beta^\infty u_2$.*

Proof. We proceed by coinduction. First, assume that $t_2 \equiv \epsilon^\omega$. By Lemma 20, u_1 is root-active. Take $u_2 \equiv u_1$. So assume $t_2 \not\equiv \epsilon^\omega$. Then by Lemmas 18, 19 there are s, u_1' with $t_1 \to_\epsilon^* s \succ u_1'$, $u_1 \to_\beta^* u_1'$ and we have one of three possibilities.

- $s \equiv \epsilon^n(x) \equiv t_2$. Then $u_1' \equiv x$ and $u_1 \to_\beta^\infty x$. So we may take $u_2 \equiv u_1'$.
- $s \equiv \epsilon^n(\lambda x.r)$ and $t_2 \equiv \epsilon^n(\lambda x.r')$ with $r \to_\epsilon^\infty r'$. Then $u_1' \equiv \lambda x.w$ with $r \succ w$. By the coinductive hypothesis there is w' with $w \to_\beta^\infty w'$ and $r' \succ w'$. So $t_2 \equiv \epsilon^n(\lambda x.r') \succ \lambda x.w'$. Also $u_1 \to_\beta^\infty \lambda x.w'$, since $u_1 \to_\beta^* \lambda x.w$ and $w \to_\beta^\infty w'$. Thus take $u_2 \equiv w'$.
- $s \equiv \epsilon^n(r_1 r_2)$ and $t_2 \equiv \epsilon^n(r_1' r_2')$ with $r_1 \to_\epsilon^\infty r_1'$ and $r_2 \to_\epsilon^\infty r_2'$. Analogous to the previous case.

Lemma 22. *If $u_1, u_2 \in \Lambda^\infty$ then: $u_1 \sim u_2$ iff there exists $t \in \Lambda^\epsilon$ such that $t \succ u_1$ and $t \succ u_2$.*

Proof. By coinduction.

Theorem 1. *If $t \sim t'$, $t \to_\beta^\infty s$ and $t' \to_\beta^\infty s'$, then there exist r, r' such that $s \to_\beta^\infty r$, $s' \to_\beta^\infty r'$ and $r \sim r'$.*

Proof. By Lemma 22 there is $t_0 \in \Lambda^\epsilon$ such that $t_0 \succ t$ and $t_0 \succ t'$. By Lemma 16 there are s_0, s_0' such that $s_0 \succ s$, $s_0' \succ s'$, $t_0 \to_\epsilon^\infty s_0$ and $t_0 \to_\epsilon^\infty s_0'$. By Lemma 13 there is r_0 such that $s_0 \to_\epsilon^\infty r_0$ and $s_0' \to_\epsilon^\infty r_0$. By Lemma 21 there are r, r' such that $r_0 \succ r$, $r_0 \succ r'$, $s \to_\beta^\infty r$ and $s' \to_\beta^\infty r'$. By Lemma 22, $r \sim r'$. See Fig. 1.

4 Confluence of Böhm Reduction

In this section we show that the relation $\to_{\beta\bot}^\infty$ of infinitary Böhm reduction is confluent. The high-level strategy of the proof resembles that of the corresponding proof in [11]. At the end of this section we also indicate how to show that our definition of infinitary Böhm reduction corresponds to the definition in [11].

Definition 4. Given the $\bot$-rules

$$t \to \bot \qquad \text{if } t \text{ is root-active and } t \not\equiv \bot$$

we define the relation $\to_{\beta\bot}$ of $\beta\bot$-*contraction* as the compatible closure of the β-rule and the $\bot$-rules. The relation $\to_{\beta\bot}^*$ of $\beta\bot$-*reduction* is the transitive-reflexive closure of $\to_{\beta\bot}$.

The relation $\to_{\beta\bot}^\infty$ of *infinitary Böhm reduction* is defined coinductively.

$$\frac{s \to_{\beta\bot}^* x}{s \to_{\beta\bot}^\infty x} \qquad \frac{s \to_{\beta\bot}^* t_1 t_2 \quad t_1 \to_{\beta\bot}^\infty t_1' \quad t_2 \to_{\beta\bot}^\infty t_2'}{s \to_{\beta\bot}^\infty t_1' t_2'} \qquad \frac{s \to_{\beta\bot}^* \lambda x.r \quad r \to_{\beta\bot}^\infty r'}{s \to_{\beta\bot}^\infty \lambda x.r'}$$

The relation $\to_\bot$ is defined coinductively.

$$\frac{s \text{ is root-active and } s \not\equiv \bot}{s \to_\bot \bot} \qquad \frac{}{x \to_\bot x} \qquad \frac{s_1 \to_\bot t_1 \quad s_2 \to_\bot t_2}{s_1 s_2 \to_\bot t_1 t_2} \qquad \frac{s \to_\bot s'}{\lambda x.s \to_\bot \lambda x.s'}$$

Lemma 23. *If $s \to_\bot s'$ and $t \to_\bot t'$ then $s[t/x] \to_\bot s'[t'/x]$.*

Proof. Coinduction with case analysis on $s \to_\bot s'$, using Lemma 5.

Lemma 24. *If $t_1 \to_\bot t_2 \to_\beta t_3$ then there exists t_1' such that $t_1 \to_\beta t_1' \to_\bot t_3$.*

Proof. Induction on $t_2 \to_\beta t_3$. The only interesting case is when $t_2 \equiv (\lambda x.s_1)s_2$ and $t_3 \equiv s_1[s_2/x]$. We exclude $t_2 \equiv \bot$, because then $t_3 \equiv \bot$. So $t_1 \equiv (\lambda x.u_1)u_2$ with $u_i \to_\bot s_i$. By Lemma 23, $u_1[u_2/x] \to_\bot s_1[s_2/x]$. Thus take $t_1' \equiv u_1[u_2/x]$.

Lemma 25. *If $t \succ u$ and u is root-active, then $t \to_\epsilon^\infty \epsilon^\omega$.*

Proof. By coinduction. If $t \equiv \epsilon^\omega$ then the claim is obvious, so suppose otherwise. Since u is root-active, $u \equiv u_1 u_2$ with $u_1 \to_\beta^\infty \lambda x.s$. Then $u_1 \to_\beta^* \lambda x.s'$ for some s'. Since $t \not\equiv \epsilon^\omega$, we have $t \equiv \epsilon^n(t_1 t_2)$ with $t_1 \succ u_1$ and $t_2 \succ u_2$. By Lemma 15 there is t_1' such that $t_1 \to_\epsilon^* t_1' \succ \lambda x.s'$. We have $t_1' \equiv \epsilon^m(\lambda x.r)$ with $r \succ s'$. Hence $t \equiv t_1 t_2 \to_1^* \epsilon^m(\lambda x.r)t_2 \to_1 \epsilon(r[t_2/x])$. By Lemma 14, $r[t_2/x] \succ s'[u_2/x]$. Since u is root-active and $u \equiv u_1 u_2 \to_\beta^\infty s'[u_2/x]$, by Lemma 4 we conclude that $s'[u_2/x]$ is root-active. By the coinductive hypothesis $r[t_2/x] \to_\epsilon^\infty \epsilon^\omega$. Therefore $t \to_\epsilon^\infty \epsilon^\omega$, by applying the last rule in the definition of $\to_\epsilon^\infty$.

Lemma 26. *If $s \to_\bot t$ then $s \sim t$.*

Proof. By coinduction.

Lemma 27. *If t is root-active and $s \to_\bot t$ or $t \to_\bot s$, then s is root-active.*

Proof. By Lemma 26, $s \sim t$. By Lemma 22 there is r with $r \succ s$ and $r \succ t$. Since t is root-active, by Lemma 25, $r \to_\epsilon^\infty \epsilon^\omega$. But $r \succ s$, so s is root-active by Lemma 20.

Lemma 28. *If $t_1 \to_\bot t_2 \to_\bot t_3$ then $t_1 \to_\bot t_3$.*

Proof. Coinduction with case analysis on $t_2 \to_\bot t_3$, using Lemma 27.

Lemma 29. *If $s \to_{\beta\bot}^* t$ then there exists r such that $s \to_\beta^* r \to_\bot t$.*

Proof. Induction on the length of $s \to_{\beta\bot}^* t$, using Lemma 24 and Lemma 28.

Lemma 30. *If $t_1 \to_\bot t_2 \to_{\beta\bot}^\infty t_3$ then $t_1 \to_{\beta\bot}^\infty t_3$.*

Proof. Coinduction with case analysis on $t_2 \to_\beta^\infty t_3$, using Lemmas 29, 24, 28.

Lemma 31. *If $s \to_{\beta\bot}^\infty t$ then there exists r such that $s \to_\beta^\infty r \to_\bot t$.*

Proof. By coinduction with case analysis on $s \to_{\beta\bot}^\infty t$, using Lemmas 29, 30.

Lemma 32. *If $t \to_\bot t_1$ and $t \to_\bot t_2$ then there exists t_3 such that $t_1 \to_\bot t_3$ and $t_2 \to_\bot t_3$.*

Proof. Coinduction with case analysis on $t \to_\bot t_1$. The only non-trivial case is when $t \equiv s_1 s_2$ is root-active and e.g. $t_1 \equiv \bot$. But since $t \to_\bot t_2$, by Lemma 27, t_2 is also root-active. Thus $t_2 \to_\bot \bot \equiv t_1$ and we take $t_3 \equiv t_1$.

Lemma 33. *If $t_1 \sim t_2$ then there is s with $t_1 \to_\bot s$ and $t_2 \to_\bot s$.*

Lemma 34. *If $t_1 \sim t_2 \to_\beta^\infty t_3$ then there is t_2' such that $t_1 \to_\beta^\infty t_2' \sim t_3$.*

Proof. By Lemma 22 there is s with $s \succ t_1$ and $s \succ t_2$. Since $s \succ t_2 \to_\beta^\infty t_3$, by Lemma 16 there is s' with $s \to_\epsilon^\infty s' \succ t_3$. Since $s \succ t_1$ and $s \to_\epsilon^\infty s'$, by Lemma 21 there is t_2' with $t_1 \to_\beta^\infty t_2'$ and $s' \succ t_2'$. Since $s \succ t_2'$ and $s' \succ t_3$, by Lemma 22 we obtain $t_2' \sim t_3$.

Lemma 35. *If $t \to_\beta^\infty s$ and s is root-active, then so is t.*

Proof. Suppose $t \to_\beta^\infty t'$ for a root-stable t'. By Theorem 1 there are s_1, s_2 with $t' \to_\beta^\infty s_2$, $s \to_\beta^\infty s_1$ and $s_1 \sim s_2$. First, we show s_2 is root-stable, from which we obtain that s_1 is root-stable – a contradiction. Without loss of generality, assume $t' \equiv t_1 t_2$, $s_2 \equiv r_1 r_2$ with $t_i \to_\beta^\infty r_i$. If $r_1 \to_\beta^\infty \lambda x.r_0$ for some r_0, then $t_1 \to_\beta^\infty \lambda x.r_0$ by Lemma 3, which would contradict the fact that t' is root-stable. Since s_2 is root-stable, $s_1 \equiv u_1 u_2$ with $u_i \sim r_i$. Because s_1 is root-active, there is w with $u_1 \to_\beta^\infty \lambda x.w$. Then by Lemma 34 there is w' with $r_1 \to_\beta^\infty \lambda x.w' \sim \lambda x.w$. This contradicts the fact that s_2 is root-stable.

Lemma 36. *If $t \to_{\beta\bot}^\infty s$ and s is root-active, then so is t.*

Proof. Follows from Lemmas 31, 27, 35.

Lemma 37. *If $t_1 \to_\beta^\infty t_2 \to_\bot t_3$ then $t_1 \to_{\beta\bot}^\infty t_3$.*

Proof. Coinduction with case analysis on $t_1 \to_\beta^\infty t_2$, using Lemma 36.

Theorem 2. *If $t \to_{\beta\bot}^\infty t_1$ and $t \to_{\beta\bot}^\infty t_2$ then there exists t_3 such that $t_1 \to_{\beta\bot}^\infty t_3$ and $t_2 \to_{\beta\bot}^\infty t_3$.*

Proof. The proof is illustrated by the following diagram.

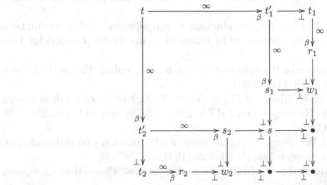

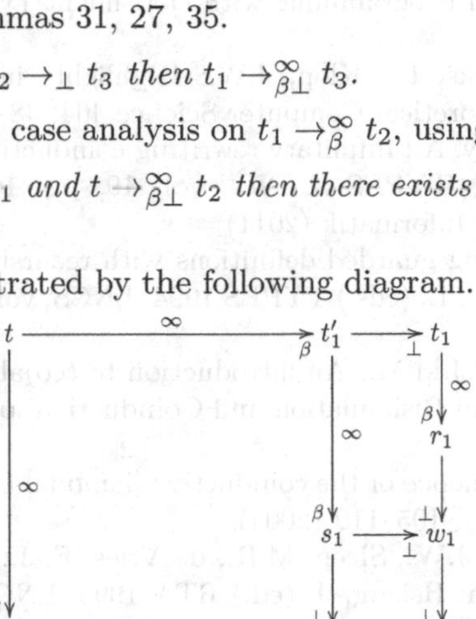

By Lemma 31 there are t_1', t_2' with $t \to_\beta^\infty t_i' \to_\bot t_i$. By Theorem 1 there are s_1, s_2 with $t_i' \to_\beta^\infty s_i$ and $s_1 \sim s_2$. By Lemma 33 there is s with $s_i \to_\bot s$. By Lemma 26, $t_i' \sim t_i$. By Lemma 34 there are r_1, r_2 with $t_i \to_\beta^\infty r_i \sim s_i$. By Lemma 33 there are w_1, w_2 with $r_i \to_\bot w_i$ and $s_i \to_\bot w_i$. The remaining squares follow from Lemma 32. The claim then follows from Lemmas 28, 37.

4.1 Equivalence with the Standard Definition

Let $\to_{\mathcal{B}}$, $\to^i_\beta$ and $\to^i_\bot$ denote infinitary Böhm reduction w.r.t. root-active terms, infinitary β-reduction and parallel $\bot$-reduction, all as defined in [11] by means of strong convergence. From [6] we have: (a) $t \to^i_\beta s$ iff $t \to^\infty_\beta s$. We show: (b) $t \to_{\mathcal{B}} s$ iff $t \to^\infty_{\beta\bot} s$. This could probably be shown by easy modification of the proof of (a) from [6]. We derive (b) from (a) using some results from [11]. Suppose $t \to_{\mathcal{B}} s$. In [11] it is shown that then there is r with $t \to^i_\beta r \to^i_\bot s$. By (a), $t \to^\infty_\beta r$. From definitions $r \to^i_\bot s$ implies $r \to_\bot s$. Then $t \to^\infty_{\beta\bot} s$ by Lemma 37. Now suppose $t \to^\infty_{\beta\bot} s$. Then by Lemma 31 there is r with $t \to^\infty_\beta r \to_\bot s$. By (a), $t \to^i_\beta r$. From definitions $r \to^i_\bot s$. Then $t \to_{\mathcal{B}} s$.

References

1. Abel, A., Pientka, B.: Wellfounded recursion with copatterns: a unified approach to termination and productivity. In: Morrisett, G., Uustalu, T. (eds.) ICFP, pp. 185–196. ACM (2013)
2. Barendregt, H., Klop, J.W.: Applications of infinitary lambda calculus. Information and Computation 207(5), 559–582 (2009)
3. Berarducci, A.: Infinite lambda-calculus and non-sensible models. In: Ursini, A., Aglianò, P. (eds.) Logic and Algebra. Lecture Notes in Pure and Applied Mathematics Series, vol. 180, pp. 339–378 (1996)
4. Chlipala, A.: Certified Programming with Dependent Types, ch. 5. The MIT Press (2013)
5. Endrullis, J., Hendriks, D., Klop, J.W.: Highlights in infinitary rewriting and lambda calculus. Theoretical Computer Science 464, 48–71 (2012)
6. Endrullis, J., Polonsky, A.: Infinitary rewriting coinductively. In: Danielsson, N.A., Nordström, B. (eds.) TYPES. LIPIcs, vol. 19, pp. 16–27. Schloss Dagstuhl - Leibniz-Zentrum fuer Informatik (2011)
7. Giménez, E.: Codifying guarded definitions with recursive schemes. In: Smith, J., Dybjer, P., Nordström, B. (eds.) TYPES 1994. LNCS, vol. 996, pp. 39–59. Springer, Heidelberg (1995)
8. Jacobs, B., Rutten, J.J.M.M.: An introduction to (co)algebras and (co)induction. In: Advanced Topics in Bisimulation and Coinduction, pp. 38–99. Cambridge University Press (2011)
9. Joachimski, F.: Confluence of the coinductive [lambda]-calculus. Theoretical Computer Science 311(1-3), 105–119 (2004)
10. Kennaway, R., Klop, J.W., Sleep, M.R., de Vries, F.-J.: Infinitary lambda calculi and Böhm models. In: Hsiang, J. (ed.) RTA 1995. LNCS, vol. 914, pp. 257–270. Springer, Heidelberg (1995)
11. Kennaway, R., Klop, J.W., Sleep, M.R., de Vries, F.-J.: Infinitary lambda calculus. Theoretical Computer Science 175(1), 93–125 (1997)
12. Rutten, J.J.M.M.: Universal coalgebra: a theory of systems. Theoretical Computer Science 249(1), 3–80 (2000)
13. Severi, P., de Vries, F.-J.: An extensional Böhm model. In: Tison, S. (ed.) RTA 2002. LNCS, vol. 2378, pp. 159–173. Springer, Heidelberg (2002)
14. Terese: Term Rewriting Systems. Cambridge Tracts in Theoretical Computer Science, vol. 55. Cambridge University Press (2003)

An Implicit Characterization of the Polynomial-Time Decidable Sets by Cons-Free Rewriting

Daniel de Carvalho and Jakob Grue Simonsen

Department of Computer Science, University of Copenhagen (DIKU)
Njalsgade 128-132, DK-2300 Copenhagen S, Denmark
{daniel.carvalho,simonsen}@diku.dk

Abstract. We define the class of *constrained cons-free* rewriting systems and show that this class characterizes P, the set of languages decidable in polynomial time on a deterministic Turing machine. The main novelty of the characterization is that it allows very liberal properties of term rewriting, in particular non-deterministic evaluation: no reduction strategy is enforced, and systems are allowed to be non-confluent.

We present a class of constructor term rewriting systems that characterizes the complexity class P—the set of languages decidable in polynomial time on a deterministic Turing machine. The class is an analogue of similar classes in functional programming that use *cons-freeness*–the inability of a program to construct new compound data during its evaluation–to characterize a range of complexity classes, including L and P [1, 2], and for higher-order programs $PSPACE$ and hierarchies of exponential space and time classes [3]. The primary novelty is that while previous work has crucially utilized the deterministic evaluation (in particular, call-by-value) and typing disciplines usually found in functional programming languages, we allow for the full rewriting relation to be used, and we allow non-orthogonal systems.

The ability to use non-orthogonal and non-confluent systems means that we do not have access to standard results on orthogonality such as normalization or finite developments of sets of redexes, and we cannot appeal to results connecting deterministic Turing machines to confluent rewriting [4], or to functional programming without overlapping function declarations [1, 3, 5]. These are the main reasons that our proofs are substantially more difficult than similar work by Bonfante showing that introducing non-determinism to a cons-free functional language characterizes P [2].

Related Work

The original impetus for devising languages or calculi characterizing complexity classes was the seminal work of Bellantoni and Cook [6] who introduced a scheme of constrained recursion in function declarations in applicative languages, called *safe recursion*, later followed by similar constraints, *tiered* or *ramified recursion*

G. Dowek (ed.): RTA-TLCA 2014, LNCS 8560, pp. 179–193, 2014.

[7, 5]. Roughly, the idea of this approach is to partition the arguments of every function into "normal" and "safe" variables, where only normal variables are used for recursion. Our approach contains no such constraints. Other approaches have used type systems, typically based on variants of linear logic [8–10]; in contrast, we employ no type system, but enforce a simple syntactic criterion to constrain copying.

Much effort has been directed towards performing *polynomial complexity analysis* in term rewriting, that is, devising methods to automatically infer that specific TRSs have polynomial runtime or derivational complexity. This work has almost invariably considered analogues of call-by-value semantics, e.g. innermost evaluation; in this vein of research, several reduction orders have been defined such that TRSs are compatible with the orders iff they have polynomial runtime complexity [11, 12]. The main difference with our work is that we do not necessarily enforce polynomial runtime complexity, but use a form of memoization to ensure that our class of systems can be evaluated in polynomial time on a Turing machine. For full rewriting with no constraints on reduction strategy, Avanzini and Moser [4] have shown that a *confluent* constructor rewriting system characterizes a language in P iff it has polynomial runtime complexity, that is, if the maximal reduction lengths starting from appropriately formed terms are polynomially bounded. Most research in this vein has focused on *functional* complexity classes, whereas we only consider the case of decision problems; we believe our results can be extended to the function classes, but with some difficulty as input constructors may not be used as output constructors in cons-free systems.

The restriction to cons-free systems was originally developed in functional programming by Jones [1], [3] inspired by similar work by Goerdt in recursion-theoretic settings [13, 14], and leading to similar characterizations in other language paradigms [15], [2]. The primary difference between this work and ours is that we do not consider a particular reduction order, and work in a completely untyped setting, that is, the standard liberal setting of term rewriting; the cost of this freedom is that we need to enforce technical demands on our class of systems leading to *constrained* cons-free systems, rather than merely cons-free ones.

1 Constrained Cons-Free Term Rewriting Systems

We presuppose basic knowledge about rewriting, corresponding to the introductory chapters of [16]. Throughout the text, we assume a denumerable set $\mathcal{X}$ of *variables*.

Let Σ be a signature (i.e., a function from a set $\mathcal{F}$ of *function symbols* to $\mathbb{N}$ which associates with every $f \in \mathcal{F}$ its arity $ar(f)$); we then denote by $\mathcal{T}(\Sigma)$ the set of terms built from Σ and $\mathcal{X}$. The set of ground terms over Σ is denoted by $\mathcal{T}_0(\Sigma)$. By abuse of language, if $\mathcal{F}_0$ is a set of function symbols, we will write also $\mathcal{T}(\mathcal{F}_0)$ instead of $\mathcal{T}(\Sigma_{|\mathcal{F}_0})$ (or $\mathcal{T}_0(\mathcal{F}_0)$ instead of $\mathcal{T}_0(\Sigma_{|\mathcal{F}_0})$). The set of positions in a term t is denoted by $Pos(t)$ (a position q is said to be below the position

p if $p \leq q$): if p is a position in a term t, then p determines the subterm $t_{|p}$ of t occurring at position p and the symbol $t(p)$ occurring in t at p. If s and t are terms, we write $s \trianglelefteq t$ if s is a subterm of t; note that $s \trianglelefteq t$ iff $(\exists p \in Pos(t))s = t_{|p}$. For any term t, we denote by $Oc(t)$ the set of variables occurring in t, that is, $Oc(t) = \{x \in \mathcal{X}; (\exists p \in Pos(t))t(p) = x\}$ and, for any $x \in \mathcal{X}$, by $Oc(x,t)$ the number of occurrences of x in t, that is, $Oc(x,t) = Card(\{p \in Pos(t); t(p) = x\})$.

A *constructor TRS* is a term rewriting system (TRS) in which the set of function symbols $\mathcal{F}$ is partitioned into a set $\mathcal{D}$ of *defined* function symbols and a set $\mathcal{C}$ of *constructors*, such that for every rewrite rule $(l,r) \in \mathcal{R}$, the left-hand side l has the form $f(t_1,\ldots,t_n)$ with $f \in \mathcal{D}$ and $t_1,\ldots,t_n \in \mathcal{T}(\mathcal{C})$, the set of terms built from variables and constructors.

We introduce *cons-free TRS* that corresponds essentially to the functional programming language called "F+ro" (ro for "read-only") in [17].

Definition 1. *A* cons-free TRS *is a finite constructor TRS such that, for every rewrite rule* (l,r), *for any* $c(u_1,\ldots,u_n) \trianglelefteq r$ *such that* c *is a constructor and* $n > 0$, *we have* $c(u_1,\ldots,u_n) \trianglelefteq l$.

The functional programming languages considered in [3] and in [17] have a call-by-value semantics, and proofs generally assume terminating programs; in contrast, terms in (cons-free) term rewriting systems may be subjected to different reduction strategies, are not necessarily terminating, and terms may have more than one normal form. To obviate technical problems due to these facts, we restrict the class of term rewriting systems to the *constrained* cons-free term rewriting systems.

Definition 2. *A cons-free TRS R is said to be* constrained *if there exists some subset* $\mathcal{A} \subseteq \mathcal{D}$ *such that, for any rule* $(f(c_1,\ldots,c_q),r) \in \mathcal{R}$ *and for any* $x \in \mathcal{X} \cap \{c_1,\ldots,c_q\}$, *we have:*

- $(\forall p,p' \in Pos(r))(r(p) = x \Rightarrow (p' < p \Rightarrow r(p') \in \mathcal{A}))$
- *and* $f \in \mathcal{A} \Rightarrow Oc(x,r) \leq 1$.

Every variable occurring just below the root symbol of a left-hand side of a rule occur only below defined symbols of a certain kind that do not allow for non-linear recursion. Note that duplication may occur in constrained cons-free TRSs, both for variables that occur "deep" in a left-hand side (i.e., below constructor symbols), and for variables occurring just below the root of defined symbols not in the special subset $\mathcal{A} \subseteq \mathcal{D}$. E.g., if $f/1, g/2 \in \mathcal{D}$ and $c/1 \in \mathcal{C}$, the TRS $\{f(c(x)) \to g(x,x), g(c(x),y) \to y, g(c(x),c(y)) \to g(x,y)\}$ is constrained cons-free (set $\mathcal{A} = \{g\}$ or $\mathcal{A} = \{f,g\}$).

In Sections 2 and Section 3, we will prove some properties of cons-free term, respectively constrained cons-free TRSs that will allow for efficient simulation on Turing machines. The main aim of the two sections is to prove Proposition 1, respectively Corollary 1.

2 Computation in Cons-Free TRS

In this section, we introduce a class of "generalized terms", and we show that any reduction sequence in a cons-free TRS from a ground term to a ground constructor term can be simulated by some "innermost" reduction sequence of such "generalized terms" (i.e. some sequence of $\triangleright$-reductions).

We are given a cons-free TRS with $\mathcal{R}$ the set of rules, $\mathcal{D}$ the set of defined function symbols, and $\mathcal{C}$ the set of constructors. Moreover, for any $m \in \mathbb{N}$, we denote by $\mathcal{D}_m$ the set of defined function symbols of arity m and by $\mathcal{C}_m$ the set of constructors of arity m. We first set notations used in the remainder of the paper.

Notations: As usual, the reflexive transitive closure of a relation E is denoted by E^*. Throughout the text, $A \rightharpoonup B$ refers to the type of partial maps with domain A and co-domain B. If $f : A \rightharpoonup B$, we denote by $dom(f)$ the set of $x \in A$ such that $f(x)$ is defined and by $im(f)$ the set $f(dom(f)) = \{f(x) : x \in dom(f)\}$.

For any $t \in \mathcal{T}(\mathcal{D} \cup \mathcal{C})$, we denote by $|t|$ the *size* of t, i.e., $|x| = |c| = 1$ for all variables x and nullary $c \in \mathcal{D}_0 \cup \mathcal{C}_0$, and $|f(s_1, \ldots, s_m)| = 1 + \sum_{i=1}^{m} |s_i|$ for $f \in \mathcal{D}_m \cup \mathcal{C}_m$.

Let $u, v, t \in \mathcal{T}(\mathcal{D} \cup \mathcal{C})$. We denote by $\mathbf{Seq}(u, v)$ the set of (finite) reduction sequences from u to v and we set $\mathbf{Seq}(u, -) = \bigcup_{v \in \mathcal{T}(\mathcal{D} \cup \mathcal{C})} \mathbf{Seq}(u, v)$. If $\rho_1 \in \mathbf{Seq}(t, u)$ and $\rho_2 \in \mathbf{Seq}(u, v)$, then we denote by $(\rho_1;\ \rho_2)$ the reduction sequence from t to v consisting in ρ_1 followed by ρ_2.

For any reduction step $\rho : t \rightarrow_{C_0[],(l,r)} u$, for any occurrence $\langle v|C[]\rangle$ of v in $t = C[v]$, we denote by $\langle v|C[]\rangle \setminus \rho$ the set of descendants of $\langle v|C[]\rangle$ in u after ρ.

We denote by $\mathcal{U}_0$ the set of ground terms that may be written as $C[t_1, \ldots, t_n]$ for some $n \in \mathbb{N}$ where $C[\cdot, \ldots, \cdot]$ is an n-hole context over $\mathcal{D}$, and $t_1, \ldots, t_n$ are ground terms over $\mathcal{C}$. Notice that, if $v \in \mathcal{U}_0$ and $v \rightarrow u$ in some cons-free TRS, then $u \in \mathcal{U}_0$.

For any $i \in \mathbb{N}$, we denote by Φ_i the set of i-hole contexts obtained by substituting exactly i distinct occurrences of constants in an element of $\mathcal{U}_0$ such that, for any hole, the unique path from the root to the hole passes through only elements of $\mathcal{D}$.

Recall that a semi-ring is an algebraic structure $(R, \cdot, +)$ satisfying the standard ring axioms with the exceptions that every element need not have a $+$-inverse. Recall further that a semi-module is an algebraic structure satisfying the usual module axioms over a commutative semi-ring. We denote by $\mathbf{2}$ the semi-ring with exactly two elements 0 and 1, where $1 + 1 = 1$. Let $\mathcal{E}$ be some set. We denote by $\mathbf{2}\langle\mathcal{E}\rangle$ the free $\mathbf{2}$-semi-module on $\mathcal{E}$. For any $V \in \mathbf{2}\langle\mathcal{E}\rangle$, we denote by *Supp*$(V)$ the unique $\mathcal{F} \subseteq \mathcal{E}$ such that $V = \sum_{v \in \mathcal{F}} v$. If $\mathcal{F} = \{v\}$, then we still denote by v the vector $\sum_{v \in \mathcal{F}} v \in \mathbf{2}\langle\mathcal{E}\rangle$.

We will use the notation $\mathbf{2}\langle\mathcal{E}\rangle$ either with $\mathcal{E} = \mathcal{T}_0(\mathcal{C})$ or $\mathcal{E} = \Delta$, the set of "generalized terms" defined just below. In those cases, an element of $\mathbf{2}\langle\mathcal{E}\rangle$ may be thought of as a "formal sum" of (generalized) terms, and *Supp*(V) as the set of (generalized) terms occurring in the sum. A benefit of considering formal sums instead of finite sets is that it allows to painlessly identify a term with the singleton containing this term. In later developments, we shall use the sum to track the possible reducts of subterms, i.e. each summand will correspond to a possible reduct.

Definition 3. *For any $i \in \mathbb{N}$, we define Δ_i by induction on i:*

- $\Delta_0 = \mathcal{T}_0(\mathcal{C})$;
- $\Delta_{i+1} = \Delta_i \cup \left(\bigcup_{m\in\mathbb{N}}\{f(U_1,\ldots,U_m);\ f \in \mathcal{D}_m \text{ and } U_1,\ldots,U_m \in \mathit{2}\langle\Delta_i\rangle\}\right)$.

We set $\Delta = \bigcup_{i\in\mathbb{N}} \Delta_i$. For any $u \in \Delta$, we set level$(u) = \min\{i \in \mathbb{N};\ u \in \Delta_i\}$.

Thus, e.g., if $\mathcal{D} = \{f/1, g/2\}$ and $\mathcal{C}' = \{s/1, n/0\}$, then $f(s(n)+s(s(n))) \in \Delta_1$ and $g(f(s(n)) + f(s(s(n))), s(n) + s(s(n))) \in \Delta_2$. Note further that $\mathcal{U}_0 \subseteq \Delta$ and every term on the form $C[c_1,\ldots,c_m]$, where $C[\cdot,\ldots,\cdot]$ is an m-hole context over $\mathcal{D}$ and $C_1,\ldots,C_m \in \mathbf{2}\langle\mathcal{T}_0(\mathcal{C})\rangle$, is an element of Δ.

Now, we want to define a notion of reduction on $\mathbf{2}\langle\Delta\rangle$: we will denote this reduction by $\rhd$. First, we define an auxiliary binary relation $\rhd_\Delta$. For any $r \in \mathcal{T}(\mathcal{D}\cup\mathcal{C}_0)$, we homomorphically extend the notation r^φ with $\varphi : \mathcal{X} \rightharpoonup \mathcal{T}(\mathcal{D}\cup\mathcal{C}_0)$ to any $\varphi : \mathcal{X} \rightharpoonup \mathbf{2}\langle\mathcal{T}_0(\mathcal{C})\rangle$ such that $Oc(r) \subseteq dom(\varphi)$: instead of having $r^\varphi \in \mathcal{T}(\mathcal{D}\cup\mathcal{C}_0)$, we have $r^\varphi \in \mathbf{2}\langle\Delta\rangle$.

Definition 4. *We define the relation $\rhd_\Delta \subseteq (\Delta_1 \setminus \Delta_0) \times \mathit{2}\langle\Delta\rangle$ as follows: $u \rhd_\Delta V$ if, and only if, there exist $q \in \mathbb{N}$, $f \in \mathcal{D}_q$, $(f(c_1,\ldots,c_q), r) \in \mathcal{R}$, $V_1,\ldots,V_q \in \mathit{2}\langle\mathcal{T}_0(\mathcal{C})\rangle$ and $\varphi : \mathcal{X} \rightharpoonup \mathit{2}\langle\mathcal{T}_0(\mathcal{C})\rangle$ such that, for any $j \in \{1,\ldots,q\}$, we have*

- *$Oc(c_j) \subseteq dom(\varphi)$ and $(c_j \notin \mathcal{X} \Rightarrow (\forall x \in Oc(c_j))\varphi(x) \in \mathcal{T}_0(\mathcal{C}))$;*
- *and $V_j = c_j{}^\varphi$*

and $u = f(V_1,\ldots,V_q)$ and $V = r^\varphi$.

If $u \rhd_\Delta V$, then U may be replaced by V inside a one-hole generalized context $C[]$; this gives rise to a reduction step $C[u] \rhd_{C[]} C[V]$. We also write $C[u] \rhd_{C[]} C[0]$, i.e. whenever u is erased. However, in this last case, we will not count this step when we define the length of $\rhd$-reductions (see Definition 6). The set of one-hone generalized contexts is denoted by Θ_1 and is defined by setting $\Theta_1 = \bigcup_{i\in\mathbb{N}} \Delta_i^\square$, where $\Delta_i^\square$ is defined by induction on i:

- $\Delta_0^\square = \{U + \square;\ U \in \mathbf{2}\langle\Delta\rangle\}$;
- $$\Delta_{i+1}^\square = \bigcup_{m\in\mathbb{N}} \left\{ \begin{array}{c} U+ \\ f(U_1,\ldots,U_m) \end{array};\ \begin{array}{l} U \in \mathbf{2}\langle\Delta\rangle, f \in \mathcal{D}_m \text{ and} \\ (\exists j \in \{1,\ldots,m\})(U_j \in \Delta_i^\square \text{ and} \\ U_1,\ldots,U_{j-1},U_{j+1},\ldots,U_m \in \mathbf{2}\langle\Delta\rangle) \end{array} \right\}.$$

More generally, we can define the set Θ_i of i-hole generalized contexts: if $i = 0$, then $\Theta_i = \mathbf{2}\langle\Delta\rangle$; if $i = 1$, then Θ_i is already defined. Now, for $i > 1$, if $D[] \in \Theta_i$, then $D[] = C[f(U_1,\ldots U_m)]$ with $C[] \in \Theta_1$ and $U_1 \in \Theta_{i_1},\ldots,U_m \in \Theta_{i_m}$, $i_1 + \ldots + i_m = i$, so the several holes have to be in the same summand.

Definition 5. *Let $C[] \in \Theta_1$. We define the binary relation $\rhd_{C[]}$ on $\mathit{2}\langle\Delta\rangle$ as follows: for any $U, U' \in \mathit{2}\langle\Delta\rangle$, we have $U \rhd_{C[]} U'$ if, and only if, there exist $u \in \Delta$ and $V \in \mathit{2}\langle\Delta\rangle$ such that $U = C[u]$, $U' = C[V]$ and ($u \rhd_\Delta V$ or $V = 0$).*

Then we define the binary relation $\rhd$ on $\mathbf{2}\langle\Delta\rangle$ by writing (as usual) $U \rhd V$ if, and only if, there exists $C[] \in \Theta_1$ such that $U \rhd_{C[]} V$.

For any generalized term $t \in \Delta$, we denote by $\|t\|$ the maximum number of distict summands occurring anywhere in t. In particular, if $t \in \Delta_0 = \mathcal{T}_0(\mathcal{C})$, we have $\|t\| = 1$. We generalize this notation to any element U of $\mathbf{2}\langle\Delta\rangle$ by setting $\|U\| = \begin{cases} 0 & \text{if } U = 0; \\ \max\{\|u\|;\ u \in \mathit{Supp}(U)\} & \text{otherwise.} \end{cases}$

For any $k \in \mathbb{N}$, we denote by $\mathbf{2}\langle\Delta\rangle_k$ the set $\{U \in \mathbf{2}\langle\Delta\rangle;\ \|U\| \leq k\}$ and by $\rhd_k$ the restriction of the binary relation $\rhd$ to $\mathbf{2}\langle\Delta\rangle_k$, i.e. $\mathbf{2}\langle\Delta\rangle_k = \rhd_{|\mathbf{2}\langle\Delta\rangle_k \times \mathbf{2}\langle\Delta\rangle_k}$. For any $k \in \mathbb{N}$, the relation $\rhd_k^*$ enjoys the following properties:

- For any $U, V \in \mathbf{2}\langle\Delta\rangle_k$, we have $U + V \rhd_k^* V$.
- Let $q \in \mathbb{N}$. Let $W_1, \ldots, W_q, V_1, \ldots, V_q \in \mathbf{2}\langle\Delta\rangle_k$ such that $W_1 \rhd_k^* V_1, \ldots, W_q \rhd_k^* V_q$. Then we have $\sum_{j=1}^q W_j \rhd_k^* \sum_{j=1}^q V_j$. Moreover, for any $C[] \in \Theta_q$ such that $C[W_1, \ldots, W_q] \in \mathbf{2}\langle\Delta\rangle_k$, we have $C[W_1, \ldots, W_q] \rhd_k^* C[V_1, \ldots, V_q]$.

Definition 6. *Let $k \in \mathbb{N}$ and let $U, V \in \mathbf{2}\langle\Delta\rangle$.*

We denote by $\mathbf{Seq}_\Delta(U, V)$ (resp. $\mathbf{Seq}_{\Delta,k}(U, V)$) the set of finite sequences $(U_1, \ldots, U_n) \in \mathbf{2}\langle\Delta\rangle^{<\infty}$ such that $U = U_1$, $V = U_n$ and, for any $i \in \{1, \ldots, n-1\}$, we have $U_i \rhd U_{i+1}$ (resp. $U_i \rhd_k U_{i+1}$).

For any $(U_1, \ldots, U_n) \in \mathbf{Seq}_\Delta(U, V)$, we denote by $\mathit{length}_\Delta((U_1, \ldots, U_n))$ the integer $\mathit{Card}\left(\left\{ i \in \{1, \ldots, n-1\};\ \begin{array}{l} (\exists C[] \in \Theta_1, u \in \Delta_1 \setminus \Delta_0, V \in \mathbf{2}\langle\Delta\rangle) \\ (u \rhd_\Delta V, U_i = C[u] \text{ and } U_{i+1} = C[V]) \end{array} \right\}\right)$.

Definition 7. *For any $(\rho, v, C[]) \in \mathbf{Seq} \times \mathcal{U}_0 \times \Phi_1$ such that $\rho \in \mathbf{Seq}(C[v], -)$, we define $\mathcal{R}(\rho, v, C[]) \subseteq \mathbf{Seq}(v, -)$ by induction on $\mathit{length}(\rho)$ as follows:*

- *if $\mathit{length}(\rho) = 0$, then $\mathcal{R}(\rho, v, C[]) = \{id_v\}$;*
- *if $\rho = C_0[u] \to_{C_0[],(l,r)} C_0[u'];\ \rho_0$ with $C_0[] = C[C'[]]$, then $\mathcal{R}(\rho, v, C[]) = \{(v \to_{C'[],(l,r)} C'[u'];\ \rho_0');\ \rho_0' \in \mathcal{R}(\rho_0, C'[u'], C[])\}$;*
- *if $\rho = C_0[u] \to_{C_0[],(l,r)} C_0[u'];\ \rho_0$ and there is no $C'[] \in \Phi_1$ such that $C_0[] = C[C'[]]$, then $\mathcal{R}(\rho, v, C[])$ is the set*

$$\{id_v\} \cup \left(\bigcup_{\substack{C''[] \in \Phi_1 \\ \langle v | C''[] \rangle \in \langle v | C[] \rangle \setminus C_0[u] \to_{C_0[],(l,r)} C_0[u']}} \mathcal{R}(\rho_0, v, C''[]) \right)$$

and we set $\mathcal{N}(\rho, v, C[]) = \{c \in \mathcal{T}_0(\mathcal{C});\ \mathcal{R}(\rho, v, C[])) \cap \mathbf{Seq}(v, c) \neq \emptyset\}$.

In other words, $\mathcal{N}(\rho, v, C[])$ is the set of constructor terms that are descendants of the occurrence $\langle v | C[] \rangle$ of v in $C[v]$ during the reduction ρ. Notice that in the case ρ is an innermost reduction sequence, the set $\mathcal{R}(\rho, v, C[])$ is a singleton.

Lemma 1. *Let $m \in \mathbb{N}$. Let $E[] \in \Phi_m$. Let $u_1, \ldots, u_m \in \mathcal{U}_0$. Let $c \in \mathcal{T}_0(\mathcal{C})$. Let $\rho \in \mathbf{Seq}(E[u_1, \ldots, u_m], c)$. For any $l \in \{1, \ldots, m\}$, let $U_l \in \mathbf{2}\langle\Delta\rangle$ such that $\mathcal{N}(\rho, u_l, E[u_1, \ldots, u_{l-1}, \Box, u_{l+1}, \ldots, u_m]) \subseteq \mathit{Supp}(U_l)$. Then $E[U_1, \ldots, U_m] \rhd^* c$.*

Definition 8. *Let $t \in \mathcal{U}_0$. For any $u \in \Delta$, we define the relation $t \downarrow u$ by induction on level(u) as follows:*

- *if $u \in \Delta_0$, then $t \downarrow u$ if, and only if, $t \to^* u$;*
- *if $u = f(V_1, \dots, V_q) \in \Delta_{i+1} \setminus \Delta_i$, then $t \downarrow u$ if, and only if, there exist $v_1, \dots, v_q \in \mathcal{U}_0$ such that $t \to^* f(v_1, \dots, v_q)$ and, for any $j \in \{1, \dots, q\}$, for any $v \in Supp(V_j)$, $v_j \downarrow v$.*

This relation is extended to the relation $\downarrow \subseteq \mathcal{U}_0 \times 2\langle\Delta\rangle$ defined by: $t \downarrow U$ if, and only if, for any $u \in Supp(U)$, $t \downarrow u$.

Notice that, for any $t, u \in \mathcal{U}_0$, we have $t \downarrow u$ if, and only if, $t \to^* u$.

Lemma 2. *Let $C[] \in \Theta_1$. Let $t \in \mathcal{U}_0$, $U, V \in 2\langle\Delta\rangle$ such that $t \downarrow U$ and $U \rhd_{C[]} V$. Then $t \downarrow V$.*

Proposition 1. *Let $t \in \mathcal{U}_0$, $c \in \mathcal{T}_0(\mathcal{C})$. We have $t \to^* c$ if, and only if, $t \rhd^* c$.*

Proof: Assume that $t \to^* c$. We have $t \in \Phi_0$ and $\mathbf{Seq}(t, c) \neq \emptyset$. Therefore, by Lemma 1, we have $t \rhd^* c$.

Conversely, we prove, by induction on n and applying Lemma 2, that, for any $n \in \mathbb{N}$, for any $U_0, \dots, U_n \in \mathbf{2}\langle\Delta\rangle$ such that $t = U_0 \rhd U_1 \dots U_{n-1} \rhd U_n$, we have $t \downarrow U_n$.

Example 1. Consider the following (constrained) cons-free TRS: the set $\mathcal{D}$ is $\{k/1, h/2, p/1\}$ and the set $\mathcal{C}$ is $\{c/2, n/0, \mathsf{true}/0, \mathsf{false}/0\}$ with the following rewrite rules:

- $p(c(x, c(y, z))) \to x$
- $p(c(x, c(y, z))) \to y$
- $h(x, \mathsf{false}) \to x$
- $k(x) \to h(x, x)$

We have $k(p(c(\mathsf{true}, c(\mathsf{false}, n)))) \to h(p(c(\mathsf{true}, c(\mathsf{false}, n))), p(c(\mathsf{true}, c(\mathsf{false}, n))))$ $\to^* h(\mathsf{true}, \mathsf{false}) \to \mathsf{true}$ (notice that there is no innermost reduction sequence from $k(p(c(\mathsf{true}, c(\mathsf{false}, n))))$ to true). Now, we have

$$\begin{aligned} k(p(c(\mathsf{true}, c(\mathsf{false}, n)))) &\rhd_2 k(p(c(\mathsf{true}, c(\mathsf{false}, n))) + \mathsf{false}) \\ &\rhd_2 k(\mathsf{true} + \mathsf{false}) \\ &\rhd_2 h(\mathsf{true} + \mathsf{false}, \mathsf{true} + \mathsf{false}) \\ &\rhd_2^* h(\mathsf{true}, \mathsf{false}) \\ &\rhd_2 \mathsf{true} \end{aligned}$$

3 Computation in Constrained Cons-Free TRS

In this section, we show that, for any constrained cons-free TRS, it is enough to consider $\rhd$-reduction sequences $(U_i)_{i \in \mathbb{N}}$ of elements of $\mathbf{2}\langle\Delta\rangle_K$ (i.e. $\rhd_K$-reduction sequences) for some integer K depending only on the TRS.

We are given a constrained cons-free TRS and we set $\mathcal{B} = \mathcal{D} \setminus \mathcal{A}$ and as the TRS is finite, we let $K \geq 1$ be an integer such that, for any $(f(c_1, \dots, c_q), r) \in \mathcal{R}$ for any $x \in \mathcal{X} \cap \{c_1, \dots, c_q\}$, $Oc(x, r) \leq K$.

Definition 9. *Let $i \in \mathbb{N}$. For any $U \in \mathcal{2}\langle \Delta_i \rangle$, we define $U^* \in \mathcal{2}\langle \Delta_i \rangle \cap \mathcal{2}\langle \Delta \rangle_K$ by induction on i:*

- *$i = 0$: we set $U^* = U$;*
- *$i > 0$, $U = f(U_1, \ldots, U_q) \in \Delta_i \setminus \Delta_{i-1}$: we set*

$$U^* = \sum_{\substack{W_1, \ldots, W_q \in \mathcal{2}\langle \Delta_{i-1} \rangle \\ Supp(W_j) \subseteq Supp({U_j}^*) \\ Card(Supp(W_j)) = \min\{Z, Card(Supp({U_j}^*))\} \\ for\ j \in \{1, \ldots, q\}}} f(W_1, \ldots, W_q),$$

where $Z = \begin{cases} 1 & if\ f \in \mathcal{A}; \\ K & if\ f \in \mathcal{B}. \end{cases}$

- *$i > 0$, $U \in \mathcal{2}\langle \Delta_i \rangle \setminus (\Delta_i \cup \mathcal{2}\langle \Delta_{i-1} \rangle)$: we set $U^* = \sum_{u \in Supp(U)} u^*$.*

Definition 10. *Let $q \in \mathbb{N}$. Let $c_1, \ldots, c_q \in \mathcal{T}(\mathcal{C})$. Let $\varphi : \mathcal{X} \rightharpoonup \mathcal{2}\langle \mathcal{T}_0(\mathcal{C}) \rangle$. Assume that, for any $l \in \{1, \ldots, q\}$, we have $Oc(c_l) \subseteq dom(\varphi)$ and $(c_l \notin \mathcal{X} \Rightarrow {c_l}^\varphi \in \mathcal{T}_0(\mathcal{C}))$. Let $W_1, \ldots, W_q \in \mathcal{2}\langle \mathcal{T}_0(\mathcal{C}) \rangle$ such that, for any $l \in \{1, \ldots, q\}$, $(c_l \notin \mathcal{X} \Rightarrow W_l = {c_l}^\varphi)$ and $Supp(W_l) \subseteq Supp({c_l}^\varphi)$. Then we denote by $\varphi_{(W_1, \ldots, W_q)}$ the partial function $\mathcal{X} \rightharpoonup \mathcal{2}\langle \mathcal{T}_0(\mathcal{C}) \rangle$ such that, for any $x \in \mathcal{X}$, we have $\varphi_{(W_1, \ldots, W_q)}(x) = \begin{cases} W_l & if\ x = c_l; \\ \varphi(x) & otherwise. \end{cases}$*

Lemma 3. *Let $(f(c_1, \ldots, c_q), r) \in \mathcal{R}$. Let $\varphi : \mathcal{X} \rightharpoonup \mathcal{2}\langle \mathcal{T}_0(\mathcal{C}) \rangle$. Assume that, for any $j \in \{1, \ldots, q\}$, we have $Oc(c_j) \subseteq dom(\varphi)$ and $(c_j \notin \mathcal{X} \Rightarrow {c_j}^\varphi \in \mathcal{T}_0(\mathcal{C}))$. Then $\sum_{W_1 \in \mathcal{W}_1, \ldots, W_q \in \mathcal{W}_q} r^{\varphi(W_1, \ldots, W_q)} \rhd_K^* (r^\varphi)^*$, where, for any $j \in \{1, \ldots, q\}$, $\mathcal{W}_j$ is the following subset of $\mathcal{2}\langle \mathcal{T}_0(\mathcal{C}) \rangle$:*

- *$\{{c_j}^\varphi\}$ in the case $c_j \notin \mathcal{X}$;*
- *$\left\{ W \in \mathcal{2}\langle \mathcal{T}_0(\mathcal{C}) \rangle; \begin{array}{l} Supp(W) \subseteq Supp({c_j}^\varphi)\ and \\ Card(Supp(W)) = \min\{Oc(c_j, r), Card(Supp({c_j}^\varphi))\} \end{array} \right\}$ in the case $c_j \in \mathcal{X}$.*

Proposition 2. *Let $i \in \mathbb{N}$. Let $C[] \in \Delta_i^\square$. For any $U, V \in \mathcal{2}\langle \Delta \rangle$ such that $U \rhd_{C[]} V$, we have $U^* \rhd_K^* V^*$.*

Proof: The proof is by induction on i.

- If $i = 0$, then $C[] = U_0 + \square$ for some $U_0 \in \mathbf{2}\langle \Delta \rangle$; we distinguish between two cases:
  - $V = U_0$ and there exists $u \in \Delta$ such that $U = U_0 + u$: in this case, we have $\|U^*\|, \|{U_0}^*\| \leq K$. Thus we have $U^* = {U_0}^* + u^* \rhd_K^* {U_0}^* = V^*$.
  - $U = U_0 + f(c_1, \ldots, c_q)^\varphi$, $V = U_0 + r^\varphi$ with $f \in \mathcal{D}_q$, $c_1, \ldots, c_q \in \mathcal{T}(\mathcal{C})$, $\varphi : \mathcal{X} \rightharpoonup \mathbf{2}\langle \mathcal{T}_0(\mathcal{C}) \rangle$ such that, for any $j \in \{1, \ldots, q\}$, $c_j \notin \mathcal{X} \Rightarrow {c_j}^\varphi \in \mathcal{T}_0(\mathcal{C})$: For any $j \in \{1, \ldots, q\}$, we define $\mathcal{W}_j$ as in Lemma 3 and we define ${\mathcal{W}'}_j$ as follows:
    - if $c_j \notin \mathcal{X}$, then ${\mathcal{W}'}_j = \{{c_j}^\varphi\}$;

* if $c_j \in \mathcal{X}$, then $\mathcal{W}'_j$ is the set

$$\left\{ W \in \mathbf{2}\langle \mathcal{T}_0(\mathcal{C})\rangle;\ \begin{array}{l} \mathit{Supp}(W) \subseteq \mathit{Supp}({c_j}^{\varphi}) \text{ and} \\ \mathit{Card}(\mathit{Supp}(W)) = \min\{Z, \mathit{Card}(\mathit{Supp}({c_j}^{\varphi}))\} \end{array} \right\}$$

with $Z = \begin{cases} 1 & \text{if } f \in \mathcal{A}; \\ K & \text{if } f \in \mathcal{B}. \end{cases}$

We have

$$\begin{aligned} (f(c_1,\ldots,c_q)^{\varphi})^* &= \sum_{W_1 \in \mathcal{W}'_1,\ldots,W_q \in \mathcal{W}'_q} f({c_1}^{\varphi(W_1,\ldots,W_q)},\ldots,{c_q}^{\varphi(W_1,\ldots,W_q)}) \\ &\rhd_K^* \sum_{W_1 \in \mathcal{W}_1,\ldots,W_q \in \mathcal{W}_q} f(c_1,\ldots,c_q)^{\varphi(W_1,\ldots,W_q)} \\ &\rhd_K^* \sum_{W_1 \in \mathcal{W}_1,\ldots,W_q \in \mathcal{W}_q} r^{\varphi(W_1,\ldots,W_q)} \\ &\rhd_K^* (r^{\varphi})^* \text{ (by Lemma 3).} \end{aligned}$$

We have $U_0 \in \mathbf{2}\langle \Delta\rangle_K$, so $U^* = {U_0}^* + (f(c_1,\ldots,c_q)^{\varphi})^* \rhd_K^* {U_0}^* + (r^{\varphi})^* = V^*$.

– If $i > 0$, $U = C[U_0]$, $V = C[V_0]$, $C[] = U' + f(U_1,\ldots,U_m)$, $k \in \{1,\ldots,m\}$, $U_k \in \Delta_{i-1}^{\square}$ and $U_0 \rhd_{\square} V_0$, then, by the induction hypothesis, $U_k[{U_0}^*] \rhd_K^* U_k[{V_0}^*]$. Let $V_1,\ldots,V_m \in \mathbf{2}\langle \Delta\rangle$ such that
  - for any $j \in \{1,\ldots,m\} \setminus \{k\}$, we have

$$\mathit{Supp}(V_j) \subseteq \mathit{Supp}({U_j}^*) \text{ and } \mathit{Card}(\mathit{Supp}(V_j)) = \min\{1, \mathit{Card}(\mathit{Supp}({U_j}^*))\}$$

  - and $\mathit{Supp}(V_k) \subseteq \mathit{Supp}(U_k[V_0]^*)$ and

$$\mathit{Card}(\mathit{Supp}(V_k)) = \begin{cases} \min\{1, \mathit{Card}(\mathit{Supp}(U_k[V_0]^*))\} & \text{if } f \in \mathcal{A}; \\ \min\{K, \mathit{Card}(\mathit{Supp}(U_k[V_0]^*))\} & \text{if } f \in \mathcal{B}. \end{cases}$$

There exists $V \in \mathbf{2}\langle \Delta\rangle$ such that $\mathit{Supp}(V) \subseteq \mathit{Supp}(U_k[U_0])$, $\mathit{Card}(\mathit{Supp}(V)) \leq \mathit{Card}(\mathit{Supp}(V_k))$ and $V \rhd_K^* V_k$, and hence $f(V_1,\ldots,V_{k-1},V,V_{k+1},\ldots,V_m) \rhd_K^* f(V_1,\ldots,V_k)$. We obtain

$$\begin{aligned} & f(U_1,\ldots,U_{k-1},U_k[U_0],U_{k+1},\ldots,U_m)^* \\ \rhd_K^* \ & f(U_1,\ldots,U_{k-1},U_k[V_0],U_{k+1},\ldots,U_m)^*; \end{aligned}$$

moreover we have $\|U'^*\| \leq K$, hence

$$\begin{aligned} U^* &= U'^* + f(U_1,\ldots,U_{k-1},U_k[U_0],U_{k+1},\ldots,U_m)^* \\ &\rhd_K^* U'^* + f(U_1,\ldots,U_{k-1},U_k[V_0],U_{k+1},\ldots,U_m)^* \\ &= V^* \end{aligned}$$

Corollary 1. *Let $q \in \mathbb{N}$. Let $C_1,\ldots,C_q \in \mathbf{2}\langle \mathcal{T}_0(\mathcal{C})\rangle$ such that $\mathit{Card}(\mathit{Supp}(C_1))$, $\ldots$, $\mathit{Card}(\mathit{Supp}(C_q)) \leq K$, let $c \in \mathcal{T}_0(\mathcal{C})$ and $f \in \mathcal{D}_q$. We have $f(C_1,\ldots,C_q) \rhd^* c$ if, and only if, $f(C_1,\ldots,C_q) \rhd_K^* c$.*

Proof: Apply Proposition 2, noticing that $f(C_1,\ldots,C_q)^* = f(C_1,\ldots,C_q)$ and $c^* = c$.

4 A Polynomial Time Algorithm

In this section we describe a polynomial time algorithm that computes the constructor terms obtained by $\rhd_K$-reduction sequences in a constrained cons-free TRS. Assume that we are given a constrained cons-free TRS. We assume that $\mathcal{R} = \{(l_1, r_1), \ldots, (l_R, r_R)\}$ and $\bigcup_{j=1}^{R} Oc(l_j) = \{x_1, \ldots, x_V\}$. We set $\mathcal{T}_\mathcal{R} = \bigcup\{t \in \mathcal{T}(\mathcal{D} \cup \mathcal{C});\ (\exists j \in \{1, \ldots, R\})(t \unlhd l_j \text{ or } t \unlhd r_j)\}$. We set $A = \max\{ar(f);\ f \in \mathcal{D} \cup \mathcal{C}\}$, $O = \max\{1, \max\{Oc(f, r_j);\ f \in \mathcal{D} \text{ and } j \in \{1, \ldots, R\}\}\}$, $Q = Card(\mathcal{D})$ and $S = Card(\mathcal{C}_0)$. For any $c_0 \in \mathcal{T}_0(\mathcal{C})$, we set $\mathcal{I}(c_0) = \{c \in \mathcal{T}_0(\mathcal{C});\ c \in \mathcal{C}_0 \text{ or } c \unlhd c_0\}$.

Remark 1. For any $c_0 \in \mathcal{T}_0(\mathcal{C})$, we have $Card(\mathcal{I}(c_0)) = S + |c_0|$.

Definition 11. *For any $c_0 \in \mathcal{T}_0(\mathcal{C})$, we set $\mathcal{V}(c_0) = \bigcup_{i \in \mathbb{N}} \mathcal{V}_i(c_0)$, where $\mathcal{V}_i(c_0)$ is a subset of $\{U \in \mathcal{2}\langle \Delta_i \rangle;\ \|U\| \leq K\}$ defined by induction on i:*

- $\mathcal{V}_0(c_0) = \{U \in \mathcal{2}\langle \mathcal{T}_0(\mathcal{C}) \rangle;\ Card(Supp(U)) \leq K \text{ and } Supp(U) \subseteq \mathcal{I}(c_0)\}$
- $\mathcal{V}_{i+1}(c_0) = \mathcal{V}_i(c_0) \cup \bigcup_{m=1}^{A} \{f(V_1, \ldots, V_m);\ f \in \mathcal{D}_m \text{ and } V_1, \ldots, V_m \in \mathcal{V}_i(c_0)\}$

Given $c_0 \in \mathcal{T}_0(\mathcal{C})$, the algorithm will compute, for every element u of $\mathcal{V}_1(c_0) \setminus \mathcal{V}_0(c_0)$, the set of constructor terms c such that $u \rhd_K^* c$. In particular, by Proposition 1 and Corollary 1, if K is large enough, then, for every $f \in \mathcal{D}_m$ and every $c_1, \ldots, c_m \in \mathcal{I}(c_0)$, it will return exactly all the constructor terms c such that $f(c_1, \ldots, c_m) \to^* c$.

Remark 2. For any $c_0 \in \mathcal{T}_0(\mathcal{C})$, we have $Card(\mathcal{V}_0(c_0)) \leq Card(\mathcal{I}(c_0))^{K+1}$, hence $Card(\mathcal{V}_1(c_0) \setminus \mathcal{V}_0(c_0)) \leq Q \cdot Card(\mathcal{I}(c_0))^{A \cdot (K+1)}$.

Definition 12. *For any $c_0 \in \mathcal{T}_0(\mathcal{C})$, for any $i \in \mathbb{N} \setminus \{0\}$, for any $V \in \mathcal{V}_i(c_0) \setminus \mathcal{V}_{i-1}(c_0)$, we define, by induction on i,* the leftmost-innermost redex $\langle U | E[] \rangle$ *of V with $U \in \mathcal{V}_1(c_0) \setminus \mathcal{V}_0(c_0)$ and $E[] \in \Theta_1$:*

- *if $i = 1$, then the leftmost-innermost redex of V is $\langle V | \Box \rangle$;*
- *if $i > 1$ and $V = f(V_1, \ldots, V_m)$, then the leftmost-innermost redex of V is $\langle W | f(V_1, \ldots, V_{j-1}, C[], V_{j+1}, \ldots, V_m) \rangle$, where $j = \min\{k \in \{1, \ldots, m\};\ V_k \notin \mathcal{V}_0\}$ and $\langle W | C[] \rangle$ is the leftmost-innermost redex of V_j.*

From now M will be an integer and E the subset $\{1, \ldots, M\}$ of $\mathbb{N}$; L will be a function $E \to \mathcal{D} \cup \mathcal{C} \cup \{x_1, \ldots, x_V, \perp\}$, $Succ$ will be a partial function $E \rightharpoonup (E \cup \{\perp\})^{\{1, \ldots, A\}}$ and $Comp$ will be a partial function $E \rightharpoonup (E \cup \{\perp\})^{\{1, \ldots, K\}}$.

Definition 13. *For any $t \in \mathcal{T}(\mathcal{D} \cup \mathcal{C})$, for any $n \in E$, we define, by induction on t, $F_\mathcal{T}(t, n) \in \{0, 1\}$ as follows: $F_\mathcal{T}(f(t_1, \ldots, t_m), n) = 1$ if, and only if, $n \in dom(Succ)$, $L(n) = f$ and $F_\mathcal{T}(t_1, Succ(n)(1)) = \ldots = F_\mathcal{T}(t_m, Succ(n)(m)) = 1$.*

Notice that $F_\mathcal{T}(t, n) = F_\mathcal{T}(t', n) = 1 \Rightarrow t = t'$, hence we can define a partial function $[\![\cdot]\!]_\mathcal{T} : E \rightharpoonup \mathcal{T}(\mathcal{D} \cup \mathcal{C})$ by setting $[\![n]\!]_\mathcal{T} = t$ if, and only if, $F_\mathcal{T}(t, n) = 1$. In the same way, we define a partial function $[\![\cdot]\!]_{\mathcal{V}, c_0} : E \rightharpoonup \mathcal{V}(c_0)$ for any $c_0 \in \mathcal{T}_0(\mathcal{C})$:

Definition 14. *Let $c_0 \in \mathcal{T}_0(\mathcal{C})$. For any $i \in \mathbb{N}$, for any $V \in \mathcal{V}_i(c_0)$, for any $n \in E$, we define, by induction on i, $F_{\mathcal{V}, c_0}(V, n) \in \{0, 1\}$:*

- *if* $i = 0$, *then* $F_{\mathcal{V},c_0}(V, n) = 1$ *if, and only if,* $[\![n]\!]_{\mathcal{T}} = V$ *or the following holds:* $L(n) = \bot$, $n \in dom(Comp)$ *and* $\sum_{\substack{k \in \{1,\ldots,K\} \\ Comp(n)(k) \neq \bot}} [\![Comp(n)(k)]\!]_{\mathcal{T}} = V$;
- *if* $i > 0$ *and* $V = f(V_1, \ldots, V_m) \notin \mathcal{V}_{i-1}$, *then* $F_{\mathcal{V},c_0}(V, [\![n]\!]_{\mathcal{V},c_0}) = 1$ *if, and only if,* $n \in dom(Succ)$, $L(n) = f$ *and* $[\![Succ(n)(1)]\!]_{\mathcal{V},c_0} = V_1$, $\ldots$, $[\![Succ(n)(ar(f))]\!]_{\mathcal{V},c_0} = V_{ar(f)}$.

Since, for any $c_0 \in \mathcal{T}_0(\mathcal{C})$, we have $F_{\mathcal{V},c_0}(V, v) = F_{\mathcal{V},c_0}(V', n) = 1 \Rightarrow V = V$, we can define a partial function $[\![\cdot]\!]_{\mathcal{V},c_0} : E \rightharpoonup \mathcal{V}(c_0)$ by setting $[\![n]\!]_{\mathcal{V},c_0} = V$ if, and only if, $F_{\mathcal{V},c_0}(V, n) = 1$.

In the two following definitions, we restrict the partial functions $[\![\cdot]\!]_{\mathcal{T}}$ and $[\![\cdot]\!]_{\mathcal{V},c_0}$ to elements of E that *unshare* (hence the symbol U) defined symbol functions.

Definition 15. *For any* $n \in dom([\![\cdot]\!]_{\mathcal{T}})$, *we define* $Reach_{\mathcal{T}}(n) \subseteq E$ *by induction on* $[\![n]\!]_{\mathcal{T}}$: *if* $L(n) \notin \mathcal{D}$, *then* $Reach_{\mathcal{T}}(n) = \emptyset$; *if* $L(n) \in \mathcal{D}$, *then* $Reach_{\mathcal{T}}(n) = \{n\} \cup \bigcup_{j=1}^{ar(L(n))} Reach_{\mathcal{T}}(Succ(n)(j))$.

We set $U_{\mathcal{T}} = \{n \in dom([\![\cdot]\!]_{\mathcal{T}});\ (\forall m, m' \in Reach_{\mathcal{T}}(n))(L(m) = L(m') \Rightarrow (m \neq m' \text{ or } L(m) \notin \mathcal{D}))\}$ *and* $[\![\cdot]\!]_{\mathcal{T},U} = [\![\cdot]\!]_{\mathcal{T}}|_{U_{\mathcal{T}}}$.

Definition 16. *For any* $c_0 \in \mathcal{T}_0(\mathcal{C})$, *for any* $i \in \mathbb{N}$, *for any* $V \in \mathcal{V}_i(c_0)$, *for any* $n \in E$ *such that* $[\![n]\!]_{\mathcal{V},c_0} = V$, *we define* $Reach_{\mathcal{V},c_0}(n)$ *by induction on* i:

- *if* $i = 0$ *and* $L(n) = \bot$, *then*

$$Reach_{\mathcal{V},c_0}(n) = \bigcup_{\substack{k \in \{1,\ldots,K\} \\ Comp(n)(k) \neq \bot}} Reach_{\mathcal{T}}(Comp(n)(k))$$

- *if* $i = 0$ *and* $L(n) \neq \bot$, *then* $Reach_{\mathcal{V},c_0}(n) = Reach_{\mathcal{T}}(n)$
- *if* $i > 0$, *then* $Reach_{\mathcal{V},c_0}(n) = \{n\} \cup \bigcup_{j=1}^{ar(L(n))} Reach_{\mathcal{V},c_0}(Succ(n)(j))$

For any $c_0 \in \mathcal{T}_0(\mathcal{C})$, *we denote by* $U_{\mathcal{V}}(c_0)$ *the set of* $n \in dom([\![\cdot]\!]_{\mathcal{V},c_0})$ *such that* $(\forall m, m' \in Reach_{\mathcal{V},c_0}(n))(L(m) = L(m') \Rightarrow (m \neq m' \text{ or } L(m) \notin \mathcal{D}))$ *and we set* $[\![\cdot]\!]_{\mathcal{V},c_0,U} = [\![\cdot]\!]_{\mathcal{V},c_0}|_{U_{\mathcal{V}}(c_0)}$.

From now, we assume that, for any $c_0 \in \mathcal{T}_0(\mathcal{C})$, we are given a bijection $inp(c_0) : \begin{array}{ccc} \{1, \ldots, Inp\text{-}Max\} & \to & \mathcal{V}_1(c_0) \setminus \mathcal{V}_0(c_0) \\ i & \mapsto & [\![i]\!]_{\mathcal{V},c_0} \end{array}$. By Remark 2, we can assume that $Inp\text{-}Max \leq Q \cdot Card(\mathcal{I}(c_0))^{A\cdot(K+1)}$.

The algorithm begins with a procedure Inst-Init() which performs the following one: for any $c_0 \in \mathcal{T}_0(\mathcal{C})$, after the execution of the procedure, we have

- $\{[\![Inp(i)]\!]_{\mathcal{V},c_0};\ 1 \leq i \leq Inp\text{-}Max\} = \mathcal{V}_1(c_0) \setminus \mathcal{V}_0(c_0)$
- and, for any $i \in \{1, \ldots, Inp\text{-}Max\}$, we have $\{[\![Inst(i)(s)]\!]_{\mathcal{V},c_0,U};\ 1 \leq s \leq R\} \setminus \{\bot\} = \{V \in \mathbf{2}\langle\Delta\rangle;\ [\![Inp(i)]\!]_{\mathcal{V},c_0} \rhd_{\Delta} V\}$.

Here we used the following crucial property of cons-free term rewriting systems: whenever we perform a reduction step $u \rhd_{\Delta} V$ with $u \in \mathcal{V}_1(c_0) \setminus \mathcal{V}_0(c_0)$, we have $V \in \mathcal{V}(c_0)$ (and not only in $\mathbf{2}\langle\Delta\rangle$).

The algorithm calls the procedure $\mathsf{Inf\text{-}Inp}(i,j)$ with $i,j \in \{1,\ldots,\mathit{Inp\text{-}Max}\}$. This procedure performs the following one: for any $c_0 \in \mathcal{T}_0(\mathcal{C})$, if there exist $q \in \mathbb{N}$, $f \in \mathcal{D}_q$ and $C_1,\ldots,C_q,C'_1,\ldots,C'_q \in \mathbf{2}\langle\mathcal{T}_0(\mathcal{C})\rangle$ such that

- $[\![\mathit{Inp}(i)]\!]_{\mathcal{V},c_0} = f(C'_1,\ldots,C'_q)$,
- $[\![\mathit{Inp}(j)]\!]_{\mathcal{V},c_0} = f(C_1,\ldots,C_q)$
- and $\mathit{Supp}(C'_1) \subseteq \mathit{Supp}(C_1)$, ..., $\mathit{Supp}(C'_q) \subseteq \mathit{Supp}(C_q)$,

then the procedure $\mathsf{Inf\text{-}Inp}(i,j)$ returns true; otherwise it returns false.

Definition 17. *Let $c_0 \in \mathcal{T}_0(\mathcal{C})$. For any $i \in \mathbb{N}$, we define some subset $\Psi_i(c_0)$ of $\Delta_i^{\square}$ by induction on i as follows: $\Psi_0(c_0) = \{\square\}$ and*

$$\Psi_{i+1}(c_0) = \bigcup_{m\in\mathbb{N}} \left\{ f(V_1,\ldots,V_m);\ \begin{array}{l} f \in \mathcal{D}_m \text{ and } (\exists j \in \{1,\ldots,m\}) \\ (V_j \in \Psi_i(c_0) \text{ and} \\ V_1,\ldots,V_{j-1},V_{j+1},\ldots,V_m \in \mathcal{V}(c_0)) \end{array} \right\}.$$

We set $\Psi(c_0) = \bigcup_{i\in\mathbb{N}} \Psi_i(c_0)$.

Definition 18. *Let $c_0 \in \mathcal{T}_0(\mathcal{C})$. Let $Y : \mathcal{V}_1(c_0) \setminus \mathcal{V}_0(c_0) \to \{\mathsf{true},\mathsf{false}\}^{\mathcal{I}}$. For any $C[] \in \Psi(c_0)$, we define the binary relation on $\mathcal{V}(c_0)$ as follows: $V \rhd_{Y,C[]} V'$ if, and only if, there exist $u \in \mathcal{V}_1(c_0) \setminus \mathcal{V}_0(c_0)$ and $V_0 \in \mathcal{V}_0(c_0)$ such that $\mathit{Supp}(V_0) \subseteq \{c \in \mathcal{I}(c_0);\ Y(u)(c) = \mathsf{true}\}$, $V = C[u]$ and $V' = C[V_0]$.*

We define the binary relation $\rhd_Y$ on $\mathcal{V}(c_0)$ as follows: $V \rhd_Y V'$ if, and only if, there exists $W \in \mathbf{2}\langle\mathcal{T}_0(\mathcal{C})\rangle$ such that

- *$\mathit{Card}(\mathit{Supp}(W)) \leq K$,*
- *for any $w \in \mathit{Supp}(W)$, $Y(U)(w) = \mathsf{true}$*
- *and $V' = E[W]$,*

where $\langle U|E[]\rangle$ is the leftmost-innermost redex of V.

The algorithm uses a procedure $\mathsf{Computation}$, which has the following properties: Let $c_0 \in \mathcal{T}_0(\mathcal{C})$. Let $Q \leq O$. Let $V \in \mathcal{V}_Q(c_0)$. Let $n \in E$ such that $[\![n]\!]_{\mathcal{V},c_0,U} = V$. Let $Y : \mathcal{V}_1(c_0) \setminus \mathcal{V}_0(c_0) \to \{\mathsf{true},\mathsf{false}\}^{\mathcal{I}(c_0)}$ such that, for any $V' \in \mathcal{V}_1(c_0) \setminus \mathcal{V}_0(c_0)$, for any $c \in \mathcal{I}(c_0)$, $Y(V')(c) = \mathsf{true}$ if, and only if, $\mathit{Val}(\mathit{inp}(c_0)^{-1}(V'))(c) = \mathsf{true}$. After the execution of the procedure $\mathsf{Computation}(n)$, we have:

- apart from M and D, which increased, and apart from *Result*, no value of any global variable changed;
- the increasing of D is bound by $(\mathit{Card}(\mathcal{I}(c_0)) + 1)^{Q\cdot K} \cdot K$;
- for any $c \in \mathcal{I}(c_0)$, there exists $j \in \{1,\ldots,D\}$ such that $\mathit{Result}(j) = c$ if, and only if, there exists $V' \in \mathcal{V}_0(c_0)$ such that $c \in \mathit{Supp}(V')$ and $V \rhd_Y^* V'$.

The execution time of the procedure $\mathsf{Computation}$ is polynomial in the size of c_0.

The key-point to notice is that the execution time of the algorithm is in $\mathcal{O}(|c_0|^H)$ for some constant H is that the size of the table *Val* is $\mathit{Card}(\mathit{Inp\text{-}Max}) \times \mathit{Card}(\mathcal{I}(c_0)) \leq (Q \cdot \mathit{Card}(\mathcal{I}(c_0))^{A\cdot(K+1)}) \times (S + |c_0|)$. Hence, for any $c_0 \in \mathcal{T}_0(\mathcal{C})$,

```
Inst-Init(); change := true;
while change do
    change := false;
    for i := 1 to Inp-Max do
        D := 0; for s := 1 to R do Computation(Inst(i)(s)); od;
        for o := 1 to Inp-Max do
            if Inf-Inp(i, o) then
                        for j := 1 to D do
                            if Val(o)(Result(j)) ≠ false
                                then change := true;
                                    Val(o)(Result(j)) := true;
                            fi;
                        od;
            fi;
        od; od; od;
```

Fig. 1. The algorithm

for any $m \in \mathbb{N}$, for any $f \in \mathcal{D}_m$, for any $C_1, \ldots, C_m \in \mathbf{2}\langle \mathcal{T}_0(\mathcal{C})\rangle$ such that $Card(Supp(C_1)), \ldots, Card(Supp(C_m)) \leq K$, for any $c \in \{c' \in \mathcal{T}(\mathcal{C});\ c' \trianglelefteq c_0 \text{ or } c' \in \mathcal{C}_0\}$, the problem of deciding whether $f(C_1, \ldots, C_m) \rhd^*_K c$ holds is solvable in time polynomial in the size of c_0.

5 Characterizing P

Let Γ be the signature {one/1, zero/1, nil/0}. For each $t = f_1(f_2(\cdots f_n(\text{nil}))) \in \mathcal{T}_0(\Gamma)$, we define the string $\langle t\rangle$ to be $\langle f_1\rangle\langle f_2\rangle \cdots \langle f_n\rangle$ where $\langle\text{one}\rangle$ ='1' and $\langle\text{zero}\rangle$ ='0'. Clearly, $\mathcal{T}_0(\Gamma)$ is in bijective correspondence with $\{0,1\}^{<\infty}$ under $\langle\cdot\rangle$.

Jones [3] considers (deterministic) *cons-free functional programs.* Now, the following lemma holds:

Lemma 4. *Any (deterministic) cons-free functional program taking only zeroth-order data and involving only terminating functions can be simulated by an orthogonal cons-free TRS.*

Proof: Given a (deterministic) cons-free functional program p taking only zeroth-order data, we consider the following cons-free TRS: for any declaration of the form $f\, x_1 \ldots x_n = e^f$ in p, we have the rewrite rule $f(x_1, \ldots, x_n) \to (e^f)^*$, where $(e^f)^*$ is defined by induction on e^f: for instance, if $e^f = \mathsf{if}\ e_1\, e_2\, e_3$, then $(e^f)^* = \mathsf{if}({e_1}^*, {e_2}^*, {e_3}^*)$; moreover we have the rewrite rules $\mathsf{if}(\mathsf{true}, x, y) \to x$ and $\mathsf{if}(\mathsf{false}, x, y) \to y$.

As the language of [3] involves only a single function declaration per function name, and all left-hand sides of such declaration have the form $f(x_1, \ldots, x_n)$ (for distinct $x_1, \ldots, x_n$), it is straightforward that the cons-free TRS we obtained is orthogonal. The operational semantics in [3] is essentially call-by-value and can be straightforwardly simulated by innermost reduction steps (the exceptions are whenever we have expressions of the form if e_1 e_2 e_3: following the evaluation of e_1, either e_2 or e_3 will not be evaluated). Hence, if $f\, c_1 \ldots c_n$ evaluates to some normal form in the functional program, then t reduces to the same normal form in the corresponding TRS. Conversely, as orthogonal TRSs are confluent (hence each term has at most one normal form), and the functions are terminating, if $f(c_1, \ldots, c_n)$ reduces to some normal form c in the TRS, then $f\, c_1 \ldots c_n$ evaluates to the value c in the functional program.

Theorem 1. *Let $L \subseteq \{0,1\}^{<\infty}$. Then, $L \in P$ if, and only if, there exists a constrained cons-free TRS over some signature $\mathcal{F} = \mathcal{D} \cup \mathcal{C}$ such that (i) $\Gamma \subseteq \mathcal{C}$, and there is $f \in \mathcal{D}$ and* true $\in \mathcal{C}_0$ *such that, for any $t \in \mathcal{T}_0(\Gamma)$, we have $f(t) \to^*$* true *if, and only if, $\langle t \rangle \in L$.*

Proof: Corollary 24.2.4 of [17] (or Theorem 6.12 of [3] in the case $k = 0$) shows that we can simulate any polynomial-time Turing Machine by a (deterministic) cons-free (called *read-only* in [17]) functional program taking only zeroth-order data (Note that cons-free in the above setting is slightly stronger than our notion: No constructors are allowed in the right-hand side of function declarations). This simulation involves only terminating functions, hence, by Lemma 4, any polynomial-time Turing Machine can be simulated by an orthogonal cons-free TRS.

The cons-free term rewriting system R obtained from a functional program is not necessarily constrained. To obtain a constrained system, we do the following for each function declaration def $f(x_1, \ldots, x_n) = e^f$ (where the function body e^f is an expression in the functional language): Let $f(x_1, \ldots, x_n) \to r$ be the corresponding cons-free rule. For every such rule, let $\{x_1, \ldots, x_n\}$ be the set of variables that occur immediately beneath the defined symbol at the root of the left-hand side. Choose a set $\{y_1, \ldots, y_n\}$ of distinct variables, and let M be the set of all n-tuples $w = (s_1, \ldots, s_n)$ where s_i (for $1 \leq i \leq n$) is either zero(y_i), one(y_i), or nil. Then replace the rule $f(x_1, \ldots, x_n) \to r$ by the $|M|$ rules on the form $f(s_1, \ldots, s_n) \to r[s_1/x_1, \ldots, s_n/x_n]$, where $(s_1, \ldots, s_n) \in M$ and $r[s_1/x_1, \ldots, s_n/x_n]$ denotes the obvious substitution. Observe that (i) each of the new rules is left-linear if the original rule was, and (ii) that the only overlaps between these rules occur when the left-hand sides are equal. Thus, as R was orthogonal, so is R', and it is clearly constrained as no variable in a left-hand side occurs immediately below the defined symbol at the root.

It is obvious that, for any terms t and t' such that $t \to t'$ in R', we have $t \to t'$ in R: indeed if $t \to_{(l,r)} t'$ and (l, r) is not a rule of R, then there exists a unique rule (l_0, r_0) of R such that (l, r) is obtained from (l_0, r_0); we have $t \to_{(l_0,r_0)} t'$. Reciprocally, if t and t' are two terms such that $t = C[f(t_1, \ldots, t_m)^\sigma] \to_{(f(t_1,\ldots,t_m),r)} C[r^\sigma] = t'$ is a innermost reduction step in R, then ${t_1}^\sigma$, ..., ${t_m}^\sigma$ are constructor terms, hence there exists a rule (l, r) in R' such that $t \to_{(l,r)} t'$. Now, since R is confluent and the functions are terminating, for any term t and any constructor term c, we have $t \to^* c$ in R if, and only if, t reduces to c in R by some innermost strategy.

To see that every constrained, cons-free TRS can be suitably simulated by a polynomial-time Turing machine, let $K \geq 1$ be an integer such that, for any rule $(f(c_1, \ldots, c_q), r)$, for any $x \in \mathcal{X} \cap \{c_1, \ldots, c_q\}$, $Oc(x, r) \leq K$. By Proposition 1 and Corollary 1, we have $f(t) \to^* \mathsf{true}$ if, and only if, $f(t) \rhd_K^* \mathsf{true}$. And the previous section showed that the problem of deciding whether $f(t) \rhd_K^* \mathsf{true}$ holds is solvable in time polynomial in the size of t.

References

1. Jones, N.D.: Logspace and ptime characterized by programming languages. Theor. Comput. Sci. 228(1-2), 151–174 (1999)
2. Bonfante, G.: Some programming languages for LOGSPACE and PTIME. In: Johnson, M., Vene, V. (eds.) AMAST 2006. LNCS, vol. 4019, pp. 66–80. Springer, Heidelberg (2006)
3. Jones, N.D.: The expressive power of higher-order types or, life without cons. J. Funct. Program. 11(1), 5–94 (2001)
4. Avanzini, M., Moser, G.: Closing the gap between runtime complexity and polytime computability. In: Lynch, C. (ed.) RTA. LIPIcs, vol. 6, pp. 33–48. Schloss Dagstuhl - Leibniz-Zentrum fuer Informatik (2010)
5. Hofmann, M.: Type systems for polynomial-time computation. Habilitationsschrift (1999)
6. Bellantoni, S., Cook, S.A.: A new recursion-theoretic characterization of the polytime functions. Computational Complexity 2, 97–110 (1992)
7. Bellantoni, S.J., Niggl, K.H., Schwichtenberg, H.: Higher type recursion, ramification and polynomial time. Ann. Pure Appl. Logic 104(1-3), 17–30 (2000)
8. Baillot, P., Lago, U.D.: Higher-Order Interpretations and Program Complexity. In: Cégielski, P., Durand, A. (eds.) Computer Science Logic (CSL 2012). Leibniz International Proceedings in Informatics (LIPIcs), vol. 16, pp. 62–76. Schloss Dagstuhl–Leibniz-Zentrum fuer Informatik, Dagstuhl (2012)
9. Baillot, P.: From proof-nets to linear logic type systems for polynomial time computing. In: Ronchi Della Rocca, S. (ed.) TLCA 2007. LNCS, vol. 4583, pp. 2–7. Springer, Heidelberg (2007)
10. Baillot, P., Gaboardi, M., Mogbil, V.: A polytime functional language from light linear logic. In: Gordon, A.D. (ed.) ESOP 2010. LNCS, vol. 6012, pp. 104–124. Springer, Heidelberg (2010)
11. Avanzini, M., Eguchi, N., Moser, G.: A new order-theoretic characterisation of the polytime computable functions. In: Jhala, R., Igarashi, A. (eds.) APLAS 2012. LNCS, vol. 7705, pp. 280–295. Springer, Heidelberg (2012)
12. Avanzini, M., Moser, G.: Polynomial path orders. Logical Methods in Computer Science 9(4) (2013)
13. Goerdt, A.: Characterizing complexity classes by general recursive definitions in higher types. Inf. Comput. 101(2), 202–218 (1992)
14. Goerdt, A.: Characterizing complexity classes by higher type primitive recursive definitions. Theor. Comput. Sci. 100(1), 45–66 (1992)
15. Kristiansen, L., Niggl, K.H.: On the computational complexity of imperative programming languages. Theor. Comput. Sci. 318(1-2), 139–161 (2004)
16. Terese (ed.): Term Rewriting Systems. Cambridge Tracts in Theoretical Computer Science, vol. 55. Cambridge University Press (2003)
17. Jones, N.D.: Computability and complexity: from a programming perspective. The MIT Press (1997)

Preciseness of Subtyping on Intersection and Union Types*

Mariangiola Dezani-Ciancaglini[1,**] and Silvia Ghilezan[2,***]

[1] Dipartimento di Informatica, Università di Torino, Italy
dezani@di.unito.it

[2] Faculty of Technical Sciences, University of Novi Sad, Serbia
gsilvia@uns.ac.rs

Abstract. The notion of subtyping has gained an important role both in theoretical and applicative domains: in lambda and concurrent calculi as well as in programming languages. The soundness and the completeness, together referred to as the preciseness of subtyping, can be considered from two different points of view: denotational and operational. The former preciseness is based on the denotation of a type which is a mathematical object that describes the meaning of the type in accordance with the denotations of other expressions from the language. The latter preciseness has been recently developed with respect to type safety, i.e. the safe replacement of a term of a smaller type when a term of a bigger type is expected.

We propose a technique for formalising and proving operational preciseness of the subtyping relation in the setting of a concurrent lambda calculus with intersection and union types. The key feature is the link between typings and the operational semantics. We then prove soundness and completeness getting that the subtyping relation of this calculus enjoys both denotational and operational preciseness.

1 Introduction

Preciseness, Soundness and Completeness of Subtyping. A subtyping relation is a pre-order (reflexive and transitive relation) on types that validates the principle: if σ is a subtype of τ (notation $\sigma \leq \tau$), then a term of type σ may be provided whenever a term of type τ is needed; see Pierce [20] (Chapter 15) and Harper [14] (Chapter 23).

The soundness and the completeness, together referred to as the preciseness of subtyping, can be considered from two different points of view: denotational and operational.

* Partly supported by COST IC1201, DART bilateral project between Italy and Serbia.

** Partly supported by MIUR PRIN Project CINA Prot. 2010LHT4KM and Torino University/Compagnia San Paolo Project SALT.

*** Partly supported by the projects ON174026 and III44006 of the Ministry of Education and Science, Serbia.

G. Dowek (ed.): RTA-TLCA 2014, LNCS 8560, pp. 194–207, 2014.

Denotational Preciseness: A usual approach to preciseness of subtyping for a calculus is to consider the interpretation of a type σ (notation $[\![\sigma]\!]$) to be a set that describes the meaning of the type in accordance with the denotations of the terms of the calculus, in general a subset of the domain of a model of the calculus. Then a subtyping relation is denotationally sound when $\sigma \leq \tau$ implies $[\![\sigma]\!] \subseteq [\![\tau]\!]$ and denotationally complete when $[\![\sigma]\!] \subseteq [\![\tau]\!]$ implies $\sigma \leq \tau$. This well-established powerful technique is applied to the pure λ-calculus with arrow and intersection types by Barendregt et al. [3], to a call-by-value λ-calculus with arrow, intersection and union types by van Bakel et al. [1] and by Ishihara and Kurata [16], and to a wide class of calculi with arrow, union and pair types by Vouillon [25].

Operational Preciseness: Blackburn et al. [4] bring a new operational approach to preciseness of subtyping and apply it to subtyping iso-recursive types. They assume a multi-step reduction $M \longrightarrow^* N$ between terms (including error), standard typing judgements $\Gamma \vdash M : \sigma$ and evaluation contexts C.

A subtyping relation is operationally sound when $\sigma \leq \tau$ implies that if $x : \tau \vdash C[x] : \rho$ (for some ρ) and $\vdash M : \sigma$, then $C[M] \not\longrightarrow^* \mathsf{error}$, for all C and M. A subtyping relation is operationally complete when $\sigma \not\leq \tau$ implies that $x : \tau \vdash C[x] : \rho$ and $\vdash M : \sigma$ and $C[M] \longrightarrow^* \mathsf{error}$, for some ρ, C and M.

Semantic subtyping of Frisch et al. [12,13] supports both notions of preciseness for a typed calculus with arrow, intersection, union and negation types. Each type is interpreted as the set of values having that type. The subtyping relation is defined semantically rather than syntactically and the typing algorithms are directly derived from semantics. In this way denotational preciseness is obtained by construction. On the other hand, operational preciseness holds as well, because of the presence of a type case operator in the calculus.

The Concurrent λ-Calculus. Dezani et al. [9] develop the concurrent $\lambda_{+\|}$-calculus, which is an enriched λ-calculus with a demonic non-deterministic choice operator $+$ [19], and an angelic parallel operator $\|$ [5]. Their type system embodies arrow, intersection and union types and a subtyping relation which enables the construction of filter models. The choice of the subtyping relation on types is crucial, since it determines the structure of the set of filters and provides a denotational semantics which ensures soundness and completeness of the type assignment.

Main Contributions. In this paper, we adapt the ideas of Blackburn et al. [4] to the setting of the concurrent $\lambda_{+\|}$-calculus with intersection and union types of [9]. We propose a technique for formalising and proving denotational and operational preciseness of subtyping, operational completeness being the real novelty. The key feature is the link between typings and the operational semantics. For the denotational preciseness we interpret a type as the set of terms having that type. For the operational preciseness we take the view that well-typed terms always evaluate to values. In this calculus applicative contexts are enough. Lastly, we can make soundness and completeness more operational by asking that some applications converge instead of being typable. To sum up, our definition of operational preciseness becomes:

A subtyping $\leq$ is operationally precise when $\sigma \leq \tau$ if and only if there are no closed terms M, N such that MP converges for all closed terms P of type τ and N has type σ and MN diverges.

Overview of the Paper. Section 2 presents the syntax and the operational semantics of the $\lambda_{+\|}$-calculus. Section 3 presents the type system with intersection and union types for the $\lambda_{+\|}$-calculus. These sections just review some definitions and results of [9]. Denotational and operational preciseness are shown in Section 4, where a crucial role is played by the construction of terms which fully characterise the functional behaviour of types. Section 5 discusses related and further work.

2 The Calculus and Its Operational Semantics

This section revisits the syntax of the λ-calculus with a non-deterministic choice operator $+$ and a parallel operator $\|$, which was introduced and developed in [9]. The obtained calculus is dubbed $\lambda_{+\|}$-calculus. There are two sorts of variables, namely the set Vn of call-by-name variables, denoted by x, y, z, and the set Vv of call-by-value variables, denoted by v, w. The symbol χ will range over the set $\mathsf{Vn} \cup \mathsf{Vv}$. The terms of the concurrent λ-calculus are defined by the following grammar

$$M ::= x \mid v \mid (\lambda x.M) \mid (\lambda v.M) \mid (MM) \mid (M + M) \mid (M\|M).$$

The set of terms is denoted by $\Lambda_{+\|}$. For any $M \in \Lambda_{+\|}$, $FV(M)$ denotes the set of free variables of M; $\Lambda^0_{+\|}$ is the set of terms M such that $FV(M) = \emptyset$.

Notation. As usual for pure λ-calculus, we omit parentheses by assuming that application associates to the left and λ-abstraction associates to the right. Moreover, application and abstraction have precedence over $+$ and $\|$, e.g. $MN + P$ stands for $((MN) + P)$ and $\lambda x.M + N$ for $((\lambda x.M) + N)$. The operator $\|$ takes precedence over $+$: for example $M\|P + Q$ is short for $((M\|P) + Q)$. As usual $\lambda\chi_1.\cdots\chi_n.M$ is short for $\lambda\chi_1.\cdots\lambda\chi_n.M$.

We will abbreviate some λ-terms as follows

$$\mathbf{I} = \lambda x.x \qquad \mathbf{K} = \lambda xy.x \qquad \mathbf{O} = \lambda xy.y$$
$$\Delta = \lambda x.xx \qquad \Omega = \Delta\Delta \qquad \mathbf{Y} = \lambda y.(\lambda x.y(xx))(\lambda x.y(xx)).$$

In pure λ-calculus values are either value variables or λ-abstractions [21]. Here we need to distinguish between partial and total values: the difference concerns the parallel operator. In fact we require both M and N to be total values to ensure that $M\|N$ is a total value, while in general it suffices that either M or N is a value to have that $M\|N$ is a value. As it is clear from the next definition, a value is either a total or a partial value.

Definition 1. *We define the set* Val *of* values *according to the grammar*

$$V ::= v \mid \lambda x.M \mid \lambda v.M \mid V\|M \mid M\|V$$

and the set TVal *of* total values *as the subset of* Val

$$W ::= v \mid \lambda x.M \mid \lambda v.M \mid W \| W.$$

A value V *is* partial *if and only if* $V \notin \mathsf{TVal}$.

We now introduce a reduction relation which is intended to formalise the expected behaviour of a machine which evaluates in a synchronous way parallel compositions, until a value is produced. Partial values can be further evaluated (rule $(\|_{app})$), and this is essential for applications of a call-by-value abstraction (rule $(\beta_v\|)$).

Definition 2. *The reduction relation* $\longrightarrow$ *is the least binary relation over* $\Lambda^0_{+\|}$ *such that*

$$(\beta)\ (\lambda x.M)N \longrightarrow M[N/x] \qquad (\beta_v)\ \frac{W \in \mathsf{TVal}}{(\lambda v.M)W \longrightarrow M[W/v]}$$

$$(\mu_v)\ \frac{N \longrightarrow N' \quad N \notin \mathsf{Val}}{(\lambda v.M)N \longrightarrow (\lambda v.M)N'} \qquad (\beta_v\|)\ \frac{V \longrightarrow V' \quad V \in \mathsf{Val}}{(\lambda v.M)V \longrightarrow M[V/v]\|(\lambda v.M)V'}$$

$$(\nu)\ \frac{M \longrightarrow M' \quad M \notin \mathsf{Val} \bigcup \mathsf{Par}}{MN \longrightarrow M'N} \qquad (\|_{app})\ (M\|N)L \longrightarrow ML\|NL$$

$$(\|_s)\ \frac{M \longrightarrow M' \quad N \longrightarrow N'}{M\|N \longrightarrow M'\|N'} \qquad (\|_a)\ \frac{M \longrightarrow M' \quad W \in \mathsf{TVal}}{M\|W \longrightarrow M'\|W,\ W\|M \longrightarrow W\|M'}$$

$$(+_L)\ M+N \longrightarrow M \qquad (+_R)\ M+N \longrightarrow N$$

We denote by $\longrightarrow^*$ *the reflexive and transitive closure of* $\longrightarrow$.

In rule (ν) we use Par defined by

$$\mathsf{Par} = \{M\|N \mid M,N \in \Lambda_{+\|}\}.$$

2.1 Rule $(\beta_v\|)$

We need to justify the rule $(\beta_v\|)$. Let us consider the context $C[\,] = (\lambda v.vv)[\,]\Omega\mathbf{I}$, the value $V = \mathbf{I}\|(\mathbf{K}+\mathbf{O})$, and the total values $W_1 = \mathbf{I}\|\mathbf{K}$, $W_2 = \mathbf{I}\|\mathbf{O}$. Then $V \longrightarrow W_1$ and $V \longrightarrow W_2$. Now considering $\|$ associative to spare parentheses, and writing $\stackrel{n}{\longrightarrow}$ when rules (β_v), $(\|_{app})$, $(+_L)$ or $(+_R)$ are applied n times:

$$\begin{aligned}
C[W_1] &\longrightarrow (\mathbf{I}\|\mathbf{K})(\mathbf{I}\|\mathbf{K})\Omega\mathbf{I}\\
&\stackrel{3}{\longrightarrow} (\mathbf{I}(\mathbf{I}\|\mathbf{K})\Omega\mathbf{I})\|(\mathbf{K}(\mathbf{I}\|\mathbf{K})\Omega\mathbf{I})\\
&\stackrel{2}{\longrightarrow} ((\mathbf{I}\|\mathbf{K})\Omega\mathbf{I})\|((\lambda y.(\mathbf{I}\|\mathbf{K}))\Omega\mathbf{I})\\
&\stackrel{2}{\longrightarrow} ((\mathbf{I}\Omega\|\mathbf{K}\Omega)\mathbf{I})\|((\mathbf{I}\|\mathbf{K})\mathbf{I})\\
&\stackrel{2}{\longrightarrow} (\mathbf{I}\Omega\mathbf{I})\|(\mathbf{K}\Omega\mathbf{I})\|(\mathbf{I}\mathbf{I})\|(\mathbf{K}\mathbf{I})\\
&\stackrel{4}{\longrightarrow} (\Omega\mathbf{I})\|((\lambda y.\Omega)\mathbf{I})\|\mathbf{I}\|(\lambda y.\mathbf{I})\\
&\longrightarrow (\Omega\mathbf{I})\|\Omega\|\mathbf{I}\|(\lambda y.\mathbf{I})
\end{aligned}$$

which is a value, and it is not hard to see that this is the only reduction out of $C[W_1]$ according to the rules given in Definition 2. Similarly,

$$\begin{aligned} C[W_2] &\longrightarrow (\mathbf{I}\|\mathbf{O})(\mathbf{I}\|\mathbf{O})\Omega\mathbf{I} \\ &\longrightarrow^* (\Omega\mathbf{I})\|\mathbf{I}\|(\Omega\mathbf{I}) \end{aligned}$$

and again this is the only reduction out of $C[W_2]$. But now consider the following reduction of $C[V]$

$$\begin{aligned} C[V] &\longrightarrow (\mathbf{I}\|(\mathbf{K}+\mathbf{O}))(\mathbf{I}\|(\mathbf{K}+\mathbf{O}))\Omega\mathbf{I} \\ &\xrightarrow{3} (\mathbf{I}(\mathbf{I}\|(\mathbf{K}+\mathbf{O}))\Omega\mathbf{I})\|((\mathbf{K}+\mathbf{O})(\mathbf{I}\|(\mathbf{K}+\mathbf{O}))\Omega\mathbf{I}) \\ &\xrightarrow{2} ((\mathbf{I}\|(\mathbf{K}+\mathbf{O}))\Omega\mathbf{I})\|(\mathbf{O}(\mathbf{I}\|(\mathbf{K}+\mathbf{O}))\Omega\mathbf{I}) && \text{choosing } \mathbf{O} \ldots \\ &\xrightarrow{4} (\mathbf{I}\Omega\mathbf{I})\|((\mathbf{K}+\mathbf{O})\Omega\mathbf{I})\|(\Omega\mathbf{I}) \\ &\xrightarrow{2} (\Omega\mathbf{I})\|(\mathbf{K}\Omega\mathbf{I})\|(\Omega\mathbf{I}) && \ldots \text{ choosing } \mathbf{K} \\ &\xrightarrow{2} (\Omega\mathbf{I})\|\Omega\|(\Omega\mathbf{I}) \end{aligned}$$

and from $(\Omega\mathbf{I})\|\Omega\|(\Omega\mathbf{I})$ we will never reach a value.

The problem of designing the β-contraction rule for call-by-value is that, given a value V, we cannot decide whether it has been computed enough to perform the reduction step $(\lambda v.M)V \longrightarrow M[V/v]$, or if it is necessary to reduce V further, before contracting the outermost β-redex. We cannot reduce V as long as possible, since this could not terminate. In the meantime, $M[V/v]$ can diverge, while $M[V'/v]$ can converge for all V' which are reducts of V, as shown by the previous example. On the other hand, any effective description of the operational semantics calls for a definition of a recursive one step reduction relation.

Now the solution we propose is to distinguish two cases: if V is a total value, then the standard call-by-value β-contraction rule applies (rule (β_v)). If, instead, V can be reduced further, to compute $(\lambda v.M)V$ we want to "take the best" between the terms $M[V'/v]$, for any V' such that $V \longrightarrow^* V'$. We realise this by evaluating in parallel $M[V/v]$ and $(\lambda v.M)V'$ for any V' such that $V \longrightarrow V'$ (rule $(\beta_v\|)$).

The previous example also shows that there are values V_0, V_1 and V_2 such that $V_0\|(V_1 + V_2)$ and $(V_0\|V_1) + (V_0\|V_2)$ have different behaviours in some context. Indeed, $(\lambda v.vv)(V_0\|(V_1 + V_2))$ can reduce to

$$(V_0\|(V_1 + V_2))(V_0\|(V_1 + V_2)), \tag{1}$$

while $(\lambda v.vv)((V_0\|V_1) + (V_0\|V_2))$ can reduce either to $(V_0\|V_1)(V_0\|V_1)$ or to $(V_0\|V_2)(V_0\|V_2)$, but never to (1). Note that in the present context call-by-name and call-by-value implement run-time-choice and call-time-choice, respectively (see [17]).

2.2 Convergence

In order to define convergence (Definition 4) it is useful to consider reduction trees of closed terms and their bars.

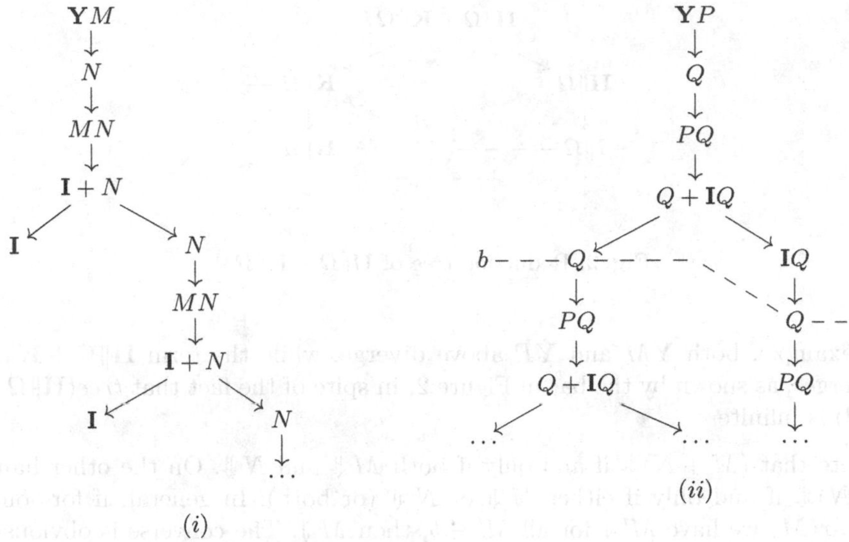

Fig. 1. (*i*) Reduction tree of $\mathbf{Y}M$. (*ii*) Reduction tree of $\mathbf{Y}P$

Definition 3. *Let* $M \in \Lambda^0_{+\|}$.

1. *tree*(M) *is the (unordered) reduction tree of* M.
2. *A* bar *of tree*(M) *is a subset of the nodes of tree*(M) *such that each maximal path intersects the bar at exactly one node.*

Inspecting the reduction rules, we see that *tree*(M) is a finitely branching tree for all $M \in \Lambda^0_{+\|}$. This implies by König's Lemma that if we cut *tree*(M) at a fixed height we obtain a finite tree. This does not contradict the fact that a term may have infinite reduction paths. For example, let us consider the infinite reduction tree of $\mathbf{Y}M$ with $M = \lambda x.(\mathbf{I}+x)$, which is shown in Figure 1(i), where $N = (\lambda x.M(xx))(\lambda x.M(xx))$. Admittedly, the set of nodes in *tree*($\mathbf{Y}M$) which are labeled by $\mathbf{I}$ is infinite, but it is not a bar. Indeed the infinite path in this tree does not have any node in such set and every bar of *tree*($\mathbf{Y}M$) must contain exactly one node of this path.

A bar is always relative to a tree and cannot be identified with the set of the labels of its nodes. For example *tree*($\mathbf{Y}P$) with $P = \lambda x.x + \mathbf{I}x$ has the shape shown in Figure 1(ii), where $Q = (\lambda x.P(xx))(\lambda x.P(xx))$. Now the indicated set of nodes b is a bar whose set of labels is the singleton $\{Q\}$. But the set containing a single node labeled by Q is not a bar of this tree.

We now define the convergence predicate. A term is *convergent* if and only if all reduction paths will eventually reach a value.

Definition 4. *Let* $M \in \Lambda^0_{+\|}$, *then* M converges *(notation* $M\Downarrow$*) if there is a bar b in tree*(M) *such that each node of b is labelled by a value. A term in* $\Lambda^0_{+\|}$ diverges *if it does not converge.*

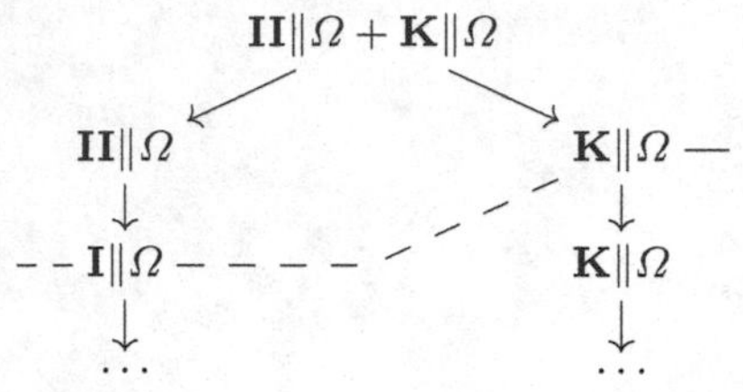

Fig. 2. Reduction tree of $\mathbf{II}\|\Omega + \mathbf{K}\|\Omega$

For example, both $\mathbf{Y}M$ and $\mathbf{Y}P$ above diverge, while the term $\mathbf{II}\|\Omega + \mathbf{K}\|\Omega$ converges, as shown by the bar in Figure 2, in spite of the fact that $tree(\mathbf{II}\|\Omega + \mathbf{K}\|\Omega)$ is infinite.

Note that $(M + N) \Downarrow$ if and only if both $M \Downarrow$ and $N \Downarrow$. On the other hand $(M\|N) \Downarrow$ if and only if either $M \Downarrow$ or $N \Downarrow$ (or both). In general, if for some $b \in bar(M)$ we have $M' \Downarrow$ for all $M' \in b$, then $M \Downarrow$. The converse is obviously true.

3 Types and Typing Rules

We consider the type system introduced in [9]. This type system has intersection and union types, dually reflecting the conjunctive and disjunctive operational semantics of $\|$ and $+$. The only atomic type is the universal type ω. Note that ω carries no information, but it is not a unit type, since all terms have type ω, see the typing rule (ω) in Definition 7. The type syntax is then as follows

$$\sigma ::= \omega \mid \sigma \to \sigma \mid \sigma \wedge \sigma \mid \sigma \vee \sigma$$

and we call *Type* the resulting set. In writing types, we assume that $\wedge$ and $\vee$ take precedence over $\to$.

The subtyping takes into account:

1. the meaning of ω as universal type;
2. the conjunctive and disjunctive meanings of intersection and union, respectively;
3. the meaning of arrow as functional type constructor.

Definition 5. *Let $\sigma \leq \tau$ be the smallest pre-order on types such that*

1. *$\langle Type, \leq\rangle$ is a distributive lattice, in which $\wedge$ is the meet, $\vee$ is the join and ω is the top;*
2. *the arrow satisfies*
   (a) $\sigma \to \omega \leq \omega \to \omega$;
   (b) $(\sigma \to \rho) \wedge (\sigma \to \tau) \leq \sigma \to \rho \wedge \tau$;
   (c) $\sigma \geq \sigma', \tau \leq \tau' \Rightarrow \sigma \to \tau \leq \sigma' \to \tau'$.

Two types σ and τ are equivalent *(notation $\sigma = \tau$) if $\sigma \leq \tau$ and $\tau \leq \sigma$.*

Axiom *2a* tells that each function maps an arbitrary term in a term.

The subtyping of the previous definition without the axioms on type ω (i.e, $\sigma \leq \omega$ for all σ and axiom *2a*) coincides with the implication in the minimal relevant logic $\mathbf{B}^+$ [22] and with the semantic subtyping [12,13] restricted to the present type constructors, but for the absence of rule $(\sigma \to \rho) \wedge (\tau \to \rho) \leq \sigma \vee \tau \to \rho$. This is proved in [10]. We will show at the end of this section that this rule is unsound for $\lambda_{+\|}$.

A special role is played by coprime types. A type σ is *coprime* if and only if

$$\sigma \leq \tau \vee \rho \text{ implies } \sigma \leq \tau \text{ or } \sigma \leq \rho$$

for any τ, ρ. Let *CType* be the set of coprime types different from ω. Observe that, because of distributivity, coprime types are closed under $\wedge$. Since $\langle Type, \leq\rangle$ is the free distributive lattice satisfying the arrow axioms, each type is the join of a finite number of coprime types. To see this, it suffices to define the following mapping Θ from types to finite sets of coprime types:

$$\begin{aligned}
\Theta(\omega) &= \{\omega\}\\
\Theta(\sigma \to \tau) &= \{\sigma \to \tau\}\\
\Theta(\sigma \wedge \tau) &= \{\sigma' \wedge \tau' \mid \sigma' \in \Theta(\sigma) \;\&\; \tau' \in \Theta(\tau)\}\\
\Theta(\sigma \vee \tau) &= \Theta(\sigma) \cup \Theta(\tau).
\end{aligned}$$

If $\Theta(\sigma) = \{\sigma_1, \ldots, \sigma_n\}$, it is easy to verify that σ_i is coprime for each i and $\sigma = \sigma_1 \vee \cdots \vee \sigma_n$.

To introduce the type assignment system we start with the notion of basis. We state that only coprime types different from ω can be assumed for call-by-value variables. This agrees with the fact that values cannot be choices and with the correspondence between choices and union types.

Definition 6. *A* basis $\Gamma : (\mathsf{Vn} \to Type) \cap (\mathsf{Vv} \to CType)$ *is a mapping such that $\Gamma(x) = \omega$ for all x but a finite subset of* Vn *and $\Gamma(v) = \omega \to \omega$ for all v but a finite subset of* Vv.

The notation $\Gamma, \chi : \sigma$ is a shorthand for the function $\Gamma'(\chi') = \sigma$ if $\chi' = \chi$, $\Gamma(\chi')$ otherwise.

Definition 7. *The axioms and rules of the type assignment system are given in Figure 3.*

To help the understanding of rule $(\to \mathrm{I}_v)$, we consider the following example. Let W_1, W_2 be total values such that $\vdash W_i : \sigma_i$ $(i = 1, 2)$ for some coprime types σ_1, σ_2. Clearly this implies $\vdash W_1 + W_2 : \sigma_1 \vee \sigma_2$ by rule $(+\mathrm{I})$. Consider $(\lambda v.M)(W_1 + W_2)$: it reduces to $M[W_1/v]$ and $M[W_2/v]$. Therefore $v : \sigma_i \vdash M : \tau$ for $i = 1, 2$ suffices to assure that $(\lambda v.M)$ has type $\sigma_1 \vee \sigma_2 \to \tau$.

An inversion lemma can be easily proved by induction on derivations and by cases on the last applied rule. In particular we need the inversion of rule $(\to E)$.

$$(\mathrm{Ax})\ \Gamma \vdash \chi : \Gamma(\chi) \qquad\qquad (\omega)\ \Gamma \vdash M : \omega$$

$$(\to \mathrm{I}_n)\ \frac{\Gamma, x : \sigma \vdash M : \tau}{\Gamma \vdash \lambda x.M : \sigma \to \tau} \qquad (\to \mathrm{I}_v)\ \frac{\Gamma, v : \sigma' \vdash M : \tau \quad \forall \sigma' \in \Theta(\sigma)}{\Gamma \vdash \lambda v.M : \sigma \to \tau}$$

$$(\to \mathrm{E})\ \frac{\Gamma \vdash M : \sigma \to \tau \quad \Gamma \vdash N : \sigma}{\Gamma \vdash MN : \tau}$$

$$(\wedge \mathrm{I})\ \frac{\Gamma \vdash M : \sigma \qquad \Gamma \vdash M : \tau}{\Gamma \vdash M : \sigma \wedge \tau} \qquad (\leq)\ \frac{\Gamma \vdash M : \sigma \quad \sigma \leq \tau}{\Gamma \vdash M : \tau}$$

$$(+\mathrm{I})\ \frac{\Gamma \vdash M : \sigma \qquad \Gamma \vdash N : \tau}{\Gamma \vdash M + N : \sigma \vee \tau} \qquad (\| \mathrm{I})\ \frac{\Gamma \vdash M : \sigma \qquad \Gamma \vdash N : \tau}{\Gamma \vdash M \| N : \sigma \wedge \tau}$$

Fig. 3. Type assigmment system

Lemma 1. *If $\Gamma \vdash MN : \sigma$, then $\Gamma \vdash M : \tau \to \sigma$ and $\Gamma \vdash N : \tau$ for some τ.*

As expected (and shown in [9]) this type assignment system enjoys subject reduction.

Theorem 1 (Subject Reduction). *If $\Gamma \vdash M : \sigma$ and $M \longrightarrow^* N$, then $\Gamma \vdash N : \sigma$.*

The main property of the present type assignment system is that convergence implies typability by $\omega \to \omega$ and vice versa. Therefore this type completely characterises terms whose meaning is a function, even if not a unique one. For a proof see [9].

Theorem 2. *A closed term is convergent if and only if it has type $\omega \to \omega$.*

It is easy to verify that each type is either a subtype of $\omega \to \omega$ or it is equivalent to ω. Therefore Theorem 2 can be rephrased as follows:
A closed term is convergent if and only if it has a type not equivalent to ω.
As an immediate consequence we get:

Corollary 1. *A closed term is divergent if and only if it has only types equivalent to ω.*

Note that the subtyping $(\sigma \to \rho) \wedge (\tau \to \rho) \leq \sigma \vee \tau \to \rho$ is unsound, since we would lose subject reduction. In fact we can derive

$$\vdash \lambda x.x\mathbf{I}\Omega \| x\Omega\mathbf{I} : (\sigma \to \rho) \wedge (\tau \to \rho) \text{ and } \vdash \mathbf{K} + \mathbf{O} : \sigma \vee \tau$$

where $\sigma = \rho \to \omega \to \rho$, $\tau = \omega \to \rho \to \rho$ and $\rho = \omega \to \omega$. From these judgments using $(\sigma \to \rho) \wedge (\tau \to \rho) \leq \sigma \vee \tau \to \rho$ we could conclude

$$\vdash (\lambda x.x\mathbf{I}\Omega \| x\Omega\mathbf{I})(\mathbf{K} + \mathbf{O}) : \rho$$

Since

$$\begin{aligned}(\lambda x.x\mathbf{I}\Omega \| x\Omega\mathbf{I})(\mathbf{K} + \mathbf{O}) &\longrightarrow (\mathbf{K} + \mathbf{O})\mathbf{I}\Omega \| (\mathbf{K} + \mathbf{O})\Omega\mathbf{I}) \\ &\longrightarrow \mathbf{O}\mathbf{I}\Omega \| \mathbf{K}\Omega\mathbf{I} \longrightarrow \Omega \| \Omega\end{aligned}$$

subject reduction would imply $\vdash \Omega \| \Omega : \rho$, which cannot be derived by Corollary 1, since $\Omega \| \Omega$ diverges. As a matter of fact, $\mathbf{K} + \mathbf{O}$ has type $\sigma \vee \tau$, but it has neither type σ nor type τ.

4 Preciseness

We are now able to specialise the definitions of soundness, completeness and preciseness to the calculus and the subtyping of the present paper.

To define denotational preciseness we need to interpret types. Following a common practice [15], we take the set of closed terms having type σ as the interpretation of σ. So the denotational preciseness particularises in our setting as follows.

Definition 8 (Denotational Preciseness). *The subtyping $\leq$ is* denotationally precise *when $\sigma \leq \tau$ if and only if*

$$\{M \mid \vdash M : \sigma\} \subseteq \{M \mid \vdash M : \tau\}.$$

The left-to-right implication is denotational soundness and the right-to-left implication is denotational completeness.

Frisch et al. [12,13] interpret a type as the set of values (instead of the set of terms) having that type. With this interpretation our subtyping is denotationally incomplete, since the set of all values is the meaning of both ω and $\omega \to \omega$.

For operational preciseness we take convergency as the test of safety.

Definition 9 (Operational Preciseness)

1. *The subtyping $\leq$ is* sound *if $\sigma \leq \tau$ implies that for all closed terms M whenever MP converges for all closed terms P of type τ, then MN converges also for all closed terms N of type σ.*
2. *The subtyping $\leq$ is* complete *if $\sigma \not\leq \tau$ implies that there are closed terms M, N such that MP converges for all closed terms P of type τ and N has type σ and MN diverges.*
3. *The subtyping $\leq$ is* precise *if it is both sound and complete.*

In order to show preciseness of our subtyping we define for each type σ a closed term R_σ which is the "worst" (with respect to convergence) term of type σ and a closed term T_σ which applied to an arbitrary closed term M reduces to the call-by-name identity if and only if M has type σ.

Definition 10. *The* characteristic terms R_σ *and the* test terms T_σ *are defined by simultaneous induction on σ:*

$$\begin{array}{ll}
\mathsf{R}_\omega = \Omega; & \mathsf{T}_\omega = \lambda xy.y; \\
\mathsf{R}_{\sigma\to\tau} = \lambda x.(\mathsf{T}_\sigma x)\,\mathsf{R}_\tau; & \mathsf{T}_{\sigma\to\tau} = \lambda v.\mathsf{T}_\tau(v\,\mathsf{R}_\sigma); \\
\mathsf{R}_{\sigma\wedge\tau} = \mathsf{R}_\sigma \,\|\, \mathsf{R}_\tau; & \mathsf{T}_{\sigma\wedge\tau} = \lambda x.(\mathsf{T}_\sigma\,x + \mathsf{T}_\tau\,x); \\
\mathsf{R}_{\sigma\vee\tau} = \mathsf{R}_\sigma + \mathsf{R}_\tau. & \mathsf{T}_{\sigma\vee\tau} = \lambda v.(\mathsf{T}_\sigma\,v \,\|\, \mathsf{T}_\tau\,v) \textit{ where } \sigma\vee\tau \neq \omega.
\end{array}$$

For example $\mathsf{R}_{\omega\to\omega} = \lambda x.\Omega$, $\mathsf{T}_{\omega\to\omega} = \lambda v.\mathbf{I}$.
More interestingly $\mathsf{R}_{(\omega\to\omega)\to\omega\to\omega} = \lambda x.((\lambda v.\mathbf{I})x)(\lambda y.\Omega)$ applied to a term returns $\lambda y.\Omega$ only if the term reduces to a value. Similarly

$$\mathsf{T}_{(\omega\to\omega)\to\omega\to\omega} = \lambda v.(\lambda v'.\mathbf{I})(v(\lambda x.\Omega))$$

applied to a term which reduces to a value, first applies this term to $\lambda x.\Omega$, and then reduces to $\mathbf{I}$ only if the result of this application reduces to a value too.

In the previous definition we start from the term Ω, but we can replace safely Ω by any divergent term.

Notice the different use of call-by-value variables in the definition of $\mathsf{T}_{\sigma\to\tau}$ and $\mathsf{T}_{\sigma\vee\tau}$: the term $\mathsf{T}_{\sigma\to\tau}$ must check that its argument has type $\sigma \to \tau$ which implies that it has to be convergent, see Theorem 2. On the other hand the argument of $\mathsf{T}_{\sigma\vee\tau}$ may reduce to a sum $P + Q$ having type $\sigma \vee \tau$ because P has type σ and Q has type τ, but neither σ nor τ can be deduced for $P + Q$. Therefore it is essential that it is evaluated before the application in parallel of T_σ and T_τ.

The types which can be deduced for R_σ and T_σ are meaningful for their operational behaviour. In fact one can verify by induction on types that:

- R_σ has exactly the types greater than or equal to σ;
- T_σ has type $\tau \to \rho \to \rho$ only if $\tau \leq \sigma$.

To establish the main result of this paper we use the types of characteristic terms and the discriminability power of test terms. For a proof see [9].

Theorem 3. *1.* R_σ *has type* τ *if and only if* $\sigma \leq \tau$.
2. Let M *be a closed term. Then* $\mathsf{T}_\sigma M$ *converges if and only if* M *has type* σ.

We can then conclude by showing soundness, completeness and preciseness of the current subtyping for the given calculus.

Theorem 4 (Denotational Preciseness). *The subtyping* $\leq$ *is denotationally precise for the* $\lambda_{+\|}$*-calculus.*

Proof. *Soundess* follows immediately by the typing rule ($\leq$), which implies
$$\{M \mid \vdash M : \sigma\} \subseteq \{M \mid \vdash M : \tau\}$$
whenever $\sigma \leq \tau$. For *completeness* it is enough to notice that if $\sigma \not\leq \tau$, then the characteristic term R_σ belongs to $[\![\sigma]\!]$, but it does not belong to $[\![\tau]\!]$ by Theorem 3(1).

Theorem 5 (Operational Preciseness). *The subtyping* $\leq$ *is operationally precise for the* $\lambda_{+\|}$*-calculus.*

Proof. For *soundness* if MN converges for all N of type τ, then $M\mathsf{R}_\tau$ converges. By Theorem 2 this implies that $M\mathsf{R}_\tau$ has type $\omega \to \omega$. Lemma 1 gives

$$\vdash M : \rho \to \omega \to \omega \tag{2}$$

and $\vdash \mathsf{R}_\tau : \rho$ for some ρ. Theorem 3(1) implies $\tau \leq \rho$. If $\vdash N : \sigma$ and $\sigma \leq \tau$ we derive

$$\vdash N : \rho \tag{3}$$

by rule $(\leq)$. Applying rule $(\to E)$ to (2) and (3) we get

$$\vdash MN : \omega \to \omega.$$

We conclude that MN converges for all N of type σ by Theorem 2.

For *completeness* if $\sigma \not\leq \tau$ we can take $M = \mathsf{T}_\tau$ and $N = \mathsf{R}_\sigma$. In fact by Theorem 3(2) $\mathsf{T}_\tau N$ converges for all N of type τ. Since $\vdash \mathsf{R}_\sigma : \rho$ implies $\sigma \leq \rho$ by Theorem 3(1) and $\sigma \not\leq \tau$ by hypothesis, we have that $\mathsf{T}_\tau \mathsf{R}_\sigma$ diverges by Theorem 3(2).

5 Related Work and Conclusion

There is a large literature on intersection and union types, a very interesting paper is [11], where the reader can also find references to the main publications. We will only consider works dealing with completeness of subtyping.

Barendregt et al. [3] interpret a type as the set of filters containing that type and get denotational preciseness for the restriction of our subtyping to arrow and intersection types.

Barbanera et al. [2] consider the subtyping obtained by adding to the axioms and rules of Definition 5 the axioms

$$\omega \leq \omega \to \omega \qquad\qquad (\sigma \to \rho) \wedge (\tau \to \rho) \leq \sigma \vee \tau \to \rho$$

and the rule

$$\mathbf{PP}(\sigma) \text{ implies } \sigma \to \tau \vee \rho \leq (\sigma \to \tau) \vee (\sigma \to \rho)$$

where $\mathbf{PP}(\sigma)$ is the predicate defined by $\mathbf{PP}(\omega) = \mathsf{true}$, $\mathbf{PP}(\tau \to \rho) = \mathbf{PP}(\rho)$, $\mathbf{PP}(\tau \wedge \rho) = \mathbf{PP}(\tau)$ & $\mathbf{PP}(\rho)$ and $\mathbf{PP}(\tau \vee \rho) = \mathsf{false}$. Types are interpreted as subsets of domains of λ-models and different conditions for denotational soundness are discussed. Our completeness results imply that the subtyping of [2] is neither denotationally nor operationally sound for the $\lambda_{+\|}$-calculus.

The subtyping considered in [1,16] is that of Definition 5 plus the axiom $(\sigma \to \rho) \wedge (\tau \to \rho) \leq \sigma \vee \tau \to \rho$. A type is interpreted as the set of filters over types which contain it and denotational preciseness is easily shown. In the same papers a λ-model (whose domain is the set of filters over types) of the call-by-value λ-calculus is built.

Vouillon [25] interprets types as sets of terms built out of functions, pairs and constants. These sets of terms must be closed with respect to indistinguishable terms, and they can only contain terms which are safe (do not produce error) and do not always diverge. The subtyping which is shown to be denotationally precise in [25] coincides with our subtyping when both relations are restricted to arrow and union types.

As already said in the introduction, to the best of our knowledge operational preciseness of subtyping was first introduced in [4]. In that paper Blackburn et al. showed that the subtyping of iso-recursive types with the Amber rule is not complete and they propose a new subtyping rule which gives preciseness.

Chen et al. [8] consider operational preciseness for session calculi. This is interesting since for these calculi also the denotational preciseness was never studied, to our knowledge. We conjecture that denotational preciseness holds for the synchronous and asynchronous subtypings as defined in [8].

Semantic subtyping is always studied for a rich set of types including arrow, intersection, union and negation type constructors. Various calculi have been considered: a λ-calculus with pairs and pattern matching in [12,13], a π-calculus with a patterned input in [6] and a session calculus with internal and external choices and typed input in [7]. For all these subtypings denotational preciseness holds by construction and operational preciseness holds by the discriminating power of the calculi.

In the present paper we have shown operational and denotational preciseness of the subtyping relation for the $\lambda_{+\|}$-calculus [9].

Operational completeness requires that all empty (i.e. not inhabited) types are less than all inhabited types. This makes unfeasible an operationally complete subtyping for the pure λ-calculus, both in the case of polymorphic types [18] and of intersection and union types. In fact inhabitation is undecidable for polymorphic types being equivalent to derivability in second order logic, while [24] shows undecidability of inhabitation for intersection types, which implies undecidability of inhabitation for intersection and union types.

An interesting open problem we plan to study is an extension of λ-calculus enjoying operational preciseness for the decidable subtypings between polymorphic types discussed in [18,23].

Acknowledgments. The authors gratefully thank the anonymous referees for their numerous constructive remarks.

References

1. van Bakel, S., Dezani-Ciancaglini, M., de' Liguoro, U., Motohama, Y.: The Minimal Relevant Logic and the Call-by-Value Lambda Calculus. Technical Report TR-ARP-05-2000, The Australian National University (2000)
2. Barbanera, F., Dezani-Ciancaglini, M., de' Liguoro, U.: Intersection and Union types: Syntax and Semantics. Information and Computation 119, 202–230 (1995)
3. Barendregt, H., Coppo, M., Dezani-Ciancaglini, M.: A Filter Lambda Model and the Completeness of Type Assignment. Journal of Symbolic Logic 48(4), 931–940 (1983)
4. Blackburn, J., Hernandez, I., Ligatti, J., Nachtigal, M.: Completely Subtyping Iso-recursive Types. Technical Report CSE-071012, University of South Florida (2012)
5. Boudol, G.: Lambda-calculi for (Strict) Parallel Functions. Information and Computation 108(1), 51–127 (1994)

6. Castagna, G., De Nicola, R., Varacca, D.: Semantic Subtyping for the Pi-calculus. Theoretical Computer Science 398(1-3), 217–242 (2008)
7. Castagna, G., Dezani-Ciancaglini, M., Giachino, E., Padovani, L.: Foundations of Session Types. In: López-Fraguas, F.J. (ed.) PPDP, pp. 219–230. ACM (2009)
8. Chen, T.-C., Dezani-Ciancaglini, M., Yoshida, N.: On the Preciseness of Subtyping in Session Types (2014), `http://www.di.unito.it/~dezani/papers/cdy14.pdf`
9. Dezani-Ciancaglini, M., de' Liguoro, U., Piperno, A.: A Filter Model for Concurrent lambda-Calculus. SIAM Journal on Computing 27(5), 1376–1419 (1998)
10. Dezani-Ciancaglini, M., Frisch, A., Giovannetti, E., Motohama, Y.: The Relevance of Semantic Subtyping. In: van Bakel, S. (ed.) ITRS. ENTCS, vol. 70(1), pp. 88–105 (2002)
11. Dunfield, J.: Elaborating Intersection and Union Types. In: Thiemann, P., Findler, R.B. (eds.) ICFP, pp. 17–28. ACM (2012)
12. Frisch, A., Castagna, G., Benzaken, V.: Semantic Subtyping. In: Plotkin, G. (ed.) LICS, pp. 137–146. IEEE (2002)
13. Frisch, A., Castagna, G., Benzaken, V.: Semantic Subtyping: Dealing set-theoretically with Function, Union, Intersection, and Negation Types. Journal of ACM 55(4) (2008)
14. Harper, R.: Practical Foundations for Programming Languages. Cambridge University Press (2013)
15. Hindley, J.R.: The Completeness Theorem for Typing Lambda-Terms. Theoretical Computer Science 22, 1–17 (1983)
16. Ishihara, H., Kurata, T.: Completeness of Intersection and Union Type Assignment Systems for Call-by-value Lambda-models. Theoretical Computer Science 272(1-2), 197–221 (2002)
17. de' Liguoro, U., Piperno, A.: Non Deterministic Extensions of Untyped Lambda-Calculus. Information and Computation 122(2), 149–177 (1995)
18. Mitchell, J.C.: Polymorphic Type Inference and Containment. Information and Computation 76(2/3), 211–249 (1988)
19. Ong, C.-H.L.: Non-Determinism in a Functional Setting. In: Vardi, M.Y. (ed.) LICS, pp. 275–286. IEEE (1993)
20. Pierce, B.C.: Types and Programming Languages. MIT Press (2002)
21. Plotkin, G.D.: Call-by-name, Call-by-value and the λ-calculus. Theoretical Computer Science 1(2), 125–159 (1975)
22. Routley, R., Meyer, R.K.: The Semantics of Entailment III. Journal of Philosophical Logic 1, 192–208 (1972)
23. Tiuryn, J., Urzyczyn, P.: The Subtyping Problem for Second-Order Types is Undecidable. Information and Computation 179(1), 1–18 (2002)
24. Urzyczyn, P.: The Emptiness Problem for Intersection Types. Journal of Symbolic Logic 64(3), 1195–1215 (1999)
25. Vouillon, J.: Subtyping Union Types. In: Marcinkowski, J., Tarlecki, A. (eds.) CSL 2004. LNCS, vol. 3210, pp. 415–429. Springer, Heidelberg (2004)

Abstract Datatypes for Real Numbers in Type Theory

Martín Hötzel Escardó[1] and Alex Simpson[2]

[1] School of Computer Science, University of Birmingham, UK
[2] LFCS, School of Informatics, University of Edinburgh, UK

Abstract. We propose an abstract datatype for a closed interval of real numbers to type theory, providing a representation-independent approach to programming with real numbers. The abstract datatype requires only function types and a natural numbers type for its formulation, and so can be added to any type theory that extends Gödel's System T. Our main result establishes that programming with the abstract datatype is equivalent in power to programming intensionally with representations of real numbers. We also consider representing arbitrary real numbers using a mantissa-exponent representation in which the mantissa is taken from the abstract interval.

1 Introduction

Exact real-number computation uses infinite representations of real numbers to compute exactly with them, avoiding round-off errors [16,2,3]. In practice, such representations can be implemented as streams or functions, allowing any computable (and hence *a fortiori* continuous) function to be programmed.

This approach of programming with representations of real numbers has drawbacks from the programmer's perspective. Great care must be taken to ensure that different representations of the same real number are treated equivalently. Furthermore, a programmer ought to be able to program with real numbers without knowing how they are represented, leading to more transparent programs, and also allowing the underlying implementation of real-number computation to be changed, e.g., to improve efficiency. In short, the programmer would like to program with an *abstract datatype of real numbers.*

Various interfaces for an abstract datatype for real numbers have been investigated in the context of typed functional programming languages based on PCF, e.g., [6,7,10,8,1,9], making essential use of the presence of general recursion. In this paper, we consider the more general scenario of typed functional programming with primitive recursion. This generality has the advantage that it can be seen as a common core both to standard functional programming languages with general recursion (ML, Haskell, etc.), and also to the type theories used in dependently-typed programming languages such as Agda [4], and proof assistants such as Coq [13], in which all functions are total.

To maximize generality, we keep the type theory in this paper as simple as possible. Our base calculus is just simply-typed λ-calculus with a base type of

G. Dowek (ed.): RTA-TLCA 2014, LNCS 8560, pp. 208–223, 2014.

natural numbers, otherwise known as Gödel's System T. To this, we add a new type constant I, which acts as an abstract datatype for the interval $[-1, 1]$ of real numbers, together with an associated interface of basic operations. Our main result (Theorem 1) establishes that programming with the abstract datatype is equivalent in power to programming intensionally with representations of reals.

The development in this paper builds closely on our LICS 2001 paper [11], where we gave a category-theoretic universal property for the interval $[-1, 1]$. The interface we provide for the type I is based directly on the universal property defined there. In [11], the definability power of the universal property was already explored, to some extent, via a class of *primitive interval functions* on $[-1, 1]$, named by analogy to the primitive recursive functions. The role of a crucial doubling function was identified, relative to which all continuous functions on $[-1, 1]$ were shown to be primitive-interval definable relative to oracles $\mathbb{N} \to \mathbb{N}$.

The new departure of the present paper is to exploit these ideas in a type-theoretic context. The cumbersome definition of primitive interval functions is replaced by a very simple interface for the abstract datatype I (Sect. 3). The role of the doubling function is again crucial, with its independence from the other constants of the interface now being established by a logical relations argument (proof of Prop. 4). And the completeness of the interface once doubling is added (Theorem 1) is now established relative to the setting at hand (System T computability) rather than relative to oracles (Sect. 4). In addition, we show that the type theoretic framework provides a natural context for proving equalities between functions on reals (based on the equalities in Fig. 2), and for programming on the full real line $\mathbb{R}$ via a mantissa-exponent representation (Sect. 5).

2 Real-Number Computation in System T

In this section, we recall how exact real-number computation is rendered possible by choosing an appropriate representation of real numbers. A natural first attempt would be to represent real numbers using streams or functions to implement one of the standard digit representations (decimal, binary, etc.). For example, a real number in $[0, 1]$ would be represented via a binary expansion as an infinite sequence of 0s and 1s. As is well known (see, e.g., [5,6,7]), however, such representations makes it impossible to compute even simple functions (on the interval) such as binary average on real numbers. The technical limitation here is that there is no continuous function $\{0,1\}^\omega \times \{0,1\}^\omega \to \{0,1\}^\omega$ that given sequences α, β as input, representing $x, y \in [0, 1]$ respectively, returns a representation of $\frac{x+y}{2}$ as result. In general, it is impossible to return even a single output digit without examining all (infinitely many) input digits.

This problem is avoided by choosing a different representation. To be appropriate for computation, any representation must be *computably admissible* in the sense of [15]. Each of the examples below is a computably admissible representation of real numbers, in the interval $[-1, 1]$, using streams:

```
type I = [Int]   -- Represents [-1,1] in binary using digits -1,0,1.
minusOne, one :: I
minusOne  = repeat (-1)
one       = repeat 1
type J = [Int]   -- Represents [-n,n] in binary using digits |d| <= n
divideBy :: Int -> J -> I
divideBy n (a:b:x) = let d = 2*a+b
                     in if d < -n then -1 : divideBy n (d+2*n:x)
                   else if d >  n then  1 : divideBy n (d-2*n:x)
                                  else  0 : divideBy n (d:x)
mid :: I -> I -> I
mid x y = divideBy 2 (zipWith (+) x y)
bigMid :: [I] -> I
bigMid = (divideBy 4).bigMid'
 where bigMid'((a:b:x):(c:y):zs) = 2*a+b+c : bigMid'((mid x y):zs)
affine :: I -> I -> I -> I
affine a b x = bigMid [h d | d <- x]
 where h (-1) = a
       h  0   = mid a b
       h  1   = b
```

Fig. 1. Haskell programs using signed binary notation

$$q_0 : q_1 : q_2 : q_3 : q_4 : q_5 : \ldots$$

of discrete data.

1. *Fast Cauchy sequences:* Require q_i to be rational numbers in $[-1, 1]$ such that $|q_{i+1} - q_i| \leq 2^{-i}$ for all i. The stream represents the real number $\lim_i q_i$.
2. *Signed binary:* Require $q_i \in \{-1, 0, 1\}$. The stream represents the real number $\sum_{i \geq 0} 2^{-(i+1)} q_i$.

Many other variations are possible. Crucially, all computably admissible representations are computably interconvertible. For representations used in practice, the conversions can be defined in System T.

Both to illustrate the style of programming that is required with such representations, and for later reference, Fig. 1 presents some simple functions on real numbers, in Haskell, using signed binary notation. The code defines a type I for the interval $[-1, 1]$, and constants one and minusOne of type I, for the streams 1:1:1:1:1:... and -1:-1:-1:-1:-1:..., which represent 1 and -1 respectively. The function mid represents the binary average function, for which we use a convenient algebraic notation:

$$x \oplus y = \frac{x+y}{2} \; .$$

The function `bigMid` maps infinite streams of real numbers to real numbers, and represents the function $\mathrm{M}\colon [-1,1]^\omega \to [-1,1]$ defined by

$$\mathrm{M}((x_n)_n) \;=\; \sum_{n\geq 0} \frac{x_n}{2^{n+1}} \;.$$

Finally, `affine` represents $\mathrm{aff}\colon [-1,1] \to [-1,1] \to [-1,1] \to [-1,1]$ defined by

$$\mathrm{aff}\, x\, y\, z \;=\; \frac{(1-z)\,x + (1+z)\,y}{2} \;.$$

The type `J` and function `divideBy` just provide auxiliary machinery.

At this point, the selection of example functions in Fig. 1 will appear peculiar. The reasons behind the choice will be clarified in Sect. 3.

Although presented in Haskell, the above algorithms can be formalized in almost any type theory containing a type of natural numbers and function types. Moreover, the recursive structure of the algorithms is tame enough to be formulated using primitive recursion. Thus a natural basic type theory for studying this approach to real number computation is Gödel's System T, see, e.g., [12], which is simply-typed λ-calculus with a natural numbers type with associated primitive recursion operator. Since this type theory will form the basis of the rest of the paper we now review it in some detail.

Types (we include product types for convenience in Sect. 5) are given by:

$$\sigma \;::=\; \mathsf{N} \mid \sigma \times \tau \mid \sigma \to \tau \;.$$

These have a set-theoretic semantics with types being interpreted by their set-theoretic counterparts:

$$[\![\mathsf{N}]\!] = \mathbb{N} \qquad [\![\sigma \times \tau]\!] = [\![\sigma]\!] \times [\![\tau]\!] \qquad [\![\sigma \to \tau]\!] = [\![\sigma]\!] \to [\![\tau]\!] \;.$$

We use standard notation for terms of the simply-typed λ-calculus; e.g., we write $\Gamma \vdash t\colon \tau$ to mean that term t has type τ in type context Γ. The constants associated with the type N are:

$$0 : \mathsf{N} \qquad \mathsf{s} : \mathsf{N} \to \mathsf{N} \qquad \mathsf{primrec}_\sigma : \sigma \to (\sigma \to \mathsf{N} \to \sigma) \to \mathsf{N} \to \sigma$$

with semantics defined by:

$$[\![0]\!] \;=\; 0 \qquad\qquad [\![s]\!]\; n \;=\; n+1$$
$$[\![\mathsf{primrec}]\!]\; x\; f\; 0 \;=\; x \qquad [\![\mathsf{primrec}]\!]\; x\; f\; (n+1) \;=\; f\;([\![\mathsf{primrec}]\!]\; x\; f\; n)\; n \;.$$

We have two main interests in System T. The first is that it serves as a basic functional programing language, for which, the standard strongly normalizing and confluent β-reduction relation is used. The second is that System T serves as the basis of a formal system for reasoning about equality between functions. For this, we introduce axioms and rules for deriving *typed equations* of the form $\Gamma \vdash t = u : \sigma$ between terms t, u such that $\Gamma \vdash t : \sigma$ and $\Gamma \vdash u : \sigma$. These rules include the usual ones asserting that equality is a typed congruence relation. Also, whenever $\Gamma \vdash t : \sigma$ and t β-reduces to u, we have an equation $\Gamma \vdash t = u : \sigma$. Finally, we add extensionality equations, which comprise the usual η-equalities for product and function types, together with the rule below, which asserts the uniqueness of the primrec iterator.

$$\frac{\Gamma \vdash t\ u\ v\ 0 = u : \sigma \qquad \Gamma, x{:}\mathsf{N} \vdash t\ u\ v\ (\mathsf{s}(x)) \ = \ v\ u\ (t\ u\ v\ x) : \sigma}{\Gamma \vdash t\ u\ v \ = \ \mathsf{primrec}\ u\ v : \mathsf{N} \to \sigma}$$

(The types of the component terms are not stated explicitly since they can be inferred from the context.)

The term language of System T, in the version we are considering, can be interpreted in any cartesian-closed category with natural numbers object; and our equational rules are sound and complete with respect to such interpretations.

Exact-real-number computation can be carried out in System T by encoding any of the usual computably admissible representation of $[-1,1]$ using the type $\mathsf{N} \to \mathsf{N}$, and these representations are all interconvertible in System T.

We examine just the case of signed binary in detail. We consider a function $\alpha\colon \mathbb{N} \to \mathbb{N}$ as encoding the signed binary stream

$$((\alpha(0) \bmod 3) - 1) : ((\alpha(1) \bmod 3) - 1) : ((\alpha(2) \bmod 3) - 1) : ((\alpha(3) \bmod 3) - 1) : \ldots$$

All the Haskell programs in Fig. 1 are then routinely translatable into System T terms of appropriate type. (Just a little effort is needed to translate the general recursion into uses of primrec.)

Consider the function real: $(\mathbb{N} \to \mathbb{N}) \to [-1,1]$ defined by

$$\mathrm{real}(\alpha) \ = \ \sum_{i \geq 0} 2^{-(i+1)} \left((\alpha(i) \bmod 3) - 1\right) .$$

We say that a function $f\colon [-1,1]^k \to [-1,1]$ is *T-representable* if there exists a closed term $t\colon (\mathsf{N} \to \mathsf{N})^k \to \mathsf{N} \to \mathsf{N}$ making the following diagram commute.

$$\begin{array}{ccc}
(\mathbb{N} \to \mathbb{N})^k & \xrightarrow{\ [\![t]\!]\ } & (\mathbb{N} \to \mathbb{N}) \\
\Big\downarrow{\scriptstyle \mathrm{real}^k} & & \Big\downarrow{\scriptstyle \mathrm{real}} \\
[-1,1]^k & \xrightarrow[\ f\ ]{} & [-1,1]
\end{array}$$

Since the vertical maps are topological quotients relative to the product (Baire space) topology on $\mathbb{N} \to \mathbb{N}$, and every T-definable $[\![t]\!]$ is continuous, it follows that every T-representable function f is continuous.

The programs in Fig. 1, when recast as System T terms, provide examples of representations of functions on reals. But programming in this style has the disadvantages discussed in Sect. 1. Accordingly, we now turn to the alternative approach of defining an abstract datatype for real numbers.

3 System I

We add a new type constant I to our base type theory to act as an abstract datatype for the closed interval $[-1, 1]$ of real numbers. In the case of System T, we call the resulting extension System I; it has types

$$\sigma ::= \mathsf{N} \mid \mathsf{I} \mid \sigma \times \tau \mid \sigma \to \tau ,$$

and we extend the set-theoretic semantics with the clause

$$[\![\mathsf{I}]\!] = [-1, 1] .$$

The interface for the type will roughly implement the idea that the closed interval is determined as the free convex set on 2 generators (-1 and 1) with respect to affine maps. However, since the notion of *convexity* requires a pre-existing interval for its formulation, we replace convexity with the existence of *iterated midpoints* and we replace affineness with preservation of (iterated) midpoints, following [11].

The term language is generated by adding the following new typed constants. We pair each with its denotational interpretation in order to specify its intended meaning. In doing so, we make use of the functions defined in Sect. 2.

$$\begin{aligned} \mathsf{1} &: \mathsf{I} & [\![\mathsf{1}]\!] &= 1 \\ \mathsf{-1} &: \mathsf{I} & [\![\mathsf{-1}]\!] &= -1 \\ \mathsf{m} &: \mathsf{I} \times \mathsf{I} \to \mathsf{I} & [\![\mathsf{m}]\!](x, y) &= x \oplus y \\ \mathsf{M} &: (\mathsf{N} \to \mathsf{I}) \to \mathsf{I} & [\![\mathsf{M}]\!] &= \mathrm{M} \\ \mathsf{aff} &: \mathsf{I} \to \mathsf{I} \to \mathsf{I} \to \mathsf{I} & [\![\mathsf{aff}]\!] &= \mathrm{aff} \end{aligned}$$

We have here adopted a convention that we shall continue to follow of using sans-serif for λ-calculus constants and defined terms and using the same symbol in roman the corresponding mathematical function.

Figure 2 presents equational axioms and rules extending those for System T. A simple consequence of the equational rules for midpoints and iterated midpoints is that

$$x, y : I \vdash \mathsf{m}(x, y) = \mathsf{M}(x, y, y, y, y, \dots) : \mathsf{I} ,$$

where we write $(x, y, y, y, y, \dots)$ as a convenient shorthand for the System I term $\mathsf{primrec}\,(x)\,(\lambda z : \mathsf{I}.\, \lambda n : \mathsf{N}.\, y)$, i.e., the function that is x at 0, and y at every natural number > 0. (Henceforth, we shall adopt other similar notational shorthands, without discussion.) Thus the constant m is redundant, and could be removed from the system. We include it, however, since the equations are more perspicuous with m included as basic. In fact, m is used frequently in the sequel, and we adopt the more suggestive notation $t \oplus u$ in preference to $\mathsf{m}(t, u)$.

(m) Midpoint equations.

$$\Gamma \vdash \mathsf{m}(t,t) = t : \mathsf{I} \qquad \Gamma \vdash \mathsf{m}(t,u) = \mathsf{m}(u,t) : \mathsf{I}$$

$$\Gamma \vdash \mathsf{m}(\mathsf{m}(t,u), \mathsf{m}(v,w)) = \mathsf{m}(\mathsf{m}(t,v), \mathsf{m}(u,w)) : \mathsf{I}$$

(M) Iterated midpoint equations.

$$\Gamma \vdash \mathsf{M}(t) = \mathsf{m}(t(0), \mathsf{M}(\lambda i{:}\mathsf{N}.\, t(i+1))) : \mathsf{I} \qquad \frac{\Gamma, i{:}\mathsf{N} \vdash t(i) = \mathsf{m}(u(i), t(i+1)) : \mathsf{I}}{\Gamma \vdash t(0) = \mathsf{M}(u) : \mathsf{I}}$$

(a) Equations for aff.

$$\Gamma \vdash \mathsf{aff}\, t\, u\, (-1) = t : \mathsf{I} \qquad \Gamma \vdash \mathsf{aff}\, t\, u\, 1 = u : \mathsf{I}$$

$$\Gamma \vdash \mathsf{aff}\, t\, u\, (\mathsf{m}(v,w)) = \mathsf{m}(\mathsf{aff}\ t\, u\, v,\ \mathsf{aff}\, t\, u\, w) : \mathsf{I}$$

$$\frac{\Gamma, x{:}\mathsf{I}, y{:}\mathsf{I} \vdash f\,(\mathsf{m}(x,y)) = \mathsf{m}(f(x),\ f(y)) : \mathsf{I}}{\Gamma \vdash f = \mathsf{aff}\,(f(-1))\,(f(1)) : \mathsf{I} \to \mathsf{I}}$$

(C) Cancellation

$$\frac{\Gamma \vdash \mathsf{m}(t,v) = \mathsf{m}(u,v) : \mathsf{I}}{\Gamma \vdash t = u : \mathsf{I}}$$

(E) Joint I-epimorphicity of $\mathsf{m}(\cdot\,, 1)$ and $\mathsf{m}(\cdot\,, -1)$.

$$\frac{\Gamma, x{:}I \vdash f(\mathsf{m}(x,1)) = g(\mathsf{m}(x,1)) : \mathsf{I} \quad \Gamma, x{:}I \vdash f(\mathsf{m}(x,-1)) = g(\mathsf{m}(x,-1)) : \mathsf{I}}{\Gamma \vdash f = g : \mathsf{I} \to \mathsf{I}}$$

Fig. 2. Equations for System I

We now develop some simple programming in System I, to explore its power as a programming language for defining real numbers, and functions on them.

$$\begin{aligned} 0 &:= (-1) \oplus 1 \\ -x &:= \mathsf{aff}\ 1\ (-1)\ x \\ xy &:= \mathsf{aff}\ (-x)\ x\ y \\ \frac{1}{3} &:= \mathsf{M}(1, -1, 1, -1, 1, -1, 1, -1, \ldots) \end{aligned}$$

More generally, any rational number is definable using M applied to an eventually periodic sequence of 1s and (-1)s. Even more generally, any real number with a System-T-definable binary expansion is definable.

Proposition 1. *The following equalities are derivable from the axioms and rules in Fig. 2 (without using (M), (C) and (E)).*

$$\begin{aligned} --x &= x & (x\,y)\,z &= x\,(y\,z) \\ x \oplus -x &= 0 & x\,(-y) &= -(x\,y) \\ x0 &= 0 & x\,(y \oplus z) &= (x\,y) \oplus (x\,z) \\ x\,y &= y\,x \end{aligned}$$

So far, we have seen that the type I supports the arithmetic of multiplication and average, together with its expected equational properties. We now look at possibilities for defining functions that arise in analysis. Suppose we have a function f defined by a power series

$$f(x) = \sum_{n \geq 0} a_n\, x^n$$

where $a_n \in [-1, 1]$. Then

$$\frac{1}{2} f\left(\frac{x}{2}\right) = \underset{n}{\mathrm{M}}\; a_n x^n \qquad \text{(which abbreviates } \mathrm{M}(\lambda n.\, a_n x^n)\text{).}$$

As a consequence, using the arithmetic defined above, the following are all definable in System I.

$$\begin{aligned} \frac{1}{2-x} &:= \underset{n}{\mathsf{M}}\, x^n \\ \frac{1}{2}\exp\left(\frac{x}{2}\right) &:= \underset{n}{\mathsf{M}}\, \frac{x^n}{n!} \\ \frac{1}{2}\cos\left(\frac{x}{2}\right) &:= \underset{n}{\mathsf{M}}\, (1 - \mathsf{parity}(n))\,(-1)^{\frac{n}{2}}\, \frac{x^n}{n!} \end{aligned}$$

These (and other similar) examples cover many functions from analysis, in versions with very particular scalings. We shall return to the issue of scaling below.

All functions defined above are continuous and smooth on $[-1, 1]$. System I is also powerful enough to define non-smooth functions. We present two examples, exhibiting different degrees of non-smoothness. Define:

$$\begin{aligned} \mathsf{times}^*(x, y) &:= \mathsf{aff}\ (-1)\ x\ y \\ \mathsf{sq}^*(x) &:= \mathsf{times}^*(x, x) \\ \mathsf{g}(x) &:= \mathsf{times}^*\left(\frac{7}{9}, \mathsf{sq}^*(-\mathsf{sq}^*(-x))\right) \\ \mathsf{h}(x) &:= \underset{i}{\mathsf{M}}\ \mathsf{g}^{3(i+1)}(x) \\ \mathsf{H}(x) &:= \underset{i}{\mathsf{M}}\ (\mathsf{g}^{3(i+1)}(x))^2 \end{aligned}$$

Here times^* and sq^* are so named because they encode multiplication and square if the endpoints of the interval are renamed from $[-1, 1]$ to $[0, 1]$. Keeping to our convention that the interval is $[-1, 1]$ the function g defined by g above is

$$g(x) = \frac{1}{9}x^4 - \frac{4}{9}x^3 - \frac{2}{9}x^2 + \frac{4}{3}x$$

which satisfies $g(0) = 0$ and $g'(0) = \frac{4}{3}$. Hence, $(g^n)'(0) = (\frac{4}{3})^n$. This leads to the result below.

Proposition 2. *1. The function defined by* h *has derivative* ∞ *at* 0.
2. The function defined by H *has derivative* ∞ *when* 0 *is approached from above, and derivative* $-\infty$ *when* 0 *is approached from below.*

Since I is an abstract datatype, to compute with system I terms, we must give the datatype an implementation. In Sect. 2, we have implicitly discussed one such implementation in System T: the type $\mathsf{N} \to \mathsf{N}$ implements I, and System T versions of the programs in Fig. 1 implement the functions in the interface. Given this implementation, Prop. 3 below is immediate. We say that a function $f\colon [-1,1]^k \to [-1,1]$ is *I-definable* if there exists a closed System I term $u\colon \mathsf{I}^k \to \mathsf{I}$ such that $[\![u]\!] = f$.

Proposition 3. *Every I-definable function is T-representable.*

The converse, however, does not hold. We use square brackets for the truncation function $[\,\cdot\,] : \mathbb{R} \to [-1,1]$ defined by:

$$[x] \;:=\; \min(1, \max(-1, x)) \; .$$

We write dbl for the function $x \mapsto [2x] : [-1,1] \to [-1,1]$.

Proposition 4. *The function* dbl *is T-representable but not I-definable.*

The non-definability of dbl shows that System I is, as already hinted above, limited in its capacity for rescaling the interval.

We end the section with the proof of Prop. 4. Using the notation of Fig. 1, a Haskell program computing dbl is:

```
dbl :: I -> I                         dbl (0:x) = x
dbl (1:1:x) = one                     dbl ((-1):(-1):x) = minusOne
dbl (1:0:x) = 1:(dbl (1:x))           dbl ((-1):0:x) = (-1):(dbl ((-1):x))
dbl (1:(-1):x) = 1:x                  dbl ((-1):1:x) = (-1):x
```

This is easily converted into a System T term, showing that dbl is T-representable.

The non-definability proof uses logical relations. For every type τ we define a binary relation $\Delta_\tau \subseteq [\![\tau]\!] \times [\![\tau]\!]$ by:

$$\begin{aligned}
\Delta_{\mathsf{N}}(m,n) &\iff m = n \\
\Delta_{\mathsf{I}}(x,y) &\iff \text{if } x \in \{-1,1\} \text{ or } y \in \{-1,1\} \text{ then } x = y \\
\Delta_{\sigma\to\tau}(f,g) &\iff \forall x,y \in [\![\sigma]\!].\ \Delta_\sigma(x,y) \text{ implies } \Delta_\tau(f(x),g(y)) \\
\Delta_{\sigma\times\tau}((x,x'),(y,y')) &\iff \Delta_\sigma(x,y) \text{ and } \Delta_\tau(x',y')
\end{aligned}$$

Lemma 1. *For every System I constant* $c\colon \tau$, *it holds that* $\Delta_\tau([\![c]\!],[\![c]\!])$.

Proof. We consider two cases. To show that $\Delta_{(\mathsf{N}\to\mathsf{I})\to\mathsf{I}}(\mathrm{M},\mathrm{M})$, suppose $\Delta_{\mathsf{N}\to\mathsf{I}}(f,f')$. Then, for all n, we have $\Delta_{\mathsf{I}}(f(n),f'(n))$. We must show that if $\mathrm{M}(f)\in\{-1,1\}$ or $\mathrm{M}(f')\in\{-1,1\}$ then $\mathrm{M}(f)=\mathrm{M}(f')$. We consider just the case that $\mathrm{M}(f)=-1$ (the others are similar). If $\mathrm{M}(f)=-1$ then $f(n)=-1$, for all $n\geq 0$. Since $\Delta_{\mathsf{I}}(f(n),f'(n))$, we have $f'(n)=-1$, for all $n\geq 0$. Thus $\mathrm{M}(f')=-1=\mathrm{M}(f)$. We have thus shown that $\Delta_{\mathsf{I}}(\mathrm{M}(f),\mathrm{M}(f'))$ as required.

To show that $\Delta_{\mathsf{I}\to\mathsf{I}\to\mathsf{I}\to\mathsf{I}}(\mathrm{aff},\mathrm{aff})$, suppose

$$\Delta_{\mathsf{I}}(x,x') \quad\text{and}\quad \Delta_{\mathsf{I}}(y,y') \quad\text{and}\quad \Delta_{\mathsf{I}}(z,z')\ . \tag{1}$$

We must show that if $\mathrm{aff}\,x\,y\,z\in\{-1,1\}$ or $\mathrm{aff}\,x'\,y'\,z'\in\{-1,1\}$ then $\mathrm{aff}\,x\,y\,z=\mathrm{aff}\,x'\,y'\,z'$. Suppose, without loss of generality, that $\mathrm{aff}\,x\,y\,z=-1$, i.e., $((1-z)\,x+(1+z)\,y)/2=-1$. Then there are three possible cases: (i) $x=z=-1$; (ii) $y=-1$ and $z=1$; (iii) $x=y=-1$. In each case, by (1), the corresponding equations hold for x',y',z'. Thus indeed $\mathrm{aff}\,x'\,y'\,z'=-1=\mathrm{aff}\,x\,y\,z$. □

Lemma 2. *For every closed System I term $t\colon\tau$, it holds that $\Delta_\tau([\![t]\!],[\![t]\!])$.*

Proof. This is an immediate consequence of the previous lemma, by the fundamental lemma of logical relations. □

Proposition 5. *If $f\colon[-1,1]\to[-1,1]$ is I-definable and $f(x)\in\{-1,1\}$ for some $x\in(-1,1)$ then f is a constant function.*

Proof. Let $x\in(-1,1)$ be such that $f(x)\in\{-1,1\}$. Consider any $y\in(-1,1)$. Then $\Delta_{\mathsf{I}}(x,y)$. By Lemma 2, $\Delta_{\mathsf{I}\to\mathsf{I}}(f,f)$. Thus $\Delta_{\mathsf{I}}(f(x),f(y))$, whence $f(x)=f(y)$. Thus f is constant on $(-1,1)$, hence on $[-1,1]$ since continuous. □

The non-definability statement of Proposition 4 is an immediate consequence, as are many other non-definability results. For example, $\cos(x)$ and $\cos(\frac{x}{2})$ are not I-definable, even though $\frac{1}{2}\cos(\frac{x}{2})$ is (see above).

4 System II

We address the weakness identified above in the obvious way. System II ("double I") is obtained by adding dbl to System I.

$$\mathsf{dbl}:\ \mathsf{I}\to\mathsf{I} \qquad\qquad [\![\mathsf{dbl}]\!]=\mathrm{dbl}\ .$$

The equations from Fig. 2 are then augmented with:

(d) Equations for dbl:

$$\Gamma\vdash\mathsf{dbl}(\mathsf{m}(1,\,\mathsf{m}(1,\,t)))=1:\mathsf{I} \qquad \Gamma\vdash\mathsf{dbl}(\mathsf{m}(-1,\,\mathsf{m}(-1,\,t)))=-1:\mathsf{I}$$
$$\Gamma\vdash\mathsf{dbl}(\mathsf{m}(0,\,t))=t:\mathsf{I}\ .$$

Proposition 6. *1. Using (m), (a) and (E) only,* dbl *is the unique (up to provable equality) term of type* $\mathsf{I}\to\mathsf{I}$ *for which equations (d) hold.*

2. *Using (m), (a) and (d) only, cancellation (C) is a consequence.*

Using dbl, we can define, in System II, several useful functions (using the square bracket truncation notation from Sect. 3).

$$\begin{aligned}[x+y] &:= \mathsf{dbl}(x \oplus y) & \mathsf{max}(0,x) &:= [[x-1]+1] \\ x \ominus y &:= x \oplus (-y) & \mathsf{max}(x,y) &:= \mathsf{dbl}\left(\left[\frac{x}{2} + \mathsf{max}\,(0, y \ominus x)\right]\right) \\ [x-y] &:= \mathsf{dbl}(x \ominus y) & \mathsf{min}(x,y) &:= -\mathsf{max}(-x,-y) \\ & & |x| &:= \mathsf{max}(-x,x)\end{aligned}$$

Question 1. Are $\mathsf{max}(0,x)$, $\mathsf{max}(x,y)$ and $|x|$ definable in System I?

(The logical relation used in the proof of Prop. 4 does not help here.)

Having defined truncated versions of arithmetic functions, a very useful way of combining functions is by taking limits of Cauchy sequences. For *fast* Cauchy sequences (see Sect. 2), a limit-finding function $\mathsf{fastlim}\colon (\mathsf{N} \to \mathsf{I}) \to \mathsf{I}$ is definable:

$$\mathsf{fastlim} := \lambda f\!:\!\mathsf{N} \to \mathsf{I}.\ \mathsf{dbl}\left(\underset{n}{\mathsf{M}}\ \mathsf{dbl}^{n+1}(f(n+1) \ominus f(n))\right).$$

We write fastlim for the function $(\mathbb{N} \to [-1,1]) \to [-1,1]$ defined by $\mathsf{fastlim}$.

Lemma 3. *Let $(x_n)_n$ be a sequence from $[-1,1]$. If $|x_{n+1} - x_n| \leq 2^{-n}$, for all n, then* fastlim$(n \mapsto x_n)$ *is the limit of the (fast) Cauchy sequence $(x_n)_n$.*

Of course, fastlim$(n \mapsto x_n)$ always returns a value, even if $(x_n)_n$ is non-convergent. Also if $(x_n)_n$ converges, but too slowly, then fastlim$(n \mapsto x_n)$ need not be the limit value.

We have seen that dbl is representable in System T. Thus the System T implementation of System I extends to an implementation of System II. Naturally, we say that $f\colon [-1,1]^k \to [-1,1]$ is *II-definable* if there exists a closed System II term $u\colon \mathsf{I}^k \to \mathsf{I}$ such that $[\![u]\!] = f$. Proposition 7 below is immediate.

Proposition 7. *Every II-definable function is T-representable.*

The main result of the paper is the converse.

Theorem 1. *Every T-representable function is II-definable.*

The rest of this section is devoted to the proof of Theorem 1. We need some auxiliary definitions.

Define $\mathsf{glue}\colon (\mathsf{I} \to \mathsf{I})^2 \to \mathsf{I} \to \mathsf{I}$ by

$$\mathsf{glue} := \lambda f\, g\, x.\,\mathsf{dbl}\left(\mathsf{dbl}\left(\left(f\left(\mathsf{dbl}\left[x+\frac{1}{2}\right]\right) \oplus g\left(\mathsf{dbl}\left[x-\frac{1}{2}\right]\right)\right) \ominus \frac{1}{2}\,f(1)\right)\right).$$

which, whenever $f(1) = g(-1)$, satisfies

$$\mathsf{glue}\ f\ g\ x = \begin{cases} f(2x+1) & \text{if } -1 \leq x \leq 0 \\ g(2x-1) & \text{if } 0 \leq x \leq 1\,. \end{cases}$$

Next, for every $k \geq 1$, we define a System II term:

$$\mathsf{appr}_k \colon \mathsf{N} \to ((\mathsf{N} \to \mathsf{N})^k \to \mathsf{I}) \to (\mathsf{I}^k \to \mathsf{I})$$

The base case appr_1 is defined by primitive recursion on N to satisfy:

$$\begin{aligned} \mathsf{appr}_1\ 0\ h &= \mathsf{aff}\ (h(\overline{-1}))\ (h(\overline{1})) \quad \text{(where } \overline{-1} \text{ and } \overline{1} \text{ represent } -1 \text{ and } 1) \\ \mathsf{appr}_1\ (n+1)\ h &= \mathsf{glue}\ (\mathsf{appr}_1\ n\ (\lambda x.\, h(x \oplus (-1))))\ (\mathsf{appr}_1\ n\ (\lambda x.\, h(x \oplus 1)))\ . \end{aligned}$$

Given appr_k, the term appr_{k+1} is given by

$$\mathsf{appr}_{k+1}\ n\ h\ x_0\ x_1\ \ldots\ x_k \ = \ \mathrm{appr}_1\ n\ (\lambda y_0.\, \mathsf{appr}_k\ n\ (h\ y_0)\ x_1\ \ldots\ x_k)\ x_0\ .$$

Let appr_k be the denotation of appr_k. If $h\colon (\mathbb{N} \to \mathbb{N})^k \to [-1,1]$ is real-extensional then the application $\mathrm{appr}_k\ n\ h$ produces a piecewise multilinear approximation to the function h, with the argument types changed from $\mathbb{N} \to \mathbb{N}$ to $[-1,1]$.

More precisely, the $\mathrm{appr}_k\ n$ function uses k-tuples of values from the set

$$\mathbb{Q}_n := \{q_n^i \mid 0 \leq i \leq 2^n\} \quad \text{where } q_n^i := \frac{i}{2^{n-1}} - 1$$

to form a lattice of $(2^n+1)^k$ rational partition points in $[-1,1]^k$. The application $\mathrm{appr}_k\ n\ h$ then results in a function $[-1,1]^k \to [-1,1]$ that agrees with h at the partition points, and is (separately) affine in each coordinate between partition points. It is also affine in the h argument. The lemma below formalises this.

Lemma 4. *If $h\colon (\mathbb{N} \to \mathbb{N})^k \to [-1,1]$ represents $f\colon [-1,1]^k \to [-1,1]$ then:*

1. *For all $r_0, \ldots, r_{k-1} \in \mathbb{Q}_n$ we have:*

$$\mathrm{appr}_k\ n\ h\ r_0\ \ldots\ r_{k-1} = f\ r_0\ \ldots\ r_{k-1}$$

2. *For $0 \leq j < k$, $0 \leq i < 2^n$, and $0 \leq \lambda \leq 1$*

$$\begin{aligned} \mathrm{appr}_k\ n\ h\ x_0\ \ldots\ x_{j-1}\ \left(\tfrac{i+\lambda}{2^{n-1}} - 1\right)\ x_{j+1}\ \ldots\ x_{k-1} &= \\ (1-\lambda)\, \mathrm{appr}_k\ n\ h\ x_0\ \ldots\ x_{j-1}\ q_n^i\ x_{j+1}\ \ldots\ x_{k-1} & \\ +\, \lambda\, \mathrm{appr}_k\ n\ h\ x_0\ \ldots\ x_{j-1}\ q_n^{i+1}\ x_{j+1}\ \ldots\ x_{k-1} & \end{aligned}$$

Note that $\left(\frac{i+\lambda}{2^{n-1}} - 1\right) = ((1-\lambda)\, q_n^i + \lambda\, q_n^{i+1})$.

Also, if $h_1, h_2\colon (\mathbb{N} \to \mathbb{N})^k \to [-1,1]$ are real-extensional then

3. *For $0 \leq \lambda \leq 1$, we have:*

$$\begin{aligned} &\mathrm{appr}_k\ n\ ((1-\lambda)h_1 + \lambda h_2)\ x_0\ \ldots\ x_{k-1} = \\ &\quad (1-\lambda)\, \mathrm{appr}_k\ n\ h_1\ x_0\ \ldots\ x_{k-1} + \lambda\, \mathrm{appr}_k\ n\ h_2\ x_0\ \ldots\ x_{k-1} \end{aligned}$$

In fact, under the conditions of the lemma, $\lambda n.\, \mathsf{appr}_k\ n\ h$ is a sequence of functions $[-1,1]^k \to [-1,1]$ that converges pointwise, and hence uniformly, to h. All that remains to be done is to extract a fast-converging subsequence, since then h can be defined using the fastlim function. In order to get a handle on the rate of convergence, we exploit the following classic fact [14]. (For $\alpha\colon \mathbb{N} \to \mathbb{N}$, and $k \in \mathbb{N}$, we write $\alpha \restriction_k$ for the sequence $\alpha(0), \ldots, \alpha(k-1) \in \mathbb{N}^k$.)

Lemma 5 (Definable modulus of uniform continuity). *Suppose we have a closed System T term*

$$t\colon (\mathsf{N} \to \mathsf{N}) \to (\mathsf{N} \to \mathsf{N})$$

Then there exists a closed System T term

$$U_t\colon \mathsf{N} \to \mathsf{N}$$

satisfying: for all $e \geq 0$, and for all $\beta, \gamma\colon \mathbb{N} \to \{0,1,2\}$ such that $\beta \restriction_{U_t(e)} = \gamma \restriction_{U_t(e)}$, it holds that $[\![t]\!](\beta) \restriction_e = [\![t]\!](\gamma) \restriction_e$.

We now complete the proof of Theorem 1. Suppose $t\colon (\mathsf{N} \to \mathsf{N})^k \to \mathsf{N} \to \mathsf{N}$ is a closed term that T-represents $f\colon [-1,1]^k \to [-1,1]$. Let U_t be a uniform modulus for continuity for t on $\mathbb{N} \to \{0,1,2\}$, as given by Lemma 5. Let $g_n\colon [-1,1]^k \to [-1,1]$ be defined by:

$$\mathsf{appr}_k\,(U_t(n+1))\,(\mathsf{real} \circ t)\ :\ \mathsf{I}^k \to \mathsf{I}\ .$$

Then for all $x_1, \ldots, x_k \in [-1,1]$

$$|f(x_1, \ldots, x_n) - g_n(x_1, \ldots, x_n)|\ \leq\ 2^{-n}\ .$$

Therefore the term below II-defines f (where real is the easily defined system I term of type $(\mathsf{N} \to \mathsf{N}) \to \mathsf{I}$ implementing the function real from Section 2).

$$\lambda\, x_0\ \ldots\ x_{k-1}.\, \mathsf{fastlim}\,(\lambda n.\, \mathsf{appr}_k\,(U_t(n+1))\,(\mathsf{real} \circ t)\, x_0\ \ldots\ x_{k-1})\ .$$

5 Mantissa-Exponent Representation

There are many ways of extending signed binary to represent the full real line. Typically, one represents real number by a pair $\langle \alpha, z \rangle$ where $\alpha \in \{-1,0,1\}^\omega$, is the signed binary representation of $\mathrm{real}(\alpha) \in [-1,1]$ and $z \in \mathbb{Z}$. One natural option is for $\langle \alpha, z \rangle$ to represent the real number $z + \mathrm{real}(\alpha)$, thus treating z as an *offset*. Another is to use α as a *mantissa* and z as an *exponent*, giving the real number $2^z\, \mathrm{real}(\alpha)$. Again, both representations are intertranslatable.

In Systems I and II, a variation on such representations is available. Instead of using signed binary to represent a number in $[-1,1]$, it is natural to use the type I itself. Thus we can encode real numbers in Systems I and II, using the type $\mathsf{I} \times \mathsf{Z}$, where we write Z as an alternative notation for N to emphasise that the type is being used to encode all (including negative) integers (and we shall adopt similar suggestive notation for manipulation of integers). Curiously, under this approach, even the most basic functions cannot be programmed using the offset representation, so we are forced to use mantissa-exponent. Thus a term $\langle t, u\rangle : \mathsf{I}\times\mathsf{Z}$, represents the real number $2^{[\![u]\!]}\,[\![t]\!]$, where we mildly abuse notation to give u an interpretation $[\![u]\!] \in \mathbb{Z}$. We call this representation *semi-extensional*, since it combines a continuous value t, which is extensional, with a discrete scaling u, which is intensional. Although representations of real numbers are not unique, the continuous part is determined once the scaling is fixed.

It is straightforward to extend our main definability result to a characterisation of functions on $\mathbb{R}$ definable in System II. We say that a function $f : \mathbb{R}^k \to \mathbb{R}$ is T-representable, if there exists a System T term

$$t\colon ((\mathsf{N}\to\mathsf{N})\times\mathsf{Z})^k \to (\mathsf{N}\to\mathsf{N})\times\mathsf{Z}$$

that computes f under mantissa-exponent representation. And we say that f is I (resp. II)-representable if there exists a System I (resp. II) term

$$t\colon (\mathsf{I}\times\mathsf{Z})^k \to \mathsf{I}\times\mathsf{Z}$$

that computes f under mantissa-exponent representation.

Theorem 2. *A function $f : \mathbb{R}^k \to \mathbb{R}$ is T-representable if and only if it is II-representable.*

This result is essentially just an N-indexed version of Theorem 1. We omit the proof for space reasons.

Curiously, we do not know whether dbl is necessary for Theorem 2.

Question 2. Is every II-representable function $f : \mathbb{R}^k \to \mathbb{R}$ also I-representable?

A positive answer may sound implausible. But we now show that surprisingly many functions on real numbers can be defined in System I. At the same time, we show that reasoning about equality between functions on $\mathbb{R}$ can be reduced to equational reasoning in System I.

Equivalence between representations is given by the smallest equivalence relation on $[-1, 1] \times \mathbb{Z}$ satisfying

$$\langle x, m\rangle \sim \left(\frac{x}{2},\, m+1\right) .$$

Indeed, this equivalence relation is defined explicitly by

$$\langle x, m\rangle \;\sim\; \langle y, n\rangle \quad\Longleftrightarrow\quad \frac{x}{2^{\max(m,n)-m}} = \frac{y}{2^{\max(m,n)-n}} ,$$

where the right-hand-side is an equality expressible in System I.

Proposition 8. *The relation $\sim$ is provably an equivalence relation in System I.*

The intended formulation of the proposition is that the transitivity (symmetry and reflexivity being trivial) of $\sim$ is a derivable inference rule in System I. The proof makes essential use of cancellation (C) from Fig. 2.

The basic arithmetic operations on $\mathbb{R}$ are definable in System I.

$$\begin{aligned}
\mathbf{0} &:= \langle 0, 0\rangle \\
\mathbf{1} &:= \langle 1, 0\rangle \\
-\langle x, m\rangle &:= \langle -x, m\rangle \\
\langle x, m\rangle + \langle y, n\rangle &:= \left\langle \frac{x}{2^{\max(m,n)-m}} \oplus \frac{y}{2^{\max(m,n)-n}}, \max(m,n)+1 \right\rangle \\
\langle x, m\rangle \times \langle y, n\rangle &:= \langle x\, y, m+n\rangle
\end{aligned}$$

It is provable in System I that the above operations respect $\sim$. (Once again, by this, we mean that the inference rule expressing this property is derivable.) Also, the usual equations for the arithmetic operations are provable (commutativity, associativity, distributivity, etc.).

Since every rational number is System I definable, it follows that polynomials with rational coefficients are I-representable. We now show that we can also define limits of fast Cauchy sequences, as long the Cauchy sequences come with a witness to their speed of convergence.

Suppose we have a sequence $(\mathbf{x}_i)_i$ given by $\mathbf{x}_{(-)} : \mathsf{N} \to \mathsf{I} \times \mathsf{Z}$, such that the inequalities $|\mathbf{x}_{i+1} - \mathbf{x}_i| \leq 2^{-(i+1)}$ are witnessed by $d_{(-)} : \mathsf{N} \to \mathsf{I}$ satisfying

$$\mathbf{x}_{i+1} - \mathbf{x}_i \sim \langle d_i, -(i+1)\rangle \ .$$

Then we define

$$\lim_i \mathbf{x}_i := \mathbf{x}_0 + \langle \mathop{\mathsf{M}}_i d_i, 0\rangle \ .$$

Given the definability of rational polynomials and Cauchy limits, it is not implausible that a positive answer to Question 2 might be modelled on a constructive proof of the Stone-Weierstrass theorem. But this needs further investigation.

Another direction to explore is how much analysis can be developed using the mantissa-exponent representation of real numbers with the mantissa taken from our abstract datatype I. It would be interesting to explore this both using just the equational logic of Systems I and II, and also in the richer context of dependent type theory.

Acknowledgements. We would like to thank Jeremy Avigad, Ulrich Kohlenbach, Yitong Li, John Longley and the anonymous referees for helpful suggestions.

References

1. Bauer, A., Escardó, M.H., Simpson, A.K.: Comparing functional paradigms for exact real-number computation. In: Widmayer, P., Triguero, F., Morales, R., Hennessy, M., Eidenbenz, S., Conejo, R. (eds.) ICALP 2002. LNCS, vol. 2380, pp. 488–500. Springer, Heidelberg (2002)
2. Boehm, H.J.: Constructive real interpretation of numerical programs. SIGPLAN Notices 22(7), 214–221 (1987)

3. Boehm, H.J., Cartwright, R.: Exact real arithmetic: Formulating real numbers as functions. In: Turner, D. (ed.) Research Topics in Functional Programming, pp. 43–64. Addison-Wesley (1990)
4. Bove, A., Dybjer, P.: Dependent types at work. In: Bove, A., Barbosa, L.S., Pardo, A., Pinto, J.S. (eds.) LerNet 2008. LNCS, vol. 5520, pp. 57–99. Springer, Heidelberg (2009)
5. Brouwer, L.E.J.: Besitzt jede reelle Zahl eine Dezimalbruchentwicklung? Math. Ann. 83, 201–210 (1920)
6. Di-Gianantonio, P.: A Functional Approach to Computability on Real Numbers. PhD thesis, Università Degli Studi di Pisa, Dipartamento di Informatica (1993)
7. Di-Gianantonio, P.: Real number computability and domain theory. Information and Computation 127(1), 11–25 (1996)
8. Edalat, A., Escardó, M.H.: Integration in Real PCF. In: Proceedings of the Eleventh Annual IEEE Symposium on Logic In Computer Science, New Brunswick, New Jersey, USA, pp. 382–393 (1996)
9. Di Gianantonio, P., Edalat, A.: A language for differentiable functions. In: Pfenning, F. (ed.) FoSSaCS 2013. LNCS, vol. 7794, pp. 337–352. Springer, Heidelberg (2013)
10. Escardó, M.H.: PCF extended with real numbers. Theoret. Comput. Sci. 162(1), 79–115 (1996)
11. Escardó, M.H., Simpson, A.: A universal characterization of the closed Euclidean interval. In: Proceedings of the 16th Annual IEEE Symposium on Logic in Computer Science, pp. 115–128. IEEE Computer Society (2001)
12. Kohlenbach, U.: Applied Proof Theory: Proof Interpretations and their Use in Mathematics. Monographs in Mathematics. Springer (2008)
13. The Coq development team. The Coq proof assistant reference manual. LogiCal Project, Version 8.0. (2004)
14. Troelstra, A.S.: Some models for intuitionistic finite type arithmetic with fan functional. J. Symbolic Logic 42(2), 194–202 (1977)
15. Weihrauch, K.: Computable analysis. Springer (2000)
16. Wiedmer, E.: Computing with infinite objects. Theoret. Comput. Sci. 10, 133–155 (1980)

Self Types for Dependently Typed Lambda Encodings

Peng Fu and Aaron Stump

Computer Science, The University of Iowa, Iowa City, Iowa, USA

Abstract. We revisit lambda encodings of data, proposing new solutions to several old problems, in particular dependent elimination with lambda encodings. We start with a type-assignment form of the Calculus of Constructions, restricted recursive definitions and Miquel's implicit product. We add a type construct $\iota x.T$, called a *self type*, which allows T to refer to the subject of typing. We show how the resulting System **S** with this novel form of dependency supports dependent elimination with lambda encodings, including induction principles. Strong normalization of **S** is established by defining an erasure from **S** to a version of $\mathbf{F}_\omega$ with positive recursive type definitions, which we analyze. We also prove type preservation for **S**.

1 Introduction

Modern type-theoretic tools Coq and Agda extend a typed lambda calculus with a rich notion of primitive datatypes. Both tools build on established foundational concepts, but the interactions of these, particularly with datatypes and recursion, often leads to unexpected problems. For example, it is well-known that type preservation does not hold in Coq, due to the treatment of coinductive types [14]. Arbitrary nesting of coinductive and inductive types is not supported by the current version of Agda, leading to new proposals like co-patterns [2]. And new issues are discovered with disturbing frequency; e.g., an unexpected incompatibility of extensional consequences of Homotopy Type Theory with both Coq and Agda was discovered in December, 2013 [21].

The above issues all are related to the datatype system, which must determine what are the legal inductive/coinductive datatypes, in the presence of indexing, dependency, and generalized induction (allowing functional arguments to constructors). And for formal study of the type theory – either on paper [23], or in a proof assistant [5] – one must formalize the datatype system, which can be daunting, even in very capable hands (cf. Section 2 of [6]).

Fortunately, an alternative to primitive datatypes exists: lambda encodings, like the well-known Church and Scott encodings [7,10]. Utilizing the core typed lambda calculus for representing data means that no datatype system is needed at all, greatly simplifying the formal theory. We focus here just on inductive types, since in extensions of System **F**, coinductive types can be reduced to inductive ones [12].

Several problems historically prevented lambda encodings from being adopted in practical type theories. Scott encodings are efficient but do not inherently

G. Dowek (ed.): RTA-TLCA 2014, LNCS 8560, pp. 224–239, 2014.

provide a form of iteration or recursion. Church encodings inherently provide iteration, and are typable in System **F**. Due to strong normalization of System **F** [15], they are thus suitable for use in a total (impredicative) type theory, but:

1. The predecessor of n takes $O(n)$ time to compute instead of constant time.
2. We cannot prove $0 \neq 1$ with the usual definition of $\neq$.
3. Induction is not derivable [13].

These issues motivated the development of the Calculus of Inductive Constructions (cf. [22]). Problem (1) is best known but has a surprisingly underappreciated solution: if we accept positive recursive definitions (which preserve normalization), then we can use Parigot numerals, which are like Church numerals but based on recursors not iterators [20]. Normal forms of Parigot numerals are exponential in size, but a reasonable term-graph implementation should be able to keep them linear via sharing. The other three problems have remained unsolved.

In this paper, we propose solutions to problems (2) and (3). For problem (2) we propose to change the definition of falsehood from explosion ($\forall X.X$, everything is true) to equational inconsistency ($\forall X.\Pi x : X.\Pi y : X.x =_X y$, everything is equal for any type). We point out that $0 \neq 1$ is derivable with this notion. Our main contribution is for problem (3). We adapt **CC** to support dependent elimination with Church or Parigot encodings, using a novel type construct called *self types*, $\iota x.T$, to express dependency of a type on its subject. This allows deriving induction principles in a total type theory, and we believe it is the missing piece of the puzzle for dependent typing of pure lambda calculus.

We summarize the main technical points of this paper:

- System **S**, which enables us to encode Church and Parigot data and derive induction principles for these data.
- We prove strong normalization of **S** by erasure to a version of $\mathbf{F}_\omega$ with positive recursive type definitions. We prove strong normalization of this version of $\mathbf{F}_\omega$ by adapting a standard argument.
- Type preservation for **S** is proved by extending Barendregt's method [4] to handle implicit products and making use of a confluence argument.

Detailed arguments omitted here may be found in an extended version [11].

2 Overview of System S

System **S** extends a type-assignment formulation of the Calculus of Constructions (**CC**) [9]. We allow global recursive definitions in a form we call a *closure*:

$$\{(x_i : S_i) \mapsto t_i\}_{i \in N} \cup \{(X_i : \kappa_i) \mapsto T_i\}_{i \in M}$$

The x_i are term variables which cannot appear in the terms t_i, but can appear in the types T_i. And N, M are nonempty index set. Occurrences in types are used to express dependency, and are crucial for our approach. Erasure to $\mathbf{F}_\omega$ with positive recursive definitions will drop all such occurrences. The X_i are type variables that can appear positively in the T_i or at erased positions (explained later).

The essential new construct is the self type $\iota x.T$. Note that this is different from self typing in the object-oriented (OO) literature, where the central problem has been to allow self-application while still validating natural record-subtyping rules [19,1]. Typing the self parameter of an object's methods appears different from allowing a type to refer to its subject, though Hickey proposes a type-theoretic encoding of objects based on very dependent function types $\{f \,|\, x : A \to B\}$, where the range B can depend on both x and values of the function f itself [16]. The self types we propose appear to be simpler.

2.1 Induction Principle

Let us take a closer look at the difficulties of deriving an induction principle for Church numerals in **CC**, and then consider our solutions. In **CC** à la Curry, let $\mathsf{Nat} := \forall X.(X \to X) \to X \to X$. One can obtain a notion of *indexed iterator* by $\mathsf{It} := \lambda x.\lambda f.\lambda a.x\ f\ a$ and $\mathsf{It} : \forall X.\Pi x : \mathsf{Nat}.(X \to X) \to X \to X$. Thus we have $\mathsf{It}\ \bar{n} =_\beta \lambda f.\lambda a.\bar{n}\ f\ a =_\beta \lambda f.\lambda a.\underbrace{f(f(f...(f}_{n}\ a)...))$. One may want to know if we can obtain a finer version, namely, the induction principle-Ind such that:

$$\mathsf{Ind} : \forall P : \mathsf{Nat} \to *.\Pi x : \mathsf{Nat}.(\Pi y : \mathsf{Nat}.(Py \to P(\mathsf{S}y))) \to P\,\bar{0} \to P\ x$$

Let us try to construct such Ind. First observe the following beta-equalities and typings:

$$\mathsf{Ind}\ \bar{0} =_\beta \lambda f.\lambda a.a$$
$$\mathsf{Ind}\ \bar{0} : (\Pi y : \mathsf{Nat}.(Py \to P(\mathsf{S}y))) \to P\,\bar{0} \to P\,\bar{0}$$
$$\mathsf{Ind}\ \bar{n} =_\beta \lambda f.\lambda a.\underbrace{f\ \overline{n-1}(...f\ \bar{1}\ (f\ \bar{0}\ a))}_{n>0}$$
$$\mathsf{Ind}\ \bar{n} : (\Pi y : \mathsf{Nat}.(Py \to P(\mathsf{S}y))) \to P\,\bar{0} \to P\,\bar{n}$$
$$\text{with } f : \Pi y : \mathsf{Nat}.(Py \to P(\mathsf{S}y)), a : P\,\bar{0}$$

These equalities suggest that $\mathsf{Ind} := \lambda x.\lambda f.\lambda a.x\ f\ a$, using Parigot numerals [20]:

$$\bar{0} := \lambda s.\lambda z.z$$
$$\bar{n} := \lambda s.\lambda z.s\ \overline{n-1}\ (\overline{n-1}\ s\ z)$$

Each numeral corresponds to its terminating recursor.

Now, let us try to type these lambda numerals. It is reasonable to assign $s : \Pi y : \mathsf{Nat}.(P\ y \to P(\mathsf{S}\ y))$ and $z : P\ \bar{0}$. Thus we have the following typing relations:

$$\bar{0} : \Pi y : \mathsf{Nat}.(P\ y \to P(\mathsf{S}\ y)) \to P\,\bar{0} \to P\,\bar{0}$$
$$\bar{1} : \Pi y : \mathsf{Nat}.(P\ y \to P(\mathsf{S}\ y)) \to P\,\bar{0} \to P\,\bar{1}$$
$$\bar{n} : \Pi y : \mathsf{Nat}.(P\ y \to P(\mathsf{S}\ y)) \to P\,\bar{0} \to P\,\bar{n}$$

So we want to define Nat to be something like:

$$\forall P : \mathsf{Nat} \to *.\Pi y : \mathsf{Nat}.(P\ y \to P(\mathsf{S}\ y)) \to P\,\bar{0} \to P\,\bar{n} \text{ for any } \bar{n}.$$

Two problems arise with this scheme of encoding. The first problem involves recursiveness. The definiens of Nat contains Nat, S and $\bar{0}$, while the type of S

is $\mathsf{Nat} \to \mathsf{Nat}$ and the type of $\bar{0}$ is Nat. So the typing of Nat will be mutually recursive. Observe that the recursive occurrences of Nat are all at the type-annotated positions; i.e., the right side of the ":".

Note that the subdata of $\bar{n}$ is responsible for one recursive occurrence of Nat, namely, $\Pi y : \mathsf{Nat}$. If one never computes with the subdata, then these numerals will behave just like Church numerals. This inspires us to use Miquel's implicit product [18]. In this case, we want to redefine Nat to be something like:

$\forall P : \mathsf{Nat} \to *.\forall y : \mathsf{Nat}.(P\ y \to P(\mathsf{S}\ y)) \to P\ \bar{0} \to P\ \bar{n}$ for any $\bar{n}$.

Here $\forall y : \mathsf{Nat}$ is the implicit product. Now our notion of numerals are exactly Church numerals instead of Parigot numerals. Even better, this definition of Nat can be erased to $\mathbf{F}_\omega$. Since $\mathbf{F}_\omega$'s types do not have dependency on terms, $P : \mathsf{Nat} \to *$ will get erased to $P : *$. It is known that one can also erase the implicit product [3]. The erasure of Nat will be $\Pi P : *.(P \to P) \to P \to P$, which is the definition of Nat in $\mathbf{F}_\omega$.

The second problem is about quantification. We want to define a type Nat for any $\bar{n}$, but right now what we really have is one Nat for each numeral $\bar{n}$. We solve this problem by introducing a new type construct $\iota x.T$ called a *self type*. This allows us to make this definition (for Church-encoded naturals):

$\mathsf{Nat} := \iota x.\forall P : \mathsf{Nat} \to *.\forall y : \mathsf{Nat}.(P\ y \to P(\mathsf{S}\ y)) \to P\ \bar{0} \to P\ x$

We require that the self type can only be instantiated/generalized by its own subject, so we add the following two rules:

$$\frac{\Gamma \vdash t : [t/x]T}{\Gamma \vdash t : \iota x.T}\ \mathit{selfGen} \qquad \frac{\Gamma \vdash t : \iota x.T}{\Gamma \vdash t : [t/x]T}\ \mathit{selfInst}$$

We have the following inferences[1]:

$$\frac{\bar{n} : \forall P : \mathsf{Nat} \to *.\forall y : \mathsf{Nat}.(P\ y \to P(\mathsf{S}\ y)) \to P\ \bar{0} \to P\ \bar{n}}{\bar{n} : \iota x.\forall P : \mathsf{Nat} \to *.\forall y : \mathsf{Nat}.(P\ y \to P(\mathsf{S}\ y)) \to P\ \bar{0} \to P\ x}$$

2.2 The Notion of Contradiction

In **CC** à la Curry, it is customary to use $\forall X : *.X$ as the notion of contradiction, since an inhabitant of the type $\forall X : *.X$ will inhabit any type, so the law of explosion is subsumed by the type $\forall X : *.X$. However, this notion of contradiction is too strong to be useful. Let $t =_A t'$ denote $\forall C : A \to *.C\ t \to C\ t'$ with $t, t' : A$. Then $0 =_{\mathsf{Nat}} 1$ can be expanded to $\forall C : \mathsf{Nat} \to *.C\ 0 \to C\ 1$ (0 is Leibniz equals to 1). One can not derive a proof for $(\forall C : \mathsf{Nat} \to *.C\ 0 \to C\ 1) \to \forall X : *.X$, because the erasure of $(\forall C : \mathsf{Nat} \to *.C\ 0 \to C\ 1) \to \forall X : *.X$ in System **F** would be $(\forall C : *.C \to C) \to \forall X : *.X$, and we know that $\forall C : *.C \to C$ is inhabited. So the inhabitation of $(\forall C : \mathsf{Nat} \to *.C\ 0 \to C\ 1) \to \forall X : *.X$ will imply the inhabitation of $\forall X : *.X$ in System **F**, which does not hold. If we take Leibniz equality and use $\forall X : *.X$ as contradiction, then we can not prove any negative results about equality.

[1] The double bar means that the converse of the inference also holds.

On the other hand, an equational theory is considered inconsistent if $a = b$ for all term a and b. So we propose to use $\forall A : *.\Pi x : A.\Pi y : A.x =_A y$ as the notion of contradiction in **CC**. We first want to make sure it is uninhabited. The way to argue that is first assume it is inhabited by t. Since **CC** is strongly normalizing, the normal form of t must be of the form[2] $[\lambda A : *.]\lambda x[: A].\lambda y[: A].[\lambda C : A \to *].\lambda z[: C\ x].n$ for some normal term n with type $C\ y$, but we know that there is no combination of x, y, z to make a term of type $C\ y$. So the type $\forall A : *.\Pi x : A.\Pi y : A.\forall C : A \to *.Cx \to Cy$ is uninhabited. We can then prove the following theorem[3]:

Theorem 1. *$0 = 1 \to \bot$ is inhabited in* **CC**, *where* $\bot := \forall A : *.\Pi x : A.\Pi y : A.\forall C : A \to *.C\ x \to C\ y$, $0 := \lambda s.\lambda z.z$, $1 := \lambda s.\lambda z.s\ z$.

Once $\bot$ is derived, one can not distinguish the domain of individuals. Note that this notion of contradiction does not subsume law of explosion.

3 System S

We use gray boxes in this section to highlight the terms, types and rules that are not in $\mathbf{F}_\omega$ with positive recusive definitions[4].

3.1 Syntax

$$
\begin{array}{lll}
\textit{Terms } t & ::= & x \mid \lambda x.t \mid tt' \\
\textit{Types } T & ::= & X \mid \forall X : \kappa.T \mid \Pi x : T_1.T_2 \mid \boxed{\forall x : T_1.T_2} \mid \\
 & & \boxed{\iota x.T} \mid \boxed{T\ t} \mid \lambda X.T \mid \boxed{\lambda x.T} \mid T_1 T_2 \\
\textit{Kinds } \kappa & ::= & * \mid \boxed{\Pi x : T.\kappa} \mid \Pi X : \kappa'.\kappa \\
\textit{Context } \Gamma & ::= & \cdot \mid \Gamma, x : T \mid \Gamma, X : \kappa \mid \Gamma, \mu \\
\textit{Closure } \mu & ::= & \{(x_i : S_i) \mapsto t_i\}_{i \in N} \cup \{(X_i : \kappa_i) \mapsto T_i\}_{i \in M}
\end{array}
$$

Closures. For $\{(x_i : S_i) \mapsto t_i\}_{i\in N}$, we mean the term variable x_i of type S_i is defined to be t_i for some $i \in N$; similarly for $\{(X_i : \kappa_i) \mapsto T_i\}_{i\in M}$.

Legal Positions for Recursion in Closures. For $\{(x_i : S_i) \mapsto t_i\}_{i\in N}$, we do not allow any recursive (or mutually recursive) definitions. For $\{(X_i : \kappa_i) \mapsto T_i\}_{i\in M}$, we only allow singly recursive type definitions, but not mutually recursive ones. This is not a fundamental limitation of the approach; it is just for simplicity of the normalization argument. The recursive occurrences of type variables can only be at positive or erased positions. Erased positions, following the erasure function we will see in Section 5.1, are those in kinds or in the types for $\forall$-bound variables.

Variable Restrictions for Closures. Let $\mathrm{FV}(e)$ denote the set of free term variables in expression e (either term, type, or kind), and let $\mathrm{FVar}(T)$ denote the

[2] We use square brackets [ ] to show annotations that are not present in the inhabiting lambda term in Curry-style System **F**.

[3] Coq code for this is in the extended version.

[4] Full specification of $\mathbf{F}_\omega$ with positive recursive definitions is in the extended version.

set of free type variables in type T. Then for $\{(x_i : S_i) \mapsto t_i\}_{i\in N} \cup \{(X_i : \kappa_i) \mapsto T_i\}_{i\in M}$, we make the simplifying assumption that for any $1 \leq i \leq n$, $\mathrm{FV}(t_i) = \emptyset$. Also, for any $1 \leq i \leq m$, we require $\mathrm{FV}(T_i) \subseteq \mathrm{dom}(\mu)$, and $\mathrm{FVar}(\mathrm{T_i}) \subseteq \{X_i\}$. All our examples below satisfy these conditions.

3.2 Kinding and Typing

Some remarks on the typing and kinding rules:

Notation for Accessing Closures. $(t_i : S_i) \in \mu$ means $(x_i : S_i) \mapsto t_i \in \mu$ and $(T_i : \kappa_i) \in \mu$ means $(X_i : \kappa_i) \mapsto T_i \in \mu$. Also, $x_i \mapsto t_i \in \mu$ means $(x_i : S_i) \mapsto t_i \in \mu$ for some S_i and $X_i \mapsto T_i \in \mu$ means $(X_i : \kappa_i) \mapsto T_i \in \mu$ for some κ_i.

Well-Formed Annotated Closures. $\Gamma \vdash \mu\ \mathsf{ok}$ stands for $\{\Gamma, \mu \vdash t_j : T_j\}_{(t_j:T_j)\in\mu}$ and $\{\Gamma, \mu \vdash T_j : \kappa_j\}_{(T_j:\kappa_j)\in\mu}$. In other words, the defining expressions in closures must be typable with respect to the context and the entire closure.

Notation for Equivalence. $\cong$ is the congruence closure of $\to_\beta$.

Self Type Formation. Typing and kinding do not depend on well-formedness of the context, so the self type formation rule *self* is not circular.

Well-formed Contexts $\boxed{\Gamma \vdash \mathsf{wf}}$

$$\frac{}{\cdot \vdash \mathsf{wf}} \quad \frac{\Gamma \vdash \mathsf{wf} \quad \Gamma \vdash T : *}{\Gamma, x : T \vdash \mathsf{wf}} \quad \frac{\Gamma \vdash \mathsf{wf} \quad \Gamma \vdash \kappa : \Box}{\Gamma, X : \kappa \vdash \mathsf{wf}} \quad \frac{\Gamma \vdash \mathsf{wf} \quad \Gamma \vdash \mu\ \mathsf{ok}}{\Gamma, \mu \vdash \mathsf{wf}}$$

Well-Formed Kinds $\boxed{\Gamma \vdash \kappa : \Box}$

$$\frac{}{\Gamma \vdash * : \Box} \quad \frac{\Gamma, X : \kappa' \vdash \kappa : \Box \quad \Gamma \vdash \kappa' : \Box}{\Gamma \vdash \Pi X : \kappa'.\kappa : \Box} \quad \frac{\Gamma, x : T \vdash \kappa : \Box \quad \Gamma \vdash T : *}{\Gamma \vdash \Pi x : T.\kappa : \Box}$$

Kinding $\boxed{\Gamma \vdash T : \kappa}$

$$\frac{(X : \kappa) \in \Gamma}{\Gamma \vdash X : \kappa} \qquad \frac{\Gamma \vdash T : \kappa \quad \Gamma \vdash \kappa \cong \kappa' \quad \Gamma \vdash \kappa' : \Box}{\Gamma \vdash T : \kappa'}$$

$$\frac{\Gamma \vdash T_1 : * \quad \Gamma, x : T_1 \vdash T_2 : *}{\Gamma \vdash \Pi x : T_1.T_2 : *} \qquad \frac{\Gamma, X : \kappa \vdash T : * \quad \Gamma \vdash \kappa : \Box}{\Gamma \vdash \forall X : \kappa.T : *}$$

$$\frac{\Gamma, x : T_1 \vdash T_2 : * \quad \Gamma \vdash T_1 : *}{\Gamma \vdash \forall x : T_1.T_2 : *} \qquad \frac{\Gamma, x : \iota x.T \vdash T : *}{\Gamma \vdash \iota x.T : *}\ \mathit{Self}$$

$$\frac{\Gamma, X : \kappa \vdash T : \kappa' \quad \Gamma \vdash \kappa : \Box}{\Gamma \vdash \lambda X.T : \Pi X : \kappa.\kappa'} \qquad \frac{\Gamma, x : T' \vdash T : \kappa \quad \Gamma \vdash T' : *}{\Gamma \vdash \lambda x.T : \Pi x : T'.\kappa}$$

$$\frac{\Gamma \vdash S : \Pi x : T.\kappa \quad \Gamma \vdash t : T}{\Gamma \vdash S\ t : [t/x]\kappa} \qquad \frac{\Gamma \vdash S : \Pi X : \kappa'.\kappa \quad \Gamma \vdash T : \kappa'}{\Gamma \vdash S\ T : [T/X]\kappa}$$

Typing $\boxed{\Gamma \vdash t : T}$

$$\frac{\Gamma \vdash t : T_1 \quad \Gamma \vdash T_1 \cong T_2 \quad \Gamma \vdash T_2 : *}{\Gamma \vdash t : T_2}\ \mathit{Conv} \qquad \frac{(x : T) \in \Gamma}{\Gamma \vdash x : T}\ \mathit{Var}$$

$$\frac{\Gamma \vdash t : [t/x]T \quad \Gamma \vdash \iota x.T : *}{\Gamma \vdash t : \iota x.T}\ \mathit{SelfGen} \qquad \frac{\Gamma \vdash t : \iota x.T}{\Gamma \vdash t : [t/x]T}\ \mathit{SelfInst}$$

$$\frac{\Gamma, x : T_1 \vdash t : T_2 \quad \Gamma \vdash T_1 : * \quad x \notin \mathrm{FV}(t)}{\Gamma \vdash t : \forall x : T_1.T_2}\ \mathit{Indx} \qquad \frac{\Gamma \vdash t : \forall x : T_1.T_2 \quad \Gamma \vdash t' : T_1}{\Gamma \vdash t : [t'/x]T_2}\ \mathit{Dex}$$

$$\frac{\Gamma \vdash t : \Pi x : T_1.T_2 \quad \Gamma \vdash t' : T_1}{\Gamma \vdash tt' : [t'/x]T_2}\ \mathit{App} \qquad \frac{\Gamma, X : \kappa \vdash t : T \quad \Gamma \vdash \kappa : \Box}{\Gamma \vdash t : \forall X : \kappa.T}\ \mathit{Poly}$$

$$\frac{\Gamma \vdash t : \forall X : \kappa.T \quad \Gamma \vdash T' : \kappa}{\Gamma \vdash t : [T'/X]T}\ \mathit{Inst} \qquad \frac{\Gamma, x : T_1 \vdash t : T_2 \quad \Gamma \vdash T_1 : *}{\Gamma \vdash \lambda x.t : \Pi x : T_1.T_2}\ \mathit{Func}$$

Reductions $\boxed{\Gamma \vdash t \to_\beta t'}$, $\boxed{\Gamma \vdash T \to_\beta T'}$

$$\frac{(x \mapsto t) \in \Gamma}{\Gamma \vdash x \to_\beta t} \qquad \frac{}{\Gamma \vdash (\lambda x.t)t' \to_\beta [t'/x]t} \qquad \frac{(X \mapsto T) \in \Gamma}{\Gamma \vdash X \to_\beta T}$$

$$\frac{}{\Gamma \vdash (\lambda x.T)t \to_\beta [t/x]T} \qquad \frac{}{\Gamma \vdash (\lambda X.T)T' \to_\beta [T'/X]T}$$

4 Lambda Encodings in S

Now let us see some concrete examples of lambda encoding in **S**. For convenience, we write $T \to T'$ for $\Pi x : T.T'$ with $x \notin \mathrm{FV}(T')$, and similarly for kinds.

4.1 Natural Numbers

Definition 1 (Church Numerals). *Let μ_c be the following closure:*
$(\mathsf{Nat} : *) \mapsto \iota x.\forall C : \mathsf{Nat} \to *.(\forall n : \mathsf{Nat}.C\ n \to C\ (\mathsf{S}\ n)) \to C\ 0 \to C\ x$
$(\mathsf{S} : \mathsf{Nat} \to \mathsf{Nat}) \mapsto \lambda n.\lambda s.\lambda z.s\ (n\ s\ z)$
$(0 : \mathsf{Nat}) \mapsto \lambda s.\lambda z.z$

With $s : \forall n : \mathsf{Nat}.C\ n \to C\ (\mathsf{S}\ n), z : C\ 0, n : \mathsf{Nat}$, we have $\mu_c \vdash \mathsf{wf}$ (using *selfGen* and *selfInst* rules). Also note that the μ_c satisfies the constraints on recursive definitions. Similarly, if we choose to use explicit product, then we can define Parigot numerals.

Definition 2 (Parigot Numerals). *Let μ_p be the following closure:*
$(\mathsf{Nat} : *) \mapsto \iota x.\forall C : \mathsf{Nat} \to *.(\Pi n : \mathsf{Nat}.C\ n \to C\ (\mathsf{S}\ n)) \to C\ 0 \to C\ x$
$(\mathsf{S} : \mathsf{Nat} \to \mathsf{Nat}) \mapsto \lambda n.\lambda s.\lambda z.s\ n\ (n\ s\ z)$
$(0 : \mathsf{Nat}) \mapsto \lambda s.\lambda z.z$

Note that the recursive occurences of Nat in Parigot numerals are at positive positions. The rest of the examples are about Church numerals, but a similar development can be carried out with Parigot numerals.

Theorem 2 (Induction Principle)
$\mu_c \vdash \mathsf{Ind} : \forall C : \mathsf{Nat} \to *.(\forall n : \mathsf{Nat}.C\ n \to C\ (\mathsf{S}\ n)) \to C\ 0 \to \Pi n : \mathsf{Nat}.C\ n$
where $\mathsf{Ind} := \lambda s.\lambda z.\lambda n.n\ s\ z$
with $s : \forall n : \mathsf{Nat}.C\ n \to C\ (\mathsf{S}\ n), z : C\ 0, n : \mathsf{Nat}$.

Proof. Let $\Gamma = \mu_c, C : \mathsf{Nat} \to *, s : \forall n : \mathsf{Nat}.C\ n \to C\ (\mathsf{S}\ n), z : C\ 0, n : \mathsf{Nat}$. Since $n : \mathsf{Nat}$, by *selfInst*, $n : \forall C : \mathsf{Nat} \to *.(\forall y : \mathsf{Nat}.C\ y \to C\ (\mathsf{S}\ y)) \to C\ 0 \to C\ n$. Thus $n\ s\ z : C\ n$.

It is worth noting that it is really the definition of Nat and the *selfInst* rule that give us the induction principle, which is not derivable in **CC** [8].

Definition 3 (Addition). $m + n := \mathsf{Ind}\ \mathsf{S}\ n\ m$

One can check that $\mu_c \vdash + : \mathsf{Nat} \to \mathsf{Nat} \to \mathsf{Nat}$ by instantiating the C in the type of Ind by $\lambda y.\mathsf{Nat}$, then the type of Ind is $(\mathsf{Nat} \to \mathsf{Nat}) \to \mathsf{Nat} \to (\mathsf{Nat} \to \mathsf{Nat})$.

Definition 4 (Leibniz's Equality). $\mathsf{Eq} := \lambda A[: *].\lambda x[: A].\lambda y[: A].\forall C : A \to *.C\ x \to C\ y$.

Note that we use $x =_A y$ to denote $\mathsf{Eq}\ A\ x\ y$. We often write $t = t'$ when the type is clear. One can check that if $\vdash A : *$ and $\vdash x, y : A$, then $\vdash x =_A y : *$.

Theorem 3. $\mu_c \vdash \Pi x : \mathsf{Nat}.x + 0 =_{\mathsf{Nat}} x$

Proof. We prove this by induction. We instantiate C in the type of Ind with $\lambda n.(n + 0) =_{\mathsf{Nat}} n$. So by beta reduction at type level, we have $(\forall n : \mathsf{Nat}.(n + 0 =_{\mathsf{Nat}} n) \to ((\mathsf{S}\ n) + 0 =_{\mathsf{Nat}} \mathsf{S}\ n)) \to 0 + 0 =_{\mathsf{Nat}} 0 \to \Pi n : \mathsf{Nat}.n + 0 =_{\mathsf{Nat}} n$. So for the base case, we need to show $0 + 0 =_{\mathsf{Nat}} 0$, which is easy. For the step case, we assume $n + 0 =_{\mathsf{Nat}} n$ (Induction Hypothesis), and want to show $(\mathsf{S}\ n) + 0 =_{\mathsf{Nat}} \mathsf{S}\ n$. Since $(\mathsf{S}\ n) + 0 \to_\beta \mathsf{S}\ (n\ \mathsf{S}\ 0) =_\beta \mathsf{S}(n + 0)$, by congruence on the induction hypothesis, we have $(\mathsf{S}\ n) + 0 =_{\mathsf{Nat}} \mathsf{S}\ n$. Thus $\Pi x : \mathsf{Nat}.x + 0 =_{\mathsf{Nat}} x$.

The above theorem is provable inside **S**. It shows how to inhabit the type $\Pi x : \mathsf{Nat}.x + 0 =_{\mathsf{Nat}} x$ given μ_c, using Ind.

4.2 Vector Encoding

Definition 5 (Vector). *Let* μ_v *be the following definitions:*
$(\mathsf{vec} : * \to \mathsf{Nat} \to *) \mapsto$
$\quad \lambda U : *.\lambda n : \mathsf{Nat}.\ \iota x\ .\forall C :\ \Pi p : \mathsf{Nat}.\mathsf{vec}\ U\ p \to *\ .$
$\quad (\Pi m : \mathsf{Nat}.\Pi u : U.\forall y : \mathsf{vec}\ U\ m.(C\ m\ y\ \to C\ (\mathsf{S}\ m)\ (\mathsf{cons}\ m\ u\ y)))$
$\quad \to C\ 0\ \mathsf{nil} \to C\ n\ \ x$
$(\mathsf{nil} : \forall U : *.\mathsf{vec}\ U\ 0) \mapsto \lambda y.\lambda x.x$
$(\mathsf{cons} : \Pi n : \mathsf{Nat}.\forall U : *.U \to \mathsf{vec}\, U\, n \to \mathsf{vec}\, U\, (\mathsf{S}\, n)) \mapsto \lambda n.\lambda v.\lambda l.\lambda y.\lambda x.y\, n\, v\, (l\, y\, x)$
where $n : \mathsf{Nat}, v : U, l : \mathsf{vec}\ U\ n, y : \Pi m : \mathsf{Nat}.\Pi u : U.\forall z : \mathsf{vec}\ U\ m.(C\ m\ z \to C\ (\mathsf{S}\ m)\ (\mathsf{cons}\ m\ u\ z)), x : C\ 0\ \mathsf{nil}$.

Typing: It is easy to see that nil is typable to $\forall U : *.\mathsf{vec}\ U\ 0$. Now we show how cons is typable to $\Pi n : \mathsf{Nat}.\forall U : *.U \to \mathsf{vec}\ U\ n \to \mathsf{vec}\ U\ (\mathsf{S}\ n)$. We can see that $l\ y\ x : C\ n\ l$ (using *selfinst* on l). After the instantiation with l, the type of $y\ n\ v$ is $C\ n\ l \to C\ (\mathsf{S}\ n)\ (\mathsf{cons}\ n\ v\ l)$. So $y\ n\ v\ (l\ y\ x) : C\ (\mathsf{S}\ n)\ (\mathsf{cons}\ n\ v\ l)$. So $\lambda y.\lambda x.y\ n\ v\ (l\ y\ x) : \Pi C : (\mathsf{Nat} \to \mathsf{vec}\ U\ p \to *).(\Pi m : \mathsf{Nat}.\Pi u : U.\forall y : \mathsf{vec}\ U\ m.(C\ m\ y \to C\ (\mathsf{S}\ m)\ (\mathsf{cons}\ m\ u\ y))) \to C\ 0\ \mathsf{nil} \to C\ (\mathsf{S}\ n)\ (\lambda y.\lambda x.y\ n\ v\ (l\ y\ x))$. So by *selfGen*, we have $\lambda y.\lambda x.y\ n\ v\ (l\ y\ x) : \mathsf{vec}\ U(\mathsf{S}\ n)$. Thus $\mathsf{cons} : \Pi n : \mathsf{Nat}.\forall U : *.U \to \mathsf{vec}\ U\ n \to \mathsf{vec}\ U\ (\mathsf{S}\ n)$.

Definition 6 (Induction Principle for Vector)
$\mu_v \vdash \mathsf{Ind} :$
$\forall U : *.\Pi n : \mathsf{Nat}.\forall C : \mathsf{Nat} \to \mathsf{vec}\ U\ p \to *.$
$(\Pi m : \mathsf{Nat}.\Pi u : U.\forall y : \mathsf{vec}\ U\ m.(C\ m\ y \to C\ (\mathsf{S}\ m)\ (\mathsf{cons}\ m\ u\ y)))$
$\to C\ 0\ \mathsf{nil} \to \Pi x : \mathsf{vec}\ U\ n.(C\ n\ x)$
where $\mathsf{Ind} := \lambda n.\lambda s.\lambda z.\lambda x.x\ s\ z$
$n : \mathsf{Nat}, s : \forall C : (\Pi p : \mathsf{Nat}.\mathsf{vec}\ U\ p \to *).(\Pi m : \mathsf{Nat}.\Pi u : U.\forall y : \mathsf{vec}\ U\ m.(C\ m\ y \to C\ (\mathsf{S}\ m)\ (\mathsf{cons}\ m\ u\ y))), z : C\ 0\ \mathsf{nil}, x : \mathsf{vec}\ U\ n.$

Definition 7 (Append)
$\mu_v \vdash \mathsf{app} : \forall U : *.\Pi n_1 : \mathsf{Nat}.\Pi n_2 : \mathsf{Nat}.\mathsf{vec}\ U\ n_1 \to \mathsf{vec}\ U\ n_2 \to \mathsf{vec}\ U\ (n_1 + n_2)$
where $\mathsf{app} := \lambda n_1.\lambda n_2.\lambda l_1.\lambda l_2.(\mathsf{Ind}\ n_1)\ (\lambda n.\lambda x.\lambda v.\mathsf{cons}\ (n + n_2)\ x\ v)\ l_2\ l_1$.

Typing: We want to show $\mathsf{app} : \forall U : *.\Pi n_1 : \mathsf{Nat}.\Pi n_2 : \mathsf{Nat}.\mathsf{vec}\ U\ n_1 \to \mathsf{vec}\ U\ n_2 \to \mathsf{vec}\ U\ (n_1 + n_2)$. Observe that $\lambda n.\lambda x.\lambda v.\mathsf{cons}(n + n_2)\ x\ v : \Pi n : \mathsf{Nat}.\Pi x : U.\mathsf{vec}\ U\ (n + n_2) \to \mathsf{vec}\ U\ (n + n_2 + 1)$. We instantiate $C := \lambda y.(\lambda x.\mathsf{vec}\ U\ (y + n_2))$, where x free over $\mathsf{vec}\ U\ (y + n_2)$, in $\mathsf{Ind}\ n_1$. By beta reductions, we get $\mathsf{Ind}\ n_1 : (\Pi m : \mathsf{Nat}.\Pi u : U.\forall y : \mathsf{vec}\ U\ m.(\mathsf{vec}\ U\ (m + n_2) \to \mathsf{vec}\ U\ ((\mathsf{S}\ m) + n_2)) \to \mathsf{vec}\ U\ (0 + n_2) \to \Pi x : \mathsf{vec}\ U\ n_1.\mathsf{vec}\ U\ (n_1 + n_2)$. So $(\mathsf{Ind}\ n_1)\ (\lambda n.\lambda x.\lambda v.\mathsf{cons}(n+n_2)\ x\ v) : \mathsf{vec}\ U\ (0+n_2) \to \Pi x : \mathsf{vec}\ U\ n_1.\mathsf{vec}\ U\ (n_1+n_2)$. We assume $l_1 : \mathsf{vec}\ U\ n_1, l_2 : \mathsf{vec}\ U\ n_2$. Thus $(\mathsf{Ind}\ n_1)\ (\lambda n.\lambda x.\lambda v.\mathsf{cons}(n+n_2)\ x\ v)\ l_2\ l_1 : \mathsf{vec}\ U\ (n_1 + n_2)$.

5 Metatheory

We first outline the erasure from **S** to $\mathbf{F}_\omega$ with positive recursive definitions. Then we conclude strong normalization for **S** by the strong normalization of $\mathbf{F}_\omega$ with positive recursive definitions. The strong normalization proof is an extension of the method describes in [17]. We also prove type preservation for **S**, which involves *confluence analysis* (Section 5.2) and *morph analysis* (Section 5.3). All omitted proofs may be found in the extended version [11].

5.1 Strong Normalization

We prove strong normalization of **S** through the strong normalization of $\mathbf{F}_\omega$ with positive recursive definitions. We first define the syntax for $\mathbf{F}_\omega$ with positive recursive definitions. We work with kind-annotated types to avoid the interpretation for ill-formed types like $\lambda X.X \to \lambda X.X$.

Definition 8 (Syntax for $\mathbf{F}_\omega$ with positive definitions)

Terms $t ::= x \mid \lambda x.t \mid tt'$
Kinds $\kappa ::= * \mid \kappa' \to \kappa$
Types $T^\kappa ::= X^\kappa \mid (\forall X^\kappa.T^*)^* \mid (T_1^* \to T_2^*)^* \mid (\lambda X^{\kappa_1}.T^{\kappa_2})^{\kappa_1 \to \kappa_2} \mid (T_1^{\kappa_1 \to \kappa_2} T_2^{\kappa_1})^{\kappa_2}$
Context $\Gamma ::= \cdot \mid \Gamma, x : T^\kappa \mid \Gamma, \mu$
Definitions $\mu ::= \{(x_i : S_i^\kappa) \mapsto t_i\}_{i \in N} \cup \{X_i^\kappa \mapsto T_i^\kappa\}_{i \in M}$
Term definitions $\rho ::= \{x_i \mapsto t_i\}_{i \in N}$

Note that for every $x \mapsto t, X^\kappa \mapsto T^\kappa \in \mu$, we require $\mathrm{FV}(t) = \emptyset$ and $\mathrm{FVar}(T^\kappa) \subseteq \{X^\kappa\}$; and the X^κ can only occur at the positive position in T^κ, no mutually recusive definitions are allowed. We elide the typing rules for space reason.

Definition 9 (Erasure for kinds). *We define a function F which maps kinds in* $\mathbf{S}$ *to kinds in* $\mathbf{F}_\omega$ *with positive definitions.*

$F(*) := *$
$F(\Pi x : T.\kappa) := F(\kappa)$
$F(\Pi X : \kappa'.\kappa) := F(\kappa') \to F(\kappa)$

Definition 10 (Erasure relation). *We define a relation $\Gamma \vdash T \rhd T'^\kappa$ (intuitively, it means that type T can be erased to T'^κ under the context Γ), where T, Γ are types and context in* $\mathbf{S}$*, T'^κ is a type in* $\mathbf{F}_\omega$ *with positive definitions.*

$$\frac{F(\kappa') = \kappa \quad (X : \kappa') \in \Gamma}{\Gamma \vdash X \rhd X^\kappa} \qquad \frac{\Gamma \vdash T \rhd T_1^\kappa}{\Gamma \vdash \iota x.T \rhd T_1^\kappa}$$

$$\frac{\Gamma, X : \kappa \vdash T \rhd T_1^*}{\Gamma \vdash \forall X : \kappa.T \rhd (\forall X^{F(\kappa)}.T_1^*)^*} \qquad \frac{\Gamma \vdash T_1 \rhd T_a^* \quad \Gamma \vdash T_2 \rhd T_b^*}{\Gamma \vdash \Pi x : T_1.T_2 \rhd (T_a^* \to T_b^*)^*}$$

$$\frac{\Gamma \vdash T_2 \rhd T^\kappa}{\Gamma \vdash \forall x : T_1.T_2 \rhd T^\kappa} \qquad \frac{\Gamma \vdash T_1 \rhd T_a^{\kappa_1 \to \kappa_2} \quad \Gamma \vdash T_b^{\kappa_1}}{\Gamma \vdash T_1 T_2 \rhd (T_a^{\kappa_1 \to \kappa_2} T_b^{\kappa_1})^{\kappa_2}}$$

$$\frac{\Gamma, X : \kappa \vdash T \rhd T_a^{\kappa'}}{\Gamma \vdash \lambda X.T \rhd (\lambda X^{F(\kappa)}.T_a^{\kappa'})^{\kappa \to \kappa'}} \qquad \frac{\Gamma \vdash T \rhd T_1^\kappa}{\Gamma \vdash T\ t \rhd T_1^\kappa}$$

$$\frac{\Gamma \vdash T \rhd T_1^\kappa}{\Gamma \vdash \lambda x.T \rhd T_1^\kappa}$$

Definition 11 (Erasure for Context). *We define relation $\Gamma \rhd \Gamma'$ inductively.*

$$\frac{\Gamma \vdash T \rhd T_a^{F(\kappa)} \quad \Gamma \rhd \Gamma'}{\Gamma, (X : \kappa) \mapsto T \rhd \Gamma', X^{F(\kappa)} \mapsto T_a^{F(\kappa)}} \qquad \frac{\Gamma \vdash \Gamma'}{\Gamma, X : \kappa \rhd \Gamma'} \qquad \overline{\cdot \rhd \cdot}$$

$$\frac{\Gamma \vdash T \rhd T_a^\kappa \quad \Gamma \rhd \Gamma'}{\Gamma, (x : T) \mapsto t \rhd \Gamma', x : T_a^\kappa \mapsto t} \qquad \frac{\Gamma \vdash T \rhd T_a^\kappa \quad \Gamma \rhd \Gamma'}{\Gamma, x : T \rhd \Gamma', x : T_a^\kappa}$$

Theorem 4 (Erasure Theorem)

1. *If $\Gamma \vdash T : \kappa$, then there exists a $T_a^{F(\kappa)}$ such that $\Gamma \vdash T \triangleright T_a^{F(\kappa)}$.*
2. *If $\Gamma \vdash t : T$ and $\Gamma \vdash \mathsf{wf}$, then there exist T_a^* and Γ' such that $\Gamma \vdash T \triangleright T_a^*$, $\Gamma \triangleright \Gamma'$ and $\Gamma' \vdash t : T_a^*$.*

Now that we obtained an erasure from **S** to $\mathbf{F}_\omega$ with positive definitions. We continue to show that the latter is strongly normalizing. The development below is in $\mathbf{F}_\omega$ with positive definitions. Let $\mathfrak{R}_\rho$ be the set of all reducibility candidates[5]. Let σ be a mapping between type variable of kind κ to element of $\rho[\![\kappa]\!]$.

Definition 12

- $\rho[\![*]\!] := \mathfrak{R}_\rho$.
- $\rho[\![\kappa \to \kappa']\!] := \{f \mid \forall a \in \rho[\![\kappa]\!], f(a) \in \rho[\![\kappa']\!]\}$.
- $\rho[\![X^\kappa]\!]_\sigma := \sigma(X^\kappa)$.
- $\rho[\![(T_1^* \to T_2^*)^*]\!]_\sigma := \{t \mid \forall u. \in \rho[\![T_1^*]\!]_\sigma, tu \in \rho[\![T_2^*]\!]_\sigma\}$.
- $\rho[\![(\forall X^\kappa.T^*)^*]\!]_\sigma := \bigcap_{f \in \rho[\![\kappa]\!]} \rho[\![T^*]\!]_{\sigma[f/X]}$.
- $\rho[\![(\lambda X^{\kappa'}.T^\kappa)^{\kappa' \to \kappa}]\!]_\sigma := f$ *where f is the map $a \mapsto \rho[\![T^\kappa]\!]_{\sigma[a/X]}$ for any $a \in \rho[\![\kappa']\!]$.*
- $\rho[\![(T_1^{\kappa' \to \kappa} T_2^{\kappa'})^\kappa]\!]_\sigma := \rho[\![T_1^{\kappa' \to \kappa}]\!]_\sigma(\rho[\![T_2^{\kappa'}]\!]_\sigma)$.

Let $|\cdot|$ be a function that retrieves all the term definitions from the context Γ.

Definition 13. *Let $\rho = |\Gamma|$, and* $\mathrm{FVar}(\Gamma)$ *be the set of free type variables in Γ. We define $\sigma \in \rho[\![\Gamma]\!]$ if $\sigma(X^\kappa) \in \rho[\![\kappa]\!]$ for undefined variable X^κ; and $\sigma(X^\kappa) = \mathrm{lfp}(b \mapsto \rho[\![T^\kappa]\!]_{\sigma[b/X^\kappa]})$ for $b \in \rho[\![\kappa]\!]$ if $X^\kappa \mapsto T^\kappa \in \Gamma$.*

Note that the least fix point operation in $\mathrm{lfp}(b \mapsto \rho[\![T^\kappa]\!]_{\sigma[b/X^\kappa]})$ is defined since we can extend the complete lattice of reducibility candidate to complete lattice $(\rho[\![\kappa]\!], \subseteq_\kappa, \cap_\kappa)$.

Definition 14. *Let $\rho = |\Gamma|$ and $\sigma \in \rho[\![\Gamma]\!]$. We define the relation $\delta \in \rho[\![\Gamma]\!]$ inductively:*

$$\frac{}{\cdot \in \rho[\![\cdot]\!]} \quad \frac{\delta \in \rho[\![\Gamma]\!] \quad t \in \rho[\![T^\kappa]\!]_\sigma}{\delta[t/x] \in \rho[\![\Gamma, x : T^\kappa]\!]} \quad \frac{\delta \in \rho[\![\Gamma]\!]}{\delta \in \rho[\![\Gamma, (x : T^\kappa) \mapsto t]\!]}$$

Theorem 5 (Soundness theorem[6]). *Let $\rho = |\Gamma|$. If $\Gamma \vdash t : T^*$ and $\Gamma \vdash \mathsf{wf}$, then for any $\sigma, \delta \in \rho[\![\Gamma]\!]$, we have $\delta t \in \rho[\![T^*]\!]_\sigma$, with $\rho[\![T^*]\!]_\sigma \in \mathfrak{R}_\rho$.*

Theorem 4 and 5 imply all the typable term in **S** is strongly normalizing.

---

[5] The notion of reducibility candidate here slightly extends the standard one [15] to handle definitional reduction: $\rho \vdash x \to_\beta t$, where $x \mapsto t \in \rho$. So it is parametrized by ρ.

[6] Please note that since we are in Curry style assignment, the infinite reduction sequence in term will not be thrown away.

5.2 Confluence Analysis

The complications of proving type preservation are due to several rules which are not syntax-directed. To prove type preservation, one needs to ensure that if $\Pi x : T.T'$ can be transformed to $\Pi x : T_1.T_2$, then it must be the case that T can be transformed to T_1 and T' can be transformed to T_2. This is why we need to show confluence for type-level reduction. We first observe that the *selfGen* rule and *selfInst* rule are mutually inverse, and model the change of self type by the following reduction relation.

Definition 15
$\Gamma \vdash T_1 \to_\iota T_2$ *if* $T_1 \equiv \iota x.T'$[7] *and* $T_2 \equiv [t/x]T'$ *for some fix term* t.

Note that $\to_\iota$ models the *selfInst* rule, $\to_\iota^{-1}$ models the *selfGen* rule. Importantly, the notion of ι-reduction does not include congruence; that is, we do not allow reduction rules like if $T \to_\iota T'$, then $\lambda x.T \to_\iota \lambda x.T'$. The purpose of ι-reduction is to emulate the typing rule *selfInst* and *selfGen*.

We first show confluence of $\to_\beta$ by applying the standard Tait-Martin Löf method, and then apply Hindley-Rossen's commutativity theorem to show $\to_\iota$ commutes with $\to_\beta$. We use $\to^*$ to denote the reflexive symmetric transitive closure of $\to$.

Lemma 1. $\to_\beta$ *is confluent.*

Definition 16 (Commutativity). *Let* $\to_1, \to_2$ *be two notions of reduction. Then* $\to_1$ *commutes with* $\to_2$ *iff* $\leftarrow_1 \cdot \to_2 \subseteq \to_1 \cdot \leftarrow_2$.

Proposition 1. *Let* $\to_1, \to_2$ *be two notions of reduction. Suppose both* $\to_1$ *and* $\to_2$ *are confluent, and* $\to_1^*$ *commutes with* $\to_2^*$. *Then* $\to_1 \cup \to_2$ *is confluent.*

Lemma 2. $\to_\beta$ *commutes with* $\to_\iota$. *Thus* $\to_{\beta,\iota}$ *is confluent, where* $\to_{\beta,\iota} = \to_\beta \cup \to_\iota$.

Theorem 6 (ι-elimination). *If* $\Gamma \vdash \Pi x : T_1.T_2 =_{\beta,\iota} \Pi x : T_1'.T_2'$, *then* $\Gamma \vdash T_1 =_\beta T_1'$ *and* $\Gamma \vdash T_2 =_\beta T_2'$.

Proof. If $\Gamma \vdash \Pi x : T_1.T_2 =_{\beta,\iota} \Pi x : T_1'.T_2'$, then by the confluence of $\to_{\beta,\iota}$, there exists a T such that $\Gamma \vdash \Pi x : T_1.T_2 \to_{\iota,\beta}^* T$ and $\Gamma \vdash \Pi x : T_1'.T_2' \to_{\iota,\beta}^* T$. Since all the reductions on $\Pi x : T_1.T_2$ preserve the structure of the dependent type, one will never have a chance to use $\to_\iota$-reduction, thus $\Gamma \vdash \Pi x : T_1.T_2 \to_\beta^* T$ and $\Gamma \vdash \Pi x : T_1'.T_2' \to_\beta^* T$. So T must be of the form $\Pi x : T_3.T_4$. And $\Gamma \vdash T_1 \to_\beta^* T_3$, $\Gamma \vdash T_1' \to_\beta^* T_3$, $\Gamma \vdash T_2 \to_\beta^* T_4$ and $\Gamma \vdash T_2' \to_\beta^* T_4$. Finally, we have $\Gamma \vdash T_1 =_\beta T_1'$ and $\Gamma \vdash T_2 =_\beta T_2'$.

[7] We use $\equiv$ to mean syntactic identity.

5.3 Morph Analysis

The methods of the previous section are not suitable for dealing with implicit polymorphism, since as a reduction relation, polymorphic instantiation is not confluent. For example, $\forall X : \kappa.X$ can be instantiated either to T or to $T \to T$. The only known syntactic method (to our knowledge) to deal with preservation proof for Curry-style System **F** is Barendregt's method [4]. We will extend his method to handle the instantiation of $\forall x : T.T'$.

Definition 17 (Morphing Relations)

- $([\Gamma], T_1) \to_i ([\Gamma], T_2)$ *if* $T_1 \equiv \forall X : \kappa.T'$ *and* $T_2 \equiv [T/X]T'$ *for some* T *such that* $\Gamma \vdash T : \kappa$.
- $([\Gamma, X : \kappa], T_1) \to_g ([\Gamma], T_2)$ *if* $T_2 \equiv \forall X : \kappa.T_1$ *and* $\Gamma \vdash \kappa : \Box$.
- $([\Gamma], T_1) \to_I ([\Gamma], T_2)$ *if* $T_1 \equiv \forall x : T.T'$ *and* $T_2 \equiv [t/x]T'$ *for some* t *such that* $\Gamma \vdash t : T$.
- $([\Gamma, x : T], T_1) \to_G ([\Gamma], T_2)$ *if* $T_2 \equiv \forall x : T.T_1$ *and* $\Gamma \vdash T : *$.

Intuitively, $([\Gamma], T_1) \to ([\Gamma'], T_2)$ means T_1 can be transformed to T_2 with a change of context from Γ to Γ'. One can view morphing relations as a way to model typing rules which are not syntax-directed. Note that morphing relations are not intended to be viewed as rewrite relation. Instead of proving confluence for these morphing relations, we try to use substitutions to *summarize* the effects of a sequence of morphing relations. Before we do that, first we "lift" $=_{\beta,\iota}$ to a form of morphing relation.

Definition 18. $([\Gamma], T) =_{\beta,\iota} ([\Gamma], T')$ *if* $\Gamma \vdash T =_{\beta,\iota} T'$ *and* $\Gamma \vdash T : *$ *and* $\Gamma \vdash T' : *$.

The best way to understand the E, G mappings below is through understanding Lemmas 4 and 5. They give concrete demonstrations of how to *summarize* a sequence of morphing relations.

Definition 19

$E(\forall X : \kappa.T) := E(T)$	$E(X) := X$	$E(\Pi x : T_1.T_2) := \Pi x : T_1.T_2$
$E(\lambda X.T) := \lambda X.T$	$E(T_1 T_2) := T_1 T_2$	$E(\forall x : T'.T) := \forall x : T'.T$
$E(\iota x.T) := \iota x.T$	$E(T\ t) := T\ t$	$E(\lambda x.T) := \lambda x.T$

Definition 20

$G(\forall X : \kappa.T) := \forall X : \kappa.T$	$G(X) := X$	$G(\Pi x : T_1.T_2) := \Pi x : T_1.T_2$
$G(\lambda X.T) := \lambda X.T$	$G(T_1 T_2) := T_1 T_2$	$G(\forall x : T'.T) := G(T)$
$G(\iota x.T) := \iota x.T$	$G(T\ t) := T\ t$	$G(\lambda x.T) := \lambda x.T$

Lemma 3. $E([T'/X]T) \equiv [T''/X]E(T)$ *for some* T''; $G([t/x]T) \equiv [t/x]G(T)$.

Proof. By induction on the structure of T.

Lemma 4. *If* $([\Gamma], T) \to^*_{i,g} ([\Gamma'], T')$, *then there exists a type substitution* σ *such that* $\sigma E(T) \equiv E(T')$.

Proof. It suffices to consider $([\Gamma],T)\to_{i,g}([\Gamma'],T')$. If $T'\equiv\forall X:\kappa.T$ and $\Gamma=\Gamma',X:\kappa$, then $E(T')\equiv E(T)$. If $T\equiv\forall X:\kappa.T_1$ and $T'\equiv[T''/X]T_1$ and $\Gamma=\Gamma'$, then $E(T)\equiv E(T_1)$. By Lemma 3, we know $E(T')\equiv E([T''/X]T_1)\equiv[T_2/X]E(T_1)$ for some T_2.

Lemma 5. *If* $([\Gamma],T)\to^*_{I,G}([\Gamma'],T')$*, then there exists a term substitution* δ *such that* $\delta G(T)\equiv G(T')$.

Proof. It suffices to consider $([\Gamma],T)\to_{I,G}([\Gamma'],T')$. If $T'\equiv\forall x:T_1.T$ and $\Gamma=\Gamma',x:T_1$, then $G(T')\equiv G(T)$. If $T\equiv\forall x:T_2.T_1$ and $T'\equiv[t/x]T_1$ and $\Gamma=\Gamma'$, then $E(T)\equiv E(T_1)$. By Lemma 3, we know $E(T')\equiv E([t/x]T_1)\equiv[t/x]E(T_1)$.

Lemma 6. *If* $([\Gamma],\Pi x:T_1.T_2)\to^*_{i,g}([\Gamma'],\Pi x:T_1'.T_2')$*, then there exists a type substitution* σ *such that* $\sigma(\Pi x:T_1.T_2)\equiv\Pi x:T_1'.T_2'$.

Proof. By Lemma 4.

Lemma 7. *If* $([\Gamma],\Pi x:T_1.T_2)\to^*_{I,G}([\Gamma'],\Pi x:T_1'.T_2')$*, then there exists a term substitution* δ *such that* $\delta(\Pi x:T_1.T_2)\equiv\Pi x:T_1'.T_2'$.

Proof. By Lemma 5.

Let $\to^*_{\iota,\beta,i,g,I,G}$ denote $(\to_{i,g,I,G}\cup=_{\iota,\beta})^*$. Let $\to_{\iota,\beta,i,g,I,G}$ denote $\to_{i,g,I,G}\cup=_{\iota,\beta}$. The goal of confluence analysis and morph analysis is to establish the following *compatibility* theorem.

Theorem 7 (Compatibility). *If* $([\Gamma],\Pi x:T_1.T_2)\to^*_{\iota,\beta,i,g,I,G}([\Gamma'],\Pi x:T_1'.T_2')$*, then there exists a mixed substitution*[8] ϕ *such that* $([\Gamma],\phi(\Pi x:T_1.T_2))=_{\iota,\beta}([\Gamma],\Pi x:T_1'.T_2')$. *Thus* $\Gamma\vdash\phi T_1=_\beta T_1'$ *and* $\Gamma\vdash\phi T_2=_\beta T_2'$ *(by Theorem 6).*

Proof. By Lemma 7 and 6, making use of the fact that if $\Gamma\vdash t=_{\iota,\beta}t'$, then for any mixed substitution ϕ, we have $\Gamma\vdash\phi t=_{\iota,\beta}\phi t'$.

Theorem 8 (Type Preservation). *If* $\Gamma\vdash t:T$ *and* $\Gamma\vdash t\to_\beta t'$ *and* $\Gamma\vdash\mathrm{wf}$, *then* $\Gamma\vdash t':T$.

6 $0\neq 1$ in S

The proof of $0\neq 1$ follows the same method as in Theorem 1, while emptiness of $\bot$ needs the erasure and preservation theorems. Notice that in this section, by $a=b$, we mean $\forall C:A\to *.C\ a\to C\ b$ with $a,b:A$.

Definition 21. $\bot:=\forall A:*.\forall x:A.\forall y:A.x=y$.

Theorem 9. *There is no term* t *such that* $\mu_c\vdash t:\bot$

[8] A substitution that contains both term substitution and type substitution.

Proof. Suppose $\mu_c \vdash t : \bot$. By the erasure theorem (Theorem 4) in Section 5.1, we have $F(\mu_c) \vdash t : \forall A : *.\forall C : *.C \to C$ in $\mathbf{F}_\omega$. We know that $\forall A : *.\forall C : *.C \to C$ is the singleton type[9], which is inhabited by $\lambda z.z$. This means $t \to_\beta^* \lambda z.z$ (the term reductions of $\mathbf{F}_\omega$ with let-bindings are the same as $\mathbf{S}$) and $\mu_c \vdash \lambda z.z : \bot$ in $\mathbf{S}$ (by type preservation, Theorem 8). Let $\Gamma = \mu_c, A : *, x : A, y : A, C : A \to *, z : C\ x$. Then we would have $\Gamma \vdash z : C\ y$. So by inversion, we have $\Gamma \vdash C\ x \to_{\iota,\beta,i,g,I,G}^* C\ y$, which means $\Gamma \vdash C\ x \to_\beta^* C\ y$. We know this is impossible by confluence of $\to_\beta$.

Theorem 10. $\mu_c \vdash 0 = 1 \to \bot$.

Proof. This proof follows the method in Theorem 1. Let $\Gamma = \mu_c, a : (\forall B : \mathsf{Nat} \to *.B\ 0 \to B\ 1), A : *, x : A, y : A, C : A \to *, c : C\ x$. We want to construct a term of type $C\ y$. Let $F := \lambda n[: \mathsf{Nat}].n\ [\lambda p : \mathsf{Nat}.A]\ (\lambda q[: A].y)x$, and note that $F : \mathsf{Nat} \to A$. We know that $F\ 0 =_\beta x$ and $F\ 1 =_\beta y$. So we can indeed convert the type of c from $C\ x$ to $C\ (F\ 0)$. And then we instantiate the B in $\forall B : \mathsf{Nat} \to *.B\ 0 \to B\ 1$ with $\lambda x[: \mathsf{Nat}].C\ (F\ x)$. So we have $C\ (F\ 0) \to C\ (F\ 1)$ as the type of a. So $a\ c : C\ (F\ 1)$, which means $a\ c : C\ y$. So we have just shown how to inhabit $0 = 1 \to \bot$ in $\mathbf{S}$.

7 Conclusion

We have revisited lambda encodings in type theory, and shown how a new self type construct $\iota x.T$ supports dependent eliminations with lambda encodings, including induction principles. We considered System $\mathbf{S}$, which incorporates self types together with implicit products and a restricted version of global positive recursive definition. The corresponding induction principles for Church- and Parigot-encoded datatypes are derivable in $\mathbf{S}$. By changing the notion of contradiction from explosion to equational inconsistency, we are able to show $0 \neq 1$ in both $\mathbf{CC}$ and $\mathbf{S}$. We proved type preservation, which is nontrivial for $\mathbf{S}$ since several rules are not syntax-directed. We also defined an erasure from $\mathbf{S}$ to $\mathbf{F}_\omega$ with positive definitions, and proved strong normalization of $\mathbf{S}$ by showing strong normalization of $\mathbf{F}_\omega$ with positive definitions. Future work includes further explorations of dependently typed lambda encodings for practical type theory. In particular, we would like to implement our system and carry out some case studies. Last but not least, we want to thank anonymous reviewers for their helpful comments.

References

1. Abadi, M., Cardelli, L.: A Theory of Primitive Objects - Second-Order Systems. In: European Symposium on Programming (ESOP), pp. 1–25 (1994)
2. Abel, A., Pientka, B.: Wellfounded recursion with copatterns: a unified approach to termination and productivity. In: Morrisett, G., Uustalu, T. (eds.) International Conference on Functional Programming (ICFP), pp. 185–196 (2013)

[9] Note that we are dealing with Curry-style $\mathbf{F}_\omega$.

3. Ahn, K.Y., Sheard, T., Fiore, M., Pitts, A.M.: System f_i. In: Hasegawa, M. (ed.) TLCA 2013. LNCS, vol. 7941, pp. 15–30. Springer, Heidelberg (2013)
4. Barendregt, H.: Lambda calculi with types, handbook of logic in computer science. In: Background: Computational Structures, vol. 2 (1993)
5. Barras, B.: Sets in coq, coq in sets. Journal of Formalized Reasoning 3(1) (2010)
6. Capretta, V.: General recursion via coinductive types. Logical Methods in Computer Science 1(2) (2005)
7. Church, A.: The Calculi of Lambda Conversion (AM-6) (Annals of Mathematics Studies) (1985)
8. Coquand, T.: Metamathematical investigations of a calculus of constructions. Technical Report RR-1088, INRIA (September 1989)
9. Coquand, T., Huet, G.: The calculus of constructions. Inf. Comput. 76(2-3), 95–120 (1988)
10. Curry, H.B., Hindley, J.R., Seldin, J.P.: Combinatory Logic, vol. II (1972)
11. Fu, P., Stump, A.: Self Types for Dependently Typed Lambda Encodings (2014), Extended version available from `http://homepage.cs.uiowa.edu/~pfu/document/papers/rta-tlca.pdf`
12. Geuvers, H.: Inductive and Coinductive Types with Iteration and Recursion. In: Nordstrom, B., Petersson, K., Plotkin, G. (eds.) Informal Proceedings of the 1992 Workshop on Types for Proofs and Programs, pp. 183–207 (1994)
13. Geuvers, H.: Induction Is Not Derivable in Second Order Dependent Type Theory. In: Abramsky, S. (ed.) TLCA 2001. LNCS, vol. 2044, pp. 166–181. Springer, Heidelberg (2001)
14. Gimenez, E.: Un calcul de constructions infinies et son application a la verification de systemes communicants. PhD thesis (1996)
15. Girard, J.-Y.: Interprétation fonctionnelle et élimination des coupures de l'arithmétique d'ordre supérieur (1972)
16. Hickey, J.: Formal objects in type theory using very dependent types. In: Bruce, K. (ed.) In Foundations of Object Oriented Languages (FOOL) 3 (1996)
17. Mendler, P.: Inductive definition in type theory. Technical report, Cornell University (1987)
18. Miquel, A.: Le Calcul des Constructions implicite: syntaxe et sémantique. PhD thesis, PhD thesis, Université Paris 7 (2001)
19. Odersky, M., Cremet, V., Röckl, C., Zenger, M.: A Nominal Theory of Objects with Dependent Types. In: Cardelli, L. (ed.) ECOOP 2003. LNCS, vol. 2743, pp. 201–224. Springer, Heidelberg (2003)
20. Parigot, M.: Programming with Proofs: A Second Order Type Theory. In: Ganzinger, H. (ed.) ESOP 1988. LNCS, vol. 300, pp. 145–159. Springer, Heidelberg (1988)
21. Schepler, D.: Bijective function implies equal types is provably inconsistent with functional extensionality in coq. Message to the Coq Club mailing list (December 12, 2013)
22. Werner, B.: A Normalization Proof for an Impredicative Type System with Large Elimination over Integers. In: Nordström, B., Petersson, K., Plotkin, G. (eds.) International Workshop on Types for Proofs and Programs (TYPES), pp. 341–357 (1992)
23. Werner, B.: Une théorie des constructions inductives. PhD thesis, Université Paris VII (1994)

First-Order Formative Rules*

Carsten Fuhs[1] and Cynthia Kop[2]

[1] University College London, Dept. of Computer Science, London WC1E 6BT, UK
[2] University of Innsbruck, Institute of Computer Science, 6020 Innsbruck, Austria

Abstract. This paper discusses the method of *formative rules* for first-order term rewriting, which was previously defined for a higher-order setting. Dual to the well-known *usable rules*, formative rules allow dropping some of the term constraints that need to be solved during a termination proof. Compared to the higher-order definition, the first-order setting allows for significant improvements of the technique.

1 Introduction

In [12,13] C. Kop and F. van Raamsdonk introduce the notion of *formative rules*. The technique is similar to the method of *usable rules* [1,9,10], which is commonly used in termination proofs, but has different strengths and weaknesses.

Since, by [15], the more common *first-order* style of term rewriting, both with and without types, can be seen as a subclass of the formalism of [13], this result immediately applies to first-order rewriting. In an untyped setting, we will, however, lose some of its strength, as sorts play a relevant role in formative rules.

On the other hand, by omitting the complicating aspects of higher-order term rewriting (such as λ-abstraction and "collapsing" rules $l \to x \cdot y$) we also gain possibilities not present in the original setting; both things which *have not* been done, as the higher-order dependency pair framework [11] is still rather limited, and things which *cannot* be done, at least with current theory. Therefore, in this paper, we will redefine the method for (many-sorted) first-order term rewriting.

New compared to [13], we will integrate formative rules into the dependency pair framework [7], which is the basis of most contemporary termination provers for first-order term rewriting. Within this framework, formative rules are used either as a stand-alone processor or with reduction pairs, and can be coupled with usable rules and argument filterings. We also formulate a semantic characterisation of formative rules, to enable future generalisations of the definition. Aside from this, we present a (new) way to weaken the detrimental effect of collapsing rules.

This paper is organised as follows. After the preliminaries in Section 2, a first definition of formative rules is given and then generalised in Section 3. Section 4 shows various ways to use formative rules in the dependency pair framework.

* Support by EPSRC & the Austrian Science Fund (FWF) international project I963.

G. Dowek (ed.): RTA-TLCA 2014, LNCS 8560, pp. 240–256, 2014.

Section 5 gives an alternative way to deal with collapsing rules. In Section 6 we consider innermost termination, Section 7 describes implementation and experiments, and in Section 8 we point out possible future work and conclude. All proofs and an improved formative rules approximation are provided in [5].

2 Preliminaries

We consider *many-sorted term rewriting*: term rewriting with *sorts*, basic types. While sorts are not usually considered in studies of first-order term rewrite systems (TRSs) and for instance the Termination Problems Data Base[1] does not include them (for first-order TRSs),[2] they are a natural addition; in typical applications there is little need to allow untypable terms like `3+apple`. Even when no sorts are present, a standard TRS can be seen as a many-sorted TRS with only one sort.[3]

Many-Sorted TRSs. We assume given a non-empty set $\mathcal{S}$ of *sorts*; these are typically things like `Nat` or `Bool`, or (for representing unsorted systems) $\mathcal{S}$ might be the set with a single sort $\{\mathtt{o}\}$. A *sort declaration* is a sequence $[\kappa_1 \times \ldots \times \kappa_n] \Rightarrow \iota$ where ι and all κ_i are sorts. A sort declaration $[] \Rightarrow \iota$ is just denoted ι.

A *many-sorted signature* is a set Σ of function symbols f, each equipped with a sort declaration σ, notation $f : \sigma \in \Sigma$. Fixing a many-sorted signature Σ and an infinite set $\mathcal{V}$ of sorted variables, the set of *terms* consists of those expressions s over Σ and $\mathcal{V}$ for which we can derive $s : \iota$ for some sort ι, using the clauses:

$$x : \iota \text{ if } x : \iota \in \mathcal{V}$$
$$f(s_1, \ldots, s_n) : \iota \text{ if } f : [\kappa_1 \times \ldots \times \kappa_n] \Rightarrow \iota \in \Sigma \text{ and } s_1 : \kappa_1, \ \ldots, \ s_n : \kappa_n$$

We often denote $f(s_1, \ldots, s_n)$ as just $f(\boldsymbol{s})$. Clearly, every term has a unique sort. Let $\mathit{Var}(s)$ be the set of all variables occurring in a term s. A term s is *linear* if every variable in $\mathit{Var}(s)$ occurs only once in s. A term t is a *subterm* of another term s, notation $s \trianglerighteq t$, if either $s = t$ or $s = f(s_1, \ldots, s_n)$ and some $s_i \trianglerighteq t$. A *substitution* γ is a mapping from variables to terms of the same sort; the application $s\gamma$ of a substitution γ on a term s is s with each $x \in \mathsf{domain}(\gamma)$ replaced by $\gamma(x)$.

A *rule* is a pair $\ell \to r$ of terms with the same sort such that ℓ is not a variable.[4] A rule is *left-linear* if ℓ is linear, and *collapsing* if r is a variable. Given a set of rules $\mathcal{R}$, the *reduction relation* $\to_{\mathcal{R}}$ is given by: $\ell\gamma \to_{\mathcal{R}} r\gamma$ if $\ell \to r \in \mathcal{R}$ and γ a

[1] More information on the *TPDB*: `http://termination-portal.org/wiki/TPDB`

[2] This may also be due to the fact that currently most termination tools for first-order rewriting only make very limited use of the additional information carried by types.

[3] However, the method of this paper is stronger given more sorts. We may be able to (temporarily) infer richer sorts, however. We will say more about this in Section 6.

[4] Often also $\mathit{Var}(r) \subseteq \mathit{Var}(\ell)$ is required. However, we use *filtered* rules $\overline{\pi}(\ell) \to \overline{\pi}(r)$ later, where the restriction is inconvenient. As a rule is non-terminating if $\mathit{Var}(r) \not\subseteq \mathit{Var}(\ell)$, as usual we forbid such rules in the input $\mathcal{R}$ and in dependency pair problems.

substitution; $f(\ldots, s_i, \ldots) \to_{\mathcal{R}} f(\ldots, s_i', \ldots)$ if $s_i \to_{\mathcal{R}} s_i'$. A term s is in *normal form* if there is no t such that $s \to_{\mathcal{R}} t$.

The relation $\to_{\mathcal{R}}^*$ is the transitive-reflexive closure of $\to_{\mathcal{R}}$. If there is a rule $f(\boldsymbol{l}) \to r \in \mathcal{R}$ we say that f is a *defined symbol*; otherwise f is a *constructor*.

A *many-sorted term rewrite system (MTRS)* is a pair $(\Sigma, \mathcal{R})$ with signature Σ and a set $\mathcal{R}$ of rules $\ell \to r$ with $Var(r) \subseteq Var(\ell)$. A term s is *terminating* if there is no infinite reduction $s \to_{\mathcal{R}} t_1 \to_{\mathcal{R}} t_2 \ldots$ An MTRS is terminating if all terms are.

Example 1. An example of a many-sorted TRS $(\Sigma, \mathcal{R})$ with more than one sort is the following system, which uses lists, natural numbers and a RESULT sort:

$$\begin{array}{rlrlrl}
\mathtt{0} : & \mathtt{NAT} & \mathtt{Cons} : & [\mathtt{NAT} \times \mathtt{LIST}] \Rightarrow \mathtt{LIST} & \mathtt{Run} : & [\mathtt{LIST}] \Rightarrow \mathtt{RESULT} \\
\mathtt{S} : & [\mathtt{NAT}] \Rightarrow \mathtt{NAT} & \mathtt{Ack} : & [\mathtt{NAT} \times \mathtt{NAT}] \Rightarrow \mathtt{NAT} & \mathtt{Return} : & [\mathtt{NAT}] \Rightarrow \mathtt{RESULT} \\
\mathtt{Nil} : & \mathtt{LIST} & \mathtt{Big} : & [\mathtt{NAT} \times \mathtt{LIST}] \Rightarrow \mathtt{NAT} & \mathtt{Rnd} : & [\mathtt{NAT}] \Rightarrow \mathtt{NAT} \\
\mathtt{Err} : & \mathtt{RESULT} & \mathtt{Upd} : & [\mathtt{LIST}] \Rightarrow \mathtt{LIST} & &
\end{array}$$

$$\begin{array}{llll}
1. & \mathtt{Rnd}(x) \to x & 6. & \mathtt{Big}(x, \mathtt{Nil}) \to x \\
2. & \mathtt{Rnd}(\mathtt{S}(x)) \to \mathtt{Rnd}(x) & 7. & \mathtt{Big}(x, \mathtt{Cons}(y, z)) \to \mathtt{Big}(\mathtt{Ack}(x, y), \mathtt{Upd}(z)) \\
3. & \mathtt{Upd}(\mathtt{Nil}) \to \mathtt{Nil} & 8. & \mathtt{Upd}(\mathtt{Cons}(x, y)) \to \mathtt{Cons}(\mathtt{Rnd}(x), \mathtt{Upd}(y)) \\
4. & \mathtt{Run}(\mathtt{Nil}) \to \mathtt{Err} & 9. & \mathtt{Run}(\mathtt{Cons}(x, y)) \to \mathtt{Return}(\mathtt{Big}(x, y)) \\
5. & \mathtt{Ack}(\mathtt{0}, y) \to \mathtt{S}(y) & 10. & \mathtt{Ack}(\mathtt{S}(x), y) \to \mathtt{Ack}(x, \mathtt{S}(y)) \\
 & & 11. & \mathtt{Ack}(\mathtt{S}(x), \mathtt{S}(y)) \to \mathtt{Ack}(x, \mathtt{Ack}(\mathtt{S}(x), y))
\end{array}$$

Run(*lst*) calculates a potentially very large number, depending on the elements of *lst* and some randomness. We have chosen this example because it will help to demonstrate the various aspects of formative rules, without being too long.

The Dependency Pair Framework. As a basis to study termination, we will use the *dependency pair (DP) framework* [7], adapted to include sorts.

Given an MTRS $(\Sigma, \mathcal{R})$, let $\Sigma^\sharp = \Sigma \cup \{f^\sharp : [\iota_1 \times \ldots \times \iota_n] \Rightarrow \mathtt{dpsort} \mid f : [\iota_1 \times \ldots \times \iota_n] \Rightarrow \kappa \in \Sigma \wedge f$ a defined symbol of $\mathcal{R}\}$, where dpsort is a fresh sort. The set $\mathsf{DP}(\mathcal{R})$ of *dependency pairs (DPs)* of $\mathcal{R}$ consists of all rules of the form $f^\sharp(l_1, \ldots, l_n) \to g^\sharp(r_1, \ldots, r_m)$ where $f(\boldsymbol{l}) \to r \in \mathcal{R}$ and $r \trianglerighteq g(\boldsymbol{r})$ with g a defined symbol.

Example 2. The dependency pairs of the system in Example 1 are:

$$\begin{array}{rr}
\mathtt{Rnd}^\sharp(\mathtt{S}(x)) \to \mathtt{Rnd}^\sharp(x) & \mathtt{Big}^\sharp(x, \mathtt{Cons}(y, z)) \to \mathtt{Big}^\sharp(\mathtt{Ack}(x, y), \mathtt{Upd}(z)) \\
\mathtt{Upd}^\sharp(\mathtt{Cons}(x, y)) \to \mathtt{Rnd}^\sharp(x) & \mathtt{Big}^\sharp(x, \mathtt{Cons}(y, z)) \to \mathtt{Ack}^\sharp(x, y) \\
\mathtt{Upd}^\sharp(\mathtt{Cons}(x, y)) \to \mathtt{Upd}^\sharp(y) & \mathtt{Big}^\sharp(x, \mathtt{Cons}(y, z)) \to \mathtt{Upd}^\sharp(z) \\
\mathtt{Run}^\sharp(\mathtt{Cons}(x, y)) \to \mathtt{Big}^\sharp(x, y) & \mathtt{Ack}^\sharp(\mathtt{S}(x), \mathtt{S}(y)) \to \mathtt{Ack}^\sharp(x, \mathtt{Ack}(\mathtt{S}(x), y)) \\
\mathtt{Ack}^\sharp(\mathtt{S}(x), y) \to \mathtt{Ack}^\sharp(x, \mathtt{S}(y)) & \mathtt{Ack}^\sharp(\mathtt{S}(x), \mathtt{S}(y)) \to \mathtt{Ack}^\sharp(\mathtt{S}(x), y)
\end{array}$$

For sets $\mathcal{P}$ and $\mathcal{R}$ of rules, an infinite $(\mathcal{P}, \mathcal{R})$-chain is a sequence $[(\ell_i \to r_i, \gamma_i) \mid i \in \mathbb{N}]$ where each $\ell_i \to r_i \in \mathcal{P}$ and γ_i is a substitution such that $r_i\gamma_i \to_{\mathcal{R}}^* \ell_{i+1}\gamma_{i+1}$. This chain is *minimal* if each $r_i\gamma_i$ is terminating with respect to $\to_{\mathcal{R}}$.

Theorem 3. *(following [1,7,9,10]) An MTRS $(\Sigma, \mathcal{R})$ is terminating if and only if there is no infinite minimal $(\mathsf{DP}(\mathcal{R}), \mathcal{R})$-chain.*

A *DP problem* is a triple $(\mathcal{P}, \mathcal{R}, f)$ with $\mathcal{P}$ and $\mathcal{R}$ sets of rules and $f \in \{\mathsf{m}, \mathsf{a}\}$ (denoting $\{\mathsf{minimal}, \mathsf{arbitrary}\}$).[5] A DP problem $(\mathcal{P}, \mathcal{R}, f)$ is *finite* if there is no infinite $(\mathcal{P}, \mathcal{R})$-chain, which is minimal if $f = \mathsf{m}$. A *DP processor* is a function which maps a DP problem to a set of DP problems. A processor *proc* is *sound* if, for all DP problems A: if all $B \in proc(A)$ are finite, then A is finite.

The goal of the DP framework is, starting with a set $D = \{(\mathsf{DP}(\mathcal{R}), \mathcal{R}, \mathsf{m})\}$, to reduce D to $\emptyset$ using sound processors. Then we may conclude termination of the initial MTRS $(\Sigma, \mathcal{R})$.[6] Various common processors use a *reduction pair*, a pair $(\succsim, \succ)$ of a monotonic, stable (closed under substitutions) quasi-ordering $\succsim$ on terms and a well-founded, stable ordering $\succ$ compatible with $\succsim$ (i.e., $\succ \cdot \succsim \subseteq \succ$).

Theorem 4. *(following [1,7,9,10]) Let $(\succsim, \succ)$ be a reduction pair. The processor which maps a DP problem $(\mathcal{P}, \mathcal{R}, f)$ to the following result is sound:*

- $\{(\mathcal{P} \setminus \mathcal{P}^{\succ}, \mathcal{R}, f)\}$ *if:*
  - $\ell \succ r$ *for* $\ell \to r \in \mathcal{P}^{\succ}$ *and* $\ell \succsim r$ *for* $\ell \to r \in \mathcal{P} \setminus \mathcal{P}^{\succ}$ *(with* $\mathcal{P}^{\succ} \subseteq \mathcal{P}$*);*
  - $\ell \succsim r$ *for* $\ell \to r \in \mathcal{R}$.
- $\{(\mathcal{P}, \mathcal{R}, f)\}$ *otherwise*

Here, we must orient all elements of $\mathcal{R}$ with $\succsim$. As there are many processors which remove elements from $\mathcal{P}$ and few which remove from $\mathcal{R}$, this may give many constraints. *Usable rules*, often combined with *argument filterings*, address this:

Definition 5. *(following [9,10]) Let Σ be a signature and $\mathcal{R}$ a set of rules. An* argument filtering *is a function that maps each $f : [\iota_1 \times \ldots \times \iota_n] \Rightarrow \kappa$ to a set $\{i_1, \ldots, i_k\} \subseteq \{1, \ldots, n\}$.*[7] *The* usable rules *of a term t with respect to an argument filtering π are defined as the smallest set $UR(t, \mathcal{R}, \pi) \subseteq \mathcal{R}$ such that:*

- *if $\mathcal{R}$ is not finitely branching (i.e. there are terms with infinitely many direct reducts), then $UR(t, \mathcal{R}, \pi) = \mathcal{R}$;*
- *if $t = f(t_1, \ldots, t_n)$, then $UR(t_i, \mathcal{R}, \pi) \subseteq UR(t, \mathcal{R}, \pi)$ for all $i \in \pi(f)$;*
- *if $t = f(t_1, \ldots, t_n)$, then $\{\ell \to r \in \mathcal{R} \mid \ell = f(\ldots)\} \subseteq UR(t, \mathcal{R}, \pi)$;*
- *if $\ell \to r \in UR(t, \mathcal{R}, \pi)$, then $UR(r, \mathcal{R}, \pi) \subseteq UR(t, \mathcal{R}, \pi)$.*

For a set of rules $\mathcal{P}$, we define $UR(\mathcal{P}, \mathcal{R}, \pi) = \bigcup_{s \to t \in \mathcal{P}} UR(t, \mathcal{R}, \pi)$.

Argument filterings π are used to disregard arguments of certain function symbols. Given π, let $f_\pi : [\iota_{i_1} \times \ldots \times \iota_{i_k}] \Rightarrow \kappa$ be a fresh function symbol for all f with $\pi(f) = \{i_1, \ldots, i_k\}$ and $i_1 < \ldots < i_k$, and define $\overline{\pi}(x) = x$ for x a variable, and $\overline{\pi}(f(s_1, \ldots, s_n)) = f_\pi(\overline{\pi}(s_{i_1}), \ldots, \overline{\pi}(s_{i_k}))$ if $\pi(f) = \{i_1, \ldots, i_k\}$ and $i_1 < \ldots < i_k$. For a set of rules $\mathcal{R}$, let $\overline{\pi}(\mathcal{R}) = \{\overline{\pi}(l) \to \overline{\pi}(r) \mid l \to r \in \mathcal{R}\}$. The idea of usable rules is to only consider rules relevant to the pairs in $\mathcal{P}$ after applying $\overline{\pi}$.

---

[5] Here we do not modify the signature $\Sigma^\sharp$ of a DP problem, so we leave $\Sigma^\sharp$ implicit.

[6] The full DP framework [7] can also be used for proofs of *non-termination*. Indeed, by [7, Lemma 2], all processors introduced in this paper (except Theorem 17 for innermost rewriting) are "complete" and may be applied in a non-termination proof.

[7] Usual definitions of argument filterings also allow $\pi(f) = i$, giving $\overline{\pi}(f(\boldsymbol{s})) = \overline{\pi}(s_i)$, but for usable rules, $\pi(f) = i$ is treated the same as $\pi(f) = \{i\}$, cf. [9, Section 4].

Combining usable rules, argument filterings and reduction pairs, we obtain:

Theorem 6. *([9,10]) Let $(\succsim, \succ)$ be a reduction pair and π an argument filtering. The processor which maps a DP problem $(\mathcal{P}, \mathcal{R}, f)$ to the following result is sound:*

- $\{(\mathcal{P} \setminus \mathcal{P}^{\succ}, \mathcal{R}, \mathsf{m})\}$ *if* $f = \mathsf{m}$ *and:*
  - $\overline{\pi}(\ell) \succ \overline{\pi}(r)$ *for* $\ell \to r \in \mathcal{P}^{\succ}$ *and* $\overline{\pi}(\ell) \succsim \overline{\pi}(r)$ *for* $\ell \to r \in \mathcal{P} \setminus \mathcal{P}^{\succ}$*;*
  - $\overline{\pi}(\ell) \succsim \overline{\pi}(r)$ *for* $\ell \to r \in \mathit{UR}(\mathcal{P}, \mathcal{R}, \pi) \cup \mathcal{C}_{\epsilon}$*,* *where* $\mathcal{C}_{\epsilon} = \{\mathsf{c}_{\iota}(x, y) \to x, \mathsf{c}_{\iota}(x, y) \to y \mid$ *all sorts* $\iota\}$.
- $\{(\mathcal{P}, \mathcal{R}, f)\}$ *otherwise*

We define $\mathit{UR}(\mathcal{P}, \mathcal{R})$ as $\mathit{UR}(\mathcal{P}, \mathcal{R}, \pi_{\mathcal{T}})$, where $\pi_{\mathcal{T}}$ is the *trivial filtering*: $\pi_{\mathcal{T}}(f) = \{1, \ldots, n\}$ for $f : [\iota_1 \times \ldots \times \iota_n] \Rightarrow \kappa \in \Sigma$. Then Theorem 6 is exactly the standard reduction pair processor, but with constraints on $\mathit{UR}(\mathcal{P}, \mathcal{R}) \cup \mathcal{C}_{\epsilon}$ instead of $\mathcal{R}$. We could also use a processor which maps $(\mathcal{P}, \mathcal{R}, \mathsf{m})$ to $\{(\mathcal{P}, \mathit{UR}(\mathcal{P}, \mathcal{R}) \cup \mathcal{C}_{\epsilon}, \mathsf{a})\}$, but as this loses the minimality flag, it is usually not a good idea (various processors need this flag, including usable rules!) and can only be done once.

3 Formative Rules

Where *usable rules* [1,9,10] are defined primarily by the right-hand sides of $\mathcal{P}$ and $\mathcal{R}$, the *formative rules* discussed here are defined by the left-hand sides. This has consequences; most importantly, we cannot handle non-left-linear rules very well.

We fix a signature Σ. A term $s : \iota$ *has shape* f with $f : [\boldsymbol{\kappa}] \Rightarrow \iota \in \Sigma$ if either $s = f(r_1, \ldots, r_n)$, or s is a variable of sort ι. That is, there exists some γ with $s\gamma = f(\ldots)$: one can *specialise* s to have f as its root symbol.

Definition 7. *Let $\mathcal{R}$ be a set of rules. The* basic formative rules *of a term t are defined as the smallest set $\mathit{FR}_{\mathsf{base}}(t, \mathcal{R}) \subseteq \mathcal{R}$ such that:*

- *if t is not linear, then $\mathit{FR}_{\mathsf{base}}(t, \mathcal{R}) = \mathcal{R}$;*
- *if $t = f(t_1, \ldots, t_n)$, then $\mathit{FR}_{\mathsf{base}}(t_i, \mathcal{R}) \subseteq \mathit{FR}_{\mathsf{base}}(t, \mathcal{R})$;*
- *if $t = f(t_1, \ldots, t_n)$, then $\{\ell \to r \in \mathcal{R} \mid r$ has shape $f\} \subseteq \mathit{FR}_{\mathsf{base}}(t, \mathcal{R})$;*
- *if $\ell \to r \in \mathit{FR}_{\mathsf{base}}(t, \mathcal{R})$, then $\mathit{FR}_{\mathsf{base}}(\ell, \mathcal{R}) \subseteq \mathit{FR}_{\mathsf{base}}(t, \mathcal{R})$.*

For rules $\mathcal{P}$, let $\mathit{FR}_{\mathsf{base}}(\mathcal{P}, \mathcal{R}) = \bigcup_{s \to t \in \mathcal{P}} \mathit{FR}_{\mathsf{base}}(s, \mathcal{R})$. Note that $\mathit{FR}_{\mathsf{base}}(x, \mathcal{R}) = \emptyset$.

Note the strong symmetry with Definition 5. We have omitted the argument filtering π here, because the definitions are simpler without it. In Section 4 we will see how we can add argument filterings back in without changing the definition.

Example 8. In the system from Example 1, consider $\mathcal{P} = \{\mathtt{Big}^{\sharp}(x, \mathtt{Cons}(y, z)) \to \mathtt{Big}^{\sharp}(\mathtt{Ack}(x, y), \mathtt{Upd}(z))\}$. The symbols in the left-hand side are just $\mathtt{Big}^{\sharp}$ (which has sort $\mathtt{dpsort}$, which is not used in $\mathcal{R}$) and $\mathtt{Cons}$. Thus, $\mathit{FR}_{\mathsf{base}}(\mathcal{P}, \mathcal{R}) = \{8\}$.

Intuitively, the formative rules of a dependency pair $\ell \to r$ are those rules which might contribute to creating the pattern ℓ. In Example 8, to reduce a term $\mathtt{Big}^\sharp(\mathtt{Ack}(\mathtt{S}(\mathtt{0}),\mathtt{0}),\mathtt{Upd}(\mathtt{Cons}(\mathtt{0},\mathtt{Nil})))$ to an instance of $\mathtt{Big}^\sharp(x,\mathtt{Cons}(y,z))$, a single step with the Upd rule 8 gives $\mathtt{Big}^\sharp(\mathtt{Ack}(\mathtt{S}(\mathtt{0}),\mathtt{0}),\mathtt{Cons}(\mathtt{Rnd}(\mathtt{0}),\mathtt{Upd}(\mathtt{Nil})))$; we need not reduce the Ack() or Rnd() subterms for this. To create a *non*-linear pattern, any rule could contribute, as a step deep inside a term may be needed.

Example 9. Consider $\Sigma = \{\mathtt{a},\mathtt{b} : \mathtt{A}, \mathtt{f}^\sharp : [\mathtt{B} \times \mathtt{B}] \Rightarrow \mathtt{dpsort}, \mathtt{h} : [\mathtt{A}] \Rightarrow \mathtt{B}\}$, $\mathcal{R} = \{\mathtt{a} \to \mathtt{b}\}$ and $\mathcal{P} = \{\mathtt{f}^\sharp(x,x) \to \mathtt{f}^\sharp(\mathtt{h}(\mathtt{a}),\mathtt{h}(\mathtt{b}))\}$. Without the linearity restriction, $\mathit{FR}_{\mathsf{base}}(\mathcal{P},\mathcal{R})$ would be $\emptyset$, as dpsort does not occur in the rules and $\mathit{FR}_{\mathsf{base}}(x,\mathcal{R}) = \emptyset$. But there is no infinite $(\mathcal{P},\emptyset)$-chain, while we do have an infinite $(\mathcal{P},\mathcal{R})$-chain, with $\gamma_i = [x := \mathtt{h}(\mathtt{b})]$ for all i. The $\mathtt{a} \to \mathtt{b}$ rule is needed to make $\mathtt{h}(\mathtt{a})$ and $\mathtt{h}(\mathtt{b})$ equal. Note that this happens even though the sort of x does not occur in $\mathcal{R}$!

Thus, as we will see, in an infinite $(\mathcal{P},\mathcal{R})$-chain we can limit interest to rules in $\mathit{FR}_{\mathsf{base}}(\mathcal{P},\mathcal{R})$. We call these *basic* formative rules because while they demonstrate the concept, in practice we would typically use more advanced extensions of the idea. For instance, following the *TCap* idea of [8, Definition 11], a rule $l \to f(\mathtt{0})$ does not need to be a formative rule of $f(\mathtt{S}(x)) \to r$ if $\mathtt{0}$ is a constructor.

To use formative rules with DPs, we will show that any $(\mathcal{P},\mathcal{R})$-chain can be altered so that the $r_i\gamma_i \to_{\mathcal{R}}^* \ell_{i+1}\gamma_{i+1}$ reduction has a very specific form (which uses only formative rules of ℓ_{i+1}). To this end, we consider *formative reductions.* A formative reduction is a reduction where, essentially, a rewriting step is only done if it is needed to obtain a result of the right form.

Definition 10 (Formative Reduction). *For a term ℓ, substitution γ and term s, we say $s \to_{\mathcal{R}}^* \ell\gamma$ by a* formative ℓ-reduction *if one of the following holds:*

1. *ℓ is non-linear;*
2. *ℓ is a variable and $s = \ell\gamma$;*
3. *$\ell = f(l_1,\dots,l_n)$ and $s = f(s_1,\dots,s_n)$ and each $s_i \to_{\mathcal{R}}^* l_i\gamma$ by a formative l_i-reduction;*
4. *$\ell = f(l_1,\dots,l_n)$ and there are a rule $\ell' \to r' \in \mathcal{R}$ and a substitution δ such that $s \to_{\mathcal{R}}^* \ell'\delta$ by a formative ℓ'-reduction and $r'\delta = f(t_1,\dots,t_n)$ and each $t_i \to_{\mathcal{R}}^* l_i\gamma$ by a formative l_i-reduction.*

Point 2 is the key: a reduction $s \to_{\mathcal{R}}^* x\gamma$ must be postponed. Formative reductions are the base of a semantic definition of formative rules:

Definition 11. *A function FR that maps a term ℓ and a set of rules $\mathcal{R}$ to a set $\mathit{FR}(\ell,\mathcal{R}) \subseteq \mathcal{R}$ is a* formative rules approximation *if for all s and γ: if $s \to_{\mathcal{R}}^* \ell\gamma$ by a formative ℓ-reduction, then this reduction uses only rules in $\mathit{FR}(\ell,\mathcal{R})$.*

Given a formative rules approximation FR, let $\mathit{FR}(\mathcal{P},\mathcal{R}) = \bigcup_{s \to t \in \mathcal{P}} \mathit{FR}(s,\mathcal{R})$.

As might be expected, $\mathit{FR}_{\mathsf{base}}$ is indeed a formative rules approximation:

Lemma 12. *A formative ℓ-reduction $s \to_{\mathcal{R}}^* \ell\gamma$ uses only rules in $\mathit{FR}_{\mathsf{base}}(\ell,\mathcal{R})$.*

Proof. By induction on the definition of a formative ℓ-reduction. If ℓ is non-linear, then $FR_{\mathsf{base}}(\ell,\mathcal{R}) = \mathcal{R}$, so this is clear. If $s = \ell\gamma$ then no rules play a part.

If $s = f(s_1,\dots,s_n)$ and $\ell = f(l_1,\dots,l_n)$ and each $s_i \to_{\mathcal{R}}^* l_i\gamma$ by a formative l_i-reduction, then by the induction hypothesis each formative l_i-reduction $s_i \to_{\mathcal{R}}^* l_i\gamma$ uses only rules in $FR_{\mathsf{base}}(l_i,\mathcal{R})$. Observing that by definition $FR_{\mathsf{base}}(l_i,\mathcal{R}) \subseteq FR_{\mathsf{base}}(\ell,\mathcal{R})$, we see that all steps of the reduction use rules in $FR_{\mathsf{base}}(\ell,\mathcal{R})$.

If $s \to_{\mathcal{R}}^* \ell'\delta \to_{\mathcal{R}} r'\delta = f(t_1,\dots,t_n) \to_{\mathcal{R}}^* f(l_1,\dots,l_n)\gamma = \ell\gamma$, then by the same reasoning the reduction $r'\delta \to_{\mathcal{R}}^* \ell\gamma$ uses only formative rules of ℓ, and by the induction hypothesis $s \to_{\mathcal{R}}^* \ell'\delta$ uses only formative rules of ℓ'. Noting that r' obviously has the same sort as ℓ, and either r' is a variable or a term $f(r'_1,\dots,r'_n)$, we see that r' has shape f, so $\ell' \to r' \in FR_{\mathsf{base}}(\ell,\mathcal{R})$. Therefore $FR_{\mathsf{base}}(\ell',\mathcal{R}) \subseteq FR_{\mathsf{base}}(\ell,\mathcal{R})$, so all rules in the reduction are formative rules of ℓ. □

In the following, we will assume a fixed formative rules approximation *FR*. The relevance of formative rules is clear from their definition: if we can prove that a $(\mathcal{P},\mathcal{R})$-chain can be altered to use formative reductions in the $\to_{\mathcal{R}}$ steps, then we can drop all non-formative rules from a DP problem.

The key result in this paper is the following technical lemma, which allows us to alter a reduction $s \to_{\mathcal{R}}^* \ell\gamma$ to a formative reduction (by changing γ):

Lemma 13. *If $s \to_{\mathcal{R}}^* \ell\gamma$ for some terms s,ℓ and a substitution γ on domain $Var(\ell)$, then there is a substitution δ on the same domain such that $s \to_{FR(\ell,\mathcal{R})}^* \ell\delta$ by a formative ℓ-reduction.*

Proof. For non-linear ℓ this is clear, choosing $\delta := \gamma$. So let ℓ be a linear term. By definition of *FR*, it suffices to see that $s \to_{\mathcal{R}}^* \ell\delta$ by a formative ℓ-reduction. This follows from the following claim: *If $s \mathrel{+\!\!\!\!\!\!\!\to}_{\mathcal{R}}^{k} \ell\gamma$ for some k, term s, linear term ℓ and substitution γ on domain $Var(\ell)$, then there is a substitution δ on $Var(\ell)$ such that $s \to_{\mathcal{R}}^* \ell\delta$ by a formative ℓ-reduction, and each $\delta(x) \mathrel{+\!\!\!\!\!\!\!\to}_{\mathcal{R}}^{k} \gamma(x)$.*

Here, the parallel reduction relation $\mathrel{+\!\!\!\!\!\!\!\to}_{\mathcal{R}}$ is defined by: $x \mathrel{+\!\!\!\!\!\!\!\to}_{\mathcal{R}} x$; $\ell\gamma \mathrel{+\!\!\!\!\!\!\!\to}_{\mathcal{R}} r\gamma$ for $\ell \to r \in \mathcal{R}$; if $s_i \mathrel{+\!\!\!\!\!\!\!\to}_{\mathcal{R}} t_i$ for $1 \le i \le n$, then $f(s_1,\dots,s_n) \mathrel{+\!\!\!\!\!\!\!\to}_{\mathcal{R}} f(t_1,\dots,t_n)$. The notation $\mathrel{+\!\!\!\!\!\!\!\to}_{\mathcal{R}}^{k}$ indicates k *or fewer* successive $\mathrel{+\!\!\!\!\!\!\!\to}_{\mathcal{R}}$ steps. Note that $\mathrel{+\!\!\!\!\!\!\!\to}_{\mathcal{R}}$ is reflexive, and if each $s_i \mathrel{+\!\!\!\!\!\!\!\to}_{\mathcal{R}}^{N_i} t_i$, then $f(\boldsymbol{s}) \mathrel{+\!\!\!\!\!\!\!\to}_{\mathcal{R}}^{\max(N_1,\dots,N_n)} f(\boldsymbol{t})$.

We prove the claim by induction first on k, second on the size of ℓ.

If ℓ is a variable we are immediately done, choosing $\delta := [\ell := s]$.

Otherwise, let $\ell = f(l_1,\dots,l_n)$ and $\gamma = \gamma_1 \cup \dots \cup \gamma_n$ such that all γ_i have disjoint domains and each $l_i\gamma_i = l_i\gamma$; this is possible due to linearity.

First suppose the reduction $s \mathrel{+\!\!\!\!\!\!\!\to}_{\mathcal{R}}^{k} \ell\gamma$ uses no topmost steps. Thus, we can write $s = f(s_1,\dots,s_n)$ and each $s_i \mathrel{+\!\!\!\!\!\!\!\to}_{\mathcal{R}}^{k} l_i\gamma$. By the second induction hypothesis we can find $\delta_1,\dots,\delta_n$ such that each $s_i \to_{\mathcal{R}}^* l_i\delta_i$ by a formative l_i-reduction and each $\delta_i(x) \mathrel{+\!\!\!\!\!\!\!\to}_{\mathcal{R}}^{k} \gamma_i(x)$. Choose $\delta := \delta_1 \cup \dots \cup \delta_n$; this is well-defined by the assumption on the disjoint domains. Then $s \to_{\mathcal{R}}^* \ell\delta$ by a formative ℓ-reduction.

Alternatively, a topmost step was done, which cannot be parallel with other steps: $s \mathrel{+\!\!\!\!\!\!\!\to}_{\mathcal{R}}^{m} \ell'\gamma' \to_{\mathcal{R}} r'\gamma' \mathrel{+\!\!\!\!\!\!\!\to}_{\mathcal{R}}^{k-m-1} \ell\gamma$ for some $\ell' \to r' \in \mathcal{R}$ and substitution γ'; we can safely assume that $r'\gamma' \mathrel{+\!\!\!\!\!\!\!\to}_{\mathcal{R}}^{k-m-1} \ell\gamma$ does not use topmost steps (otherwise we could just choose a later step). Since $m < k$, the first induction hypothesis provides δ' such that $s \rightarrow_{\mathcal{R}}^* \ell'\delta'$ by a formative ℓ'-reduction and each $\delta'(x) \dashrightarrow\!\!\!\!\!+\!\!\!\!\rightarrow_{\mathcal{R}}{}^m \gamma'(x)$. But then also $r'\delta' \dashrightarrow_{\mathcal{R}}{}^m r'\gamma'$. Since $r'\gamma' \dashrightarrow_{\mathcal{R}}{}^{k-m-1} \ell\gamma$, we have that $r'\delta' \dashrightarrow_{\mathcal{R}}{}^{k-1} \ell\gamma$. Thus, by the first induction hypothesis, there is δ such that $r'\delta' \rightarrow_{\mathcal{R}}^* \ell\delta$ by a formative ℓ-reduction, and each $\delta(x) \dashrightarrow_{\mathcal{R}}{}^{k-1} \gamma(x)$.

We are done if the full reduction $s \rightarrow_{\mathcal{R}}^* \ell'\delta' \rightarrow_{\mathcal{R}} r'\delta' \rightarrow_{\mathcal{R}}^* \ell\delta$ is ℓ-formative; this is easy with induction on the number of topmost steps in the second part. □

Lemma 13 lays the foundation for all theorems in this paper. To start:

Theorem 14. *$(\Sigma, \mathcal{R})$ is non-terminating if and only if there is an infinite minimal formative* $(\mathsf{DP}(\mathcal{R}), FR(\mathsf{DP}(\mathcal{R}), \mathcal{R}))$*-chain. Here, a chain* $[(\ell_i \to r_i, \gamma_i) \mid i \in \mathbb{N}]$ *is* formative *if always* $r_i\gamma_i \rightarrow_{FR(\ell_{i+1},\mathcal{R})}^* \ell_{i+1}\gamma_{i+1}$ *by a formative* ℓ_{i+1}*-reduction.*

Proof Sketch: Construct an infinite $(\mathsf{DP}(\mathcal{R}), \mathcal{R})$-chain following the usual proof, but when choosing γ_{i+1}, use Lemma 13 to guarantee that $r_i\gamma_i \rightarrow_{FR(\ell_{i+1},\mathcal{R})}^* \ell_{i+1}\gamma_{i+1}$ by a formative ℓ_{i+1}-reduction. □

Note that this theorem extends the standard dependency pairs result (Theorem 3) by limiting interest to chains with formative reductions.

Example 15. The system from Example 1 is terminating iff there is no infinite minimal formative $(\mathcal{P}, Q)$-chain, where $\mathcal{P} = \mathsf{DP}(\mathcal{R})$ from Example 2 and $Q = \{1, 2, 3, 5, 6, 7, 8, 10, 11\}$. Rules 4 and 9 have right-hand sides headed by symbols `Err` and `Return` which do not occur in the left-hand sides of DP or its formative rules.

4 Formative Rules in the Dependency Pair Framework

Theorem 14 provides a basis for using DPs with formative rules to prove termination: instead of proving that there is no infinite minimal $(\mathsf{DP}(\mathcal{R}), \mathcal{R})$-chain, it suffices if there is no infinite minimal formative $(\mathsf{DP}(\mathcal{R}), FR(\mathsf{DP}(\mathcal{R}), \mathcal{R}))$-chain. So in the DP framework, we can start with the set $\{(\mathsf{DP}(\mathcal{R}), FR(\mathsf{DP}(\mathcal{R}), \mathcal{R}), \mathsf{m})\}$ instead of $\{(\mathsf{DP}(\mathcal{R}), \mathcal{R}, \mathsf{m})\}$, as we did in Example 15. We thus obtain a similar improvement to Dershowitz' refinement [3] in that it yields a smaller *initial* DP problem: by [3], we can reduce the initial set $\mathsf{DP}(\mathcal{R})$; by Theorem 14 we can reduce the initial set $\mathcal{R}$. However, there (currently) is no way to keep track of the information that we only need to consider formative chains. Despite this, we can define several processors. All of them are based on this consequence of Lemma 13:

Lemma 16. *If there is a $(\mathcal{P}, \mathcal{R})$-chain $[(\ell_i \to r_i, \gamma_i) \mid i \in \mathbb{N}]$, then there are δ_i for $i \in \mathbb{N}$ such that $[(\ell_i \to r_i, \delta_i) \mid i \in \mathbb{N}]$ is a formative $(\mathcal{P}, FR(\mathcal{P}, \mathcal{R}))$-chain.*

Proof. Given $[(\ell_i \to r_i, \gamma_i) \mid i \in \mathbb{N}]$ we construct the formative chain as follows. Let $\delta_1 := \gamma_1$. For given i, suppose δ_i is a substitution such that $\delta_i \rightarrow_{\mathcal{R}}^* \gamma_i$, so still $r_i\delta_i \rightarrow_{\mathcal{R}}^* \ell_{i+1}\gamma_{i+1}$. Use Lemma 13 to find δ_{i+1} such that $r_i\delta_i \rightarrow_{FR(\ell_{i+1},\mathcal{R})}^* \ell_{i+1}\delta_{i+1}$ by a formative ℓ_{i+1}-reduction, and moreover $\delta_{i+1} \rightarrow_{\mathcal{R}}^* \gamma_{i+1}$. □

This lemma for instance allows us to remove all non-formative rules from a DP problem. To this end, we use the following processor:

Theorem 17. *The DP processor which maps a DP problem* $(\mathcal{P}, \mathcal{R}, f)$ *to the set* $\{(\mathcal{P}, FR(\mathcal{P}, \mathcal{R}), \mathsf{a})\}$ *is sound.*

Proof Sketch: This follows immediately from Lemma 16. □

Example 18. Let $Q = FR_{\mathsf{base}}(\mathsf{DP}(\mathcal{R}), \mathcal{R})$ from Example 15, and let $\mathcal{P} = \{\mathtt{Big}^\sharp(x, \mathtt{Cons}(y, z)) \to \mathtt{Big}^\sharp(\mathtt{Ack}(x, y), \mathtt{Upd}(z))\}$ as in Example 8. If, during a termination proof with dependency pairs, we encounter a DP problem $(\mathcal{P}, Q, \mathsf{m})$, we can soundly replace it by $(\mathcal{P}, T, \mathsf{a})$, where $T = FR_{\mathsf{base}}(\mathcal{P}, Q) = \{8\}$.

Thus, we can (permanently) remove all non-formative rules from a dependency pair problem. This processor has a clear downside, however: given a problem $(\mathcal{P}, \mathcal{R}, \mathsf{m})$, we lose minimality. This m flag is very convenient to have, as several processors require it (such as reduction pairs with usable rules from Theorem 6).

Could we preserve minimality? Unfortunately, the answer is no. By modifying a chain to use formative reductions, we may lose the property that each $r_i\gamma_i$ is terminating. This happens for instance for $(\mathcal{P}, \mathcal{R}, \mathsf{m})$, where $\mathcal{P} = \{\mathtt{g}^\sharp(x) \to \mathtt{h}^\sharp(\mathtt{f}(x)),\ \mathtt{h}^\sharp(\mathtt{c}) \to \mathtt{g}^\sharp(\mathtt{a})\}$ and $\mathcal{R} = \{\mathtt{a} \to \mathtt{b},\ \mathtt{f}(x) \to \mathtt{c},\ \mathtt{f}(\mathtt{a}) \to \mathtt{f}(\mathtt{a})\}$. Here, $FR_{\mathsf{base}}(\mathcal{P}, \mathcal{R}) = \{\mathtt{f}(x) \to \mathtt{c},\ \mathtt{f}(\mathtt{a}) \to \mathtt{f}(\mathtt{a})\}$. While there is an infinite minimal $(\mathcal{P}, \mathcal{R})$-chain, the only infinite $(\mathcal{P}, FR_{\mathsf{base}}(\mathcal{P}, \mathcal{R}))$-chain is non-minimal.

Fortunately, there *is* an easy way to use formative rules without losing any information: by using them in a reduction pair, as we typically do for usable rules. In fact, although usable and formative rules seem to be opposites, there is no reason why we should use either one or the other; we can combine them. Considering also argument filterings, we find the following extension of Theorem 6.

Theorem 19. *Let* $(\succsim, \succ)$ *be a reduction pair and* π *an argument filtering. The processor which maps* $(\mathcal{P}, \mathcal{R}, f)$ *to the following result is sound:*

- $\{(\mathcal{P} \setminus \mathcal{P}^{\succ}, \mathcal{R}, f)\}$ *if:*
  - $\overline{\pi}(\ell) \succ \overline{\pi}(r)$ *for* $\ell \to r \in \mathcal{P}^{\succ}$ *and* $\overline{\pi}(\ell) \succsim \overline{\pi}(r)$ *for* $\ell \to r \in \mathcal{P} \setminus \mathcal{P}^{\succ}$;
  - $u \succsim v$ *for* $u \to v \in FR(\overline{\pi}(\mathcal{P}), \overline{\pi}(U))$,
    where $U = \mathcal{R}$ *if* $f = \mathsf{a}$ *and* $U = UR(\mathcal{P}, \mathcal{R}, \pi) \cup \mathcal{C}_\epsilon$ *if* $f = \mathsf{m}$;
- $\{(\mathcal{P}, \mathcal{R}, f)\}$ *otherwise.*

Proof Sketch: Given an infinite $(\mathcal{P}, \mathcal{R})$-chain, we use argument filterings and maybe usable rules to obtain a $(\overline{\pi}(\mathcal{P}), \overline{\pi}(U))$-chain which uses the same dependency pairs infinitely often (as in [9]); using Lemma 16 we turn this chain formative. □

Note that we use the argument filtering here in a slightly different way than for usable rules: rather than including π in the definition of FR and requiring that $\overline{\pi}(\ell) \succsim \overline{\pi}(r)$ for $\ell \to r \in FR(\mathcal{P}, \mathcal{R}, \pi)$, we simply use $FR(\overline{\pi}(\mathcal{P}), \overline{\pi}(\mathcal{R}))$. For space reasons, we give additional semantic and syntactic definitions of formative rules with respect to an argument filtering in the technical report [5, Appendix C].

Example 20. To handle $(\mathcal{P}, Q, \mathsf{m})$ from Example 18, we can alternatively use a reduction pair. Using the trivial argument filtering, with a polynomial interpretation with $\mathtt{Big}^\sharp(x,y) = x+y$, $\mathtt{Ack}(x,y) = 0$, $\mathtt{Upd}(x) = x$ and $\mathtt{Cons}(x,y) = y+1$, all constraints are oriented, and we may remove the only element of $\mathcal{P}$.

Note that we *could* have handled this example without using formative rules; Ack and Rnd can be oriented with an extension of $\succsim$, or we might use an argument filtering with $\pi(\mathtt{Big}^\sharp) = \{2\}$. Both objections could be cancelled by adding extra rules, but we kept the example short, as it suffices to illustrate the method.

Discussion. It is worth noting the parallels between formative and usable rules. To start, their definitions are very similar; although we did not present the semantic definition of usable rules from [16] (which is only used for *innermost* termination), the syntactic definitions are almost symmetric. Also the usage corresponds: in both cases, we lose minimality when using the direct rule removing processor, but can safely use the restriction in a reduction pair (with argument filterings).

There are also differences, however. The transformations used to turn a chain usable or formative are very different, with the usable rules transformation (which we did not discuss) encoding subterms whose root is not usable, while the formative rules transformation is simply a matter of postponing reduction steps.

Due to this difference, usable rules are useful only for a finitely branching system (which is standard, as all finite MTRSs are finitely branching); formative rules are useful mostly for left-linear systems (also usual, especially in MTRSs originating from functional programming, but typically seen as a larger restriction). Usable rules introduce the extra $\mathcal{C}_\epsilon$ rules, while formative rules are all included in the original rules. But for formative rules, even definitions extending FR_{base}, necessarily all collapsing rules are included, which has no parallel in usable rules; the parallel of collapsing rules would be rules $x \to r$, which are not permitted.

To use formative rules without losing minimality information, an alternative to Theorem 17 allows us to permanently delete rules. The trick is to add a new component to DP problems, as for higher-order rewriting in [11, Ch. 7]. A DP problem becomes a tuple $(\mathcal{P}, \mathcal{R}, f_1, f_2)$, with $f_1 \in \{\mathsf{m}, \mathsf{a}\}$ and $f_2 \in \{\mathsf{form}, \mathsf{arbitrary}\}$, and is finite if there is no infinite $(\mathcal{P}, \mathcal{R})$-chain which is minimal if $f_1 = \mathsf{m}$, and formative if $f_2 = \mathsf{form}$. By Theorem 14, $\mathcal{R}$ is terminating iff $(\mathsf{DP}(\mathcal{R}), \mathcal{R}, \mathsf{m}, \mathsf{form})$ is finite.

Theorem 21. *In the extended DP framework, the processor which maps* $(\mathcal{P}, \mathcal{R}, f_1, f_2)$ *to* $\{(\mathcal{P}, FR(\mathcal{P}, \mathcal{R}), f_1, f_2)\}$ *if* $f_2 = \mathsf{form}$ *and* $\{(\mathcal{P}, \mathcal{R}, f_1, f_2)\}$ *otherwise, is sound.*

Proof: This follows immediately from Lemma 12. □

The downside of changing the DP framework in this way is that we have to revisit all existing DP processors to see how they interact with the formative flag. In many cases, we can simply pass the flag on unmodified (i.e. if $proc((\mathcal{P}, \mathcal{R}, f_1)) = A$, then $proc'((\mathcal{P}, \mathcal{R}, f_1, f_2)) = \{(\mathcal{P}', \mathcal{R}', f_1', f_2) \mid (\mathcal{P}', \mathcal{R}', f_1') \in A\})$. This is for example the case for processors with reduction pairs (like the one in Theorem 19), the dependency graph and the subterm criterion. Other processors would have to be checked individually, or reset the flag to arbitrary by default.

Given how long the dependency pair framework has existed (and how many processors have been defined, see e.g. [16]), and that the formative flag clashes with the component for innermost rewriting (see Section 6), it is unlikely that many tool programmers will make the effort for a single rule-removing processor.

5 Handling the Collapsing Rules Problem

A great weakness of the formative rules method is the matter of collapsing rules. Whenever the left-hand side of a dependency pair or formative rule has a symbol $f : [\iota] \Rightarrow \kappa$, all collapsing rules of sort κ are formative. And then all *their* formative rules are also formative. Thus, this often leads to the inclusion of all rules of a given sort. In particular for systems with only one sort (such as all first-order benchmarks in the Termination Problems Data Base), this is problematic.

For this reason, we will consider a new notion, building on the idea of formative rules and reductions. This notion is based on the observation that it might suffice to include *composite rules* rather than the formative rules of all collapsing rules. To illustrate the idea, assume given a uni-sorted system with rules $\mathtt{a} \to \mathtt{f}(\mathtt{b})$ and $\mathtt{f}(x) \to x$. $FR_{\mathsf{base}}(\mathtt{c})$ includes $\mathtt{f}(x) \to x$, so also $\mathtt{a} \to \mathtt{f}(\mathtt{b})$. But a term $\mathtt{f}(\mathtt{b})$ does not reduce to $\mathtt{c}$. So intuitively, we should not really need to include the first rule.

Instead of including the formative rules of all collapsing rules, we might imagine a system where we *combine* rules with collapsing rules that could follow them. In the example above, this gives $\mathcal{R} = \{\mathtt{a} \to \mathtt{f}(\mathtt{b}),\ \mathtt{a} \to \mathtt{b},\ \mathtt{f}(x) \to x\}$. Now we might consider an alternative definition of formative rules, where we still need to include the collapsing rule $\mathtt{f}(x) \to x$, but no longer need to have $\mathtt{a} \to \mathtt{f}(\mathtt{b})$.

To make this idea formal, we first consider how rules can be combined. In the following, we consider systems with *only one sort*; this is needed for the definition to be well-defined, but can always be achieved by replacing all sorts by $\mathtt{o}$.

Definition 22 (Combining Rules). *Given an MTRS $(\Sigma, \mathcal{R})$, let $A := \{f(\boldsymbol{x}) \to x_i \mid f : [\iota_1 \times \ldots \times \iota_n] \Rightarrow \kappa \in \Sigma \wedge 1 \leq i \leq n\}$ and $B := \{\ell \to p \mid \ell \to r \in \mathcal{R} \wedge r \trianglerighteq p\}$. Let $X \subseteq A \cup B$ be the smallest set such that $\mathcal{R} \subseteq X$ and for all $\ell \to r \in X$:*

a. if r is a variable, $\ell \trianglerighteq f(l_1, \ldots, l_n)$ and $l_i \trianglerighteq r$, then $f(x_1, \ldots, x_n) \to x_i \in X$;
b. if $r = f(r_1, \ldots, r_n)$ and $f(x_1, \ldots, x_n) \to x_i \in X$, then $\ell \to r_i \in X$.

Let $Cl := A \cap X$ and $NC = \{\ell \to r \in X \mid r$ not a variable$\}$. Let $A_{\mathcal{R}} := Cl \cup NC$.

It is easy to see that $\to_{\mathcal{R}}^*$ is included in $\to_{A_{\mathcal{R}}}^*$: all non-collapsing rules of $\mathcal{R}$ are in NC, and all collapsing rules are obtained as a concatenation of steps in Cl.

Example 23. Consider an unsorted version of Example 1. Then for $(\mathcal{P}, Q)$ as in Example 18, we have $U := UR(\mathcal{P}, Q) = \{1, 2, 3, 5, 8, 10, 11\}$. Unfortunately, only (3) is not formative, as the two `Rnd` rules cause inclusion of all rules in $FR_{\mathsf{base}}(\mathtt{S}(x), U)$. Let us instead calculate X, which we do as an iterative procedure starting from $\mathcal{R}$. In the following, $C \Rightarrow D_1, \ldots, D_n$ should be read as: "by requirement a, rule C enforces inclusion of each D_i", and $C, D \Rightarrow E$ similarly refers to requirement b.

$2, 1 \Rightarrow 12$	$5, 13 \Rightarrow 14$	$10, 15 \Rightarrow 16$	$16, 13 \Rightarrow 18$	$17, 15 \Rightarrow 19$
$12 \Rightarrow 1, 13$	$14 \Rightarrow 15$	$11, 15 \Rightarrow 17$	$18 \Rightarrow 15, 13$	$19 \Rightarrow 15, 13$

12. $\mathtt{Rnd}(\mathtt{S}(x)) \to x$
13. $\mathtt{S}(x) \to x$
14. $\mathtt{Ack}(\mathtt{0}, y) \to y$
15. $\mathtt{Ack}(x, y) \to y$
16. $\mathtt{Ack}(\mathtt{S}(x), y) \to \mathtt{S}(y)$
17. $\mathtt{Ack}(\mathtt{S}(x), \mathtt{S}(y)) \to \mathtt{Ack}(\mathtt{S}(x), y)$
18. $\mathtt{Ack}(\mathtt{S}(x), y) \to y$
19. $\mathtt{Ack}(\mathtt{S}(x), \mathtt{S}(y)) \to y$

Now $Cl = \{1, 13, 15\}$ and $NC = \{2, 3, 5, 8, 10, 11, 16, 17\}$, and $A_U = Cl \cup NC$.

Although combining a system $\mathcal{R}$ into $A_{\mathcal{R}}$ may create significantly more rules, the result is not necessarily harder to handle. For many standard reduction pairs, like RPO or linear polynomials over $\mathbb{N}$, we have: if $s \succsim x$ where $x \in Var(s)$ occurs exactly once, then $f(\ldots, t, \ldots) \succsim t$ for any t with $s \trianglerighteq t \trianglerighteq x$. For such a reduction pair, $A_{\mathcal{R}}$ can be oriented whenever $\mathcal{R}$ can be (if $\mathcal{R}$ is left-linear).

$A_{\mathcal{R}}$ has the advantage that we never need to follow a non-collapsing rule $l \to f(\boldsymbol{r})$ by a collapsing step. This is essential to use the following definition:

Definition 24. *Let A be a set of rules. The* split-formative rules *of a term t are defined as the smallest set $SR(t, A) \subseteq A$ such that:*

- *if t is not linear, then $SR(t, A) = A$;*
- *all collapsing rules in A are included in $SR(t, A)$;* (boxed)
- *if $t = f(t_1, \ldots, t_n)$, then $SR(t_i, A) \subseteq SR(t, A)$;*
- *if $t = f(t_1, \ldots, t_n)$, then $\{\ell \to r \in A \mid r$ has the form $f(\ldots)\} \subseteq SR(t, A)$;*
- *if $\ell \to r \in SR(t, A)$ and r is not a variable (boxed), then $SR(\ell, A) \subseteq SR(t, A)$.*

For a set of rules $\mathcal{P}$, we define $SR(\mathcal{P}, A) = \bigcup_{s \to t \in \mathcal{P}} SR(s, A)$.

Definition 24 is an alternative definition of formative rules, where collapsing rules have a smaller effect (differences to Definition 7 are highlighted (boxed)). SR is *not* a formative rules approximation, as shown by the a-formative reduction $\mathtt{f}(\mathtt{a}) \to_{\mathcal{R}} \mathtt{g}(\mathtt{a}) \to_{\mathcal{R}} \mathtt{a}$ with $\mathcal{R} = \{\mathtt{f}(x) \to \mathtt{g}(x),\ \mathtt{g}(x) \to x\}$ but $SR(\mathtt{a}, \mathcal{R}) = \{\mathtt{g}(x) \to x\}$. However, given the relation between $\mathcal{R}$ and $A_{\mathcal{R}}$, we find a similar result to Lemma 12:

Lemma 25. *Let $(\Sigma, \mathcal{R})$ be an MTRS. If $s \to_{\mathcal{R}}^* \ell\gamma$ by a formative ℓ-reduction, then $s \to_{SR(\ell, A_{\mathcal{R}})}^* \ell\gamma$ by a formative ℓ-reduction.*

Unlike Lemma 12, the altered reduction might be different. We also do not have that $SR(\mathcal{P}, A_{\mathcal{R}}) \subseteq \mathcal{R}$. Nevertheless, by this lemma we can use split-formative rules in reduction pair processors with formative rules, such as Theorem 19.

Proof Sketch: The original reduction $s \rightarrow_{\mathcal{R}}^* \ell\gamma$ gives rise to a formative reduction over $A_{\mathcal{R}}$, simply replacing collapsing steps by a sequence of rules in *Cl*. So, we assume given a formative ℓ-reduction over $A_{\mathcal{R}}$, and prove with induction first on the number of non-collapsing steps in the reduction, second on the length of the reduction, third on the size of s, that $s \rightarrow_{SR(\ell,A_{\mathcal{R}})}^* \ell\gamma$ by a formative ℓ-reduction.

This is mostly easy with the induction hypotheses; note that if a root-rule in *NC* is followed by a rule in *Cl*, there can be no internal $\rightarrow_{\mathcal{R}}^*$ reduction in between (as this would not be a formative reduction); combining a rule in *NC* with a rule in *Cl* gives either a rule in *NC* (and a continuation with the second induction hypothesis) or a sequence of rules in *Cl* (and the first induction hypothesis). □

Note that this method unfortunately does not transpose directly to the higher-order setting, where collapsing rules may have more complex forms. We also had to give up sort differentiation, as otherwise we might not be able to flatten a rule $f(g(x)) \rightarrow x$ into $f(x) \rightarrow x$, $g(x) \rightarrow x$. This is not such a great problem, as reduction pairs typically do not care about sorts, and we circumvented the main reason why sorts are important for formative rules. We have the following result:

Theorem 26. *Let* $(\succsim, \succ)$ *be a reduction pair and* π *an argument filtering. The processor which maps a DP problem* $(\mathcal{P}, \mathcal{R}, f)$ *to the following result is sound:*

- $\{(\mathcal{P} \setminus \mathcal{P}^{\succ}, \mathcal{R}, f)\}$ *if:*
  - $\overline{\pi}(\ell) \succ \overline{\pi}(r)$ *for* $\ell \rightarrow r \in \mathcal{P}^{\succ}$ *and* $\overline{\pi}(\ell) \succsim \overline{\pi}(r)$ *for* $\ell \rightarrow r \in \mathcal{P} \setminus \mathcal{P}^{\succ}$;
  - $u \succsim v$ *for* $u \rightarrow v \in SR(\overline{\pi}(\mathcal{P}), A_{\overline{\pi}(U)})$, *and* $Var(t) \subseteq Var(s)$ *for* $s \rightarrow t \in \overline{\pi}(U)$, *where* $U = \mathcal{R}$ *if* $f = \mathsf{a}$ *and* $U = UR(\mathcal{P}, \mathcal{R}, \pi) \cup \mathcal{C}_{\epsilon}$ *if* $f = \mathsf{m}$;
- $\{(\mathcal{P}, \mathcal{R}, f)\}$ *otherwise.*

Proof Sketch: Like Theorem 19, but using Lemma 25 to alter the created formative $(\overline{\pi}(\mathcal{P}), \overline{\pi}(U))$-chain to a split-formative $(\overline{\pi}(\mathcal{P}), SR(\overline{\pi}(\mathcal{P}), A_{\overline{\pi}(U)}))$-chain. □

Example 27. Following Example 23, $SR(\mathtt{Big}^\sharp(x, \mathtt{Cons}(y,z)) \rightarrow \mathtt{Big}^\sharp(\mathtt{Ack}(x,y), \mathtt{Upd}(z)), A_U) = Cl \cup \{8\}$, and Theorem 26 gives an easily orientable problem.

6 Formative Rules for Innermost Termination

So far, we have considered only *full termination.* A very common related query is *innermost termination*; that is, termination of $\rightarrow_{\mathcal{R}}^{\mathsf{in}}$, defined by:

- $f(\boldsymbol{l})\gamma \rightarrow_{\mathcal{R}}^{\mathsf{in}} r\gamma$ if $f(\boldsymbol{l}) \rightarrow r \in \mathcal{R}$, γ a substitution and all $l_i\gamma$ in normal form;
- $f(s_1, \dots, s_i, \dots, s_n) \rightarrow_{\mathcal{R}}^{\mathsf{in}} f(s_1, \dots, s_i', \dots, s_n)$ if $s_i \rightarrow_{\mathcal{R}}^{\mathsf{in}} s_i'$.

The innermost reduction relation is often used in for instance program analysis.

An innermost strategy can be included in the dependency pair framework by adding the innermost flag [9] to DP problems (or, more generally, a component $\mathcal{Q}$ [7] which indicates that when reducing any term with $\rightarrow_{\mathcal{P}}$ or $\rightarrow_{\mathcal{R}}$, its strict subterms must be normal with respect to $\mathcal{Q}$). Usable rules are more viable for innermost than normal termination: we do not need minimality, the $\mathcal{C}_{\epsilon}$ rules do not need to be handled by the reduction pair, and we can define a sound processor that maps $(\mathcal{P}, \mathcal{R}, f, \mathsf{innermost})$ to $\{(\mathcal{P}, \mathit{UR}(\mathcal{P}, \mathcal{R}), f, \mathsf{innermost})\}$.

This is not the case for formative rules. Innermost reductions eagerly evaluate arguments, yet formative reductions postpone evaluations as long as possible. In a way, these are exact opposites. Thus, it should not be surprising that formative rules are *weaker* for innermost termination than for full termination. Theorem 14 has no counterpart for $\to_{\mathcal{R}}^{\mathsf{in}}$; for innermost termination we must start the DP framework with $(\mathcal{P}, \mathcal{R}, \mathsf{m}, \mathsf{innermost})$, not with $(\mathcal{P}, \mathit{FR}(\mathcal{P}, \mathcal{R}), \mathsf{m}, \mathsf{innermost})$. Theorem 17 is only sound if the innermost flag is removed: $(\mathcal{P}, \mathcal{R}, f, \mathsf{innermost})$ is mapped to $\{(\mathcal{P}, \mathit{FR}(\mathcal{P}, \mathcal{R}), \mathsf{a}, \mathsf{arbitrary})\}$. Still, we *can* safely use formative rules with reduction pairs. For example, we obtain this variation of Theorem 19:

Theorem 28. *Let* $(\succsim, \succ)$ *be a reduction pair and* π *an argument filtering. The processor which maps a DP problem* $(\mathcal{P}, \mathcal{R}, f_1, f_2)$ *to the following result is sound:*

- $\{(\mathcal{P} \setminus \mathcal{P}^{\succ}, \mathcal{R}, f_1, f_2)\}$ *if:*
  - $\overline{\pi}(\ell) \succ \overline{\pi}(r)$ *for* $\ell \to r \in \mathcal{P}^{\succ}$ *and* $\overline{\pi}(\ell) \succsim \overline{\pi}(r)$ *for* $\ell \to r \in \mathcal{P} \setminus \mathcal{P}^{\succ}$*;*
  - $u \succsim v$ *for* $u \to v \in \mathit{FR}(\overline{\pi}(\mathcal{P}), \overline{\pi}(U))$*, where* U *is:* $\mathit{UR}(\mathcal{P}, \mathcal{R}, \pi)$ *if* $f_2 = \mathsf{innermost}$*; otherwise* $\mathit{UR}(\mathcal{P}, \mathcal{R}, \pi) \cup \mathcal{C}_\epsilon$ *if* $f_1 = \mathsf{m}$*; otherwise* $\mathcal{R}$*.*
- $\{(\mathcal{P}, \mathcal{R}, f_1, f_2)\}$ *otherwise.*

Proof Sketch: The proof of Theorem 19 still applies; we just ignore that the given chain might be innermost (aside from getting more convenient usable rules). □

Theorem 26 extends to innermost termination in a similar way.

Conveniently, innermost termination is *persistent* [4], so modifying Σ does not alter innermost termination behaviour, as long as all rules stay well-sorted. In practice, we could infer a typing with as many different sorts as possible, and get stronger formative-rules-with-reduction-pair processors. With the *innermost switch processor* [16, Thm. 3.14], which in cases can set the innermost flag on a DP problem, we could also often use this trick even for proving full termination.

In Section 4, we used the extra flag f_2 as the formative flag. It is not contradictory to use f_2 in both ways, allowing $f_2 \in \{\mathsf{arbitrary}, \mathsf{form}, \mathsf{innermost}\}$, since it is very unlikely for a $(\mathcal{P}, \mathcal{R})$-chain to be both formative and innermost at once! When using both extensions of the DP framework together, termination provers (human or computer) will, however, sometimes have to make a choice which flag to add.

7 Implementation and Experiments

We have performed a preliminary implementation of formative rules in the termination tool AProVE [6]. Our automation borrows from the usable rules of [8] (see [5, Appendices B+D]) and uses a constraint encoding [2] for a combined search for argument filterings and corresponding formative rules. While we did not find any termination proofs for examples from the TPDB where none were known before, our experiments show that formative rules do improve the power of reduction pairs for widely used term orders (e.g., polynomial orders [14]). For more information, see also: `http://aprove.informatik.rwth-aachen.de/eval/Formative`.

For instance, we experimented with a configuration where we applied dependency pairs, and then alternatingly dependency graph decomposition and reduction pairs with linear polynomials and coefficients ≤ 3. On the TRS Standard category of the TPDB (v8.0.7) with 1493 examples, this configuration (without formative rules, but with usable rules w.r.t. an argument filter) shows termination of 579 examples within a timeout of 60 seconds (on an Intel Xeon 5140 at 2.33 GHz). With additional formative rules, our implementation of Theorem 19 proved termination of 6 additional TRSs. (We did, however, lose 4 examples to timeouts, which we believe are due in part to the currently unoptimised implementation.)

The split-formative rules from Theorem 26 are not a subset of $\mathcal{R}$, in contrast to the usable rules. Thus, it is *a priori* not clear how to combine their encodings w.r.t. an argument filtering, and we conducted experiments using only the standard usable rules. Without formative rules, 532 examples are proved terminating. In contrast, adding either the formative rules of Theorem 19 or the split-formative rules of Theorem 26 we solved 6 additional examples each (where Theorem 19 and Theorem 26 each had 1 example the other could not solve), losing 1 to timeouts.

Finally, we experimented with the improved dependency pair transformation based on Theorem 14, which drops non-formative rules from $\mathcal{R}$. We applied DPs as the first technique on the 1403 TRSs from TRS Standard with at least one DP. This reduced the number of rules in the initial DP problem for 618 of these TRSs, without any search problems and without sacrificing minimality.

Thus, our current impression is that while formative rules are not the next "killer technique", they nonetheless provide additional power to widely-used orders in an elegant way and reduce the number of term constraints to be solved in a termination proof. The examples from the TPDB are all untyped, and we believe that formative rules may have a greater impact in a typed first-order setting.

8 Conclusions

In this paper, we have simplified the notion of formative rules from [13] to the first-order setting, and integrated it in the dependency pair framework. We did so by means of *formative reductions*, which allows us to obtain a semantic definition of formative rules (more extensive syntactic definitions are discussed in [5]).

We have defined three processors to use formative rules in the standard dependency pair framework for full termination: one is a processor to permanently remove rules, the other two combine formative rules with a reduction pair.

We also discussed how to strengthen the method by adding a new flag to the framework – although doing so might require too many changes to existing processors and strategies to be considered worthwhile – and how we can still use the technique in the innermost case, and even profit from the innermost setting.

Related Work. In the first-order DP framework two processors stand out as relevant to formative rules. The first is, of course, usable rules; see Section 4 for a detailed discussion. The second is the *dependency graph*, which determines whether any two dependency pairs can follow each other in a $(\mathcal{P}, \mathcal{R})$-chain, and uses this information to eliminate elements of $\mathcal{P}$, or to split $\mathcal{P}$ in multiple parts.

In state-of-the-art implementations of the dependency graph (see e.g. [16]), both left- and right-hand side of dependency pairs are considered to see whether a pair can be preceded or followed by another pair. Therefore it seems quite surprising that the same mirroring was not previously tried for usable rules.

Formative rules *have* been previously defined, for higher-order term rewriting, in [13], which introduces a limited DP framework, with formative rules (but not formative reductions) included in the definition of a chain: we simply impose the restriction that always $r_i\gamma_i \rightarrow^*_{FR(\mathcal{P},\mathcal{R})} \ell_{i+1}\gamma_{i+1}$. This gives a reduction pair processor which considers only formative rules, although it cannot be combined with usable rules and argument filterings. The authors do not yet consider rule removing processors, but if they did, Theorem 21 would also go through.

In the second author's PhD thesis [11], a more complete higher-order DP framework is considered. Here, we do see formative reductions, and a variation of Lemma 13 which, however, requires that s is terminating: the proof style used here does not go through there due to β-reduction. Consequently, Lemma 16 does not go through in the higher-order setting, and there is no counterpart to Theorems 17 or 19. We *do*, however, have Theorem 21. Furthermore, the results of Section 5 are entirely new to this paper, and do not apply in the higher-order setting, where rules might also have a form $l \rightarrow x \cdot s_1 \cdots s_n$ (with x a variable).

Future Work. In the future, it would be interesting to look back at higher-order rewriting, and see whether we can obtain some form of Lemma 16 after all. Alternatively, we might be able to use the specific form of formative chains to obtain formative (and usable) rules w.r.t. an argument filtering.

In the first-order setting, we might turn our attention to non-left-linear rules. Here, we could think for instance of renaming apart some of these variables; a rule $f(x, x) \rightarrow g(x, x)$ could become any of $f(x, y) \rightarrow g(x, y)$, $f(x, y) \rightarrow g(y, x)$, ...

References

1. Arts, T., Giesl, J.: Termination of term rewriting using dependency pairs. Theoretical Computer Science 236(1-2), 133–178 (2000)
2. Codish, M., Giesl, J., Schneider-Kamp, P., Thiemann, R.: SAT solving for termination proofs with recursive path orders and dependency pairs. Journal of Automated Reasoning 49(1), 53–93 (2012)
3. Dershowitz, N.: Termination by abstraction. In: Demoen, B., Lifschitz, V. (eds.) ICLP 2004. LNCS, vol. 3132, pp. 1–18. Springer, Heidelberg (2004)
4. Fuhs, C., Giesl, J., Parting, M., Schneider-Kamp, P., Swiderski, S.: Proving termination by dependency pairs and inductive theorem proving. Journal of Automated Reasoning 47(2), 133–160 (2011)
5. Fuhs, C., Kop, C.: First-order formative rules. Technical Report arXiv:1404.7695 [cs.LO] (2014), `http://arxiv.org/abs/1404.7695`

6. Giesl, J., Brockschmidt, M., Emmes, F., Frohn, F., Fuhs, C., Otto, C., Plücker, M., Schneider-Kamp, P., Ströder, T., Swiderski, S., Thiemann, R.: Proving termination of programs automatically with AProVE. In: Proc. IJCAR (to appear, 2014)
7. Giesl, J., Thiemann, R., Schneider-Kamp, P.: The dependency pair framework: Combining techniques for automated termination proofs. In: Baader, F., Voronkov, A. (eds.) LPAR 2004. LNCS (LNAI), vol. 3452, pp. 301–331. Springer, Heidelberg (2005)
8. Giesl, J., Thiemann, R., Schneider-Kamp, P.: Proving and disproving termination of higher-order functions. In: Gramlich, B. (ed.) FroCos 2005. LNCS (LNAI), vol. 3717, pp. 216–231. Springer, Heidelberg (2005)
9. Giesl, J., Thiemann, R., Schneider-Kamp, P., Falke, S.: Mechanizing and improving dependency pairs. Journal of Automated Reasoning 37(3), 155–203 (2006)
10. Hirokawa, N., Middeldorp, A.: Tyrolean termination tool: Techniques and features. Information and Computation 205(4), 474–511 (2007)
11. Kop, C.: Higher Order Termination. PhD thesis, Vrije Univ. Amsterdam (2012)
12. Kop, C., van Raamsdonk, F.: Higher order dependency pairs for algebraic functional systems. In: Schmidt-Schauß, M. (ed.) RTA 2011. LIPIcs, vol. 10, pp. 203–218. Dagstuhl Publishing (2011)
13. Kop, C., van Raamsdonk, F.: Dynamic dependency pairs for algebraic functional systems. Logical Methods in Computer Science 8(2) (2012)
14. Lankford, D.: On proving term rewriting systems are Noetherian. Technical Report MTP-3, Louisiana Technical University (1979)
15. Tannen, V., Gallier, G.H.: Polymorphic rewriting conserves algebraic strong normalization. Theoretical Computer Science 83(1), 3–28 (1991)
16. Thiemann, R.: The DP framework for proving termination of term rewriting. PhD thesis, RWTH Aachen (2007)

Automated Complexity Analysis Based on Context-Sensitive Rewriting*

Nao Hirokawa[1] and Georg Moser[2]

[1] School of Information Science, JAIST, Japan
hirokawa@jaist.ac.jp

[2] Institute of Computer Science, University of Innsbruck, Austria
georg.moser@uibk.ac.at

Abstract. In this paper we present a simple technique for analysing the runtime complexity of rewrite systems. In complexity analysis many techniques are based on reduction orders. We show how the monotonicity condition for orders can be weakened by using the notion of context-sensitive rewriting. The presented technique is very easy to implement, even in a modular setting, and has been integrated in the Tyrolean Complexity Tool. We provide ample experimental data for assessing the viability of our method.

Keywords: term rewriting, complexity analysis, context-sensitive rewriting, automation.

1 Introduction

This paper is concerned with runtime complexity analysis of term rewrite systems. In recent years the field of complexity analysis of rewrite systems has been dramatically revived. Nowadays this area provides a wide range of different techniques to analyse the time complexity of rewrite systems, fully automatically. Techniques range from *direct methods*, like *polynomial interpretations*, *matrix interpretations* of *polynomial path orders* (e.g. [1–3]) to *transformation techniques*, like adaptions of the *dependency pair method* [4, 5] or modular techniques [6, 7]. See [8] for an overview of complexity analysis methods for term rewrite systems. Furthermore the connection between (runtime) complexity analysis and *implicit computational complexity* [9] is by now well-understood, cf. [10]. Despite this wealth of results, very simple examples cannot be handled and in particular the enormous power of today's termination provers for rewrite systems is still far beyond the ability of today's complexity analysers. Modern termination provers provide termination or non-termination certificates for upto 90 % of the problems *Termination Problem Database* (*TPDB* for short), while with respect to

* This research is partly supported by JSPS KAKENHI Grant Number 25730004 and FWF (Austrian Science Fund) project I 963-N15.

G. Dowek (ed.): RTA-TLCA 2014, LNCS 8560, pp. 257–271, 2014.

automated polynomial runtime complexity analysis of rewrite systems, we see a success rate of 38 %.[1]

Consider Example 1 below. The example encodes division in a natural way. It can be analysed with the techniques mentioned above, but the optimal linear bound on the runtime complexity is not attainable.

Example 1. Consider the following TRS $\mathcal{R}_{div}$[2]

$$\begin{array}{llll} 1: & x - 0 \to x & 3: & 0 \div \mathsf{s}(y) \to 0 \\ 2: & \mathsf{s}(x) - \mathsf{s}(y) \to x - y & 4: & \mathsf{s}(x) \div \mathsf{s}(y) \to \mathsf{s}((x - y) \div \mathsf{s}(y)) . \end{array}$$

The example also clarifies a difference between the *derivational* and the *runtime complexity* of rewrite systems. The *derivational complexity function* with respect to a terminating TRS relates the maximal derivation height to the size of the initial term, cf. [12,13]. On the other hand the *runtime complexity function* with respect to a terminating TRS restricts the derivational complexity function so that only basic terms are considered as starting terms. Here basic terms refer to terms that contain a defined symbol only at root. This terminology was suggested in [4]. Related notions have been studied in [1,14]. It is easy to see that the derivational complexity with respect to $\mathcal{R}_{div}$ bounded from below by an exponential function, while the runtime complexity is linear. Essentially this is due to the fact that, in the computation of division, no contraction is ever required below the second argument. Furthermore, dependency on the first argument is linear.

An inspection of the motivating example in the context of runtime complexity analysis reveals that direct methods are not applicable as the monotonicity constraints are too restrictive. While monotonicity is no longer an issue for transformation techniques, neither the *weak dependency pair method* [4] nor the *dependency tuple method* [5] can deduce the linear (innermost) runtime complexity, as essential constraints cannot be met. In this paper we extend the applicability of direct techniques for complexity results by showing how the monotonicity constraints can be significantly weakened through the employ of *usable replacement maps*, which govern those argument position actually used in rewriting. As usable replacement maps are not computable in general, we provide sufficiently expressive approximations of usable replacement maps. More generally, we show how notions from context-sensitive rewriting can be made applicable in the context of complexity analysis.

This paper is structured as follows. In the next section we cover basics. In Section 3 we define usable replacement maps. In Section 4 we provide experimental data that verifies that the proposed technique makes a difference in practice. Furthermore, in Section 5 we present related work and conclude in Section 6.

---

[1] We base the comparison on last year's run of TERMCOMP, where we consider the categories *TRS Standard* and *Runtime Complexity – Innermost Rewriting*. Note that for termination YES and NO answers have been counted.

[2] This is Example 3.1 in Arts and Giesl's collection of TRSs [11].

2 Runtime Complexity Analysis Based on Matrix Interpretations

We assume familiarity with term rewriting [13,15] but briefly review basic concepts and notations from term rewriting, context-sensitive rewriting and recall matrix interpretations. In particular we will adapt triangular matrix interpretations for runtime complexity analysis.

Let $\mathcal{V}$ denote a countably infinite set of variables and $\mathcal{F}$ a signature, such that $\mathcal{F}$ contains at least one constant. The set of terms over $\mathcal{F}$ and $\mathcal{V}$ is denoted by $\mathcal{T}(\mathcal{F},\mathcal{V})$. The *set of positions* $\mathcal{P}\mathsf{os}(t)$ of a term t is defined as usual. We write $\mathcal{P}\mathsf{os}_{\mathcal{G}}(t) \subseteq \mathcal{P}\mathsf{os}(t)$ for the set of positions of subterms whose root symbol is contained in $\mathcal{G} \subseteq \mathcal{F}$. The subterm of t at position p is denoted as $t|_p$, and $t[u]_p$ denotes the term that is obtained from t by replacing the subterm at p by u. The subterm relation is denoted as $\trianglelefteq$. $\mathcal{V}\mathsf{ar}(t)$ denotes the set of variables occurring in a term t. The *size* $|t|$ of a term is defined as the number of symbols in t.

A *term rewrite system* (*TRS*) $\mathcal{R}$ over $\mathcal{T}(\mathcal{F},\mathcal{V})$ is a *finite* set of rewrite rules $l \to r$. In the sequel, $\mathcal{R}$ always denotes a TRS. The rewrite relation is denoted as $\to_{\mathcal{R}}$ and we use the standard notations for its transitive and reflexive closure. We simply write $\to$ for $\to_{\mathcal{R}}$ if $\mathcal{R}$ is clear from context. Let s and t be terms. If exactly n steps are performed to rewrite s to t, we write $s \to^n t$. With $\mathsf{NF}(\mathcal{R})$ we denote the set of all normal forms of a term rewrite system $\mathcal{R}$. The *innermost rewrite relation* $\xrightarrow{i}_{\mathcal{R}}$ of a TRS $\mathcal{R}$ is defined on terms as follows: $s \xrightarrow{i}_{\mathcal{R}} t$ if there exist a rewrite rule $l \to r \in \mathcal{R}$, a context C, and a substitution σ such that $s = C[l\sigma]$, $t = C[r\sigma]$, and all proper subterms of $l\sigma$ are normal forms of $\mathcal{R}$. *Defined symbols* of $\mathcal{R}$ are symbols appearing at root in left-hand sides of $\mathcal{R}$. The set of defined function symbols is denoted as $\mathcal{D}$, while the *constructor symbols* $\mathcal{F} \setminus \mathcal{D}$ are collected in $\mathcal{C}$. We call a term $t = f(t_1,\ldots,t_n)$ *basic* or *constructor based* if $f \in \mathcal{D}$ and $t_i \in \mathcal{T}(\mathcal{C},\mathcal{V})$ for all $1 \leqslant i \leqslant n$. The set of all basic terms are denoted by $\mathcal{T}_{\mathsf{b}}$. We call a TRS *(innermost) terminating* if no infinite (innermost) rewrite sequence exists.

A replacement map μ is a function with $\mu(f) \subseteq \{1,\ldots,n\}$ for all n-ary functions with $n \geqslant 1$ [16]. The set $\mathcal{P}\mathsf{os}_\mu(t)$ of *μ-replacing positions* in t is defined as follows:

$$\mathcal{P}\mathsf{os}_\mu(t) := \begin{cases} \{\epsilon\} & \text{if } t \text{ is a variable },\\ \{\epsilon\} \cup \{ip \mid i \in \mu(f) \text{ and } p \in \mathcal{P}\mathsf{os}_\mu(t_i)\} & \text{if } t = f(t_1,\ldots,t_n)\,. \end{cases}$$

A *μ-step* $s \xrightarrow{\mu} t$ is a rewrite step $s \to t$ whose rewrite position is in $\mathcal{P}\mathsf{os}_\mu(s)$. The set of all non-$\mu$-replacing positions in t is denoted by $\overline{\mathcal{P}\mathsf{os}_\mu}(t)$; namely, $\overline{\mathcal{P}\mathsf{os}_\mu}(t) := \mathcal{P}\mathsf{os}(t) \setminus \mathcal{P}\mathsf{os}_\mu(t)$.

A *monotone $\mathcal{F}$-algebra* is a pair $(\mathcal{A},\succ)$ where $\mathcal{A}$ is an $\mathcal{F}$-algebra and $\succ$ is a proper order such that for every function symbol $f \in \mathcal{F}$, $f_{\mathcal{A}}$ is strictly monotone in all coordinates with respect to $\succ$. A (monotone) $\mathcal{F}$-algebra $(\mathcal{A},\succ)$ is called *well-founded* if $\succ$ is well-founded. Any monotone $\mathcal{F}$-algebra $(\mathcal{A},R)$ induces a binary relation $R_{\mathcal{A}}$ on terms: define $s\ R_{\mathcal{A}}\ t$ if $[\alpha]_{\mathcal{A}}(s)\ R\ [\alpha]_{\mathcal{A}}(t)$ for all assignments α. We say $\mathcal{A}$ is *compatible* with a TRS $\mathcal{R}$ if $\mathcal{R} \subseteq R_{\mathcal{A}}$. Let μ denote a replacement map. Then we call a well-founded algebra $(\mathcal{A}, \succ)$ *μ-monotone* if for every function symbol $f \in \mathcal{F}$, $f_{\mathcal{A}}$ is strictly monotone *on* $\mu(f)$, i.e. $f_{\mathcal{A}}$ is strictly monotone with respect to every argument position in $\mu(f)$. Similarly a relation R is called μ-monotone if it is strictly monotone on $\mu(f)$ for all $f \in \mathcal{F}$. Let $\mathcal{R}$ be a TRS compatible with a μ-monotone relation R. Then clearly any μ-step $s \xrightarrow{\mu} t$ implies $s\,R\,t$.

We recall the concept of *matrix interpretations* on natural numbers (see [17] but compare also [18]). Let $\mathcal{F}$ denote a signature. We fix a dimension $d \in \mathbb{N}$ and use the set $\mathbb{N}^d$ as the carrier of an algebra $\mathcal{A}$, together with the following extension of the natural order $>$ on $\mathbb{N}$: $(x_1, x_2, \ldots, x_d) > (y_1, y_2, \ldots, y_d) :\Longleftrightarrow x_1 > y_1 \wedge x_2 \geqslant y_2 \wedge \ldots \wedge x_d \geqslant y_d$. Let μ be a replacement map. For each n-ary function symbol f, we choose as an interpretation a linear function of the following shape:

$$f_{\mathcal{A}} \colon (\boldsymbol{v}_1, \ldots, \boldsymbol{v}_n) \mapsto F_1\boldsymbol{v}_1 + \cdots + F_n\boldsymbol{v}_n + \boldsymbol{f}\ ,$$

where $\boldsymbol{v}_1, \ldots, \boldsymbol{v}_n$ are (column) vectors of variables, $F_1, \ldots, F_n$ are matrices (each of size $d \times d$), and $\boldsymbol{f}$ is a vector over $\mathbb{N}$. Moreover, suppose for any $i \in \mu(f)$ the top left entry $(F_i)_{1,1}$ is positive. Then it is easy to see that the algebra $\mathcal{A}$ forms a μ-monotone well-founded algebra Let $\mathcal{A}$ be a matrix interpretation, let α_0 denote the assignment mapping any variable to $\mathbf{0}$, i.e. $\alpha_0(x) = \mathbf{0}$ for all $x \in \mathcal{V}$, and let t be a term. In the following we write $[t]$, $[t]_j$ as an abbreviation for $[\alpha_0]_{\mathcal{A}}(t)$, or $([\alpha_0]_{\mathcal{A}}(t))_j$ $(1 \leqslant j \leqslant d)$, respectively, if the algebra $\mathcal{A}$ is clear from the context.

The *derivation height* of a term s with respect to a well-founded, finitely branching relation $\to$ is defined as: $\mathsf{dh}(s, \to) = \max\{n \mid \exists t\ s \to^n t\}$.

Definition 2. *We define the* runtime complexity function $\mathsf{rc}_{\mathcal{R}}(n)$ *and the* innermost runtime complexity function $\mathsf{rc}^{\mathsf{i}}_{\mathcal{R}}(n)$ *as follows:*

$$\mathsf{rc}_{\mathcal{R}}(n) := \max\{\mathsf{dh}(t, \to_{\mathcal{R}}) \mid t \textit{ is basic and } |t| \leqslant n\}$$
$$\mathsf{rc}^{\mathsf{i}}_{\mathcal{R}}(n) := \max\{\mathsf{dh}(t, \xrightarrow{\mathsf{i}}_{\mathcal{R}}) \mid t \textit{ is basic and } |t| \leqslant n\}\ .$$

We may say the (innermost) runtime complexity of $\mathcal{R}$ is *linear*, *quadratic*, or *polynomial* if there exists a (linear, quadratic) polynomial $p(n)$ such that $\mathsf{rc}^{(\mathsf{i})}_{\mathcal{R}}(n) \leqslant p(n)$ for sufficiently large n.

Note that $\mathsf{dh}(t, \succ)$ is undefined, if the relation $\succ$ is not well-founded or not finitely branching. In fact compatibility of a constructor TRS with the polynomial path order $>_{\mathsf{pop*}}$ ([3]) induces polynomial innermost runtime complexity, whereas $\mathsf{f}(x) >_{\mathsf{pop*}} \mathsf{g}^n(x) >_{\mathsf{pop*}} \cdots >_{\mathsf{pop*}} \cdots >_{\mathsf{pop*}} \mathsf{g}^2(x) >_{\mathsf{pop*}} \mathsf{g}(x) >_{\mathsf{pop*}} x$ holds for all $n \in \mathbb{N}$, when precedence $\mathsf{f} > \mathsf{g}$ is used. Hence $\mathsf{dh}(t, >_{\mathsf{pop*}})$ is undefined, while the order $>_{\mathsf{pop*}}$ can still be employed in complexity analysis. Let R be a binary relation over terms, let $\succ$ be a proper order on terms, and let G denote a mapping associating a term with a natural number. Then $\succ$ is G-*collapsible on R* if $\mathsf{G}(s) > \mathsf{G}(t)$, whenever $s\,R\,t$ and $s \succ t$ holds. An order $\succ$ is *collapsible (on R)*, if there is a mapping G such that $\succ$ is G-collapsible (on R).

Lemma 3. *Let R be a finitely branching and well-founded relation. Further, let $\succ$ be a* G*-collapsible order with $R \subseteq \succ$. Then* $\mathsf{dh}(t, R) \leqslant \mathsf{G}(t)$ *holds for all terms t.*

If a TRS $\mathcal{R}$ and a μ-monotone matrix interpretation $\mathcal{A}$ are compatible, $\mathsf{G}(t)$ can be given by $[t]_1$. In order to estimate derivational or runtime complexity, one needs to associate $[t]_1$ to $|t|$. For this sake we define degrees of matrix interpretations.

Definition 4. *A matrix interpretation is of* (basic) degree k *if there is a constant* c *such that* $[t]_i \leqslant c \cdot |t|^k$ *for all (basic) terms* t *and* i*, respectively.*

An *upper triangular complexity matrix* is a matrix M in $\mathbb{N}^{d \times d}$ such that we have $M_{j,k} = 0$ for all $1 \leqslant k < j \leqslant d$, and $M_{j,j} \leqslant 1$ for all $1 \leqslant j \leqslant d$. We say that a ($\mu$-)monotone well-founded algebra $\mathcal{A}$ is a *triangular matrix interpretation* (*TMI* for short) if $\mathcal{A}$ is a matrix interpretation (over $\mathbb{N}$) and all matrices employed are of upper triangular complexity form. The following result can be easily distilled from the literature, cf. [19,20].

Theorem 5. *Let* $\mathcal{A}$ *be a TMI and let* M *denote the component-wise maximum of all matrices occurring in* $\mathcal{A}$*. Further, let* k *denote the number of ones occurring along the diagonal of* M*. Then,* $\succ_{\mathcal{A}}$ *is* $\mathsf{O}(n^k)$*-collapsible.*

In order to cope with runtime complexity, a similar idea to restricted polynomial interpretations (see [1]) can be integrated to triangular matrix interpretations. We call $\mathcal{A}$ a *restricted matrix interpretation* (*RMI* for short) if $\mathcal{A}$ is a matrix interpretation, but for each constructor symbol $f \in \mathcal{C}$, the interpretation $f_{\mathcal{A}}$ of f employs upper triangular complexity matrices, only. The following theorem is obtained from the combination of existing results [1,2,19].

Theorem 6. *Let* $\mathcal{A}$ *be an RMI and let* t *be a basic term. Further, let* M *denote the component-wise maximum of all matrices used for the interpretation of constructor symbols, and let* k *denote the number of ones occurring along the diagonal of* M*. Then* $\mathcal{A}$ *is of basic degree* k*. Furthermore, if* M *is the unit matrix then* $\mathcal{A}$ *is of basic degree* 1.

It is not difficult to see that Theorem 6 also holds for RMIs based on *lower* triangular complexity matrices. We refrain from given the formal details, but rather exemplify the definition below.

Example 7. Consider the TRS $\mathcal{R}_{sum}$[3]

1:	$\mathsf{sum}(\mathsf{0}) \to \mathsf{0}$	3:	$\mathsf{sum1}(\mathsf{0}) \to \mathsf{0}$
2:	$\mathsf{sum}(\mathsf{s}(x)) \to \mathsf{sum}(x) + \mathsf{s}(x)$	4:	$\mathsf{sum1}(\mathsf{s}(x)) \to \mathsf{s}(\mathsf{sum1}(x) + (x + x))$

where, sum and sum1 are defined symbols, and 0, s, and + are constructor symbols. Consider the 2-dimensional RMI $\mathcal{A}$ (based on lower triangular complexity matrices) with

[3] The TRS is Example 2.17 in Steinbach and Kühler's collection of TRSs [21].

$$
\begin{aligned}
\mathsf{0}_{\mathcal{A}} &= \begin{pmatrix}0\\1\end{pmatrix} & \mathsf{sum}_{\mathcal{A}}(\boldsymbol{x}) &= \begin{pmatrix}1\,2\\1\,3\end{pmatrix}\boldsymbol{x} + \begin{pmatrix}0\\1\end{pmatrix} \\
\mathsf{s}_{\mathcal{A}}(\boldsymbol{x}) &= \begin{pmatrix}1\,0\\1\,1\end{pmatrix}\boldsymbol{x} + \begin{pmatrix}2\\2\end{pmatrix} & \mathsf{sum1}_{\mathcal{A}}(\boldsymbol{x}) &= \begin{pmatrix}1\,2\\0\,3\end{pmatrix}\boldsymbol{x} \\
+_{\mathcal{A}}(\boldsymbol{x},\boldsymbol{y}) &= \begin{pmatrix}1\,0\\0\,0\end{pmatrix}\boldsymbol{x} + \begin{pmatrix}1\,0\\0\,0\end{pmatrix}\boldsymbol{y}\,.
\end{aligned}
$$

The rules in $\mathcal{R}_{sum}$ are interpreted and ordered as follows.

$$
\begin{aligned}
&1\colon \begin{pmatrix}2\\4\end{pmatrix} > \begin{pmatrix}0\\1\end{pmatrix} && 3\colon \begin{pmatrix}2\\3\end{pmatrix} > \begin{pmatrix}0\\1\end{pmatrix} \\
&2\colon \begin{pmatrix}3\,2\\4\,3\end{pmatrix}\boldsymbol{x} + \begin{pmatrix}6\\9\end{pmatrix} > \begin{pmatrix}2\,2\\0\,0\end{pmatrix}\boldsymbol{x} + \begin{pmatrix}2\\0\end{pmatrix} && 4\colon \begin{pmatrix}3\,2\\3\,3\end{pmatrix}\boldsymbol{x} + \begin{pmatrix}6\\6\end{pmatrix} > \begin{pmatrix}3\,2\\3\,2\end{pmatrix}\boldsymbol{x} + \begin{pmatrix}2\\2\end{pmatrix}\,.
\end{aligned}
$$

Therefore, $\mathcal{R}_{sum} \subseteq >_{\mathcal{A}}$ holds. By an application of Theorem 6 we conclude that the runtime complexity is *quadratic*. As we see later, there is a tighter bound.

3 Usable Replacement Maps

Unfortunately, there is no RMI compatible with the TRS of our running example (Example 1). The reason is that the monotonicity requirement of matrix interpretations is too severe for complexity analysis. Inspired by the idea of Fernández [22], we show how context-sensitive rewriting is used in complexity analysis. Here we briefly explain our idea. Let $\mathbf{n}$ denote the numeral $s^n(\mathsf{0})$. Consider the derivation from $\mathbf{4} \div \mathbf{2}$:

$$\underline{\mathbf{4} \div \mathbf{2}} \to \mathsf{s}((\underline{\mathbf{3} - \mathbf{1}}) \div \mathbf{2}) \to \mathsf{s}((\underline{\mathbf{2} - \mathsf{0}}) \div \mathbf{2}) \to \mathsf{s}(\underline{\mathbf{2} \div \mathbf{2}}) \to \cdots$$

where redexes are underlined. Observe that e.g. any second argument of $\div$ is never rewritten. More precisely, any derivation from a basic term consists of only μ-steps with the replacement map μ: $\mu(\mathsf{s}) = \mu(\div) = \{1\}$ and $\mu(-) = \varnothing$.

Recall that $\mathcal{P}\mathsf{os}_\mu(t)$ denotes the set of μ-replacing positions in t and $\overline{\mathcal{P}\mathsf{os}_\mu}(t) = \mathcal{P}\mathsf{os}(t) \setminus \mathcal{P}\mathsf{os}_\mu(t)$. Further, a term t is a *μ-replacing term* with respect to a TRS $\mathcal{R}$ if $p \in \overline{\mathcal{P}\mathsf{os}_\mu}(t)$ implies $t|_p \in \mathsf{NF}(\mathcal{R})$. The set of all μ-replacing terms is denoted by $\mathcal{T}(\mu)$. Below $[L](\to^*)$ denotes the set $\{t \mid s \to^* t \text{ for some } s \in L\}$.

The above observation is cast in the following definition, borrowed from [23]. Usable replacement maps satisfy a desired property for runtime complexity analysis, as detailed in this section.

Definition 8. *Let $\to$ denote a binary relation. A replacement map μ is called a* usable replacement map *with respect to $\to$ and the set of starting terms $\mathcal{T}_\mathsf{b}$ if $[\mathcal{T}_\mathsf{b}](\to^*) \subseteq \mathcal{T}(\mu)$.*

The main result of this section is the definition of suitable *approximations* of usable replacement maps. For that we adapt the cap-function ICAP suitably, cf. [24]. Let μ be a replacement map. Clearly the function μ is representable as set of ordered pairs (f, i). Below we often confuse the notation of μ as a function or as a set.

Definition 9. *Let $\mathcal{R}$ be a TRS and let μ be a replacement map. We define the operator $\Upsilon^{\mathcal{R}}$ as follows:*

$$\Upsilon^{\mathcal{R}}(\mu) := \{(f,i) \mid l \to C[f(r_1,\ldots,r_n)] \in \mathcal{R} \text{ and } \mathsf{CAP}^l_\mu(r_i) \neq r_i\} .$$

Here $\mathsf{CAP}^s_\mu(t)$ is inductively defined on t as follows:

$$\mathsf{CAP}^s_\mu(t) = \begin{cases} t & \text{if } t = s|_p \text{ for some } p \in \overline{\mathcal{P}\mathsf{os}_\mu}(s) , \\ u & \text{if } t = f(t_1,\ldots,t_n) \text{ and } u \text{ and } l \text{ unify for no } l \to r \in \mathcal{R} , \\ y & \text{otherwise} , \end{cases}$$

where, $u = f(\mathsf{CAP}^s_\mu(t_1),\ldots,\mathsf{CAP}^s_\mu(t_n))$, y is a fresh variable, and we assume that $\mathcal{V}\mathsf{ar}(l) \cap \mathcal{V}\mathsf{ar}(u) = \varnothing$ holds.

We define the *approximated* innermost usable replacement map $\mu_{\mathsf{i}}^{\mathcal{R}}$ as follows $\mu_{\mathsf{i}}^{\mathcal{R}} := \Upsilon^{\mathcal{R}}(\varnothing)$ and let the *approximated* usable replacement map $\mu_{\mathsf{f}}^{\mathcal{R}}$ denote the least fixed point of $\Upsilon^{\mathcal{R}}$. The existence of $\Upsilon^{\mathcal{R}}$ follows from the monotonicity of $\Upsilon^{\mathcal{R}}$. If $\mathcal{R}$ is clear from context, we simple write μ_{i}, μ_{f}, and Υ, respectively. In the remainder of the section we establish that μ_{i} and μ_{f} constitute usable replacement maps for $\xrightarrow{\mathsf{i}}$ and $\to$ respectively. Suppose $s \in \mathcal{T}(\mu)$: observe that the function $\mathsf{CAP}^s_\mu(t)$ replaces a subterm u of t by a fresh variable if $u\sigma$ is a redex for some $s\sigma \in \mathcal{T}(\mu)$. This is exemplified below.

Example 10. Consider the TRS $\mathcal{R}_{div}$. Let $l \to r$ be rule 4, namely, $l = \mathsf{s}(x) \div \mathsf{s}(y)$ and $r = \mathsf{s}((x-y) \div \mathsf{s}(y))$. Suppose $\mu(f) = \varnothing$ for all functions f and let w and z be fresh variables. The next table summarises $\mathsf{CAP}^l_\mu(t)$ for each proper subterm t in r. To see the computation process, we also indicate the term u in Definition 9.

t	x	y	$x-y$	$\mathsf{s}(y)$	$(x-y) \div \mathsf{s}(y)$
$\mathsf{CAP}^l_\mu(t)$	x	y	w	$\mathsf{s}(y)$	z
u	—	—	$x-y$	$\mathsf{s}(y)$	$w \div \mathsf{s}(y)$

By underlining proper subterms t in r such that $\mathsf{CAP}^l_\mu(t) \neq t$, we have

$$\mathsf{s}(\underline{\underline{(x-y)} \div \mathsf{s}(y)})$$

which indicates $(\mathsf{s},1), (\div,1) \in \Upsilon(\mu)$.

The next lemma clarifies the rôle played by the cap function $\mathsf{CAP}^s_\mu(t)$.

Lemma 11. *Let s and t be terms, and σ a substitution such that $s\sigma \in \mathcal{T}(\mu)$ and $\mathsf{CAP}^s_\mu(t) = t$. Then $t\sigma \in \mathsf{NF}(\mathcal{R})$.*

Proof. We use induction on t. Suppose $s\sigma \in \mathcal{T}(\mu)$ and $\mathsf{CAP}^s_\mu(t) = t$. If $t = s|_p$ for some $p \in \overline{\mathcal{P}\mathsf{os}_\mu}(s)$ then $t\sigma = (s\sigma)|_p \in \mathsf{NF}$ follows by definition of $\mathcal{T}(\mu)$.

We can assume that $t = f(t_1,\ldots,t_n)$. Assume otherwise that $t = x \in \mathcal{V}$, then $\mathsf{CAP}^s_\mu(x) = x$ entails that $x\sigma$ occurs at a non-μ-replacing position in $s\sigma$. Hence $x\sigma \in \mathsf{NF}$ follows from $s\sigma \in \mathcal{T}(\mu)$. Moreover, by assumption we have:

1. $\mathsf{CAP}^s_\mu(t_i) = t_i$ for each i, and
2. there is no rule $l \to r \in \mathcal{R}$ such that t and l unify.

Due to 2) $l\sigma$ is not reducible at the root, and the induction hypothesis yields $t_i\sigma \in \mathsf{NF}$ because of 1). Therefore, we obtain $t\sigma \in \mathsf{NF}$. □

For a smooth inductive proof of the key lemma, Lemma 14, we develop an alternative characterisation of the set of μ-replacing terms $\mathcal{T}(\mu)$.

Definition 12. *The set* $\{(f, i) \mid f(t_1, \ldots, t_n) \trianglelefteq t$ *and* $t_i \notin \mathsf{NF}(\mathcal{R})\}$ *is denoted by* $\upsilon(t)$.

The next lemma shows that the set of μ-replacing terms $\mathcal{T}(\mu)$ can be characterised through the above definition.

Lemma 13. $\mathcal{T}(\mu) = \{t \mid \upsilon(t) \subseteq \mu\}$.

Proof. For the inclusion from left to right, let $t \in \mathcal{T}(\mu)$ and let $(f, i) \in \upsilon(t)$. We show $(f, i) \in \mu$. By Definition 12 there is a position $p \in \mathcal{P}\mathsf{os}(t)$ with $t|_p = f(t_1, \ldots, t_n)$ and $t|_{pi} \notin \mathsf{NF}$. Thus $pi \in \mathcal{P}\mathsf{os}_\mu(t)$ and $i \in \mathcal{P}\mathsf{os}_\mu(t|_p)$. Hence $(f, i) \in \mu$ is concluded.

Next we consider the opposite direction $\{t \mid \upsilon(t) \subseteq \mu\} \subseteq \mathcal{T}(\mu)$. Let t be a minimal counter-example such that $\upsilon(t) \subseteq \mu$ and $t \notin \mathcal{T}(\mu)$. One can write $t = f(t_1, \ldots, t_n)$. Then, there exists a position $p \in \overline{\mathcal{P}\mathsf{os}_\mu}(t)$ such that $t|_p \notin \mathsf{NF}$. Because $\epsilon \notin \overline{\mathcal{P}\mathsf{os}_\mu}(t)$ by definition, $p = iq$ with $i \in \mathbb{N}$. As $iq \in \overline{\mathcal{P}\mathsf{os}_\mu}(t)$ one of $(f, i) \notin \mu$ or $q \in \overline{\mathcal{P}\mathsf{os}_\mu}(t|_i)$ must hold. Consider the first alternative. Then by Definition 12, $(f, i) \in \upsilon(t) \subseteq \mu$ and we obtain a contradiction. Now, consider the second alternative. Note that $t|_{iq} \notin \mathsf{NF}$ implies $t|_i \notin \mathsf{NF}$. In conjunction with $q \in \overline{\mathcal{P}\mathsf{os}_\mu}(t|_i)$ this yields that t_i is counter-example which is smaller than t. This contradicts the definition of t. □

The next lemma about the operator Υ is a key for the main theorem. Note that every subterm of a μ-replacing term is a μ-replacing term.

Lemma 14. *If* $l \to r \in \mathcal{R}$ *and* $l\sigma \in \mathcal{T}(\mu)$ *then* $r\sigma \in \mathcal{T}(\mu \cup \Upsilon(\mu))$.

Proof. Let $l \to r \in \mathcal{R}$ and suppose $l\sigma \in \mathcal{T}(\mu)$. By Lemma 13 we have

$$\mathcal{T}(\mu) = \{t \mid \upsilon(t) \subseteq \mu\} \qquad \mathcal{T}(\mu \cup \Upsilon(\mu)) = \{t \mid \upsilon(t) \subseteq \mu \cup \Upsilon(\mu)\}\ .$$

Hence it is sufficient to show $\upsilon(r\sigma) \subseteq \mu \cup \Upsilon(\mu)$. Let $(f, i) \in \upsilon(r\sigma)$. There is $p \in \mathcal{P}\mathsf{os}(r\sigma)$ with $r\sigma|_p = f(t_1, \ldots, t_n)$ and $t_i \notin \mathsf{NF}$. If p is below some variable position of r, $r\sigma|_p$ is a subterm of $l\sigma$, and thus $\upsilon(r\sigma|_p) \subseteq \upsilon(l\sigma) \subseteq \mu$. Otherwise, p is a non-variable position of r. We may write $r|_p = f(r_1, \ldots, r_n)$ and $r_i\sigma = t_i \notin \mathsf{NF}$. Due to Lemma 11 we obtain $\mathsf{CAP}^l_\mu(r_i) \neq r_i$. Therefore, $(f, i) \in \Upsilon(\mu)$. □

Lemma 15. *For the approximated usable replacement maps* $\mathcal{T}(\mu_\mathsf{i})$ *and* $\mathcal{T}(\mu_\mathsf{f})$, *the following implications hold:*

1. *If* $s \in \mathcal{T}(\mu_\mathsf{i})$ *and* $s \xrightarrow{\mathsf{i}} t$ *then* $t \in \mathcal{T}(\mu_\mathsf{i})$.
2. *If* $s \in \mathcal{T}(\mu_\mathsf{f})$ *and* $s \to t$ *then* $t \in \mathcal{T}(\mu_\mathsf{f})$.

Proof. We show property 1). Suppose $s \in \mathcal{T}(\mu_\mathsf{i})$ and $s \xrightarrow{\mathsf{i}} t$ is a rewrite step at p. Due to the definition of innermost rewriting, we have $s|_p \in \mathcal{T}(\varnothing)$. Hence, $t|_p \in \mathcal{T}(\mu_\mathsf{i})$ is obtained by Lemma 14. Because $s \in \mathcal{T}(\mu_\mathsf{i})$ we have $p \in \mathcal{P}\mathsf{os}_{\mu_\mathsf{i}}(s)$. Hence due to $t|_p \in \mathcal{T}(\mu_\mathsf{i})$ we conclude $t = s[t|_p]_p \in \mathcal{T}(\mu_\mathsf{i})$ due to the above remark. The proof of 2) proceeds along the same pattern and is left to the reader. □

We arrive at the main result of this section.

Theorem 16. *The inclusions* $[\mathcal{T}(\varnothing)](\xrightarrow{\mathsf{i}}{}^*_\mathcal{R}) \subseteq \mathcal{T}(\mu_\mathsf{i})$ *and* $[\mathcal{T}(\varnothing)](\to^*_\mathcal{R}) \subseteq \mathcal{T}(\mu_\mathsf{f})$ *hold. In particular* μ_i *and* μ_f *constitute usable replacement maps for* $\xrightarrow{\mathsf{i}}$ *and* $\to$, *respectively.*

Proof. We focus on the second part of the theorem, where we have to prove that $t \in \mathcal{T}(\mu_\mathsf{f})$, whenever there exists $s \in \mathcal{T}(\varnothing)$ such that $s \to^*_\mathcal{R} t$. As $\mathcal{T}(\varnothing) \subseteq \mathcal{T}(\mu_\mathsf{f})$ this follows directly from Lemma 15.

Note that $\mathcal{T}(\varnothing)$ is the set of all argument normalised terms. Therefore, $\mathcal{T}_\mathsf{b} \subseteq \mathcal{T}(\varnothing)$. Hence the second half of the theorem follows. □

Given a TRS $\mathcal{R}$ we write $\xrightarrow{\mu_\mathsf{i}}$ for the μ_i-step relation of $\mathcal{R}$, and $\xrightarrow{\mu_\mathsf{f}}$ for the μ_f-step relation. The following corollary to Theorem 16 is immediate.

Corollary 17. *We have* $\mathsf{dh}(t, \xrightarrow{\mathsf{i}}_\mathcal{R}) \leqslant \mathsf{dh}(t, \xrightarrow{\mu_\mathsf{i}})$ *and* $\mathsf{dh}(t, \to_\mathcal{R}) = \mathsf{dh}(t, \xrightarrow{\mu_\mathsf{f}})$ *for all terminating terms* $t \in \mathcal{T}_\mathsf{b}$.

An advantage of the use of context-sensitive rewriting is that the compatibility requirement of monotone algebra in termination or complexity analysis is relaxed to μ-monotone algebra. We illustrate its use in the next examples.

Example 18. Recall the TRS $\mathcal{R}_{div}$ given in Example 1 above. The usable replacement maps are as follows:

$$\mu_\mathsf{i}(-) = \varnothing \quad \mu_\mathsf{i}(\mathsf{s}) = \mu_\mathsf{i}(\div) = \{1\} \qquad \mu_\mathsf{f}(\mathsf{s}) = \mu_\mathsf{f}(-) = \mu_\mathsf{f}(\div) = \{1\}\ .$$

Consider the 1-dimensional RMI $\mathcal{A}$ (i.e. linear polynomial interpretations) with $0_\mathcal{A} = 1$, $\mathsf{s}_\mathcal{A}(x) = x + 2$, $-_\mathcal{A}(x, y) = x + 1$, and $\div_\mathcal{A}(x, y) = 3x$; $\mathcal{A}$ is strictly μ_i-monotone and μ_f-monotone. The rules in $\mathcal{R}_{div}$ are interpreted and ordered as follows.

$$\begin{aligned} 1&: \quad x + 1 > x & 3&: \quad 3 > 1 \\ 2&: \quad x + 3 > x + 1 & 4&: \quad 3x + 6 > 3x + 5\ . \end{aligned}$$

Therefore, $\mathcal{R}_{div} \subseteq >_\mathcal{A}$ holds. Applying Theorem 6 in the context of usable replacement maps, we conclude that the (innermost) runtime complexity is *linear*, which is optimal.

Example 19. Recall the TRS $\mathcal{R}_{sum}$ of Example 7. The usable replacement map for full rewriting is as follows:

$$\mu_{\mathsf{f}}(\mathsf{sum}) = \mu_{\mathsf{f}}(\mathsf{sum1}) = \varnothing \qquad \mu_{\mathsf{f}}(\mathsf{s}) = \mu_{\mathsf{f}}(+) = \{1\} .$$

Consider the 1-dimensional RMI $\mathcal{A}$ with

$$0_{\mathcal{A}} = 2 \quad \mathsf{s}_{\mathcal{A}}(x) = x + 2 \quad +_{\mathcal{A}}(x,y) = x \quad \mathsf{sum}_{\mathcal{A}}(x) = 2x \quad \mathsf{sum1}(x,y) = 2x .$$

which is strictly μ_{f}-monotone. The rules in $\mathcal{R}_{div}$ are interpreted and ordered as follows.

$$\begin{array}{llll} 1: & 4 > 2 & 3: & 4 > 2 \\ 2: & 2x + 4 > 2x & 4: & 2x + 4 > 2x + 2 . \end{array}$$

Therefore, $\mathcal{R}_{div} \subseteq >_{\mathcal{A}}$ holds. By an application of Theorem 6 we conclude that the runtime complexity is *linear*, which is optimal.

We cast the observations in the example into another corollary to Theorem 16.

Corollary 20. *Let $\mathcal{R}$ be a TRS and let $\mathcal{A}$ be a d-degree μ_{i}-monotone (or μ_{f}-monotone) RMI compatible with $\mathcal{R}$. Then the (innermost) runtime complexity function $\mathsf{rc}^{(\mathsf{i})}_{\mathcal{R}}$ with respect to $\mathcal{R}$ is bounded by a d-degree polynomial.*

Proof. It suffices to consider the case for full rewriting. Let s, t be terms such that $s \rightarrow_{\mathcal{R}} t$. By the theorem, we have $s \xrightarrow{\mu_{\mathsf{f}}} t$. Furthermore, by assumption $\mathcal{R} \subseteq \succ_{\mathcal{A}}$ and for any $f \in \mathcal{F}$, $f_{\mathcal{A}}$ is strictly monotone on all $\mu_{\mathsf{f}}(f)$. Thus $s \succ_{\mathcal{A}} t$ follows. Finally, the corollary follows by application of Theorem 6. □

4 Experiments

The usable replacement map method has been incorporated into the *Tyrolean Complexity Tool* $\mathsf{T_CT}$ [25]. We note that the established method can easily combined with existing modular frameworks and the implementation in $\mathsf{T_CT}$ makes (essential) use of this. In this section we present an experimental evaluation of the technique based on version 8.0.6 of the *Termination Problems Database* (*TPDB* for short). We consider TRSs without theory annotation, where the runtime complexity analysis is non-trivial, that is the set of basic terms is infinite. This testbed comprises 1249 TRSs.

All experiments were conducted on a machine that is identical to the official competition server (8 AMD Opteron® 885 dual-core processors with 2.8GHz, 8x8 GB memory). As timeout we use 60 seconds. The complete experimental data can be found at `http://cl-informatik.uibk.ac.at/software/tct/experiments/rtatlca14`, where also the testbed employed is detailed.

Table 1 summarises the experimental results of the use of usable replacement maps for full and innermost runtime complexity analysis. The tests employ

Table 1. Experimental results I (one- to three-dimensional RMIs)

	full		*innermost*	
result	RMI(1-3),(−)	RMI(1-3), (+)	RMI(1-3), (−)	RMI(1-3), (+)
$O(n)$	103	134	104	140
$O(n^2)$	174	209	174	226
$O(n^k)$	183	225	183	247
timeout (60s)	113	109	113	117

Table 2. Experimental results II (overall effect)

	full		*innermost*	
result	T_CT (−)	T_CT (+)	T_CT (−)	T_CT (+)
$O(1)$	92	100	326	325
$O(n)$	408	421	500	508
$O(n^2)$	423	428	554	555
$O(n^3)$	424	429	564	565
$O(n^k)$	426	431	568	569
timeout (60s)	714	711	617	615

one- to three-dimensional RMIs.[4] The tests clearly indicate the power of the established technique. This power is not only in the absolute number of examples, but more importantly in the precision of the analysis. We note that our tests only make use of the simplest notion of RMIs for runtime complexity analysis cf. [26]. This is a rather mundane method, more sophisticated methods have been reported in [2,19]. However, to assess the power of the established technique this restriction is insignificant.

In Table 2 we present the overall power obtained for the automated runtime complexity analysis. Here we test the current version of T_CT using a strategy that avoids the new method, in contrast to its standard strategy. It is to be expected that the effect of the proposed technique is smaller than in Table 1. This is due to the presence of transformation techniques, like the *weak dependency pair method* [4] or the *dependency tuple method* [5] and the use of a modular framework [7]. While the usable argument method is still effective for weak dependency pairs, as it may lighten the *weight gap constraint*, the dependency tuple method allows to remove all monotonicity constraints. Note that the dependency tuple method is only applicable for innermost runtime complexity.

[4] Note that matrix interpretations, that is the test "RMI(1-3)", cannot discern between innermost versus full rewriting. Hence the practical differences noted in the table are coincidental.

Despite these theoretical facts, the method has a significant impact on full and innermost rewriting. While the overall effect is a lot smaller than for direct techniques, we emphasise that the method allows to win some examples that can be handled with linear rather than with quadratic complexity. The alert reader may wonder, why there is a positive effect at all in the innermost case, as transformation techniques do not require monotonicity. This is due to the case that even for innermost runtime complexity analysis, direct methods are currently not superseded by transformational techniques.

5 Related Work

Usable replacement maps Usable replacement maps were originally introduced by Fernández [22] for proving termination of *innermost rewriting*. This notion is already applicable for analysing innermost runtime complexity. More precisely, we link Theorem 16 to Fernández' work. In [22] an application of context-sensitive rewriting for innermost termination has been established.

Proposition 21 ([22]). *A TRS $\mathcal{R}$ is innermost terminating if $\xrightarrow{\mu_{\mathsf{i}}}$ is terminating.*

Proof. We show the contraposition. If $\mathcal{R}$ is not innermost terminating, there is an infinite sequence $t_0 \xrightarrow{\mathsf{i}} t_1 \xrightarrow{\mathsf{i}} t_2 \xrightarrow{\mathsf{i}} \cdots$, where $t_0 \in \mathcal{T}(\varnothing)$. From Theorem 16 and Lemma 15 we obtain $t_0 \xrightarrow{\mu_{\mathsf{i}}} t_1 \xrightarrow{\mu_{\mathsf{i}}} t_2 \xrightarrow{\mu_{\mathsf{i}}} \cdots$. Hence, $\xrightarrow{\mu_{\mathsf{i}}}$ is not terminating. □

We lifted Fernández' notion to full rewriting, exploiting the cap function ICAP in [24]. Realisation of a fixed point calculation of usable replacement maps is considered as a primary result of this paper. Note that Proposition 21 does not generalise to full termination, even if one replaces the innermost replacement map μ_{i}, by the replacement map μ_{f}.

Example 22. Consider the famous Toyama's example $\mathcal{R}$

$$\mathsf{f}(\mathsf{a},\mathsf{b},x) \to \mathsf{f}(x,x,x) \qquad \mathsf{g}(x,y) \to x \qquad \mathsf{g}(x,y) \to y\ .$$

The replacement map μ_{f} is empty. Thus, the algebra $\mathcal{A}$ over $\mathbb{N}$

$$\mathsf{f}_{\mathcal{A}}(x,y,z) = \max\{x-y,0\} \qquad \mathsf{g}_{\mathcal{A}}(x,y) = x+y+1 \qquad \mathsf{a}_{\mathcal{A}} = 1 \qquad \mathsf{b}_{\mathcal{A}} = 0\ .$$

is μ_{f}-monotone and we have $\mathcal{R} \subseteq {>_{\mathcal{A}}}$. However, we should not conclude termination of $\mathcal{R}$, because $\mathsf{f}(\mathsf{a},\mathsf{b},\mathsf{g}(\mathsf{a},\mathsf{b}))$ is non-terminating.

Cap functions In [22] usable replacement maps for innermost rewriting are defined as $\{(f,i) \mid l \to C[f(r_1,\ldots,r_n)] \in \mathcal{R}$ and $r_i|_p \ntriangleleft l$, for some $p \in \mathcal{P}\mathsf{os}_{\mathcal{D}}(r)\}$. This and our definition μ_{i} coincide if the following cap function is used during the computation of $\Upsilon^{\mathcal{R}}$:

$$\mathsf{CAP}^{s}_{\mu}(t) = \begin{cases} t & \text{if } t \lhd s\ , \\ u & \text{if } t = f(t_1,\ldots,t_n) \text{ and } f \in \mathcal{C}\ , \\ y & \text{otherwise}\ , \end{cases}$$

where, $u = f(\mathsf{CAP}^s_\mu(t_1), \ldots, \mathsf{CAP}^s_\mu(t_n))$ and y is a fresh variable. In the light of this reformulation one can (easily) verify that the usable replacement map μ_i in Section 3 is always a subset of the above set.

There exists a cap function for context-sensitive rewriting, introduced by Alarcón et al. [27] for termination analysis. Their definition is the following:

$$\mathsf{CAP}^s_\mu(t) = \begin{cases} t & \text{if } t \text{ is a variable} , \\ u & \text{if } t = f(t_1, \ldots, t_n) \text{ and } u \text{ and } l \text{ unify for no } l \to r \in \mathcal{R} , \\ y & \text{otherwise} , \end{cases}$$

where, y is a fresh variable and $u = f(u_1, \ldots, u_n)$. Each u_i stands for $\mathsf{CAP}^s_\mu(t_i)$ if $i \in \mu(f)$, and t_i otherwise. This definition cannot be used for calculation of usable replacement maps: It is designed for exploiting a given replacement map μ to *ignore* potential rewrite positions, while our cap function is aimed at *detecting* potential reducible positions to build a usable replacement map.

Dependency Pairs. Weak dependency pairs [4] and dependency tuples [5] are transformational approaches that split a rewrite relation into two relations. This split allows us to weaken the monotonicity condition. Although these approaches exploit dependencies of defined symbols, they do not analyse how variables in rewrite rules or dependency pairs (tuples) are instantiated in rewriting.[5] As a consequence, the transformations do not resolve the problem of variable duplication addressed in the introduction. We emphasise that usable replacement maps and dependency pairs (tuples) are complementary and the combination is beneficial, as seen in Section 4. A similar observation holds for techniques employing modularity.

6 Conclusion

In this paper we have defined the notion of usable replacement maps. It is a straightforward observation that only usable arguments need to be considered for monotonicity conditions. In a nutshell, we have shown how monotonicity conditions for orders can be weakened by using the notion of context-sensitive rewriting.

The presented technique is very easy to implement and has been integrated in the Tyrolean Complexity Tool. Above we have provided ample experimental data for assessing the viability of our method. The positive experimental evaluation, even in the innermost case, is somewhat surprising. One might have assumed that transformation techniques, as for example the dependency tuple method introduced in [5] supersede direct methods and thus refrain us from concerns about monotonicity. Our experiments clearly show that this is not the case. We emphasise that the here proposed method directly extends to modular frameworks and our implementation in $\mathsf{T_CT}$ makes essential use of this fact.

Apart from its practical value the proposed technique allows to incorporate features of complexity analysis of functional programs into rewrite systems. We consider a reformulation of our motivating example in an ML-like language,

[5] Approximation techniques for dependency graphs may be considered an exception.

```
minus : (nat, nat) -> nat
minus (m, n) = match m with
               | 0    -> 0
               | S m' -> match n with
                         | 0    -> m
                         | S n' -> minus (m', n');

quot : (nat, nat) -> nat
quot (m, n) = match m with
              | 0    -> 0
              | S m' -> match n with
                        | 0  -> 0
                        | S n' ->
                              (quot (minus (m', n'), n)) + 1;
```

Fig. 1. Division in RaML

cf. Figure 1. This functional program is subject to the analysis of the RaML-prototype, developed by Hoffmann et al. [28]. The prototype is based on an amortised resource analysis that employs a potential-based type system. Application of the method on the example yields the optimal linear bound on the innermost runtime complexity. Inspection of the complexity proof reveals that the method assigns zero potential to the second argument of minus and div, which is related to the fact that these arguments can be safely ignored in our setting, cf. Example 18. However, the potential-based method depends on the presence of types as detailed in [29]. We emphasise that the usable arguments method allows a similar fine-grained control for the runtime complexity analysis, even without the introduction of types.

Acknowledgments. We would like to thank the anonymous reviewers for their valuable comments that greatly helped in improving the presentation.

References

1. Bonfante, G., Cichon, A., Marion, J.Y., Touzet, H.: Algorithms with polynomial interpretation termination proof. JFP 11(1), 33–53 (2001)
2. Middeldorp, A., Moser, G., Neurauter, F., Waldmann, J., Zankl, H.: Joint spectral radius theory for automated complexity analysis of rewrite systems. In: Winkler, F. (ed.) CAI 2011. LNCS, vol. 6742, pp. 1–20. Springer, Heidelberg (2011)
3. Avanzini, M., Moser, G.: Polynomial path orders. LMCS 9(4) (2013)
4. Hirokawa, N., Moser, G.: Automated complexity analysis based on the dependency pair method. In: Armando, A., Baumgartner, P., Dowek, G. (eds.) IJCAR 2008. LNCS (LNAI), vol. 5195, pp. 364–379. Springer, Heidelberg (2008)
5. Noschinski, L., Emmes, F., Giesl, J.: Analyzing innermost runtime complexity of term rewriting by dependency pairs. JAR 51(1), 27–56 (2013)
6. Zankl, H., Korp, M.: Modular complexity analysis via relative complexity. LMCS 10(1:19), 1–33 (2014)

7. Avanzini, M., Moser, G.: A combination framework for complexity. In: Proc. 24th RTA. LIPIcs, vol. 21, pp. 55–70 (2013)
8. Moser, G.: Proof Theory at Work: Complexity Analysis of Term Rewrite Systems. CoRR abs/0907.5527 (2009) Habilitation Thesis.
9. Baillot, P., Marion, J.Y., Rocca, S.R.D.: Guest editorial: Special issue on implicit computational complexity. TOCL 10(4) (2009)
10. Avanzini, M., Moser, G.: Closing the gap between runtime complexity and polytime computability. In: Proc. 21st RTA. LIPIcs, vol. 6, pp. 33–48 (2010)
11. Arts, T., Giesl, J.: A collection of examples for termination of term rewriting using dependency pairs. Technical Report AIB-2001-09, RWTH Aachen (2001)
12. Hofbauer, D., Lautemann, C.: Termination proofs and the length of derivations. In: Dershowitz, N. (ed.) RTA 1989. LNCS, vol. 355, pp. 167–177. Springer, Heidelberg (1989)
13. Baader, F., Nipkow, T.: Term Rewriting and All That. Cambridge University Press (1998)
14. Choppy, C., Kaplan, S., Soria, M.: Complexity analysis of term-rewriting systems. TCS 67(2-3), 261–282 (1989)
15. TeReSe: Term Rewriting Systems. Cambridge Tracks in Theoretical Computer Science, vol. 55. Cambridge University Press (2003)
16. Lucas, S.: Context-sensitive rewriting strategies. IC 178(1), 294–343 (2002)
17. Endrullis, J., Waldmann, J., Zantema, H.: Matrix interpretations for proving termination of term rewriting. JAR 40(3), 195–220 (2008)
18. Hofbauer, D., Waldmann, J.: Termination of string rewriting with matrix interpretations. In: Pfenning, F. (ed.) RTA 2006. LNCS, vol. 4098, pp. 328–342. Springer, Heidelberg (2006)
19. Neurauter, F., Zankl, H., Middeldorp, A.: Revisiting matrix interpretations for polynomial derivational complexity of term rewriting. In: Fermüller, C.G., Voronkov, A. (eds.) LPAR-17. LNCS, vol. 6397, pp. 550–564. Springer, Heidelberg (2010)
20. Waldmann, J.: Polynomially bounded matrix interpretations. In: Proc. 21st RTA. LIPIcs, vol. 6, pp. 357–372 (2010)
21. Steinbach, J., Kühler, U.: Check your ordering – termination proofs and open problems. Technical Report SR-90-25, Universität Kaiserslautern (1990)
22. Fernández, M.L.: Relaxing monotonicity for innermost termination. Information Processing Letters 93(1), 117–123 (2005)
23. Avanzini, M.: Verifying Polytime Computability Automatically. PhD thesis, University of Innsbruck (2013)
24. Giesl, J., Thiemann, R., Schneider-Kamp, P.: Proving and disproving termination of higher-order functions. In: Gramlich, B. (ed.) FroCos 2005. LNCS (LNAI), vol. 3717, pp. 216–231. Springer, Heidelberg (2005)
25. Avanzini, M., Moser, G.: Tyrolean Complexity Tool: Features and usage. In: Proc. 24th RTA. LIPIcs, vol. 21, pp. 71–80 (2013)
26. Moser, G., Schnabl, A., Waldmann, J.: Complexity analysis of term rewriting based on matrix and context dependent interpretations. In: Proc. 28th FSTTCS. LIPIcs, vol. 2, pp. 304–315 (2008)
27. Alarcón, B., Gutiérrez, R., Lucas, S.: Context-sensitive dependency pairs. IC 208(8), 922–968 (2010)
28. Hoffmann, J., Aehlig, K., Hofmann, M.: Resource aware ML. In: Madhusudan, P., Seshia, S.A. (eds.) CAV 2012. LNCS, vol. 7358, pp. 781–786. Springer, Heidelberg (2012)
29. Hofmann, M., Moser, G.: Amortised resource analysis and typed polynomial interpretations. In: Dowek, G. (ed.) RTA-TLCA 2014. LNCS, vol. 8560, pp. 272–287. Springer, Heidelberg (2014)

Amortised Resource Analysis and Typed Polynomial Interpretations*

Martin Hofmann[1] and Georg Moser[2]

[1] Institute of Computer Science, LMU Munich, Germany
hofmann@ifi.lmu.de

[2] Institute of Computer Science, University of Innsbruck, Austria
georg.moser@uibk.ac.at

Abstract. We introduce a novel resource analysis for typed term rewrite systems based on a potential-based type system. This type system gives rise to polynomial bounds on the innermost runtime complexity. We relate the thus obtained amortised resource analysis to polynomial interpretations and obtain the perhaps surprising result that whenever a rewrite system $\mathcal{R}$ can be well-typed, then there exists a polynomial interpretation that orients $\mathcal{R}$. For this we adequately adapt the standard notion of polynomial interpretations to the typed setting.

Keywords: term rewriting, types, amortised resource analysis, complexity of rewriting, polynomial interpretations.

1 Introduction

In recent years there have been several approaches to the automated analysis of the complexity of programs. Without hope for completeness, we mention work by Albert et al. [1] that underlies COSTA, an automated tool for the resource analysis of Java programs. Related work, targeting C programs, has been reported by Alias et al. [2]. In Zuleger et al. [3] further approaches for the runtime complexity analysis of C programs is reported, incorporated into LOOPUS. Noschinski et al. [4] study runtime complexity analysis of rewrite systems, which has been incorporated in AProVE. Finally, the RaML prototype [5] provides an automated potential-based resource analysis for various resource bounds of functional programs and $\mathsf{T_CT}$ [6] is one of the most powerful tools for complexity analysis of rewrite systems.

Despite the abundance in the literature on complexity analysis of programs, almost no comparison results are known that relate the sophisticated methods developed. Indeed a precise comparison proves often difficult. Consider Example 1; $\mathcal{R}_{\mathsf{que}}$ encodes an efficient implementation of a queue in functional programming. A queue is represented as a pair of two lists $\mathsf{que}(f, r)$, encoding the initial part f and the reversal of the remainder r. The invariant of the algorithm is that the first list never becomes empty, which is achieved by reversing r if necessary. Should the invariant ever be violated, an exception (err_head or err_tail) is raised.

* This research is partly supported by FWF (Austrian Science Fund) project P25781.

G. Dowek (ed.): RTA-TLCA 2014, LNCS 8560, pp. 272–286, 2014.

Example 1. Consider the following term rewrite system (TRS for short) $\mathcal{R}_{\mathsf{que}}$, encoding a variant of an example by Okasaki [7, Section 5.2].

1: $\mathsf{chk}(\mathsf{que}(\mathsf{nil}, r)) \to \mathsf{que}(\mathsf{rev}(r), \mathsf{nil})$	7: $\mathsf{enq}(0) \to \mathsf{que}(\mathsf{nil}, \mathsf{nil})$
2: $\mathsf{chk}(\mathsf{que}(x \sharp xs, r)) \to \mathsf{que}(x \sharp xs, r)$	8: $\mathsf{rev}'(\mathsf{nil}, ys) \to ys$
3: $\mathsf{tl}(\mathsf{que}(x \sharp f, r)) \to \mathsf{chk}(\mathsf{que}(f, r))$	9: $\mathsf{rev}(xs) \to \mathsf{rev}'(xs, \mathsf{nil})$
4: $\mathsf{snoc}(\mathsf{que}(f, r), x) \to \mathsf{chk}(\mathsf{que}(f, x \sharp r))$	10: $\mathsf{hd}(\mathsf{que}(x \sharp f, r)) \to x$
5: $\mathsf{rev}'(x \sharp xs, ys) \to \mathsf{rev}'(xs, x \sharp ys)$	11: $\mathsf{hd}(\mathsf{que}(\mathsf{nil}, r)) \to \mathsf{err\_head}$
6: $\mathsf{enq}(\mathsf{s}(n)) \to \mathsf{snoc}(\mathsf{enq}(n), n)$	12: $\mathsf{tl}(\mathsf{que}(\mathsf{nil}, r)) \to \mathsf{err\_tail}$.

We exemplify the physicist's method of amortised analysis [8]. We assign to every queue $\mathsf{que}(f, r)$ the length of r as *potential*. Then the amortised cost for each operation is constant, as the costly reversal operation is only executed if the potential can pay for the operation, cf. [7]. Thus, based on an amortised analysis, we deduce the optimal linear runtime complexity for $\mathcal{R}$. Let us attempt to apply the interpretation method instead. Termination proofs by interpretations are well-established and can be traced back to work by Turing [9]. It is straightforward to restrict *polynomial interpretations* [10] so that compatibility with a TRS $\mathcal{R}$ induces polynomial runtime complexity of $\mathcal{R}$, cf. [11]. Such polynomial interpretations are called *restricted*. However, we can see that no restricted polynomial interpretation can exist that is compatible with $\mathcal{R}_{\mathsf{que}}$. The constraints induced by $\mathcal{R}_{\mathsf{que}}$ imply that the function snoc has to be interpreted by a linear polynomial. Thus an exponential interpretation is required for enqueuing (enq). Looking more closely at the different proofs, we observe the following. While in the amortised analysis the potential of a queue $\mathsf{que}(f, r)$ depends only on the remainder r, the interpretation of que has to be monotone in both arguments by definition. This difference induces that snoc is assigned a *strongly linear* potential in the amortised analysis, while only a *linear* interpretation is possible for snoc.

Still it is possible to relate amortised analysis to polynomial interpretations if we base our investigation on many-sorted (or typed) TRSs and make suitable use of the concept of *annotated types* originally introduced in [12]. We note that Example 1 is also subject to other techniques like quasi-interpretations [13] and can also be handled fully automatically in AProVE or $\mathsf{T_CT}$. However, our interest in the example stems from the fact that it shows a separation between amortised analysis and restricted polynomial interpretations.

We establish a novel innermost runtime complexity analysis for typed constructor rewrite systems $\mathcal{R}$. This complexity analysis is based on a potential-based amortised analysis incorporated into a type system. From the annotated type of a term its derivation height with respect to innermost rewriting can be read off, inducing polynomial bounds on the runtime complexity with respect to $\mathcal{R}$ (see Theorem 12). The correctness proof of the obtained bound rests on an operational big-step semantics decorated with counters for the derivation height of the evaluated terms. We complement this big-step semantics with a similar decorated small-step semantics and prove equivalence between these semantics.

Furthermore we establish a second soundness result based on the small-step semantics (see Theorem 20). Exploiting the small-step semantics we prove our main result that from the well-typing of $\mathcal{R}$ we can read off a typed polynomial interpretation that orients $\mathcal{R}$ (see Theorem 23).

While the type system exhibited is inspired by Hoffmann et al. [14] we generalise their use of annotated types to arbitrary (data) types. Furthermore the introduced small-step semantics (and our main result) directly establish that any well-typed TRS is terminating, cf. [15]. As a corollary to our main result, we obtain that the physicist's method of amortised analysis conceptually amounts to the interpretation method, if we allow for the following changes: (i) every term bears a potential, not only values, (ii) polynomial interpretations are defined over annotated types, and (iii) compatibility is replaced by orientability.

Our study is purely theoretic, and we have not (yet) an implementation of the provided techniques. However, automation is straightforward and seems to yield fairly precise bounds on the runtime complexity. Furthermore, we have restricted our study to typed (constructor) TRSs. In the conclusion we sketch application of the established results to innermost runtime complexity analysis of untyped TRSs.

This paper is structured as follows. In the next section we cover basics. In Section 3 we provide our first soundness result. In Section 4 we establish our second soundness result. Our main result will be stated and proved in Section 5. Finally, we conclude in Section 6. Due to space limitations some proofs are only sketched, or have been completely omitted. The reader is kindly referred to the extended version of this paper [16].

2 Typed Term Rewrite Systems

Let $\mathcal{C}$ denote a finite, non-empty set of *constructor symbols* and $\mathcal{D}$ a finite set of *defined function symbols*. Let S be a finite set of (data) types. A family $(X_A)_{A \in S}$ of sets is called *S-typed* and denoted as X. Let $\mathcal{V}$ denote an S-typed set of *variables*, such that the $\mathcal{V}_A$ are pairwise disjoint. In the following, variables will be denoted by x, y, ..., possibly extended by subscripts.

Following [17], a *type declaration* is of the form $[A_1 \times \cdots \times A_n] \to C$, where A_i and C are types. Type declarations serve as input-output specifications for function symbols. We write A instead of $[] \to A$. A *signature* $\mathcal{F}$ (with respect to the set of types S) is a mapping from $\mathcal{C} \cup \mathcal{D}$ to type declarations. We often write $f\colon [A_1 \times \cdots \times A_n] \to C$, if $\mathcal{F}(f) = [A_1 \times \cdots \times A_n] \to C$ and refer to a type declaration as a type, if no confusion can arise.

We define the S-typed set of terms $\mathcal{T}(\mathcal{D} \cup \mathcal{C}, \mathcal{V})$ (or $\mathcal{T}$ for short): (i) for each $A \in S$: $\mathcal{V}_A \subseteq \mathcal{T}_A$, (ii) for $f \in \mathcal{C} \cup \mathcal{D}$ such that $\mathcal{F}(f) = [A_1, \ldots, A_n] \to C$ and $t_i \in \mathcal{T}_{A_i}$, we have $f(t_1, \ldots, t_n) \in \mathcal{T}_C$. Type assertions are denoted $t\colon A$. Terms of type A will sometimes be referred to as instances of A: a term of list type, is simply called a list. If $t \in \mathcal{T}(\mathcal{C}, \varnothing)$ then t is called a *ground constructor term* or a *value*. The set of values is denoted $\mathcal{T}(\mathcal{C})$. The (S-typed) set of variables of a term t is denoted $\mathcal{V}\mathsf{ar}(t)$. The root of t is denoted $\mathsf{rt}(t)$ and the size of t, that is the number of symbols in t, is denoted $|t|$. In the following terms are denoted by $s, t, u, v, \ldots$, possibly extended by subscripts. Furthermore, we use v (possibly extended by subscripts) to denote values.

A *substitution* σ is a mapping from variables to terms that respects types. Substitutions are denoted as sets of assignments: $\sigma = \{x_1 \mapsto t_1, \ldots, x_n \mapsto t_n\}$. We write $\mathsf{dom}(\sigma)$ ($\mathsf{rg}(\sigma)$) to denote the domain (range) of σ. Let σ be a substitution and V be a set of variables; $\sigma \restriction V$ denotes the restriction of the domain of σ to V. The substitution τ is called an *extension* of substitution σ if $\tau \restriction \mathsf{dom}(\sigma) = \sigma$. Let σ, τ be substitutions such that $\mathsf{dom}(\sigma) \cap \mathsf{dom}(\tau) = \varnothing$. Then we denote the (disjoint) union of σ and τ as $\sigma \uplus \tau$. We call a substitution σ *normalised* if all terms in the range of σ are values. In the following all considered substitutions will be normalised.

A *typing context* is a mapping from variables $\mathcal{V}$ to types. Type contexts are denoted by upper-case Greek letters. Let Γ be a context and let t be a term. The typing relation $\Gamma \vdash t\colon A$ expresses that based on context Γ, t has type A (with respect to the signature $\mathcal{F}$). The typing rules that define the typing relation are given in Figure 2, where we forget the annotations. In the sequel we sometimes make use of an abbreviated notation for sequences of terms $\boldsymbol{t} := t_1, \ldots, t_n$.

A *typed rewrite rule* is a pair $l \to r$ of terms, such that (i) the types of l and r coincide, (ii) $\mathsf{rt}(l) \in \mathcal{D}$, and (iii) $\mathcal{V}\mathsf{ar}(l) \supseteq \mathcal{V}\mathsf{ar}(r)$. An S-typed *term rewrite system* (*TRS* for short) over the signature $\mathcal{F}$ is a finite set of typed rewrite rules. We define the *innermost rewrite relation* $\xrightarrow{\text{i}}_{\mathcal{R}}$ for typed TRSs $\mathcal{R}$. For well-typed terms s and t, $s \xrightarrow{\text{i}}_{\mathcal{R}} t$ holds, if there exists a context C, a normalised substitution σ and a rewrite rule $l \to r \in \mathcal{R}$ such that $s = C[l\sigma]$ and $t = C[r\sigma]$. In the sequel we are only concerned with *innermost* rewriting. A TRS is *orthogonal* if it is left-linear and non-overlapping [10,18]. A TRS is *completely defined* if all ground normal-forms are values. These notions naturally extend to typed TRS. In particular note that an orthogonal typed TRS is confluent. Let s and t be terms, such that t is in normal-form. Then an *(innermost) derivation* $D\colon s \xrightarrow{\text{i}}{}^{*}_{\mathcal{R}} t$ with respect to a TRS $\mathcal{R}$ is a finite sequence of rewrite steps. The *derivation height* of a term s with respect to a well-founded, finitely branching relation $\to$ is defined as: $\mathsf{dh}(s, \to) = \max\{n \mid \exists t\ s \to^n t\}$. A term $t = f(t_1, \ldots, t_k)$ is called *basic* if f is defined, and all $t_i \in \mathcal{T}(\mathcal{C}, \mathcal{V})$.

Definition 2. *We define the* runtime complexity *(with respect to $\mathcal{R}$):* $\mathsf{rc}_{\mathcal{R}}(n) := \max\{\mathsf{dh}(t, \xrightarrow{\text{i}}_{\mathcal{R}}) \mid t \text{ is basic and } |t| \leqslant n\}$.

We study *typed constructor* TRSs $\mathcal{R}$, that is, for each rule $f(l_1, \ldots, l_n) \to r$ we have that the arguments l_i are constructor terms. Furthermore, we restrict to *completely defined* and *orthogonal* systems. These restrictions are natural in the context of functional programming. If no confusion can arise from this, we simply call $\mathcal{R}$ a TRS. $\mathcal{F}$ denotes the signature underlying $\mathcal{R}$. In the sequel, $\mathcal{R}$ and $\mathcal{F}$ are kept fixed.

Example 3 (continued from Example 1). Consider the TRS $\mathcal{R}_{\mathsf{que}}$ and let $S = \{\mathsf{Nat}, \mathsf{List}, \mathsf{Q}\}$, where Nat, List, and Q represent the type of natural numbers, lists over natural numbers, and queues respectively. Then $\mathcal{R}_{\mathsf{que}}$ is an S-typed

$$\frac{x\sigma = v}{\sigma \stackrel{0}{\vdash} x \Rightarrow v} \qquad \frac{c \in \mathcal{C} \quad x_1\sigma = v_1 \quad \cdots \quad x_n\sigma = v_n}{\sigma \stackrel{0}{\vdash} c(x_1,\dots,x_n) \Rightarrow c(v_1,\dots,v_n)}$$

$$\frac{f(l_1,\dots,l_n) \to r \in \mathcal{R} \quad \exists \tau\, \forall i\colon x_i\sigma = l_i\tau \quad \sigma \uplus \tau \stackrel{m}{\vdash} r \Rightarrow v}{\sigma \stackrel{m+1}{\vdash} f(x_1,\dots,x_n) \Rightarrow v}$$

$$\frac{\text{all } x_i \text{ are fresh} \quad \sigma \uplus \rho \stackrel{m_0}{\vdash} f(x_1,\dots,x_n) \Rightarrow v \quad \sigma \stackrel{m_1}{\vdash} t_1 \Rightarrow v_1 \quad \cdots \quad \sigma \stackrel{m_n}{\vdash} t_n \Rightarrow v_n \quad m = \sum_{i=0}^{n} m_i}{\sigma \stackrel{m}{\vdash} f(t_1,\dots,t_n) \Rightarrow v}$$

Here $\rho := \{x_1 \mapsto v_1, \dots, x_n \mapsto v_n\}$. Recall that σ, τ, and ρ are normalised.

Fig. 1. Operational Big-Step Semantics

TRSs over signature $\mathcal{F}$. We exemplify the signature of some constructors: $\mathsf{0}\colon \mathsf{Nat}$, $\mathsf{s}\colon [\mathsf{Nat}] \to \mathsf{Nat}$, $\mathsf{nil}\colon \mathsf{List}$, $\sharp\colon [\mathsf{Nat} \times \mathsf{List}] \to \mathsf{List}$, $\mathsf{que}\colon [\mathsf{List} \times \mathsf{List}] \to \mathsf{Q}$. Finally, consider $\mathsf{snoc}\colon [\mathsf{Q} \times \mathsf{Nat}] \to \mathsf{Q}$.

As $\mathcal{R}$ is completely defined, any derivation ends in a value. In connection with innermost rewriting this yields a *call-by-value* strategy. Furthermore, as $\mathcal{R}$ is non-overlapping any innermost derivation is determined modulo the order in which parallel redexes are contracted. This allows us to recast innermost rewriting into an operational big-step semantics instrumented with resource counters, cf. Figure 1. Its definition is instrumental in the proof of our first soundness theorem. The semantics resembles similar definitions given in the literature on amortised resource analysis (see for example [14, 19, 20]).

Proposition 4. *Let f be a defined function symbol of arity n and σ a substitution. Then $\sigma \stackrel{m}{\vdash} f(x_1,\dots,x_n) \Rightarrow v$ holds iff $\mathsf{dh}(f(x_1\sigma,\dots,x_n\sigma), \xrightarrow{\mathrm{i}}_{\mathcal{R}}) = m$.*

Proof. In the proof of the direction form left to right, we show the stronger statement that $\sigma \stackrel{m}{\vdash} t \Rightarrow v$ implies $\mathsf{dh}(t\sigma, \xrightarrow{\mathrm{i}}_{\mathcal{R}}) = m$ by induction on the derivation of $\sigma \stackrel{m}{\vdash} t \Rightarrow v$. For the opposite direction, we show that if $\mathsf{dh}(t\sigma, \xrightarrow{\mathrm{i}}_{\mathcal{R}}) = m$, then $\sigma \stackrel{m}{\vdash} t \Rightarrow v$ by induction on the length of the derivation $D\colon t\sigma \xrightarrow{\mathrm{i}}{}^{*}_{\mathcal{R}} v$. □

3 Annotated Types

Let S be a set of types. We call a type $A \in S$ *annotated*, if A is decorated with resource annotations. These annotations will allow us to read off the potential of a well-typed term t from the annotations.

Definition 5. *An* annotated type $A^{\mathbf{p}}$, *is a pair consisting of a type $A \in S$ and a vector $\mathbf{p} = (p_1,\dots,p_k)$ over non-negative rational numbers, typically natural numbers. The vector $\mathbf{p}$ is called* resource annotation.

Resource annotations are denoted by $\boldsymbol{p}$, $\boldsymbol{q}$, $\boldsymbol{u}$, $\boldsymbol{v}$, ..., possibly extended by subscripts and we write $\mathcal{A}$ for the set of such annotations. For resource annotations (p) of length 1 we write p. We will see that a resource annotation does not change its meaning if zeroes are appended at the end, so, conceptually, we can identify $()$ with (0) (and also with 0). If $\boldsymbol{p} = (p_1, \ldots, p_k)$ we write $k = |\boldsymbol{p}|$ and $\max \boldsymbol{p} = \max_i p_i$. We define the notations $\boldsymbol{p} \leqslant \boldsymbol{q}$ and $\boldsymbol{p} + \boldsymbol{q}$ and $\lambda \boldsymbol{p}$ for $\lambda \geqslant 0$ component-wise, filling up with 0s if needed. So, for example $(1,2) \leqslant (3,4,5)$ and $(1,2) + (3,4,5) = (4,6,5)$.

Furthermore, we recall the additive shift [14] given by $\lhd(p_1, \ldots, p_k) = (p_1 + p_2, p_2 + p_3, \ldots, p_{k-1} + p_k, p_k)$. We also define the interleaving $\boldsymbol{p} ||| \boldsymbol{q}$ by $(p_1, q_1, p_2, q_2, \ldots, p_k, q_k)$ where, as before the shorter of the two vectors is padded with 0s. Finally, we use the notation $\Diamond \boldsymbol{p} = p_1$ for the first entry of an annotation vector. If no confusion can arise, we refer to annotated types simply as types. In contrast to Hoffmann et al. [14, 21], we generalise the concept of annotated types to arbitrary (data) types. In [14] only list types, in [21] list and tree types have been annotated.

Definition 6. *Let $\mathcal{F}$ be a signature. Suppose $\mathcal{F}(f) = [A_1 \times \cdots \times A_n] \to C$, such that the A_i $(i = 1, \ldots, n)$ and C are types. Consider the annotated types $A_i^{\boldsymbol{u}_i}$ and $A^{\boldsymbol{v}}$. Then an* annotated type declaration *for f is a type declaration over annotated types, decorated with $p \in \mathbb{N}$: $[A_1^{\boldsymbol{u}_1} \times \cdots \times A_n^{\boldsymbol{u}_n}] \xrightarrow{p} C^{\boldsymbol{v}}$. The set of annotated type declarations is denoted $\mathcal{F}_{\mathsf{pol}}$.*

We lift signatures to *annotated signatures* $\mathcal{F} \colon \mathcal{C} \cup \mathcal{D} \to (\mathcal{P}(\mathcal{F}_{\mathsf{pol}}) \setminus \varnothing)$ by mapping a function symbol to a non-empty set of annotated type declarations. Hence for any function symbol f we allow multiple types. If f has result type C, then for each annotation $C^{\boldsymbol{q}}$ there should exist exactly one declaration of the form $[A_1^{\boldsymbol{p}_1} \times \cdots \times A_n^{\boldsymbol{p}_n}] \xrightarrow{p} C^{\boldsymbol{q}}$ in $\mathcal{F}(c)$. Moreover, constructor annotations are to satisfy the *superposition principle*: If a constructor c admits the annotations $[A_1^{\boldsymbol{p}_1} \times \cdots \times A_n^{\boldsymbol{p}_n}] \xrightarrow{p} C^{\boldsymbol{q}}$ and $[A_1^{\boldsymbol{p}'_1} \times \cdots \times A_n^{\boldsymbol{p}'_n}] \xrightarrow{p'} C^{\boldsymbol{q}'}$ then it also has the annotations $[A_1^{\lambda \boldsymbol{p}_1} \times \cdots \times A_n^{\lambda \boldsymbol{p}_n}] \xrightarrow{\lambda p} C^{\lambda \boldsymbol{q}}$ $(\lambda \geqslant 0)$ and $[A_1^{\boldsymbol{p}_1 + \boldsymbol{p}'_1} \times \cdots \times A_n^{\boldsymbol{p}_n + \boldsymbol{p}'_n}] \xrightarrow{p+p'} C^{\boldsymbol{q}+\boldsymbol{q}'}$.

Note that, in view of superposition and uniqueness, the annotations of a given constructor are uniquely determined once we fix the annotated types for result annotations of the form $(0, \ldots, 0, 1)$ (remember the implicit filling up with 0s). An annotated signature $\mathcal{F}$ is simply called signature, where we sometimes write $f \colon [A_1 \times \cdots \times A_n] \xrightarrow{p} C$ instead of $[A_1 \times \cdots \times A_n] \xrightarrow{p} C \in \mathcal{F}(f)$. Note that the A_i $(i = 1, \ldots, n)$ and C denote *annotated* types.

Example 7 (continued from Example 3). In order to extend $\mathcal{F}$ to an annotated signature we can set $\mathcal{F}(\mathsf{0}) := \{[] \xrightarrow{0} \mathsf{Nat}^{\boldsymbol{p}} \mid \boldsymbol{p} \in \mathcal{A}\}$ and $\mathcal{F}(\mathsf{s}) := \{[\mathsf{Nat}^{\lhd(\boldsymbol{p})}] \xrightarrow{\Diamond \boldsymbol{p}} \mathsf{Nat}^{\boldsymbol{p}} \mid \boldsymbol{p} \in \mathcal{A}\}$. Furthermore, we set $\mathcal{F}(\mathsf{nil}) := \{[] \xrightarrow{0} \mathsf{List}^{\boldsymbol{p}} \mid \boldsymbol{p} \in \mathcal{A}\}$ and $\mathcal{F}(\sharp) := \{[\mathsf{Nat}^0 \times \mathsf{List}^{\lhd(\boldsymbol{p})}] \xrightarrow{\Diamond \boldsymbol{p}} \mathsf{List}^{\boldsymbol{p}} \mid \boldsymbol{p} \in \mathcal{A}\}$ and $\mathcal{F}(\mathsf{que}) := \{[\mathsf{List}^{\boldsymbol{p}} \times \mathsf{List}^{\boldsymbol{q}}] \xrightarrow{0} \mathsf{Q}^{\boldsymbol{p} ||| \boldsymbol{q}} \mid \boldsymbol{p}, \boldsymbol{q} \in \mathcal{A}\}$. In particular, we have the typings $\sharp \colon [\mathsf{Nat}^0 \times \mathsf{List}^7] \xrightarrow{7} \mathsf{List}^7$ and $\sharp \colon [\mathsf{Nat}^0 \times \mathsf{List}^{(10,7)}] \xrightarrow{3} \mathsf{List}^{(3,7)}$ and $\mathsf{que} \colon [\mathsf{List}^1 \times \mathsf{List}^3] \xrightarrow{0} \mathsf{Q}^{(1,3)}$.

We omit annotations for the defined symbols and refer to Example 13 for a complete signature with a different annotation for the constructor symbol que.

The next definition introduces the notion of the potential of a value.

Definition 8. *Let $v = c(v_1, \ldots, v_n) \in \mathcal{T}(\mathcal{C})$ and let C be an annotated type. The* potential *of v under C, written $\Phi(v{:}\, C)$, is defined recursively by $\Phi(v{:}\, C) := p + \Phi(v_1{:}\, A_1) + \cdots + \Phi(v_n{:}\, A_n)$ when $[A_1 \times \cdots \times A_n] \xrightarrow{p} C \in \mathcal{F}(c)$.*

Note that by assumption the declaration in $\mathcal{F}(c)$ is unique.

Example 9 (continued from Example 7). It is easy to see that for any term t of type Nat, we have $\Phi(t{:}\, \mathsf{Nat}^0) = 0$ and $\Phi(t{:}\, \mathsf{Nat}^\lambda) = \lambda t$.

If l is a list then $\Phi(l{:}\, \mathsf{List}^{(p,q)}) = p \cdot |l| + q \cdot \binom{|l|}{2}$. where $|l|$ denotes the length of l, that is the number of $\sharp$ in l. More generally, we have $\Phi(l{:}\, \mathsf{List}^{\boldsymbol{p}}) = \sum_i p_i \binom{|l|}{i}$. Finally, if $\mathsf{que}(l, k)$ has type Q then $\Phi(\mathsf{que}(l,k){:}\, \mathsf{Q}^{\boldsymbol{p}|||\boldsymbol{q}}) = \Phi(l{:}\, \mathsf{List}^{\boldsymbol{p}}) + \Phi(k{:}\, \mathsf{List}^{\boldsymbol{q}})$.

The *sharing relation* $\curlyvee(A^{\boldsymbol{p}} \mid A^{\boldsymbol{p_1}}, A^{\boldsymbol{p_2}})$ holds if $\boldsymbol{p_1} + \boldsymbol{p_2} = \boldsymbol{p}$. The *subtype relation* is defined as follows: $A^{\boldsymbol{p}} <: B^{\boldsymbol{q}}$, if $A = B$ and $\boldsymbol{p} \geqslant \boldsymbol{q}$.

Lemma 10. *If $\curlyvee(A^{\boldsymbol{p}} \mid A^{\boldsymbol{p_1}}, A^{\boldsymbol{p_2}})$ then $\Phi(v{:}\, A^{\boldsymbol{p}}) = \Phi(v{:}\, A^{\boldsymbol{p_1}}) + \Phi(v{:}\, A^{\boldsymbol{p_2}})$ holds for any value of type A. If $A^{\boldsymbol{p}} <: B^{\boldsymbol{q}}$ then $\Phi(v{:}\, A^{\boldsymbol{p}}) \geqslant \Phi(v{:}\, B^{\boldsymbol{q}})$ again for any $v{:}\, A$.*

Proof. The proof of the first claim is by induction on the structure of v. We note that by superposition together with uniqueness the additivity property propagates to the argument types. For example, if we have the annotations $\mathsf{s} : [\mathsf{Nat}^2] \xrightarrow{4} \mathsf{Nat}^3$ and $\mathsf{s} : [\mathsf{Nat}^4] \xrightarrow{6} \mathsf{Nat}^5$ and $\mathsf{s} : [\mathsf{Nat}^x] \xrightarrow{10} \mathsf{Nat}^y$ then we can conclude $x = 6$, $y = 8$, for this annotation must be present by superposition and there can only be one by uniqueness.

The second claim follows from the first one and nonnegativity of potentials. □

The set of typing rules for TRSs $\mathcal{R}$ are given in Figure 2. Observe that the type system employs the assumption that $\mathcal{R}$ is left-linear. In a nutshell, the method works as follows: Let Γ be a typing context and let us consider the typing judgement $\Gamma \overset{p}{\vdash} t{:}\, A$ derivable from the type rules. Then p is an upper-bound to the amortised cost required for reducing t to a value. The derivation height of $t\sigma$ (with respect to innermost rewriting) is bound by the difference in the potential before and after the evaluation plus p. Thus if the sum of the potential of the arguments of $t\sigma$ is in $\mathsf{O}(n^k)$, where n is the size of the arguments, then the runtime complexity of $\mathcal{R}$ lies in $\mathsf{O}(n^k)$.

Recall that any rewrite rule $l \to r \in \mathcal{R}$ can be written as $f(l_1, \ldots, l_n) \to r$ with $l_i \in \mathcal{T}(\mathcal{C}, \mathcal{V})$. We introduce *well-typed* TRSs.

Definition 11. *Let $f(l_1, \ldots, l_n) \to r$ be a rewrite rule in $\mathcal{R}$ and let $\mathcal{V}\mathsf{ar}(f(\boldsymbol{l})) = \{y_1, \ldots, y_\ell\}$. Then $f \in \mathcal{D}$ is* well-typed *wrt. $\mathcal{F}$, if we obtain*

$$y_1{:}\, B_1, \ldots, y_\ell{:}\, B_\ell \overset{p-1+\sum_{i=1}^{n} k_i}{\vdash} r{:}\, C\ , \qquad (1)$$

$$\frac{f \in \mathcal{C} \cup \mathcal{D} \quad [A_1^{u_1} \times \cdots \times A_n^{u_n}] \xrightarrow{p} C^v \in \mathcal{F}(f)}{x_1 \colon A_1^{u_1}, \ldots, x_n \colon A_n^{u_n} \overset{p}{\vdash} f(x_1, \ldots, x_n) \colon C^v} \qquad \frac{\Gamma \overset{p}{\vdash} t \colon C \quad p' \geqslant p}{\Gamma \overset{p'}{\vdash} t \colon C}$$

$$\frac{\begin{array}{c}\text{all } x_i \text{ are fresh} \qquad\qquad p = \sum_{i=0}^{n} p_i \\ x_1 \colon A_1, \ldots, x_n \colon A_n \overset{p_0}{\vdash} f(x_1, \ldots, x_n) \colon C \quad \Gamma_1 \overset{p_1}{\vdash} t_1 \colon A_1 \; \cdots \; \Gamma_n \overset{p_n}{\vdash} t_n \colon A_n\end{array}}{\Gamma_1, \ldots, \Gamma_n \overset{p}{\vdash} f(t_1, \ldots, t_n) \colon C}$$

$$\frac{\Gamma \overset{p}{\vdash} t \colon C}{\Gamma, x \colon A \overset{p}{\vdash} t \colon C} \qquad \frac{\Gamma, x \colon A_1, y \colon A_2 \overset{p}{\vdash} t[x, y] \colon C \quad \curlyvee(A \mid A_1, A_2) \quad x, y \text{ are fresh}}{\Gamma, z \colon A \overset{p}{\vdash} t[z, z] \colon C}$$

$$\frac{\Gamma, x \colon B \overset{p}{\vdash} t \colon C \quad A <: B}{\Gamma, x \colon A \overset{p}{\vdash} t \colon C} \qquad \frac{}{x \colon A \overset{0}{\vdash} x \colon A} \qquad \frac{\Gamma \overset{p}{\vdash} t \colon D \quad D <: C}{\Gamma \overset{p}{\vdash} t \colon C}$$

Fig. 2. Type System for Rewrite Systems

for all $[A_1 \times \cdots \times A_n] \xrightarrow{p} C \in \mathcal{F}(f)$*, for all types* B_j $(j \in \{1, \ldots, \ell\})$*, and all costs* k_i*, such that* $y_1 \colon B_1, \ldots, y_\ell \colon B_\ell \overset{k_i}{\vdash} l_i \colon A_i$ *is derivable. A TRS* $\mathcal{R}$ *over* $\mathcal{F}$ *is* well-typed *if any defined* f *is well-typed.*

Let Γ be a typing context and let σ be a substitution. We call σ *well-typed (with respect to* Γ*)* if for all $x \in \mathsf{dom}(\Gamma)$, $x\sigma$ is of type $\Gamma(x)$. We extend the definition of potential to substitutions σ and typing contexts Γ. Suppose σ is well-typed with respect to Γ. Then $\Phi(\sigma \colon \Gamma) := \sum_{x \in \mathsf{dom}(\Gamma)} \Phi(x\sigma \colon \Gamma(x))$. We establish our first soundness result.

Theorem 12. *Let* $\mathcal{R}$ *and* σ *be well-typed. Suppose* $\Gamma \overset{p}{\vdash} t \colon A$ *and* $\sigma \overset{m}{\vdash} t \Rightarrow v$*. Then* $\Phi(\sigma \colon \Gamma) - \Phi(v \colon A) + p \geqslant m$*.*

Proof. Let Π be the proof deriving $\sigma \overset{m}{\vdash} t \Rightarrow v$ and let Ξ be the proof of $\Gamma \overset{p}{\vdash} t \colon A$. The proof of the theorem proceeds by main-induction on the length of Π and by side-induction on the length of Ξ.

We exemplify the pattern of the proof on one case. We employ the notation from Figure 1. Suppose the last rule in Π has the form

$$\frac{\sigma \uplus \rho \overset{m_0}{\vdash} f(x_1, \ldots, x_n) \Rightarrow v \quad \sigma \overset{m_1}{\vdash} t_1 \Rightarrow v_1 \quad \cdots \quad \sigma \overset{m_n}{\vdash} t_n \Rightarrow v_n}{\sigma \overset{m}{\vdash} f(t_1, \ldots, t_n) \Rightarrow v} ,$$

where $m = \sum_{i=0}^{n} m_i$. W.l.o.g. we can assume that t is linear. Otherwise we would consider the case, where Ξ ends with the type rule for sharing. Thus, we assume the last rule in the type inference Ξ is of the following form.

$$\frac{\overbrace{x_1 \colon A_1, \ldots, x_n \colon A_n}^{=:\Delta} \overset{p_0}{\vdash} f(\boldsymbol{x}) \colon C \quad \Gamma_1 \overset{p_1}{\vdash} t_1 \colon A_1 \quad \cdots \quad \Gamma_n \overset{p_n}{\vdash} t_n \colon A_n}{\Gamma_1, \ldots, \Gamma_n \overset{p}{\vdash} f(t_1, \ldots, t_n) \colon C} ,$$

such that $p = \sum_{i=0}^{n} p_i$. By induction hypothesis: $\Phi(\sigma\colon \Gamma_i) - \Phi(v_i\colon A_i) + p_i \geqslant m_i$ for all $i = 1, \ldots, n$. Hence (i) $\sum_{i=1}^{n} \Phi(\sigma\colon \Gamma_i) - \sum_{i=1}^{n} \Phi(v_i\colon A_i) + \sum_{i=1}^{n} p_i \geqslant \sum_{i=1}^{n} m_i$ Again by induction hypothesis we obtain: (ii) $\Phi(\sigma \uplus \rho\colon \Delta) - \Phi(v\colon C) + p_0 \geqslant m_0$ Now $\Phi(\sigma\colon \Gamma) = \sum_{i=1}^{n} \Phi(\sigma\colon \Gamma_i)$ and $\Phi(\sigma \uplus \rho\colon \Delta) = \Phi(\rho\colon \Delta) = \sum_{i=1}^{n} \Phi(v_i\colon A_i)$. By (i) and (ii), we obtain

$$\begin{aligned}\Phi(\sigma\colon \Gamma) + \sum_{i=0}^{n} p_i &= \sum_{i=1}^{n} \Phi(\sigma\colon \Gamma_i) + \sum_{i=1}^{n} p_i + p_0 \\ &\geqslant \sum_{i=1}^{n} \Phi(v_i\colon A_i) + \sum_{i=1}^{n} m_i + p_0 \geqslant \Phi(v\colon C) + \sum_{i=0}^{n} m_i ,\end{aligned}$$

and thus $\Phi(\sigma\colon \Gamma) - \Phi(v\colon C) + p \geqslant m$. □

Example 13 (continued from Example 1). We extend our example signature with annotated typings for the defined functions as follows.

$$\begin{array}{lll}\mathsf{chk}\colon [\mathsf{Q}^{(0,1)}] \xrightarrow{3} \mathsf{Q}^{(0,1)} & \mathsf{tl}\colon [\mathsf{Q}^{(0,1)}] \xrightarrow{4} \mathsf{Q}^{(0,1)} & \mathsf{hd}\colon [\mathsf{Q}^{(0,1)}] \xrightarrow{1} \mathsf{Nat}^0 \\ \mathsf{rev}'\colon [\mathsf{List}^1 \times \mathsf{List}^0] \xrightarrow{1} \mathsf{List}^0 & \mathsf{rev}\colon [\mathsf{List}^1] \xrightarrow{2} \mathsf{List}^0 & \\ \mathsf{snoc}\colon [\mathsf{Q}^{(0,1)} \times \mathsf{Nat}^0] \xrightarrow{5} \mathsf{Q}^{(0,1)} & \mathsf{enq}\colon [\mathsf{Nat}^6] \xrightarrow{1} \mathsf{Q}^{(0,1)} , & \end{array}$$

where the annotations of the constructors are as in Example 7, with the exception of que: $[\mathsf{List}^0 \times \mathsf{List}^1] \xrightarrow{0} \mathsf{Q}^{(0,1)}$. It is not difficult to verify that $\mathcal{R}_{\mathsf{que}}$ is well-typed wrt. $\mathcal{F}$. We show in detail that enq is well-typed. Consider rule 6. First, we observe that 6 resource units become available for the recursive call, as $n\colon \mathsf{Nat}^6 \mid\!\frac{6}{}\, \mathsf{s}(n)\colon \mathsf{Nat}^6$ is derivable. Second, we have the following partial type derivation; missing parts are easy to fill in.

$$\dfrac{\dfrac{q\colon \mathsf{Q}^{(0,1)}, m\colon \mathsf{Nat}^0 \mid\!\frac{5}{}\, \mathsf{snoc}(q,m)\colon \mathsf{Q}^{(0,1)} \qquad n_1\colon \mathsf{Nat}^6 \mid\!\frac{1}{}\, \mathsf{enq}(n_1)\colon \mathsf{Q}^{(0,1)} \qquad n_2\colon \mathsf{Nat}^0 \mid\!\frac{0}{}\, n_2\colon \mathsf{Nat}^0}{n_1\colon \mathsf{Nat}^6, n_2\colon \mathsf{Nat}^0 \mid\!\frac{6}{}\, \mathsf{snoc}(\mathsf{enq}(n_1), n_2)\colon \mathsf{Q}^{(0,1)}}}{n\colon \mathsf{Nat}^6 \mid\!\frac{6}{}\, \mathsf{snoc}(\mathsf{enq}(n), n)\colon \mathsf{Q}^{(0,1)}}$$

Considering rule 7, we see that $n\colon \mathsf{Nat}^6 \mid\!\frac{0}{}\, \mathsf{que}(\mathsf{nil},\mathsf{nil})\colon \mathsf{Q}^{(0,1)}$ is derivable. Thus enq is well-typed and we conclude optimal linear runtime complexity of $\mathcal{R}_{\mathsf{que}}$.

Polynomial bounds Note that if the type annotations are chosen such that for each type A we have $\Phi(v\colon A) \in \mathsf{O}(n^k)$ for $n = |v|$ then $\mathsf{rc}_{\mathcal{R}}(n) \in \mathsf{O}(n^k)$ as well. The following proposition gives a sufficient condition as to when this is the case and in particular subsumes the type system in [14].

Theorem 14. *Suppose that for each constructor c with $[A_1^{\boldsymbol{u}_1} \times \cdots \times A_n^{\boldsymbol{u}_n}] \xrightarrow{p} C^{\boldsymbol{w}} \in \mathcal{F}(c)$, there exists $\boldsymbol{r}_i \in \mathcal{A}$ such that $\boldsymbol{u}_i \leqslant \boldsymbol{w} + \boldsymbol{r}_i$ where $\max \boldsymbol{r}_i \leqslant \max \boldsymbol{w} =: r$ and $p \leqslant r$ with $|\boldsymbol{r}_i| < |\boldsymbol{w}| =: k$. Then $\Phi(v\colon C^{\boldsymbol{w}}) \leqslant r|v|^k$.*

$$\frac{x\sigma = v}{\vdash^{0} \langle x, \sigma\rangle \to \langle v, \sigma\rangle} \qquad \frac{c \in \mathcal{C} \quad x_1\sigma = v_1 \quad \cdots \quad x_n\sigma = v_n}{\vdash^{0} \langle c(x_1,\ldots,x_n), \sigma\rangle \to \langle c(v_1,\ldots,v_n), \sigma\rangle}$$

$$\frac{\forall i\colon v_i \text{ is a value} \quad \rho = \{x_1 \mapsto v_1, \ldots, x_n \mapsto v_n\} \quad f \text{ is defined and all } x_i \text{ are fresh}}{\vdash^{0} \langle f(v_1,\ldots,v_n), \sigma\rangle \to \langle f(x_1,\ldots,x_n), \sigma \uplus \rho\rangle}$$

$$\frac{f(l_1,\ldots,l_n) \to r \in \mathcal{R} \quad \forall i\colon x_i\sigma = l_i\tau}{\vdash^{1} \langle f(x_1,\ldots,x_n), \sigma\rangle \to \langle r, \sigma \uplus \tau\rangle} \qquad \frac{\vdash^{1} \langle t_i, \sigma\rangle \to \langle u, \sigma'\rangle}{\vdash^{1} \langle f(\ldots,t_i,\ldots), \sigma\rangle \to \langle f(\ldots,u,\ldots), \sigma'\rangle}$$

Note that the substitutions σ, σ', τ, and ρ are normalised.

Fig. 3. Operational Small-Step Semantics

Proof. The proof is by induction on the size of v. Note that, if $k = 0$ then $\Phi(v\colon C^{w}) = 0$. Otherwise, we have $\Phi(c(v_1,\ldots,v_n)\colon C^{w}) \leqslant r + \Phi(v_1\colon A_1^{w+r_1}) + \cdots + \Phi(v_n\colon A_n^{w+r_n}) \leqslant r(1 + |v_1|^k + |v_1|^{k-1} + \cdots + |v_n|^k + |v_n|^{k-1})$ by application of the induction hypothesis in conjunction with Lemma 10. The latter quantity can be bounded by $r(1 + |v_1| + \cdots + |v_n|)^k = r|v|^k$ due to the multinomial theorem. □

We note that our running example satisfies the premise to the proposition. In concrete cases more precise bounds than those given by Theorem 14 can be computed as has been done in Example 9. The next example clarifies that potentials are not restricted to polynomials.

Example 15. Consider that we annotate the constructors for natural numbers as $\mathsf{0}\colon [] \xrightarrow{0} \mathsf{Nat}^{p}$ and $\mathsf{s}\colon [\mathsf{Nat}^{2p}] \xrightarrow{\lozenge p} \mathsf{Nat}^{p}$. We then have, for example, $\Phi(t\colon \mathsf{Nat}^{1}) = 2^{t+1} - 1$.

As mentioned in the introduction, foundational issues are our main concern. However, the potential-based method detailed above are susceptible to automation. One conceives the resource annotations as variables and encodes the constraints of the typing rules in Figure 2 over these resource variables.

4 Small-Step Semantics

The big-step semantics, the type system, and Theorem 12 provide a potential-based resource analysis for typed TRSs that yields polynomial bounds. However, Theorem 12 is not directly applicable, if we want to link this analysis to the interpretation method. We recast the method and present a small-step semantics, used in our second soundness result (Theorem 20 below), cf. Figure 3. As the big-step semantics, the small-step semantics is decorated with counters for the derivation height of the evaluated terms. Its definition is instrumental in the proof of our second soundness theorem.

The transitive closure of the judgement $\vdash^{m} \langle s, \sigma\rangle \to \langle t, \tau\rangle$ is defined as follows:

– $\vdash^{m} \langle s,\sigma\rangle \twoheadrightarrow \langle t,\tau\rangle$ if $\vdash^{m} \langle s,\sigma\rangle \to \langle t,\tau\rangle$ ($m \in \{0,1\}$)
– $\vdash^{m_1+m_2} \langle s,\sigma\rangle \twoheadrightarrow \langle u,\rho\rangle$ if $\vdash^{m_1} \langle s,\sigma\rangle \to \langle t,\tau\rangle$ and $\vdash^{m_2} \langle t,\tau\rangle \twoheadrightarrow \langle u,\rho\rangle$.

The next lemma proves the equivalence of big-step and small-step semantics.

Lemma 16. *Let σ be a substitution, let t be a term, $\mathcal{V}\mathsf{ar}(t) \subseteq \mathsf{dom}(\sigma)$, and let v be a value. Then $\sigma \vdash^{m} t \Rightarrow v$ if and only if $\vdash^{m} \langle t,\sigma\rangle \twoheadrightarrow \langle v,\sigma'\rangle$, where σ' is an extension of σ.*

Proof. Let Π be the proof deriving $\sigma \vdash^{m} t \Rightarrow v$ and let D denote the sequence of reductions that make up $\vdash^{m} \langle t,\sigma\rangle \twoheadrightarrow \langle v,\sigma'\rangle$.

One proves the direction left-to-right by induction on the length of Π. We observe that if $\sigma \vdash^{m} t \Rightarrow v$ and if σ' is an extension of σ, then $\sigma' \vdash^{m} t \Rightarrow v$. Furthermore the sizes of the derivations of the corresponding judgements are the same. This follows by straightforward inductive argument.

In proof of the lemma, we consider one, significant case. We employ the notation from Figure 3. Suppose the last rule in Π has the form

$$\frac{\sigma \uplus \rho \vdash^{m_0} f(x_1,\dots,x_n) \Rightarrow v \quad \sigma \vdash^{m_1} t_1 \Rightarrow v_1 \quad \cdots \quad \sigma \vdash^{m_n} t_n \Rightarrow v_n}{\sigma \vdash^{m} f(t_1,\dots,t_n) \Rightarrow v} ,$$

where $t = f(t_1,\dots,t_n)$ and $m = \sum_{i=0}^{n} m_i$. By induction hypothesis we have for all $i = 1,\dots,n$: $\vdash^{m_i} \langle t_1,\sigma_{i-1}\rangle \twoheadrightarrow \langle v_1,\sigma_i\rangle$, where we set $\sigma_0 = \sigma$ and note that all σ_i are extensions of σ. From $\vdash^{0} \langle f(v_1,\dots,v_n),\sigma_n\rangle \to \langle f(x_1,\dots,x_n),\sigma_n \uplus \rho\rangle$ we obtain:

$$\vdash^{\sum_{i=1}^{n} m_i} \langle f(t_1,\dots,t_n),\sigma\rangle \twoheadrightarrow \langle f(x_1,\dots,x_n),\sigma_n \uplus \rho\rangle .$$

Furthermore, by the above and the induction hypothesis there exists a substitution σ' such that $\vdash^{m_0} \langle f(x_1,\dots,x_n),\sigma_n \uplus \rho\rangle \twoheadrightarrow \langle v,\sigma'\rangle$ where σ' extends $\sigma_n \uplus \rho$ (and thus also σ as $\mathsf{dom}(\sigma_n) \cap \mathsf{dom}(\rho) = \varnothing$). From above, we obtain $\vdash^{m} \langle t,\sigma\rangle \twoheadrightarrow \langle v,\sigma'\rangle$.

The direction from right to left follows by induction on the sum of the size of the proofs of the single-step execution in D. The proof is based on the observation that if $\vdash^{m} \langle s,\sigma\rangle \to \langle t,\sigma'\rangle$, $m \in \mathbb{N}$, then σ' extends σ and $s\sigma = s\sigma'$. □

We extend the notion of potential (cf. Definition 8) to ground terms. Recall that by assumption the declaration in $\mathcal{F}(f)$ is unique.

Definition 17. *Let $t = f(t_1,\dots,t_n) \in \mathcal{T}(\mathcal{D} \cup \mathcal{C})$ and let $[A_1 \times \cdots \times A_n] \xrightarrow{p} C \in \mathcal{F}(f)$. Then the* potential *of t is defined as follows: $\Phi(t\colon C) := p + \Phi(t_1\colon A_1) + \cdots + \Phi(t_n\colon A_n)$.*

Example 18 (continued from Example 13). Recall the type of chk. Let $q = \mathsf{que}(f,r)$ be a queue. We obtain $\Phi(\mathsf{chk}(q)\colon \mathsf{Q}^{(0,1)}) = 3 + \Phi(q\colon \mathsf{Q}^{(0,1)}) = 3 + \Phi(f\colon \mathsf{List}^0) + \Phi(r\colon \mathsf{List}^1) = 3 + |r|$.

Lemma 19. *Let $\mathcal{R}$ and σ be well-typed. Suppose $\Gamma \vdash^{p} t\colon A$. Then we have $\Phi(\sigma\colon \Gamma) + p \geqslant \Phi(t\sigma\colon A)$.*

Proof. Let Ξ denote the proof of $\Gamma \overset{p}{\vdash} t\colon A$. Then the lemma follows by induction on Ξ. □

We obtain our second soundness result.

Theorem 20. *Let $\mathcal{R}$ and σ be well-typed. If $\Gamma \overset{p}{\vdash} t\colon A$ and $\overset{m}{\vdash} \langle t,\sigma\rangle \twoheadrightarrow \langle u,\sigma'\rangle$, then $\Phi(\sigma\colon\Gamma) - \Phi(u\sigma'\colon A) + p \geqslant m$. Thus if for all ground basic terms t and types A: $\Phi(t\colon A) \in \mathsf{O}(n^k)$, where $n = |t|$, then $\mathsf{rc}_{\mathcal{R}}(n) \in \mathsf{O}(n^k)$.*

Proof. Let D denote the derivation of $\overset{m}{\vdash} \langle t,\sigma\rangle \twoheadrightarrow \langle u,\sigma'\rangle$ and let Ξ denote the proof of $\Gamma \overset{p}{\vdash} t\colon A$. The proof proceeds by main induction on the size of D and by side induction on the length of Ξ.

We exemplify the pattern of the proof on one case. Let Π denote the proof of the judgement $\overset{m_1}{\vdash} \langle t,\sigma\rangle \to \langle w,\gamma\rangle$ where $\overset{m_2}{\vdash} \langle w,\gamma\rangle \twoheadrightarrow \langle u,\sigma'\rangle$ and $m = m_1 + m_2$. Suppose Π has the form

$$\frac{f(l_1,\ldots,l_n) \to r \in \mathcal{R} \quad \forall i\colon x_i\sigma = l_i\tau}{\overset{1}{\vdash} \langle f(x_1,\ldots,x_n),\sigma\rangle \to \langle r, \sigma \uplus \tau\rangle} \;.$$

Then $t = f(x_1,\ldots,x_n)$ and $f(x_1,\ldots,x_n)\sigma = f(l_1,\ldots,l_n)\tau$. Let $\mathcal{V}\mathsf{ar}(f(\boldsymbol{l})) = \{y_1,\ldots,y_\ell\}$ and let $\mathcal{V}\mathsf{ar}(l_i) = \{y_{i1},\ldots,y_{il_i}\}$ for $i \in \{1,\ldots,n\}$. As $\mathcal{R}$ is left-linear we have $\mathcal{V}\mathsf{ar}(f(l_1,\ldots,l_n)) = \biguplus_{i=1}^{n} \mathcal{V}\mathsf{ar}(l_i)$. We set $\Gamma = x_1\colon A_1,\ldots,x_n\colon A_n$. By the assumption $\Gamma \overset{p}{\vdash} t\colon A$ and well-typedness of $\mathcal{R}$ we obtain

$$\overbrace{y_1\colon B_1,\ldots,y_\ell\colon B_\ell}^{=:\Delta} \overset{p-1+\sum_{i=1}^{n} k_i}{\vdash} r\colon A \;, \tag{2}$$

similar to (1). We have $\Phi(\sigma\colon\Gamma) + p = \sum_{i=1}^{n} (k_i + \Phi(y_{i1}\tau\colon B_{i1}) + \cdots + \Phi(y_{il_i}\tau\colon B_{il_i})) + p = \Phi(\tau\colon\Delta) + \sum_{i=1}^{n} k_i + (p-1) + 1$. The first equality follows by an inspection on the cases for the constructors. Furthermore note that $r\tau = r(\sigma \uplus \tau)$, as $\mathsf{dom}(\sigma) \cap \mathsf{dom}(\tau) = \varnothing$. The theorem follows by application of the main induction hypothesis on r in conjunction with the typing judgement (2). □

5 Typed Polynomial Interpretations

We adapt the concept of polynomial interpretation to typed TRSs. For that we suppose a mapping $[\![\cdot]\!]$ that assigns to every *annotated* type C a subset of the natural numbers, whose elements are ordered with $>$ in the standard way. The set $[\![C]\!]$ is called the *interpretation* of C.

Definition 21. *An* interpretation γ of function symbols *is a mapping from function symbols and types to functions over $\mathbb{N}$. Consider a function symbol f and an annotated type C such that $[A_1 \times \cdots \times A_n] \xrightarrow{p} C \in \mathcal{F}(f)$. Then the interpretation $\gamma(f,C)\colon [\![A_1]\!] \times \cdots \times [\![A_n]\!] \to [\![C]\!]$ of f is defined as: $\gamma(f,C)(x_1,\ldots,x_n) := x_1 + \cdots + x_n + p$.*

Note that by assumption the declaration in $\mathcal{F}(f)$ is unique and thus $\gamma(f, C)$ is unique. Interpretations of function symbols naturally extend to interpretations on ground terms. $[\![f(t_1,\dots,t_n){:}\,C]\!]^\gamma := \gamma(f,C)([\![t_1{:}\,A_1]\!]^\gamma,\dots,[\![t_n{:}\,A_n]\!]^\gamma)$. Let $\mathcal{R}$ be well-typed and let the interpretation γ of function symbols in $\mathcal{F}$ be induced by the well-typing of $\mathcal{R}$ as in Definition 21. Then by construction $[\![t{:}\,A]\!]^\gamma = \Phi(t{:}\,A)$.

Example 22 (continued from Example 13). We obtain the following definitions of the interpretation of function symbols γ. We start with the constructor symbols.

$$\gamma(\mathsf{0},\mathsf{Nat}^p) = 0 \qquad \gamma(\mathsf{s},\mathsf{Nat}^p)(x) = x + p \qquad \gamma(\mathsf{err\_head},\mathsf{Nat}^p) = 0$$
$$\gamma(\mathsf{nil},\mathsf{List}^q) = 0 \qquad \gamma(\sharp,\mathsf{List}^q)(x,y) = x + y + q \quad \gamma(\mathsf{err\_tail},\mathsf{Q}^{(0,1)}) = 0$$
$$\gamma(\mathsf{que},\mathsf{Q}^{(0,1)})(x,y) = x + y ,$$

where $p, q \in \mathbb{N}$. Similarly the definition of γ for defined symbols follows from the signature detailed in Example 13. Then for any rule $l \to r \in \mathcal{R}_{\mathsf{que}}$ and any substitution σ, we obtain $[\![l\sigma]\!]^\gamma > [\![r\sigma]\!]^\gamma$. We show this for rule 1.

$$\begin{aligned}[\![\mathsf{chk}(\mathsf{que}(\mathsf{nil}, r\sigma)){:}\,\mathsf{Q}^{(0,1)}]\!]^\gamma = [\![r\sigma{:}\,\mathsf{List}^1]\!]^\gamma + 3 > 0 \\ = [\![\mathsf{rev}(r\sigma){:}\,\mathsf{List}^0]\!]^\gamma + [\![\mathsf{nil}{:}\,\mathsf{List}^1]\!]^\gamma \\ = [\![\mathsf{que}(\mathsf{rev}(r\sigma),\mathsf{nil}){:}\,\mathsf{Q}^{(0,1)}]\!]^\gamma .\end{aligned}$$

Orientability of $\mathcal{R}_{\mathsf{que}}$ with the above given interpretation implies the optimal linear innermost runtime complexity.

We lift the standard order $>$ on the interpretation domain $\mathbb{N}$ to an order on terms. Let s and t be terms of type A. Then $s > t$ if for all well-typed substitutions σ we have $[\![s\sigma{:}\,A]\!]^\gamma > [\![t\sigma{:}\,A]\!]^\gamma$.

Theorem 23. *Let $\mathcal{R}$ be well-typed, constructor TRS over signature $\mathcal{F}$ and let the interpretation of function symbols γ be induced by the type system. Then $l > r$ for any rule $l \to r \in \mathcal{R}$. Thus if for all ground basic terms t and types A: $[\![t{:}\,A]\!]^\gamma \in \mathsf{O}(n^k)$, where $n = |t|$, then $\mathsf{rc}_{\mathcal{R}}(n) \in \mathsf{O}(n^k)$.*

Proof. Let $l = f(l_1,\dots,l_n)$ and let $x_1,\dots,x_n$ be fresh variables. Suppose further $[A_1 \times \cdots \times A_n] \xrightarrow{p} C \in \mathcal{F}(f)$. As $\mathcal{R}$ is well-typed we have

$$\overbrace{x_1{:}\,A_1,\dots,x_n{:}\,A_n}^{=:\Gamma} \overset{p}{\vdash} f(x_1,\dots,x_n){:}\,C ,$$

for $p \in \mathbb{N}$. Now suppose that τ denotes any well-typed substitution for the rule $l \to r$. In the standard way, we extend τ to a well-typed substitution σ such that $l\tau = f(x_1,\dots,x_n)\sigma$. By definition of the small-step semantics, we obtain $\overset{1}{\vdash} \langle f(x_1,\dots,x_n),\sigma\rangle \to \langle r, \sigma \uplus \tau\rangle$. Then by Theorem 20, $\Phi(\sigma{:}\,\Gamma) + p > \Phi(r(\sigma \uplus \tau){:}\,C)$ and by definitions, we have:

$$\Phi(l\tau{:}\,C) = \Phi(f(x_1\sigma,\dots,x_n\sigma){:}\,C) = \sum_{i=1}^{n} \Phi(x_i\sigma{:}\,A_i) + p = \Phi(\sigma{:}\,\Gamma) + p .$$

Furthermore, observe that $r(\sigma \uplus \tau) = r\tau$ as $\mathsf{dom}(\sigma) \cap \mathsf{dom}(\tau) = \varnothing$. In sum, we obtain $\Phi(l\tau{:}\, C) > \Phi(r\tau{:}\, C)$, from which we conclude $[\![l\tau{:}\, C]\!]^{\gamma} > [\![r\tau{:}\, C]\!]^{\gamma}$. As τ was chosen arbitrarily, we obtain $\mathcal{R} \subseteq >$. $\square$

We say that an interpretation *orients* a typed TRS $\mathcal{R}$, if $\mathcal{R} \subseteq >$. As an immediate consequence of the theorem, we obtain the following corollary.

Corollary 24. *Let $\mathcal{R}$ be a well-typed and constructor TRS. Then there exists a typed polynomial interpretation over $\mathbb{N}$ that orients $\mathcal{R}$.*

At the end of Section 3 we have remarked on the automatabilty of the obtained amortised analysis. Observe that Theorem 23 gives rise to a related but conceptually different implementation. Instead of encoding the constraints of the typing rules in Figure 2 one directly encodes the orientability constraints for each rule, cf. [22].

6 Conclusion

This paper is concerned with the connection between amortised resource analysis, originally introduced for functional programs, and polynomial interpretations, which are frequently used in complexity and termination analysis of rewrite systems.

In order to study this connection we have established a novel resource analysis for typed term rewrite systems based on a potential-based type system. This type system gives rise to polynomial bounds for innermost runtime complexity. A key observation is that the classical notion of potential can be altered so that not only values but any term is assigned a potential. Ie. the potential function Φ is conceivable as an interpretation. Based on this observation we have shown that well-typedness of a TRSs $\mathcal{R}$ induces a typed polynomial interpretation that orients $\mathcal{R}$.

Apart from clarifying the connection between amortised resource analysis and polynomial interpretation our results induce two new methods for the innermost runtime complexity of typed TRSs. If we restrict the length of the resource annotations, standard techniques for type inference performed on our type system yield linear constraints that can be solved with a linear constraint solver. On the other hand considering the synthesis of typed polynomial interpretations on fixes abstract polynomials and feeds the obtained constraints into an SMT solver. A prototype is in preparation.

We emphasise that these methods are not restricted to typed TRSs, as our cost model gives rise to a *persistent* property. A property is called persistent if for any typed TRS $\mathcal{R}$ the property holds iff it holds for the corresponding untyped TRS $\mathcal{R}'$. While termination is in general not persistent [18], it is not difficult to see that runtime complexity is persistent. This is due to the restricted set of starting terms. Thus the proposed techniques directly give rise to novel methods of automated runtime complexity analysis. In future work we will clarify whether the established results extend to the multivariate amortised resource analysis presented in [20].

References

1. Albert, E., Arenas, P., Genaim, S., Puebla, G.: Closed-form upper bounds in static cost analysis. JAR 46 (2011)
2. Alias, C., Darte, A., Feautrier, P., Gonnord, L.: Multi-dimensional rankings, program termination, and complexity bounds of flowchart programs. In: Cousot, R., Martel, M. (eds.) SAS 2010. LNCS, vol. 6337, pp. 117–133. Springer, Heidelberg (2010)
3. Zuleger, F., Gulwani, S., Sinn, M., Veith, H.: Bound analysis of imperative programs with the size-change abstraction. In: Yahav, E. (ed.) Static Analysis. LNCS, vol. 6887, pp. 280–297. Springer, Heidelberg (2011)
4. Noschinski, L., Emmes, F., Giesl, J.: Analyzing innermost runtime complexity of term rewriting by dependency pairs. JAR 51, 27–56 (2013)
5. Hoffmann, J., Aehlig, K., Hofmann, M.: Resource aware ML. In: Madhusudan, P., Seshia, S.A. (eds.) CAV 2012. LNCS, vol. 7358, pp. 781–786. Springer, Heidelberg (2012)
6. Avanzini, M., Moser, G.: Tyrolean complexity tool: Features and usage. In: Proc. 24th RTA. LIPIcs, vol. 21, pp. 71–80 (2013)
7. Okasaki, C.: Purely functional data structures. Cambridge University Press (1999)
8. Tarjan, R.: Amortized computational complexity. SIAM J. Alg. Disc. Math. 6, 306–318 (1985)
9. Turing, A.: Checking a large routine. In: Report of a Conference on High Speed Automatic Calculating Machines, pp. 67–69. Cambridge University (1949)
10. Baader, F., Nipkow, T.: Term Rewriting and All That. Cambridge University Press (1998)
11. Bonfante, G., Cichon, A., Marion, J.Y., Touzet, H.: Algorithms with polynomial interpretation termination proof. JFP 11, 33–53 (2001)
12. Hofmann, M., Jost, S.: Static prediction of heap space usage for first-order functional programs. In: Proc. 30th POPL, pp. 185–197. ACM (2003)
13. Bonfante, G., Marion, J.Y., Moyen, J.Y.: Quasi-interpretations a way to control resources. TCS 412, 2776–2796 (2011)
14. Hoffmann, J., Hofmann, M.: Amortized resource analysis with polynomial potential. In: Gordon, A.D. (ed.) ESOP 2010. LNCS, vol. 6012, pp. 287–306. Springer, Heidelberg (2010)
15. Hoffmann, J., Hofmann, M.: Amortized resource analysis with polymorphic recursion and partial big-step operational semantics. In: Ueda, K. (ed.) APLAS 2010. LNCS, vol. 6461, pp. 172–187. Springer, Heidelberg (2010)
16. Hofmann, M., Moser, G.: Amortised resource analysis and typed polynomial interpretations (extended version). CoRR, cs.LO (2014), `http://arxiv.org/abs/1402.1922`
17. Jouannaud, J.P., Rubio, A.: The higher-order recursive path ordering. In: Proc. 14th LICS, pp. 402–411. IEEE Computer Society (1999)
18. TeReSe: Term Rewriting Systems. Cambridge Tracks in Theoretical Computer Science, vol. 55. Cambridge University Press (2003)
19. Jost, S., Loidl, H.-W., Hammond, K., Scaife, N., Hofmann, M.: "Carbon Credits" for resource-bounded computations using amortised analysis. In: Cavalcanti, A., Dams, D.R. (eds.) FM 2009. LNCS, vol. 5850, pp. 354–369. Springer, Heidelberg (2009)
20. Hoffmann, J., Aehlig, K., Hofmann, M.: Multivariate amortized resource analysis. TOPLAS 34, 14 (2012)
21. Hoffmann, J.: Types with Potential: Polynomial Resource Bounds via Automatic Amortized Analysis. PhD thesis, Ludwig-Maximilians-Universiät München (2011)
22. Contejean, E., Marché, C., Tomás, A.P., Urbain, X.: Mechanically proving termination using polynomial interpretations. JAR 34, 325–363 (2005)

Confluence by Critical Pair Analysis

Jiaxiang Liu[1,2,3], Nachum Dershowitz[4], and Jean-Pierre Jouannaud[1,3]

[1] School of Software, Tsinghua University, Tsinghua Nat. Lab. for IST, Beijing, China
[2] Key Lab. for Information System Security, Ministry of Education, Beijing, China
[3] École Polytechnique, Palaiseau, France
[4] Tel Aviv University, Tel Aviv, Israel

Abstract. Knuth and Bendix showed that confluence of a terminating first-order rewrite system can be reduced to the joinability of its finitely many critical pairs. We show that this is still true of a rewrite system $R_T \cup R_{NT}$ such that R_T is terminating and R_{NT} is a left-linear, rank non-increasing, possibly non-terminating rewrite system. Confluence can then be reduced to the joinability of the critical pairs of R_T and to the existence of decreasing diagrams for the critical pairs of R_T inside R_{NT} as well as for the rigid parallel critical pairs of R_{NT}.

1 Introduction

Rewriting is a non-deterministic rule-based mechanism for describing intentional computations. Confluence is the property expressing that the associated extensional relation is functional. It is well-known that confluence of a set of rewrite rules is undecidable. There are two main methods for showing confluence of a binary relation: the first applies to terminating relations [8] and is the basis of the Knuth-Bendix test, reducing confluence to the *joinability* of its so-called *critical pairs* obtained by unifying left-hand sides of rules at subterms [7]. Based on the Hindley-Rosen Lemma, the second applies to non-terminating relations [9] and is the basis of Tait's confluence proof for the pure λ-calculus. Reduction to critical pairs is also possible under strong linearity assumptions [3], although practice favors orthogonal (left-linear, critical pair free) systems for which there are no pairs. It is our ambition to develop a critical-pair criterion capturing both situations together.

Problem. Van Oostrom succeeded in capturing both confluence methods within a single framework thanks to the notion of *decreasing diagram* of a *labelled abstract relation* [12]. In [5], the method is applied to concrete rewrite relations on terms, opening the way to an analysis of non-terminating rewrite relations in terms of the joinability of their critical pairs. The idea is to split the set of rules into a set R_T of terminating rules and a set R_{NT} of non-terminating ones. While left-linearity is required from R_{NT} as shown by simple examples, it is not from R_T. This problem has however escaped efforts so far.

Contributions. We deliver the first true generalization of the Knuth-Bendix test to rewrite systems made of two subsets, R_T of terminating rules and R_{NT} of possibly non-terminating, rank non-increasing, left-linear rules. Confluence is reduced – via decreasing diagrams – to joinability of the finitely many *critical pairs* of rules in R_T within rules

G. Dowek (ed.): RTA-TLCA 2014, LNCS 8560, pp. 287–302, 2014.

in $R_T \cup R_{NT}$ and the finitely many *rigid parallel critical pairs* of rules in R_{NT} within rules in $R_T \cup R_{NT}$. The result is obtained thanks to a new notion, *sub-rewriting*, which appears as the key to glue together many concepts that appeared before in the study of termination and confluence of union systems, namely: caps and aliens, rank non-increasing rewrites, parallel rewriting, decreasing diagrams, stable terms, and constructor-lifting rules. This culminates with the solution of an old open problem raised by Huet who exhibited a critical pair free, non-terminating, non-confluent system [3]. We show that the computation of critical pairs should then involve unification over infinite rational trees, and then, indeed, Huet's example is no longer critical-pair free.

Organization. Sections 4 and 5 are devoted to the main result, its proof, and extension to Huet's open problem. Relevant literature is analyzed in Sect. 6.

2 Term Algebras

Given a *signature* $\mathcal{F}$ of *function symbols* and a denumerable set $\mathcal{X}$ of *variables*, $\mathcal{T}(\mathcal{F}, \mathcal{X})$ denotes the set of *terms* built up from $\mathcal{F}$ and $\mathcal{X}$. Terms are identified with finite labelled trees as usual. *Positions* are strings of positive integers, identifying the empty string Λ with the root position. We use "·" for concatenation of positions, or sets thereof. We assume a set of variables $\mathcal{Y}$ disjoint from $\mathcal{X}$ and a bijective mapping ξ from the set of positions to $\mathcal{Y}$. We use $\mathcal{FP}os(t)$ to denote the set of non-variable positions of t, $t(p)$ for the function symbol at position p in t, $t|_p$ for the *subterm* of t at position p, and $t[u]_p$ for the result of replacing $t|_p$ with u at position p in t. We may omit the position p, writing $t[u]$ for simplicity and calling $t[\cdot]$ a *context*. We use $\geq$ for the partial order on positions (further from the root is bigger), $p\#q$ for incomparable positions p, q, called *disjoint*. The order on positions is extended to sets as follows: $P \geq Q$ (resp. $P > Q$) if $(\forall p \in P)(\exists q \in max(Q))\, p \geq q$ (resp. $p > q$), where $max(P)$ is the set of maximal positions in P. We use p for the singleton set $\{p\}$. We write $u[v_1, \ldots, v_n]_Q$ for $u[v_1]_{q_1} \cdots [v_n]_{q_n}$ if $Q = \{q_i\}_1^n$. By $\mathcal{V}ar(t)$ we mean the set of variables occurring in t. We say that t is *linear* if no variable occurs more than once in t.

Substitutions are mappings from variables to terms, called *variable substitutions* when mapping variables onto variables, and *variable renamings* when also bijective. We denote by $\sigma_{|X}$ the restriction of σ to a subset X of variables. We use Greek letters for substitutions and postfix notation for their application. The strict *subsumption order* $\gtrdot$ on terms (resp. substitutions) associated with the quasi-order $s \trianglerighteq t$ (resp. $\sigma \trianglerighteq \tau$) iff $s = t\theta$ (resp. $\sigma = \tau\theta$) for some substitution θ, is well-founded. Given terms s, t, computing the substitution σ whenever it exists such that $t = s\sigma$ (resp. $t\sigma = s\sigma$) is called *matching* (resp. unification) and σ is called a *match* (resp. *unifier*). Two unifiable terms s, t have a unique (up to variable renaming) *most general unifier* $mgu(s, t)$, which is the smallest with respect to subsumption. The result remains true when unifying terms $s, t_1, \ldots, t_n$ at a set of disjoint positions $\{p_i\}_1^n$ such that $s|_{p_1}\sigma = t_1\sigma \wedge \ldots \wedge s|_{p_n}\sigma = t_n\sigma$, of which the previous result is a particular case when $n = 1$ and $p_1 = \Lambda$.

Given $F \subseteq \mathcal{F}$, a term t is F-headed if $t(\Lambda) \in F$. The notion extends to substitutions.

3 Rewriting

Our goal is to reduce the Church-Rosser property of the union of a terminating rewrite relation R_T and a non-terminating relation R_{NT} to that of finitely many critical pairs. The particular case where R_{NT} is empty was carried out by Knuth and Bendix and is based on Newman's result stating that a terminating relation is Church-Rosser provided its local peaks are joinable. The other particular case, where R_T is empty, was considered by Huet and is based on Hindley's result stating that a (non-terminating) relation is Church-Rosser provided its local peaks are joinable in at most one step from each side. The general case requires using both, which has been made possible by van Oostrom, who introduced labelled relations and decreasing diagrams to replace joinability.

Definition 1. *A* rewrite rule *is a pair of terms, written $l \to r$, whose* left-hand side *l is not a variable and whose* right-hand side *r satisfies $\mathcal{V}ar(r) \subseteq \mathcal{V}ar(l)$. A rewrite system R is a set of rewrite rules. A rewrite system is* left-linear *(resp.* linear*) if for every rule $l \to r$, the left-hand side l is a linear term (resp. l and r are linear terms).*

Definition 2. *A term u* rewrites in parallel *to v at a set $P = \{p_i\}_1^n$ of pairwise disjoint positions, written $u \Rightarrow_{l\to r}^{P} v$, if $(\forall p_i \in P)\ u|_{p_i} = l\sigma_i$ and $v = u[r\sigma_1, \ldots, r\sigma_n]_P$. The term $l\sigma_i$ is a* redex*. We may omit P or replace it by a property that it satisfies.*

We call our notion of parallel rewriting *rigid*. It departs from the literature [3,1] by imposing the use of a *single* rule. Rewriting extends naturally to lists of terms of the same length, hence to substitutions of the same domain. Rewriting *terminates* if there exists no infinite sequence of rewriting issuing from an arbitrary term.

Plain rewriting is obtained as the particular case of parallel rewriting when $n = 1$. We then also write $u \to_{l\to r}^{p} v$. As a consequence, most of the following definitions will be given for parallel rewriting, while also applying to plain rewriting.

Consider two parallel rewrites issuing from the same term u with possibly different rules, say $u \Rightarrow_{l\to r}^{P} v$ and $u \Rightarrow_{g\to d}^{Q} w$. Following Huet [3], we distinguish three cases,

$P \# Q$, that is, $(\forall p \in P\ \forall q \in Q)\ p \# q$, (disjoint case)
$P = \{p\}, Q > p \cdot \mathcal{FP}os(l)$, (ancestor case)
$P = \{p\}, Q \subseteq p \cdot \mathcal{FP}os(l)$, (critical case)

all other cases being a combination of the above three.

Definition 3 (Rigid parallel critical pairs). *Given a rule $l \to r$, a set $P = \{p_i \in \mathcal{FP}os(l)\}_1^n$ of disjoint positions and n copies $\{g_i \to d_i\}_1^n$ of a rule $g \to d$ sharing no variable among themselves nor with $l \to r$, such that σ is a most general unifier of the terms $l, g_1, \ldots, g_n$ at P. Then $l\sigma$ is the* overlap *and $\langle r\sigma, l\sigma[d_1\sigma, \ldots, d_n\sigma]_P\rangle$ the* rigid (parallel) critical pair *of $\{g_i \to d_i\}_1^n$ on $l \to r$ at P (a* critical pair *if $n = 1$).*

Definition 4. *A* labelled rewrite relation *is a pair made of a rewrite relation $\to$ and a mapping from rewrite steps to a set of labels $\mathcal{L}$ equipped with a partial quasi-order $\trianglerighteq$ whose strict part $\triangleright$ is well-founded. We write $u \Rightarrow_{R}^{P,m} v$ for a parallel rewrite step from u to v at positions P with label m and rewrite system R. Indexes P, m, R may be omitted. We also write $\alpha \triangleright l$ (resp. $l \triangleright \alpha$) if $m \triangleright l$ (resp. $l \triangleright m$) for all m in the multiset α.*

Given an arbitrary (possibly labelled) rewrite step $\longrightarrow^{l}$, we denote its projection on terms by $\longrightarrow$, its inverse by ${}^{l}\!\longleftarrow$, its reflexive closure by $\Longrightarrow^{l}$, its symmetric closure by $\longleftrightarrow^{l}$, its reflexive and transitive closure by $\twoheadrightarrow^{\alpha}$ for some word α on the alphabet of labels, and its reflexive, symmetric, transitive closure, called *conversion*, by $\twoheadleftarrow\!\!\twoheadrightarrow^{\alpha}$. We sometimes consider the word α to be a multiset. Given u, $\{v \mid u \twoheadrightarrow v\}$ is the set of reducts of u. We say that a reduct of u is *reachable* from u.

The triple v, u, w is said to be a *local peak* if $v\ {}^{l}\!\longleftarrow u \longrightarrow^{m} w$, a *peak* if $v\ {}^{\alpha}\!\twoheadleftarrow u \twoheadrightarrow^{\beta} w$, a *joinability diagram* if $v \twoheadrightarrow^{\alpha} u\ {}^{\beta}\!\twoheadleftarrow w$. The local peak $v \underset{l\to r}{\overset{p,m}{\longleftarrow}} u \underset{g\to d}{\overset{q,n}{\longrightarrow}} w$ is a *disjoint, critical, ancestor* local peak if $p\#q, q \in p{\cdot}\mathcal{FP}os(l), q > p{\cdot}\mathcal{FP}os(l)$, respectively. The pair v, w is *convertible* if $v \twoheadleftarrow\!\!\twoheadrightarrow^{\alpha} w$, *divergent* if $v\ {}^{\alpha}\!\twoheadleftarrow u \twoheadrightarrow^{\beta} w$ for some u, and *joinable* if $v \twoheadrightarrow^{\alpha} t\ {}^{\beta}\!\twoheadleftarrow w$ for some t. The relation $\longrightarrow$ is *locally confluent* (resp. *confluent, Church-Rosser*) if every local peak (resp. divergent pair, convertible pair) is joinable.

Decreasing Diagrams. Given a rewrite relation $\longrightarrow$ on terms, we first consider specific conversions made of a local peak and an associated conversion called a *local diagram* and recall the important subclass of van Oostrom's decreasing diagrams and their main property: a relation all whose local diagrams are decreasing enjoys the Church-Rosser property, hence confluence. Decreasing diagrams were introduced in [12], where it is shown that they imply confluence. Van Oostrom's most general form of decreasing diagrams is discussed in [5].

Definition 5 (Local diagrams). *A* local diagram D *is a conversion made of a* local peak $D_{peak} = v \longleftarrow u \longrightarrow w$ *and a* conversion $D_{conv} = v \twoheadleftarrow\!\!\twoheadrightarrow u$. *We call* diagram rewriting *the rewrite relation* $\Rightarrow_{\mathcal{D}}$ *on conversions associated with a set* $\mathcal{D}$ *of local diagrams, in which a local peak is replaced by one of its associated conversions:*

$$P\, D_{peak}\, Q \Rightarrow_{\mathcal{D}} P\, D_{conv}\, Q \text{ for some } D \in \mathcal{D}$$

Definition 6 (Decreasing diagrams [12]). *A local diagram D with peak $v\ {}^{l}\!\longleftarrow u \longrightarrow^{m} w$ is* decreasing *if $D_{conv} = v \twoheadrightarrow^{\alpha} s \Longrightarrow^{m} s' \twoheadrightarrow^{\delta} {}^{\delta'}\!\twoheadleftarrow t' \ {}^{l}\!\Longleftarrow t\ {}^{\beta}\!\twoheadleftarrow w$, with labels in α (resp. β) strictly smaller than l (resp. m), and labels in δ, δ' strictly smaller than l or m. The rewrites $v \twoheadrightarrow^{\alpha} s$ and $t\ {}^{\beta}\!\twoheadleftarrow w$, $s \Longrightarrow^{m} s'$ and $t' \ {}^{l}\!\Longleftarrow t$, $s' \twoheadrightarrow^{\delta} {}^{\delta'}\!\twoheadleftarrow t'$ are called the* side steps, facing steps, *and* middle steps *of the diagram, respectively. A decreasing diagram D is* stable *if $C[D\gamma]$ is decreasing for arbitrary context $C[\cdot]$ and substitution γ.*

Theorem 1 ([5]). *The relation $\Rightarrow_{\mathcal{D}}$ terminates for any set $\mathcal{D}$ of decreasing diagrams.*

Corollary 1. *Assume that $T \subseteq \mathcal{T}(\mathcal{F}, \mathcal{X})$ and $\mathcal{D}$ is a set of decreasing diagrams in T such that T is closed under $\Rightarrow_{\mathcal{D}}$. Then the restriction of $\longrightarrow$ to T is Church-Rosser if every local peak in T has a decreasing diagram in $\mathcal{D}$.*

This simple corollary of Theorem 1 implies van Oostrom decreasing diagram theorem by taking $T = \mathcal{T}(\mathcal{F}, \mathcal{X})$. With a different choice of the set T, it will be the basis of our main Church-Rosser result to come.

Layering. From now on, we assume two signatures F_T and F_{NT} satisfying
(A1) $F_T \cap F_{NT} = \varnothing$.
and proceed by slicing terms into homogeneous subparts, following definitions in [4].

Definition 7. *A term* $s \in \mathcal{T}(F_T \cup F_{NT}, \mathcal{X})$ *is* homogeneous *if it belongs to* $\mathcal{T}(F_T, \mathcal{X})$ *or to* $\mathcal{T}(F_{NT}, \mathcal{X})$*; otherwise it is* heterogeneous.

Thanks to assumption (A1), a heterogeneous term can be uniquely decomposed (w.r.t. $\mathcal{Y}$ and ξ introduced in Section 2) into a topmost homogeneous part, its *cap*, and a multiset of remaining subterms, its *aliens*, headed by symbols of the other signature.

Definition 8 (Cap, aliens). *Let* $t \in \mathcal{T}(F_T \cup F_{NT}, \mathcal{X})$*. An* alien *of* t *is a maximal non-variable subterm of* t *whose head does not belong to the signature of* t*'s head. We use* $\mathcal{AP}os(t)$ *for its set of pairwise disjoint* alien positions, $\mathcal{A}(t)$ *for its list of aliens from left to right, and* $\mathcal{CP}os(t) = \{p \in \mathcal{P}os(t) \mid p \not\geq \mathcal{AP}os(t)\}$ *for its set of* cap positions. *We define the* cap $\bar{t}$ *and* alien substitution $\overline{\gamma}_t$ *of* t *as follows: (i)* $\mathcal{P}os(\bar{t}) = \mathcal{CP}os(t) \cup \mathcal{AP}os(t)$*; (ii)* $(\forall p \in \mathcal{CP}os(t)), \bar{t}(p) = t(p)$*; (iii)* $(\forall p \in \mathcal{AP}os(t)), \bar{t}(p) = \xi(p)$ *and* $\overline{\gamma}_t(\xi(p)) = t|_p$*. The* rank *of* t*, denoted* $rk(t)$*, is 1 plus the maximal rank of its aliens.*

Fact. Given $t \in \mathcal{T}(F_T \cup F_{NT}, \mathcal{X})$, then $t = \bar{t}\overline{\gamma}_t$.

Example 1. Let $F_T = \{G\}, F_{NT} = \{F, 0, 1\}, t = F(G(0,1,1), G(0,1,x), G(0,1,1))$. Then t has cap $F(y_1, y_2, y_3)$ and aliens $G(0,1,1)$ and $G(0,1,x)$. $G(0,1,1)$ has cap $G(y_1, y_2, y_3)$ and homogeneous aliens 0 and 1, while $G(0,1,x)$ has cap $G(y_1, y_2, x)$ and same set of homogeneous aliens. Hence, the rank of t is 3.

4 From Church-Rosser to Critical Pairs

Definition 9. *A rewrite rule* $l \to r$ *is* rank non-increasing *iff for all rewrites* $u \to_{l \to r} v$, $rk(u) \geq rk(v)$*. A rewrite system is rank non-increasing iff all its rules are.*

From now on, we assume we are given two rewrite systems R_T and R_{NT} satisfying:
(A2) R_T is a terminating rewrite system in $\mathcal{T}(F_T, \mathcal{X})$;
(A3) R_{NT} is a set of rank non-increasing, left-linear rules $f(s) \to g(t)$ s.t. $f, g \in F_{NT}$, $s, t \in \mathcal{T}(F_T \cup F_{NT}, \mathcal{X})$;
(A4) if $g \to d \in R_T$ overlaps $l \to r \in R_{NT}$ at $p \in \mathcal{FP}os(l)$, then $l|_p \in \mathcal{T}(F_T, \mathcal{X})$.

Our goal is to show that $R_T \cup R_{NT}$ is Church-Rosser provided its critical pairs have appropriate decreasing diagrams.

Strategy. Since R_T and R_{NT} are both rank non-increasing, by assumption for the latter and homogeneity assumption of its rules for the former, we shall prove our result by induction on the rank of terms. To this end, we introduce the set $\mathcal{T}_n(F_T \cup F_{NT}, \mathcal{X})$ of terms of rank at most n. Since rewriting is rank non-increasing, $\mathcal{T}_n(F_T \cup F_{NT}, \mathcal{X})$ is closed under diagram rewriting. This is why we adopted this restricted form of decreasing diagrams rather than the more general form studied in [5].

We say that two terms in $\mathcal{T}_n(F_T \cup F_{NT}, \mathcal{X})$ are n*-*$(R_T \cup R_{NT})$*-convertible* (in short, n*-convertible*) if their conversion involves terms in $\mathcal{T}_n(F_T \cup F_{NT}, \mathcal{X})$ only. We shall assume that n-$(R_T \cup R_{NT})$-convertible terms are joinable, and show that $(n+1)$-$(R_T \cup R_{NT})$-convertible terms are joinable as well by exhibiting decreasing diagrams for all their local peaks, using Corollary 1.

Since R_{NT} may have non-linear right-hand sides, we classically use parallel rewriting with R_{NT} rules to enable the existence of decreasing diagrams for ancestor peaks in case R_{NT} is below R_{NT}. The main difficulty, however, has to do with ancestor peaks $v \;{}_{R_{NT}}^{\;q}\!\!\leftarrow u \rightarrow_{R_T}^{p} w$ for which R_{NT} is below R_T. Due to non-left-linearity of the rules in R_T, the classical diagram for such peaks, $v \twoheadrightarrow_{R_{NT}} s \rightarrow_{R_T}^{p} t \twoheadleftarrow_{R_{NT}} w$, can hardly be made decreasing in case $s \rightarrow_{R_T}^{p} t$ must be a facing step and $v \twoheadrightarrow_{R_{NT}} s$ side steps with labels identical to that of the top R_{NT}-step. A way out is to group them together as a single facing step from v to t. To this end, we introduce a specific rewriting relation:

Definition 10 (Sub-rewriting). *A term* u sub-rewrites *to* v *at* $p \in \mathcal{P}os(u)$ *with* $l \rightarrow r$ *in* R_T*, written* $u \rightarrow_{R_{T_{sub}}}^{p} v$ *if the following conditions hold: (i)* $\mathcal{FP}os(l) \subseteq \mathcal{CP}os(u|_p)$*; (ii)* $u\,(\rightarrow_{R_T \cup R_{NT}}^{\geq p \cdot \mathcal{AP}os(u|_p)})^{*}\, w = u[l\sigma]_p$ *; (iii)* $v = u[r\sigma]_p$*.*

Condition (ii) allows *arbitrary rewriting* in $\mathcal{A}(u|_p)$ until an R_T-redex is obtained. Thanks to assumptions (A1–3), these aliens remain aliens along the derivation from u to w, implying (i). Condition (i) will however be needed later when relaxing assumptions (A1) and (A3). Note also that the cap of $w|_p$ may collapse in the last step, in which case $v|_p$ becomes F_{NT}-headed.

A Hierarchy of Decompositions. Sub-rewriting needs another notion of cap for F_T-headed terms. Let ζ_n be a bijective mapping from $\mathcal{Y} \cup \mathcal{X}$ to n-$(R_T \cup R_{NT})$-convertibility classes of terms in $\mathcal{T}(F_T \cup F_{NT}, \mathcal{X})$, which is the identity on $\mathcal{X}$. The rank of a term being at least one, 0-$(R_T \cup R_{NT})$-convertibility does not identify any two different terms; hence ζ_0 is a bijection from $\mathcal{Y} \cup \mathcal{X}$ to $\mathcal{T}(F_T \cup F_{NT}, \mathcal{X})$. Similarly we denote by ζ_∞ a bijective mapping from $\mathcal{Y} \cup \mathcal{X}$ to $(R_T \cup R_{NT})$-convertibility classes, abbreviated as ζ.

Definition 11 (Hat). *The* hat at rank n *of a term* $t \in \mathcal{T}(F_T \cup F_{NT}, \mathcal{X})$ *is the term* $\widehat{t^n}$ *defined as: if* t *is* F_{NT}*-headed,* $\widehat{t^n} = \zeta_n^{-1}(t)$*; otherwise,* $(\forall p \in \mathcal{CP}os(t))\ \widehat{t^n}(p) = \bar{t}(p)$ *and* $(\forall p \in \mathcal{AP}os(t))\ \widehat{t^n}(p) = \zeta_n^{-1}(t|_p)$*.*

Since n-$(R_T \cup R_{NT})$-convertibility is an infinite hierarchy of equivalences identifying more and more terms, given t, $\widehat{t^n}$ is an infinite sequence of terms, each of them being an instance of the previous one, which is stable from some index n_t. We use $\widehat{t}$ for $\widehat{t^\infty}$.

Lemma 1. *Let* $t \in \mathcal{T}(F_T \cup F_{NT}, \mathcal{X})$ *and* $m \geq n \geq 0$*. Then* $\widehat{t} \trianglerighteq \widehat{t^m} \trianglerighteq \widehat{t^n} \trianglerighteq \bar{t}$*.*

The associated variable substitution from $\widehat{t^n}$ to $\widehat{t^m}$ is $\xi_{n,m}$, omitting m when infinite.

Note that $\xi_{n,m}$ does not actually depend on the term t, but only on the m- and n-convertibility classes. Also, $\widehat{t^0}$ corresponds to the case where identical terms only are identified by ζ_0^{-1}, while $\widehat{t}$ corresponds to the case where any two $(R_T \cup R_{NT})$-convertible terms are identified by ζ^{-1}. In the literature, $\widehat{t^0}$ is usually called a hat (or a cap!).

Example 2. Let $F_{NT} = \{F\}$, $F_T = \{G, 0, 1\}$ and $R_T = \{1 \rightarrow 0\}$. Then, $G(F(1,0,x), F(1,0,x), 1) \rightarrow_{1\rightarrow 0}^{2\cdot 1} G(F(1,0,x), F(0,0,x), 1)$. 0-hats of these terms are $G(y, y, 1)$ and $G(y, y', 1)$, respectively. Their 1-hats are the same as their 0-hats, since their aliens have rank 2, hence cannot be 1-convertible. On the other hand, their $(i \geq 2)$-hats are $G(y, y, 1)$ and $G(y, y, 1)$, since $F(1, 0, x)$ and $F(0, 0, x)$ are 2-convertible.

The following lemmas are standard, with $\zeta_t = \zeta_{0|\mathcal{V}ar(\widehat{t^0})}$.

Lemma 2. *Let* $t \in \mathcal{T}(F_T \cup F_{NT}, \mathcal{X})$. *Then* $t = \widehat{t}^0 \zeta_t$.

Lemma 3. *Let* $u \rightarrow^p_{R_T} v$, $p \in \mathcal{CP}os(u)$. *Then* $\widehat{u}^0 \rightarrow^p_{R_T} \widehat{v}^0$ *and* $(\forall y \in \mathcal{V}ar(\widehat{v}^0))\, \zeta_u(y) = \zeta_v(y)$.

Lemma 4. *Let* $u(\Lambda) \in F_T$ *and* $u \rightarrow^p_{R_T \cup R_{NT}} v$ *at* $p \geq \mathcal{AP}os(u)$. *Then* $\mathcal{CP}os(u) = \mathcal{CP}os(v)$, $(\forall q \in \mathcal{CP}os(u))\, u(q) = v(q)$, $\mathcal{AP}os(u) = \mathcal{AP}os(v)$, $(\forall q \in \mathcal{AP}os(u))\, u|_q \Rightarrow_{R_T \cup R_{NT}} v|_q$.

Key properties of sub-rewriting are the following:

Lemma 5. *Let* u *be an* F_T*-headed term of rank* $n+1$ *s.t.* $u \rightarrow^{\geq \mathcal{AP}os(u)}_{R_T \cup R_{NT}} v$. *Then,* $(\forall i \geq n) \widehat{u}^i = \widehat{v}^i$.

Proof. Rules in R_{NT} being F_{NT}-headed, $\mathcal{AP}os(u) = \mathcal{AP}os(v)$, and rewriting in aliens does not change $\mathcal{CP}os(u)$. It does not change $(i \geq n)$-convertibility either, hence the statement. □

Lemma 6. *Let* u *of rank* $n+1$, $p \in \mathcal{CP}os(u)$, *and* $u \rightarrow^p_{R_{T\,sub}} v$. *Then,* $(\forall i \geq n)\, \widehat{u}^i \rightarrow^p_{R_T} \widehat{v}^i$.

Proof. By definition of sub-rewriting, we get $u(\rightarrow^{\geq \mathcal{AP}os(u)}_{R_T \cup R_{NT}})^* w \rightarrow^p_{l \rightarrow r \in R_T} v$, therefore $w|_p = l\sigma$ for some substitution σ and $v = w[r\sigma]_p$. Let $i \geq n$.

By Lemma 3, $\widehat{w}^0 \rightarrow^p_{l \rightarrow r} \widehat{v}^0$. By repeated applications of Lemma 4, $\mathcal{CP}os(u) = \mathcal{CP}os(w)$, $(\forall q \in \mathcal{CP}os(u))\, u(q) = w(q)$, and $\mathcal{A}(u)$ rewrites to $\mathcal{A}(w)$; hence aliens in $\mathcal{A}(u)$ are n-convertible iff the corresponding aliens in $\mathcal{A}(w)$ are n-convertible. By definition 11, we get $\widehat{u}^n = \widehat{w}^n$.

Putting things together, $\widehat{u}^i = \widehat{u}^n \xi_{n,i} = \widehat{w}^n \xi_{n,i} = \widehat{w}^0 \xi_{0,n} \xi_{n,i} \rightarrow \widehat{v}^0 \xi_{0,n} \xi_{n,i} = \widehat{v}^i$. □

Definition 12 (Rewrite root). *The root of a rewrite* $u \rightarrow^p_{R_{T\,sub}} v$ *is the minimal position, written* $\widehat{p}$, *such that* $(\forall q : p \geq q \geq \widehat{p})\, u(q) \in F_T$.

Note that $u|_p$ is a subterm of $u|_{\widehat{p}}$. By monotony of rewriting:

Corollary 2. *Let* $u \rightarrow^p_{R_{T\,sub}} v$. *Then* $\widehat{u|_{\widehat{p}}} \rightarrow_{R_T} \widehat{v|_{\widehat{p}}}$.

Main Result. We assume from here on that rules are indexed, those in R_T by 0, and those in R_{NT} by (non-zero) natural numbers, making R_{NT} into a disjoint union $\{R_i\}_{i \in I}$ where $I \subseteq i > 0$. Having a strictly smaller index for R_T rules is no harm nor necessity.

Our relations, parallel rewriting with R_{NT} and sub-rewriting with R_T, are labelled by triples made of the rank of the rewritten term first, the index of the rule used, and – approximately – the hat of the considered redex, ordered by the well-founded order $\rhd := (>, >, \rightarrow^+_{R_T})_{lex}$. More precisely,

$u \Rightarrow^P_{R_{i>0}} v$ is given label $\langle k, i, \_ \rangle$, where $k = max\{rk(u|_{p_i})\}_{p_i \in P}$;

$u \rightarrow^q_{R_{T\,sub}} v$ is given label $\langle k, 0, \widehat{u|_{q'}} \rangle$, where $k = rk(u|_q)$ and q' is the root $\widehat{q}$ of q.

The third component of an R_{NT}-rewrite is never used. Decreasing diagrams for critical pairs need be stable and satisfy a *variable condition* introduced by Toyama, see also [1]:

Definition 13. *The R_{NT} rigid critical peak $v \leftarrow^{\Lambda} u \Rightarrow^{Q} w$ (resp. rigid critical pair (v, w)) is* naturally decreasing *if it has a stable decreasing diagram in which:*

(i) step $s \Rightarrow^{Q'} s'$ facing $u \Rightarrow w$ uses the same rule and satisfies $\mathcal{V}ar(s'|_{Q'}) \subseteq \mathcal{V}ar(u|_{Q})$;

(ii) step $t \Rightarrow t'$ facing $u \rightarrow v$ uses the same rule.

Note the variable condition is automatically satisfied for an overlapping at the root.

Definition 14. *The R_{NT}-R_T critical peak $v \leftarrow^{\Lambda}_{R_{NT}} u \rightarrow^{q}_{R_T} w$ (resp. critical pair (v, w)) is* naturally decreasing *if it has a stable decreasing diagram whose step $t \Rightarrow^{P} t'$ facing $u \rightarrow v$ uses the same rule.*

Theorem 2 (Church-Rosser unions). *A rewrite union $R_T \cup R_{NT}$ satisfying: (A1–4), R_{NT}-R_T critical pairs are naturally decreasing, R_{NT} rigid critical pairs are naturally decreasing, is Church-Rosser iff its R_T critical pairs are joinable in R_T.*

Proof. While the "only if" direction is trivial, we are going to prove the "if" direction.

Since $\rightarrow_{R_T \cup R_{NT}} \subseteq \rightarrow_{R_{T\,sub}} \cup \Rightarrow_{R_{NT}}$ and $(\rightarrow_{R_{T\,sub}} \cup \Rightarrow_{R_{NT}})^* = (\rightarrow_{R_T \cup R_{NT}})^*$, $R_T \cup R_{NT}$ is Church-Rosser iff $\rightarrow_{R_{T\,sub}} \cup \Rightarrow_{R_{NT}}$ is. By induction on the rank, we therefore show that every local peak $v\ ({}_{R_{T\,sub}}\!\leftarrow \cup \Leftarrow_{R_{NT}})\ u\ (\rightarrow_{R_{T\,sub}} \cup \Rightarrow_{R_{NT}})\ w$, where $rk(u) = n+1$, enjoys a decreasing diagram, implying confluence on terms of rank $n+1$ by Corollary 1.

The proof is divided into three parts according to the considered local peak. Each key case is described by a picture to ease the reading, in which $\rightarrow$, $\rightarrow$ and $\rightarrow$ are used for plain steps with R_T, $R_{T\,sub}$ and $R_T \cup R_{NT}$, respectively, while $\rightarrow$ is used for parallel (sometimes plain) steps with R_{NT}. Every omitted case is symmetric to some considered case, or is easily solved by induction in case all rewrites take place in the aliens of u.

1) Consider a local peak $v \Leftarrow^{P,\langle k,i,\_\rangle}_{R_{NT}} u \Rightarrow^{Q,\langle m,j,\_\rangle}_{R_{NT}} w$. Following [1], we carry out first the particular case of a root peak, for which a rule $l \rightarrow r \in R_i$ applies at the root of u

(a) Root case. Although our labelling technique is different from [1], with ranks playing a prominent role here, the proof can be adapted without difficulty, as described in Fig. 1. Let $Q_1 := \{q \in Q \mid q \in \mathcal{FP}os(l)\}$. We first split the parallel rewrite from u to w into two successive parallel steps, at positions in Q_1 first, then at positions in $Q_2 = Q \setminus Q_1$. Note that the peak is specialized into ancestor peak when $Q_1 = \varnothing$. The inner part of the figure uses the fact that l unifies at Q_1 with some R_{NT} rule, yielding a rigid critical peak (v', u', w') of which the peak $(v, u, w'\sigma)$ is a σ-instance. By assumption, (v', w') has a stable diagram which is instantiated by σ in the figure. Since $Q_1 \cup Q_2$ are pairwise disjoint positions and $Q_2 > \mathcal{FP}os(w')$, by left-linearity of R_{NT}, $w'\sigma \Rightarrow^{Q_2}_{R_j} w'\sigma' = w$. Now, we can push that parallel rewrite from $w'\sigma$ to $s'\sigma$ as indicated, using stability and monotony of rewriting, thereby making ancestor redexes commute.

Finally, Toyama's variable condition ensures that Q'_1 and Q'_2 are disjoint sets of positions; hence $s\sigma$ rewrites to $s'\sigma'$ in one parallel step with the same j-rule as $u \Rightarrow w$. The obtained diagram is decreasing as a consequence of stability of the rigid critical pair diagram and rank non-increasingness of rewrites.

(b) For the general case, we proceed again as in [1]. For every position $p \in min(P \cup Q)$, the peak $v \Leftarrow^{P,\langle k,i,\_\rangle}_{R_{NT}} u \Rightarrow^{Q,\langle m,j,\_\rangle}_{R_{NT}} w$ induces a root-peak $v|_p \Leftarrow^{P',\langle k',i,\_\rangle}_{R_{NT}} u|_p \Rightarrow^{Q',\langle m',j,\_\rangle}_{R_{NT}} w|_p$. As just shown, root-peaks have decreasing diagrams; hence, for each p, we have a decreasing diagram between $v|_p$ and $w|_p$. Notice

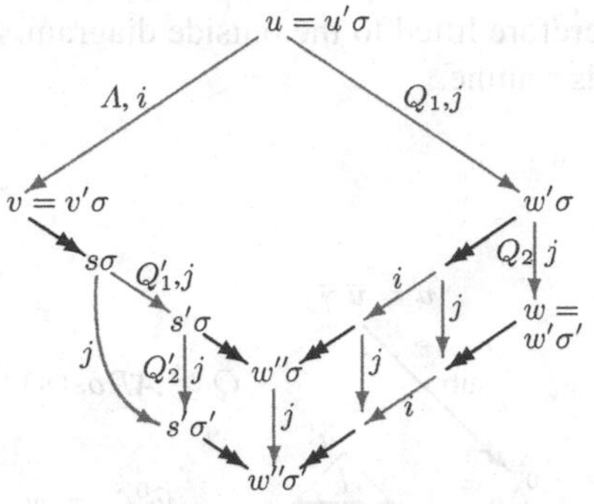

Fig. 1. R_{NT} root peak

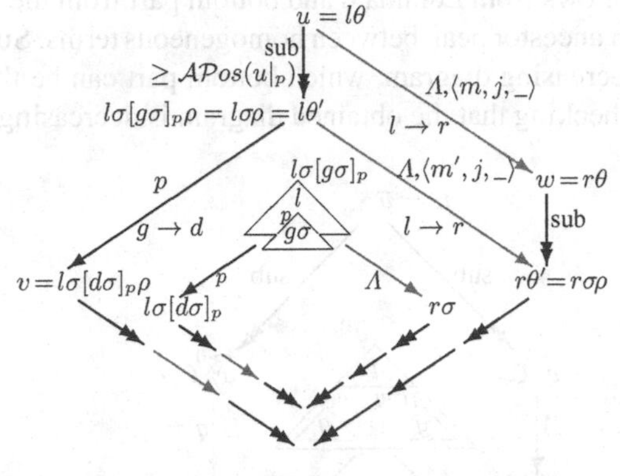

Fig. 2. R_{NT} above R_T critical peak

that in the decreasing diagram we have shown, each facing step – if it exists – uses the same rule as that one it faces. Since positions in $min(P \cup Q)$ are pairwise disjoint, these decreasing diagrams combine into a single decreasing diagram: in particular, the facing steps $\Rightarrow_{R_{NT}}^{\langle m',j,\_\rangle}$ (resp. $\Leftarrow_{R_{NT}}^{\langle k',i,\_\rangle}$) yield the facing step $\Rightarrow_{R_{NT}}^{\langle m,j,\_\rangle}$ (resp. $\Leftarrow_{R_{NT}}^{\langle k,i,\_\rangle}$).

2) Consider a local peak $v \;{}_{R_{T_{sub}}}^{p,\langle k,0,\widehat{u|_{\hat{p}}}\rangle}\!\!\leftarrow u \rightarrow_{R_{T_{sub}}}^{q,\langle m,0,\widehat{u|_{\hat{q}}}\rangle} w$. We denote by $l \rightarrow r$ and $g \rightarrow d$ the R_T-rules applied from u to v at p and u to w at q, respectively. We discuss cases depending on $p, \hat{p}, q, \hat{q}$, instead of only p, q as usual.

(a) Disjoint case: $p\#q$. The usual commutation lemma yields $v \rightarrow_{R_{T_{sub}}}^{q,\langle m,0,\widehat{v|_{\hat{q}}}\rangle} t \;{}_{R_{T_{sub}}}^{p,\langle k,0,\widehat{w|_{\hat{p}}}\rangle}\!\!\leftarrow w$ for some t. It is decreasing easily by Corollary 2 or Lemma 5, decided by $\hat{p}, \hat{q}$.

(b) Root ancestor case: $\hat{q} > p$. By Definition 12, $m < k$; hence $q \geq \mathcal{AP}os(u|_{\hat{p}})$. This case is thus similar to the R_T above R_{NT} ancestor case considered later, pictured at Fig. 4.

(c) Ancestor case: $\hat{q} = \hat{p}$; hence $k = m$, with $q > p \cdot \mathcal{FP}os(l)$. This is the usual ancestor case, within a given layer. The proof is depicted in Fig. 3, simplified by taking $p = \Lambda$.

Using Definition 8 and Lemma 2, then, by Definition 10, the rewrite from $u = \overline{u}\,\overline{\gamma}_u$, to $v = \widehat{v}^0\zeta_v$ (resp. $w = \widehat{w}^0\zeta_w$) factors out through $v' = \widehat{v'}^0\zeta_{v'}$ (resp., $w' = \widehat{w'}^0\zeta_{w'}$). By Lemma 3, ζ_v and $\zeta_{v'}$ coincide on $\mathcal{V}ar(\widehat{v'}^0)$, and so do ζ_w and $\zeta_{w'}$ on $\mathcal{V}ar(\widehat{w'}^0)$. By Lemma 4, $\mathcal{A}(u)$ rewrites to both $\mathcal{A}(v')$ and $\mathcal{A}(w')$, hence each alien in $\mathcal{A}(v)$ and $\mathcal{A}(w)$ originates from some in $\mathcal{A}(u)$. It follows that the aliens in $\mathcal{A}(v)$ and $\mathcal{A}(w)$ originating from the same one in $\mathcal{A}(u)$ are n-convertible. For each $y \in \mathcal{V}ar(\widehat{v}^n) \cup \mathcal{V}ar(\widehat{w}^n)$, we choose all aliens of v and w which belong to the n-convertibility class $\zeta^n(y)$, and apply induction hypothesis to get a common reduct t_y of them, mapping y to t_y to construct the substitution $\zeta_{v\downarrow_w^n}$. Letting v_n be the term $\widehat{v}^n\zeta_{v\downarrow_w^n}$, v rewrites to v_n. Similarly, w rewrites to w_n. This technique, which we call **equalization**, of equalizing all n-convertible aliens to construct $\zeta_{v\downarrow_w^n}$ is somewhat crucial in our proof. The last three steps follow from the inner ancestor diagram between hats of u, v, w, which upper part follows from Lemma 6 and bottom part from the fact that $q > p \cdot \mathcal{FP}os(l)$, resulting in an ancestor peak between homogeneous terms. Such an ancestor peak has an easy stable decreasing diagram, which bottom part can be therefore lifted to the outside diagram. Checking that the obtained diagram is decreasing is routine.

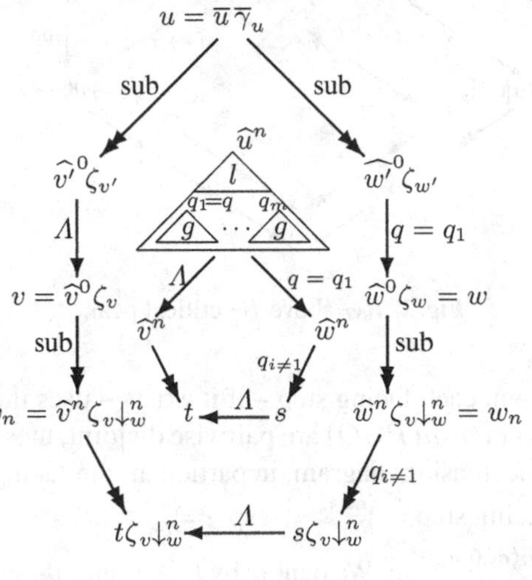

Fig. 3. R_T ancestor peak

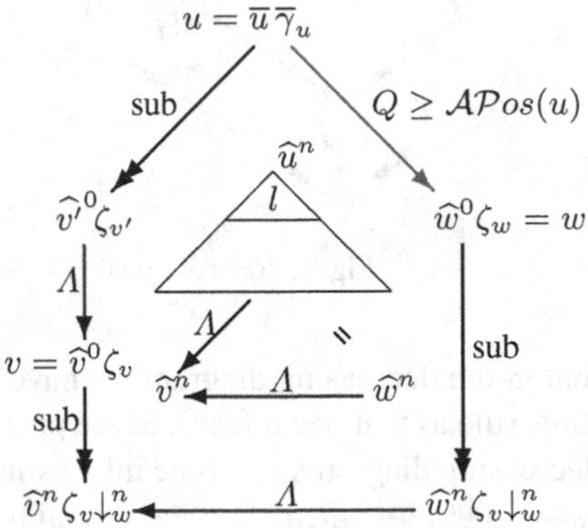

Fig. 4. R_T above R_{NT} ancestor peak

(d) Critical case: $\hat{q} = \hat{p}$; hence $k = m$, with $q \in \mathcal{FP}os(l)$. This is the usual critical case, happening necessarily within same layer. The proof works as in Case (2c), except that the inner diagram is now of a critical peak. Since the R_T critical peak has a joinability diagram by assumption, thanks to stability of rewriting, it can be lifted to the outer diagram, yielding a decreasing diagram for the starting peak.

3) Consider a local peak $v \; {}_{R_{T_{sub}}}\xleftarrow{p,\langle k,0,\widehat{u|_p}\rangle} u \Rightarrow^{Q,\langle m,j,\_\rangle}_{R_{NT}} w$. There are three cases.

(a) Disjoint case: $p\#Q$. We get the usual commuting diagram with two facing steps.

(b) Ancestor case. There are two sub-cases: (α) $p > Q$; hence $m > k$. Since R_{NT} is left-linear, then $v \Rightarrow^{\langle m',j,\_\rangle}_{R_{NT}} t \twoheadleftarrow^{\langle k,0,?\rangle}_{R_{T_{sub}}} w$ for some t and $m' \leq m$, being a clearly decreasing diagram. (β) $p < Q$. This case is a little bit more delicate, since the R_T-rule $l \to r$ used at position p may be non-left-linear. We use equalization as for Case (2c), depicted in Fig. 4 in the particular case where $p = \Lambda$ for simplicity. The main difference with Case (2c) is that the R_{NT}-step must occur in an alien; hence $\widehat{w}^n = \widehat{u}^n$, which somewhat simplifies the figure.

(c) Critical case. By assumption (A1-3), $Q = \{q_i\}_i$ and $p \in q_i \cdot \mathcal{FP}os(l)$ for some q_i. The proof is depicted at Fig. 2 with $Q = \{\Lambda\}$ for simplicity, implying a unique redex for that parallel rewrite at the top. Note that the R_T- and R_{NT}-redexes must have different ranks, hence $m > k$.

By assumption, $u = l\theta \Rightarrow^{\Lambda}_{l\to r} r\theta = w$ and $u(\to^{\geq \mathcal{AP}os(u|_p)}_{R_T \cup R_{NT}})^* u[g\theta]_p \to^{p}_{g\to d} v$ for some substitution θ (assuming l and g are renamed apart). The key of the proof is the fact that $u[g\theta]_p = l\theta'$ for some substitution θ' such that $\theta \twoheadrightarrow \theta'$. By assumption (A4), if o is a variable position in g and $p \cdot o \in \mathcal{FP}os(l)$, then $l|_{p \cdot o} \in \mathcal{T}(F_T, \mathcal{X})$. This indeed ensures that the sub-rewrites from u to v cannot occur at positions in $\mathcal{FPos}(l)$, therefore ensuring the fact $u[g\theta]_p = l\theta'$ since l is linear. It follows that $l\theta'$ rewrites to $r\theta'$ at the root, and to v at $p \in \mathcal{FPos}(l)$, which proves the existence of a critical pair of R_T inside R_{NT}. The rest of the proof is routine, the lifting part being ensured by stability.

To conclude, we simply remark that any two $(R_T \cup R_{NT})$-convertible terms are n-$(R_T \cup R_{NT})$-convertible for some n possibly strictly larger than their respective ranks. □

5 Relaxing Assumptions

One must understand that there is no room for relaxing the conditions on R_T and little for R_{NT}. Left-linearity is mandatory, rank non-increasingness as well, and the fact that left-hand sides are headed by symbols which do not belong to F_T serves avoiding critical pairs of R_{NT} inside R_T. This does not forbid left-hand sides to stretch over possibly several layers, making our result very different from known modularity results. Therefore, the only potential relaxations apply to the right-hand sides of R_{NT}-rules, which need not be headed by F_{NT}-symbols, as we assumed to make the proof more comfortable. We will allow them to be headed by some symbols from F_T.

From now on, we replace our assumption (A1) by the following: Let $F_C = F_T \cap F_{NT}$ be the set of *constructor symbols* s.t. no rule in $R_T \cup R_{NT}$ can have an F_C-headed left-hand side. We use $F_{T\setminus C}$ and $F_{NT\setminus C}$ as shorthand for $F_T \setminus F_C$ and $F_{NT} \setminus F_C$, respectively.

Terms in $\mathcal{T}(F_C, \mathcal{X})$ are *constructor terms*, trivial ones if in $\mathcal{X}$. The definitions of rank, cap and alien for terms headed by $F_{T\setminus C}$- or $F_{NT\setminus C}$-symbols are as before with respect to F_T and F_{NT}, respectively. An F_C-headed term has its cap and aliens defined with respect to F_C, and its rank is the maximal rank of its aliens, which are headed in $F_{T\setminus C}$ or $F_{NT\setminus C}$. The rank of a homogeneous constructor term is therefore 0, which explains why we started with rank 1 before.

Definition 15. *We introduce names for three important categories of terms:*
– type 1: $F_{NT\setminus C}$-headed terms have a variable as cap and themselves as alien;
– type 2: terms u whose cap $\overline{u} \in \mathcal{T}(F_C, \mathcal{Y})$ and aliens are all $F_{NT\setminus C}$-headed;
– type 3: $F_{T\setminus C}$-headed terms whose cap $u \in \mathcal{T}(F_T, \mathcal{X} \cup \mathcal{Y})$, and aliens are $F_{NT\setminus C}$-headed.

We also modify our assumption (A3), which becomes:
(A3) R_{NT} is a left-linear, rank non-increasing rewrite system whose rules have the form $f(\boldsymbol{l}) \to r$, $f \in F_{NT\setminus C}$, $\boldsymbol{l} \in \mathcal{T}(F_T \cup F_{NT}, \mathcal{X})$, r is a term of type 2.
Previous assumption (A3) is a particular case of the new one when r has type 1 $\subseteq$ type 2.

The proof structure of Theorem 2 depends on layering and labelling. Allowing constructor lifting rules in R_{NT} invalidates Lemmas 5, 6 used to control the label's third component of R_T-sub-rewriting steps, since R_{NT}-rewrites in aliens may now modify the cap of an F_T-headed term. Our strategy is to modify the notion of hat and get analogs of Lemmas 5, 6, making the whole proof work by changing the third component of the label of an R_T-sub-rewriting step. Following [4], the idea is to *estimate* the constructors which can pop up at the head of a given $F_{NT\setminus C}$-headed term, by rewriting it until stabilization.

From here on, we assume the Church-Rosser property for n-convertible terms of rank up to n. Being fixed throughout this section, the rank n will often be left implicit.

Finite Constructor Lifting

Definition 16. *A derivation $s \twoheadrightarrow u$, where s : type 1 and u : type 2 \ type 1, is said to be* constructor lifting. *$R_T \cup R_{NT}$ is a* finite constructor lifting *rewrite system if* $(\forall s :$ *type* $1)\ \exists n_s \geq 0$ *s.t. for all constructor lifting derivation* $s \twoheadrightarrow u$, $|\overline{u}| \leq n_s$.

Definition 17 (Stable terms). *A term whose multiset M of aliens only contains $F_{NT\setminus C}$-headed terms of rank at most n, is* stable *if M is stable. A multiset M of $F_{NT\setminus C}$-headed terms of rank at most n is* stable *if (i) reducts of terms in M are $F_{NT\setminus C}$-headed; (ii) any two convertible terms in M are equal.*

Example 3. Let $R_T = \{G(x,x,y) \to y, G(x,y,x) \to y, G(y,x,x) \to y, 1 \to 0\}$, $R_{NT} = \{F(0,1,x) \to F(x,x,x), F(1,0,x) \to F(x,x,x), F(0,0,x) \to F(x,x,x)\}$. Then, $u = G(F(0,1,G(0,0,0)), F(0,0,0), F(1,0,0))$ is not stable since its aliens are all convertible but different. But u rewrites to stable $G(F(0,0,0), F(0,0,0), F(0,0,0))$.

From rank non-increasingness and the Church-Rosser assumption, we get:

Lemma 7. *Let u a stable term of type 1 s.t. $u \twoheadrightarrow v$. Then v is a stable term of type 1.*

Lemma 8. *Let u a stable term whose aliens are of rank up to n. Then,* $(\forall i \leq n)\ \widehat{u}^i = \widehat{u}^0$.

Lemma 9 (Stabilization). *A term s of type 1, 2, 3 whose aliens have rank up to n has a stable term t such that $\widehat{t}^n = \widehat{s}^n\,\theta$ for some constructor substitution θ which depends only on the aliens of s.*

Proof. Let M be a multiset of type 1 terms, and $u \in M$. By assumption (A3), the set of constructor positions on top can only increase along a derivation from u. Being bounded, it has a maximum. Let v be such a reduct. If v is of type 1, then it is stable. Otherwise, we still needto equalize its convertible aliens, using the Church-Rosser property of terms of rank up to n, and we are done. Applying this procedure to all terms in M, we are left equalizing as above the convertible stable terms which are stable by Lemma 7. Taking now a type 2/3 term, we apply the procedure to its multiset of aliens, all of which have type 1. The relationship between the hats of s and t is clear: θ is generated by constructor lifting, which is the same for equivalent aliens, hence for equal aliens. □

Lemma 10 (Structure). *Let s be a term of type 1,2,3 whose aliens have rank up to n, and u, v be two stable terms obtained from s by stabilization. Then,* $(\forall i \leq n)\ \widehat{u}^i = \widehat{v}^i$.

Proof. Let $p \in \mathcal{AP}os(s)$. By stabilization $u|_p$ and $v|_p$ are convertible stable terms of type 2. By Church-Rosser assumption $u|_p \twoheadrightarrow t \twoheadleftarrow v|_p$. Since constructors cannot be rewritten, $u|_p$ and $v|_p$ must have the same constructor cap, thus u, v have the same cap. Since they are stable, two convertible aliens of u (resp., v) must be equal, hence u, v have the same 0-hat. We conclude by Lemma 8. □

Definition 18 (Estimated hat). *Let u be a term of type 1,2,3 whose aliens have rank up to n and v a stable term obtained from u by stabilization. The* estimated hat $\overset{\triangle n}{u_v}$ *of u w.r.t. v is the term $\widehat{v}^n$.*

By Lemma 10, the choice of v has no impact on $\overset{\triangle n}{u_v}$, hence the short notation $\overset{\triangle}{u}$.

Lemma 11 (Alien rewriting). *Let u, v be terms of type 3 whose aliens are of rank up to n, such that $u \rightarrow^{\geq \mathcal{APos}(u)}_{R_T \cup R_{NT}} v$. Then $\overset{\triangle}{u} = \overset{\triangle}{v}$.*

Proof. Follows from Lemmas 9 and 10: any stable term for v is a stable term for u. □

Lemma 12. *Let u be a term of type 3 whose aliens have rank up to n, s.t. $u \rightarrow^{p}_{R_{T\,sub}} v$ with $p \in \mathcal{CPos}(u)$. Then $\overset{\triangle}{u} \rightarrow_{R_T} \overset{\triangle}{v}$.*

Proof. By definition of sub-rewriting $u \twoheadrightarrow^{\geq \mathcal{APos}(u)} w \rightarrow^{p}_{R_T} v$. By Lemma 11, $\overset{\triangle}{u} = \overset{\triangle}{w}$. By Lemma 6, $\widehat{w}^n \rightarrow^{p}_{R_T} \widehat{v}^n$, and aliens of v are aliens of w. Let now w', v' be stable terms obtained from w, v by stabilization, hence $\widehat{w'}^n = \widehat{w}^n \theta_w$ and $\widehat{v'}^n = \widehat{v}^n \theta_v$ by Lemma 9, where θ_v, θ_w depend only on the aliens of v, w, respectively; hence θ_v and θ_w coincide on $\mathcal{V}ar(\widehat{v}^n) \subseteq \mathcal{V}ar(\widehat{w}^n)$ and $\widehat{v'}^n = \widehat{v}^n \theta_w$. We conclude by stability of rewriting and definition of estimated hats. □

Theorem 3. *Theorem 2 holds with finite constructor lifting.*

Proof. Same as for Theorem 2, with the exception of the crucial sub-rewriting cases, which are marginally modified by using stabilization instead of equalization of terms. □

Infinite Constructor Lifting. It is easy to see that the only difficult case in the main proof is the elimination of sub-rewriting critical peaks. Consider the critical peak $v \; {}_{l\rightarrow r}\!\overset{\Lambda}{\leftarrow} v' \twoheadleftarrow^{\geq \mathcal{APos}(u)}_{R_T \cup R_{NT}} u \twoheadrightarrow^{\geq \mathcal{APos}(u)}_{R_T \cup R_{NT}} w' \rightarrow^{p}_{g \rightarrow d} w$, $p \in \mathcal{FPos}(l)$ and $l \rightarrow r, g \rightarrow d \in R_T$. To obtain a term instance of l whose subterm at position p is an instance of g, v' and w' must be equalized into a term s whose hat rewrites at Λ with $l \rightarrow r$ and at p with $g \rightarrow d$ to the hats of the corresponding equalizations of v and w. The heart of the problem lies therefore in equalization which constructs here a solution in the signature of F_T to F_T-unification problems associated with critical pairs by rewriting in $R_T \cup R_{NT}$. Hence,

Theorem 4. *With new assumption (A3), Theorem 2 holds if R_T critical pairs modulo $R_T \cup R_{NT}$ are joinable in R_T.*

Because sub-rewriting can only equalize aliens, $R_T \cup R_{NT}$-unification sole purpose is to solve *occurs-check* failures that occur in the plain unification problem $l|_p = g$.

Definition 19. *Let $l \rightarrow r$ and $g \rightarrow d$ be two rules in R_T s.t. g Prolog unifies with l at position $p \in \mathcal{FPos}(l)$. Let $\bigwedge_i x_i = s_i \wedge \bigwedge_j y_j = t_j$ be a dag solved form returned by Prolog unification, where $\bigwedge_i x_i = s_i$ is the finite substitution part, and $\bigwedge_j y_j = t_j$ the occurs-check part. Let now σ be the substitution $\{x_i \mapsto s_i\}_i$ and $\tau = \{y_j \mapsto t_j\}_j$. Then $\langle r\sigma, l\sigma[d\sigma]_p \rangle$ is a Prolog critical pair of R_T, constrained by the occurs checks $y_j = t_j$.*

If the critical pairs obtained by Prolog unification are joinable in R_T constrained by the occurs-check equations, then the Church-Rosser property is satisfied:

Conjecture 1. With new assumption (A3), Theorem 2 holds if R_T critical pairs are joinable in R_T and Prolog critical pairs of R_T are joinable in R_T modulo their occurs checks.

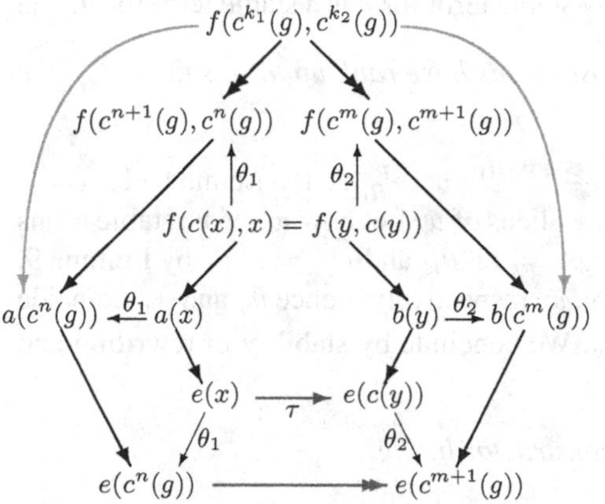

Example 4 (Variation of Huet's example [3]). Let

$$R_T = \{\, f(c(x), x) \to a(x),\ f(y, c(y)) \to b(y),\ a(x) \to e(x),\ b(y) \to e(c(y))\,\},$$

$$R_{NT} = \{\, g \to c(g)\,\}.$$

Then the unification problem $f(c(x), x) = f(y, c(y))$ results in an empty substitution and the occurs-check equations $\tau = \{x = c(y), y = c(x)\}$. The critical pair $\langle a(x), b(y)\rangle$ is then joinable by $a(x) \to e(x) = e(c(y)) \leftarrow b(y)$, as exemplified in the figure, where $\theta_1 = \{x \mapsto c^n(g)\}, \theta_2 = \{y \mapsto c^m(g)\}$.

Fig. 5. Variation of Huet's example

The idea is shown in Fig. 5. Note that the red bottom steps operate on aliens, hence have a small rank, making the whole joinability diagram decreasing. We have no clear formulation of the converse yet. Confluence is indeed satisfied if the occurs check is unsolvable, that is, when there exists no $F_{NT\backslash C}$-headed substitution θ of the y_j's such that $y_j\theta \twoheadleftarrow\!\!\twoheadrightarrow_{R_T \cup R_{NT}} t_j\theta$. We suspect this condition can be reinforced as $y_j\theta \twoheadrightarrow_{R_T \cup R_{NT}} t_j\theta$, possibly leading to interesting sufficient conditions for unsolvability of occurs checks.

6 Related Work

In [5], it is shown that confluence can be characterized by the existence of decreasing diagrams for the critical pairs in $R_T \cup R_{NT}$ provided all rules are linear (an assumption that was forgotten [but used] for R_T, as pointed out to the third author by Aart Middeldorp). This is a particular case of a recent result of Felgenhauer [1] showing that R_{NT} is confluent if rules are left-linear and parallel critical pairs have decreasing diagrams with respect to rule indexes used as labels. When F_T is empty, all terms have rank 1, hence our labels for non-linear rules reduce to his. A difference is that we assume R_{NT}-rules to be non-collapsing. One could argue that R_{NT} collapsing rules can be moved to R_T, but this answer is not satisfactory for two different reasons: the resulting change of labels may affect the search for decreasing diagrams, and it can also impact condition (A1). A second difference is that we use rigid parallel rewriting, which yields exponentially fewer parallel critical pairs than when allowing parallel steps with different rules of a given index (which we could have done too). The price to pay – having less flexibility for finding decreasing diagrams – should not make a difference in practice.

A very recent result of Klein and Hirokawa, generalizing [2], extends Knuth and Bendix's critical pair test to relatively terminating systems [6]. It is an extension in the sense that it boils down to it when $R_{NT} = \varnothing$. Otherwise, it requires computing critical pairs of R_T *modulo* a confluent R_{NT}, hence modifies the critical pair test for the subset of terminating rules. Further, it requires proving relative termination (termination of $\twoheadrightarrow_{R_{NT}} \rightarrow_{R_T} \twoheadrightarrow_{R_{NT}}$), complete unification modulo R_{NT}, and absence of critical pairs between R_T and R_{NT}, all tests implemented in CSI[http://dx.doi.org/10.1007/978-3-642-22438-6_38] – to our surprise! This is used to detect that Huet's example is non-confluent.

Theorem 2 can be seen as a modularity theorem to some extent, since rewriting a term in $\mathcal{T}(F_T, \mathcal{X})$ can only involve R_T rules. But left-hand sides of R_{NT} rules may have F_T-symbols. That is why we need to compute critical pairs of R_T inside R_{NT}. Our proof uses many concepts and techniques inherited from previous work on modularity, such as the decomposition of terms (caps and aliens, hats and estimated caps [10]). We have not tried using van Oostrom's notion of cap, in which aliens must have maximal rank [13], nor the method developed by Klein and Hirokawa for studying the Church-Rosser property of *disjoint* rewrite relations on terms [6], which we could do by considering cap rewriting with R_T-rules and alien rewriting with all rules. This remains to be done.

7 Conclusion

Decreasing diagrams opened the way for generalizing Knuth and Bendix's critical-pair test for confluence to non-terminating systems, re-igniting these questions. Our results answer important open questions, in particular by allowing both non-left-linear and non-terminating rules. While combining many existing as well as new techniques, our proof has proved quite robust. Two technical questions have been left open: having collapsing rules in R_{NT}, following [1], and eliminating assumption (A4).

A major theoretical question is whether layering requires assumption (A1). Our proof is based on two key properties, layering and the absence of overlaps of R_{NT} inside R_T. Currently, (A1) serves both purposes. The question is however open whether the latter property is sufficient to define some form of layering, as we suspect.

We end up with our long term goal, applying this technique in practice. The need for showing the Church-Rosser property of mixed terminating and non-terminating rewrite computations arises in at least two areas, first-order and higher-order. The development of sophisticated type theories with complex elimination rules requires proving Church-Rosser *before* strong-normalization and type preservation, directly on untyped terms. Unfortunately, besides being collapsing, β-reduction is also rank-increasing in the presence of another signature. We therefore need to develop another notion of rank that would apply to pure λ-calculus, a question related to the previous one.

Transformation valuation is a static analysis that tries to verify that an optimizer is semantics preserving by constructing a *value graph* for both programs and showing their equivalence by rewriting techniques [11]. Here, the user has a good feeling of which subset of rules is a candidate for R_{NT}. Where this is not the case, work is of course needed to find good splits automatically. Implementers are invited to lead the way.

Acknowledgements. Work supported by NSFC grants 61272002, 91218302, 973 Program 2010CB328003 and Nat. Key Tech. R&D Program SQ2012BAJY4052 of China.

References

1. Felgenhauer, B.: Rule labeling for confluence of left-linear term rewrite systems. In: IWC, pp. 23–27 (2013)
2. Hirokawa, N., Middeldorp, A.: Decreasing diagrams and relative termination. J. Autom. Reasoning 47, 481–501 (2011)
3. Huet, G.P.: Confluent reductions: Abstract properties and applications to term rewriting systems: Abstract properties and applications to term rewriting systems. J. ACM 27, 797–821 (1980)
4. Jouannaud, J.P., Toyama, Y.: Modular Church-Rosser modulo: The complete picture. Int. J. Software and Informatics 2, 61–75 (2008)
5. Jouannaud, J.P., van Oostrom, V.: Diagrammatic confluence and completion. In: Albers, S., Marchetti-Spaccamela, A., Matias, Y., Nikoletseas, S., Thomas, W. (eds.) ICALP 2009, Part II. LNCS, vol. 5556, pp. 212–222. Springer, Heidelberg (2009)
6. Klein, D., Hirokawa, N.: Confluence of non-left-linear TRSs via relative termination. In: Bjørner, N., Voronkov, A. (eds.) LPAR-18 2012. LNCS, vol. 7180, pp. 258–273. Springer, Heidelberg (2012)
7. Knuth, D.E., Bendix, P.B.: Simple word problems in universal algebras. In: Leech, J. (ed.) Computational Problems in Abstract Algebra, Elsevier (1970)
8. Newman, M.H.A.: On theories with a combinatorial definition of 'equivalence'. Ann. Math. 43, 223–243 (1942)
9. Rosen, B.K.: Tree-manipulating systems and Church-Rosser theorems. J. ACM 20, 160–187 (1973)
10. Toyama, Y.: On the Church-Rosser property for the direct sum of term rewriting systems. J. ACM 34, 128–143 (1987)
11. Tristan, J.B., Govereau, P., Morrisett, G.: Evaluating value-graph translation validation for LLVM. In: Proceedings of ACM SIGPLAN Conference PLDI. ACM, New York (2011)
12. van Oostrom, V.: Confluence by decreasing diagrams. Theor. Comput. Sci. 126, 259–280 (1994)
13. van Oostrom, V.: Modularity of confluence. In: Armando, A., Baumgartner, P., Dowek, G. (eds.) IJCAR 2008. LNCS (LNAI), vol. 5195, pp. 348–363. Springer, Heidelberg (2008)

Proof Terms for Infinitary Rewriting

Carlos Lombardi[1,2,3], Alejandro Ríos[1,3], and Roel de Vrijer[4]

[1] Universidad Nacional de Quilmes, Argentina
[2] PPS (Université Paris-Diderot and CNRS), France
[3] Universidad de Buenos Aires, Argentina
[4] VU University Amsterdam, The Netherlands

Abstract. We generalize the notion of proof term to the realm of transfinite reduction. Proof terms represent reductions in the first-order term format, thereby facilitating their formal analysis. Transfinite reductions can be faithfully represented as infinitary proof terms, unique up to infinitary associativity. We use proof terms to define equivalence of transfinite reductions on the basis of permutation equations. A proof of the compression property via proof terms is presented, which establishes permutation equivalence between the original and the compressed reductions.

Keywords: infinitary rewriting, proof terms, permutation equivalence.

1 Introduction

We study infinitary left-linear rewriting, based on the notion of strong convergence, as developed in [5]. Proof terms denoting reductions for finitary rewriting have been introduced in [11], Chap. 8. The use of proof terms marks a shift of attention from reduction *as a relation* between terms, towards reductions *as objects* by themselves. Proof terms represent reductions in the first-order term format, thereby facilitating their formal analysis. For example, proof-theoretic analysis, equational reasoning, and even rewriting techniques, can now be applied in the study of properties of rewriting such as standardisation and the like. This use of proof terms in finitary rewriting has been well developed. See [11], Chaps. 8 and 9.

The main objective of our work is to generalize the notion of proof term to the realm of infinite reduction, and to use this to define and study the notion of permutation equivalence for transfinite reduction. The representation of transfinite reductions by proof terms can be exploited to yield transparent proofs of results like standardisation and compression. As a matter of fact, strong versions of these theorems: the standardised or compressed reductions thus obtained are permutation equivalent to the original ones. The compression proof will be presented here, standardisation is beyond the scope of this paper.

We will prove in Section 5 that any transfinite reduction can be faithfully represented as an infinitary, so-called *stepwise* proof term. Moreover, the original reduction can be reconstructed from the stepwise proof term: the representation is injective. The proof terms formalism is broader, however. Apart from sequential reduction, also parallel and multi-step reduction can be represented.

G. Dowek (ed.): RTA-TLCA 2014, LNCS 8560, pp. 303–318, 2014.

An alternative, *co-inductive* approach to the study of infinitary rewriting, was proposed in [3]. Proof objects emerge there as witnesses in the co-inductive characterisation of the reduction relation. Their focus, however, is on techniques for proving properties of the reduction *relation*, rather than the fine structure of the space of transfinite reductions, which is our primary interest. We use *inductive* techniques. Although a term can be infinite, the distance of each symbol to its root is finite, allowing for inductive reasoning about occurrences in a term. Likewise, transfinite induction can be used to reason about infinite reduction sequences, of which the length can always be expressed as an ordinal.

An extended version of the present article, containing full details of the definitions and proofs, is available as [8].

2 Preliminaries

We assume acquaintance with term rewriting, both finitary and infinitary, and with ordinal numbers. As background sources on rewriting we mention [1], [11], [5], [7]. The following survey fixes some notations and terminology.

Ordinals. Properties of countable ordinals (cf. e.g. [10]) are extensively used in this work. Especially we will make use of the following cofinality result.

Lemma 1. *Let α be a countable limit ordinal. Then there exists a sequence of ordinals $\langle \alpha_i \rangle_{i<\omega}$ such that $0 < \alpha_i < \alpha$ for all $i < \omega$, and $\alpha = \sum_{i<\omega} \alpha_i$.*

Positions, terms. A *position* is a finite sequence of elements of $\mathbb{N}_{>0}$. We will use p, q, r for positions. The length of a position p is called its *depth*, denoted $|p|$. The empty position is denoted by ϵ. Concatenation of positions is denoted by an infix dot, i.e. $p \cdot q$. The dot will be mostly omitted. The *prefix order* on positions is denoted by $\leq$, i.e. $p \leq q$ iff $q = pq'$.

A *tree domain* (cf. [2]) is a set of positions P, such that $P \neq \emptyset$, P is prefix-closed, and $pj \in P$ and $i < j$ imply $pi \in P$. Note that a tree domain can be infinite. However, the depth of any position in a tree domain is finite. A *signature* Σ is a function from a set of symbols to $\mathbb{N}_{\geq 0}$; the value of this function for a symbol f is called its *arity*, denoted as $ar(f)$. We write $f/m \in \Sigma$ to indicate $ar(f) = m$. We assume a countably infinite set $\mathtt{Var}$ of *variables*. An *infinitary* (i.e. finite or infinite) *term* over Σ is a pair $\langle P, F \rangle$ where P is a tree domain, and $F : P \to (\Sigma \cup \mathtt{Var})$ verifying the following: if $p \in P$, then $pi \in P$ iff $F(p) \in \Sigma$ and $ar(F(p)) \geq i$. We denote the set of infinitary terms over Σ as $Ter^{\infty}(\Sigma)$. If $t = \langle P, F \rangle$, then we define $\mathtt{pos}(t) := P$ and $t(p) := F(p)$; we say that t is *finite* iff $\mathtt{pos}(t)$ is. If f is a unary symbol, then f^{ω} denotes the term $f(f(f \ldots))$.

Given t, u terms and p a position in t, we let $t|_p$ denote the *subterm* of t at p, and $t[u]_p$ the term that results by replacing that subterm of t by u.

The *distance* between two terms t and u is defined as follows: $d(t, u) := 0$ if $t = u$; otherwise, $d(t, u) := 2^{-|p|}$ where t and u coincide on q if $|q| < |p|$, and $t(p) \neq u(p)$. Given this notion of distance, the *limit* of any Cauchy-convergent sequence of terms is defined.

Term rewriting systems, reduction. A *rule* over a signature Σ is a pair $\langle l, h \rangle$ of terms over Σ such that $l \notin \mathtt{Var}$, l is a finite term, and all the variables occurring in h occur also in l. We denote the rule $\langle l, h \rangle$ as $\mu : l \to h$; here μ is the *name* of the rule, a symbol from a signature disjoint with Σ. A *term rewriting system* (TRS) is a pair $T = \langle \Sigma, R \rangle$, where Σ is a signature and R a set of rules over Σ. We only consider *left-linear* TRSs, i.e., if $\langle l, h \rangle \in R$ then l has no multiple occurrences of the same variable. If the rules are used for *infinitary rewriting*, as in this paper, then T is also called an *infinitary TRS* (iTRS).

A *reduction step* over a TRS $T = \langle \Sigma, R \rangle$ is a triple $a = \langle t, p, \mu \rangle$ where t is a term, $p \in \mathbf{pos}(t)$, and $\mu : l \to h \in R$, iff $t|_p = \sigma l$ for some substitution σ. We define the source term, target term, activity position and depth of $a = \langle t, p, \mu \rangle$ as follows: $src(a) := t$, $tgt(a) := t[\sigma h]_p$, $\mathbf{rpos}(a) := p$, $d(a) := |p|$.

A *reduction sequence* is: either Id_t, the *empty reduction sequence for* the term t, or else a non-empty sequence of reduction steps $\delta := \langle \delta[\alpha] \rangle_{\alpha < \beta}$, where $\beta > 0$ and the following conditions are met: (1) for all α such that $\alpha + 1 < \beta$, $src(\delta[\alpha+1]) = tgt(\delta[\alpha])$; and for all limit ordinals $\beta_0 < \beta$: (2.a) the sequence $\langle tgt(\delta[\alpha]) \rangle_{\alpha < \beta_0}$ has a limit, (2.b) that limit coincides with $src(\delta[\beta_0])$, and (2.c) for all $n < \omega$, there exists $\beta' < \beta_0$ such that $d(\delta[\alpha]) > n$ if $\beta' < \alpha < \beta_0$. We say that a reduction sequence δ is *strongly convergent* if either $\delta = \mathsf{Id}_t$ for some term t, or else $\delta = \langle \delta[\alpha] \rangle_{\alpha < \beta}$, and either β is a successor ordinal, or else β is a limit ordinal and conditions (2.a) and (2.c) hold for β as well. By *convergence*, we will always mean strong convergence.

The symbols $\to$, $\twoheadrightarrow$ and $\twoheadrightarrow\!\!\!\!\!\rightarrow$ denote the one-step, finite rewrite, and convergent infinitary rewrite relations respectively.

We will denote the step $a = \langle t, p, \mu \rangle$ as $t \xrightarrow{a} u$, or $t \xrightarrow{\langle p, \mu \rangle} u$ where $u = tgt(a)$. Analogously for $\twoheadrightarrow$ and $\twoheadrightarrow\!\!\!\!\!\rightarrow$. We indicate the TRS below the arrow if needed.

WN^∞ is the property of having an infinitary normal form and UN^∞ of unicity of normal forms. SN^∞ is more complicated, but boils down to the absence of non-convergent reduction sequences. All orthogonal iTRSs are UN^∞, see [11], Chap. 13 or [7].

3 Finitary and Infinitary Proof Terms

The idea motivating the definition and application of proof terms, as they were introduced in [11], Chap. 8, is to denote reductions of a TRS T as *terms* over an extended signature. For each reduction rule $\rho : l \to r$ of T a *rule symbol* is introduced, which we will also denote by ρ. The arity of the rule symbol ρ equals the number of different variables occurring in l. So e.g., the signature of proof terms for a TRS $T = \langle \Sigma, R \rangle$ with rules $\mu : f(x) \to g(x)$, $\rho : h(m(x), m(y)) \to k(x)$ and $\nu : g(x) \to k(x)$ adds to Σ rule symbols $\mu/1$, $\rho/2$ and $\nu/1$.

We describe some valid proof terms along with the T-reductions they denote. Proof terms with only one occurrence of a rule symbol denote single reduction steps: $\mu(a) : f(a) \to g(a)$ and $g(\rho(a,b)) : g(h(m(a), m(b))) \to g(k(a))$. With more occurrences of rule symbols, they denote multi-steps, like $h(\mu(a), \mu(b)) : h(f(a), f(b)) \twoheadrightarrow h(g(a), g(b))$ and $\rho(\mu(a), b) : h(m(f(a)), m(b)) \twoheadrightarrow k(g(a))$. A proof term ψ without rule-symbol occurrences denotes an empty step. It will also be denoted by the symbol 1_ψ, or just briefly 1.

The beginning and end terms of the corresponding reductions are called the source and target of the proof term. For the proof terms considered so far, they can be obtained via rewriting in two companion TRSs, denoted as SRC and TGT respectively. For each rule symbol $\rho : l \to r$, SRC includes a rule $\rho(x_1, \ldots, x_m) \to l[x_1, \ldots, x_m]$ and TGT a rule $\rho(x_1, \ldots, x_m) \to r[x_1, \ldots, x_m]$. Source and target of a proof term are its normal forms in SRC and TGT, respectively. It is easy to see that SRC and TGT are orthogonal and terminating.

To complete the definition of the set of finitary proof terms, we add a new binary function symbol $\cdot$ (written infix), expressing concatenation, or composition, of reductions. There is a restriction on term formation, though: $\psi \cdot \phi$ is a valid proof term only if $tgt(\psi) = src(\phi)$. Computing source and target for proof terms containing the symbol $\cdot$ is made possible by schematically adding rewrite rules $\psi \cdot \phi \to \psi$ to SRC and $\psi \cdot \phi \to \phi$ to TGT.

Just to give a simple example, the proof term $f(\mu(a)) \cdot f(\nu(a))$ denotes the two-step reduction $f(f(a)) \to f(g(a)) \to f(k(a))$. The same reduction is represented by the proof term $f(\mu(a) \cdot \nu(a))$.

Proof terms without the symbol $\cdot$ are called *multi-steps*, with *one-step* proof terms, multi-steps with precisely one rule-symbol occurrence, as a special case.

3.1 Infinitary Proof Terms

We extend the concept of proof terms to the setting of infinitary rewriting. *Infinitary multi-steps* are finite *or infinite* terms over the signature extended with rule symbols. A multi-step may now contain infinitely many rule symbol occurrences, as e.g. in the proof term $\mu^\omega : f^\omega \to g^\omega$. As in the finite case, infinitary multi-steps with just one rule-symbol occurrence are called *one-steps*.

Source $src(\psi)$ and target $tgt(\psi)$ for an infinitary multi-step ψ can again be defined using the TRSs SRC and TGT, now considered as *infinitary* TRSs, as the respective infinitary normal forms of ψ. Of course there are the questions of existence and uniqueness. First note that UN^∞ holds for both SRC and TGT, since they are orthogonal iTRSs. It is also not hard to verify WN^∞ for SRC. For TGT, however, we have WN^∞ only if the object TRS does not include collapsing rules. We conclude that the source of an infinitary multi-step ψ is always uniquely defined. The target is only defined if ψ is WN^∞, but if so, it is also unique. If ψ is not WN^∞ for TGT, then we say that $tgt(\psi)$ is undefined.

The set of redexes in $src(\psi)$ corresponding to the rule symbol occurrences in ψ admits at least one convergent development (respectively, all developments are convergent) precisely if ψ is WN^∞ (respectively SN^∞) in the TRS TGT. An infinitary multi-step is called *convergent*, if its target can be computed.

Composition is expressed, as before, by the binary symbol $\cdot$ of which a proof term may now contain an infinite number of occurrences. Not all terms over the thus extended signature are valid proof terms though, but only those that can be constructed starting from the infinitary multi-steps by the following three inductive clauses.

First, closure under application of function or rule symbols: if $\psi_1, \ldots, \psi_n$ are proof-terms, then so are $f(\psi_1, \ldots, \psi_n)$ and $\mu(\psi_1, \ldots, \psi_n)$.

Secondly, *binary composition*: if ψ, ϕ are proof terms, then so is $\psi \cdot \phi$, provided that $tgt(\psi) = src(\phi)$. This presupposes convergence of ψ. The proof term $\psi \cdot \phi$ is convergent iff ϕ is.

Thirdly, *infinite composition*: the term corresponding to the figure is a proof term, if $\psi_0, \psi_1, \psi_2, \ldots$ are, provided that for each $i < \omega$ we have convergence of ψ_i and $tgt(\psi_i) = src(\psi_{i+1})$. A linear rendering would be $\psi_0 \cdot (\psi_1 \cdot (\psi_2 \cdot \ldots))$. We use $\cdot_{i<\omega}\ \psi_i$ as shorthand for this proof term.

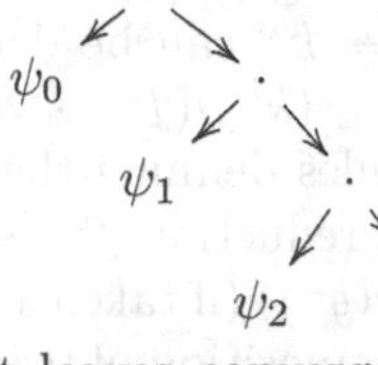

For $\psi = \cdot_{i<\omega}\ \psi_i$, we define $src(\psi) = src(\psi_0)$. That leaves convergence for $\cdot_{i<\omega}\ \psi_i$ to be defined, and the target, if it exists. For this we need the notion of *minimal activity depth* of a proof term ψ, notation $mind(\psi)$. This can be defined as the minimal depth of a rule symbol occurrence, where dot occurrences are not counted when computing the depth of a symbol. (If ψ does not have any rule symbol occurrence, then we call ψ *trivial* and define $mind(\psi) = \omega$.) For example: if $\psi = \mu(a) \cdot \nu(a) : f(a) \to g(a) \to k(a)$, then $mind(\psi) = 0$. This is the correct value since ψ denotes a reduction including contractions at the root.

Now we define $\psi = \cdot_{i<\omega}\ \psi_i$ to be convergent if the sequence $\langle mind(\psi_i) \rangle_{i<\omega}$ tends to infinity, and if so, $tgt(\psi)$ is defined as the limit of the sequence $tgt(\psi_i)$.

In [8] infinitary proof terms are defined in layers corresponding to ordinal numbers. We implicitly refer to these layers when applying induction on proof terms. Let $\nu(\psi)$ denote the ordinal layer for ψ. We have $\nu(\psi_1 \cdot \psi_2) = \nu(\psi_1) + \nu(\psi_2) + 1$, and $\nu(\cdot_{i<\omega}\ \psi_i) = \Sigma_{i<\omega}\, \nu(\psi_i)$. Hence $\nu(\psi_i) < \nu(\psi_1 \cdot \psi_2)$ for $i = 1, 2$, $\nu(\psi_0 \cdot \ldots \cdot \psi_n) < \nu(\cdot_{i<\omega}\ \psi_i)$ for all $n < \omega$, and $\nu(\psi)$ is a limit ordinal iff ψ is an infinite composition.

By the above definition of proof terms an infinite composition is also a binary composition: $\cdot_{i<\omega}\ \psi_i = \psi_0 \cdot (\cdot_{i<\omega}\ \psi_{i+1})$. However, since proof terms have unique layers, in the layered definition the proof term is constructed only once, as infinite composition, and with layer a limit ordinal. We still have unique constructibility.

3.2 Examples

The definition of infinitary proof terms extends that of the finitary ones so that the examples given earlier are infinitary proof terms as well.

Let us show some infinite examples. Let $\mu : f(x) \to g(x)$ and $\nu : g(x) \to k(x)$. Consider $f^\omega \twoheadrightarrow g^\omega$ taken as the simultaneous reduction of all the μ redexes present in the term f^ω. Such a simultaneous reduction can be denoted by the infinitary multi-step $\psi_1 := \mu^\omega$. To compute $src(\psi_1) = f^\omega$ and $tgt(\psi_1) = g^\omega$ we observe that the "companion TRSs" SRC and TGT include the rules $\mu(x) \to f(x)$ and $\mu(x) \to g(x)$ respectively, leading to the convergent reduction sequences $\mu^\omega \to f(\mu^\omega) \to f^2(\mu^\omega) \twoheadrightarrow f^\omega$ and $\mu^\omega \to g(\mu^\omega) \to g^2(\mu^\omega) \twoheadrightarrow g^\omega$.

Consider the reduction sequence $f^\omega \to g(f^\omega) \to g^2(f^\omega) \twoheadrightarrow g^\omega$ having length ω. The i-th step of this sequence, namely $g^i(f^\omega) \to g^{i+1}(f^\omega)$, can be described by the proof term $g^i(\mu(f^\omega))$. It is straightforward to check that the sequence formed by these proof terms verifies the conditions of the infinitary composition rule, and that the depth of the denoted activity tends to infinity. Therefore $\psi_2 := \cdot_{i<\omega}\ g^i(\mu(f^\omega))$ is a valid and convergent proof term; we observe $src(\psi_2) = f^\omega$. In order to obtain $tgt(\psi_2) = g^\omega$, it is enough to observe that the sequence of targets of each $g^i(\mu(f^\omega))$, namely $g(f^\omega), g^2(f^\omega), \ldots$, converges to that term. Analogously, the reduction sequence $f^\omega \to g(f^\omega) \to k(f^\omega) \to k(g(f^\omega)) \to k^2(f^\omega) \twoheadrightarrow k^\omega$ can be denoted by either $\psi_3 := \cdot_{i<\omega}\ (k^i(\mu(f^\omega)) \cdot k^i(\nu(f^\omega)))$ or $\psi_4 := \cdot_{i<\omega}\ (k^i(\mu(f^\omega) \cdot \nu(f^\omega)))$.

The rules defining the set of proof terms can be combined in different ways. E.g., the reduction $f^\omega \twoheadrightarrow g^\omega \twoheadrightarrow k^\omega$, can be denoted by either $\cdot_{i<\omega}\ g^i(\mu(f^\omega)) \cdot \cdot_{i<\omega}\ k^i(\nu(g^\omega))$ (if taken as a sequence having length $\omega \times 2$) or $\mu^\omega \cdot \nu^\omega$ (if considered as the composition of two infinite simultaneous reductions). The reduction $f^\omega \to f(g(f^\omega)) \to f(k(f^\omega)) \to f(k(g(f^\omega))) \to f(k^2(f^\omega)) \twoheadrightarrow f(k^\omega) \to g(k^\omega)$ can be denoted by $\cdot_{i<\omega}\ f(k^i(\mu(f^\omega) \cdot \nu(f^\omega))) \cdot \mu(k^\omega)$.

Particularly, infinite composition can be combined with itself. Let us consider a reduction sequence having length ω^2, and ϕ_{ij} be a proof term denoting its $\omega * i + j$-th step, so that for each $i < \omega$, $\cdot_{j<\omega}\ \phi_{ij}$ denotes the subsequence including the steps from the $\omega * i$-th up to the $\omega * (i+1)$-th. Then $\cdot_{i<\omega} \cdot_{j<\omega}\ \phi_{ij}$ is a proof term denoting the entire reduction sequence. By iteration of this pattern, proof terms can be built denoting reduction sequences of any countable ordinal length. This claim is proved in Sec. 5.

4 Permutation Equivalence

Two proof terms can be the result of arranging the same contraction activity in different ways, regarding parallelism/nesting degree, sequential order, and/or localisation of contractions. Such proof terms should be recognised as being *permutation equivalent.*

In this section we give the definition of equivalence between infinitary proof terms. We do this by extending *permutation equivalence*, as it is defined in [11] Sec. 8.3, to the infinitary setting. Permutation equivalence, which will be denoted $\approx$ henceforth, is defined there for finitary proof terms as the congruence generated by the following six basic equivalences.

$$
\begin{array}{lrcl}
(\mathsf{IdLeft}) & 1 \cdot \psi & \approx & \psi \\
(\mathsf{IdRight}) & \psi \cdot 1 & \approx & \psi \\
(\mathsf{Assoc}) & \psi \cdot (\phi \cdot \chi) & \approx & (\psi \cdot \phi) \cdot \chi \\
(\mathsf{Struct}) & f(\psi_1, \ldots, \psi_m) \cdot f(\phi_1, \ldots, \phi_m) & \approx & f(\psi_1 \cdot \phi_1, \ldots, \psi_m \cdot \phi_m) \\
(\mathsf{OutIn}) & \mu(\psi_1, \ldots, \psi_m) & \approx & \mu(s_1, \ldots, s_m) \cdot r[\psi_1, \ldots, \psi_m] \\
(\mathsf{InOut}) & \mu(\psi_1, \ldots, \psi_m) & \approx & l[\psi_1, \ldots, \psi_m] \cdot \mu(t_1, \ldots, t_m)
\end{array}
$$

where $\mu : l \to r$, $s_i = src(\psi_i)$ and $t_i = tgt(\psi_i)$ in (InOut) and (OutIn).

In the present, infinitary setting, three elements must be added in order to obtain an adequate characterisation of infinitary permutation equivalence. Two of these elements are related with infinitary composition.

First, we add one basic equation, the analogue of (Struct) for infinite composition:

$$(\mathsf{InfStruct})\ \cdot_{i<\omega} f(\psi_i^1,\ldots,\psi_i^m) \approx f(\cdot_{i<\omega}\psi_i^1,\ldots,\cdot_{i<\omega}\psi_i^m)$$

Secondly, we need an infinitary equational logic rule in order to ensure that $\approx$ behaves as a congruence also w.r.t. infinite composition:

$$\frac{\psi_i \approx \phi_i \quad \text{for all } i<\omega}{\cdot_{i<\omega}\psi_i \approx \cdot_{i<\omega}\phi_i}\ \mathsf{InfComp}$$

Lastly, we need a limit rule, by which permutation equivalence of two convergent reductions can be concluded, roughly, from permutation equivalence up to any finite depth. In the following Lim-rule this is implemented by requiring that for any k the reductions can be factorised into prefixes that coincide and "tails" for which all activity occurs at depth greater than k.

$$\frac{\left.\begin{array}{ll}\psi \approx_1 \chi_k \cdot \psi'_k & mind(\psi'_k) > k\\ \phi \approx_1 \chi_k \cdot \phi'_k & mind(\phi'_k) > k\end{array}\right\}\ \text{for all } k<\omega}{\psi \approx \phi}\ \mathsf{Lim}$$

Here $\approx_1$ is the congruence generated by the six basic equations, (InfStruct) and InfComp, but excluding the Lim-rule itself. Variations of the Lim-rule can be considered; we will briefly discuss this point in the conclusions.

The equational logic that we use to define permutation equivalence is infinitary in two ways. It involves the infinite proof terms, and two of the rules, InfComp and Lim, require an infinite number of premises. In order to obtain a sound (transfinite) induction principle to reason about infinitary permutation equivalence, its definition can be *layered* by ordinals, analogously to the definition of the set of proof terms. For details we refer to [8].

In the remainder of this section, we will give some examples of infinitary permutation equivalence. Let $\mu : f(x) \to g(x)$, $\nu : g(x) \to k(x)$, $\sigma : k(x) \to m(x)$, $\rho : h(x,y) \to x$, $\pi : a \to b$, $\tau : i(x) \to x$, and $\kappa : f(x) \to g(f(x))$.

Consider $\psi = \psi_1 \cdot \psi_2$ where $\psi_1 = \cdot_{i<\omega}\, g^i(\mu(f^\omega))$ and $\psi_2 := \cdot_{i<\omega}\, k^i(\nu(g^\omega))$, and $\phi = \cdot_{i<\omega}\chi_i$ where $\chi_i = k^i(\mu(f^\omega)\cdot\nu(f^\omega))$. These proof terms denote different sequentialisations of the same contraction activity, namely, the transformation of each occurrence of f in f^ω to g and subsequently to k, by means of the μ- and ν-rules respectively.

Using the augmented congruence, including the Lim rule, the assertion $\psi \approx \phi$ can be justified as follows. Let $n<\omega$. Then

$$\begin{array}{llll}\psi & \approx_1 \mu(f^\omega)\cdot g(\psi_1)\cdot\nu(g^\omega)\cdot k(\psi_2) & \approx_1 & \mu(f^\omega)\cdot\nu(f^\omega)\cdot k(\psi_1)\cdot k(\psi_2)\\ & \approx_1 \chi_0 \cdot k(\psi) & \approx_1 & \chi_0 \cdot k(\chi_0 \cdot k(\psi))\\ & \approx_1 \chi_0 \cdot k(\chi_0)\cdot k^2(\psi) & = & \chi_0\cdot\chi_1\cdot k^2(\psi)\\ & \cdots & \approx_1 & \chi_0\cdot\chi_1\cdot\ldots\cdot\chi_n\cdot k^{n+1}(\psi)\end{array}$$

by applying (InfStruct), (InOut), (OutIn), (Struct), repeating this for the copy of ψ in $k(\psi)$ followed by (Struct), and so on. For ϕ we have $\phi \approx_1 \chi_0\cdot\chi_1\cdot\ldots\chi_n\cdot\cdot_{i<\omega}\chi_{n+1+i}$. Hence Lim yields $\psi \approx \phi$.

We point out the infinitary aspects of this transformation. An *infinite* number of ν-steps must be permuted, each one over an *infinite* number of μ-steps. Permuting a ν-step requires equations (InOut) and (OutIn). The equation (InfStruct) is used to extract the pattern $g(\Box)$ of ν, obtaining $g(\psi_1)$, thereby allowing permutation of a ν step over an infinite number of μ-steps, using (InOut) and (OutIn) just once. The infinite number of ν-permutations take place at ever-bigger depths. This makes it possible to apply Lim.

4.1 The Role of the InfComp Rule

We consider the following variation of the first permutation equivalence example discussed: let $\psi = \psi_1 \cdot \psi_2$ where $\psi_1 = \cdot_{i<\omega} k^i(\mu(f^\omega)) \cdot k^i(\nu(f^\omega))$ and $\psi_2 = \cdot_{i<\omega} m^i(\sigma(k^\omega))$, and $\phi = \cdot_{i<\omega} \chi_i$ where $\chi_i = m^i(\mu(f^\omega) \cdot \nu(f^\omega) \cdot \sigma(f^\omega))$. Let $\psi_1' := \cdot_{i<\omega} k^{i+1}(\mu(f^\omega)) \cdot k^{i+1}(\nu(f^\omega))$. The proof terms ψ and ϕ denote the transformation of f^ω to m^ω, where at each depth level the rules μ, ν and σ are applied successively.

The idea of a judgement allowing to assert $\psi \approx \phi$ is similar to that of the previous example: in order to transform ψ into ϕ, ψ_1' must be transformed into $k(\psi_1)$, so that the first σ step can be permuted. But in this case, for any $i < \omega$, the (Struct) equation gives $k^{i+1}(\mu(f^\omega)) \cdot k^{i+1}(\nu(f^\omega)) \approx_1 k(k^i(\mu(f^\omega)) \cdot k^i(\nu(f^\omega)))$. Consequently, InfComp yields $\psi_1' \approx_1 \cdot_{i<\omega} k(k^i(\mu(f^\omega)) \cdot k^i(\nu(f^\omega)))$. Therefore,

$$\begin{aligned}
\psi &\approx_1 \mu(f^\omega) \cdot \nu(f^\omega) \cdot \psi_1' \cdot \sigma(k^\omega) \cdot \cdot_{i<\omega} m^{i+1}(\sigma(k^\omega)) \\
&\approx_1 \mu(f^\omega) \cdot \nu(f^\omega) \cdot \left(\cdot_{i<\omega} k(k^i(\mu(f^\omega)) \cdot k^i(\nu(f^\omega)))\right) \cdot \sigma(k^\omega) \cdot \cdot_{i<\omega} m^{i+1}(\sigma(k^\omega)) \\
&\approx_1 \mu(f^\omega) \cdot \nu(f^\omega) \cdot k\left(\cdot_{i<\omega} k^i(\mu(f^\omega)) \cdot k^i(\nu(f^\omega))\right) \cdot \sigma(k^\omega) \cdot m(\cdot_{i<\omega} m^i(\sigma(k^\omega))) \\
&= \mu(f^\omega) \cdot \nu(f^\omega) \cdot k(\psi_1) \cdot \sigma(k^\omega) \cdot m(\psi_2) \\
&\approx_1 \mu(f^\omega) \cdot \nu(f^\omega) \cdot \sigma(f^\omega) \cdot m(\psi_1) \cdot m(\psi_2) \\
&\approx_1 \chi_0 \cdot m(\psi)
\end{aligned}$$

by (Assoc), the previous argument, (InfStruct), (InOut) followed by (OutIn), and (Struct) respectively. An iteration of this argument yields $\psi \approx_1 \chi_0 \cdot \ldots \cdot \chi_n \cdot m^{n+1}(\psi)$ for any $n < \omega$. From this result, it is easy to conclude $\psi \approx \phi$ by resorting to the Lim rule.

4.2 Infinitary Erasure

As for the finitary setting, this characterisation of permutation equivalence models adequately the phenomenon of *erasure* of some contraction activity by step permutation.A simple example follows: by applying twice the (OutIn) equation, we obtain $\rho(\mu(a), \tau^\omega) \approx \rho(f(a), i^\omega) \cdot \mu(a) \approx \rho(\mu(a), i^\omega)$.

Now consider $\psi = f(\pi) \cdot \cdot_{i<\omega} g^i(\kappa(b))$ and $\phi = \cdot_{i<\omega} g^i(\kappa(a))$. These proof terms are sequential descriptions of the reduction sequences $f(a) \to f(b) \to g(f(b)) \to g^2(f(b)) \twoheadrightarrow g^\omega$ and $f(a) \to g(f(a)) \to g^2(f(a)) \twoheadrightarrow g^\omega$ respectively. Observe that the π step in the former sequence can be permuted with each κ step in turn, yielding $f(a) \to g(f(a)) \to \ldots \to g^n(f(a)) \to g^n(f(b)) \twoheadrightarrow g^\omega$. The latter sequence can be seen as the result of taking the limit of this sequence of step permutations.

This example shows the existence of a form of erasure unique to infinitary rewriting: some contraction activity (in this case the π step) is erased as the result of a limit of step permutations. We call this phenomenon *infinitary erasure.*

According to this view, ψ and ϕ denote equivalent contraction activities, and they therefore should be stated as permutation equivalent. Observe that

$$\begin{aligned}\psi &\approx_1 f(\pi) \cdot \kappa(b) \cdot g(\cdot_{i<\omega}\, g^i(\kappa(b))) \\ &\approx_1 \kappa(a) \cdot g(f(\pi)) \cdot g(\kappa(b)) \cdot g^2(\cdot_{i<\omega}\, g^i(\kappa(b))) \\ &\approx_1 \kappa(a) \cdot g(\kappa(a)) \cdot g^2(f(\pi)) \cdot g^2(\kappa(b)) \cdot g^3(\cdot_{i<\omega}\, g^i(\kappa(b))) \ \ldots\end{aligned}$$

where the first component of each successive infinitary concatenation is extracted by (Assoc), (Struct) and (InfStruct), and each inversion involves (Struct) (except for the first one), (InOut) and (OutIn). Consequently, for any $n < \omega$, we can obtain $\psi \approx_1 \kappa(a) \cdot \ldots \cdot g^n(\kappa(a)) \cdot g^{n+1}(f(\pi)) \cdot g^{n+1}(\cdot_{i<\omega}\, g^i(\kappa(b)))$. On the other hand, it is straightforward to get $\phi \approx_1 \kappa(a) \cdot \ldots \cdot g^n(\kappa(a)) \cdot g^{n+1}(\phi)$. Hence, the Lim rule yields $\psi \approx \phi$.

This example adequately describes infinitary erasure: the equivalence between ψ and ϕ is obtained by a limit argument.

5 Denotation

In this section we show that *any* reduction sequence can be faithfully denoted by a proof term. Since also simultaneous reduction can be expressed by proof terms, in the form of infinitary multi-steps, not every proof term uniquely denotes a reduction sequence. However, we will define a subclass of so-called *stepwise* proof terms, which do have unique denotations. We will show that any reduction can be represented by a *stepwise* proof term, which is unique up to an equivalence which we baptized "rebracketing", and which results from applying associativity of composition in this infinitary setting.

5.1 Existence

A *stepwise proof term* is inductively defined as: either a one-step proof term, or a binary or infinite concatenation of stepwise proof terms. A *stepwise-or-nil proof term* is either a stepwise proof term or a term in $Ter^\infty(\Sigma)$.

Given a stepwise-or-nil proof term ψ, we define the *number of steps* and the *α-th step* of ψ, notation $ns(\psi)$ and $\psi[\alpha]$ respectively. The definitions are straightforward. For number of steps: $ns(\psi) := 0$ if $\psi \in Ter^\infty(\Sigma)$, $ns(\psi) := 1$ if ψ is a one-step, $ns(\psi_1 \cdot \psi_2) := ns(\psi_1) + ns(\psi_2)$, $ns(\cdot_{i<\omega}\, \psi_i) := \Sigma_{i<\omega} ns(\psi_i)$. Moreover, *$\alpha$-th step* is defined for $\alpha < ns(\psi)$ by: $\psi[0] := \psi$ if ψ is a one-step, $\cdot_{i<\omega}\, \psi_i[\alpha] := \psi_k[\gamma]$ where $\alpha = \Sigma_{i<k} ns(\psi_i) + \gamma$ and $\gamma < ns(\psi_k)^\dagger$, and analogously for binary concatenation.

Observe that the steps of stepwise proof terms are one-step proof terms. A stepwise proof term is a representation of the sequence of its steps, organised

† Properties of ordinal infinitary sum imply uniqueness of k and γ here.

by means of binary and infinite compositions. In this way, one-step proof terms denote reduction steps, while stepwise proof terms denote reduction sequences.

Properties that support this view of stepwise proof terms can be proved by analysing their form. Specifically, if ψ is a stepwise proof term: $src(\psi) = src(\psi[0])$, $tgt(\psi) = tgt(\psi[\alpha])$ if $ns(\psi) = \alpha + 1$, $tgt(\psi) = \lim_{\alpha \to ns(\psi)} tgt(\psi[\alpha])$ if $ns(\psi)$ is a limit ordinal, $src(\psi[\alpha + 1]) = tgt(\psi[\alpha])$ if $\alpha + 1 < ns(\psi)$, and $mind(\psi) = min\{d(\psi[\alpha]) \mid \alpha < ns(\psi)\}$.

We formalise the relation between stepwise proof terms and reduction sequences as follows: a stepwise-or-nil proof term ψ *denotes* a reduction sequence δ iff $ns(\psi) = \mathsf{length}(\delta)$, $src(\psi) = src(\delta)$, and for all $\alpha < ns(\psi)$, $\psi[\alpha]$ coincides in source term, redex position and rule (symbol) with $\delta[\alpha]$. This relation preserves convergence: if ψ denotes δ, then ψ is convergent if and only if δ is, cf. [8].

In this setting, the existence requirement amounts to the following:

Proposition 1. *Let δ be a reduction sequence having countable length. Then there exists a stepwise-or-nil proof term ψ such that ψ denotes δ.*

Proof. By induction on $\mathsf{length}(\delta)$. If $\mathsf{length}(\delta) = 0$, i.e. $\delta = \mathsf{Id}_t$, then just take $\psi := t$. If $\mathsf{length}(\delta) = 1$, then replacing the pattern of the contracted redex in the source term by the corresponding rule symbol yields a one-step proof term denoting δ. If $\mathsf{length}(\delta)$ is a successor (resp. limit) ordinal, then it is easy to split δ in two (resp. ω, cf. Lem. 1) parts, so that IH can be applied on each part and then the results joined by binary (resp. infinite) concatenation.

5.2 Uniqueness

We say that two stepwise-or-nil proof terms ψ and ϕ are *denotationally equivalent* iff they contain exactly the same sequence of steps, organised in possibly different ways. Formally, $\psi \equiv \phi$ iff $ns(\psi) = ns(\phi)$ and either $ns(\psi) = 0$ and $\psi = \phi$, or $ns(\psi) > 0$ and $\psi[\alpha] = \phi[\alpha]$ if $\alpha < ns(\psi)$. It is straightforward to verify that two stepwise-or-nil proof terms are denotationally equivalent iff they denote the same reduction sequence.

We define *rebracketing equivalence*, notation $\approx_{()}$, as the congruence generated using (Assoc) as the only basic equation, and including InfComp and Lim as equational logic rules.[‡] The corresponding base rebracketing equivalence, notation $\approx_{(1)}$, is defined analogously but not including Lim as a rule. In the premises of Lim for $\approx_{()}$, the occurrences of $\approx_1$ must be replaced by $\approx_{(1)}$.

We claim that the denotation of reduction sequences by using proof terms is unique modulo rebracketing. To verify this claim, we show that the relations $\equiv$ and $\approx_{()}$ coincide.

Lemma 2. *Let ψ be a stepwise proof term, and α an ordinal verifying $0 < \alpha < ns(\psi)$. Then there exist stepwise proof terms ϕ, χ such that $\psi \approx_{(1)} \phi \cdot \chi$ and $ns(\phi) = \alpha$. Moreover, $\nu(\phi) < \nu(\psi)$ and $\nu(\chi) \leq \nu(\psi)$.*

---

‡ This is part of what in [11], Table 8.1, are called the *reduction identities*, adapted to the infinitary setting.

Proof. Easy induction on ψ. Cf. the stated properties about ν in Sec. 3.1.

Proposition 2. *Let ψ, ϕ be stepwise-or-nil proof terms such that $\psi \approx_{()} \phi$. Then $\psi \equiv \phi$.*

Proof. Induction on the judgement $\psi \approx_{()} \phi$, observing the rule used to conclude it. We give some details w.r.t. the Lim rule. We prove $ns(\psi) = ns(\phi)$ by contradiction. Assume wlog that $ns(\phi) > ns(\psi)$, implying that the step $\phi[ns(\psi)]$ exists. Let $n := d(\phi[ns(\psi)])$, observe $n < \omega$. Then the premises of the Lim rule yield $\psi \approx_{(1)} \chi \cdot \psi'$ and $\phi \approx_{(1)} \chi \cdot \phi'$, where $mind(\psi') > n$ and $mind(\phi') > n$. In turn, $\psi \approx_{(1)} \chi \cdot \psi'$ implies $ns(\psi) \geq ns(\chi)$, while $\phi \approx_{(1)} \chi \cdot \phi'$, IH, and $mind(\phi') > n = d(\phi[ns(\psi)])$, imply $\phi[ns(\psi)] = \chi[ns(\psi)]$, then $ns(\chi) > ns(\psi)$; hence, a contradiction. Let $\gamma < ns(\psi)$, and consider $\psi \approx_{(1)} \chi \cdot \psi'$ and $\phi \approx_{(1)} \chi \cdot \phi'$ where $mind(\psi') > d(\psi[\gamma])$ and $mind(\phi') > d(\psi[\gamma])$. Then IH on these premises and the definition of $\equiv$ imply that $\psi[\gamma] = (\chi \cdot \psi')[\gamma]$ and $\phi[\gamma] = (\chi \cdot \phi')[\gamma]$. In turn, the conditions on $mind(\psi')$ imply that $\gamma < ns(\chi)$, so that $\psi[\gamma] = \phi[\gamma] = \chi[\gamma]$.

Proposition 3. *Let ψ, ϕ be convergent stepwise-or-nil proof terms such that $\psi \equiv \phi$. Then $\psi \approx_{()} \phi$.*

Proof. We proceed by induction on $\langle \psi, \phi \rangle$. The interesting cases are when ψ is either a binary or an infinite concatenation.

Assume $\psi = \psi_1 \cdot \psi_2$. It is easy to see that also ϕ must be of this form: $\phi = \phi_1 \cdot \phi_2$. If $ns(\psi_1) = ns(\phi_1)$, then $\psi_i \equiv \phi_i$ for $i = 1, 2$, so that IH and then congruence suffice to conclude. Assume $ns(\psi_1) < ns(\phi_1)$. Then Lem. 2 implies $\phi_1 \approx_{()} \chi_1 \cdot \chi_2$ where $ns(\chi_1) = ns(\psi_1)$, and therefore $\phi \approx_{()} (\chi_1 \cdot \chi_2) \cdot \phi_2 \approx_{()} \chi_1 \cdot (\chi_2 \cdot \phi_2)$. Hence hypotheses and Prop. 2 imply $\psi_1 \cdot \psi_2 = \psi \equiv \phi \equiv \chi_1 \cdot (\chi_2 \cdot \phi_2)$. From this, since $ns(\chi_1) = ns(\psi_1)$, it is easy to show $\psi_1 \equiv \chi_1$, and also $\psi_2 \equiv \chi_2 \cdot \phi_2$. The result then follows by applying IH twice, congruence and transitivity. Finally, if $ns(\psi_1) > ns(\phi_1)$, an analogous argument yields $\chi_1 \cdot (\chi_2 \cdot \psi_2) \equiv \psi \equiv \phi = \phi_1 \cdot \phi_2$ where $ns(\chi_1) = ns(\phi_1)$, and hence $\chi_1 \equiv \phi_1$ and $\chi_2 \cdot \psi_2 \equiv \phi_2$. Again using $\psi_1 \approx_{()} \chi_1 \cdot \chi_2$, it is now easy to obtain $\nu(\chi_1) < \nu(\psi)$ and $\nu(\chi_2 \cdot \psi_2) \leq \nu(\psi)$. Hence the result follows again by IH, congruence and transitivity.

Assume $\psi = \cdot_{i<\omega} \psi_i$. Then also ϕ is an infinite concatenation: $\phi = \cdot_{i<\omega} \phi_i$. By recombining the steps of ψ, a proof term $\chi = \cdot_{i<\omega} \chi_i$ can be built satisfying: *(i)* $\chi \equiv \psi$, *(ii)* $ns(\chi_i) = ns(\phi_i)$ for all $i < \omega$, and *(iii)* for all $p < \omega$, $\chi_0 \cdot \ldots \cdot \chi_p \approx_{(1)} \psi_0 \cdot \ldots \cdot \psi_j \cdot \xi$, where $ns(\chi_0 \cdot \ldots \cdot \chi_p) = ns(\psi_0 \cdot \ldots \cdot \psi_j) + \alpha$, $\alpha < ns(\psi_{j+1})$, $\psi_{j+1} \approx_{(1)} \xi \cdot \xi'$ and $ns(\xi) = \alpha$; cf. Lem. 2. E.g. if $ns(\phi_0) = ns(\psi_0) + \alpha$ such that $\alpha < ns(\psi_1)$, then we define $\chi_0 := \psi_0 \cdot \xi_1$ and χ_1 such that $\chi_1[0] = \xi_2[0]$, where $\psi_1 \approx_{(1)} \xi_1 \cdot \xi_2$ and $ns(\xi_1) = \alpha$.

Let $n < \omega$, p such that $mind(\chi_i) > n$ if $i > p$, and j, ξ, ξ' as in condition *(iii)* above. Let us call $\xi_n := \chi_0 \cdot \ldots \cdot \chi_p$, $\psi'_n := \xi' \cdot \cdot_{i<\omega} \psi_{j+2+i}$, and $\chi'_n := \cdot_{i<\omega} \chi_{p+1+i}$. Then $\xi_n \approx_{(1)} \psi_0 \cdot \ldots \cdot \psi_j \cdot \xi$, implying $\psi \approx_{(1)} \xi_n \cdot \psi'_n$. Prop. 2 yields $\xi_n \equiv \psi_0 \cdot \ldots \cdot \psi_j \cdot \xi$, from this and *(i)* it is easy to obtain $\chi'_n \equiv \psi'_n$, and therefore $mind(\psi'_n) = mind(\chi'_n) > n$. Hence we can obtain $\psi \approx_{()} \chi$ by applying the Lim equation.

Moreover, *(i)* and $\psi \equiv \phi$ imply $\phi \equiv \chi$. From this and considering *(ii)*, it is easy to obtain $\chi_i \equiv \phi_i$ for all i. Furthermore, for any $i < \omega$ it can also be proved that $\nu(\chi_i) < \nu(\psi)$, so that we can apply IH to obtain $\chi_i \approx_{()} \phi_i$.

We conclude by building the following rebracketing equivalence judgement.

$$\dfrac{\dfrac{\begin{matrix} & \psi \approx_{(1)} \xi_n \cdot \psi'_n & \\ \cdots & \chi \approx_{(1)} \xi_n \cdot \chi'_n & \cdots \end{matrix}}{\psi \approx_{()} \chi}\,\mathsf{Lim} \qquad \dfrac{\cdots \quad \dfrac{\ldots IH \ldots}{\chi_n \approx_{()} \phi_n} \quad \cdots}{\chi \approx_{()} \phi}\,\mathsf{InfComp}}{\psi \approx_{()} \phi}$$

6 Compression

The compression lemma, [5,11,6] states that the full power of (strongly) convergent reduction can be achieved by reductions having length at most ω. Formally: if $t \xrightarrow{\delta}\!\!\!\twoheadrightarrow u$, then there exists a convergent reduction $t \xrightarrow{\gamma}\!\!\!\twoheadrightarrow u$, such that $\mathsf{length}(\gamma) \leq \omega$. In [5] a more precise statement is given: for orthogonal TRSs, γ can be chosen such that it is Lévy-equivalent (cf. [4]) to δ.

We present the compression result for arbitrary proof terms. Moreover, it establishes permutation equivalence between the original and the compressed proof terms. Via the faithful representation of reduction sequences as stepwise proof terms, our compression result extends to the original reductions as well.

As in [6], we do not require orthogonality for compression (although our general assumption of left-linearity is crucial here). However, as far as we know the combination of not assuming orthogonality and yet obtaining equivalence between original and compressed form is new.

The compression proof uses as a key technical result, that any proof term can be *factorised* into a leading part denoting *finite* contraction activity, composed with a tail denoting activity at *arbitrary depths.* That is, for any proof term ψ and $n < \omega$, we obtain two proof terms χ and ϕ, such that $\psi \approx_1 \chi \cdot \phi$, χ is a finite stepwise-or-nil proof term, and $mind(\phi) > n$.

Collapsing rules make factorisation non-trivial for infinitary multi-steps. We need the notion of a *collapsing chain* for an infinitary multi-step ψ: a sequence of positions $\langle p_i \rangle_{i \leq n}$ (resp. $\langle p_i \rangle_{i < \omega}$), such that for all $i < n$ $(i < \omega)$, $\psi(p_i)$: $l[x_1, \ldots, x_m] \to x_j$ is a collapsing rule symbol, and $p_{i+1} = p_i\ j$. We consider particularly chains *starting at* ϵ, i.e. such that $p_0 = \epsilon$.

Lemma 3. *Let ψ be an infinitary multi-step such that an infinite collapsing chain for ψ starting at ϵ exists. Then ψ is not TGT-weakly normalising.*

Proof. It is enough to verify that if $\psi \xrightarrow{\delta}\!\!\!\twoheadrightarrow \phi$ by using the TGT companion TRS, then ϕ includes an infinite collapsing chain starting at ϵ, implying that $\phi(\epsilon)$ is a (collapsing) rule symbol, and therefore it is not a TGT-normal form.

Lemma 4. *Let ψ be an infinitary multi-step and n, $1 < n < \omega$, such that there is no collapsing chain of length n for ψ, starting at ϵ. Then there exists a TGT-reduction sequence δ, such that $\psi \xrightarrow[TGT]{\delta}\!\!\!\twoheadrightarrow \phi$, $\mathsf{length}(\delta) < n$, $d(\delta[i]) = 0$ for all $i < \mathsf{length}(\delta)$, and $\phi(\epsilon) \in \Sigma$.*

Proof. By induction on n. If $n = 2$, then either $\psi(\epsilon) \in \Sigma$ so that we conclude immediately, or $\psi(\epsilon) = \mu$ where $\mu : l[x_1, \ldots, x_m] \to f(t_1, \ldots, t_k)$, and then $\psi \xrightarrow[TGT]{\langle\epsilon,\mu\rangle} f(t'_1, \ldots, t'_k)$, thus we conclude. Assume $n = n' + 1$. Then $\psi(\epsilon)$ not being a collapsing rule symbol allows an argument similar to that of the previous case. Otherwise, i.e. if $\psi(\epsilon) = \mu$ where $\mu : l[x_1, \ldots, x_m] \to x_k$, and $\psi \xrightarrow[TGT]{\langle\epsilon,\mu\rangle} \psi|_k$, it is not difficult to verify that a collapsing chain having length n' in $\psi|_k$ would imply the existence of a collapsing chain having length n in ψ, contradicting the lemma hypotheses. Then IH can be applied on $\psi|_k$, which suffices to conclude.

Lemma 5. *Let ψ be a convergent infinitary multi-step. Then there exist χ, ϕ such that $\psi \approx_1 \chi \cdot \phi$, χ is a finite stepwise-or-nil proof term, $d(\chi[i]) = 0$ for all $i < ns(\chi)$, and ϕ is a convergent infinitary multi-step verifying $mind(\phi) > 0$.*

Proof. Convergence of ψ implies that ψ is WN^∞ in TGT, so that Lem. 3 implies the existence of some $n < \omega$ such that ψ does not have a collapsing chain of length n starting with ϵ. Therefore, Lem 4. implies $\psi \xrightarrow[TGT]{\delta}\!\!\!\!\rightarrow \phi$, $\mathsf{length}(\delta) < n$, $d(\delta[i]) = 0$ for all $i < \mathsf{length}(\delta)$, and $\phi(\epsilon) \in \Sigma$. Then, it is not difficult to prove that $\psi \xrightarrow[TGT]{\delta}\!\!\!\!\rightarrow \phi$ implies $\psi \approx_1 \chi \cdot \phi$, where χ is a stepwise-or-nil proof term verifying $ns(\chi) = \mathsf{length}(\delta)$ and $d(\chi[i]) = d(\delta[i])$ if $i < \mathsf{length}(\delta)$: induction on $d(\delta[0])$ yields the property for one-steps, and then induction on $\mathsf{length}(\delta)$ suffices. Observe that $\phi(\epsilon) \in \Sigma$ implies $mind(\phi) > 0$. To verify that ϕ is convergent, the strip lemma or parallel moves lemma, valid for TGT since it is an orthogonal iTRS, yields that ψ convergent implies ϕ convergent. Thus we conclude.

To extend factorisation to any proof term we need to be able to *swap* its components. The (InOut) and (OutIn) equations allow for this swapping in simple cases, e.g. as in the following permutation equivalence judgement: $f(\mu(a)) \cdot \mu(g(a)) \approx_1 \mu(\mu(a)) \approx_1 \mu(f(a)) \cdot g(\mu(a))$, where $\mu : f(x) \to g(x)$. The following lemma shows that such an operation can be performed in a general case.

Lemma 6. *Let ψ be a finite stepwise-or-nil proof term. Then there exist two numbers $n, n' < \omega$ such that, for any convergent proof term ξ verifying $tgt(\xi) = src(\psi)$ and $mind(\xi) \geq n + n'$, a finite stepwise-or-nil proof term ψ' and a convergent proof term ξ' can be found, such that $\xi \cdot \psi \approx_1 \psi' \cdot \xi'$, $ns(\psi') = ns(\psi)$, $d(\psi'[i]) = d(\psi[i])$ for all i, and $mind(\xi') \geq mind(\xi) - n' \geq n$.*

Proof. Induction on $ns(\psi)$, the interesting case being when ψ is a one-step.

Let μ be the rule symbol occurring in ψ, and p the position of the occurrence of μ in ψ. We consider $n := |p|$ and n' to be the depth of the deepest variable occurrence in the left-hand side of μ. It is not difficult to obtain that $\psi = C[t_1, \ldots, t_{j-1}, \mu(u_1, \ldots, u_m), t_{j+1}, \ldots, t_k]$ where the depth of all the holes in C is exactly n. By carefully defining the *compressed to fixed prefix form* of a proof term for which no activity is performed at any position in a prefix-closed set, we can obtain that $\xi \approx_1 C[\xi_1, \ldots, \xi_{j-1}, l[\phi_1, \ldots, \phi_m], \xi_{j+1}, \ldots, \xi_k]$. Therefore

$$\begin{aligned}
&\xi \cdot \psi \\
&\approx_1 C[\xi_1 \cdot t_1, ..., \xi_{j-1} \cdot t_{j-1}, l[\phi_1, ..., \phi_m] \cdot \mu(u_1, ..., u_m), \xi_{j+1} \cdot t_{j+1}, ..., \xi_k \cdot t_k] \\
&\approx_1 C[\xi_1, ..., \xi_{j-1}, \mu(\phi_1, ..., \phi_m), \xi_{j+1}, ..., \xi_k] \\
&\approx_1 C[s_1 \cdot \xi_1, ..., s_{j-1} \cdot \xi_{j-1}, \mu(w_1, ..., w_m) \cdot h[\phi_1, ..., \phi_m], s_{j+1} \cdot \xi_{j+1}, ..., s_k \cdot \xi_k] \\
&\approx_1 C[s_1, ..., s_{j-1}, \mu(w_1, ..., w_m), s_{j+1}, ..., s_k] \cdot \\
&\qquad C[\xi_1, ..., \xi_{j-1}, h[\phi_1, ..., \phi_m], \xi_{j+1}, ..., \xi_k]
\end{aligned}$$

where $\mu : l \to h$, $s_i = src(\xi_i)$ and $w_i = src(\phi_i)$.

We consider $\psi' = C[s_1, \ldots, s_{j-1}, \mu(w_1, \ldots, w_m), s_{j+1}, \ldots, s_k]$ and $\xi' = C[\xi_1, \ldots, \xi_{j-1}, h[\phi_1, \ldots, \phi_m], \xi_{j+1}, \ldots, \xi_k]$. Convergence of ξ' stems from convergence of each ξ_i and each ϕ_i. Finally, a careful analysis of minimal activity depth yields the desired condition about $mind(\xi')$.

Now we can prove the general factorisation result.

Proposition 4. *Let ψ be a convergent proof term and $n < \omega$. Then there exist χ and ϕ such that $\psi \approx_1 \chi \cdot \phi$, χ is a finite stepwise-or-nil proof term, ϕ is convergent and $mind(\phi) > n$.*

Proof. We proceed by induction on ψ.

If ψ is an infinitary multi-step, then we proceed by induction on n. Lem. 5 implies $\psi \approx_1 \chi_0 \cdot f(\psi_1, \ldots, \psi_m)$. If $n = 0$ then we are done. Otherwise convergence of all ψ_i can be obtained from convergence of $f(\psi_1, \ldots, \psi_m)$ by a careful study of projections. Then IH applied to the ψ_i's yields $\psi \approx_1 \chi_0 \cdot f(\chi_1 \cdot \phi_1, \ldots, \chi_m \cdot \phi_m)$. From this proof term, a permutation equivalence argument allows to conclude.

Assume $\psi = \psi_1 \cdot \psi_2$. Then IH on ψ_2 yields $\psi \approx_1 \psi_1 \cdot \chi_2 \cdot \phi_2$. IH applies also on ψ_1 w.r.t. a sufficiently big n', yielding $\psi \approx_1 \chi_1 \cdot \phi_1 \cdot \chi_2 \cdot \phi_2$. Then Lem. 6 on $\phi_1 \cdot \chi_2$ implies $\psi \approx_1 \chi_1 \cdot \chi_2' \cdot \phi_1' \cdot \phi_2$, which suffices to conclude.

If $\psi = \cdot_{i<\omega} \psi_i$, then it is enough to observe that IH can be applied to $\psi_0 \cdot \ldots \cdot \psi_k$, where $mind(\psi_i) > n$ if $i > k$.

If $\psi = f(\psi_1, \ldots, \psi_m)$ then an easy inductive argument suffices.

Finally, if $\psi = \mu(\psi_1, \ldots, \psi_m)$, then two cases must be considered: μ is collapsing or not. In either case an easy argument suffices. If μ is not collapsing, then observe that the (Struct) equation can be extended to arbitrary contexts having a finite number of holes.

Theorem 1. *Let ψ be a convergent proof term. Then there exists some stepwise proof term ϕ verifying $\psi \approx \phi$ and $ns(\phi) \leq \omega$.*

Proof. We define the sequences of proof terms $\langle \psi_i \rangle_{i<\omega}$ and $\langle \phi_i \rangle_{i<\omega}$ as follows. We start defining $\psi_0 := \psi$. Then, for each $i < \omega$, we define ϕ_i and ψ_{i+1} to be proof terms such that $\psi_i \approx_1 \phi_i \cdot \psi_{i+1}$, ϕ_i is a finite stepwise-or-nil proof term and either $mind(\psi_{i+1}) > mind(\psi_i)$ or $mind(\psi_{i+1}) = mind(\psi_i) = \omega$; cf. Prop. 4. Observe that $mind(\psi_i) < \omega$ implies $mind(\phi_i) = mind(\psi_i)$, so in that case ϕ_i is a stepwise proof term, i.e. it is not trivial. Moreover, an easy induction on n yields $\psi \approx_1 \phi_0 \cdot \ldots \cdot \phi_n \cdot \psi_{n+1}$ for all n.

There are two cases to consider. If there is some n such that ψ_n is trivial, then it is enough to take $\phi := \phi_0 \cdot \ldots \cdot \phi_{n-1}$ for the minimal such n (or just $\phi := src(\psi)$

if $\psi_0 = \psi$ is trivial). Otherwise, it is easy to prove that $\psi \approx \cdot_{i<\omega} \phi_i$ using the Lim rule, and, moreover, ϕ_i being finite for all i implies $ns(\cdot_{i<\omega} \phi_i) \leq \omega$.

7 Conclusions and Future Work

The reasoning in the compression proof in Sec. 6 can be extended in order to obtain *standardisation* results. As noted in [6], a concept of standard reduction being adequate for infinitary rewriting should be used, leftmost-outermost reduction does not fit in this setting. We claim that it is possible to prove the existence of a *unique* standard reduction in each permutation equivalence class, using depth-leftmost standardness as defined in [6].

It seems natural to consider a variant of permutation equivalence in which the Lim-rule can be used at most once in a derivation, and only as its last step. The derivations of permutation equivalence in our examples in Sec. 4 are all of this form. By a proof-theoretic analysis using the purported standardisation result, one can show that the restricted variant is equivalent to the more general version used in this paper. We plan to present these results in a future paper.

Another area for further exploration is the comparison of permutation equivalence as defined in this paper with other notions of equivalence of reduction. Compare [11], Ch. 8 and [9], where the equivalence of several such notions is established for finitary rewriting. An obvious first candidate would be Lévy equivalence as defined in [5] and [11], Ch. 12, via projections of reductions.

Acknowledgements. We thank Eduardo Bonelli, Delia Kesner, Vincent van Oostrom and Femke van Raamsdonk for support, stimulating discussions and some useful suggestions. This work was partially funded by the French–Argentinian Laboratory in Computer Science **INFINIS**.

References

1. Baader, F., Nipkow, T.: Term Rewriting and All That. Cambridge University Press, Cambridge (1998)
2. Courcelle, B.: Fundamental properties of infinite trees. Theor. Comput. Sci. 25, 95–169 (1983)
3. Endrullis, J., Hansen, H.H., Hendriks, D., Polonsky, A., Silva, A.: A coinductive treatment of infinitary rewriting. Presented at WIR 2013, First International Workshop on Infinitary Rewriting, Eindhoven, Netherlands (June 2013)
4. Huet, G.P., Lévy, J.J.: Computations in orthogonal rewriting systems, i and ii. In: Computational Logic - Essays in Honor of Alan Robinson, pp. 395–414 (1991)
5. Kennaway, R., Klop, J.W., Sleep, M.R., de Vries, F.J.: Transfinite reductions in orthogonal term rewriting systems. Inf. Comput. 119(1), 18–38 (1995)
6. Ketema, J.: Reinterpreting compression in infinitary rewriting. In: 23rd International Conference on Rewriting Techniques and Applications, RTA. LIPIcs, vol. 15, pp. 209–224. Schloss Dagstuhl - Leibniz-Zentrum für Informatik (2012)
7. Klop, J.W., de Vrijer, R.C.: Infinitary normalization. In: We Will Show Them: Essays in Honour of Dov Gabbay, vol. 2, pp. 169–192. College Publications (2005)

8. Lombardi, C., Ríos, A., de Vrijer, R.: Proof terms for infinitary rewriting, progress report (2014), `http://arxiv.org/abs/1402.2245`
9. van Oostrom, V., de Vrijer, R.: Four equivalent equivalences of reductions. Electronic Notes in Theoretical Computer Science 70(6), 21–61 (2002)
10. Suppes, P.: Axiomatic Set Theory. D. van Nostrand, Princeton, USA (1960)
11. Terese: Term Rewriting Systems. Cambridge Tracts in Theoretical Computer Science, vol. 55. Cambridge University Press, Cambridge (2003)

Construction of Retractile Proof Structures

Roberto Maieli

Department of Mathematics and Physics, "Roma Tre" University
Largo San Leonardo Murialdo 1 – 00146 Rome, Italy
roberto.maieli@uniroma3.it

Abstract. In this work we present a paradigm of focusing proof search based on an incremental construction of *retractile* (i.e, *correct* or *sequentializable*) proof structures of the pure (units free) multiplicative and additive fragment of linear logic. The correctness of proof construction steps (or expansion steps) is ensured by means of a system of graph retraction rules; this graph rewriting system is shown to be convergent, that is, terminating and confluent. Moreover, the proposed proof construction follows an optimal (parsimonious, indeed) retraction strategy that, at each expansion step, allows to take into account (abstract) graphs that are "smaller" (w.r.t. the size) than the starting proof structures.

Keywords: linear logic, sequent calculus, focusing proofs, proof search, proof construction, proof nets, proof net retraction, graph rewriting.

1 Introduction

This work aims to make a further step towards the development of a research programme, firstly launched by Andreoli in 2001 (see [1], [2] and [3]), which points to a theoretical foundation of a *computational programming paradigm based on the construction of linear logic proofs* (LL, [8]). Naively, this paradigm relies on the following isomorphism: *proof = state* and *construction* (or *inference*) *step = state transition*. Traditionally, this paradigm is presented as an incremental (bottom-up) construction of possibly *incomplete* (i.e., open or with proper axioms) proofs of the *bipolar focusing sequent calculus* (see Sect. 2 for a brief introduction). This calculus satisfies the property that the complete (i.e., closed or with logical axioms) bipolar focusing proofs are fully representative of all closed proofs of linear logic: this correspondence is, in general, not satisfied by the polarized fragments of linear logic. Bipolarity and focusing properties ensure more compact proofs since they get rid of some irrelevant intermediate steps during proof search (or proof construction).

Now, while the view of sequent proof construction is well adapted to theorem proving, it is inadequate when we want to model some proof-theoretic intuitions behind, e.g., concurrent logic programming which requires very flexible and modular approaches. Due to their artificial sequential nature, sequent proofs are difficult to cut into modular (reusable) concurrent components.

G. Dowek (ed.): RTA-TLCA 2014, LNCS 8560, pp. 319–333, 2014.

A much more appealing solution consists of using the technology offered by *proof nets* of linear logic or, more precisely, some forms of de-sequentialized (geometrical, indeed) proof structures in which the composition operation is simply given by (possibly, constrained) juxtaposition, obeying some correctness criteria. Actually, the proof net construction, as well the proof net cut reduction, can be performed in parallel (concurrently), but despite the cut reduction, there may not exist executable (i.e., sequentializable) construction steps: in other words, construction steps must satisfy a, possibly efficient, correctness criterion. Here, a proof net is a particular "open" proof structure, called *transitory net* (see Sect. 3), that is incrementally built bottom-up by juxtaposing, via construction steps, simple proof structures or modules, called *bipoles*. Roughly, bipoles correspond to Prolog-like *methods* of Logic Programming Languages: the *head* is represented by a multiple trigger (i.e., a multiset of positive atoms) and the *body* is represented by a layer of negative connectives with negative atoms. We say that a construction step is correct (that is, a *transaction*) when it preserves, after juxtaposition, the property of being a transitory net: that is the case when the given abstract transitory structure *retracts* (after a finite sequence of rewriting steps) to an elementary collapsed graph (i.e., single node with only pending edges). Each retraction step consists of a simple (local) graph deformation or graph rewriting. The resulting rewriting system is shown to be convergent (i.e., terminating and confluent), moreover, it preserves, step by step, the property of being a transitory structure (see Theorem 1 and Lemma 1 in Sect. 3.1). Transitory nets (i.e., retractile structures) correspond to derivations of the focusing bipolar sequent calculus (Sect. 4, Theorem 2).

The first retraction algorithm for checking correctness of the proof structures of the pure multiplicative fragment of linear logic (MLL), was given by Danos in his Thesis ([6]); the complexity of this algorithm was later shown to be linear, in the size of the given proof structure, by Guerrini in [10]. Then, the retraction criterion was extended, respectively, by the author, in [14], to the pure multiplicative and additive (MALL) proof nets with *boolean weights* and then by Fouqueré and Mogbil, in [7], to polarized multiplicative and exponential proof structures.

Traditionally, concerning proof nets of linear logic, the main interest on the retraction system is oriented to study the complexity of correctness criteria or cut reduction. Here, our (original) point of view is rather to exploit retraction systems for incrementally building (correct) proof structures. Indeed, the convergence of our retraction system allows to focus on particular retraction strategies that turn out to be optimal (in the graph size) w.r.t. the problem of incrementally constructing transitory nets. Actually, checking correctness of an expanded proof structure is a task which may involve visiting (i.e., retracting) a large portion of the so obtained net: some good bound for these task would be welcome. Here, we show that checking correctness (retraction) of a MALL transitory net, after a construction attempt, is a task that can be performed by restricting to some "minimal" (i.e., already partially retracted) transitory nets. The reason is that some subgraphs of the given transitory net will not play an active role in the construction process, since they are already correct and encapsulated (i.e., border free): so, their retraction can be performed regardless of the construction process (that is the main content of Corollary 1, in Sect. 3.2).

Finally, we give in Sect. 5 a comparing with some related works concerning:

1. analogous attempts to give a theoretical foundation of computational programming paradigms based on the construction of proofs of intuitionistic or linear logic (notably, some works of Pfenning and co-authors, [4], and some works of Miller and co-authors, [15] and [5]);
2. alternative syntaxes for additive-multiplicative proof structures (mainly, those ones given, respectively, by Girard [8] and Hughes–van Glabbeek in [11]).

2 Construction of Bipolar Focusing Proofs

In this section we give a brief presentation of the *bipolar focusing sequent calculus* introduced by Andreoli; more technical details can be found in [1]. We start with the basic notions of the *MALL fragment* of LL, without units and Mix rule. We arbitrarily assume literals $a, a^\perp, b, b^\perp, ...$ with a polarity: *negative* for atoms and *positive* for their duals. A *formula* is built from literals by means of the two groups of connectives:
– *negative*, $⅋$ ("par") and & ("with");
– *positive*, $\otimes$ ("tensor") and $\oplus$ ("plus").
A proof is then built by the following rules of the *MALL sequent calculus*:

$$\frac{}{A, A^\perp}\,id \qquad \frac{\Gamma, A \quad \Delta, A^\perp}{\Gamma, \Delta}\,cut \qquad \frac{\Gamma, A \quad \Delta, B}{\Gamma, \Delta, A \otimes B}\,\otimes \qquad \frac{\Gamma, A, B}{\Gamma, A ⅋ B}\,⅋ \qquad \frac{\Gamma, A \quad \Gamma, B}{\Gamma, A \& B}\,\& \qquad \frac{\Gamma, A_i}{\Gamma, A_1 \oplus_i A_2}\,\oplus_{i=1,2}$$

The *bipolar focusing sequent calculus* is a refinement of the previous one, based on the crucial properties of *focusing* and *bipolarity* (see, also, [12]). The *focusing property* states that, in proof search (or proof construction), we can build (bottom up) a sequent proof by alternating clusters of negative inferences with clusters of positive ones. As consequence of this bipolar alternation we get more compact proofs in which we get rid of the most part of the bureaucracy hidden in sequential proofs (as, for instance, irrelevant permutations of rules). Remind that, w.r.t. proof search, negative (resp., positive) connectives involve a kind of *don't care non-determinism* (resp., *true non-determinism*).

A *monopole* is a formula built on negative atoms using only the negative connectives, while a *bipole* is a formula built from monopoles and positive atoms, using only positive connectives; moreover, bipoles must contain at least one positive connective or be reduced to a positive atom, so that they are always disjoint from monopoles. Given a set $\mathcal{F}$ of bipoles, the *bipolar focusing sequent calculus* $\Sigma[\mathcal{F}]$ is a set of inferences of the form

$$\frac{\Gamma_1 \quad \dots \quad \Gamma_n}{\Gamma}\,B$$

where the conclusion Γ is a sequent made by a multiset of negative atoms and the premises $\Gamma_1, ..., \Gamma_n$ are obtained by fully focusing decomposition of some bipole

$B \in \mathcal{F}$ in the context Γ (therefore, $\Gamma_1, ..., \Gamma_n$ are mutiset of negative atoms too). More precisely, due to the presence of additives (in particular the sum $\oplus$ connective) a bipole B is naturally associated to a set of inferences $B_1, ..., B_{m+1}$, where m is the number of $\oplus$ connectives present in B. For instance, in the purely multiplicative fragment of LL (i.e, MLL), the bipole $B = a^\perp \otimes b^\perp \otimes (c \parr d) \otimes e$, where a, b, c, d and e are (negative) atoms, yields the inference below (on the left hand side), more compact than the explicit one (on the right hand side):

$$\frac{\Gamma, c, d \qquad \Delta, e}{\Gamma, \Delta, a, b}\, B \quad \Leftrightarrow \quad \dfrac{\dfrac{\dfrac{\Gamma, c, d}{\Gamma, c \parr d}\,\parr \qquad \Delta, e}{\Gamma, \Delta, (c \parr d) \otimes e}\,\otimes \qquad b, b^\perp \qquad a, a^\perp}{\Gamma, \Delta, a, b, \mathbf{a^\perp \otimes b^\perp \otimes (c \parr d) \otimes e}}\,\otimes$$

where Γ and Δ range over a multiset of negative atoms; the identity axioms $\overline{a, a^\perp}$ and $\overline{b, b^\perp}$ are omitted in the bipolar sequent proof for simplicity. Observe, the couple a and b plays the role of a *trigger* (or *multi-focus*) of the B inference; more generally, a trigger (of a bipole) is a multi-set of duals of the positive atoms occurring in the bipole. Intuitively, the main feature of the bipolar focusing sequent calculus is that its inferences are triggered by multiple focus, like in [15] and and [5]. Bipoles are clearly inspired by the *methods* used in logic programming languages: the positive layer of a bipole corresponds to the *head*, while the negative layer corresponds to the *body* of a *Prolog-like method.*

The bipolar focusing sequent calculus, with only logical axioms (*id*), has been proven in [1] to be isomorphic to the focusing sequent calculus, so that (closed) proof construction can be performed indifferently in the two systems. The main idea behind this isomorphism is the *bipolarisation technique*, that is a simple procedure that allows to transform any provable formula F of the LL sequent calculus into a *set* of bipoles, called *universal program* of the bipolar sequent calculus. In Example 1 we give an instance of (closed) bipolar focusing derivation.

Example 1. Assume the universal program $\mathcal{U} = \{B_1 = f^\perp \otimes (x \parr g \parr h \parr (d \& e)),$ $B_2 = x^\perp \otimes (a \& b)$, $B_3 = g^\perp \otimes ((a^\perp \oplus b^\perp) \otimes c^\perp)$ a $B_4 = h^\perp \otimes c \otimes (d^\perp \oplus e^\perp)\}$.
Each bipole induces a non empty set of bipolar inferences as follows:

– both bipoles B_1 and B_2 induce a single inference

$$\frac{\Gamma, x, g, h, d \qquad \Gamma, x, g, h, e}{\Gamma, f}\, B_1 \qquad \text{resp.,} \qquad \frac{\Gamma, a \qquad \Gamma, b}{\Gamma, x}\, B_2$$

– while both bipoles B_3 and B_4 induce two inferences

$$\frac{\Gamma}{\Gamma, g, a, c}\, B_3' \quad \text{and} \quad \frac{\Gamma}{\Gamma, g, b, c}\, B_3'' \quad \text{resp.,} \quad \frac{\Gamma, c}{\Gamma, h, d}\, B_4' \quad \text{and} \quad \frac{\Gamma, c}{\Gamma, h, e}\, B_4''$$

Then, the resulting bipolar focusing proof Π of f if built as follows:

$$\dfrac{\dfrac{\dfrac{\dfrac{}{\mathbf{g, a, c}}\, B_3'}{a, g, \mathbf{h, d}}\, B_4' \qquad \dfrac{\dfrac{}{\mathbf{g, b, c}}\, B_3''}{b, g, \mathbf{h, d}}\, B_4'}{\mathbf{x}, g, h, d}\, B_2 \qquad \dfrac{\dfrac{\dfrac{}{\mathbf{g, a, c}}\, B_3'}{a, g, \mathbf{h, e}}\, B_4'' \qquad \dfrac{\dfrac{}{\mathbf{g, b, c}}\, B_3''}{b, g, \mathbf{h, e}}\, B_4''}{\mathbf{x}, g, h, e}\, B_2}{\mathbf{f}}\, B_1$$

Although this derivation is quite compact and abstract, it still presents some structural drawbacks like duplications of some sub-proofs. Therefore, we will move, in the next section, to more flexible proof structures.

3 Bipolar Transitory Structures

In this section we introduce the de-sequentialized version of the bipolar focusing sequent calculus, i.e. a graphical representation of *bipolar structures* (eventually correct, i.e. *bipolar nets*) which preserves only essential sequentializations.

Definition 1 (links). *Assume an infinite set* $\mathcal{L}$ *of* resource places $a, b, c, ...$ *(also* ports *or* addresses*). A* link *consists of two disjoint sets of places,* top *and* bottom, *together with a polarity,* positive *or* negative, *and s.t. a positive link must have* at least one *bottom place, while a negative link must have* exactly one *bottom place. The* border *or* frontier *of a link is the set of its top and bottom places.*

Graphically, links are represented like in Fig. 1 and distinguished by their shape: triangular for negative and round for positive links. Top (resp., bottom) places are drawn as edges incident to a vertex. We may use variables $x^p, y^p, z^p, ...$ for links with a polarity $p \in \{+, -\}$, and the compact expression $link^+$ (resp., $link^-$) for a positive (resp., negative) link. Moreover, we define some relations on the set of links; in particular, given two links, x and y, we say:

- they are *adjacent* if they have (or *share*) a common place;
- x is *just above* (resp., *just below*) y if there exists a place that is both at the bottom (resp., top) of x and at the top (resp., bottom) of y;
- they are *connected* if they belong to the transitive closure of the adjacency relation.

Definition 2 (transitory structure). *A* transitory structure *(TS) is a set* π *of links satisfying the following conditions:*

1. *if two links are one above the other, then they have opposite polarity;*
2. *if two links have a top (resp. bottom) place in common, then they must have the same polarity;*
3. *if two negative links have a top place in common, then they must share their (unique) bottom place.*

Moreover, a TS π *is called:*

- bipolar *(BTS), if any place occurring at the top of some positive link of* π *also occurs at the bottom of some negative link of* π *and vice-versa (the bottom place of any negative link also occurs at the top of some positive link);*
- negative hyperlink, *if it is a set of, at least two, negative links with same bottom place;*
- positive hyperlink, *if it is a set of connected positive links;*
- bipole, *when it contains exactly one positive link; a bipole is then called* elementary *(or* multiplicative*) when it does not contain any negative hyperlink.*

Finally, in a TS π*, the set of bottom (resp., top) places that do not occur at the top (resp., bottom) of any link of* π *is called the* bottom *(resp.,* top*)* border *or* frontier *of* π*. If the top border of* π *is empty, then* π *is called* closed. *A place shared by at least two links of the same polarity is called* (additive) multiport.

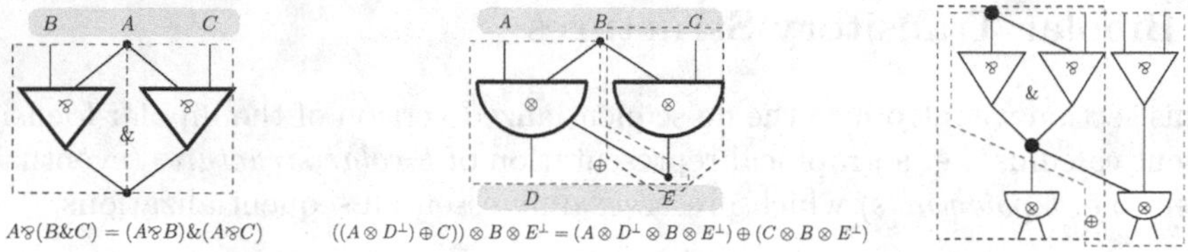

Fig. 1. Links, hyperlinks, bipoles and bipolar transitory structures

Intuitively, w.r.t. the standard syntax of proof nets of linear logic, negative (resp., positive) links correspond to generalized (i.e., n-ary) $\parr$-links (resp., $\otimes$-links). Similarly, negative (resp., positive) hyperlinks correspond, modulo distributivity and associativity of linear connectives, to generalized & (resp., $\oplus$) of negative (resp., positive) links. Instances of negative and positive hyperlinks are, respectively, given in the leftmost and middle side pictures of Fig. 1, where links are enclosed within dashed lines; graphically, these hyperlinks represent the distributive law of negative ($\parr$/&), respectively, positive ($\otimes/\oplus$) connectives. An instance of BTS is also given in the rightmost picture of Fig. 1, with two bipoles enclosed within dashed lines (bullets, $\bullet$, graphically represent multiports). Intuitively, bipoles correspond to bipolar inferences of the sequent calculus.

3.1 Retraction of Bipolar Transitory Structures

We are interested in those BTSs that *correspond* to bipolar focusing sequent proofs: these *correct* BTSs will be called *bipolar transitory nets* (BTN). In the following we will give a geometrical way to characterize BTNs: actually we will show that BTNs are those BTSs whose *abstract structures retract*, by means of sequences of rewriting rules (graph deformation steps), to special terminal *collapsed* graphs . This retraction technique was primarily exploited by Danos in his thesis ([6]), limited to the multiplicative proof structures (see *rules* R_1, R_2 *and* R_3 of Definition 5) and then extended by Maieli in [14] to the multiplicative and additive proof structures. The latter work provides a binary version of *rules* R_5 *and* R_6 of Definition 5 that only works with closed proof structures labeled by *boolean monomial weights* (see [9]). Here, we further extend these techniques, by generalizing the rules above, to weightless proof structures that are focusing, bipolar, possibly open and with n-ary links.

Definition 3 (abstract structure). *An* abstract structure *(*AS*) is a undirected graph* π^* *equipped with a set* $\mathcal{C}(\pi^*)$ *of pairs of coincident edges: two edges are* coincident *if they share at least a vertex, called* base *of the pair. Each pair has a* type $\alpha \in \{\parr, \&, C\}$ *(where* C *denotes the* additive contraction*). We call* cluster *of type* α *a tuple of edges that are pairwise pairs of* $\mathcal{C}(\pi^*)$ *with type* α*. A pair (resp., cluster) is graphically denoted by a crossing* arc *close to the base. Some* pending edges *(i.e., edges that are incident to only one node) of an AS are called* conclusions *(resp.* hypotheses*) of the AS. We call* collapsed *any acyclic AS* π^* *with at most a single node and* $\mathcal{C}(\pi^*) = \emptyset$.

Notation: a dashed edge incident to a vertex v is a compact representation of possibly several edges (with possibly clusters) incident to v; variables $a, b, c, \ldots$ denote (dashed) edges; possibly partially dotted arcs with labels $\alpha \in \{\parr, \&, C\}$ are compact representations of pairs (clusters) of type α; vertices may be denoted by naturals inside (dotted) circles ①, ②, ... A cluster of n edges, $a_1, \ldots, a_n$, with type α, is denoted by $\alpha(a_1, \ldots, a_n)$ (sometimes, simply α_n).

Definition 4 (abstraction). *We may transform (*abstract*) a given BTS π, with bottom border Γ and top border Δ, in to an AS π^* (also* abstraction of π*) with conclusions Γ and hypothesis Δ, built by applying the following* procedure*:*

1. *a link$^+$ with border $a_1, \ldots, a_n$ becomes a vertex with incident edges $a_1, \ldots, a_n$;*
2. *a link$^-$ with top places $a_1, \ldots, a_n$ and bottom place b becomes a vertex that is base for a cluster $\parr(a_1, \ldots, a_n)$ and with b as an additional incident edge;*
3. *a place (multiport) a that is bottom (resp., top) place of n links$^-$ becomes a vertex that it is base of a cluster $\&(a_1, \ldots, a_n)$ (resp., $C(a_1, \ldots, a_n)$) with n copies of a, and with an additional incident edge labeled by a;*
4. *a place (multiport) a that is top (resp., bottom) place of n links$^+$ becomes a vertex that is base of a cluster $C(a_1, \ldots, a_n)$, with n copies of a, and with an additional incident edge labeled by a;*
5. *we may compact π^* by some applications of structural retractions R_1, R_2.*

Definition 5 (retraction system). *Given an AS π^*, a* retraction step *is a replacement (also,* deformation *or* rewriting*) of a subgraph S (called,* retraction graph*) of π^* with a new graph S' (called,* retracted graph*), leading to π'^* according to one of the following rules $R_1, \ldots, R_9$.*

R_1 **(structural):** *with the condition that, like in Fig. 2, the retraction graph of π^* contains a vertex ① with only two incident edges, a and b, none of them pending; then, this graph is replaced in π'^* by a single new edge c s.t. any pair of $\mathcal{C}(\pi^*)$ containing a or b is replaced in $\mathcal{C}(\pi'^*)$ by a pair of the same type and with c at the place of a or b.*

R_2 **(structural):** *with the condition that, like in Fig. 2, the retraction graph of π^* contains two distinct vertices ① and ② with a common edge c not occurring in any pair of $\mathcal{C}(\pi^*)$; then, one of these two nodes, ① or ②, together with the edge c, does not occur in π'^*; moreover, $\mathcal{C}(\pi^*) = \mathcal{C}(\pi'^*)$.*

R_3 **(multiplicative):**[1] *with the condition that, w.r.t. the retraction graph of π^* in Fig. 2, all vertices are distinct and there exists a cluster $\parr(a_1, \ldots, a_n)$, with base in ①, whose edges, a_{n-1} and a_n are also incident to vertex ②; moreover, a_{n-1} and a_n do not occur in any pair, except the cluster $\parr(a_1, \ldots, a_n)$. Then, π'^* (resp. $\mathcal{C}(\pi'^*)$) is obtained from π^* (resp. from $\mathcal{C}(\pi^*)$) by erasing a_n (resp., by replacing $\parr(a_1, \ldots, a_n)$ with $\parr(a_1, \ldots, a_{n-1})$).*

R_4 **(associative):**[2] *with the conditions that, w.r.t. the retraction graph of π^* in the Fig. 2 (all vertices are distinct):*

1. *vertex ① is a base for the cluster $\alpha(a_1, \ldots, a_n)$;*

[1] Intuitively, this rule corresponds to the replacement of an axiom by its η-*expansion*.
[2] Intuitively, this rule corresponds to the *associativity* of, respectively, $\parr, \&$ and C.

2. vertex ② is a base for the cluster $\alpha(b_1, ..., b_m)$;
3. $\alpha \in \{\parr, \&, C\}$ and $n, m \geq 2$;
4. the only edges incident to the vertex ② are $b_1, ..., b_m, a_n$.

Then, the edge a_n (resp., vertex ②) does not occur in π'^* and both clusters, $\alpha(a_1, ..., a_n)$ and $\alpha(b_1, ..., b_m)$ of $\mathcal{C}(\pi^*)$, are replaced in $\mathcal{C}(\pi'^*)$ by an unique cluster $\alpha(a_1, ..., a_{n-1}, b_1, ..., b_m)$ with base in vertex ①.

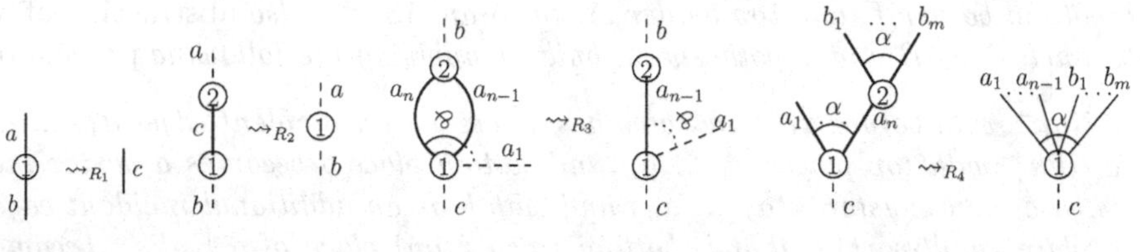

Fig. 2. *Structurals* (R_1, R_2), *multiplicative* (R_3) and *associative* (R_4) retractions

R_5 **(distributive):**[3] *with the condition that, w.r.t. the retraction graph of* π^* *in Fig. 3, all vertices are distinct and each vertex* v_i $(1 \leq i \leq n)$ *has only* a_i, b_i *and* c_i $(1 \leq i \leq n)$ *as incident edges with the following conditions:*
1. c_i is an edge occurring in the cluster $\&(c_1, ..., c_n, d)$ with base in v_h;
2. b_i is an edge occurring both in the cluster $C(b_1, ..., b_n)$, with base in vertex v_k, and in the cluster $\parr_i(a_i, b_i)$, with base in vertex v_i;
3. a_i is a non empty bundle of edges occurring in the cluster $\parr(a_i, b_i)$; moreover, each edge $e \in a_i$ must satisfy one of the following conditions:
   (a) either it is a pending edge or an edge incident to a vertex with only pending edges not labeled by any conclusion of π^*; in that case, there must exist at least such an analogous edge for each bundle $a_1, ..., a_n$;
   (b) or it must occur in a C cluster and, in that case, for each bundle $a_1, ..., a_n$, there must exist exactly one edge that occurs in this C cluster too.

   Then, π^* retracts to π'^* like in Fig. 3. Observe that edges $b_1, ..., b_n$, except one, b_i, do not occur in π'^*; similarly, the cluster $C(b_1, ..., b_n) \notin \mathcal{C}(\pi'^*)$. Moreover, new edges g and e are added to π'^* (similarly, new pairs, $\parr(b_i, g)$ and $\&(d, e)$ occur in $\mathcal{C}(\pi'^*)$ with base, respectively, in the new vertex $v_{h'}$ and $v_{h''}$).

R_6 **(semi-distributive):**[4] *with the condition that, w.r.t. the retraction graph of* π^* *in Fig. 3, all vertices are distinct and each vertex* v_i*, with* $1 \leq i \leq n$*, has only* a_i, b_i *and* c_i $(1 \leq i \leq n)$ *as incident edges with the following conditions:*
1. c_i is an edge occurring in the cluster $\&(c_1, ..., c_n, d)$ with base in v_h;
2. b_i is an edge occurring in the cluster $C(b_1, ..., b_n)$ with base in v_k;
3. a_i is, possibly, a bundle of edges occurring neither in a pair with b_i nor in a pair containing c_i.

[3] A reminiscence of the *distributivity* $(\&_{i=1}^n(a_i \parr f))\&d \dashv\vdash (\&_{i=1}^n(a_i)\parr f)\&d$ (see [14]).
[4] Reminiscence of the *semi-distributivity* $(\&_{i=1}^n(f \otimes a_i))\&d \dashv (f \otimes (\&_{i=1}^n(a_i))\&d$ ([14]).

Then, π^* retracts to π'^* like in Fig. 3. Observe, π'^* does not contain any $b_1, ..., b_n$ except one, b_i, (resp., $C(b_1, ..., b_n) \notin \mathcal{C}(\pi'^*)$). Finally, in π'^* we add a new edge g and a new vertex $v_{h'}$ (resp., a, possibly, new cluster $\&(d, g') \in \mathcal{C}(\pi'^*)$ with base $v_{h'}$).

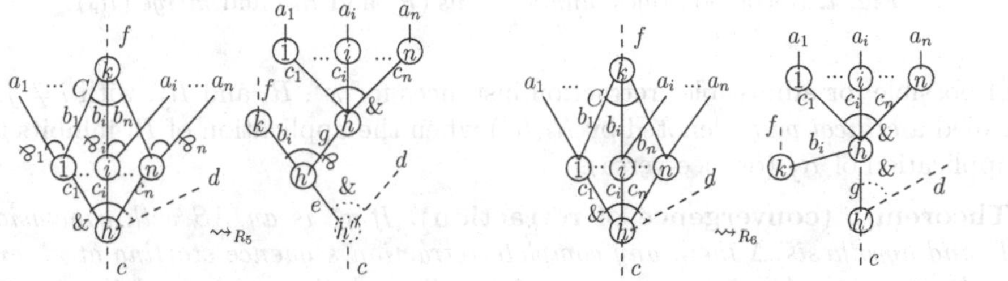

Fig. 3. Retraction rules: *distributive* (R_5) and *semi-distributive* (R_6)

R_7 **(&-annihilation):** *with the conditions that, w.r.t. the retraction graph of π^* in Fig. 4, all vertices are distinct and each a_i, with $1 \leq i \leq n$, is an edge occurring in the cluster $\&(a_1, ..., a_n)$; moreover, each a_i must belong to a collapsed subgraph of π^* non containing conclusions of π^*, with the condition that any couple a_i, a_j $(1 \leq i, j \leq n)$ cannot belong to the same collapsed graph. Then, in π'^*, each a_i will be disconnected from v_h (so, $\&(a_1, ..., a_n) \notin \mathcal{C}(\pi'^*)$).*

R_8 **($\invamp$-annihilation):** *with the condition that, w.r.t. the retraction graph of π^* in Fig. 4, all vertices are distinct and edges $a_1, ..., a_n$ occur in a cluster $\invamp(a_1, ..., a_n)$; then, π^* retracts to π'^*, like in Fig. 4, whenever d is:*

1. *either a bundle of pending edges not labeled by any conclusion of π^* and not occurring in a pair with any a_i;*
2. *or a bundle of pending edges not occurring in any pair with any a_i and e is also a bundle of pending edges with at least one of them labeled by a conclusion of π^* and none of them occurring in a pair with any a_i.*

Then, in π'^* the edge a_n will be disconnected form vertex ①; therefore, $\mathcal{C}(\pi'^*)$ will contain all the pairs of $\mathcal{C}(\pi^*)$ except those one containing a_n.

R_9 **(merge):** *with the condition that, w.r.t. the retraction graph of π^* in Fig. 4, all vertices are distinct and χ_1^* and χ_2^* are both collapsed AS made, resp., by a vertex ① and a vertex ②, with, resp., only pending edges $a_1, ..., a_{n \geq 1}$ and $b_1, ..., b_{m \geq 1}$, with b_m that is neither a conclusion nor an hypothesis of π^*. Then, π'^* is obtained by gluing χ_1^* with χ_2^* and erasing ② and b_m.*

We say that π^* *retracts to* π'^* when there exists a non empty finite sequence of retraction steps starting at π^* and terminating at π'^*; then, we say that π^* is *retractile* when there exists a $\sigma^* \neq \pi^*$ s.t. π^* retracts to σ^*. A non retractile AS is called *terminal.* A sequence of retraction steps is said *complete* when it ends with a terminal AS. An AS *collapes* when it retracts to a collapsed graph. A pair

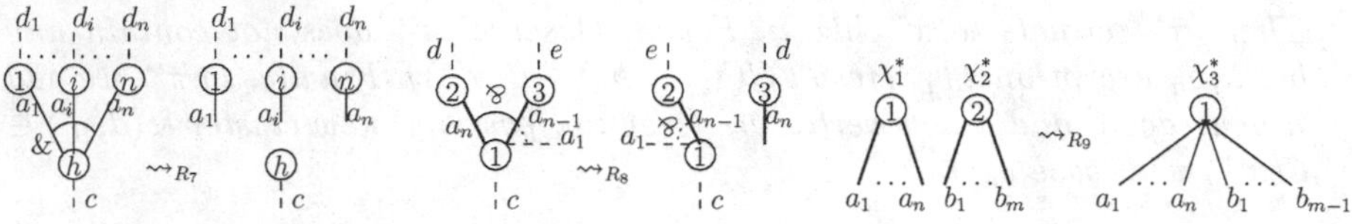

Fig. 4. Retraction rules: *annihilations* (R_7 and R_8) and *merge* (R_9)

of possible (or admissible) retraction instances for π^*, R_i and R_j, with $i \neq j$, is called a *critical pair* (denoted by $R_i|R_j$) when the application of R_i inhibits the application of R_j (or vice-versa).

Theorem 1 (convergence of retraction). *If π^* is an AS with conclusions Γ and hypothesis Δ then, any complete retraction sequence starting at π^* ends with a terminal AS χ^*; moreover, if χ^* is collapsed, then any complete retraction sequence starting at π^* ends with χ^*.*

Proof. Termination is proved by (lexicographic) induction on the *complexity degree* of π^*, that is, the triple $\langle \sharp P, \sharp N, \sharp E \rangle$, where "$\sharp P$", "$\sharp N$" and "$\sharp E$" denotes respectively the number of pairs, nodes and edges of π^*.

For the confluence, we reason, analogously, by induction on the complexity degree of the starting π^*. It is not difficult to show that for each critical pair, $R_5|R_5$, $R_5|R_8$ and $R_8|R_8$, we can find, in a few steps, an almost local confluence strategy that allows to apply the induction reasoning.

Next Lemma 1 intuitively says that abstraction commutes under retraction; it will play a crucial role in the sequentialization of BTSs (Theorem 2, Sect. 4).

Lemma 1 (abstraction). *Assume π^* is an AS that retracts to π'^* by an instance of R_i ($i = 1, ..., 9$) and assume there exists a BTS π that abstracts to π^*; then, we can find a BTS π' whose abstraction is π'^*.*

Proof. It is to show, for each $R_{i=1,...,9}$, how to locally deform some bipoles of the given BTS π in such a way to get a BTS π' whose abstraction is π'^*.

Definition 6 (bipolar transitory net). *A BTS π with bottom border Γ and top border Δ, is* correct, *that is a* bipolar transitory net *(BTN), when its abstraction π^*, with conclusions Γ and hypothesis Δ,* collapses.

Example 2. In Fig. 5 we give an instance of (closed) BTS π (*Pic.* A_1) obtained by juxtaposing bipoles $\beta_1, \beta_2, \beta_3', \beta_3'', \beta_4'$ and β_4''. Observe, π is correct (it is a BTN) since its abstraction π^* (*Pic.* A_2) collapses after few retraction steps:

1. first we get the AS of *Pic.* A_3 after some instances of distributive retraction R_5 applied to the dotted retraction graph of *Pic.* A_2;
2. then we get the AS of *Pic.* A_4 after a couple of instances of semi-distributive retractions R_6 applied to the dotted retraction graphs of *Pic.* A_3;
3. finally, we get the collapsed graph after three multiplicative retractions instances R_3 applied to the dotted retraction graphs of *Pic.* A_4 (modulo some structural retractions).

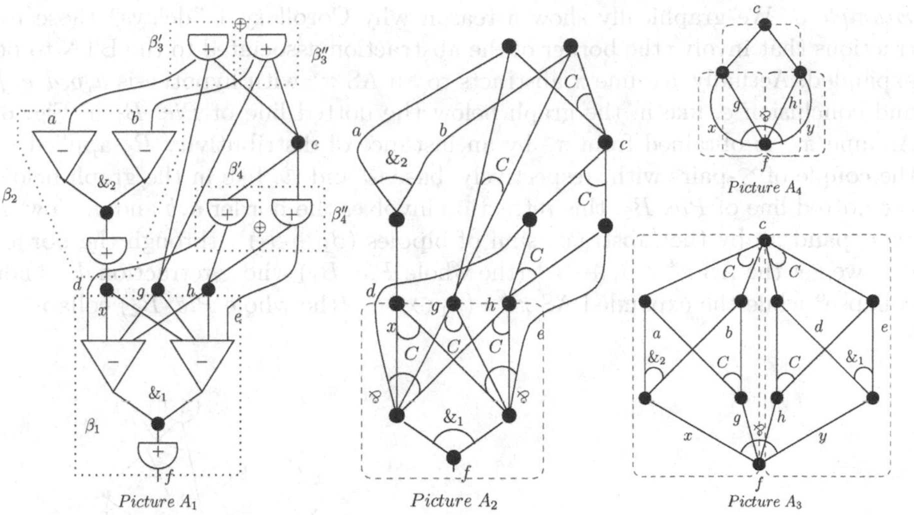

Fig. 5. Bipolar net (*Pic.* A_1) with its collapsing abstractions (*Pics.* $A_{2,3,4}$)

3.2 Construction of Transitory Nets *via* Optimal Retraction

Analogously to the construction of bipolar focusing sequent proof seen in Sect. 2, in the construction of BTNs, places are decorated by type information, that is, occurrences of negative atoms. A bipole β is viewed as an *agent* which continuously attempts to perform a bottom-up *expansion step* of the given BTN π: this step consists of adding (by a gluing operation "$\star$") a non empty cluster (a sum, indeed) of bipoles from the top border places whose types match the trigger, i.e. the bottom places, of the given bipoles. Not all construction steps are admissible. We will only consider those ones that preserve correctness by retraction. Now, checking correctness of an expansion is a task which, *a-priori*, repeatedly involves *visiting* (i.e., retracting) the whole portion of the expanded BTS. Actually, we could avoid, at each construction step, considering the whole structure built up, by e.g. taking advantage of the incremental construction in such a way to reduce the complexity of the contraction task. That is exactly the content of the next Corollary 1, immediate consequence of the Convergence Theorem 1. Intuitively, Corollary 1 allows us to incrementally pursue an optimal retraction strategy that manages, when they exist, abstract correction graphs that are strictly smaller (w.r.t. the complexity degree) than the starting ones.

Corollary 1 (optimal retraction). *Let π be a BTN (with a non empty top border) and let β a non empty cluster (a sum) of bipoles, whose bottom border matches some places of the top border of π. Assume π abstracts to π^* and assume η^* is the AS, which π^* retracts to, by only applying those retraction instances whose retraction graph does not contain pending (border) edges. Then, $(\pi \star \beta)^*$ collapses iff $\eta^* \star \beta^*$ collapses too.*

Example 3. We graphically show a reason why Corollary 1 "delays" those retractions that involve the border of the abstraction associated to the BTN to be expanded. Actually, assume π abstracts to an AS π^* with hypothesis a, b, d, e, f and conclusion c, like in the graph below the dotted line of *Pic.* B_1 in Fig. 6. Assume π'^* is obtained from π^* by an instance of distributivity R_5 applied to the couple of ⅋-pairs with, respectively, base ② and ③, like in the graph below the dotted line of *Pic.* B_2: this retraction involves the border a, b and d. Now, if we expand π^* by the (abstract) sum of bipoles $(\beta_1 \oplus \beta_2)^*$, through the border d, e, we get the AS $\pi^* \star (\beta_1 \oplus \beta_2)^*$ (the whole *Pic.* B_1) whose retraction does not collapse[5] while the expanded AS $\pi'^* \star (\beta_1 \oplus \beta_2)^*$ (the whole *Pic.* B_2) collapses.

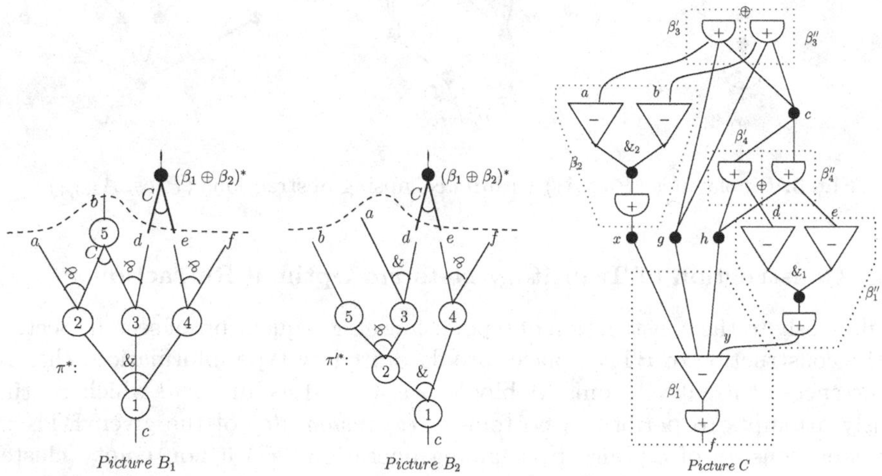

Fig. 6. Expansion steps (*Pics.* B_1, B_2) and a BTS (*Pic.* C)

4 Sequentialization of Bipolar Transitory Nets

In this section we show that correct BTSs correspond (sequentialize) to proofs of the bipolar focusing sequent calculus and vice-versa.

There exists an almost direct *correspondence* (modulo associativity and distributivity of linear connectives) between a sequential bipole B and a an additive sum of bipoles $\mathcal{B} = \{\beta_1 \oplus \cdots \oplus \beta_{n\geq 1}\}$, as follows:

1. the positive layer of B corresponds to the positive hyperlink made by the positive links of $\mathcal{B}$ connected through the border (see Definition 1);
2. the negative layer of B corresponds to the set of negative hyperlinks of $\mathcal{B}$;
3. the negative literals (i.e., atoms) of B correspond to the top places of $\mathcal{B}$ while (the dual of) the positive literals of B correspond to the bottom places of $\mathcal{B}$;

[5] It is no longer possible to apply rule R_5 since *condition 3b* of Definition 5 is violated.

4. each bipole β_i corresponds to the i-th bipolar inference induced by the sequential bipole B (see Example 1 in Sect. 2).

In general, ports (resp. multiports) correspond to a single (resp., multiple) occurrence of literals. Then, we say that a bipolar sequential proof Π with hypothesis Δ and conclusions Γ *de-sequentializes to* π, when π is a BTN with top border Δ (resp., bottom border Γ) and each instance of the i-th bipolar inference induced by $B \in \Pi$ corresponds to a bipole $\beta_i \in \pi$. The other way round, from BTNs to bipolar sequential proofs, is called *sequentialization.*

Theorem 2 ((de-)sequentialization). *A sequent proof* Π_Γ^Δ*, with conclusions* Γ *and hypothesis* Δ*,* de-sequentializes *in to a BTN* π*, with bottom places* Γ *and top places* Δ *and vice-versa (*sequentialization*).*

Proof of the de-sequentialization part: we proceed by induction on the size of Π, *via* the correspondence stated above between sequential bipoles and graphical bipoles, modulo associativity and distributivity of connectives.

Proof of the sequentialization part: it is given by induction on the complexity degree of the abstraction π^* corresponding to the given π. By the Abstraction Lemma 1, we show that at each retraction step $\pi^* \rightsquigarrow_{R_i} \pi'^*$, for $i = 1, ..., 9$, it is possible to recover a BTN π' from the retracted AS π'^*, with same border. Then, by hypothesis of induction, π' sequentializes to a proof $\Pi_{\Gamma'}^{\prime\Delta'}$ from which, finally, by deformations of Π' (i.e., permuting some bipolar inferences of Π'), we get a sequential proof Π_Γ^Δ. We reason by cases, according to R_i, with $1 \leq i \leq 9$.

Example 4. Observe, the closed bipolar net given in Example 2 (Fig. 5, *Pic.* A_1), sequentializes in to the bipolar focusing proof Π displayed at the end of Example 1; we illustrates how the sequentialization works in that case. Assume π (*Pic.* A_1, Fig. 5) abstracts to π^* (*Pic.* A_2) and assume π^* retracts to π'^* like in *Pic.* A_3, after a block of distributive retractions (without losing generality, we may treats a sequence of retractions of the same type R_5 as a single generalized retraction R_5). By Abstraction Lemma 1 we may build a BTN π' from π'^* like in *Pic.* C of Fig. 6; then, by hypothesis of induction we know that π' sequentializes to the bipolar sequent proof Π' below:

$$\dfrac{\dfrac{\dfrac{\dfrac{}{\mathbf{g,a,c}}\,B_3'}{a,g,\mathbf{h},\mathbf{d}}\,B_4' \quad \dfrac{\dfrac{}{\mathbf{g,b,c}}\,B_3''}{b,g,\mathbf{h},\mathbf{d}}\,B_4'}{\mathbf{x},g,h,d}\,B_2 \quad \dfrac{\dfrac{\dfrac{}{\mathbf{g,a,c}}\,B_3'}{a,g,\mathbf{h},\mathbf{e}}\,B_4'' \quad \dfrac{\dfrac{}{\mathbf{g,b,c}}\,B_3''}{b,g,\mathbf{h},\mathbf{e}}\,B_4''}{\mathbf{x},g,h,e}\,B_2}{\dfrac{x,g,h,\mathbf{y}}{\mathbf{f}}\,B_1'}\,B_1''$$

Clearly, π is nothing else that π' in which we replaced bipoles β_1' and β_1'' with the single bipole β_1. Since bipole β_1' (resp., β_1'') corresponds (sequentializes) to the inference B_1' (resp., B_1''), then π sequentializes to Π obtained from Π' by simply replacing the two inferences B_1' and B_1'' with the unique inference B_1 which trivially corresponds to bipole β_1.

5 Conclusions, Related and Future Works

In this work we provided:

1. a very simple syntax for open proof structures (BTSs) that allows to extend the paradigm of proof construction to the MALL fragment of LL. In particular, we set a precise correspondence, called *sequentialization* (Theorem 2) between focusing bipolar sequent proofs and correct BTSs (i.e., BTNs);
2. a convergent retraction system to check correctness of BTNs (Theorem 1);
3. an optimal strategy for incrementally building BTNs (Corollary 1).

Concerning other attempts to give a theoretical foundation of computational paradigms based on sequent proof construction, we only mention:

– some works of Pfenning and co-authors, from 2002 and later (see, e.g., [4]), which rely neither on focusing (or polarities) nor on proof nets but on softer notions of sequent calculus proofs;
– some works of Miller and co-authors which generalize focused sequent proofs to admit multiple "foci": see, e.g., [15] and [5]; the latter also provides a bijection to the unit-free proof nets of the MLL fragment, but it only discusses the possibility of a similar correspondence for larger fragments. At this moment, we are exploring a direct sequentialization from retractile transitory nets to, possibly open, multi-focus sequential calculi.

Concerning the related literature on additive proof nets, although there currently exist several satisfactory syntaxes for MALL proof structures, we briefly discuss some reasons that lead us to avoid most of them (at least in this first approach):

– *Girard*, [8]: requiring boolean (monomial) weights over proof structures is a condition that prevents certain transactional structures: take e.g. a simple BTS containing a single positive hyperlink or the rightmost BTS of Fig. 1;
– *Hughes-van Glabbeek*, [11]: similarly to the previous one, this syntax seems well adapt to take in to account only closed proof structures; actually, it has the inconvenient of allowing additive contractions only immediately below the axiom links; although this canonical form has great advantages for semantical reasons, it does not seem adapted to the composition of arbitrary modules that may require "non canonical" contractions.

Moreover, since these syntaxes make, more or less, explicit reference to graph dependencies (like *jumps*) they, *a-priori*, seem to garble the "principle of locality" required by retraction systems. Finally, as future works, we aim at investigating:

– the complexity class of the optimal BTNs construction;
– an extension of the retraction system that could preserve BTNs under the (almost local) cut reduction proposed by Laurent and Maieli in [13].

Acknowledgements. The author thanks the anonymous reviewers and Lorenzo Tortora de Falco for their useful comments and suggestions. This work was partially supported by the PRIN Research Program *Logical Methods of Information Management*. An extended version of this paper, with detailed proofs, is available at: `http://logica.uniroma3.it/~maieli/papers.html/`.

References

1. Andreoli, J.-M.: Focussing and Proof Construction. Annals of Pure and Applied Logic 107(1), 131–163 (2001)
2. Andreoli, J.-M.: Focussing proof-net construction as a middleware paradigm. In: Voronkov, A. (ed.) CADE 2002. LNCS (LNAI), vol. 2392, pp. 501–516. Springer, Heidelberg (2002)
3. Andreoli, J.-M., Mazaré, L.: Concurrent Construction of Proof-Nets. In: Baaz, M., Makowsky, J.A. (eds.) CSL 2003. LNCS, vol. 2803, pp. 29–42. Springer, Heidelberg (2003)
4. Cervesato, I., Pfenning, F., Walker, D., Watkins, K.: A concurrent logical framework II: Examples and applications. Technical Report CMU-CS-02-102, Department of Computer Science, Carnegie Mellon University (2002)
5. Chaudhuri, K., Miller, D., Saurin, A.: Canonical Sequent Proofs via Multi-Focusing. In: Ausiello, G., Karhumäki, J., Mauri, G., Ong, L. (eds.) Proc. of International Conference on Theoretical Computer Science (TCS-5). IFIP, vol. 273, pp. 383–396. Springer, Boston (2008)
6. Danos, V.: La Logique Linéaire appliquée à l'étude de divers processus de normalisation (principalment du λ-calcul). PhD Thesis, Univ. Paris VII (1990)
7. Fouqueré, C., Mogbil, V.: Rewritings for polarized multiplicative and exponential proof structures. Electronic Notes in Theoretical Computer Science 203(1), 109–121 (2008)
8. Girard, J.-Y.: Linear Logic. Theoretical Computer Science 50, 1–102 (1987)
9. Girard, J.-Y.: Proof-nets: The parallel syntax for proof-theory. In: Ursini, Agliano (eds.) Logic and Algebra (1996)
10. Guerrini, S.: A linear algorithm for MLL proof net correctness and sequentialization. Theoretical Computer Science 412(20), 1958–1978 (2011)
11. Hughes, D., van Glabbeek, R.: Proof Nets for Unit-free Multiplicative-Additive Linear Logic. In: Proc. of the 18th IEEE Logic in Computer Science. IEEE Computer Society Press, Los Alamitos (2003)
12. Laurent, O.: Polarized Proof-Nets: Proof-Nets for LC. In: Girard, J.-Y. (ed.) TLCA 1999. LNCS, vol. 1581, pp. 213–227. Springer, Heidelberg (1999)
13. Laurent, O., Maieli, R.: Cut Elimination for Monomial MALL Proof Nets. In: Proc. of the 23rd IEEE Logic in Computer Science. IEEE Computer Society Press, Los Alamitos (2008)
14. Maieli, R.: Retractile Proof Nets of the Purely Multiplicative and Additive Fragment of Linear Logic. In: Dershowitz, N., Voronkov, A. (eds.) LPAR 2007. LNCS (LNAI), vol. 4790, pp. 363–377. Springer, Heidelberg (2007)
15. Miller, D.: A multiple-conclusion specification logic. Theoretical Computer Science 165, 201–232 (1996)

Local States in String Diagrams

Paul-André Melliès*

Laboratoire Preuves, Programmes, Systèmes
CNRS, Université Paris Diderot, Paris, France

Abstract. We establish that the local state monad introduced by Plotkin and Power is a monad with graded arities in the category $[Inj, Set]$. From this, we deduce that the local state monad is associated to a graded Lawvere theory ℶ which is presented by generators and relations, depicted in the graphical language of string diagrams.

1 Introduction

In this paper, we elaborate an algebraic and graphical account of the local state monad on the category $[Inj, Set]$ of covariant presheaves on the category Inj

$$L \quad : \quad [Inj, Set] \quad \longrightarrow \quad [Inj, Set]$$

formulated ten years ago by Plotkin and Power [12] themselves inspired by seminal ideas developed by O'Hearn and Tennent [11] on the presheaf semantics of local states. Much work has been dedicated in the past decade in order to understand the algebraic nature of this specific local state monad, in particular by Power [14, 15] and by Staton [17, 18]. One main purpose of the present paper is to recast these two lines of work in the language of *monads with arities*. An immediate benefit of the reformulation is that every monad with arities comes together with a notion of *Lawvere theory with arities* formulated in [9]. By proving that the local state monad is a monad with graded arities, we are thus able to define its graded Lawvere theory מ (a letter pronounced *mem* in hebrew). The whole point of the paper is that the category מ is sufficiently simple to be presented by generators and relations easily adapted from [12, 15, 18]. The shift from finitely presentable arities to graded arities is fundamental to that purpose.

Recall that the category Inj has natural numbers $n, p, q \in \mathbb{N}$ as objects, and injective functions $f : [p] \to [q]$ as morphisms, where $[n]$ denotes the finite set $[n] = \{1, \ldots, n\}$ of cardinal n. As any presheaf category, the category $[Inj, Set]$ is cartesian closed and thus provides a model of the simply-typed λ-calculus, where every simple type is interpreted as a presheaf A. Moreover, the local state monad L defines a computational monad on the category $[Inj, Set]$ in the sense of Moggi [10]. For that reason, the category $[Inj, Set]$ together with the monad L defines an interpretation of an imperative call-by-value λ-calculus where registers may be alternatively written, read, allocated and collected, see [12] for details.

* This work has been partly supported by the ANR RECRE.

G. Dowek (ed.): RTA-TLCA 2014, LNCS 8560, pp. 334–348, 2014.

The elements M in A_n are called elements of degree n in the presheaf A. The idea underlying the model is that a program M with n registers in the language should be interpreted as an element of degree n in the presheaf A associated to the type of the program.

Now, let us recall how the monad L is defined. By convention, we suppose that every register of our language may be assigned the same finite set V of values. In that situation, the covariant presheaf LA obtained by applying the monad L on a covariant presheaf A is conveniently expressed by the formula

$$LA \quad : \quad n \mapsto \quad S^n \Rightarrow \left(\int^{p \in Inj} S^p \times A_p \times Inj(n,p) \right) \tag{1}$$

where the contravariant presheaf

$$S \quad : \quad n \quad \mapsto \quad V^n \quad : \quad Inj^{op} \quad \longrightarrow \quad Set$$

transports every number n to the set S^n of states possibly taken by n registers:

$$S^n \quad := \quad V^n \quad = \quad \left\{ (val_1, \ldots, val_n) \mid \forall i \in \{1, ..., n\}, val_i \in V \right\}.$$

Although the formula (1) may appear slightly intimidating, the intuition underlying it is easy to grasp. It simply reflects the idea that a program M of type LA with n registers behaves in the following way: first, the program M reads the state $s_{in} \in S^n$ of its n registers, then, depending on the value $s_{in} = (val_1, \ldots, val_n)$ which has been just read, the program M allocates a number $p - n$ of registers (with $p \geq n$) and returns three pieces of information to the context:

1. a state $s_{out} = (wal_1, \ldots, wal_p) \in S^p$ of the p registers,
2. a return value $M(s_{in}) \in A_p$ depending on the p registers,
3. and finally, an injective function $f : [n] \to [p]$ which tracks the n registers originally appearing in the program M among the p registers of the returned program $M(s_{in})$.

A nice aspect of the formula (1) is that it takes care of the fact that the $p - n$ registers may be allocated with different names in the memory. This is indeed the purpose of the colimit (or more precisely the coend) formula

$$\int^{p \in Inj} S^p \times A_p \times Inj(n,p) \quad = \quad \Big(\coprod_{p \in \mathbb{N}} S^p \times A_p \times Inj(n,p) \Big) / \sim \tag{2}$$

which is defined as the set of triples $(val_1, \ldots, val_p, M, f)$ in $S^p \times A_p \times Inj(n,p)$ modulo the least equivalence relation $\sim$ identifying all triples

$$(val_{h(1)}, \cdots, val_{h(p)}, M, f : [n] \to [p]) \sim (val_1, \cdots, val_q, A_h(M), h \circ f : [n] \to [q])$$

for an injection $h : [p] \to [q]$. Here, the element $A_h(M) \in A_q$ denotes the image (or pushforward) of the element $M \in A_p$ of degree p along the injection $h : [p] \to [q]$, which is typically obtained in the case of a program M by h-reindexing the names of its p registers.

2 The Global State Monad

In their work, Plotkin and Power [12] made the important observation that the local state monad L may be alternatively presented by a series of well-chosen generators and relations. One ambition of this paper is to illustrate the additional principle advocated by the author in [9] that any concise formulation such as (1) of a monad L presented by generators and relations can be derived from the existence of a class of *canonical forms* for the terms of the associated algebraic theory. This general principle was succesfully applied in [9] on the global state monad

$$T \;:\; A \;\mapsto\; S \Rightarrow (S \times A) \;:\; \mathit{Set} \;\longrightarrow\; \mathit{Set} \qquad (3)$$

induced by a finite set S of states on the category Set. For simplicity, we will suppose from now on that all the registers manipulated by the language are boolean, and thus that the set of values is equal to $V = \{true, false\}$. We will also suppose for the sake of the discussion that $S = V = \{true, false\}$. This leads to the following definition. A *mnemoid*in a cartesian category $\mathscr{C}$ is defined as an object A equipped with a binary operation lookup and a unary operation $\mathtt{update}_{\langle val \rangle}$ for each value $val \in \{true, false\}$ of the register:

$$\mathtt{lookup} : A \times A \longrightarrow A \qquad\qquad \mathtt{update}_{\langle val \rangle} : A \longrightarrow A$$

moreover satisfying three families of equations.

1. Creation lookup – update. Reading the value val of the register and then writing the very same value val in the register is like doing nothing at all. This leads to the equation below:

$$\mathtt{lookup}(\mathtt{update}_{\langle true \rangle}(term), \mathtt{update}_{\langle false \rangle}(term)) = term$$

2. Interaction update – update. Storing a value val_1 and then a value val_2 inside the register is just like storing directly the value val_2. In particular, the value val_1 is lost in the process.

$$\mathtt{update}_{\langle val_1 \rangle} \circ \mathtt{update}_{\langle val_2 \rangle} = \mathtt{update}_{\langle val_2 \rangle}$$

3. Interaction update – lookup. When one stores a value val in the register and then reads the value of the register, one gets back the value val.

$$\mathtt{update}_{\langle val \rangle} \circ \mathtt{lookup}\left[\, term(true)\,,\, term(false)\,\right] \;=\; \mathtt{update}_{\langle val \rangle}(\, term(val)\,).$$

The two operations lookup and $\mathtt{update}_{\langle val \rangle}$ of a mnemoid may be conveniently depicted in the language of string diagrams in the following way:

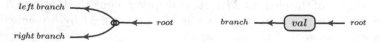

Here, the `lookup` operation is depicted as an "eye" which inspects the value of the register at the root and then branches on the left when the value is *true* and on the right when the value is *false*. The $\mathtt{update}_{\langle val \rangle}$ operation is depicted as a "cartouche" which erases the value of the register at the root and writes its own value $val \in \{true, false\}$ on the branch. The arrows on the wires indicate the direction of execution, which goes from the root to the leaves of the tree of operations. The three equations 1, 2 and $3(a, b)$ required of a mnemoid are depicted as follows in the language of string diagrams:

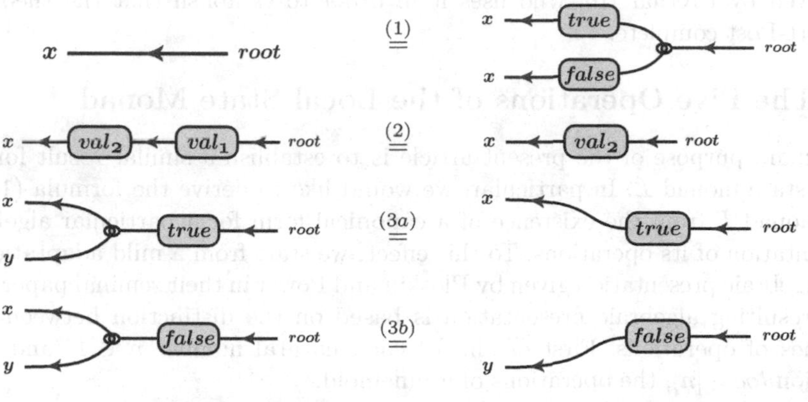

The main theorem established by Plotkin and Power [12] for the global state monad may be formulated as follows for the cartesian category $\mathscr{C} = Set$:

Theorem 1. *The category of mnemoids is equivalent to the category of algebras of the global state monad for* $S = \{true, false\}$.

The theorem is obtained in the original paper by Plotkin and Power [12] by applying the Beck theorem in order to establish the monadicity of an adjunction of interest. We advocate in [9] that a more conceptual way to obtain the same result is to deduce it from two separate facts. First of all, the global state monad is finitary. Then, the following *canonical form* theorem for mnemoids:

Theorem 2. *Every term of the theory of mnemoids with n variables* $x_1, \ldots, x_p$ *is equivalent to a term of the form*

$$\mathtt{lookup} \quad \left[\quad \mathtt{update}_{\langle val \rangle}\,(x_p)\,,\ \mathtt{update}_{\langle wal \rangle}\,(x_q) \quad \right]$$

Moreover, this canonical form is unique for a given term.

Expressed graphically, this means that every such term with n variables of the theory of mnemoids is equivalent to a unique term of the form:

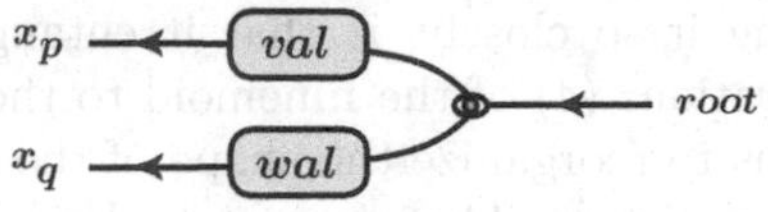

with $val, wal \in \{true, false\}$ and $p, q \in [n]$. In other words, every sequence of operations performed in the theory of mnemoids on the finite set $[n] = \{1, \ldots, n\}$

of variables is equal to a lookup operation followed on each branch *true* and *false* by an update operation and a choice of an element in $[n]$. We explain in [9] how to apply the philosophy of Lawvere in order to deduce from this canonical form theorem that the free mnemoid generated by a given set A coincides in fact with the result TA of applying the global state monad T to the set A. More conceptually, one recovers in this way the result by Plotkin and Power (Thm 1) that the category of mnemoids is equivalent to the category of algebras of the global state monad T. The existence of the canonical form was independently observed by Pretnar [16] who uses it in order to establish that the theory is Hilbert-Post complete.

3 The Five Operations of the Local State Monad

One main purpose of the present article is to establish a similar result for the local state monad L. In particular, we would like to derive the formula (1) for the monad L from the existence of a canonical form for a particular algebraic presentation of its operations. To this effect, we start from a mild adaptation of the algebraic presentation given by Plotkin and Power in their seminal paper [12]. The resulting algebraic presentation is based on the distinction between four families of operations. First of all, for each natural number $n \in \mathbb{N}$ and each location $loc \in [n]$, the operations of a mnemoid:

$$\texttt{lookup}_{\langle loc\rangle} : A_n \times A_n \longrightarrow A_n \qquad \texttt{update}_{\langle loc,val\rangle} : A_n \longrightarrow A_n \tag{4}$$

where $val \in \{true, false\}$. Note that these two operations do not alter the degree n of the elements. Then, for each natural number $n \in \mathbb{N}$, for each location $loc \in [n+1]$ and for each value $val \in \{true, false\}$, an operation

$$\texttt{fresh}_{\langle loc,val\rangle} \quad : \quad A_{n+1} \quad \longrightarrow \quad A_n \tag{5}$$

whose intuitive purpose is to allocate among n registers a fresh register at location $loc \in [n+1]$ moreover initialized with the value $val \in \{true, false\}$. Then, for each natural number $n \in \mathbb{N}$ and for each location $loc \in [n+1]$, an operation

$$\texttt{collect}_{\langle loc\rangle} \quad : \quad A_n \quad \longrightarrow \quad A_{n+1} \tag{6}$$

whose intuitive purpose is to deallocate or garbage collect the register at location $loc \in [n+1]$. Finally, for each natural number $n \in \mathbb{N}$ and for each pair of locations $loc, loc+1 \in [n]$, an operation

$$\texttt{permute}_{\langle loc,loc+1\rangle} \quad : \quad A_n \quad \longrightarrow \quad A_n \tag{7}$$

whose intuitive purpose is to permute the two registers at location loc and $loc+1$. One main conceptual difficulty of the local state monad, and also one main reason for studying it so closely, is that it entangles in a sophisticated way the read/write operations (4) of the mnemoid to the structural operations (5–6–7) whose function is to reorganize the shape of the memory by allocating, collecting or permuting registers. There seems to be here a general principle of interaction between effects and resources, which one would like to better understand.

4 A Notion of Graded Arity in $[Inj, Set]$

An apparent obstruction to the idea of canonical form in the case of the local state monad L is the fact that its definition relies on the coend formula (2). According to a naive but reasonable understanding of Lawvere's principles, there should be *exactly* one canonical form for each element of the set $(TA)_n$ and such a one-to-one correspondence seems difficult to achieve for a general covariant presheaf A because of the presence of the non-trivial equivalence relation $\sim$ in the formula (2). In order to tackle the situation, we thus need to *specialize* the formula (1) to a specific class of covariant presheaves, provided in this cas by the finite sums

$$[\,p_0\,,\;\ldots\;,p_m\,] \quad = \quad \underbrace{\langle 0\rangle + \cdots + \langle 0\rangle}_{p_0\,\text{times}} \;+\; \cdots \;+\; \underbrace{\langle m\rangle + \cdots + \langle m\rangle}_{p_m\,\text{times}}$$

of *representable* covariant presheaves

$$\langle\,k\,\rangle \;:=\; \mathbf{y}_k \quad : \quad n \mapsto Inj(k,n) \quad : \quad Inj \longrightarrow Set.$$

Note in particular that it follows from the Yoneda lemma that

$$[Inj, Set]\,(\,[p_0\,,\;\ldots\;,p_m\,]\;,\;A\,) \quad = \quad \prod_{k=0}^{m} \underbrace{A_k \times \ldots \times A_k}_{p_k\;\text{times}} \quad = \quad \prod_{k=0}^{m} A_k^{p_k}$$

for every covariant presheaf A over the category Inj. This equation justifies thinking of the presheaves $[\,p_0,\ldots,p_m\,]$ as an appropriate notion of generalized arity in the category of presheaves $[Inj, Set]$ which we call *graded arity*. Indeed, in the same way a function from $[n] = \{1,\ldots,n\}$ to a set A defines a word of length n in the alphabet A, a morphism from $[\,p_0,\ldots,p_m\,]$ to a covariant presheaf A defines a word of length $n = \sum p_k$ in the *graded* alphabet A, consisting of m words of length p_k in the alphabet A_k of elements of grade $k \in [m]$ in the covariant presheaf A. Note that the full subcategory of such arities in the category $[Inj, Set]$ is isomorphic to the free category with finite sums $\Sigma\, Inj^{op}$ generated by the category Inj^{op}. Moreover, the resulting full and faithful functor

$$i \quad : \quad \Sigma\, Inj^{op} \quad \longrightarrow \quad [Inj, Set]$$

is dense in the category $[Inj, Set]$. As such, the category ΣInj^{op} together with the functor i defines a notion of arities on the presheaf category $[Inj, Set]$ in the sense of Weber [20] who developed the notion in his study of globular operads, see also [9].

5 From a Coend Formula to a Coproduct Formula

It appears that when applied to a graded arity $A = [p_0,\ldots,p_m]$, the coend formula (2) suddenly becomes a coproduct formula. Let us briefly explain why.

The phenomenon is in fact slightly more general. Consider the category Nat with natural numbers $n \in \mathbb{N}$ as objects and no morphisms except for the identities. The object-preserving functor $\ell : Nat \longrightarrow Inj$ induces an adjunction

$$[Nat, Set] \underset{\ell^*}{\overset{\exists_\ell}{\rightleftarrows}} [Inj, Set] \qquad \bot \tag{8}$$

where the forgetful functor ℓ^* is defined by precomposition with ℓ and its left adjoint functor $\exists_\ell$ is defined by left Kan extension along ℓ. The computation of this left Kan extension is easy. Namely, given a presheaf A on the category Nat, one obtains:

$$\exists_\ell A \ : \ n \mapsto \coprod_{m \in \mathbb{N}} A_m \times Inj(m,n) \ = \ \{\, (M, f) \mid M \in A_m, \ f \in Inj(m,n) \,\}$$

Observe that the covariant presheaves of the form $\exists_\ell A$ are precisely the (possibly infinite) sums of representable presheaves over Inj. In particular, every graded arity $[\, p_0, \ldots, p_m \,]$ is of that form. Now, a simple computation shows that the coend formula (2) applied to such a covariant presheaf $\exists_\ell A$ yields a coproduct formula:

$$\int^{p \in Inj} S^p \times \Big[\coprod_{m \in \mathbb{N}} A_m \times Inj(m,p) \Big] \times Inj(n,p) \ \cong \ \coprod_{m \in \mathbb{N}} A_m \times \langle\, m, n \mid S \,\rangle$$

where the set $\langle\, m, n \mid S \,\rangle$ is defined as

$$\langle\, m, n \mid S \,\rangle \quad := \quad \int^{p \in Inj} Inj(m,p) \times Inj(n,p) \times S^p.$$

This slightly enigmatic result convinces us to study more closely the monad

$$L_{Nat} \quad := \quad \ell^* \circ L \circ \exists_\ell \quad : \quad [Nat, Set] \quad \longrightarrow \quad [Nat, Set]$$

obtained by pre and post-composing the monad L with the two components of the adjunction $\exists_\ell \dashv \ell^*$. The image of a presheaf A on Nat (also called a graded set) is thus defined as

$$L_{Nat}\, A \ : \ n \mapsto S^n \Rightarrow \coprod_{m \in \mathbb{N}} A_m \times \langle m, n \mid S \rangle. \tag{9}$$

From now on, and in order to differentiate the two local state monads, we write L_{Inj} for the local state monad L on the category $[Inj, Set]$.

6 The Category *Res* of Resource Management

The monad L_{Nat} is so simple that it deserves further analysis. In particular, remember from §3 that we are interested in clarifying the intricate interplay between read/write effects (lookup,update) and resource management (fresh, collect, permute) in the local state monad. This idea has been already explored quite far by Power [15] in his work on indexed Lawvere theories. Here, it provides us with a precious guide in our analysis. Indeed, it is folklore that the category Inj is presented (in some sense which will be later elaborated) by the `collect` and `permute` operations. This preliminary observation leads us to introduce a category Res already considered by Staton [18] whose intuitive purpose is to reflect all the resource management operations (not just `collect` and `permute` but also `fresh`) of the local state monad. By definition, the category Res has natural numbers $n \in \mathbb{N}$ as objects, and resource morphisms $[m] \to [n]$ as morphisms $m \to n$, where a resource morphism $f : [m] \to [n]$ is defined as a function $f : [m] \to [n] + \{true, false\}$ satisfying the following injectivity property: every element $k \in [n]$ has *at most* one antecedent in $[m]$. The resource morphism $g \circ f : [m] \to [n]$ obtained by composing two resource morphisms $f : [m] \to [p]$ and $g : [p] \to [n]$ is defined just as expected. Note in particular that the category Res is a subcategory of the Kleisli category induced by the exception monad $A \mapsto A + \{true, false\}$ on the category Set with natural numbers $m, n \in \mathbb{N}$ as objects and functions $[m] \to [n]$ as morphisms. Accordingly, the reader should note that there exists an object-preserving functor $\iota : Inj \to Res$ which transports every injection $f : [m] \to [n]$ of the category Inj to the function $\eta \circ f : [m] \to [n] + \{true, false\}$ defined by composing f with the unit η of the exception monad in Fin. An interesting fact to mention regarding the category Res is that there exists for every pair of numbers $m, n \in \mathbb{N}$ a one-to-one correspondence

$$\langle m, n \,|\, S \rangle \quad \cong \quad S^n \times Res(m, n). \tag{10}$$

From this follows that Formula (9) may be conveniently rewritten as

$$L_{Nat} A \;:\; n \;\mapsto\; S^n \;\Rightarrow\; S^n \times \coprod_{m \in \mathbb{N}} A_m \times Res(m, n). \tag{11}$$

7 Main Theorem

Together with the functor $\ell : Nat \to Inj$, the functor $\iota : Inj \to Res$ induces a pair of adjunctions on the associated presheaf categories:

$$[Nat, Set] \underset{\ell^*}{\overset{\exists_\ell}{\rightleftarrows}} [Inj, Set] \underset{\iota^*}{\overset{\exists_\iota}{\rightleftarrows}} [Res, Set] \tag{12}$$

(with $\exists_\ell \dashv \ell^*$ and $\exists_\iota \dashv \iota^*$, each marked $\perp$)

This pair of adjunctions $\exists_\ell \dashv \ell^*$ and $\exists_\iota \dashv \iota^*$ induces in turn a monad

$$\mathcal{B}_{Nat} \quad := \quad \ell^* \circ \iota^* \circ \exists_\iota \circ \exists_\ell \quad : \quad [Nat, Set] \quad \longrightarrow \quad [Nat, Set]$$

on the presheaf category $[Nat, Set]$. The image of a graded set A is defined as

$$\mathcal{B}_{Nat}A \ : \ n \mapsto \coprod_{m \in \mathbb{N}} A_m \times Res(m,n) \ = \ \{ (M, f) \mid M \in A_m, \ f \in Res(m,n) \}.$$

In addition to the monad $\mathcal{B}_{Nat}$, there is also a state monad

$$\mathcal{F}_{Nat}A \quad : \quad n \mapsto S^n \Rightarrow (S^n \times A_n) \quad : \quad [Nat, Set] \quad \longrightarrow \quad [Nat, Set]$$

on the presheaf category $[Nat, Set]$, simply obtained by applying the global state monad on n registers

$$T_n \quad : \quad A \quad \mapsto \quad S^n \Rightarrow (S^n \times A) \quad : \quad Set \quad \longrightarrow \quad Set$$

on each set A_n of elements of grade n in the presheaf A. Note in particular that

$$(\mathcal{F}_{Nat}A)_n \quad := \quad T_n\,(A_n).$$

The notations $\mathcal{B}_{Nat}$ and $\mathcal{F}_{Nat}$ are mnemonics for *basis monad* $\mathcal{B}_{Nat}$ and *fiber monad* $\mathcal{F}_{Nat}$. The intuition is that the basis monad $\mathcal{B}_{Nat}$ acts on the basis Nat of the covariant presheaf A by an appropriate change of basis from Nat to Res while the fiber monad $\mathcal{F}_{Nat}$ acts on each of its fibers A_n of elements of grade n. Each of the two monads $\mathcal{F}_{Nat}$ and $\mathcal{B}_{Nat}$ captures one disjoint aspect of the local state monad L_{Nat}. Intuitively, the monad $\mathcal{F}_{Nat}$ deals with the read/write operations while the monad $\mathcal{B}_{Nat}$ deals with memory management. The question is thus to understand how the two monads $\mathcal{F}_{Nat}$ and $\mathcal{B}_{Nat}$ interact. The nature of this interaction is nicely captured by the existence of a distributivity law in the sense of Beck [1] between the two monads:

Theorem 3. *The local state monad L_{Nat} is equal to the monad $\mathcal{F}_{Nat} \circ \mathcal{B}_{Nat}$ associated to a distributivity law $\lambda_{[Nat]} : \mathcal{B}_{Nat} \circ \mathcal{F}_{Nat} \Rightarrow \mathcal{F}_{Nat} \circ \mathcal{B}_{Nat}$ between the two monads $\mathcal{B}_{Nat}$ and $\mathcal{F}_{Nat}$.*

Once this decomposition of the monad L_{Nat} performed, it appears that a similar decomposition of the local state monad L_{Inj} is also possible. One recovers in this way the distributivity law noticed by Staton in [18]. Recall that the presheaf category $[Inj, Set]$ is equivalent to the category of algebras of the monad $\ell^* \circ \exists_\ell$ encountered in §5. There exists moreover a distributivity law λ between the two monads $\mathcal{F}_{Nat}$ and $\ell^* \circ \exists_\ell$. For these two reasons, the monad $\mathcal{F}_{Nat}$ extends to a monad $\mathcal{F}_{Inj}$ on the presheaf category $[Inj, Set]$ defined in just the same way:

$$(\mathcal{F}_{Inj}A)_n \quad := \quad T_n\,(A_n).$$

For the sake of comparison, it is worth mentioning here that the algebras of the monad $\mathcal{F}_{Inj}A$ coincide with the models of the indexed Lawvere theory $L_\otimes$ formulated by Power in [15]. For that reason, the distributivity law λ may be seen as an alternative but equivalent way as the functor $L_\otimes : Inj \to Law$ to "glue" together the global state monads T_n into the monad $\mathcal{F}_{Inj}$. Besides the monad $\mathcal{F}_{Inj}$ just defined on $[Inj, Set]$, one finds the monad $\mathcal{B}_{Inj} = \iota^* \circ \exists_\iota$ induced from the adjunction $\exists_\iota \dashv \iota^*$ mentioned in (12). This leads us to the following variant of Theorem 3, established this time for the local state monad L_{Inj}:

Theorem 4. *The local state monad L_{Inj} is equal to the monad $\mathcal{F}_{Inj} \circ \mathcal{B}_{Inj}$ associated to a distributivity law $\lambda_{[Inj]} : \mathcal{B}_{Inj} \circ \mathcal{F}_{Inj} \Rightarrow \mathcal{F}_{Inj} \circ \mathcal{B}_{Inj}$ between the two monads $\mathcal{B}_{Inj}$ and $\mathcal{F}_{Inj}$.*

This leads us to the main theorem of the paper:

Theorem 5. *The local state monad L_{Inj} is a monad with graded arities ΣInj^{op} on the presheaf category $[Inj, Set]$.*

The property is a direct consequence of the fact that the local state monad L_{Inj} factors as a pair of monads $\mathcal{F}_{Inj}$ and $\mathcal{B}_{Inj}$ with graded arities. Note that one establishes in the same way that the monad L_{Inj} is a monad with finitary arities, where the notion of finitary arities is defined as the full and dense subcategory $FinGrad$ of finite graded sets in $[Nat, Set]$.

8 The Graded Lawvere Theory $\beth$

One important consequence of Theorem 5 is that the monad L_{Inj} may be entirely reconstructed from its Lawvere theory $\beth$ with graded arities. This result holds for every monad with arities and thus applies in particular to the monad L_{Inj}. See [9] for details. The graded Lawvere theory $\beth$ is defined as the category with graded arities $[\,p_0, \ldots, p_k\,]$ as objects and with morphisms

$$\beth(\,[\,p_0, \ldots, p_j\,]\,,\,[\,q_0\,,\,\ldots\,,\,q_k\,]\,) = [Inj, Set]\,(\,[\,q_0\,,\,\ldots\,,\,q_k\,]\,,\,,L_{Inj}\,[\,p_0\,,\,\ldots\,,\,p_j\,]\,).$$

Note that following Lawvere's philosophy, the category $\beth$ is defined as a full subcategory of the opposite of the Kleisli category induced by the local state monad L_{Inj} on the presheaf category $[Inj, Set]$. The very last part of the paper is devoted to an algebraic presentation by generators and relations of the graded Lawvere theory $\beth$. To that purpose, we take advantage that the category $\beth$ coincides with the Lawvere theory (with finitary arities) associated to the monad L_{Nat}. The algebraic presentation is then performed in four easy steps. We start by describing in §9 the generators and relations of the fiber monad $\mathcal{F}_{Nat}$ and then carry on in §10 and §11 with a description of the generators and relations of the basis monad $\mathcal{B}_{Nat}$. We conclude in §12 by the series of equations involved in the algebraic presentation of the distributivity law $\lambda_{[Nat]}$. This concludes the algebraic presentation of the graded Lawvere theory $\beth$.

9 The Global State Monad in String Diagrams

A handy graphical notation for the update and lookup operations on the global state is to depict each location loc as a specific wire on a ribbon of registers. Typically, the lookup and update operations on the register $loc = loc_2$ for a machine with four registers $L = \{loc_1, loc_2, loc_3, loc_4\}$ are depicted as

$\texttt{lookup}_{\langle loc \rangle} =$ 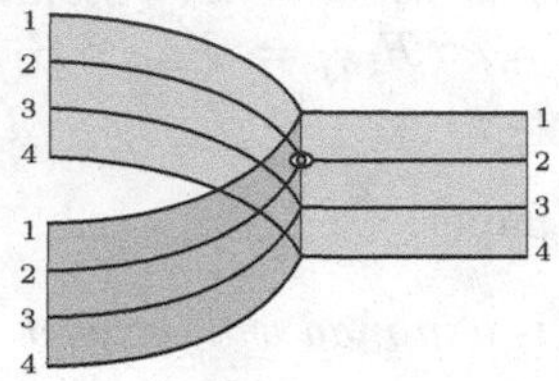

$\texttt{update}_{\langle loc,val \rangle} =$ 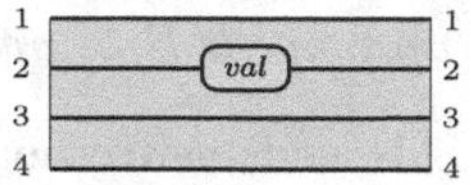

where in each case the "eye" and the "cartouche" are positioned on the register $loc = loc_2$. One recovers the three equations of mnemoids in this multi-wire setting. The first equation *creation lookup – update* is depicted as

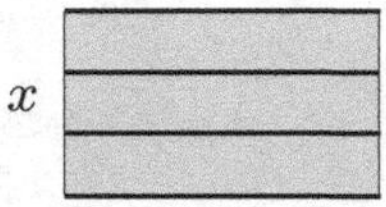

$\overset{(1)}{=}$
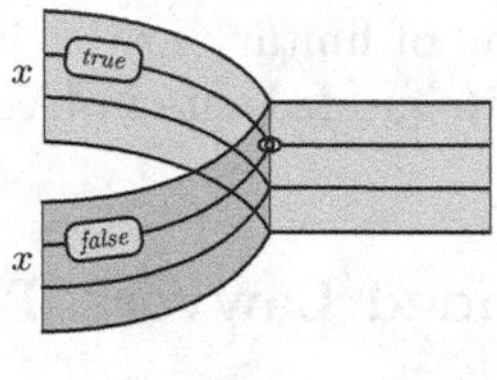

the second equation *update – update interaction* is depicted as:

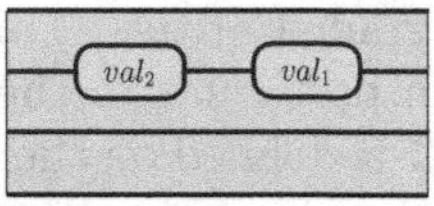

$\overset{(2)}{=}$
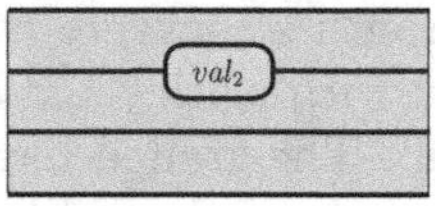

and the *true* case of the equation *update – lookup interaction* is depicted as:

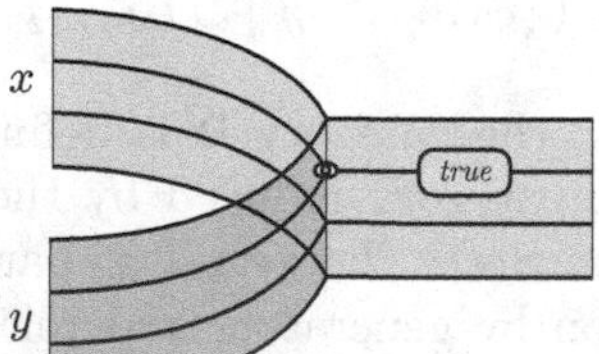

$\overset{(3a)}{=}$
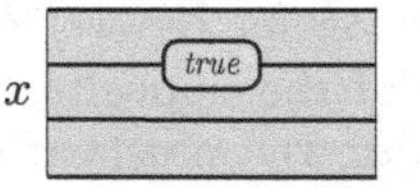

There is also a fourth equation (4) which states that two updates on *different* registers *loc* and *loc′* commute:

$$\texttt{update}_{\langle loc,val \rangle} \circ \texttt{update}_{\langle loc',val' \rangle} \overset{(4)}{=} \texttt{update}_{\langle loc',val' \rangle} \circ \texttt{update}_{\langle loc,val \rangle}$$

and is depicted in the following way:

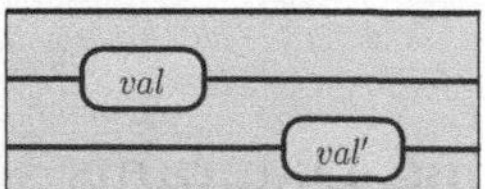

$\overset{(4)}{=}$
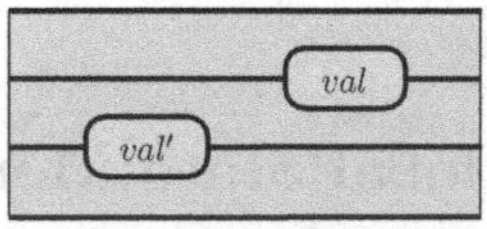

This last equation is sufficient to ensure that all the lookup/update operations applied on two different wires commute. In particular, the resulting algebraic theory for the monad T_n reflects the fact that for every two natural numbers $p, q \in \mathbb{N}$, one has $T_{p+q} = T_p \otimes T_q$ where $\otimes$ denotes the tensor product of algebraic theories (or equivalently of Lawvere theories) on the category *Set*. Hence, T_n may be seen as the n-fold tensor product of the algebraic theory T_1 of mnemoids given in §2. From this follows that the monad $\mathcal{F}$ is presented by the two families of operations `lookup` and `update` together with the four equations 1, 2, 3, 4.

10 The Action of the Category *Inj* in String Diagrams

An important ingredient of the local state monad is the action of the category Inj on the names of registers. Indeed, the very definition of the monad T relies on the equivalence relation $\sim$ between various choices of a representative $S^p \times A_p \times Inj(n,p)$ modulo an action of the category Inj on the set $[p] = \{1, \ldots, p\}$ of registers. For that reason, it is natural to introduce the notion of Inj-module $\mathscr{C}$, defined as an action $* : Inj \times \mathscr{C} \longrightarrow \mathscr{C}$ of the monoidal category $(Inj, +, 0)$ on the category $\mathscr{C}$. Lawvere observed that a monad T on a category $\mathscr{C}$ is the same thing as an action of the monoidal category $(\Delta, +, 0)$ on the category $\mathscr{C}$, where the category Δ of so-called simplices has finite numbers $p, q \in \mathbb{N}$ as objects and monotone functions $f : [p] \to [q]$ as morphisms. Similarly, an Inj-module $\mathscr{C}$ is the same thing as a category $\mathscr{C}$ equipped with a functor $D : \mathscr{C} \longrightarrow \mathscr{C}$ and two natural transformations

$$\texttt{permute} : D \circ D \longrightarrow D \circ D \qquad\qquad \texttt{collect} : Id \longrightarrow D$$

depicted as follows in the language of string diagrams:

and satisfying the familiar Yang-Baxter equation:

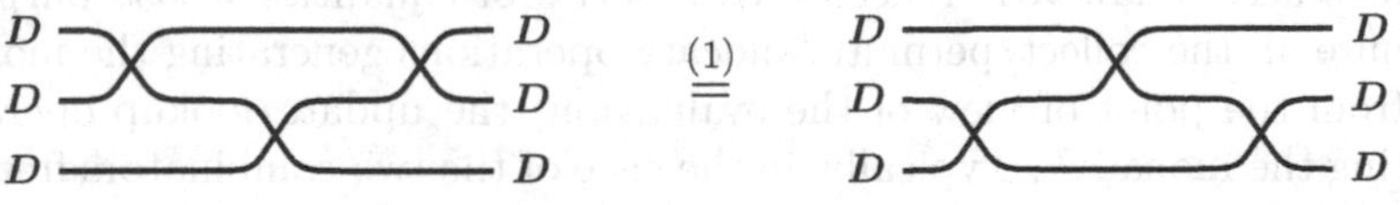

as well as the expected equation for a symmetry:

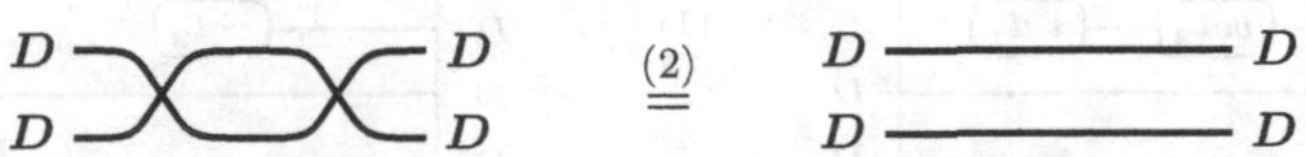

as well as two equations regulating the interaction between the permutation and the dispose combinator, the first one among them:

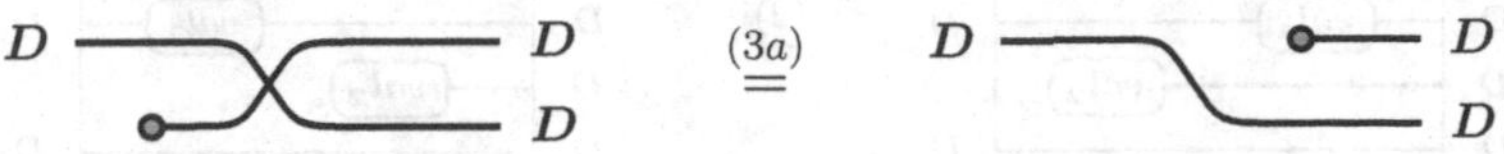

11 The Category *Res* in String Diagrams

An important instance of *Inj*-module is provided by the category *Res*. Just like the category *Inj*, the category *Res* is monoidal with tensor product $p \otimes q$ defined as the sum $p + q$. Moreover, the functor $\iota : Inj \to Res$ is monoidal in the strict sense. From this follows that *Inj* acts on the category *Res*. As a matter of fact, the category *Res* may be defined as the free *Inj*-module where the functor D is moreover equipped with a natural transformation

$$\mathtt{fresh}_{\langle val \rangle} \quad : \quad D \quad \longrightarrow \quad Id$$

for each value $val \in \{true, false\}$ and depicted as:

The two operations $\mathtt{fresh}_{\langle val \rangle}$ of allocation should satisfy a series of equations depicted below. The main equation *interaction fresh – collect* is depicted as:

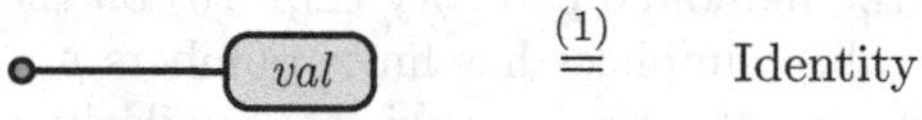

One of the two equations *interaction fresh – permutation* is depicted as:

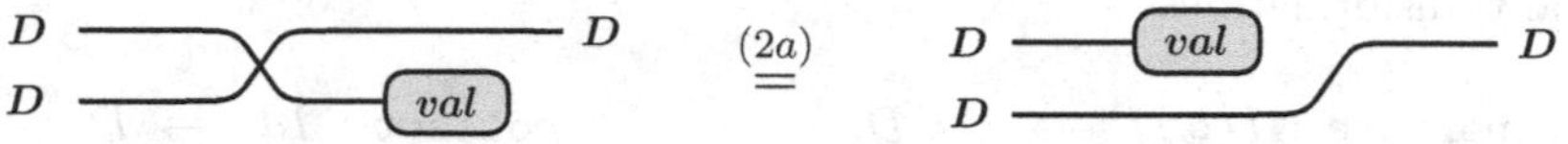

while the equation *commutation fresh – fresh* is depicted as:

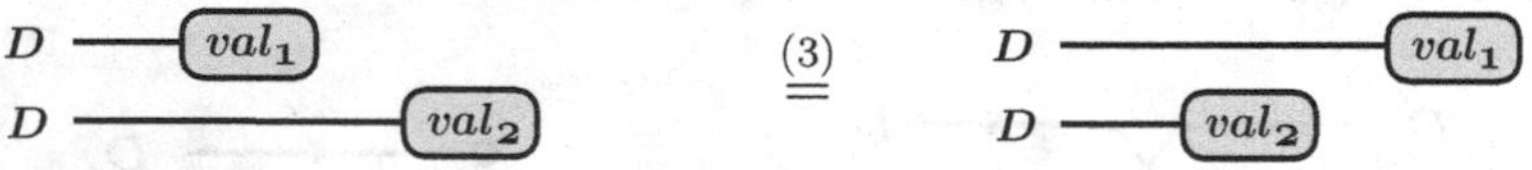

12 The Distributivity Law in String Diagrams

The distributivity law λ is reflected as a series of equalities whose purpose is to permute all the collect/permute/allocate operations generating the monad $\mathcal{B}$ *after* (from the point of view of the evaluation) the update/lookup operations generating the monad $\mathcal{F}$. Typically, in the case of the two combinators fresh and update, the first equation *interaction fresh – update* is depicted as

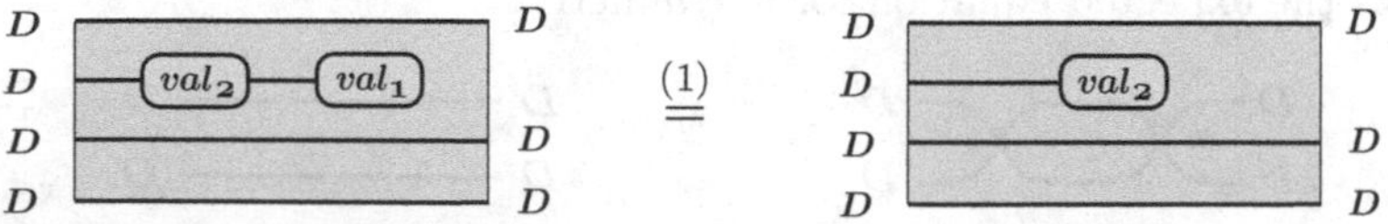

while the second equation *commutation fresh – update* is depicted as

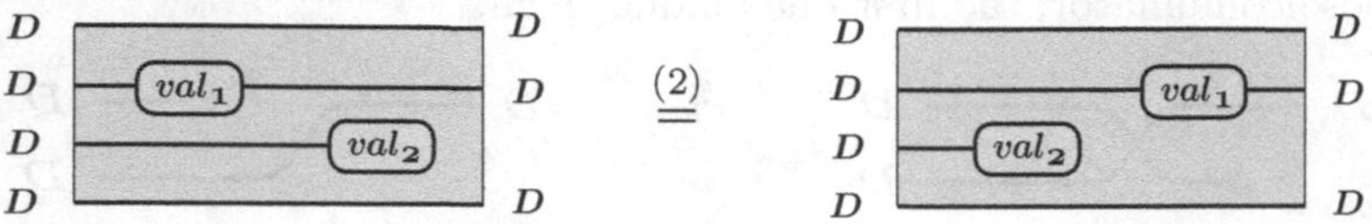

In the case of the operations fresh and lookup, the equation *commutation fresh – update* is depicted as follows:

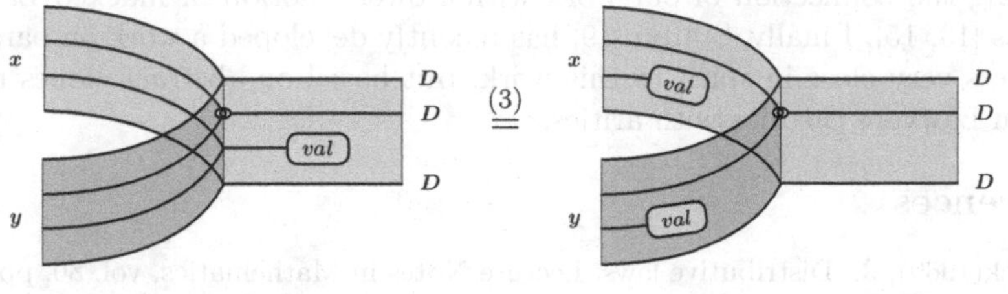

Similarly, there is an equation *interaction collect – update* depicted as

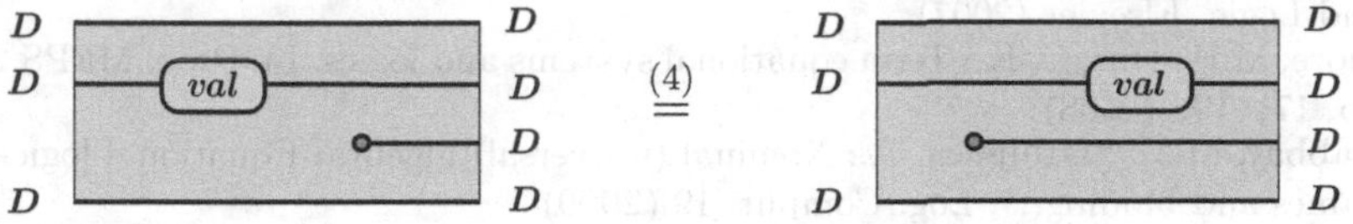

together with an equation *commutation collect – lookup*. Similar interaction and commutation equations should be then depicted for all pairs consisting of a lookup or an update operation and a permute operation. Typically, one of the two equations *interaction permute – lookup* is depicted as:

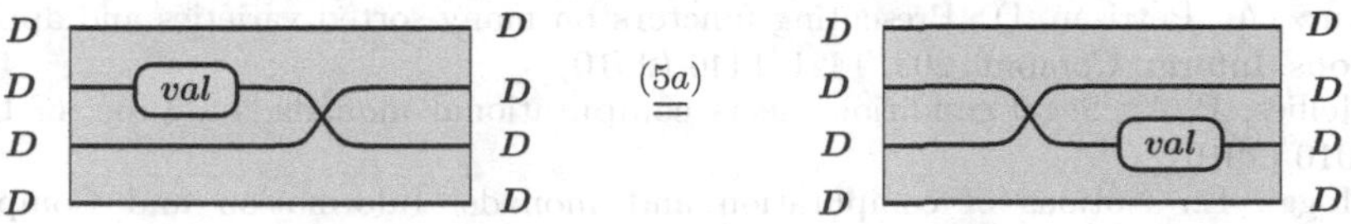

Note that the expected equation *interaction fresh – lookup* may be derived from the two equations *interaction fresh – update* and *interaction update – lookup*.

Finally, it should be mentioned that there exists a canonical form theorem extending Theorem 2 to the local state monad: informally speaking, the theorem states that every morphism of the category מ factors uniquely as a series of `lookup` operations followed by a series of `update` operations (just as in the case of Theorem 2) then followed by a series of `collect` operations followed by a series of `permute` operations followed by a series of `fresh` operations.

13 Conclusion and Related Works

Much work has been devoted in the past decade in order to understand the algebraic and combinatorial nature of the local state monad formulated by Plotkin and Power's seminal paper [12]. Besides the works by Power [13, 15] and Staton [17, 18] already mentioned, our work is close in spirit to the line of work on nominal algebraic theories developed by various authors, see in particular [2–4, 8]. A substantial work thus remains to be done in order to clarify the connection between these various notions of nominal algebraic theories and the notion formulated here of graded algebraic theory. The present paper is also tightly connected to the work by Hyland, Plotkin and Power on combining computational monads, see [5, 7]. In that respect, we are currently interested in clarifying the connection of our work with Power's notion of indexed Lawvere theories [13, 15]. Finally, Staton [19] has recently developed a work on parametric effects very close in spirit to this work, but based on abstract clones rather than on Lawvere theories with arities.

References

1. Beck(1969), J.: Distributive laws. Lecture Notes in Mathematics, vol. 80, pp. 119–140 (1969)
2. Clouston, R.A., Pitts, A.M.: Nominal equational logic. In: Computation, Meaning, and Logic. Elsevier (2007)
3. Fiore, M.P., Hur, C.-K.: Term equational systems and logics. In: Proc. MFPS 2008, pp. 171–192 (2008)
4. Gabbay, M.J., Mathijssen, A.: Nominal (universal) algebra: Equational logic with names and binding. J. Log. Comput. 19 (2009)
5. Hyland, J.M.E., Plotkin, G., Power, A.J.: Combining effects: sum and tensor. Theoretical Computer Science 357(1), 70–99 (2006)
6. Hyland, J.M.E., Levy, P.B., Plotkin, G., Power, A.J.: Combining algebraic effects with continuations. Theoretical Computer Science 375 (2007)
7. Hyland, J.M.E., Power, A.J.: Discrete lawvere theories and computational effects. Theoretical Computer Science 366, 144–162 (2006)
8. Kurz, A., Petrisan, D.: Presenting functors on many-sorted varieties and applications. Inform. Comput. 208, 1421–1446 (2010)
9. Mellies, P.-A.: Segal condition meets computational monads. In: Proc. of LICS 2010 (2010)
10. Moggi, E.: Notions of computation and monads. Information and Computation 93(1) (1991)
11. O'Hearn, P.W., Tennent, R.D.: Algol-like Languages. Progress in Theoretical Computer Science. Birkhauser, Boston (1997)
12. Plotkin, G., Power, J.: Notions of computation determine monads. In: Nielsen, M., Engberg, U. (eds.) FOSSACS 2002. LNCS, vol. 2303, p. 342. Springer, Heidelberg (2002)
13. Power, J.: Enriched lawvere theories. Theory and Applications of Categories 6(7), 83–93 (1999)
14. Power, J.: Semantics for Local Computational Effects. In: Proc. MFPS 2006. ENTCS, vol. 158, pp. 355–371 (2006)
15. Power, J.: Indexed Lawvere theories for local state. Models, Logics and Higher-Dimensional Categories: A Tribute to the Work of Mihály Makkai, pp. 213–229. American Mathematical Society
16. Pretnar, M.: The Logic and Handling of Algebraic Effects. PhD thesis, University of Edinburgh (2010)
17. Staton, S.: Two cotensors in one: Presentations of algebraic theories for local state and fresh names. In: Proc. of MFPS XXV (2009)
18. Staton, S.: Completeness for algebraic theories of local state. In: Ong, L. (ed.) FOSSACS 2010. LNCS, vol. 6014, pp. 48–63. Springer, Heidelberg (2010)
19. Staton, S.: Instances of computational effects. In: Proceedings of Twenty-Eighth Annual ACM/IEEE Symposium on Logic in Computer Science (2013)
20. Weber, M.: Familial 2-functors and parametric right adjoints. Theory and Applications of Categories 18, 665–732 (2007)

Reduction System for Extensional Lambda-mu Calculus

Koji Nakazawa and Tomoharu Nagai

Graduate School of Informatics, Kyoto University, Kyoto, Japan

Abstract. The $\Lambda\mu$-calculus is an extension of Parigot's $\lambda\mu$-calculus. For the untyped $\Lambda\mu$-calculus, Saurin proved some fundamental properties such as the standardization and the separation theorem. Nakazawa and Katsumata gave extensional models, called stream models, in which terms are represented as functions on streams. This paper introduces a conservative extension of the $\Lambda\mu$-calculus, called $\Lambda\mu_{\mathsf{cons}}$, from which the open term model is straightforwardly constructed as a stream model, and for which we can define a reduction system satisfying several fundamental properties such as confluence, subject reduction, and strong normalization.

1 Introduction

The $\lambda\mu$-calculus was originally introduced by Parigot [16] as a term assignment system for the classical natural deduction, and then a lot of studies have been devoted to the $\lambda\mu$-calculus from both sides of logic and computer science. An extension of the $\lambda\mu$-calculus was given by de Groote [6] to study continuation-passing-style translations for the calculus. As Saurin showed in [18,20], an untyped variant of this extension, called the $\Lambda\mu$-calculus, enjoys fundamental properties such as the standardization and the separation theorem. In particular, the latter does not hold for the original $\lambda\mu$-calculus as shown by David and Py [5].

For the untyped $\Lambda\mu$-calculus, Nakazawa and Katsumata [14] gave a extensional model, called *stream model*. The stream model is a simple extension of the λ-model and similar to Streicher and Reus' continuation semantics for the $\lambda\mu$-calculus [22]. The stream model naturally reflects the idea that the $\Lambda\mu$-terms represent functions on streams. Nakazawa and Katsumata showed the soundness and gave an algebraic characterization for the stream model, but they have not discussed on completeness.

Regarding types, some type assignment systems for the $\Lambda\mu$-calculus has been introduced. Pagani and Saurin [15,21] gave a type system for the $\Lambda\mu$-calculus as a stream calculus, and Gaboardi and Saurin [10] proposed its extension with recursive types. De'Liguoro [8] gave an intersection type system and filter models for the $\Lambda\mu$-calculus, based on the stream model. However, the results on the stream model in [14] have not been adapted to typed calculi.

The main results of this paper are the following: (1) An extension $\Lambda\mu_{\mathsf{cons}}$ of the $\Lambda\mu$-calculus and its reduction system are proposed. The calculus $\Lambda\mu_{\mathsf{cons}}$ induces

G. Dowek (ed.): RTA-TLCA 2014, LNCS 8560, pp. 349–363, 2014.

a term model as a stream model, and hence it is sound and complete with respect to the stream model. It is proved that the reduction on the untyped $\Lambda\mu_{\mathsf{cons}}$ is confluence. (2) A type assignment for $\Lambda\mu_{\mathsf{cons}}$ based on de'Liguoro's type system is proposed, and subject reduction and strong normalization of the reduction on the typed $\Lambda\mu_{\mathsf{cons}}$ are proved.

In Section 2 and 3, we define the equational theory and the reduction system for the untyped $\Lambda\mu_{\mathsf{cons}}$, and prove confluence. The calculus $\Lambda\mu_{\mathsf{cons}}$ explicitly contains stream expressions, and this extension is similar to the $\lambda\mu$-calculus of Streicher and Reus [22]. The reduction system proposed in this paper avoids the expansion rule in the $\Lambda\mu$-calculus, called *(fst)* in [18,21], and adopts a new rule (exp), which we can define with the new explicit stream expressions. The reduction system is confluent for whole of the untyped $\Lambda\mu_{\mathsf{cons}}$ including open terms in contrast to the $\Lambda\mu$-calculus in [21], where confluence holds for only the stream closed terms.

In Section 4 and 5, a typed variant of $\Lambda\mu_{\mathsf{cons}}$ is proposed, and subject reduction and strong normalization are proved. Following the structure of the stream model and de'Liguoro's type system [8], our type system restricts functional types to those from stream types to term types. For types of streams, we adopt (restricted forms of) recursive types for finiteness of the calculus, similarly to Gaboardi and Saurin's type system [10].

In Section 6, we discuss on the relationship with the existing calculi, such as the extended stack calculus [3], the untyped $\Lambda\mu$-calculus, and some type systems in [16,15,21,10]. In particular, we will see that the untyped $\Lambda\mu_{\mathsf{cons}}$ is conservative over the untyped $\Lambda\mu$-calculus (and hence over the λ-calculus), and inherits the separation theorem from $\Lambda\mu$.

2 $\Lambda\mu_{\mathsf{cons}}$

2.1 Definition of Untyped $\Lambda\mu_{\mathsf{cons}}$

As in [18], we adopt the notation $t\alpha$ to denote the named term $[\alpha]t$ in the original $\lambda\mu$-calculus, that can be read as a function application of t to a stream α. We use the constructors car and cdr to represent the head and the tail of a stream, respectively.

Definition 1 (Untyped $\Lambda\mu_{\mathsf{cons}}$). Suppose to have two sorts of variables: term variables, denoted by $x, y, \cdots$, and stream variables, $\alpha, \beta, \cdots$.

The *terms* and the *streams* of $\Lambda\mu_{\mathsf{cons}}$ are defined as

$$t, u ::= x \mid \lambda x.t \mid tu \mid \mu\alpha.t \mid tS \mid \mathsf{car}S \qquad S ::= \alpha \mid t :: S \mid \mathsf{cdr}S$$

The sets of terms and streams are denoted by Tm and St, respectively. Occurrences of x in $\lambda x.t$ and α in $\mu\alpha.t$ are considered to be bound. A variable occurrence which is not bound is called *free.* A term which contains no free stream variables is called *stream closed.* The size of t (and S) is defined as usual, and it is denoted by $|t|$ (and $|S|$, respectively).

The axiom schema of $\Lambda\mu_{\mathsf{cons}}$ are the following:

$$\begin{aligned}
(\lambda x.t)u &= t[x := u] && (\beta_T)\\
(\mu\alpha.t)S &= t[\alpha := S] && (\beta_S)\\
\lambda x.tx &= t && (\eta_T)\\
\mu\alpha.t\alpha &= t && (\eta_S)\\
(\mathsf{car}S) :: (\mathsf{cdr}S) &= S && (\mathsf{surj})\\
\mathsf{car}(t :: S) &= t && (\mathsf{car})\\
\mathsf{cdr}(t :: S) &= S && (\mathsf{cdr})\\
t(u :: S) &= (tu)S && (\mathsf{assoc})
\end{aligned}$$

where, t contains no free x in (η_T), and t contains no free α in (η_S). The congruence relation $=_{\Lambda\mu_{\mathsf{cons}}}$ is defined from the above axiom schema.

We write $\mathsf{cdr}^i S$ to denote the i-time application of cdr to S for $i \geq 0$, and use the abbreviation $\mathsf{cadr}^i S \equiv \mathsf{car}(\mathsf{cdr}^i\, S)$.

Example 1. 1. The usual μ-rule $(\mu\alpha.t)u = \mu\alpha.t[[]\alpha := []u\alpha]$ is admissible as follows, where the special substitution $t[[]\alpha := []u\alpha]$ recursively replaces subterm occurrences of the form $v\alpha$ in t with $(vu)\alpha$.

$$\begin{aligned}
(\mu\alpha.t)u &=_{\Lambda\mu_{\mathsf{cons}}} \mu\beta.(\mu\alpha.t)u\beta && (\eta_S)\\
&=_{\Lambda\mu_{\mathsf{cons}}} \mu\beta.t[\alpha := u :: \beta] && (\mathsf{assoc}, \beta_S)\\
&=_{\Lambda\mu_{\mathsf{cons}}} \mu\alpha.t[[]\alpha := []u\alpha] && (\mathsf{assoc}).
\end{aligned}$$

2. By a fixed-point combinator Y in the λ-calculus, the n-th function on streams is defined as

$$\mathsf{nth} \equiv Y(\lambda f.\mu\alpha.\lambda n.\mathsf{ifzero}\ n\ \mathsf{then}\ (\mathsf{car}\,\alpha)\ \mathsf{else}\ f(\mathsf{cdr}\,\alpha)(\mathsf{pred}\,n)),$$

where ifzero and pred are defined on the Church numerals, and then we have $\mathsf{nth}\,S\,\overline{k} =_{\Lambda\mu_{\mathsf{cons}}} \mathsf{cadr}^k S$, where $\overline{k}$ is the Church numeral representing k.

The calculus $\Lambda\mu_{\mathsf{cons}}$ is a natural extension of the $\Lambda\mu$-calculus, and it is also close to the $\lambda\mu$-calculus in [22], which explicitly has the expressions for continuations but no cdr operator. The main difference from these existing calculi is the surjectivity axiom (surj). We will discuss the relationship with existing calculi in Section 6.

2.2 Stream Models for Untyped $\Lambda\mu_{\mathsf{cons}}$

The stream models are defined as in [14]. We use $\overline{\lambda}$ to denote the meta-level function abstraction.

Definition 2. The set $\mathcal{S}$ is called a *stream set* on a set $\mathcal{D}$ if there is a bijective mapping $(::)$ from $\mathcal{D} \times \mathcal{S}$ to $\mathcal{S}$. For a stream set $\mathcal{S}$, the inverse of $(::)$ is denoted by $\langle \mathsf{Car}, \mathsf{Cdr} \rangle$.

Definition 3. A *stream model* consists of

a non-empty set $\mathcal{D}$ and a stream set $\mathcal{S}$ on $\mathcal{D}$,

a subset $[\mathcal{S} \to \mathcal{D}]$ of the set of functions from $\mathcal{S}$ to $\mathcal{D}$,

$\Psi : [\mathcal{S} \to \mathcal{D}] \to \mathcal{D}$ a bijective mapping,

such that the meaning function $[\![\cdot]\!]_\rho$ can be defined for any function ρ from term variables to $\mathcal{D}$ and stream variables to $\mathcal{S}$ as follows, where $d\star s$ denotes $\Psi^{-1}(d)(s)$.

$$
\begin{aligned}
[\![x]\!]_\rho &= \rho(x) & [\![\alpha]\!]_\rho &= \rho(\alpha)\\
[\![\lambda x.t]\!]_\rho &= \Psi(\overline{\lambda} s \in \mathcal{S}.[\![t]\!]_{\rho[x\mapsto \mathsf{Car}\, s]}\star(\mathsf{Cdr}\, s)) & [\![t :: S]\!]_\rho &= [\![t]\!]_\rho :: [\![S]\!]_\rho\\
[\![tu]\!]_\rho &= \Psi(\overline{\lambda} s \in \mathcal{S}.[\![t]\!]_\rho\star([\![u]\!]_\rho :: s)) & [\![\mathsf{cdr} S]\!]_\rho &= \mathsf{Cdr}[\![S]\!]_\rho\\
[\![\mu\alpha.t]\!]_\rho &= \Psi(\overline{\lambda} s \in \mathcal{S}.[\![t]\!]_{\rho[\alpha\mapsto s]}) & &\\
[\![tS]\!]_\rho &= [\![t]\!]_\rho\star[\![S]\!]_\rho & &\\
[\![\mathsf{car} S]\!]_\rho &= \mathsf{Car}[\![S]\!]_\rho & &
\end{aligned}
$$

The set $\mathsf{Tm}/=_{\Lambda\mu_{\mathsf{cons}}}$ is a stream model, which we call *open term model.*

Proposition 1 (Open term model). *Let* $[t]$ *and* $[S]$ *be the equivalence classes of* t *and* S *with respect to* $=_{\Lambda\mu_{\mathsf{cons}}}$*, and define*

$$\mathcal{D} = \{[t] \mid t \in \mathsf{Tm}\} \qquad \mathcal{S} = \{[S] \mid S \in \mathsf{St}\} \qquad [\mathcal{S} \to \mathcal{D}] = \{f_{[t]} \mid t \in \mathsf{Tm}\},$$

where $f_{[t]}$ *denotes the function* $\overline{\lambda}[S] \in \mathcal{S}.[tS]$. Ψ *is defined as* $\Psi(f_{[t]}) = [t]$. *Then, these give a stream model with the meaning function given by* $[\![t]\!]_\rho = [t\theta_\rho]$*, where* θ_ρ *is the substitution such that* $\theta_\rho(x) = u$ *for* $\rho(x) = [u]$ *and* $\theta_\rho(\alpha) = S$ *for* $\rho(\alpha) = [S]$.

Proof. Straightforward. Note that $[t\theta_\rho]$ is independent of the choice of θ_ρ.

Then, the following is easy to show.

Theorem 1 (Soundness and completeness). *For any* t *and* u*,* $t =_{\Lambda\mu_{\mathsf{cons}}} u$ *holds if and only if* $[\![t]\!]_\rho = [\![u]\!]_\rho$ *holds for any stream model and* ρ.

Some properties of the stream models are shown in [14]. One of them guarantees existence of a non-trivial stream model, which gives a semantical proof of the consistency of the equational theory of $\Lambda\mu_{\mathsf{cons}}$.

Proposition 2 ([14]). *For any pointed CPO* D*, there exists a stream model* D^S_∞ *into which* D *can be embedded.*

Corollary 1 (Consistency). *There exist closed* $\Lambda\mu_{\mathsf{cons}}$*-terms* t *and* u *such that* $t =_{\Lambda\mu_{\mathsf{cons}}} u$ *does not hold.*

3 Reduction System

3.1 Reduction for Untyped $\Lambda\mu_{\mathsf{cons}}$

Definition 4. 1. The one-step reduction $\to$ on terms and streams of $\Lambda\mu_{\mathsf{cons}}$ is the least compatible relation satisfying the following axioms.

$$
\begin{array}{rll}
(\mu\alpha.t)u \to \mu\alpha.t[\alpha := u :: \alpha] & & (\beta_T)\\
(\mu\alpha.t)S \to t[\alpha := S] & & (\beta_S)\\
\lambda x.t \to \mu\alpha.t[x := \mathsf{car}\alpha](\mathsf{cdr}\alpha) & & (\mathsf{exp})\\
t(u :: S) \to tuS & & (\mathsf{assoc})\\
\mathsf{car}(u :: S) \to u & & (\mathsf{car})\\
\mathsf{cdr}(u :: S) \to S & & (\mathsf{cdr})\\
\mu\alpha.t\alpha \to t & (\alpha \notin FV(t)) & (\eta_S)\\
(\mathsf{car}S) :: (\mathsf{cdr}S) \to S & & (\eta_{::})\\
t(\mathsf{car}S)(\mathsf{cdr}S) \to tS & & (\eta'_{::})
\end{array}
$$

Here, α in (exp) is a fresh stream variable. The relation $\to^*$ is the reflexive transitive closure of $\to$, the relation $\to^+$ is the transitive closure of $\to$, and the relation $\to^=$ is the reflexive closure of $\to$.

2. The relations $\to_{\mathsf{B}}$ and $\to_{\mathsf{E}}$ are defined as the least compatible relations satisfying the following axioms, respectively.

$\to_{\mathsf{B}}$: (β_T), (β_S), (exp), (assoc), (car), and (cdr)

$\to_{\mathsf{E}}$: (η_S), $(\eta_{::})$, $(\eta'_{::})$, (car), and (cdr)

We also use $\to^*_{\mathsf{B}}$, $\to^+_{\mathsf{E}}$, and so on.

Note that the $\to_{\mathsf{B}}$-normal forms are characterized by

$$
t ::= a \mid \mu\alpha.t \qquad\qquad a ::= x \mid \mathsf{cadr}^n\alpha \mid at \mid a(\mathsf{cdr}^n\alpha).
$$

We can easily see that the usual β- and η-rules in the λ-calculus are derivable, that is, $(\lambda x.t)u \to^* t[x := u]$ and $\lambda x.tx \to^* t$ for $x \notin FV(t)$ hold. Hence, the $\beta\eta$-reduction of the λ-calculus and the reduction of Parigot's $\lambda\mu$-calculus including the renaming and the η-rules for μ-abstractions can be simulated in $\Lambda\mu_{\mathsf{cons}}$. Furthermore, the following holds.

Proposition 3. *The equivalence closure of* $\to$ *coincides with* $=_{\Lambda\mu_{\mathsf{cons}}}$.

It is known that naïvely adding the η-rule $\lambda x.tx \to_\eta t$ to the $\lambda\mu$-calculus destroys confluence [5]. The counterexample is $t = \lambda x.(\mu\alpha.y\beta)x$, and then

$$
t \to_\eta \mu\alpha.y\beta, \qquad\qquad t \to_\beta \lambda x.\mu\alpha.y\beta.
$$

In order to recover confluence, the rule called (*fst*) in [21] has been proposed as

$$
\mu\alpha.t \to \lambda x.\mu\alpha.t[[]\alpha := []x\alpha] \qquad\qquad (\mathit{fst}).
$$

It seems natural since it means the surjectivity of the bound variable α. However, if we consider type systems, it induces a type dependent reduction for subject reduction and strong normalization.

Alternatively, by the explicit stream syntax in $\Lambda\mu_{\mathsf{cons}}$, we can define the new rule (exp), and the above critical pair is solved as

$$\lambda x.\mu\alpha.y\beta \to_{\mathsf{exp}} \mu\gamma.(\mu\alpha.y\beta)(\mathsf{cdr}\gamma) \to_{\beta_S} \mu\gamma.y\beta (= \mu\alpha.y\beta).$$

The reduction system with (exp) will be adapted to the typed $\Lambda\mu_{\mathsf{cons}}$ without any restriction of types, and subject reduction and strong normalization will be proved.

3.2 Confluence

We prove confluence of $\to$ by (1) confluence of $\to_{\mathsf{B}}$, (2) confluence of $\to_{\mathsf{E}}$, and (3) commutativity of them. In contrast to the $\Lambda\mu$-calculus [21], the result is not restricted to stream closed terms, and hence the Church-Rosser theorem directly follows from the confluence.

Proposition 4. *$\to_{\mathsf{B}}$ and $\to_{\mathsf{E}}$ are respectively confluent.*

Proof. (B) By a generalized notion of complete development, which is independently introduced in [7,11]. We define the mapping $(\cdot)^\dagger$ as follows.

$$\begin{aligned}
x^\dagger &= x & \alpha^\dagger &= \alpha\\
(\lambda x.t)^\dagger &= \mu\alpha.t^\dagger[x := \mathsf{car}\alpha](\mathsf{cdr}\alpha) & (\mathsf{cdr}(t :: S))^\dagger &= S^\dagger\\
(\mu\alpha.t)^\dagger &= \mu\alpha.t^\dagger & (\mathsf{cdr}S)^\dagger &= \mathsf{cdr}S^\dagger \quad \text{(otherwise)}\\
((\mu\alpha.t)u)^\dagger &= \mu\alpha.t^\dagger[\alpha := u^\dagger :: \alpha] & (t :: S)^\dagger &= t^\dagger :: S^\dagger\\
(tu)^\dagger &= t^\dagger u^\dagger \quad \text{(otherwise)}\\
((\mu\alpha.t)S)^\dagger &= t^\dagger[\alpha := S^\dagger]\\
((\mu\alpha.t)uS)^\dagger &= t^\dagger[\alpha := u^\dagger :: S^\dagger]\\
(t(u :: S))^\dagger &= t^\dagger u^\dagger S^\dagger \quad (t \neq \mu\text{-abst.})\\
(tS)^\dagger &= t^\dagger S^\dagger \quad \text{(otherwise)}\\
(\mathsf{car}(t :: S))^\dagger &= t^\dagger\\
(\mathsf{car}S)^\dagger &= \mathsf{car}S^\dagger \quad \text{(otherwise)}
\end{aligned}$$

Then, we can prove that $t \to_{\mathsf{B}} u$ implies $u \to^*_{\mathsf{B}} t^\dagger \to^*_{\mathsf{B}} u^\dagger$, from which the confluence follows. The only non-trivial point is that we exceptionally define $((\mu\alpha.t)uS)^\dagger = t^\dagger[\alpha := u^\dagger :: S^\dagger]$ (not $(\mu\alpha.t^\dagger[\alpha := u^\dagger :: \alpha])S^\dagger$), since we have to show that $((\mu\alpha.t)(u :: S))^\dagger \to^*_{\mathsf{B}} ((\mu\alpha.t)uS)^\dagger$, the left-hand side of which is $t^\dagger[\alpha := (u :: S)^\dagger] = t^\dagger[\alpha := u^\dagger :: S^\dagger]$.

(E) Since $\to_{\mathsf{E}}$ is clearly strongly normalizing, it is sufficient to prove local confluence. It is straightforward.

In order to prove commutativity of B and E, we consider the following restricted E-reduction.

Definition 5. The relation $\rightarrow_{\mathsf{E}^-}$ is the least compatible relation satisfying (η_S), (car), (cdr), and the restricted forms of $(\eta_{::})$ and $(\eta'_{::})$ as follows.

$$(\mathsf{car}\,\mathsf{cdr}^n\alpha) :: (\mathsf{cdr}\,\mathsf{cdr}^n\alpha) \rightarrow \mathsf{cdr}^n\alpha \qquad (\eta^-_{::})$$
$$t(\mathsf{car}\,\mathsf{cdr}^n\alpha)(\mathsf{cdr}\,\mathsf{cdr}^n\alpha) \rightarrow t(\mathsf{cdr}^n\alpha) \qquad (\eta'^-_{::})$$

The relation $\rightarrow_{\mathsf{cdr}}$ is the least compatible relation satisfying (cdr).

The relation $\rightarrow_{\mathsf{E}^-}$ is introduced to show a variant of strong commutativity, that is Lemma 2.2. Note that $t \rightarrow_{\mathsf{E}^-} t'$ does not necessarily imply $t[\alpha := S] \rightarrow_{\mathsf{E}^-} t'[\alpha := S]$ due to the restriction. Instead, we have the following lemma.

Lemma 1. *1. Any S is reduced by $\rightarrow_{\mathsf{cdr}}$ to a term of the form either $t' :: S'$ or $\mathsf{cdr}^n\alpha$ for some $n \geq 0$.*
2. For any S, there exists S' such that $S \rightarrow^*_{\mathsf{cdr}} S'$ and $(\mathsf{car}S) :: (\mathsf{cdr}S) \rightarrow^*_{\mathsf{E}^-} S'$.
3. If $t \rightarrow_{\mathsf{E}^-} t'$, then $t[\alpha := S] \rightarrow_{\mathsf{E}^-} u$ and $t'[\alpha := S] \rightarrow^*_{\mathsf{cdr}} u$ for some u.

Proof. 1. By induction on S.

2. By 1, there exists S' such that $S \rightarrow^*_{\mathsf{cdr}} S'$ and S' is of the form $t'_0 :: S'_0$ or $\mathsf{cdr}^n\alpha$. In the former case, we have $(\mathsf{car}S) :: (\mathsf{cdr}S) \rightarrow^*_{\mathsf{cdr}} (\mathsf{car}(t'_0 :: S'_0)) :: (\mathsf{cdr}(t'_0 :: S'_0)) \rightarrow^*_{\mathsf{car},\mathsf{cdr}} t'_0 :: S'_0$. In the latter case, we have $(\mathsf{car}S) :: (\mathsf{cdr}S) \rightarrow^*_{\mathsf{cdr}} (\mathsf{car}\,\mathsf{cdr}^n\alpha) :: (\mathsf{cdr}\,\mathsf{cdr}^n\alpha) \rightarrow_{\mathsf{E}^-} \mathsf{cdr}^n\alpha$.

3. By induction on $t \rightarrow_{\mathsf{E}^-} t'$. Consider the case of $\mathsf{car}\,\mathsf{cdr}^n\alpha :: \mathsf{cdr}\,\mathsf{cdr}^n\alpha \rightarrow_{\mathsf{E}^-} \mathsf{cdr}^n\alpha$ by $(\eta^-_{::})$. By 2, there exists S' such that $(\mathsf{car}\,\mathsf{cdr}^n S) :: (\mathsf{cdr}\,\mathsf{cdr}^n S) \rightarrow^*_{\mathsf{E}^-} S'$ and $(\mathsf{cdr}^n S) \rightarrow^*_{\mathsf{cdr}} S'$. The case of $\eta'^-_{::}$ is similarly proved, and the other cases are straightforward.

Lemma 2. *The following commuting diagrams hold.*

1. 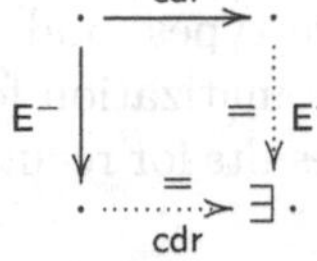

2. 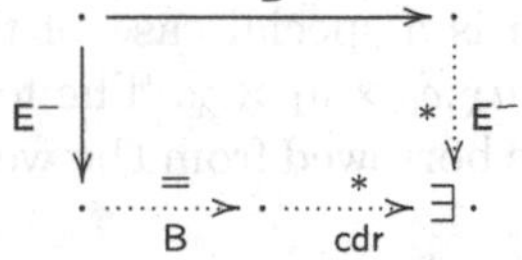

3. 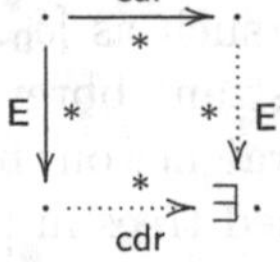

Proof. 1. By induction on the size of terms and streams.

2. By induction on the terms and streams. We only show the case of $(\mu\alpha.t)S \rightarrow_{\mathsf{B}} t[\alpha := S]$ and $(\mu\alpha.t)S \rightarrow_{\mathsf{E}^-} (\mu\alpha.t')S$ where $t \rightarrow_{\mathsf{E}^-} t'$. By Lemma 1, there exists u such that $t[\alpha := S] \rightarrow^=_{\mathsf{E}^-} u$ and $(\mu\alpha.t')S \rightarrow_{\mathsf{B}} t'[\alpha := S] \rightarrow^*_{\mathsf{cdr}} u$.

3. By induction on the length of $\rightarrow^*_{\mathsf{E}}$. By 1, $\rightarrow^*_{\mathsf{E}^-}$ and $\rightarrow^*_{\mathsf{cdr}}$ commute, so it is sufficient to consider each step of $(\eta_{::})$ and $(\eta'_{::})$ which is not restricted. It is proved since $t \rightarrow_{\mathsf{E}} t'$ implies that there exists u such that $t \rightarrow^*_{\mathsf{E}^-} u$ and $t' \rightarrow^*_{\mathsf{cdr}} u$, that is proved by Lemma 1.2.

If we consider the full E-reduction, $\rightarrow^=$ in 1 and 2 does not necessarily hold.

Proposition 5. *$\to_{\mathsf{B}}^*$ and $\to_{\mathsf{E}}^*$ commute.*

Proof. By Lemma 2.3, the left triangle in the following diagram commute. By 1 and 2 of Lemma 2, $\to_{\mathsf{B}}^*$ and $\to_{\mathsf{E}^-}^*$ commute, and hence $\to_{\mathsf{B}}^*$ and $\to_{\mathsf{E}}^*$ commute.

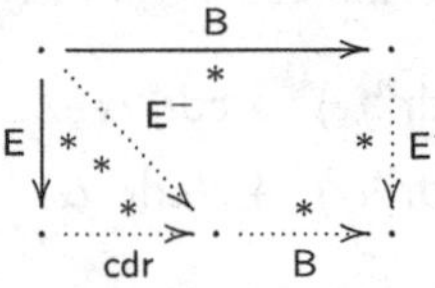

Theorem 2 (Confluence). *The reduction $\to$ is confluent.*

Proof. It follows from Proposition 4 and 5.

Corollary 2 (Church-Rosser theorem). *If $t =_{\Lambda\mu_{\mathsf{cons}}} t'$ holds, then there exists u such that $t \to^* u$ and $t' \to^* u$.*

The Church-Rosser theorem gives a syntactic proof of consistency of the equational logic of $\Lambda\mu_{\mathsf{cons}}$, since, for example, $\mu\alpha.\mathsf{car}\alpha$ and $\mu\alpha.\mathsf{cadr}\alpha$ are different normal forms.

Corollary 3 (Consistency of $\Lambda\mu_{\mathsf{cons}}$). *There exists two closed $\Lambda\mu_{\mathsf{cons}}$-terms t and u such that $t =_{\Lambda\mu_{\mathsf{cons}}} u$ does not hold.*

4 Typed $\Lambda\mu_{\mathsf{cons}}$

We will give a type assignment system for $\Lambda\mu_{\mathsf{cons}}$, inspired by de'Liguoro's type system for the $\Lambda\mu$-calculus [8], and adopting recursive types to represent types for streams like [10].

4.1 Definition of Typed $\Lambda\mu_{\mathsf{cons}}$

The types of streams will be introduced as non-empty lists of types of individual data such as $[\delta_0, \delta_1]$, which is a special case of the recursive types, and which is just an abbreviation for $\mu\chi.\delta_0 \times \delta_1 \times \chi$. The following axiomatization for the equivalence on the types are borrowed from the well-known results for recursively defined trees in [17,12,1,2].

Definition 6 (Typed $\Lambda\mu_{\mathsf{cons}}$). The types consist of two sorts, *term types* and *stream types*, which are inductively defined as

$$\delta ::= X \mid \sigma \to \delta \qquad\qquad \sigma ::= [\delta_0, \cdots, \delta_{n-1}] \mid \delta \times \sigma,$$

where X ranges over the base types, and $[\delta_0, \cdots, \delta_{n-1}]$ is a non-empty finite list of types. The relation $\sim$ on the types is defined as the least congruence relation satisfying the following.

$$\frac{}{[\delta_0, \cdots, \delta_{n-1}] \sim \delta_0 \times [\delta_1, \cdots \delta_{n-1}, \delta_0]}\ (\mathrm{Fld}) \qquad \frac{\delta_0 \times \cdots \delta_{n-1} \times \sigma \sim \sigma}{[\delta_0, \cdots, \delta_{n-1}] \sim \sigma}\ (\mathrm{Ctr})$$

Note that $\sim$ is also defined on the term types as $\sigma \to \delta \sim \sigma' \to \delta'$ if $\sigma \sim \sigma'$ and $\delta \sim \delta'$.

A term context Γ and a stream context Δ are finite lists of pairs of the form $(x : \delta)$ and $(\alpha : \sigma)$, respectively, in which each variable occurs at most once.

The typing rules of $\Lambda\mu_{\mathsf{cons}}$ are the following.

$$\frac{}{\Gamma, x : \delta \mid \Delta \vdash x : \delta} \qquad \frac{}{\Gamma \mid \Delta, \alpha : \sigma \vdash \alpha : \sigma} \qquad \frac{\Gamma \mid \Delta \vdash S : \delta \times \sigma}{\Gamma \mid \Delta \vdash \mathsf{car}\, S : \delta}$$

$$\frac{\Gamma, x : \delta \mid \Delta \vdash t : \sigma \to \delta'}{\Gamma \mid \Delta \vdash \lambda x.t : \delta \times \sigma \to \delta'} \qquad \frac{\Gamma \mid \Delta \vdash t : \delta \times \sigma \to \delta' \quad \Gamma \mid \Delta \vdash u : \delta}{\Gamma \mid \Delta \vdash tu : \sigma \to \delta'}$$

$$\frac{\Gamma \mid \Delta, \alpha : \sigma \vdash t : \delta}{\Gamma \mid \Delta \vdash \mu\alpha.t : \sigma \to \delta} \qquad \frac{\Gamma \mid \Delta \vdash t : \sigma \to \delta \quad \Gamma \mid \Delta \vdash S : \sigma}{\Gamma \mid \Delta \vdash tS : \delta}$$

$$\frac{\Gamma \mid \Delta \vdash t : \delta \quad \Gamma \mid \Delta \vdash S : \sigma}{\Gamma \mid \Delta \vdash t :: S : \delta \times \sigma} \qquad \frac{\Gamma \mid \Delta \vdash S : \delta \times \sigma}{\Gamma \mid \Delta \vdash \mathsf{cdr}, S : \sigma}$$

$$\frac{\Gamma \mid \Delta \vdash t : \delta \quad \delta \sim \delta'}{\Gamma \mid \Delta \vdash t : \delta'} \qquad \frac{\Gamma \mid \Delta \vdash S : \sigma \quad \sigma \sim \sigma'}{\Gamma \mid \Delta \vdash S : \sigma'}$$

The relation $\Gamma \mid \Delta \vdash t_1 = t_2 : \delta$ means $\Gamma \mid \Delta \vdash t_i : \delta$ $(i = 1, 2)$ and $t_1 =_{\Lambda\mu_{\mathsf{cons}}} t_2$.

We consider the restricted recursive types only for finiteness of the type system, and the choice of the equivalence $\sim$ is not essential for the following discussion. We can adopt the equivalence defined by only the fold/unfold axiom as in [10]. Indeed, the discussion in the following sections can be done in more general setting in which types of streams are represented as infinite product types, called *expanded types*. Some notions such as the stream models and the reducibility predicate for the strong normalization proof will be defined on the expanded types.

Definition 7. 1. The *expanded types* are defined by

$$\delta ::= X \mid \sigma \to \delta \qquad\qquad \sigma ::= \Pi_{i\in\mathbb{N}}\delta_i.$$

We also use the notation $\delta \times \Pi_{i\in\mathbb{N}}\delta'_i$, which is straightforwardly defined.

2. Given a stream type σ and $i \in \mathbb{N}$, we define $(\sigma)_i$ by

$$([\delta_0, \cdots, \delta_{n-1}])_i = \delta_{i \bmod n} \qquad\qquad (\delta \times \sigma)_i = \begin{cases} \delta & (i = 0) \\ (\sigma)_{i-1} & (i > 0), \end{cases}$$

where $i \bmod n$ denotes the remainder of the division of i by n. We also define the function $(\sigma)_i$ for the expanded types as $(\Pi_{j\in\mathbb{N}}\delta_j)_i = \delta_i$.

3. The *expansion* of the types is defined as follows.

$$\langle\!|X|\!\rangle = X \qquad \langle\!|\sigma \to \delta|\!\rangle = \langle\!|\sigma|\!\rangle \to \langle\!|\delta|\!\rangle \qquad \langle\!|\sigma|\!\rangle = \Pi_{i\in\mathbb{N}}\langle\!|(\sigma)_i|\!\rangle$$

Note that the relation $\sqsubset$ on expanded types defined as $\sigma, \delta \sqsubset \sigma \to \delta$ and $\delta_i \sqsubset \Pi_{i\in\mathbb{N}}\delta_i$ is a well-founded order, and we use the induction on this order.

Proposition 6. $\delta \sim \delta'$ *iff* $\langle\!\langle \delta \rangle\!\rangle = \langle\!\langle \delta' \rangle\!\rangle$.

It follows from the completeness of the axiomatization, for example, in [1].

Example 2. In [14], SCL is proposed as a combinatory calculus which is equivalent to the $\Lambda\mu$-calculus. However, some combinators of SCL are not typable in the original typed $\lambda\mu$-calculus. On the other hand, the SCL combinators are typable in $\Lambda\mu_{\mathsf{cons}}$ such as

$$(\mathsf{K}_1) \quad \cdot \mid \cdot \vdash \lambda x.\mu\alpha.x : \delta \times \sigma \to \delta$$
$$(\mathsf{W}_1) \quad \cdot \mid \cdot \vdash \lambda x.\mu\alpha.x\alpha\alpha : (\sigma \to \sigma \to \delta) \times \sigma \to \delta$$

for any term type δ and any stream type σ.

We will discuss the related typed calculi in Section 6 and in [13].

4.2 Stream Models for Typed $\Lambda\mu_{\mathsf{cons}}$

In [13], it is shown that the stream models are adapted to the typed $\Lambda\mu_{\mathsf{cons}}$, and we briefly introduce the results.

A stream model for the typed $\Lambda\mu_{\mathsf{cons}}$ consists of

- family of sets $\mathcal{A}^\delta$ and $\mathcal{A}^\sigma$ indexed by the expanded types
- an operation $(\star) : \mathcal{A}^{\sigma\to\delta} \times \mathcal{A}^\sigma \to \mathcal{A}^\delta$ for each σ and δ such that
$$\forall f, g \in \mathcal{A}^{\sigma\to\delta}.[\forall s \in \mathcal{A}^\sigma.[f \star s = g \star s] \Rightarrow f = g].$$
- a bijection $(::) : \mathcal{A}^\delta \times \mathcal{A}^\sigma \to \mathcal{A}^{\delta\times\sigma}$ for each δ and σ, the inverse of which consists of the projection functions $\langle \mathsf{Car}, \mathsf{Cdr} \rangle$.
- a meaning function $[\![\cdot]\!]$ such that $[\![\lambda x^{\delta'}.t^{\sigma\to\delta}]\!]_\rho \in \mathcal{A}^{\delta'\times\sigma\to\delta}$ and $[\![\lambda x^{\delta'}.t^{\sigma\to\delta}]\!]_\rho \star s = [\![t]\!]_{\rho[x\mapsto \mathsf{Car}(s)]} \star \mathsf{Cdr}(s)$ for any $s \in \mathcal{A}^{\delta'\times\sigma}$, and so on.

In particular, a stream model is called *full* if $\mathcal{A}^{\sigma\to\delta}$ is the whole function space from $\mathcal{A}^\sigma$ to $\mathcal{A}^\delta$ for any σ and δ, and $\mathcal{A}^\sigma$ is $\Pi_{i\in\mathbb{N}}\mathcal{A}^{(\sigma)_i}$ for any σ.

The typed $\Lambda\mu_{\mathsf{cons}}$ is sound and complete with respect to the stream model. Furthermore, we can show the following property, corresponding to Friedman's theorem [9]: the extensional equality in $\lambda^{\to}$ is characterized by an arbitrary individual full type hierarchy with infinite domains for base types. This theorem is proved by giving the logical relation on the stream models between the open term model and the full stream model.

Theorem 3 (Friedman's theorem for $\Lambda\mu_{\mathsf{cons}}$, [13]). *Suppose that a stream model $\mathcal{F}$ is full and all of $\mathcal{F}^X$ are infinite. Then, for any closed typable t and u, $t =_{\Lambda\mu_{\mathsf{cons}}} u$ holds if and only if $[\![t]\!]^{\mathcal{F}} = [\![u]\!]^{\mathcal{F}}$ holds.*

5 Reduction System for Typed $\Lambda\mu_{\mathsf{cons}}$

In this section, we show two fundamental properties of the reduction on the typed $\Lambda\mu_{\mathsf{cons}}$: subject reduction and strong normalization.

5.1 Subject Reduction

We omit the proof of the subject reduction since it is straightforwardly proved using the usual generation lemma modulo $\sim$.

Theorem 4 (Subject reduction). *If $\Gamma \mid \Delta \vdash t : \delta$ and $t \to u$ hold, then we have $\Gamma \mid \Delta \vdash u : \delta$.*

5.2 Strong Normalization

First, we prove the strong normalization of $\to_{\mathsf{B}}$ by the usual reducibility, and then extend it to the full reduction $\to$. The set of terms and streams which are strongly normalizable with respect to $\to_{\mathsf{B}}$ are denoted by $\mathsf{SN_T}$ and $\mathsf{SN_S}$, respectively. Moreover, the applicative contexts are defined as $C ::= [] \mid Ct \mid CS$, and $\mathsf{SN_C}$ is the set of the applicative contexts in which $t \in \mathsf{SN_T}$ and $S \in \mathsf{SN_S}$.

Definition 8. The predicates Red indexed by the expanded types are defined as

$\mathsf{Red}_X = \mathsf{SN_T}$,
$t \in \mathsf{Red}_{\sigma\to\delta}$ iff, for any $S \in \mathsf{Red}_\sigma$, $tS \in \mathsf{Red}_\delta$,
$S \in \mathsf{Red}_\sigma$ iff, for any $n \geq 0$, $\mathsf{cadr}^n S \in \mathsf{Red}_{(\sigma)_n}$.

For (not expanded) types, Red_δ and Red_σ mean $\mathsf{Red}_{\langle\!|\delta|\!\rangle}$ and $\mathsf{Red}_{\langle\!|\sigma|\!\rangle}$, respectively.

Note that, $S \in \mathsf{Red}_{\delta\times\sigma}$ iff $\mathsf{car}\, S \in \mathsf{Red}_\delta$ and $\mathsf{cdr}\, S \in \mathsf{Red}_\sigma$ by the definition.

Lemma 3. *1. $\mathsf{Red}_\delta \subseteq \mathsf{SN_T}$ and $\mathsf{Red}_\sigma \subseteq \mathsf{SN_S}$ hold.*
2. For any $C \in \mathsf{SN_C}$, $C[x] \in \mathsf{Red}_\delta$ and $C[\mathsf{cadr}^n \alpha] \in \mathsf{Red}_\delta$ hold.
3. $\alpha \in \mathsf{Red}_\sigma$ holds.

Proof. They are simultaneously proved by induction on the expanded types.

Lemma 4. *For any expanded types δ, σ, and any applicative context C, the following hold.*
1. For any $u \in \mathsf{SN_T}$, $C[\mu\alpha.t[\alpha := u :: \alpha]] \in \mathsf{Red}_\delta$ implies $C[(\mu\alpha.t)u] \in \mathsf{Red}_\delta$.
2. For any $S \in \mathsf{SN_S}$, $C[t[\alpha := S]] \in \mathsf{Red}_\delta$ implies $C[(\mu\alpha.t)S] \in \mathsf{Red}_\delta$.
3. $C[\mu\alpha.t[x := \mathsf{car}\alpha](\mathsf{cdr}\alpha)] \in \mathsf{Red}_\delta$ implies $C[\lambda x.t] \in \mathsf{Red}_\delta$.
4. For any $S \in \mathsf{SN_S}$, if $C[t] \in \mathsf{Red}_\delta$ implies $C[\mathsf{car}(t :: S)] \in \mathsf{Red}_\delta$.
5. For any $t \in \mathsf{SN_T}$, if $C[\mathsf{cadr}^n S] \in \mathsf{Red}_\delta$ implies $C[\mathsf{cadr}^{n+1}(t :: S)] \in \mathsf{Red}_\delta$.

Proof. We give only the proof of 2, and the others are proved similarly. In this proof, $\#t$ for $t \in \mathsf{SN_T}$ denotes the maximum length of reduction sequences from t, and $\#S$ for $S \in \mathsf{SN_S}$ is similarly defined.

First, we show that, for any $S \in \mathsf{SN_S}$, $C[t[\alpha := S]] \in \mathsf{SN_T}$ implies $C[(\mu\alpha.t)S] \in \mathsf{SN_T}$, by induction on the triple $\langle \#S, |S|, \#C[t[\alpha := S]]\rangle$ with the lexicographical order. It is sufficient to show that $u \in \mathsf{SN_T}$ for any u such that $C[(\mu\alpha.t)S] \to_{\mathsf{B}} u$.

Case $C[(\mu\alpha.t)S] \to_{\mathsf{B}} C[t[\alpha := S]]$. $C[t[\alpha := S]] \in \mathsf{SN_T}$ is the assumption.

Case $C[(\mu\alpha.t)S] \to_{\mathsf{B}} C'[(\mu\alpha.t')S]$. We have $C[t[\alpha := S]] \to_{\mathsf{B}} C'[t'[\alpha := S]]$, and hence $C'[(\mu\alpha.t')S] \in \mathsf{SN_T}$ follows from the induction hypothesis since $\#C[t[\alpha := S]] > \#C'[t'[\alpha := S]]$.

Case $C[(\mu\alpha.t)S] \to_{\mathsf{B}} C[(\mu\alpha.t)S']$. It follows from the induction hypothesis since $\#S > \#S'$.

Case $C[(\mu\alpha.t)(t_0 :: S_0)] \to_{\mathsf{B}} C[(\mu\alpha.t)t_0S_0]$ by (assoc). Since $\#(t_0 :: S_0) \geq \#S_0$ and $|t_0 :: S_0| > |S_0|$, we have $C[(\mu\alpha.t[\alpha := t_0 :: \alpha])S_0] \in \mathsf{SN}_{\mathsf{T}}$ by the induction hypothesis. Since $t_0 \in \mathsf{SN}_{\mathsf{T}}$, we have $C[(\mu\alpha.t)t_0S_0] \in \mathsf{SN}_{\mathsf{T}}$ by 1.

Secondly, the lemma is proved by induction on the expanded types. The base case is shown above, and the induction steps are straightforward.

Definition 9. $\mathsf{Red}_{\Gamma|\Delta}$ denotes the set of substitutions θ such that $\theta(x) \in \mathsf{Red}_\delta$ for any $x : \delta \in \Gamma$ and $\theta(\alpha) \in \mathsf{Red}_\sigma$ for any $\alpha : \sigma \in \Delta$.

Lemma 5. *If $\Gamma \mid \Delta \vdash t : \delta$ and $\theta \in \mathsf{Red}_{\Gamma|\Delta}$, then we have $t\theta \in \mathsf{Red}_\delta$.*

Proof. By induction on the derivation of $\Gamma \mid \Delta \vdash t : \delta$, using Lemma 4. Note that $\delta \sim \delta'$ implies $\mathsf{Red}_\delta = \mathsf{Red}_{\delta'}$.

Proposition 7. *Every typable term is in SN_{T}.*

Proof. By 2 and 3 of Lemma 3, the identity substitution θ is in $\mathsf{Red}_{\Gamma|\Delta}$ for any Γ and Δ. Hence, by Lemma 3.1 and Lemma 5, we have $t = t\theta \in \mathsf{Red}_\delta \subseteq \mathsf{SN}_{\mathsf{T}}$.

Theorem 5 (Strong normalization). *Every typable term is strongly normalizing with respect to $\to$.*

Proof. For any reduction sequence, we can postpone any E-reduction, that is, we can prove that $t \to_{\mathsf{E}} \cdot \to_{\mathsf{B}} u$ implies $t \to_{\mathsf{B}}^{+} \cdot \to_{\mathsf{E}}^{*} u$. Since $\to_{\mathsf{E}}$ is strongly normalizable, if we have an infinite sequence of $\to$, we can construct an infinite sequence of $\to_{\mathsf{B}}$, that contradicts Proposition 7.

6 Related Work

In this section, we discuss the relationship between $\Lambda\mu_{\mathsf{cons}}$ and the existing related systems such as the stack calculus in [4,3], the untyped $\Lambda\mu$-calculus in [18], Parigot's original typed $\lambda\mu$-calculus [16], Pagani and Saurin's $\Lambda_{\mathcal{S}}$ in [15], and Gaboardi and Saurin's $\Lambda_{\mathcal{S}}$ in [10].

6.1 Extended Stack Calculus

The calculus $\Lambda\mu_{\mathsf{cons}}$ can be seen as an extension of the nil-free fragment of the extended stack calculus in [3]. The stack calculus contains neither term variables, λ-abstractions, nor term applications, but they can be simulated. It is straightforward to see that the reduction of $\Lambda\mu_{\mathsf{cons}}$ is conservative over the stack calculus, that is, for terms t and u of the extended stack calculus without nil, $t \to^{*} u$ in the stack calculus if and only if $t \to^{*} u$ in $\Lambda\mu_{\mathsf{cons}}$. Moreover, our type system can be adapted to the extended stack calculus without nil in a straightforward way. The discussion in this paper on the stream models for the untyped and typed variants of $\Lambda\mu_{\mathsf{cons}}$ can be adapted to the extended stack calculus.

6.2 Untyped $\Lambda\mu$-Calculus

The calculus $\Lambda\mu_{\mathsf{cons}}$ is conservative over the $\Lambda\mu$-calculus.

Proposition 8 (Conservativity over $\Lambda\mu$). *For any $\Lambda\mu$-terms t and u, $t = u$ holds in $\Lambda\mu_{\mathsf{cons}}$ if and only if $t = u$ holds in $\Lambda\mu$.*

Corollary 4. *For any $\Lambda\mu$-terms t and u, $t = u$ holds in the $\Lambda\mu$-calculus if and only if $[\![t]\!]_\rho = [\![u]\!]_\rho$ for any stream model $\mathcal{A}$ and ρ.*

Saurin [18] proved the separation theorem of the $\Lambda\mu$-calculus. By the conservativity, $\Lambda\mu_{\mathsf{cons}}$ inherits the separation theorem from $\Lambda\mu$ for stream closed terms. The *canonical normal forms* in the $\Lambda\mu$-calculus are defined as terms which are η-normal and contain no subterm of the form either $(\lambda x.t)u$, $(\lambda x.t)\beta$, $(\mu\alpha.t)u$, or $(\mu\alpha.t)\beta$. The *stream applicative contexts* are defined as $\mathcal{C} ::= [] \mid \mathcal{C}t \mid \mathcal{C}\alpha$.

Theorem 6 (Separation theorem for $\Lambda\mu$, [18]). *Let $\Lambda\mu$-terms t_1 and t_2 be closed canonical normal forms. If $t_1 \neq t_2$ in $\Lambda\mu$, then there exists a stream applicative context $\mathcal{C}$ such that $\mathcal{C}[t_1] \to^* \lambda xy.x$ and $\mathcal{C}[t_2] \to^* \lambda xy.y$ hold in $\Lambda\mu$.*

By this theorem, the separation theorem for $\Lambda\mu_{\mathsf{cons}}$ is proved.

Theorem 7 (Separation theorem for $\Lambda\mu_{\mathsf{cons}}$). *Let $\Lambda\mu_{\mathsf{cons}}$-terms t_1 and t_2 be distinct closed normal forms. For any normal u_1 and u_2, there exists a stream applicative context $\mathcal{C}$ such that $\mathcal{C}[t_1] \to^* u_1$ and $\mathcal{C}[t_2] \to^* u_2$ hold in $\Lambda\mu_{\mathsf{cons}}$.*

In [19], Saurin also gave an interpretation of the $\Lambda\mu$-calculus (and its generalization, called stream hierarchy) with a CPS translation into the λ-calculus with surjective pairs, called λSP, and proved the completeness of the CPS translation. The term model induced from λSP is a special case of the stream models with $D = S$, so his result can be seen as the completeness of $\Lambda\mu$ with respect to the stream model.

6.3 Type Assignment for $\Lambda\mu$-Calculus

On the related type systems, more detailed discussion is found in [13].

In the typed $\Lambda\mu_{\mathsf{cons}}$, only functional types from streams to individual data are considered, inspired by the type system of de'Liguoro [8]. However, every typable term in Pagani and Saurin's $\Lambda_{\mathcal{S}}$ [15,21] is also typable in $\Lambda\mu_{\mathsf{cons}}$. Therefore, every typable term in Parigot's propositional typed $\lambda\mu$-calculus [16] is also typable in $\Lambda\mu_{\mathsf{cons}}$.

Here, we show the translation from the $\lambda\mu$-calculus to $\Lambda\mu_{\mathsf{cons}}$, which is based on the same idea of translations in Saurin [21] and van Bakel et al. [24]. They show that their type systems correspond to the image of negative translations from the classical logic to the intuitionistic logic. The following translation corresponds to the continuation-passing-style translation of Thielecke [23].

Definition 10 (Negative translation). We fix a type variable O and an arbitrary stream type θ, and we write $\neg\sigma$ for $\sigma \to O$. The *negative translation* $\overline{(\cdot)}$ from the implicational formulas to term types in $\Lambda\mu_{\mathsf{cons}}$ is defined as

$$\overline{A} = \neg A^{\bullet} \qquad p^{\bullet} = \theta \qquad (A \to B)^{\bullet} = \neg A^{\bullet} \times B^{\bullet}.$$

Proposition 9. *If $\Gamma \vdash t : A; \Delta$ holds in the propositional typed $\lambda\mu$-calculus, then $\neg\Gamma^{\bullet}; \Delta^{\bullet} \vdash t : \neg A^{\bullet}$ holds in $\Lambda\mu_{\mathsf{cons}}$.*

Hence, every typable term in either $\lambda^{\to}$, $\lambda\mu$, or $\Lambda_{\mathcal{S}}$ is typable also in $\Lambda\mu_{\mathsf{cons}}$. On the other hand, due to the recursive stream types, there is a λ-term which is typable in $\Lambda\mu_{\mathsf{cons}}$, and not typable in $\lambda^{\to}$. An example of such terms is $x(y(zw))(y(zww))$, where zw and zww can have the same type in the typed $\Lambda\mu_{\mathsf{cons}}$ under the context $z : [X] \to X, w : X$.

Gaboardi and Saurin [10] proposed another type system $\Lambda_{\mathcal{S}}$ as an extension of the type system in [15,21], equipped with the recursive types and coercion operator from streams to terms, which enables to represent functions returning streams such as cdr. We can define a translation from $\Lambda\mu_{\mathsf{cons}}$ to $\Lambda_{\mathcal{S}}$ preserving typability.

Acknowledgments. The authors would like to thank Kentaro Kikuchi, Shogo Ehara, and the anonymous referees for their helpful comments.

References

1. Amadio, R., Cardelli, L.: Subtyping recursive types. ACM Transactions on Programming Languages and Systems 15(4), 575–631 (1993)
2. Ariola, Z.M., Klop, J.W.: Equational term graph rewriting. Fundamenta Informaticae 26(3-4), 207–240 (1996)
3. Carraro, A.: The untyped stack calculus and Böhm's theorem. In: 7th Workshop on Logical and Semantic Frameworks, with Applications (LSFA 2012). Electric Proceedings in Theoretical Computer Science, vol. 113, pp. 77–92 (2012)
4. Carraro, A., Ehrhard, T., Salibra, A.: The stack calculus. In: 7th Workshop on Logical and Semantic Frameworks, with Applications (LSFA 2012). Electric Proceedings in Theoretical Computer Science, vol. 113, pp. 93–108 (2012)
5. David, R., Py, W.: $\lambda\mu$-calculus and Böhm's theorem. The Journal of Symbolic Logic 66, 407–413 (2001)
6. de Groote, P.: A CPS-translation of the $\lambda\mu$-calculus. In: Tison, S. (ed.) CAAP 1994. LNCS, vol. 787, pp. 85–99. Springer, Heidelberg (1994)
7. Dehornoy, P., van Oostrom, V.Z.: Proving confluence by monotonic single-step upperbound functions. In: Logical Models of Reasoning and Computation, LMRC 2008 (2008)
8. de'Liguoro, U.: The approximation theorem for the $\Lambda\mu$-calculus. Mathematical Structures in Computer Science (to appear)
9. Friedman, H.: Equality between functionals. In: Parikh, R. (ed.) Logic Colloquium, pp. 22–37 (1973)
10. Gaboardi, M., Saurin, A.: A foundational calculus for computing with streams. In: 12th Italian Conference on Theoretical Computer Science (2010)

11. Komori, Y., Matsuda, N., Yamakawa, F.: A simplified proof of the church-rosser theorem. Studia Logica 101(1) (2013)
12. Milner, R.: A complete inference system for a class of regular behaviours. Journal of Computer and System Sciences 28, 439–466 (1984)
13. Nakazawa, K.: Extensional models of typed lambda-mu caclulus (2014) (unpublished manuscript), http://www.fos.kuis.kyoto-u.ac.jp/~knak/papers/manuscript/typed-sm.pdf
14. Nakazawa, K., Katsumata, S.: Extensional models of untyped Lambda-mu calculus. In: Geuvers, H., de'Liguoro, U. (eds.) Proceedings Fourth Workshop on Classical Logic and Computation (CL&C 2012). Electric Proceedings in Theoretical Computer Science, vol. 97, pp. 35–47 (2012)
15. Pagani, M., Saurin, A.: Stream associative nets and $\lambda\mu$-calculus. Research Report 6431, INRIA (2008)
16. Parigot, M.: $\lambda\mu$-calculus: an algorithmic interpretation of classical natural deduction. In: Voronkov, A. (ed.) LPAR 1992. LNCS, vol. 624, pp. 190–201. Springer, Heidelberg (1992)
17. Salomaa, A.: Two complete axiom systems for the algebra of regular events. Journal of the Associations for Computing Machinery 13(1), 158–169 (1966)
18. Saurin, A.: Separation with streams in the $\Lambda\mu$-calculus. In: 20th Annual IEEE Symposium on Logic in Computer Science (LICS 2005), pp. 356–365 (2005)
19. Saurin, A.: A hierarchy for delimited continuations in call-by-name. In: Ong, L. (ed.) FOSSACS 2010. LNCS, vol. 6014, pp. 374–388. Springer, Heidelberg (2010)
20. Saurin, A.: Standardization and Böhm trees for $\Lambda\mu$-calculus. In: Blume, M., Kobayashi, N., Vidal, G. (eds.) FLOPS 2010. LNCS, vol. 6009, pp. 134–149. Springer, Heidelberg (2010)
21. Saurin, A.: Typing streams in the $\Lambda\mu$-calculus. ACM Transactions on Computational Logic 11 (2010)
22. Streicher, T., Reus, B.: Classical logic, continuation semantics and abstract machines. Journal of Functional Programming 8(6), 543–572 (1998)
23. Thielecke, H.: Categorical structure of continuation passing style. PhD thesis, University of Edinburgh (1997)
24. van Bakel, S., Barbanera, F., de'Liguoro, U.: A filter model for the $\lambda\mu$-calculus. In: Ong, L. (ed.) Typed Lambda Calculi and Applications. LNCS, vol. 6690, pp. 213–228. Springer, Heidelberg (2011)

The Structural Theory of Pure Type Systems

Cody Roux and Floris van Doorn

Carnegie Mellon University
Pittsburgh, PA 15213

Abstract. We investigate possible extensions of arbitrary given Pure Type Systems with additional sorts and rules which preserve the normalization property. In particular we identify the following interesting extensions: the disjoint union $\mathcal{P} + \mathcal{Q}$ of two PTSs $\mathcal{P}$ and $\mathcal{Q}$, the PTS $\forall\mathcal{P}.\mathcal{Q}$ which intuitively captures the "$\mathcal{Q}$-logic of $\mathcal{P}$-terms" and $\mathcal{P}_{\text{Poly}}$ which intuitively denotes the predicative polymorphism extension of $\mathcal{P}$.

These results suggest a new approach to the study of the meta-theory of PTSs, by examination of the relationships between different calculi and predicative extensions which allow more expressiveness with equivalent logical strength.

Keywords: Pure Type Systems, Type Theory, weak normalization, conservative extension, predicativity.

1 Introduction

When describing a logical system or, as is equivalent through the Curry-Howard lens, a type system, one often wishes to describe a generic situation, in which one wishes not to describe a single construct, but a *family* of constructs ranging over a set of *parameters*, which themselves are particular to the constructs being defined. This is often referred to as a *schema* in logic, as in the description of the induction rule in the usual presentation of Peano Arithmetic. This can be seen as a *meta-level* quantification: the rule is defined for all possible instance of the quantifier. It is then very natural to ask the following question: "is it possible to reify this meta-level quantification?". The immediate practical advantage to such a reification is that it now has a finite description: the meta-level quantification, which can be seen as an infinite conjunction at the object level, is now encapsulated in a single construct of the theory.

In the case that such a reification is possible, the next natural question is this: "is the resulting theory a conservative extension to the original theory?". This question can be quite tricky, and in general depends on what we mean by "reification". Is the reification of the implicit quantification over propositions in Peano Arithmetic second-order Arithmetic or ACA_0, where comprehension is restricted to first-order formulas? In the first case we have a very powerful extension to arithmetic, whereas in the second case, the extension is conservative, comforting us in the feeling that such an extension does not "add anything" to our logic (in particular, the enriched theory is consistent if and only if the original theory is).

G. Dowek (ed.): RTA-TLCA 2014, LNCS 8560, pp. 364–378, 2014.

Our first contribution is to formalize, in part, a process which allows us to perform such an enrichment. We place ourselves in the framework of Pure Type Systems (PTSs) as described by Barendregt [3]. This framework has the advantage of allowing a very fine and rich account of quantification and dependency. In this framework, there is in general no clear notion of consistency; for this reason we concentrate on normalization/cut elimination, which generally implies consistency in the frameworks in which both concepts exist (by showing that no well-typed normal proof of falsity exists).

The second observation is that there are modular constructions which allow us to combine or extend pure type systems into new systems, and identify certain transformations which preserve weak (and strong) normalization. This suggests a novel approach to describing a logical framework: first identify the components of the framework, e.g. the proof language and the term language, and the relationships between them with respect to quantification. Then use one or several of the combination methods to construct the desired framework. We identify two particularly interesting such constructs: the first takes two PTSs $\mathcal{P}$ and $\mathcal{Q}$, and forms the PTS $\forall\mathcal{P}.\mathcal{Q}$ which informally captures the "$\mathcal{Q}$-logic of $\mathcal{P}$-terms". The second takes a single PTS $\mathcal{P}$ and forms the PTS $\mathcal{P}_{\mathrm{Poly}}$ which adds predicative quantification over every sort of $\mathcal{P}$.

2 Pure Type Systems

Pure type systems are defined as a set of *type assignment systems*, parametrized by the types one is allowed to form. This is prescribed by the dependent function space formation rule, itself entirely described by a triple $(\mathcal{S}, \mathcal{A}, \mathcal{R})$ consisting of a set $s, k \in \mathcal{S}$ of *Sorts*, a set $\mathcal{A} \subseteq \mathcal{S} \times \mathcal{S}$ of *Axioms* and a set $R \subseteq \mathcal{S} \times \mathcal{S} \times \mathcal{S}$ of *Rules*.

We use these rules to assign types to terms. The untyped terms and types have the same syntax, which is given by the BNF

$$t, u, A, B \in \Lambda := s \mid x \mid \lambda x{:}A.\ t \mid t\ u \mid \Pi x{:}A.\ B\,.$$

Conversion is restricted to *β-conversion*: the equivalence relation generated by the contextual closure of the rule

$$(\lambda x : A.\ t)\ u \to_\beta t\{x \mapsto u\}\,.$$

We adopt the usual Barendregt convention for renaming variables in terms and contexts. The typing rules are standard and are given in Figure 1.

Given a PTS $\mathcal{P} = (\mathcal{S}, \mathcal{A}, \mathcal{R})$, we will write $s_1 : s_2$ for axioms (s_1, s_2) and $s_1 \overset{s_3}{\rightsquigarrow} s_2$ to denote rules (s_1, s_2, s_3). We say that t *has sort* s if there are Γ, A such that $\Gamma \vdash t : A$ and $\Gamma \vdash A : s$ in $\mathcal{P}$.

This deceptively simple framework is in fact quite expressive: it is possible to find instances of PTSs that allow the encoding of very expressive logics like higher-order arithmetic or Zermelo set-theory [16]. In general soundness of these logics can be proven by proving normalization of the corresponding PTS.

$$\textbf{leaf}\,\frac{}{\vdash} \qquad \textbf{wf}\,\frac{\Gamma \vdash A : s}{\Gamma, x : A \vdash}\; s \in \mathcal{S}, x \notin \mathrm{dom}(\Gamma)$$

$$\textbf{var}\,\frac{\Gamma, x : A, \Delta \vdash}{\Gamma, x : A, \Delta \vdash x : A} \qquad \textbf{axiom}\,\frac{\Gamma \vdash}{\Gamma \vdash s_1 : s_2}\; (s_1, s_2) \in \mathcal{A}$$

$$\textbf{prod}\,\frac{\Gamma \vdash A : s_1 \qquad \Gamma, x : A \vdash B : s_2}{\Gamma \vdash \Pi x{:}A.\ B : s_3}\; (s_1, s_2, s_3) \in \mathcal{R}$$

$$\textbf{abs}\,\frac{\Gamma, x : A \vdash t : B \qquad \Gamma \vdash \Pi x{:}A.\ B : s}{\Gamma \vdash \lambda x{:}A.\ t : \Pi x{:}A.\ B}\; s \in \mathcal{S}$$

$$\textbf{app}\,\frac{\Gamma \vdash t : \Pi x{:}A.\ B \qquad \Gamma \vdash u : A}{\Gamma \vdash t\ u : B\{x \mapsto u\}}$$

$$\textbf{conv}\,\frac{\Gamma \vdash t : A \qquad \Gamma \vdash A' : s}{\Gamma \vdash t : A'}\; A \simeq_\beta A',\ s \in \mathcal{S}$$

Fig. 1. Typing Rules for PTS

Definition 1. *Let $\mathcal{P}$ be a Pure Type System. A term is* well typed *in $\mathcal{P}$ if there is a context Γ and a type A such that*

$$\Gamma \vdash t : A.$$

The PTS $\mathcal{P}$ is weakly normalizing *(resp.* strongly normalizing*) if every well-typed term t in $\mathcal{P}$ has a normal form (resp. there is no infinite chain of reductions starting with t).*

In the remainder of the article, we use normalizing interchangeably with weakly normalizing.

We can consider a PTS to be fully described simply by the triple $(\mathcal{S}, \mathcal{A}, \mathcal{R})$. Using this fact, the class of PTSs can be seen as a *category* where the morphisms between $\mathcal{P}$ and $\mathcal{Q}$ is the set of functions $\phi\colon \mathcal{S}_\mathcal{P} \to \mathcal{S}_\mathcal{Q}$ such that $\phi(s_1) : \phi(s_2)$ whenever $s_1 : s_2$ and $\phi(s_1) \overset{\phi(s_3)}{\leadsto} \phi(s_2)$ when $s_1 \overset{s_3}{\leadsto} s_2$. Any such function induces a morphism on terms and contexts which we denote ϕ as well. We then have by simple induction, for every morphism of PTSs $\phi : \mathcal{P} \to \mathcal{Q}$ that

$$\Gamma \vdash_\mathcal{P} t : T \quad \Rightarrow \quad \phi(\Gamma) \vdash_\mathcal{Q} \phi(t) : \phi(T)\,.$$

We can now make our first non-trivial remark:

Remark 1. (Morphisms preserve non-normalization). Let $\mathcal{P}, \mathcal{Q}$ be PTSs. If $\mathcal{Q}$ is WN (resp. SN) and if there is a morphism $\phi : \mathcal{P} \to \mathcal{Q}$, then $\mathcal{P}$ is WN (resp. SN).

This can be seen simply by observing that ϕ preserves β-reduction steps:

$$t \to_\beta t' \Leftrightarrow \phi(t) \to_\beta \phi(t')\,.$$

The converse does not hold, since the terminal object in this category is not normalizing.

Now it is interesting, if not terribly useful, to observe that this category inherits rich structure from that of sets: it admits all limits and co-limits! In particular, it admits products and co-products.

It is immediate that if sorts, axioms and rules are simply restricted, then there is an inclusion morphism from the restricted system to the full system.

The terminal object of the category of PTSs is the "Martin-Löf inconsistent type theory" (see Martin-Löf [14]), which we note $* : *$, which has the unique sort $*$, the axiom $* : *$ and the rule $* \overset{*}{\rightsquigarrow} *$, and was shown to be non-normalizing by Girard [10] (see also Hurkens [11]).

Finally the fact that the co-product of PTSs preserves normalization is non-trivial, and is the object of Theorem 1. It is intuitively clear, however, that every term typed in $\mathcal{P}+\mathcal{Q}$ must be well-typed in either $\mathcal{P}$ or $\mathcal{Q}$. We will make this kind of reasoning precise, and extend it to prove the main results of this work. More generally, we show that we may track the rules which give rise to each redex and show that the subterm which contains the redex is either typable in one of the original systems, or obeys certain combinatorial commutation properties which allow such a redex to be safely eliminated.

The results we prove allow extending pure type systems with certain forms of quantifications while preserving normalization. This is the first step of a *structuralist program* to study pure type systems: rather than trying to find properties that are true of each PTS independently of the others, we study some particular pure type systems, like system F, F_ω or the ECC of Luo [13], which have "atomic" complexity and show that the systems we are interested in can be built using known transformations such as those described above. This approach is quite natural in other fields of algebra, as for example how representations of groups can be classified in terms of irreducible representations.

While we believe that this is the first time such a program has explicitly been stated, there are several instances of such an approach being used in the study of the meta-theory of pure type systems: most notably the work of Bernardy and Lasson [4] has served as inspiration for this approach.

However there have been other instances, as for example the work of Peyton-Jones and Meijer [12] who propose a particular PTS (the Calculus of Constructions) as a possible intermediate language for the `Haskell` programming language. Uncomfortable with the power of the impredicative quantification, they then define a predicative variant by duplicating certain sorts and restricting product formation. We argue that our second main theorem (Theorem 3) addresses exactly this step: the addition of predicative quantification over any given sort of the system preserves normalization.

3 Disjoint Union

To introduce the basic lemmas and techniques that will be used in the next section, we first prove that disjoint unions of PTSs (co-products in the PTS category) preserve normalization. To the knowledge of the authors, this observation has never explicitly been stated in the literature.

Theorem 1. *Suppose $\mathcal{P}$ and $\mathcal{Q}$ are two PTSs such that their respective sets of sorts are disjoint. Then the PTS $\mathcal{P}+\mathcal{Q}$ formed by*

$$\mathcal{S}_{\mathcal{P}+\mathcal{Q}} = \mathcal{S}_{\mathcal{P}} \uplus \mathcal{S}_{\mathcal{Q}} \qquad \mathcal{A}_{\mathcal{P}+\mathcal{Q}} = \mathcal{A}_{\mathcal{P}} \uplus \mathcal{A}_{\mathcal{Q}} \qquad \mathcal{R}_{\mathcal{P}+\mathcal{Q}} = \mathcal{R}_{\mathcal{P}} \uplus \mathcal{R}_{\mathcal{Q}}$$

is WN if and only if $\mathcal{P}$ and $\mathcal{Q}$ are WN.

The main difficulty for proving this theorem is to prove that by applying the conversion rule, one cannot move from one PTS to the other. It is possible to prove this directly using subject reduction. However, we prove it using techniques which can more easily generalize to the results in the following sections. We work in a modified presentation of PTSs in which we label well-typed terms with information about which rules and sorts were involved in their construction.

The labeling is similar to a number of labellings used for meta-theoretical studies of pure type systems, see e.g. Melliès and Werner [15], in which they note that it is a crucial device for building models for the Calculus of Constructions.

We want to label each variable $x : A$ with its sort. However, it is not possible in general to attribute a *unique* sort s to A. To circumvent this failure, we refine the classical result on uniqueness of types on functional PTSs in order to characterize the ways in which it may fail in the non-functional case. We define a relation $\sim_\kappa$ on $\mathcal{S}$ that will have the property that if $\Gamma \vdash A : s$ and $\Gamma \vdash A : s'$, then $s \sim_\kappa s'$.

Definition 2. *Given a PTS $(\mathcal{S}, \mathcal{A}, \mathcal{R})$, we define $\sim_\kappa \subseteq \mathcal{S} \times \mathcal{S}$ inductively:*

$$\begin{array}{lllll} & & & & s \sim_\kappa s \\ k : s & \wedge\ k' : s' & \wedge\ k \sim_\kappa k' & & \Rightarrow\ s \sim_\kappa s' \\ k_1 \sim_\kappa k_1' & \wedge\ k_2 \sim_\kappa k_2' & \wedge\ k_1 \overset{s}{\rightsquigarrow} k_2 & \wedge\ k_1' \overset{s'}{\rightsquigarrow} k_2' & \Rightarrow\ s \sim_\kappa s' \,. \end{array}$$

Note that this relation is reflexive and symmetric, but not transitive in general. However, we may easily turn $\sim_\kappa$ into an equivalence relation by taking the transitive closure $\sim_\kappa^*$. This allows us to take the equivalence classes of sorts modulo $\sim_\kappa^*$; the class of a sort s will be denoted $\overline{s}$. Similarly, for rules $r = s_1 \overset{s_3}{\rightsquigarrow} s_2$ we write $\overline{r} = \overline{s_1} \overset{\overline{s_3}}{\rightsquigarrow} \overline{s_2}$. In the rest of this document, we write $\sim$ instead of $\sim_\kappa^*$.

We notice that taking equivalence classes of sorts gives rise to a functional PTS $\mathcal{P}_{\mathsf{fun}}$ defined by

$$\overline{\mathcal{S}} = \left\{\overline{s} \mid s \in \mathcal{S}\right\} \qquad \overline{\mathcal{A}} = \left\{\overline{k} : \overline{s} \mid (k, s) \in \mathcal{A}\right\} \qquad \overline{\mathcal{R}} = \left\{\overline{r} \mid r \in \mathcal{R}\right\}.$$

It is straightforward to verify that this is indeed a functional PTS, and that there is a morphism $\phi : \mathcal{P} \to \mathcal{P}_{\mathsf{fun}}$ that sends s to $\overline{s}$.

Lemma 1. *In every PTS $\mathcal{P}$, if $\Gamma \vdash A : s, s'$, then $s \sim s'$.*
Similarly, if t has both sorts s and s' then $s \sim s'$.

Proof of Lemma 1. We only prove the first statement, the second is proven similarly. Given the morphism ϕ defined above, from the statement

$$\Gamma \vdash A : s, s'$$

we have

$$\phi(\Gamma) \vdash \phi(A) : \phi(s), \phi(s')$$

in $\mathcal{P}_{\text{fun}}$. But in functional PTSs, we have unicity of types modulo β-conversion (Barendregt [3], Lemma 5.2.21), which gives:

$$\phi(s) = \overline{s} \simeq_\beta \overline{s'} = \phi(s)$$

confluence of β-conversion gives $\overline{s} = \overline{s'}$, which is what we needed. □

This observation allows us to give an alternative version of PTSs with rule-labeled abstraction, application and products and sort-labeled variables. This system will allow extraction of sort information by straightforward induction on terms.

Definition 3. *Let $\mathcal{P}$ be a PTS. Define the $\mathcal{P}$-*labeled calculus $\overline{\mathcal{P}}$ *with the* labeled terms

$$t, u, A, B \in \Lambda_{\text{lab}} := s \mid x^{\overline{s}} \mid \lambda^{\overline{r}} x^{\overline{s}} : A.\ t \mid (t\ u)^{\overline{r}} \mid \Pi^{\overline{r}} x^{\overline{s}} : A.\ B$$

where $s \in \mathcal{S}$ and $r \in \mathcal{R}$.

We define the unlabeling $|t|$ *to be the term t in which all sort and rule labels are removed.*

We define the typing judgment $\vdash_{\text{lab}}$ as consisting of (the obvious labeling of) the rules given in Figure 1 with the following modifications:

$$\textbf{wf}\ \frac{\Gamma \vdash_{\text{lab}} A : s}{\Gamma, x^{\overline{s}} : A \vdash_{\text{lab}}}\ s \in \mathcal{S},\ x \notin \text{dom}(\Gamma)$$

$$\textbf{prod}\ \frac{\Gamma \vdash_{\text{lab}} A : s_1 \qquad \Gamma, x^{\overline{s_1}} : A \vdash_{\text{lab}} B : s_2}{\Gamma \vdash_{\text{lab}} \Pi^{\overline{r}} x^{\overline{s_1}} : A.\ B : s_3}\ r = s_1 \overset{s_3}{\rightsquigarrow} s_2 \in \mathcal{R}$$

$$\textbf{abs}\ \frac{\Gamma, x^{\overline{s_1}} : A \vdash_{\text{lab}} t : B \qquad \Gamma \vdash_{\text{lab}} \Pi^{\overline{r}} x : A.\ B : s_3}{\Gamma \vdash_{\text{lab}} \lambda^{\overline{r}} x^{\overline{s_1}} : A.\ t : \Pi^{\overline{r}} x^{\overline{s_1}} : A.\ B}\ r = s_1 \overset{s_3}{\rightsquigarrow} s_2 \in \mathcal{R}$$

$$\textbf{app}\ \frac{\Gamma \vdash_{\text{lab}} t : \Pi^{\overline{r}} x^{\overline{s_1}} : A.\ B \qquad \Gamma \vdash_{\text{lab}} u : A}{\Gamma \vdash_{\text{lab}} (t\ u)^{\overline{r}} : B\{x^{\overline{s_1}} \mapsto u\}}\ r = s_1 \overset{s_3}{\rightsquigarrow} s_2$$

$$\textbf{conv}\ \frac{\Gamma \vdash_{\text{lab}} t : A \qquad \Gamma \vdash_{\text{lab}} A : s_1 \qquad \Gamma \vdash_{\text{lab}} A' : s_2}{\Gamma \vdash_{\text{lab}} t : A'}\ |A| \simeq_\beta |A'|,\ s_1 \sim s_2$$

We remove labels in the conversion to simplify the meta-theory: if we had introduced the rule $((\lambda^{\overline{r}}x : A.\ t)\ u)^{\overline{r}} \to_\beta t\{x \mapsto u\}$ then confluence would fail on ill-typed terms, making the meta-theory more complex, and our completeness result below significantly more difficult to prove.

A *labeling* of an unlabeled term t is a labeled term $\hat{t}$ such that $|\hat{t}| = t$. We extend labeling and unlabeling to contexts in the obvious manner.

In the following section, we fix a given PTS $\mathcal{P}$. The next lemma is immediate by induction on the derivation.

Lemma 2. *Suppose that* $\Gamma \vdash_{\text{lab}} t : A$ *in* $\mathcal{P}$. *Then* $|\Gamma| \vdash |t| : |A|$.

There is a clear characterization of the sort of a well-typed term, that is on the type of its type, or simply its type if that is a top-sort.

Lemma 3. *Suppose that* $\Gamma \vdash_{\text{lab}} t : A$. *Then there is* $s \in \mathcal{S}_\mathcal{P}$ *such that either* $\Gamma \vdash_{\text{lab}} A : s$ *or* $A = s$.

Lemma 4. *The judgment* $\Gamma \vdash t : A$ *is derivable in* $\mathcal{P}$ *if and only if there is a (unique) labeling* $\widehat{\Gamma}, \hat{t}$ *and* $\widehat{A}$ *such that* $\widehat{\Gamma} \vdash_{\text{lab}} \hat{t} : \widehat{A}$ *in* $\overline{\mathcal{P}}$.

Given this theorem, we will often write t for both $|t|$ and $\hat{t}$ indistinguishably for a given well-typed term t, and $\vdash$ instead of $\vdash_{\text{lab}}$. This more explicit type-system allows us to give a straightforward proof of Theorem 1 by induction over the labeled type derivation.

Proof of Theorem 1. Suppose $\mathcal{P} + \mathcal{Q}$ is WN. The inclusions $i_1 : \mathcal{P} \to \mathcal{P} + \mathcal{Q}$ and $i_2 : \mathcal{Q} \to \mathcal{P} + \mathcal{Q}$ are morphisms, which implies that $\mathcal{P}$ and $\mathcal{Q}$ are WN.

Now suppose that $\mathcal{P}$ and $\mathcal{Q}$ are WN and let $\Gamma \vdash_{\mathcal{P}+\mathcal{Q}} t : A$. Then we have $\Gamma \vdash A : s$ or $A = s$. W.l.o.g. we may suppose that $s \in \mathcal{S}_\mathcal{P}$. Let Δ be the subset of Γ with only the type declarations of the form

$$x^{\overline{k}} : B$$

for $k \in \mathcal{S}_\mathcal{P}$. We show by induction on the typing derivation of t in $\overline{\mathcal{P} + \mathcal{Q}}$ that

$$\Delta \vdash_\mathcal{P} t : A$$

is derivable.

Now we can conclude that t is weakly normalizable, as it is typable in $\mathcal{P}$. □

4 The PTS $\forall\mathcal{P}.\mathcal{Q}$

The main result of this section is an extension of the previous one. We wish for not only $\mathcal{P}$ and $\mathcal{Q}$ to coexist, but for $\mathcal{Q}$ types to be built by quantification over $\mathcal{P}$ types.

Theorem 2. *Let* $\mathcal{P}$ *and* $\mathcal{Q}$ *be as in Theorem 1. Let* $\forall\mathcal{P}.\mathcal{Q}$ *be the PTS* $\mathcal{P} + \mathcal{Q}$ *with the additional rules:*

$$\mathcal{I} = \left\{ s \overset{k}{\rightsquigarrow} k \mid s \in \mathcal{S}_\mathcal{P},\ k \in \mathcal{S}_\mathcal{Q} \right\}.$$

Then $\forall\mathcal{P}.\mathcal{Q}$ *is WN iff* $\mathcal{P}$ *and* $\mathcal{Q}$ *are WN.*

Now suppose we call $\mathcal{P}$-types *term(-sets)* and $\mathcal{Q}$-types *propositions.* Then the rules we introduced allow us to build $\mathcal{Q}$-propositions which depend on $\mathcal{P}$-terms.

The proof of Theorem 2 is more involved than that of Theorem 1 as there are non-trivial interactions between the two systems generated by the rules in $\mathcal{I}$. Our proof is directly adapted from Bernardy and Lasson [4].

The idea is to split each term into subterms typable in $\mathcal{P}$ and *erased terms* typable in $\mathcal{Q}$. The "interaction redexes" built by the rules in $\mathcal{I}$ will be handled separately, as they strictly decrease in number after each such β-reduction, and can not duplicate $\mathcal{Q}$-redexes.

This proof bears many similarities with the Geuvers and Nederhof's [9] proof that normalization of system F_ω implies that of the Calculus of Constructions (Barendregt [3] Theorem 5.3.14), which tends to indicate that these proofs are instances of a general approach based on erasure and labeling.

Definition 4 (Erasure). *Let $\mathcal{P} = (\mathcal{S}, \mathcal{A}, \mathcal{R})$ be a PTS and $D \subseteq \mathcal{R}$ (a set of* dependencies*). Suppose in addition that D is closed under $\sim$; that is if $r \in D$ and $\overline{r} = \overline{r'}$ then $r' \in D$. The D-erasure $\lfloor t \rfloor^D$ of a term t is defined by induction on the labeled term:*

$$\begin{array}{rcll}
\lfloor s \rfloor^D & = & s & \\
\lfloor x^{\overline{s}} \rfloor^D & = & x^{\overline{s}} & \\
\lfloor \Pi^{\overline{r}} x^{\overline{s}} : t.\ u \rfloor^D & = & \lfloor u \rfloor^D & \textit{if } r \in D \\
\lfloor \Pi^{\overline{r}} x^{\overline{s}} : t.\ u \rfloor^D & = & \Pi^{\overline{r}} x^{\overline{s}} : \lfloor t \rfloor^D.\ \lfloor u \rfloor^D & \textit{otherwise} \\
\lfloor \lambda^{\overline{r}} x^{\overline{s}} : t.\ u \rfloor^D & = & \lfloor u \rfloor^D & \textit{if } r \in D \\
\lfloor \lambda^{\overline{r}} x^{\overline{s}} : t.\ u \rfloor^D & = & \lambda^{\overline{r}} x^{\overline{s}} : \lfloor t \rfloor^D.\ \lfloor u \rfloor^D & \textit{otherwise} \\
\lfloor (t\ u)^{\overline{r}} \rfloor^D & = & \lfloor t \rfloor^D & \textit{if } r \in D \\
\lfloor (t\ u)^{\overline{r}} \rfloor^D & = & (\lfloor t \rfloor^D\ \lfloor u \rfloor^D)^{\overline{r}} & \textit{otherwise.}
\end{array}$$

We sometimes omit the superscript if it is clear in the context.

Note that in general, the D-erasure of a well-typed term is not well typed, or indeed, even stable by reduction (or variable binding). However in the current case, we have enough structure to guarantee typability of erased terms. In the following section we fix $D = \mathcal{R}_\mathcal{P} \cup \mathcal{I}$.

We want to distinguish "$\mathcal{P}$-terms" from "$\mathcal{Q}$-terms."

Definition 5. *Suppose that $\Gamma \vdash_{\forall \mathcal{P}.\mathcal{Q}} t : A$. We say that t* has a sort *in $\mathcal{P}$ (resp. $\mathcal{Q}$) when there is a sort $s \in \mathcal{S}_\mathcal{P}$ (resp. $\mathcal{S}_\mathcal{Q}$) such that either $\Gamma \vdash_{\forall \mathcal{P}.\mathcal{Q}} A : s$ or $A = s$.*

By Lemma 3 we know that every well-typed term in $\forall \mathcal{P}.\mathcal{Q}$ has a sort in either $\mathcal{P}$ or $\mathcal{Q}$. Using Lemma 1 and Lemma 5 (below) we know that a term cannot have both.

Lemma 5. *Suppose $s \sim s'$ in $\forall \mathcal{P}.\mathcal{Q}$. Then $s, s' \in \mathcal{S}_\mathcal{P}$ or $s, s' \in \mathcal{S}_\mathcal{Q}$.*

We have three kind of redexes in terms, the redexes from rules in $\mathcal{P}$, from rules in $\mathcal{Q}$ and the redexes in $\mathcal{I}$.

Definition 6. *Given a term t well typed in $\forall\mathcal{P}.\mathcal{Q}$, we say that the redex at position p is a $\mathcal{P}$-redex, resp. $\mathcal{Q}$-redex, $\mathcal{I}$-redex if the redex is of shape*

$$((\lambda^{\overline{r}} x^{\overline{s}} : A.t)\ u)^{\overline{r}}$$

with $r \in \mathcal{R}_{\mathcal{P}}$, resp. $\mathcal{R}_{\mathcal{Q}}$, resp. $\mathcal{I}$.

We will sometimes write $t \to_{\mathcal{P}} t'$ if t' is obtained from t by contraction of a $\mathcal{P}$-redex (similarly for $\mathcal{Q}$- and $\mathcal{I}$-redexes).

Furthermore, conversion is preserved by erasure on well-typed terms.

Lemma 6. *Suppose t, t' are well typed in $\forall\mathcal{P}.\mathcal{Q}$ with sort in $\mathcal{Q}$. Then*

$$t \to_{\mathcal{Q}} t' \quad \Rightarrow \quad \lfloor t \rfloor \to_{\beta} \lfloor t' \rfloor\,.$$

Proof. We only treat head reduction: in that case we have $t = ((\lambda^{\overline{r}} x^{\overline{s}} : A.\, t_1)\ t_2)^{\overline{r'}}$. Well typedness gives $\overline{r} = \overline{r'}$. Again we treat the three cases:

1. $r \in \mathcal{R}_{\mathcal{P}}$. This is not possible as t has a sort in $\mathcal{Q}$.
2. $r \in \mathcal{R}_{\mathcal{Q}}$. In this case we have

$$\lfloor t \rfloor = ((\lambda^{\overline{r}} x^{\overline{s}} : \lfloor A \rfloor .\ \lfloor t_1 \rfloor)\ \lfloor t_2 \rfloor)^{\overline{r}} \to_{\beta} \lfloor t_1 \rfloor \{x^{\overline{s}} \mapsto \lfloor t_2 \rfloor\} = \lfloor t_1 \{x^{\overline{s}} \mapsto t_2\} \rfloor$$

   where the last equality is proven by simple induction over the structure of t_1.
3. $r \in \mathcal{I}$. In this case we have $x^{\overline{s}} \notin \mathrm{FV}(\lfloor t_1 \rfloor)$ This gives

$$\lfloor t \rfloor = \lfloor t_1 \rfloor = \lfloor t_1 \rfloor \{x^{\overline{s}} \mapsto \lfloor t_2 \rfloor\} = \lfloor t_1 \{x^{\overline{s}} \mapsto t_2\} \rfloor = \lfloor t' \rfloor\,.$$

□

In particular, due to confluence of β-reduction, we have

$$t \simeq_{\beta} t' \quad \Rightarrow \quad \lfloor t \rfloor \simeq_{\beta} \lfloor t' \rfloor$$

for terms with sort in $\mathcal{Q}$.

Now this allows us to show that well-typed terms in $\forall\mathcal{P}.\mathcal{Q}$ are either well typed in $\mathcal{P}$ or their erasure is well typed in $\mathcal{Q}$.

Proposition 1. *Suppose $\Gamma \vdash_{\forall\mathcal{P}.\mathcal{Q}} t : A$. We have the following:*

1. *If t has a sort in $\mathcal{P}$, then there is a subcontext Δ of Γ such that $\Delta \vdash_{\mathcal{P}} t : A$.*
2. *If t has a sort in $\mathcal{Q}$, then we have $\lfloor \Gamma \rfloor \vdash_{\mathcal{Q}} \lfloor t \rfloor : \lfloor A \rfloor$.*

Proof. First suppose t has a sort in $\mathcal{P}$. Choose Δ to be the declarations $x^k : B$ in Γ with $k \in \mathcal{S}_{\mathcal{P}}$. Then we proceed by induction on the type derivation of the labeled term t. The proof is similar to that of Theorem 1.

Now if t has a sort in $\mathcal{Q}$ we proceed similarly. The only difficult case is conversion, which is handled by appeal to Lemma 6. □

Lemma 7. *If t is well-typed term in $\forall\mathcal{P}.\mathcal{Q}$ then $\to_{\mathcal{I}}$ reductions of t are finite.*

Proof. We will show that the number of $\mathcal{I}$-abstractions will strictly decrease when contracting a $\mathcal{I}$-redex. Let $((\lambda^{\overline{r}} x^{\overline{s}} : T.t)\ u)^{\overline{r}}$ be the redex in question. We have by Inversion and Proposition 1 that

- u is of sort $s \in \mathcal{S}_{\mathcal{P}}$
- u is well typed in $\mathcal{P}$

In particular, u can contain no subterm of the form $\lambda^{\overline{r'}} x^{\overline{s'}} : A.v$ with $r' \in \mathcal{I}$. This means that no $\mathcal{I}$-abstraction can be duplicated. So the number of $\mathcal{I}$-abstractions in t must strictly decrease at each $\mathcal{I}$-reduction step, which implies termination of $\mathcal{I}$-reductions. □

The converse of Lemma 6 is not true in general, as $\mathcal{I}$-redexes can "hide" possible $\mathcal{Q}$-redexes, illustrated in the following example.

Example 1. Consider the following term:

$$t = (\lambda x^{\mathcal{P}} : A.(\lambda y^{\mathcal{Q}} : B.u_1^{\mathcal{Q}})^{\mathcal{R}_{\mathcal{Q}}})^{\mathcal{I}} u_2^{\mathcal{P}} u_3^{\mathcal{Q}}.$$

The sort labels mean that the variable/term belongs to a sort in that set, and the rule annotations means the rule belongs to that set (for clarity we didn't annotate the applications). It is possible to make explicit choices such that t is well-typed. Note that

$$\lfloor t \rfloor = (\lambda y^{\mathcal{Q}} : B.u_1)^{\mathcal{R}_{\mathcal{Q}}} u_3 \ \to_\beta \ u_1\{y \mapsto u_3\}.$$

However, t is in $\to_{\mathcal{Q}}$ normal form. The $\mathcal{Q}$-redex is hidden by the $\mathcal{I}$-redex in t. In contrast, it is not possible for $\mathcal{P}$-redexes to create $\mathcal{Q}$- or $\mathcal{I}$-redexes.

To show that terms typable in $\forall \mathcal{P}.\mathcal{Q}$ have $(\mathcal{Q} \cup \mathcal{I})$-normal forms, we need to lift reductions in the erased domain up to the richer pure type system. The crucial observation is the following:

Lemma 8. *Suppose t is well typed in $\forall \mathcal{P}.\mathcal{Q}$ with a sort in $\mathcal{Q}$ and suppose that $\lfloor t \rfloor \to_\beta v$. Then there exists a term $\tilde{v}$ such that $\lfloor \tilde{v} \rfloor = v$ and*

$$t \to_{\mathcal{I}}^* \to_{\mathcal{Q}} \tilde{v}\,.$$

Proof of Theorem 2. Suppose that t is well typed in $\forall \mathcal{P}.\mathcal{Q}$. If t has a sort in $\mathcal{P}$, then t is well typed in the PTS $\mathcal{P}$ by Proposition 1 and we are done. Otherwise, t has a sort in $\mathcal{Q}$, and we proceed as follows. We will first find a $\to_{\mathcal{QI}}$ normal form. Since $\lfloor t \rfloor$ is typable in $\mathcal{Q}$ (again by Proposition 1) and $\mathcal{Q}$ is weakly normalizing, there exists a $\mathcal{Q}$-normal form t_1 of $\lfloor t \rfloor$. By Lemma 8 we can lift every step of this reduction chain to $\forall \mathcal{P}.\mathcal{Q}$ by adding $\mathcal{I}$-reductions. This way we obtain a lift $\tilde{t}_1$ of t_1 such that $t \to_{\mathcal{QI}}^* \tilde{t}_1$. Lemma 7 tells us that we can find a $\mathcal{I}$ normal form t' of $\tilde{t}_1$. Since contracting $\mathcal{I}$-redexes doesn't change the erasure, we know that $\lfloor t' \rfloor = \lfloor \tilde{t}_1 \rfloor = t_1$. By Lemma 6 we conclude that t' is also in $\mathcal{Q}$ normal form. Hence t' is a $\to_{\mathcal{QI}}$ normal form of t.

Now we prove by induction on terms in $\to_{\mathcal{QI}}$ normal form that they have a $\to_\beta$ normal form. The only interesting case is when the term is an application which is part of a redex. Since the term is in $\to_{\mathcal{QI}}$ normal form, this must be a $\mathcal{P}$-redex. By Proposition 1 this means that the term is well-typed in $\mathcal{P}$, so it has a $\to_{\mathcal{P}}$ normal form. Since contracting $\mathcal{P}$-redexes cannot create $\mathcal{I}$- or $\mathcal{Q}$-redexes, this means the normal form is actually an $\to_\beta$ normal form. This completes the induction, which shows that t', and hence t has a $\to_\beta$ normal form. □

Additionally, Proposition 1 gives us the following logical conservativity result.

Corollary 1. *Suppose A is well typed in $\forall\mathcal{P}.\mathcal{Q}$ of type $s \in \mathcal{S}_{\mathcal{Q}}$. Suppose furthermore that A only contains subterms which have a sorts in $\mathcal{Q}$. Then we have*

$$A \text{ is inhabited in } \forall\mathcal{P}.\mathcal{Q} \quad \text{iff} \quad A \text{ is inhabited in } \mathcal{Q}.$$

5 The PTS $\mathcal{P}_{\text{Poly}}$

In this section, we show that we may extend a PTS $\mathcal{P}$ with quantification of every sort over every other sort, provided the *result* lives in a "fresh" sort. This allows internalizing quantification over free variables: in general if a term t contains a free variable x of sort s, one may instantiate x with any term u of the same type. However it is not in general possible to quantify over x. The following result shows that it is possible to safely form the term $\lambda x : T.\ t$ (if $x : T$) within the theory, by pushing the resulting term into a new sort. This is sometimes referred to as *predicative polymorphism*, *ML-style polymorphism* or *prenex polymorphism*. We feel that it is natural to try to capture such a concept in its most general form.

Additionally, such a practice seems quite useful in general for extensions of type theory with such things as size-types (See Blanqui [5] or Abel [1]) or universes [17] in order to obtain an object theory that naturally allows terms polymorphic in sizes or universes. Note that these particular extensions are impossible in the theory of *pure* type systems, but there is good hope that our approach still applies.

In the following we fix a PTS $\mathcal{P} = (\mathcal{S}, \mathcal{A}, \mathcal{R})$.

Definition 7. *Let ${s_2}^{s_1} \notin \mathcal{S}$, be a new sort for each pair of sorts $s_1, s_2 \in \mathcal{S}$. We define the PTS $\mathcal{P}_{\text{Poly}} = (\mathcal{S}_{\text{Poly}}, \mathcal{A}_{\text{Poly}}, \mathcal{R}_{\text{Poly}})$ by*

$$\mathcal{S}_{\text{Poly}} = \mathcal{S} \cup \left\{ {s_2}^{s_1} \mid s_1, s_2 \in \mathcal{S} \right\}$$

$$\mathcal{A}_{\text{Poly}} = \mathcal{A}$$

$$\mathcal{R}_{\text{Poly}} = \mathcal{R} \cup \left\{ s_1 \overset{{s_2}^{s_1}}{\rightsquigarrow} s_2 \mid s_1, s_2 \in \mathcal{S} \right\} \cup \left\{ s_1 \overset{{s_2}^{s_1}}{\rightsquigarrow} {s_2}^{s_1} \mid s_1, s_2 \in \mathcal{S} \right\}.$$

This construction also preserves normalization, by a similar argument to above.

Theorem 3. *If $\mathcal{P}$ is WN, then so is $\mathcal{P}_{\text{Poly}}$.*

Proof. (Sketch)

We identify, in the same manner as in the proof of Theorem 2, two types of redexes: those coming from a rule in $\mathcal{R}$, and those coming from the new rules, which we call $\mathcal{P}$-redexes and $\mathcal{I}$-redexes respectively.

In the same manner as before, we can show

1. Every $\mathcal{P}$-redex belongs to a subterm typable in $\mathcal{P}$.
2. Reducing a $\mathcal{I}$-redex can not create *any* redexes.

Due to this observation, and subject reduction, we may proceed as previously and reduce every $\mathcal{I}$-redex, then normalize each subterm that contains a $\mathcal{P}$-redex. Both operations normalize by the above observation, and the final term is in normal form.

6 Examples

We may verify that the PTS $\mathcal{P}^2$ given in Bernardi and Lasson [4] is a sub-PTS of $\forall\mathcal{P}.\ \mathcal{P}'$ (where $\mathcal{P}'$ is a renaming of $\mathcal{P}$ to make it disjoint from the latter), and as such, is normalizing if $\mathcal{P}$ is. The only-if direction does not follow immediately, as a sub-PTS of a non-normalizing PTS may be normalizing. In the case of that paper however, it is trivial to verify that it in fact does.

It is also easy to use the predicative polymorphism transformation to turn the simply-typed λ-calculus (STLC) into a calculus with ML-style polymorphism: define STLC to be the PTS defined by

$$\mathcal{S} = \{*, \Box\} \qquad \mathcal{A} = \{* : \Box\} \qquad \mathcal{R} = \{* \overset{*}{\rightsquigarrow} *\}\,.$$

In the PTS $\text{STLC}_{\text{Poly}}$ we can for example form the polymorphic term

$$\text{id} = \lambda X\!:\!*.\ \lambda x\!:\!X.\ x\ :\ \Pi X\!:\!*.\ X \to X\,.$$

By use of the rule $\Box \overset{*^{\Box}}{\rightsquigarrow} *$.

It is amusing to note that there is the rule $* \overset{\Box^*}{\rightsquigarrow} \Box$ in $\text{STLC}_{\text{Poly}}$, which seems to allow for the construction of dependent types. However this ability is quite restricted, in opposition to "true" dependent types as those in the $\lambda\Pi$-calculus or Martin-Löf Type Theory.

Unfortunately, we do not quite have the predicative system described by Peyton-Jones and Meijer [12], as they have the additional rules

$$*^{\Box} \overset{*^{\Box}}{\rightsquigarrow} * \qquad *^{\Box} \overset{*^{\Box}}{\rightsquigarrow} *^{\Box}\,.$$

We do not know whether such rules can be added in a general way to every PTS.

Now let us give a slightly more elaborate example in which we construct a system of interest out of more elementary systems. We wish to build a predicate logic over terms. We therefore consider an elementary PTS with a single sort $*^s$ which represents the universe of basic sets of terms, and a sort $\square^s$ with $*^s : \square^s$ which allows declaring type variables such as

$$\mathrm{Nat} : *^s, \quad \mathrm{Bool} : *^s .$$

We call TERM this PTS with two sorts, one axiom and no rules. It is easy to see that there are no possible λ-abstractions in this PTS, and so every term is trivially normalizing.

As this is a quite poor framework in which to define even first order terms, we add function spaces to be able to declare variables of a function type, e.g. $S : \mathrm{Nat} \to \mathrm{Nat}$. To do this, we add a third sort, $*^f$, to represent function spaces, along with the rules

$$*^s \overset{*^f}{\rightsquigarrow} *^s, \; *^s \overset{*^f}{\rightsquigarrow} *^f .$$

This allows us to declare variables of functional type such as $S : \mathrm{Nat} \to \mathrm{Nat}$ in a well-formed context. But this new PTS, which we call $\mathrm{TERM}_{\mathrm{ext}}$, is just a sub-PTS of $\mathrm{TERM}_{\mathrm{Poly}}$, where $*^f = *^{s*^s}$! Without any additional work, we can therefore conclude, using Theorem 3 that this PTS admits normalization.

Now we wish to reason about such terms using a propositional framework. If we choose that framework to be STLC, then we can simply form the sub-PTS of $\forall\mathrm{TERM}_{\mathrm{ext}}$. STLC obtained by adding the rules

$$*^s \overset{\square}{\rightsquigarrow} \square \quad *^s \overset{*}{\rightsquigarrow} *$$

to obtain a dependently-typed system which captures the $\forall, \Rightarrow$ fragment of (minimal, intuitionistic) first-order logic. This system admits cut elimination by Theorem 2 and normalization of STLC. Such a system was in fact described by Berardi (see Barendregt [3]) by a direct construction, and in the above reference cut elimination is derived by translation into a system with dependent types, rather than our modular approach.

We can also apply Theorem 3 a second time to obtain a system with the additional rule $\square \overset{*^\square}{\rightsquigarrow} *$. In this system, we are now able to express axiom schemas: in the context

$$\Gamma = \mathrm{Nat} : *^s, \; 0 : \mathrm{Nat}, \; S : \mathrm{Nat} \to \mathrm{Nat}$$

we have

$$\Gamma \vdash \Pi P : \mathrm{Nat} \to *. \; P\, 0 \to (\Pi n : \mathrm{Nat}\,.\, P\, n \to P\, (S\, n)) \to \Pi m : \mathrm{Nat}\,.\, P\, m \; : \; *^\square$$

which allows us to build a well-formed context with a variable *ind* of that type. This is possible without fear of losing normalization under the β-rule. Note that this fact does not help much when trying to prove meta-theoretical properties about arithmetic like consistency, which requires more elaborate cut-elimination rules.

One can iterate this construction to get an arbitrary number of "universes". However, since there is no rule $*^\square \overset{*^\square}{\rightsquigarrow} *^\square$, the resulting system is weaker than the usual Martin-Löf type theory with universes.

7 Conclusions

We have presented various operations allowing one to combine or extend arbitrary PTSs, and shown certain of these combinations to preserve normalization.

On the technical side, it is clear that sort-labeling and erasure are powerful techniques for proving properties about reduction in pure type systems, and more investigation is warranted to understand the extent of these techniques. Ideally we would like a general syntactic theorem (which depends only on combinatorial properties of $\mathcal{S}$, $\mathcal{A}$ and $\mathcal{R}$) which captures the extensions to a system (or a set of systems) that can be proven sound with this method.

Natural applications of this approach include the analysis of dependently-typed programming languages. Such languages aim to model programs and proofs using a single framework. However, the construction of a proof language and of a programming language are often at odds, as there are many features of an environment for proofs (impredicativity, normalization, irrelevance) which are not desirable for a programming environment. One approach is to compartmentalize the system into two (or more) universes, along with sometimes complex rules to guide their interaction. In particular the *Trellys* project (see e.g. Sjöberg et al. [18]) aims at exploring the consequences of such distinctions. We believe that our approach may allow a systematic study of these interactions, lightening the burden of meta-theoretical study.

There has been some effort concerning the use of dependently-typed languages to serve as a framework in which to recast, or replay proofs done in different, more complex systems. The language `Dedukti` [8], for instance, has been used to embed proofs coming from Coq [6], and HOL [2] using a suitable encoding. It is a natural question to ask whether the combination of these encodings is still coherent, or more generally under which conditions one can combine such encodings. While in general this question is quite difficult, our Theorem 1, and to a lesser extent Theorem 3 can be seen as a first step in that direction.

All the theorems in this paper can be generalized to hold with strong normalization as well, by adapting the proof to use well-known modularity results in the theory of rewrite systems. We concentrate on weak normalization, as it is sufficient to imply consistency of logical systems based on PTS.

Acknowledgments. We would like to thank Jean-Philippe Bernardy and Marc Lasson for many useful comments and corrections concerning a first draft of this paper, as well as the numerous helpful comments from the anonymous reviewers.

References

1. Abel, A.: A Polymorphic Lambda-Calculus with Sized Higher-Order Types. PhD thesis, Ludwig-Maximilians-Universität München (2006)
2. Assaf, A., Burel, G.: The Holide home page, `https://www.rocq.inria.fr/deducteam/Holide/index.html`

3. Barendregt, H.: Lambda calculi with types. In: Abramsky, S., Gabbay, D., Maibaum, T. (eds.) Handbook of Logic in Computer Science, vol. 2, Oxford University Press (1992)
4. Bernardy, J.-P., Lasson, M.: Realizability and parametricity in pure type systems. In: Hofmann, M. (ed.) FOSSACS 2011. LNCS, vol. 6604, pp. 108–122. Springer, Heidelberg (2011)
5. Blanqui, F.: A type-based termination criterion for dependently-typed higher-order rewrite systems. In: van Oostrom, V. (ed.) RTA 2004. LNCS, vol. 3091, pp. 24–39. Springer, Heidelberg (2004)
6. Boespflug, M., Burel, G.: CoqInE: Translating the calculus of inductive constructions into the lambda pi-calculus modulo. In: Pichardie, D., Weber, T. (eds.) Proceedings of the Second International Workshop on Proof Exchange for Theorem Proving. CEUR Workshop Proceedings, vol. 878. CEUR-WS.org. (2012)
7. Coquand, T.: An analysis of Girard's paradox. In: Proceedings of LICS 1986, pp. 227–236. IEEE Computer Society Press (1986)
8. Deducteam. The Dedukti Reference Manual, version 1.0 (2012)
9. Geuvers, H., Nederhof, M.: A modular proof of strong normalisation for the Calculus of Constructions. Journal of Functional Programming 1(2), 155–189 (1991)
10. Girard, J.-Y.: Interprétation fonctionelle et élimination des coupures dans l'arithmétique d'ordre supérieur. PhD thesis, Université Paris VII (1972)
11. Hurkens, A.: A Simplification of Girard's Paradox. In: Dezani-Ciancaglini, M., Plotkin, G. (eds.) TLCA 1995. LNCS, vol. 902, pp. 266–278. Springer, Heidelberg (1995)
12. Jones, S., Meijer, E.: Henk: a typed intermediate language (1997)
13. Luo, Z.E.: An Extended Calculus of Constructions. In: Proc. of LICS, pp. 385–395 (1990)
14. Martin-Löf, P.: An intuitionistic theory of types. Tech. Rep. TR 72-?, University of Stockholm (1972)
15. Melliès, P.-A., Werner, B.: A generic proof of strong normalisation for pure type systems. In: Giménez, E., Paulin-Mohring, C. (eds.) TYPES 1996. LNCS, vol. 1512, pp. 254–276. Springer, Heidelberg (1998)
16. Miquel, A.: Le Calcul des Constructions implicite: syntaxe et sémantique. PhD thesis, Université Paris 11 (2001)
17. Palmgren, E.: On universes in type theory. In: Twenty-five years of constructive type theory (Venice, 1995). Oxford Logic Guides, vol. 36, pp. 191–204. Oxford Univ. Press, New York (1998)
18. Sjöberg, V., Casinghino, C., Ahn, K.Y., Collins, N., Eades III, H.D., Fu, P., Kimmell, G., Sheard, T., Stump, A., Weirich, S.: Irrelevance, heterogeneous equality, and call-by-value dependent type systems. In: Chapman, J., Levy, P.B. (eds.) MSFP. EPTCS, vol. 76, pp. 112–162 (2012)

Applicative May- and Should-Simulation in the Call-by-Value Lambda Calculus with AMB

Manfred Schmidt-Schauß and David Sabel

Goethe University, Frankfurt, Germany
{schauss,sabel}@ki.informatik.uni-frankfurt.de

Abstract. Motivated by the question whether sound and expressive applicative similarities for program calculi with should-convergence exist, this paper investigates expressive applicative similarities for the untyped call-by-value lambda-calculus extended with McCarthy's ambiguous choice operator amb. Soundness of the applicative similarities w.r.t. contextual equivalence based on may- and should-convergence is proved by adapting Howe's method to should-convergence. As usual for nondeterministic calculi, similarity is not complete w.r.t. contextual equivalence which requires a rather complex counter example as a witness. Also the call-by-value lambda-calculus with the weaker nondeterministic construct erratic choice is analyzed and sound applicative similarities are provided. This justifies the expectation that also for more expressive and call-by-need higher-order calculi there are sound and powerful similarities for should-convergence.

1 Introduction

Our motivation for investigating program equivalences is to show correctness of program optimizations, more generally of program transformations, and also to get more knowledge of program semantics, since the induced equivalence classes can be viewed as the semantics of the program.

A foundational notion of equality of higher-order programs is contextual equivalence, which holds for two expressions s, t, if the evaluation of program $P[s]$ (may-)terminates successfully if and only if the evaluation of program $P[t]$ (may-)terminates successfully, for all programs $P[\cdot]$. Here we denote by $P[t]$ the program P, where the expression s is replaced by t. For concurrent and/or nondeterministic languages, the situation is a bit more complex, since contextual equivalence based only on successful may-termination is too weak, since it ignores paths that lead to errors, nontermination or deadlocks. There are proposals to remedy this weakness by adding another test: either a must-convergence test, where the test is that every possible evaluation is finite; another proposal is should-convergence, where the test only requests that for every (finite) reduction sequence there is always a possible may-termination. Contextual equivalence based on the combination of may- and should-convergence has been used for several extended, nondeterministic lambda calculi *e.g.* [3, 24], for process calculi and

G. Dowek (ed.): RTA-TLCA 2014, LNCS 8560, pp. 379–394, 2014.

algebras [9, 5, 23], and also for concurrent lambda calculi that model real concurrent programming languages *e.g.* Concurrent Haskell, STM Haskell and Alice ML (see [20, 25–27]).

Although contextual equivalence provides a natural notion of program equivalence, proving expressions to be contextually equivalent is usually hard, since all program contexts need to be taken into account. Establishing equivalence proofs is often easier using an applicative (bi)-similarity. For may-convergence, applicative (bi)similarity is the coinductive test consisting of evaluating the expressions to abstractions, applying them to arguments, and showing that the resulting expressions are again applicative (bi)similar.

It is known that applicative (bi)similarities in many (usually deterministic) cases are sound and complete for contextual equivalence (see *e.g.* [1, 7]). On the other hand, there are also some negative results when more expressive and complex languages are considered, *e.g.* applicative similarity (for may-convergence) is unsound in impure lambda calculi with direct storage modifications [17, 30] and also in nondeterministic languages with recursive bindings [29].

While there are several approaches for an applicative similarity for must-convergence (*e.g.* [21, 13, 12, 10]), to the best of our knowledge, no notion of applicative similarity for should-convergence has been studied. So in this paper we will make a first step to close this gap and investigate a notion of applicative similarity for should-convergence.

We choose a rather small calculus for our foundational investigation to not get sidetracked by the syntactic complexity of the calculus. Hence, we investigate the untyped call-by-value lambda calculus extended by the nondeterministic primitive `amb`. We choose McCarthy's `amb`-operator[18], since its implementation requires concurrency: `amb` $s\ t$ can be implemented by executing two concurrent threads – one evaluates s and the other one evaluates t, and the first result obtained from one of the two threads is used as the result for `amb` $s\ t$. Clearly, if both threads return a result, then the program is free to choose one of them. In a concrete implementation this will depend on the scheduling of the threads. Semantically, any (fair) scheduling must be allowed to ensure the correct implementation of `amb`. The operator `amb` is (locally) bottom-avoiding, *i.e.* speaking denotationally where $\bot$ represents diverging programs, `amb` $\bot\ s$ and `amb` $s\ \bot$ are equal to s, and for the case $s \neq \bot \neq t$ the `amb`-operator may freely choose between s and t, *i.e.* then (`amb` $s\ t) \in \{s, t\}$.

The `amb`-operator is also very expressive compared to other nondeterministic operators, *e.g.* using `amb` one can encode an erratic choice which chooses arbitrarily between its arguments, a demonic choice which is the strict variant of erratic choice and requires termination of both of its arguments before choosing between the arguments, and a parallel or. Also semantically, `amb` is challenging, since usual semantic properties do not hold for calculi with `amb`, *e.g.* nonterminating programs are not least elements *w.r.t.* the ordering of contextual semantics. A further reason for analyzing the calculus with `amb` is that it is being studied for several decades (*e.g.* [18, 2, 19, 13, 11, 10, 14]) and for the contextual equivalence with may- and must-convergence it is a long standing open question whether a

Variables: $x, x_i \in \mathcal{V}$
Expressions: $s, t \in Expr_{LCA} ::= x \mid \lambda x.s \mid (s\ t) \mid (\texttt{amb}\ s\ t)$
Values: $v, v_i \in Val ::= \lambda x.s$
Contexts: $C, C_i \in \mathbb{C}_{LCA} ::= [\cdot] \mid \lambda x.C \mid (C\ s) \mid (s\ C) \mid (\texttt{amb}\ C\ s) \mid (\texttt{amb}\ s\ C)$
Evaluation contexts: $E \in \mathbb{E} ::= [\cdot] \mid (E\ s) \mid (v\ E) \mid (\texttt{amb}\ E\ s) \mid (\texttt{amb}\ s\ E)$
Reduction rules:
(cbvbeta) $((\lambda x.s)\ (\lambda y.t)) \to s[(\lambda y.t)/x]$ where $FV(\lambda y.t) \cap BV(\lambda x.s) = \emptyset$
(ambl) $(\texttt{amb}\ (\lambda x.s)\ t) \to (\lambda x.s)$
(ambr) $(\texttt{amb}\ t\ (\lambda x.s)) \to (\lambda x.s)$
Call-by-value reduction: $\dfrac{s \to t, \text{ by (cbvbeta), (ambl) or (ambr)} \quad E \in \mathbb{E}}{E[s] \xrightarrow{LCA} E[t]}$

Fig. 1. Syntax and Operational Semantics of *LCA*

sound applicative similarity exists (see *e.g.* [10]). A negative result is provided by [14], however it requires a typed calculus and the given counterexample is no longer valid if should-convergence is used instead of must-convergence.

Results. Our main theorem (Main Theorem 3.6) states that an expressive applicative similarity is sound for a contextual equivalence defined as a conjunction of may- and should-contextual equivalence, in the untyped call-by-value calculus with amb. The proof is an adaption of Howe's method [7, 8, 22] to should-convergence. We also show that the applicative similarity is not complete *w.r.t.* contextual equivalence by providing a counter-example. We also explore and discuss other possible definitions of applicative similarity and compare them to our definition. Finally, we consider the call-by-value lambda calculus with erratic choice (which is weaker than amb) and show that the coarser applicative similarity for may- and should-convergence (called convex similarity) is sound in the calculus with choice, but unsound in the calculus with amb.

Outline. In Sect. 2 we introduce the call-by-value lambda-calculus with amb, and in Sect. 3 we define the applicative similarities for may- and should-convergence, state our main theorem, and discuss other definition of applicative similarity. The proof of the main theorem is accomplished in Sect. 4. In Sect. 5 we consider the call-by-value calculus with erratic choice and show soundness of applicative similarity for this calculus. We conclude in Sect. 6. For readability, some proofs are omitted, but they can be found in the technical report [28].

2 Call-by-Value AMB Lambda-Calculus

We introduce the call-by-value lambda-calculus with the amb-operator, and define the contextual semantics based on may- and should-convergence.

Let $\mathcal{V}$ be an infinite set of variables. The syntax of expressions and values of the calculus *LCA* is shown in Fig. 1. In $\lambda x.s$ variable x becomes bound in s. With $FV(s)$ ($BV(s)$, resp.) we denote the set of free (bound resp.) variables of expression s, which are defined as usual. If $FV(s) = \emptyset$ then s is called *closed*, otherwise s is an *open* expression. Note that values $v \in Val$ include all abstractions (also open ones). We assume the distinct variable convention to hold, *i.e.* bound names are pairwise distinct and $BV(s) \cap FV(s) = \emptyset$. This convention can always be fulfilled by applying α-renamings. Contexts $C, C_i \in \mathbb{C}_{LCA}$ (see Fig. 1) are expressions where one subexpression is replaced by a hole, denoted with $[\cdot]$. With $C[s]$ we denote the expression where in C the hole is replaced by expression s.

The reduction rules (cbvbeta), (ambl) and (ambr) and the call-by-value small-step reduction $\xrightarrow{LCA}$ are defined in Fig. 1. Call-by-value reduction applies the reduction rules inside call-by-value evaluation contexts $E \in \mathbb{E}$. With $\xrightarrow{LCA,*}$ we denote the reflexive-transitive closure of $\xrightarrow{LCA}$. The reduction is non-deterministic, *i.e.* the arguments of `amb` can be reduced non-deterministically in any sequence, and if one argument is already evaluated to an abstraction, then it is also permitted to project the `amb`-expression to this argument.

Definition 2.1 (May- and Should-Convergence). *If* $s \xrightarrow{LCA,*} \lambda x.s'$ *for some abstraction* $\lambda x.s'$*, then we say* s may-converges *and write* $s\downarrow$*, otherwise* s *is* must-divergent*, denoted as* $s\Uparrow$. *If* $s \xrightarrow{LCA,*} \lambda x.s'$ *then we also write* $s\downarrow\lambda x.s'$.

If for all s' *with* $s \xrightarrow{LCA,*} s'$*, also* $s'\downarrow$ *holds, then we say* s should-converges *and write* $s\Downarrow$*, and otherwise* s may-diverges *(denoted by* $s\uparrow$*). Note that* $s\uparrow$ *iff there is an expression* s'*, such that* $s'\Uparrow$ *and* $s \xrightarrow{LCA,*} s'$.

Definition 2.2 (Contextual Preorder & Equivalence). *For* $\xi \in \{\downarrow, \Downarrow, \uparrow, \Uparrow\}$ *the* contextual ξ-preorder $\leq_\xi$ *and* contextual ξ-equivalence *are defined as*

- $s \leq_\xi t$ *iff for all* $C \in \mathbb{C}_{LCA}$ s.t. $C[s]$ *and* $C[t]$ *are closed:* $C[s]\xi \implies C[t]\xi$.
- $s \sim_\xi t$ *iff* $s \leq_\xi t$ *and* $t \leq_\xi s$.

Contextual preorder $\leq_{LCA}$ *is defined by* $s \leq_{LCA} t$*, iff* $s \leq_\downarrow t$ *and* $s \leq_\Downarrow t$*; and* contextual equivalence $\sim_{LCA}$ *is defined by* $s \sim_{LCA} t$*, iff* $s \sim_\downarrow t$ *and* $s \sim_\Downarrow t$.

Some abbreviations for expressions that we will use in later examples are $\Omega = (\lambda x.(x\ x))\ (\lambda x.(x\ x))$, $Id = \lambda x.x$, $True = \lambda x.\lambda y.x$, $False = \lambda x.\lambda y.y$, $Y = \lambda f.(\lambda x.f\ \lambda z.(x\ x\ z))\ (\lambda x.f\ \lambda z.(x\ x\ z))$, $Top = (Y\ True)$. We will also write $\lambda x_1, x_2, \ldots, x_n.s$ abbreviating nested abstractions $\lambda x_1.\lambda x_2.\ldots\lambda x_n.s$.

The given operational semantics does not take fairness into account, *e.g.* call-by-value reduction may reduce the left argument in $\texttt{amb}\ \Omega\ Id \xrightarrow{LCA} \texttt{amb}\ \Omega\ Id$ infinitely often ignoring the right argument Id. So the bottom-avoidance of the `amb`-operator is not fully captured by our operational semantics. However, the convergence predicates may- and should-convergence and thus also the contextual semantics capture this behavior, *i.e.* if we restrict the allowed reduction sequences to fair ones (*i.e.* no redex is ignored infinitely often in an infinite reduction sequence), then the corresponding predicates for may- and should-convergence are identical to our predicates, *i.e.* should-convergence already has this kind of fairness built-in (see *e.g.* [24]). So our operational semantics is a simplification (which greatly simplifies reasoning), but all of our results also hold for an operational semantics which includes the fairness requirement.

The amb-operator is more expressive than a lot of other nondeterministic operators. E.g., amb can encode *erratic choice* which freely chooses between its two arguments and thus we will use *choice* $s\ t$ as an abbreviation for (amb $(\lambda x.s)$ $(\lambda x.t)$) Id, where x is a fresh variable. Also a *demonic choice* operator *dchoice* is expressible, which requires termination of both of its arguments before choosing between them: *dchoice* $s\ t :=$ (amb $(\lambda x, y.x)$ $(\lambda x, y.y)$) $s\ t$.

Unlike calculi with erratic or demonic choice, in LCA the inequation $s \leq_{\Downarrow} t$ implies $t \leq_{\downarrow} s$, since there is the so-called "bottom-avoiding context" which can be used to test for must-divergence using the should-convergence test. This also implies that contextual equivalence and $\sim_{\Downarrow}$ coincide.

Proposition 2.3. *$\leq_{\Downarrow} \subseteq \leq_{\Uparrow}$ and thus $\leq_{LCA} \subsetneq \sim_{\downarrow}$ as well as $\sim_{LCA} = \sim_{\Downarrow}$.*

Proof. For the context $BA :=$ (amb $((\lambda x.\lambda y.\Omega)\ [\cdot])$ Id) Id and any LCA-expression s the equivalence $BA[s]{\Downarrow} \iff s{\Uparrow}$ holds: if $s{\Uparrow}$, then the amb-expression can only evaluate to its right argument Id, and thus $BA[s]$ is should-convergent in this case. If $s{\downarrow}$, then the reduction sequence $BA[s] \xrightarrow{LCA,*}$ (amb $(\lambda y.\Omega)$ Id) $Id \xrightarrow{LCA,*} \Omega$ shows $BA[s]{\uparrow}$. Now let $s \leq_{\Downarrow} t$ and assume $s \not\leq_{\Uparrow} t$. Then there exists a context C *s.t.* $C[s], C[t]$ are closed and $C[s]{\Uparrow}$ but $C[t]{\downarrow}$. Then $BA[C[s]], BA[C[t]]$ are closed and $BA[C[s]]{\Downarrow}$ and $BA[C[t]]{\uparrow}$, which contradicts $s \leq_{\Downarrow} t$. Thus our assumption was wrong and $s \leq_{\Uparrow} t$ must hold. □

3 Applicative Similarities for LCA

In this section we define applicative similarities for may- and should-convergence in LCA. Then we present our main theorem: the applicative similarities are sound for contextual preorder. We also discuss our definitions and also consider and analyze alternative definitions of similarity. Due to its complexity, the proof of the main theorem is not included in this section, but given in the subsequent section. We use several binary relations on expressions. Sometimes the relations are defined on closed expressions only, and thus we deal with their extensions to open expressions and vice versa with the restrictions to closed expressions:

Definition 3.1. *For a binary relation η on closed LCA-expressions, η^o is the* open value-extension *on LCA: For (open) LCA-expressions s_1, s_2, the relation $s_1\ \eta^o\ s_2$ holds, if for all value-substitutions σ,* i.e. *that replace the free variables in s_1, s_2 by closed abstractions, and where $\sigma(s_1), \sigma(s_2)$ are closed, the relation $\sigma(s_1)\ \eta\ \sigma(s_2)$ holds. Conversely, for a binary relation μ on open expressions, $(\mu)^c$ is its restriction to closed expressions.*

Lemma 3.2. *Let η be a binary relation on closed expressions, and μ be a binary relation on open expressions. Then 1. $((\eta)^o)^c = \eta$, and 2. $s\ \eta^o\ t$ implies $\sigma(s)\ \eta^o\ \sigma(t)$ for any value-substitution σ, and 3. $\mu \subseteq ((\mu)^c)^o$ is equivalent to: $\forall s, t$ and all closing value-substitutions σ: $s\ \mu\ t \implies \sigma(s)\ \mu\ \sigma(t)$*

3.1 Applicative Similarities for May- and Should-Convergence

We define applicative similarity $\preccurlyeq_\downarrow$ for may-convergence and applicative similarity $\preccurlyeq_\uparrow$ for should-convergence (where in fact its negation may-divergence is used). Also mutual similarities and applicative bisimilarities are defined.

Definition 3.3. *We will define operators F_α on binary relations of closed expressions, where α is a name or a mark. The corresponding* similarity, *denoted as $\preccurlyeq_\alpha$ is the greatest fixpoint* gfp(F_α) *of F_α, and the* mutual similarity *is $\approx_\alpha := \preccurlyeq_\alpha \cap \succcurlyeq_\alpha$. If F_α is symmetric, then it is a* bisimilarity, *denoted as $\simeq_\alpha$.*

We always define monotone operators F_α, hence the greatest fixpoints exist. For closed s, t and a binary relation η on closed expressions let $LR(s, t, \eta)$ be the condition: $s{\downarrow}\lambda x.s' \implies \left(\exists \lambda x.t' \text{ with } t{\downarrow}\lambda x.t' \text{ and } s' \; \eta^o \; t'\right)$.

Definition 3.4 (Similarities for LCA). *On closed expressions we define:*

May-Similarity in LCA, $\preccurlyeq_\downarrow := \text{gfp}(F_\downarrow)$: *Let $s \; F_\downarrow(\eta) \; t$ hold iff $LR(s, t, \eta)$.*
Should-Similarity in LCA, $\preccurlyeq_\uparrow := \text{gfp}(F_\uparrow)$:
Let $s \; F_\uparrow(\eta) \; t$ hold iff $s\uparrow \implies t\uparrow$, $t \preccurlyeq_\downarrow s$ and $LR(s, t, \eta)$.
Should-Bisimilarity in LCA, $\simeq_\Downarrow := \text{gfp}(F_\Downarrow)$:
Let $s \; F_\Downarrow(\eta) \; t$ hold iff $s\uparrow \iff t\uparrow$, $LR(s, t, \eta)$, and $LR(t, s, \eta)$.

Since $\text{gfp}(F_\alpha) := \bigcup\{\eta \mid \eta \subseteq F_\alpha(\eta)\}$ by the Knaster-Tarski-Theorem on fixpoints, the following principle of coinduction holds (see *e.g.* [4, 6]):

Proposition 3.5 (Coinduction). *If a relation η on closed expressions is F_α-dense, i.e. $\eta \subseteq F_\alpha(\eta)$, then $\eta \subseteq \preccurlyeq_\alpha$, and also $(\eta)^o \subseteq (\preccurlyeq_\alpha)^o$ holds.*

We now present our main theorem, *i.e.* soundness of may- and should-similarity and also should-bisimilarity. Here we state it for the open extensions of the relations, however it also holds for the relations on closed expressions and the restriction of contextual preorders and equivalence on closed expressions.

Main Theorem 3.6 *The similarities $\preccurlyeq_\downarrow^o$ and $\preccurlyeq_\uparrow^o$ are precongruences, the mutual similarities $\approx_\downarrow^o$, $\approx_\uparrow^o$, and the bisimilarity $\simeq_\Downarrow^o$ are congruences. Moreover, the following soundness results hold:*

1. $\preccurlyeq_\downarrow^o \subset \leq_\downarrow$ *and* $\approx_\downarrow^o \subset \sim_\downarrow$.
2. $\preccurlyeq_\uparrow^o \subset \geq_{LCA}$ *and* $\approx_\uparrow^o \subset \sim_{LCA}$.
3. $\simeq_\Downarrow^o \subseteq \approx_\uparrow^o \subset \sim_{LCA}$.

We prove Main Theorem 3.6 in Sect. 4: the results for may-similarity $\preccurlyeq_\downarrow$ are standard and a sketch is given in Theorem 4.6, the full proof is given in [28, Appendix B]. The results for should-similarity $\preccurlyeq_\uparrow$ are proved in Theorems 4.14 and 4.15. For should-bisimilarity the inclusion $\simeq_\Downarrow^o \subseteq \approx_\uparrow^o$ holds, since $\simeq_\Downarrow$ is $F_\uparrow$-dense. The congruence property for $\simeq_\Downarrow$ requires a separate proof which is in [28, Appendix C]. Strictness of the inclusions will be proved by counter-examples.

3.2 Discussion on Similarities for Should-Convergence

In this section we discuss other variants of should-similarity for LCA. As we show, the first and second are unsound, the third may be a slight generalization, and the status of the fourth is unknown.

Definition 3.7. Naive Should-Similarity in LCA**,** $\preccurlyeq_{\uparrow N} := \mathrm{gfp}(\mathrm{F}_{\uparrow N})$**:**
Let $s\ F_{\uparrow N}(\eta)\ t$ *hold iff* $s\uparrow \implies t\uparrow$ *and* $LR(s,t,\eta)$*.*
Convex Should-Similarity in LCA**,** $\preccurlyeq_{\uparrow X} := \mathrm{gfp}(\mathrm{F}_{\uparrow X})$**:**
Let $s\ F_{\uparrow X}(\eta)\ t$ *hold iff* $s\uparrow \implies t\uparrow$, $t \preccurlyeq_{\downarrow} s$, *and* $t\Downarrow \implies LR(s,t,\eta)$*.*
Should-Similarity in LCA**, variant** $\preccurlyeq_{\uparrow C} := \mathrm{gfp}(\mathrm{F}_{\uparrow C})$**:**
Let $s\ F_{\uparrow C}(\eta)\ t$ *hold iff* $s\uparrow \implies t\uparrow$, $t \leq_{\downarrow} s$, *and* $LR(s,t,\eta)$*.*
Should-Similarity in LCA**, variant** $\preccurlyeq_{\uparrow'} := \mathrm{gfp}(\mathrm{F}_{\uparrow'})$**:**
Let $s\ F_{\uparrow'}(\eta)\ t$ *hold iff* $s\uparrow \implies t\uparrow$, $LR(s,t,\eta)$, *and* $LR(t,s,\eta^{-1})$*.*

Obviously, (*choice False True*) $\not\leq_{\uparrow}$ *True* using the context ([$\cdot$] *Id* Ω). This suggests the naive should-similarity $\preccurlyeq_{\uparrow N}$ which, however, is insufficient:

Lemma 3.8. $\preccurlyeq_{\uparrow N}$ *is unsound* w.r.t. $\leq_{\uparrow}$.

Proof. While $Id \preccurlyeq_{\uparrow N} \lambda x.choice\ x\ Id$ holds, we have $(Y\ (\lambda x.choice\ x\ Id)\ Id)\Downarrow$, but $(Y\ Id\ Id)\Uparrow$. Thus $\preccurlyeq_{\uparrow N}$ is not a precongruence and not sound *w.r.t.* $\leq_{\uparrow}$. □

In the definition of $\preccurlyeq_{\uparrow}$ this is the reason for the additional condition $t \preccurlyeq_{\downarrow} s$ inside $F_{\uparrow}$ (which in fact implies $s \approx_{\downarrow} t$, since $\preccurlyeq_{\uparrow} \subset \preccurlyeq_{\downarrow}$). Further generalizing the definition of $\preccurlyeq_{\uparrow}$ by requiring the recursive test to hold only if the right expression is should-convergent leads to the convex should-similarity, $\preccurlyeq_{\uparrow X}$, which is analogous to the definition of so-called (unsound) "convex similarity" in [19] for a call-by-name lambda-calculus with `amb`, but using must-convergence instead of should-convergence. However, also for LCA the similarity $\preccurlyeq_{\uparrow X}$ is unsound:

Lemma 3.9. $\preccurlyeq_{\uparrow X}$ *is unsound* w.r.t. $\leq_{\uparrow}$.

Proof. Let $s_1 :=$ `amb` $(\lambda x.\Omega)\ (\lambda x,y,z.\Omega)$ and $s_2 :=$ `amb` $s_1\ (\lambda x,y.\Omega)$. Then $s_2 \preccurlyeq_{\uparrow X} s_1$, but $s_2 \not\leq_{\uparrow} s_1$, since for the context $C :=$ (`amb` ([$\cdot$] `Id`) `Id`) `Id` we have $C[s_2] \xrightarrow{LCA,*} \Omega$ and thus $C[s_2]\uparrow$, but $C[s_1]\Downarrow$.

For calculi with only erratic or demonic choice, $\preccurlyeq_{\uparrow X}$ is sound (see Sect. 5).

A further generalization of the successful similarity $\preccurlyeq_{\uparrow}$ by replacing the $t \preccurlyeq_{\downarrow} s$ condition by $t \leq_{\downarrow} s$ leads to $\preccurlyeq_{\uparrow C}$, for which it is easy to see that $\preccurlyeq_{\uparrow} \subseteq \preccurlyeq_{\uparrow C}$, and we conjecture that it is sound, but a soundness proof would require at least a ciu-Lemma for LCA. As another strengthening of the conditions inside $F_{\uparrow N}$ we added the condition $LR(t,s,\eta^{-1})$ resulting in the should-similarity $\preccurlyeq_{\uparrow'}$ We did neither find a soundness proof for $\preccurlyeq_{\uparrow'}$, since the condition $\forall t \downarrow \lambda x.t' \exists s \downarrow \lambda x.s'$ is inappropriate for Howe's method, nor did we find a counter-example showing unsoundness, so we leave soundness of $\preccurlyeq_{\uparrow'}$ as an open question.
Our results imply that the following properties hold for $\preccurlyeq_{\uparrow'}$:

Lemma 3.10. $\preccurlyeq_{\uparrow'}\ \subseteq\ \preccurlyeq_{\downarrow}\ \subseteq\ \leq_{\downarrow}$ *and* $\simeq_{\Downarrow}\ \subseteq\ \approx_{\uparrow'}\ \subseteq\ \approx_{\downarrow}\ \subseteq\ \sim_{\downarrow}$.

Proof. The first chain of inclusions is valid, since $\preccurlyeq_{\uparrow'}$ is $F_{\downarrow}$-dense, *i.e.* $\preccurlyeq_{\uparrow'} \subseteq F_{\downarrow}(\preccurlyeq_{\uparrow'})$, and since $\preccurlyeq_{\downarrow}$ is sound for $\leq_{\downarrow}$ (Main Theorem 3.6). In the second chain, the inclusion $\simeq_{\Downarrow} \subseteq \approx_{\uparrow'}$ holds, since $\simeq_{\Downarrow} \subseteq F_{\uparrow'}(\simeq_{\Downarrow})$ and since $\simeq_{\Downarrow}$ is symmetric. The remaining inclusions follow from the first chain. □

4 Soundness Proofs for Similarity in *LCA*

4.1 Preliminaries on Howe's Method

In this section we will introduce the necessary notions to apply Howe' method for the soundness proofs of similarities *w.r.t.* contextual preorder and contextual equivalence in *LCA*. Here we employ higher order abstract syntax as *e.g.* in [7] for the proof and write $\tau(..)$ for an expression with top operator τ, which may be λ, application, or `amb`. For consistency of terminology and treatment with that in other papers such as [7], we assume that removing the top constructor λx in relations is done after a renaming. For example, $\lambda x.s\ \mu\ \lambda y.t$ is renamed to the same bound variable before further reasoning about s, t, to $\lambda z.s[z/x]\ \mu\ \lambda z.t[z/y]$ for a fresh variable z. A relation μ is *operator-respecting,* iff $s_i\ \mu\ t_i$ for $i = 1, \ldots, n$ implies $\tau(s_1, \ldots, s_n)\ \mu\ \tau(t_1, \ldots, t_n)$.In these preliminaries for Howe's method we assume that there is a preorder $\preccurlyeq$, which is a reflexive and transitive relation on closed expressions. The goal is to show that $\preccurlyeq$ is a precongruence. We then define the *Howe candidate relation* $\preccurlyeq_H$ and show its properties. Later $\preccurlyeq$ is instantiated by the may- or should-similarity or by the should-bisimilarity.

Definition 4.1. *Given a reflexive and transitive relation $\preccurlyeq$ on closed expressions, the* Howe (precongruence candidate) relation *$\preccurlyeq_H$ is a binary relation on open expressions defined inductively on the structure of the left hand expression:*

1. *If $x \preccurlyeq^o s$ then $x \preccurlyeq_H s$.*
2. *If there are expressions s, s_i, s_i'* s.t. *$\tau(s_1', \ldots, s_n') \preccurlyeq^o s$ with $s_i \preccurlyeq_H s_i'$ for $i = 1, \ldots, n$, then $\tau(s_1, \ldots, s_n) \preccurlyeq_H s$.*

Lemma 4.2. *We have $x \preccurlyeq_H s$ iff $x \preccurlyeq^o s$; and $\tau(s_1, \ldots, s_n) \preccurlyeq_H s$ iff there is some expression $\tau(s_1', \ldots, s_n') \preccurlyeq^o s$ such that $s_i \preccurlyeq_H s_i'$ for $i = 1, \ldots, n$.*

Lemma 4.3. *The following properties are proved in [28, Appendix A]:*

1. *$\preccurlyeq_H$ is reflexive.*
2. *$\preccurlyeq_H$ and $(\preccurlyeq_H)^c$ are operator-respecting.*
3. *$\preccurlyeq^o \subseteq \preccurlyeq_H$ and $\preccurlyeq \subseteq (\preccurlyeq_H)^c$.*
4. *$\preccurlyeq_H \circ \preccurlyeq^o \subseteq \preccurlyeq_H$.*
5. *$(v \preccurlyeq_H v' \wedge t \preccurlyeq_H t') \implies t[v/x] \preccurlyeq_H t'[v'/x]$ for values v, v'.*
6. *$s \preccurlyeq_H t$ implies that $\sigma(s) \preccurlyeq_H \sigma(t)$ for every value-substitution σ.*
7. *$\preccurlyeq_H \subseteq ((\preccurlyeq_H)^c)^o$.*
8. *If $(\preccurlyeq_H)^c = \preccurlyeq$, then $\preccurlyeq_H = \preccurlyeq^o$.*

9. *If s, t are closed, $s = \tau(s_1, \ldots, s_{\mathrm{ar}(\tau)})$ and $s \preccurlyeq_H t$ holds, then there are s'_i, such that $\tau(s'_1, \ldots, s'_{\mathrm{ar}(\tau)})$ is closed, $\forall i : s_i \preccurlyeq_H s'_i$ and $\tau(s'_1, \ldots, s'_{\mathrm{ar}(\tau)}) \preccurlyeq t$.*

As a general outline, the goal of Howe's method is to show that $\preccurlyeq_H \;=\; \preccurlyeq^o$, which implies that $\preccurlyeq^o$ is operator-respecting and hence it is a precongruence.

Lemma 4.4. *The relations $\preccurlyeq_\alpha$, $\preccurlyeq^o_\alpha$ from Definition 3.4 are reflexive and transitive. The relations $\simeq_\Downarrow$, and $\simeq^o_\Downarrow$ are equivalence relations.*

Lemma 4.5. $s \preccurlyeq^o_\alpha t \iff \lambda x.s \preccurlyeq^o_\alpha \lambda x.t$.

4.2 Soundness of May-Similarity

Theorem 4.6. *May-similarity behaves as expected: The similarity $\preccurlyeq_\downarrow$ for may-convergence is a precongruence on closed expressions and sound for $\leq^c_\downarrow$. Extending this on all expressions: $\preccurlyeq^o_\downarrow$ is a precongruence and sound for $\leq_\downarrow$.*

Proof (Sketch, see [28, Appendix B]). Use Howe's method. Define $\preccurlyeq_{\downarrow H}$ as an extension of $\preccurlyeq_\downarrow$ using Definition 4.1. Then show that $\preccurlyeq^c_{\downarrow H}$ satisfies the fixpoint conditions for $\preccurlyeq_\downarrow$, which implies $\preccurlyeq^c_{\downarrow H} \subseteq \preccurlyeq_\downarrow$, and so $\preccurlyeq^c_{\downarrow H} = \preccurlyeq_\downarrow$, which implies the precongruence property, and $\preccurlyeq_{\downarrow H} = \preccurlyeq^o$. □

Corollary 4.7. *The mutual similarity $\approx_\downarrow$ is a congruence and sound for $\sim^c_\downarrow$. Also $\approx^o_\downarrow$ is a congruence and sound for $\sim_\downarrow$.*

But note that $\approx_\downarrow$ is not complete using a similar example as in [15]:

Proposition 4.8. $\approx^o_\downarrow \;\neq\; \sim_\downarrow$

Proof. With $F = \lambda f.\lambda z.choice\ (\lambda x.\Omega)\ ((\lambda x_1, x_2.x_1)\ (f\ z))$ one can verify that $Y\ F\ Id$ reduces to $\lambda x_1, \ldots, x_n.\Omega$ for any $n \geq 1$. Using a context lemma for LCA, one can show that $Y\ F\ Id \sim_\downarrow Top$. However, $Top \not\preccurlyeq_\downarrow Y\ F\ Id$, since after evaluating Top to $\lambda z.(True\ Top\ z) = v_1$, we have to choose a value $\lambda x_1, \ldots, x_n.\Omega = v_2$ of $(Y\ F\ Id)$ for a fixed number n, and applying v_1 to n arguments converges, but the application of v_2 to n arguments diverges. □

4.3 Soundness of Should-Similarity

In this section we present a proof for soundness of should-similarity, *i.e.* $\preccurlyeq^o_\uparrow \;\subseteq\; \leq_{LCA}$. We first show some properties of $\preccurlyeq_\uparrow$:

Lemma 4.9. $\preccurlyeq_\uparrow \;\subseteq\; \approx_\downarrow \;\subseteq\; \sim_\downarrow$ *and* $\simeq_\Downarrow \;\subseteq\; \approx_\uparrow \;\subseteq\; \sim_\downarrow$.

Proof. The first inclusion holds, since $\preccurlyeq_\uparrow \subseteq \succcurlyeq_\downarrow$ by definition, $\preccurlyeq_\uparrow \subseteq \preccurlyeq_\downarrow$ (since $\preccurlyeq_\uparrow$ is $F_\downarrow$-dense), and $\preccurlyeq_\downarrow \subseteq \leq_\downarrow$ by Theorem 4.6. In the second chain, the inclusion $\simeq_\Downarrow \;\subseteq\; \approx_\uparrow$ holds, since $\simeq_\Downarrow$ satisfies all the conditions of $F_\uparrow$, and since $\simeq_\Downarrow$ is symmetric. The remaining inclusion follows from the first chain.

The goal in the following is to show that the candidate relation $\preccurlyeq_{\uparrow H}$ derived from $\preccurlyeq_{\uparrow}$ can be treated using Howe's method to prove its soundness. Our proof relies on the precongruence property of $\preccurlyeq_{\downarrow}^{o}$ (which is already proved in Theorem 4.6) for the transfer of may-divergence over the candidate relation.

Definition 4.10. *The candidate relation* $\preccurlyeq_{\uparrow H}$ *is defined* w.r.t. *the relation* $\preccurlyeq_{\uparrow}$.

Lemma 4.11. $\preccurlyeq_{\uparrow H} \subseteq \approx_{\downarrow}^{o}$.

Proof. To show that $s \preccurlyeq_{\uparrow H} t \implies s \approx_{\downarrow}^{o} t$, we use induction on the structure of s. In the case $s = x$ the definition of the candidate implies $x \preccurlyeq_{\uparrow}^{o} t$, which implies $x \approx_{\downarrow}^{o} t$ by Lemma 4.9. If $s = \tau(s_1, \ldots, s_n)$, there is some $\tau(t_1, \ldots, t_n) \preccurlyeq_{\uparrow}^{o} t$ with $s_i \preccurlyeq_{\uparrow H} t_i$ for all i. The induction hypothesis implies $s_i \approx_{\downarrow}^{o} t_i$ for all i, and the congruence property of $\approx_{\downarrow}^{o}$ shows $\tau(s_1, \ldots, s_n) \approx_{\downarrow}^{o} \tau(t_1, \ldots, t_n)$. Transitivity of $\approx_{\downarrow}^{o}$ and $\preccurlyeq_{\uparrow}^{o} \subseteq \approx_{\downarrow}^{o}$ now shows $s = \tau(s_1, \ldots, s_n) \approx_{\downarrow}^{o} t$. □

Proposition 4.12. *Let s, t be closed expressions, $s \preccurlyeq_{\uparrow H} t$ and $s{\downarrow}\lambda x.s'$. Then there is some $\lambda x.t'$ such that $t{\downarrow}\lambda x.t'$ and $s' \preccurlyeq_{\uparrow H} t'$.*

Proof. The proof is by induction on the length of the reduction of $s{\downarrow}\lambda x.s'$.

- If $s = \lambda x.s'$, then there is some closed $\lambda x.t'$ with $s' \preccurlyeq_{\uparrow H} t'$ and $\lambda x.t' \preccurlyeq_{\uparrow} t$. The latter implies that there is some closed $\lambda x.t''$ with $t{\downarrow}\lambda x.t''$ and $t' \preccurlyeq_{\uparrow}^{o} t''$, and so $s' \preccurlyeq_{\uparrow H} t''$ by Lemma 4.3 (4).
- Case $s = \texttt{amb}\ s_1\ s_2$, and $s{\downarrow}\lambda x.s'$. Then there is some closed expression $\texttt{amb}\ t_1\ t_2 \preccurlyeq_{\uparrow} t$ with $s_i \preccurlyeq_{\uparrow H} t_i$ for $i = 1, 2$. W.l.o.g. let $s_1{\downarrow}\lambda x.s'$. Then by induction, there is some $\lambda x.t'$ with $t_1{\downarrow}\lambda x.t'$ and $s' \preccurlyeq_{\uparrow H} t'$. Obviously, also $\texttt{amb}\ t_1\ t_2{\downarrow}\lambda x.t'$. From $\texttt{amb}\ t_1\ t_2 \preccurlyeq_{\uparrow} t$, we obtain that there is some $\lambda x.t''$ with $t{\downarrow}\lambda x.t''$ and $t' \preccurlyeq_{\uparrow}^{o} t''$, which implies $s' \preccurlyeq_{\uparrow H} t''$ by Lemma 4.3 (4).
- If $s = (s_1\ s_2)$, then there is some closed $t' = (t_1'\ t_2') \preccurlyeq_{\uparrow} t$ with $s_i \preccurlyeq_{\uparrow H} t_i'$ for $i = 1, 2$. Since $(s_1\ s_2){\downarrow}\lambda x.s'$ there is a reduction sequence $(s_1\ s_2) \xrightarrow{LCA,*} (\lambda x.s_1')\ s_2 \xrightarrow{LCA,*} (\lambda x.s_1')\ (\lambda x.s_2') \xrightarrow{LCA} s_1'[\lambda x.s_2'/x] \xrightarrow{LCA,*} \lambda x.s'$ such that $s_i{\downarrow}\lambda x.s_i'$ for $i = 1, 2$. By induction, there are expressions $\lambda x.t_i''$ with $t_i'{\downarrow}\lambda x.t_i''$ and $s_i' \preccurlyeq_{\uparrow H} t_i''$. Lemma 4.3 (5) now shows $s_1'[\lambda x.s_2'/x] \preccurlyeq_{\uparrow H} t_1''[\lambda x.t_2''/x]$. Now we can again use the induction hypothesis which shows that there is some $\lambda x.t''$ with $t_1''[\lambda x.t_2''/x]{\downarrow}\lambda x.t''$ and $s' \preccurlyeq_{\uparrow H} t''$. The relation $(t_1'\ t_2') \preccurlyeq_{\uparrow} t$ implies that $t{\downarrow}\lambda x.t_0$ with $t'' \preccurlyeq_{\uparrow}^{o} t_0$, and hence $s' \preccurlyeq_{\uparrow H} t_0$ by Lemma 4.3 (4). □

Proposition 4.13. *Let s, t be closed expressions, $s \preccurlyeq_{\uparrow H} t$ and $s{\uparrow}$. Then $t{\uparrow}$.*

Proof. The proof is by induction on the number of reductions of s to a must-divergent expression, and on the size of expressions as a second measure.

- The base case is that $s{\Uparrow}$. Then Lemma 4.11 shows $t{\Uparrow}$.
- Let $s = \texttt{amb}\ s_1\ s_2$ with $s{\uparrow}$. Then there is some closed expression $t' = \texttt{amb}\ t_1\ t_2$ with $s_i \preccurlyeq_{\uparrow H} t_i$ for $i = 1, 2$ and $\texttt{amb}\ t_1\ t_2 \preccurlyeq_{\uparrow} t$. It follows that $s_1{\uparrow}$ as well as $s_2{\uparrow}$. Applying the induction hypothesis shows that $t_1{\uparrow}$ as well as $t_2{\uparrow}$, and hence $(\texttt{amb}\ t_1\ t_2){\uparrow}$. From $\texttt{amb}\ t_1\ t_2 \preccurlyeq_{\uparrow} t$ we obtain $t{\uparrow}$.

– Let $s = (s_1\ s_2)$ with $s\uparrow$. Then there is some closed expression $t' = (t_1\ t_2) \preccurlyeq_\uparrow t$ and $s_i \preccurlyeq_{\uparrow H} t_i$ for $i = 1, 2$. There are several cases:
   1. If $(s_1\ s_2) \xrightarrow{LCA,*} (s'_1\ s_2)$ and $s'_1\Uparrow$, then $s_1\uparrow$ and by the induction hypothesis also $t_1\uparrow$, and hence $t'\uparrow$, which implies $t\uparrow$.
   2. If $(s_1\ s_2) \xrightarrow{LCA,*} (\lambda x.s'_1)\ s_2 \xrightarrow{LCA,*} (\lambda x.s'_1)\ s'_2$ and $s'_2\Uparrow$, then $s_2\uparrow$ and by induction hypothesis also $t_2\uparrow$, and hence $t'\uparrow$, which implies $t\uparrow$.
   3. If $(s_1\ s_2) \xrightarrow{LCA,*} (\lambda x.s'_1)\ s_2 \xrightarrow{LCA,*} (\lambda x.s'_1)\ (\lambda x.s'_2) \xrightarrow{LCA} s'_1[\lambda x.s'_2/x] \xrightarrow{LCA,*} s_0$ where $s_0\Uparrow$. Then $s_i\downarrow\lambda x.s'_i$ for $i = 1, 2$ and by Proposition 4.12 there are reductions $t_i\downarrow\lambda x.t'_i$ for $i = 1, 2$ with $s'_i \preccurlyeq_{\uparrow H} t'_i$. Thus $s'_1[\lambda x.s'_2/x] \preccurlyeq_{\uparrow H} t'_1[\lambda x.t'_2/x]$, and hence by the induction hypothesis $t'_1[\lambda x.t'_2/x]\uparrow$. Thus $(t_1\ t_2)\uparrow$, and now $(t_1\ t_2) \preccurlyeq_\uparrow t$ implies $t\uparrow$. □

Theorem 4.14. *The relation $\preccurlyeq_\uparrow$ is a precongruence on closed expressions and $\preccurlyeq_\uparrow^o$ is a precongruence on all expressions.*

Proof. We have $\preccurlyeq_\uparrow \subseteq \preccurlyeq_{\uparrow H}^c$ by Lemma 4.3 (3). Since $\preccurlyeq_{\uparrow H}^c$ satisfies the fixpoint conditions of $\preccurlyeq_\uparrow$ (using Propositions 4.12 and 4.13), coinduction shows that $\preccurlyeq_{\uparrow H}^c \subseteq \preccurlyeq_\uparrow$. Hence, $\preccurlyeq_{\uparrow H}^c = \preccurlyeq_\uparrow$ and also $\preccurlyeq_{\uparrow H} = \preccurlyeq_\uparrow^o$.

Theorem 4.15. *$\preccurlyeq_\uparrow^o$ is sound for $\geq_{LCA}$.*

Proof. Let $s \preccurlyeq_\uparrow^o t$, and let C be a context such that $C[s], C[t]$ are closed. First assume that $C[s]\uparrow$. Theorem 4.14 shows that $C[s] \preccurlyeq_\uparrow^o C[t]$, and so $C[t]\uparrow$. Lemma 4.9 and Theorem 4.6. imply $C[s]\downarrow \iff C[t]\downarrow$. Hence $s \geq_{LCA} t$. □

Theorem 4.16. *The similarity $\preccurlyeq_\uparrow$ is incomplete for $\geq_\Downarrow$.*

Proof. We give a counterexample (details are in [28]): Let A = choice Ω $(\lambda x.A)$, $B_0 = Top$, $B_{i+1} = \lambda x.$choice Ω B_i; and B = choice Ω (choice B_0 (choice B_1 ...)). Then $Top \simeq_\downarrow A \simeq_\downarrow B_i$ for all i and $Top \simeq_\downarrow B$. Also $B_i <_\uparrow A$ for all i. Using a context lemma for closed expressions it can be shown that $A \sim_{LCA} B$. It is easy to see that $B \preccurlyeq_\uparrow A$, but $A \not\preccurlyeq_\uparrow B$. □

Comparing s, t for $\leq_\uparrow$, the incompleteness of $\preccurlyeq_\uparrow$ cannot appear if t reduces to only finitely many abstractions.

Proposition 4.17. *Assume that s is a closed abstraction and t is a closed expression such that $s \leq_\uparrow t$ and there is a nonempty set $T := \{t_1, \ldots, t_n\}$ of closed abstractions, such that $t\downarrow\lambda x.t'$ implies $\lambda x.t' \in T$. Then there is some i with $s \leq_\uparrow t_i$.*

Proof. Suppose this is false. Then there are contexts $C_1, \ldots, C_n$, such that $C_i[s], C_i[t_i]$ are closed for all i, and for all $i = 1, \ldots, n$: $C_i[s]\uparrow$ and $C_i[t_i]\Downarrow$. The context $C = (\lambda x.\texttt{amb}\ C_1[x]\ (\texttt{amb}\ \ldots (\texttt{amb}\ C_{n-1}[x]\ C_n[x])))\ [\cdot]$ has the property: $C[s]\uparrow$, but $C[t]\Downarrow$, which is a contradiction.

Soundness of the applicative similarities implies:

Proposition 4.18. *Let s, t be closed expressions, such that for all $\lambda x.s'$: $s{\downarrow}\lambda x.s' \iff t{\downarrow}\lambda x.s'$ (the same results modulo alpha-equivalence), and $s{\uparrow} \iff t{\uparrow}$, then $s \approx_{\uparrow} t$, and hence also $s \sim_{LCA} t$.*
If s, t are open expressions, such that for all value substitutions σ, such that $\sigma(s), \sigma(t)$ are closed: $\sigma(s){\downarrow}\lambda x.s' \iff \sigma(t){\downarrow}\lambda x.s'$ (modulo alpha-equivalence), and $\sigma(s){\uparrow} \iff \sigma(t){\uparrow}$, then $s \approx_{\uparrow}^{o} t$, and hence also $s \sim_{LCA} t$.

Corollary 4.19. *Several identities obviously hold in LCA:*

$$(\lambda x.s)\ (\lambda x.t) \sim_{LCA} s[\lambda x.t/x] \qquad (\texttt{amb}\ \Omega\ s) \sim_{LCA} s \qquad (\texttt{amb}\ s\ s) \sim_{LCA} s$$
$$(\texttt{amb}\ s\ t) \sim_{LCA} (\texttt{amb}\ t\ s) \qquad \texttt{amb}\ s_1\ (\texttt{amb}\ s_2\ s_3) \sim_{LCA} \texttt{amb}\ (\texttt{amb}\ s_1\ s_2)\ s_3$$

An example that is a bit more complex is:

Example 4.20. Let $F = \lambda f.\lambda x.\texttt{amb}\ x\ (f\ x)$. We show that $Y\ F \sim Id$ using similarities. It is easy to see that for all closed abstrations r: $Id\ r{\downarrow}r$ and also $(Y\ F\ r){\downarrow}r' \implies r = r'$. Note that $(Y\ F\ r)$ has arbitrary long successful reduction sequences to r. We also have $(Id\ r) \Downarrow$ as well as $(Y\ F\ r) \Downarrow$. The simulation definitions imply $Id \simeq (Y\ F)$, and hence $Id \sim (Y\ F)$.

5 Simulations for the Call-by-Value Choice Calculus

Even though amb can simulate choice in different variants, if only (erratic or demonic) choice is permitted instead of amb, then the expressivity is different, which is reflected in different contextual equivalences. For example Ω is the smallest element if only choice is permitted, which is false in LCA. In this section we consider erratic choice only, since demonic and erratic choice can encode each other in a call-by-value calculus.

Definition 5.1 (The calculus LCC). *The calculus LCC is defined analogous to LCA with the following differences:*

- *Instead of* amb *the syntax has a binary operator* choice.
- *The hole of evaluation contexts is not inside arguments of* choice.
- *The reduction rules are (cbvbeta) and choice-reductions:*
  (choicel): $(\texttt{choice}\ s\ t) \to s$; *and (choicer)*: $(\texttt{choice}\ s\ t) \to t$.
- *Reduction* $\xrightarrow{LCC}$ *applies the reduction rules in evaluation contexts.*
- *The definitions of contextual equivalences are as for LCA.*

The general properties on similarities and the candidate relation presented in Sect. 4.1 also hold for LCC. We immediately start with the similarity definitions and use the convex variant. In abuse of notation, we use the same symbols for the relations as for LCA.

Definition 5.2. *We define simulations for LCC on closed expressions:*

May-Similarity in LCC, $\preccurlyeq_{\downarrow} := \mathrm{gfp}(F_{\downarrow})$: *Let $s\ F_{\downarrow}(\eta)\ t$ hold iff $LR(s, t, \eta)$.*

Should-Similarity in LCC, $\preccurlyeq_{\uparrow x} := \mathrm{gfp}(F_{\uparrow x})$:

Let $s\ F_{\uparrow x}(\eta)\ t$ *hold iff* $s\uparrow \implies t\uparrow$, $t \preccurlyeq_{\downarrow} s$, *and* $t\Downarrow \implies LR(s,t,\eta)$.

Doing the same using Howe's method for $\preccurlyeq_{\uparrow x}$ as for LCA shows:

Theorem 5.3. *May-similarity* $\preccurlyeq_{\downarrow}$ *in* LCC *is a precongruence and sound for the contextual may-preorder, and the mutual may-similarity* $\approx_{\downarrow}$ *is a congruence and sound for may-equivalence.*

Definition 5.4. *The candidate relation* $\preccurlyeq_{\uparrow x H}$ *is defined* w.r.t. *the relation* $\preccurlyeq_{\uparrow x}$.

Lemma 5.5. $\preccurlyeq_{\uparrow x H} \subseteq \succcurlyeq_{\downarrow}^{o}$.

Mostly, the proofs are the same as for LCA. So we only exhibit the differences.

Proposition 5.6. *Let* s,t *be closed* LCC*-expressions,* $s \preccurlyeq_{\uparrow x H} t$, $t\Downarrow$, $s{\downarrow}\lambda x.s'$. *Then there is some* $\lambda x.t'$ *such that* $t{\downarrow}\lambda x.t'$ *and* $s' \preccurlyeq_{\uparrow x H} t'$.

Proof. We work in the calculus LCC. The proof is by induction on the length of the reduction of $s{\downarrow}\lambda x.s'$. There are three cases: $s = \lambda x.s'$, $s = (\mathtt{choice}\ s_1\ s_2)$ and $s = (s_1\ s_2)$, where the first and third cases are the same as for LCA. So we only show the case for the choice-expression:
Case $s = \mathtt{choice}\ s_1\ s_2$, and $s{\downarrow}\lambda x.s'$. Then there is some closed expression $\mathtt{choice}\ t_1\ t_2 \preccurlyeq_{\uparrow x}\ t$ with $s_i \preccurlyeq_{\uparrow x H} t_i$ for $i = 1,2$. Note that $t\Downarrow$ implies $t_1\Downarrow$ and $t_2\Downarrow$. W.l.o.g. let $s_1{\downarrow}\lambda x.s'$. Then by induction, there is some $\lambda x.t'$ with $t_1{\downarrow}\lambda x.t'$ and $s' \preccurlyeq_{\uparrow x H} t'$. Obviously, also $\mathtt{choice}\ t_1\ t_2{\downarrow}\lambda x.t'$. From $\mathtt{choice}\ t_1\ t_2 \preccurlyeq_{\uparrow x}\ t$ and $t\Downarrow$, we obtain that there is some $\lambda x.t''$ with $t{\downarrow}\lambda x.t''$ and $t' \preccurlyeq_{\uparrow x}^{o} t''$, which implies $s' \preccurlyeq_{\uparrow x H} t''$ by Lemma 4.3 (4). □

Note that in the calculus LCA this proof fails, since the induction hypothesis cannot be proved for s_i, t_i.

Proposition 5.7. *Let* s,t *be closed expressions,* $s \preccurlyeq_{\uparrow x H} t$, *and* $s\uparrow$. *Then* $t\uparrow$.

Proof. The proof is by induction on the number of reductions of s to a must-divergent expression, and on the size of expressions as a second measure.
The base case is that $s\Uparrow$. Then Lemma 5.5 shows $t\Uparrow$, since $t \preccurlyeq_{\downarrow} s$ must hold, which implies $s \geq_{\downarrow} t$ and thus $s \leq_{\Uparrow} t$.
Let $s = \mathtt{choice}\ s_1\ s_2$ with $s\uparrow$, and assume that $t\Downarrow$. Then there is some closed expression $t' = \mathtt{choice}\ t_1\ t_2$ with $s_i \preccurlyeq_{\uparrow x H} t_i$ for $i = 1,2$ and $\mathtt{choice}\ t_1\ t_2 \preccurlyeq_{\uparrow x} t$. This implies $t_1\Downarrow$ and $t_2\Downarrow$. It follows that $s_1\uparrow$ or $s_2\uparrow$. Applying the induction hypothesis shows that $t_1\uparrow$ or $t_2\uparrow$, which contradicts the assumption $t\Downarrow$.

Theorem 5.8. *The relation* $\preccurlyeq_{\uparrow x}$ *in* LCC *is a precongruence on closed expressions and* $\preccurlyeq_{\uparrow x}^{o}$ *is a precongruence on all expressions.*

Proof. We already have $\preccurlyeq_{\uparrow x} \subseteq \preccurlyeq_{\uparrow x H}^{c}$ by Lemma 4.3 (3). Propositions 5.6 and 5.7 show that $(\preccurlyeq_{\uparrow x H})^c$ satisfies the fixpoint conditions of $\preccurlyeq_{\uparrow x}$ and thus coinduction shows $(\preccurlyeq_{\uparrow x H})^c \subseteq \preccurlyeq_{\uparrow x}$. Hence we have $(\preccurlyeq_{\uparrow x H})^c = \preccurlyeq_{\uparrow x}$. Lemma 4.3.(8) then shows the equation $\preccurlyeq_{\uparrow x H} = \preccurlyeq_{\uparrow x}^{o}$. □

Theorem 5.9. *$\preccurlyeq^o_{\uparrow X}$ is sound for $\geq_{LCC}$, and $\approx^o_{\uparrow X}$ is sound for $\sim_{LCC}$.*

Proof. We first show that $\preccurlyeq^o_{\uparrow X}$ is sound for $\leq_{\uparrow,LCC}$ (and thus also for $\geq_{\Downarrow,LCC}$): Let $s \preccurlyeq^o_{\uparrow X} t$, and let C be a context such that $C[s], C[t]$ are closed. First assume that $C[s]\uparrow$. Theorem 5.8 shows that $C[s] \preccurlyeq_{\uparrow X} C[t]$, and so $C[t]\uparrow$. Since $s \preccurlyeq^o_{\uparrow X} t$ also implies $t \preccurlyeq^o_{\downarrow} s$ and thus $t \leq_{\downarrow,LCC} s$, we have $\preccurlyeq^o_{\uparrow X} \subseteq \geq_{LCC}$. The second part of the theorem follows by symmetry. □

Proposition 5.10. *Let s,t be closed with $s\uparrow, t\uparrow$. Then $s \approx_{\downarrow} t \implies s \sim_{LCC} t$.*

Proof. First note that $\Omega \leq_{LCC} r$ for all r, which follows from Theorems 5.3 and 5.9. Theorem 5.9 shows that $s \approx_{\downarrow} t$, $s\uparrow, t\uparrow$ implies that $s \sim_{LCC} t$.

Note that this proposition is not valid in *LCA*.

Proposition 5.11. *Convex should-simulation $\preccurlyeq_{\uparrow X}$ is not complete for $\leq_{\uparrow,LCC}$.*

Proof. Let $s = \texttt{choice}\ \Omega\ (\lambda x.\Omega)$ and $t = \texttt{choice}\ \Omega\ \mathit{Top}$. Then $s \leq_{\uparrow,LCC} t$, as well as $t \leq_{\uparrow,LCC} s$ holds, since for every context C, if $C[s]\uparrow$, then also $C[t]\uparrow$ by selecting always the Ω in a choice-reduction, and also vice versa. However, $t \not\preccurlyeq_{\downarrow} s$ (since $\mathit{Top} \not\preccurlyeq_{\downarrow} \lambda x.\Omega$), and thus $s \preccurlyeq_{\uparrow X} t$ does not hold. □

6 Conclusion

We have shown that in the call-by-value lambda calculus with amb there exists a very expressive (an argument for this is Proposition 4.17) mutual similarity for should-convergence, which is a congruence and sound for contextual equivalence. We also showed that the used method can be transferred to the call-by-value lambda calculus with choice. This novel and encouraging result may enable further research for more expressive non-deterministic and/or concurrent calculi and languages and for call-by-need lambda calculi using the approximation techniques from *e.g.* [15, 16].

Acknowledgements We thank the anonymous reviewers for their valuable comments.

References

1. Abramsky, S.: The lazy lambda calculus. In: Turner, D.A. (ed.) Research Topics in Functional Programming, pp. 65–116. Addison-Wesley (1990)
2. Broy, M.: A theory for nondeterminism, parallelism, communication, and concurrency. Theoret. Comput. Sci. 45, 1–61 (1986)
3. Carayol, A., Hirschkoff, D., Sangiorgi, D.: On the representation of McCarthy's amb in the pi-calculus. Theoret. Comput. Sci. 330(3), 439–473 (2005)
4. Davey, B., Priestley, H.: Introduction to Lattices and Order. Cambridge University Press, Cambridge (1992)
5. Fournet, C., Gonthier, G.: A hierarchy of equivalences for asynchronous calculi. J. Log. Algebr. Program. 63(1), 131–173 (2005)

6. Gordon, A.D.: Bisimilarity as a theory of functional programming. Theoret. Comput. Sci. 228(1-2), 5–47 (1999)
7. Howe, D.: Equality in lazy computation systems. In: 4th IEEE Symp. on Logic in Computer Science, pp. 198–203 (1989)
8. Howe, D.: Proving congruence of bisimulation in functional programming languages. Inform. and Comput. 124(2), 103–112 (1996)
9. Laneve, C.: On testing equivalence: May and must testing in the join-calculus. Technical Report Technical Report UBLCS 96-04, University of Bologna (1996)
10. Lassen, S.B.: Normal form simulation for McCarthy's amb. Electr. Notes Theor. Comput. Sci. 155, 445–465 (2006)
11. Lassen, S.B., Moran, A.: Unique fixed point induction for McCarthy's amb. In: Kutyłowski, M., Wierzbicki, T., Pacholski, L. (eds.) MFCS 1999. LNCS, vol. 1672, pp. 198–208. Springer, Heidelberg (1999)
12. Lassen, S.B., Pitcher, C.S.: Similarity and bisimilarity for countable nondeterminism and higher-order functions. Electron. Notes Theor. Comput. Sci. 10 (2000)
13. Lassen, S.B.: Relational Reasoning about Functions and Nondeterminism. PhD thesis, University of Aarhus (1998)
14. Levy, P.B.: Amb breaks well-pointedness, ground amb doesn't. Electron. Notes Theor. Comput. Sci. 173(1), 221–239 (2007)
15. Mann, M.: Congruence of bisimulation in a non-deterministic call-by-need lambda calculus. Electron. Notes Theor. Comput. Sci. 128(1), 81–101 (2005)
16. Mann, M., Schmidt-Schauß, M.: Similarity implies equivalence in a class of non-deterministic call-by-need lambda calculi. Inform. and Comput. 208(3), 276–291 (2010)
17. Mason, I., Talcott, C.L.: Equivalence in functional languages with effects. J. Funct. Programming 1(3), 287–327 (1991)
18. McCarthy, J.: A Basis for a Mathematical Theory of Computation. In: Braffort, P., Hirschberg, D. (eds.) Computer Programming and Formal Systems, pp. 33–70. North-Holland, Amsterdam (1963)
19. Moran, A.K.D.: Call-by-name, call-by-need, and McCarthy's Amb. PhD thesis, Chalmers University, Sweden (1998)
20. Niehren, J., Sabel, D., Schmidt-Schauß, M., Schwinghammer, J.: Observational semantics for a concurrent lambda calculus with reference cells and futures. Electron. Notes Theor. Comput. Sci. 173, 313–337 (2007)
21. Ong, C.H.L.: Non-determinism in a functional setting. In: Vardi, M.Y. (ed.) Proc. 8th IEEE Symposium on Logic in Computer Science, pp. 275–286. IEEE Computer Society Press (1993)
22. Pitts, A.M.: Howe's method for higher-order languages. In: Sangiorgi, D., Rutten, J. (eds.) Advanced Topics in Bisimulation and Coinduction. Cambridge Tracts in Theoretical Computer Science, ch. 5, pp. 197–232. Cambridge University Press (2011)
23. Rensink, A., Vogler, W.: Fair testing. Inform. and Comput. 205(2), 125–198 (2007)
24. Sabel, D., Schmidt-Schauß, M.: A call-by-need lambda-calculus with locally bottom-avoiding choice: Context lemma and correctness of transformations. Math. Structures Comput. Sci. 18(3), 501–553 (2008)
25. Sabel, D., Schmidt-Schauß, M.: A contextual semantics for Concurrent Haskell with futures. In: Schneider-Kamp, P., Hanus, M. (eds.) Proc. 13th International ACM SIGPLAN Conference on Principles and Practice of Declarative Programming, pp. 101–112. ACM (2011)

26. Sabel, D., Schmidt-Schauß, M.: Conservative concurrency in Haskell. In: Dershowitz, N. (ed.) Proc. 27th ACM/IEEE Symposium on Logic in Computer Science (LICS 2012), pp. 561–570. IEEE Computer Society (2012)
27. Schmidt-Schauß, M., Sabel, D.: Correctness of an STM Haskell implementation. In: Morrisett, G., Uustalu, T. (eds.) Proc. 18th ACM SIGPLAN International Conference on Functional Programming, pp. 161–172. ACM (2013)
28. Schmidt-Schauß, M., Sabel, D.: Applicative may- and should-simulation in the call-by-value lambda calculus with amb. Frank report 54. Inst. f. Informatik, Goethe-University, Frankfurt (2014), `http://www.ki.informatik.uni-frankfurt.de/papers/frank/frank-54.pdf`
29. Schmidt-Schauß, M., Sabel, D., Machkasova, E.: Counterexamples to applicative simulation and extensionality in non-deterministic call-by-need lambda-calculi with letrec. Inf. Process. Lett. 111(14), 711–716 (2011)
30. Koutavas, V., Levy, P.B., Sumii, E.: Limitations of applicative bisimulation. In: Modelling, Controlling and Reasoning about State. Dagstuhl Seminar Proceedings, vol. 10351 (2010)

Implicational Relevance Logic is 2-ExpTime-Complete*

Sylvain Schmitz

ENS Cachan & INRIA, France

Abstract. We show that provability in the implicational fragment of relevance logic is complete for doubly exponential time, using reductions to and from coverability in branching vector addition systems.

Keywords: Relevance logic, branching VASS, focusing proofs, complexity.

1 Introduction

Relevance logic **R** [1, 7] provides a formalisation of 'relevant' implication: in such a system, the formula $A \to B$ indicates that the truth of A is actually useful in establishing B; an example of an *irrelevant* implication valid in classical logic would be $B \to (A \to B)$.

The pure implicational fragment $\mathbf{R}_{\to}$ of **R** was developed independently by Moh [15] in 1950 and Church [4] in 1951, and is as such the oldest of the relevance logics. Kripke already presented in 1959 a decision algorithm for provability in $\mathbf{R}_{\to}$ [10], which was later extended to larger and larger subsets of **R**, like the conjunctive-implicational fragment $\mathbf{R}_{\to,\wedge}$. Several negative results by Urquhart would however foil any hope for elementary algorithms: first in 1984 when he showed the undecidability of the full logic **R** [20]; later in 1999 with a proof that $\mathbf{R}_{\to,\wedge}$ suffers from a non primitive-recursive complexity: it is ACKERMANN-complete [22]. This left a gigantic gap for the implicational fragment $\mathbf{R}_{\to}$, between an earlier EXPSPACE lower bound [21] and the ACKERMANN upper bound shared by the variants of Kripke's procedure.

In this paper, we close this gap and show that provability in $\mathbf{R}_{\to}$ is 2-EXPTIME-complete. Our proof relies crucially on a recent result by Demri et al. [6], who show the 2-EXPTIME-completeness of the coverability problem in *branching vector addition systems with states* (BVASS). These systems form a natural generalisation of vector addition systems, and have been defined independently in a variety of contexts (see the survey [18] and Sec. 3 below), notably that of provability in *multiplicative exponential linear logic* (**MELL**, see [9]). More precisely:

- In Sec. 4, we show that so-called *expansive* BVASSs can simulate proofs in $\mathbf{R}_{\to}$ in a natural manner by exploiting the subformula property of its usual sequent calculus $L\mathbf{R}_{\to}$. We then show how to reduce reachability in

* Work supported by the ReacHard project, ANR grant 11-BS02-001-01.

G. Dowek (ed.): RTA-TLCA 2014, LNCS 8560, pp. 395–409, 2014.

expansive BVASS to coverability, thereby providing a decision procedure in doubly exponential time.
- The matching hardness proof in Sec. 5 relies on the one hand on *comprehensive* instances of the BVASS coverability problem, and on the other hand on a new *focusing* sequent calculus $F\mathbf{R}_{\rightarrow}$ for $\mathbf{R}_{\rightarrow}$.
- The reduction from $\mathbf{R}_{\rightarrow}$ provability to expansive BVASS reachability is actually a special case of a more general reduction proved in [11] for the multiplicative exponential fragment of intuitionistic contractive linear logic (**IMELLC**), i.e. **IMELL** with structural contraction. Our reduction in Sec. 4.3 from expansive reachability to coverability thus entails that **IMELLC** provability is 2-ExpTime-complete, as explained in Sec. 6.

Due to space constraints, some material is omitted and can be found in the full paper available from http://arxiv.org/abs/1402.0705.

Let us first recall the formal definition of $\mathbf{R}_{\rightarrow}$ before turning to that of BVASSs in Sec. 3.

2 The Implicational Fragment $\mathbf{R}_{\rightarrow}$

The reader will find in [7, Sec. 4] a nice overview of the decision problem for **R**, covering in particular Kripke's solution for $\mathbf{R}_{\rightarrow}$ [10] and Urquhart's lower bound argument for $\mathbf{R}_{\rightarrow,\wedge}$ [22].

2.1 A Sequent Calculus

We recall here the formal definition of $\mathbf{R}_{\rightarrow}$ as a sequent calculus $L\mathbf{R}_{\rightarrow}$ in Gentzen's style. Let $\mathcal{A}$ be a countable set of atomic propositions; we define the set of formulæ as following the abstract syntax

$$A ::= a \mid A \rightarrow A \qquad \text{(implicational formulæ)}$$

where a ranges over $\mathcal{A}$. We consider $\rightarrow$ to be right-associative, e.g. $A \rightarrow B \rightarrow C$ denotes $A \rightarrow (B \rightarrow C)$. In the following rules, we use $A, B, C, \ldots$ to denote implicational formulæ and $\Gamma, \Delta, \ldots$ to denote multisets of such formulæ; commas in e.g. 'Γ, A' and 'Γ, Δ' denote multiset unions of Γ with the singleton A and with Δ respectively; finally, a *sequent* is a pair '$\Gamma \vdash A$' stating that the succedent A is valid assuming the antecedent Γ to be relevant:

$$\frac{}{A \vdash A}\ (\mathsf{Id}) \qquad \frac{\Gamma, A, A \vdash B}{\Gamma, A \vdash B}\ (\mathsf{C})$$

$$\frac{\Gamma \vdash A \quad \Delta, B \vdash C}{\Gamma, \Delta, A \rightarrow B \vdash C}\ (\rightarrow_{\mathsf{L}}) \qquad \frac{\Gamma, A \vdash B}{\Gamma \vdash A \rightarrow B}\ (\rightarrow_{\mathsf{R}})$$

As we work with multisets, this sequent calculus includes implicitly the structural 'exchange' rule. It does however not feature the classical 'weakening' rule—which would defeat the very point of relevance—nor the 'cut' rule—which is admissible. A visible consequence of this definition is that the calculus enjoys the *subformula property*: all the formulæ in rule premises are subformulæ of the formulæ appearing in the corresponding consequences.

2.2 Decidability and Complexity

With hindsight, the decision procedure of Kripke [10] for provability in the implicational fragment of relevance logic can be seen as a precursor for many later algorithms that rely on the existence of a *well quasi ordering* (wqo) for their termination [17]. This decision procedure can be understood as an application of Dickson's Lemma to prove the finiteness of 'irredundant' proof trees for the target sequent $\vdash A$. Furthermore, combinatorial analyses of Dickson's Lemma as e.g. in [8] provide explicit upper bounds on the size of those irredundant proofs, in the form of the Ackermann function in the size of A, yielding an ACKERMANN upper bound for $\mathbf{R}_{\rightarrow}$ provability, as shown by Urquhart [22].

Regarding lower bounds, Urquhart in [21, Sec. 9] explains how to derive EXPSPACE-hardness for $\mathbf{R}_{\rightarrow}$, using model-theoretic techniques to reduce from the word problem for finitely presented commutative semigroups [14].

2.3 Strict λ-Calculus

The implicational fragment $\mathbf{R}_{\rightarrow}$ is in bijection with the typing rules of the simply typed λI-calculus, where abstracted terms $\lambda x.t$ are well-formed only if x appears free in t; see [5, Sec. 9F]. This means that $\mathbf{R}_{\rightarrow}$ provability can be restated as the *type inhabitation* problem for the simply typed λI-calculus. Our complexity results should then be contrasted with the PSPACE-completeness of the same problem for the simply typed λ-calculus [19].

3 Branching VASS

Branching vector addition systems with states (BVASS) have been independently defined in several contexts; see [18] for a survey.

3.1 Formal Definitions

Given d in $\mathbb{N}$, we write '$\bar{0}$' for the null vector in $\mathbb{N}^d$, and for $0 < i \leq d$, '$\bar{e}_i$' for the unit vector in $\mathbb{N}^d$ with 1 on coordinate i and 0 everywhere else. Let $U_d \stackrel{\text{def}}{=} \{\bar{e}_i, -\bar{e}_i \mid 0 < i \leq d\}$. Syntactically, an *ordinary BVASS* is a tuple $\mathcal{B} = \langle Q, d, T_u, T_s \rangle$ where Q is a finite set of *states*, d is a *dimension* in $\mathbb{N}$, and $T_u \subseteq Q \times U_d \times Q$ and $T_s \subseteq Q^3$ are respectively finite sets of *unary* and *split* rules. We denote unary rules $(q, \bar{u}, q_1)$ in T_u with $\bar{u}$ in U_d by '$q \xrightarrow{\bar{u}} q_1$' and split rules (q, q_1, q_2) in T_s by '$q \rightarrow q_1 + q_2$'.

We define the semantics of an ordinary BVASS through a deduction system over *configurations* $(q, \bar{v})$ in $Q \times \mathbb{N}^d$:

$$\frac{q, \bar{v}}{q_1, \bar{v} + \bar{e}_i}\,(\mathsf{incr}) \qquad \frac{q, \bar{v} + \bar{e}_i}{q_1, \bar{v}}\,(\mathsf{decr}) \qquad \frac{q, \bar{v}_1 + \bar{v}_2}{q_1, \bar{v}_1 \quad q_2, \bar{v}_2}\,(\mathsf{split})$$

respectively for unary rules $q \xrightarrow{\bar{e}_i} q_1$ and $q \xrightarrow{-\bar{e}_i} q_1$ in T_u and a split rule $q \rightarrow q_1 + q_2$ in T_s; in (split) '+' denotes component-wise addition in $\mathbb{N}^d$. Such a deduction system can be employed either *top-down* or *bottom-up* depending on the decision problem at hand (as with tree automata); the top-down direction will correspond in a natural way to goal-directed *proof search* in the sequent calculus of Sec. 2.

Ordinary BVASSs are a slight restriction over BVASSs, which would in general allow any vector in $\mathbb{Z}^d$ in unary rules. Because they often lead to more readable proofs, we only employ ordinary BVASSs in this paper. This is at no loss of generality, since one can build an ordinary BVASS 'equivalent' to a given BVASS in logarithmic space, where equivalence should be understood relative to the reachability and coverability problems; see [18] for details.

Reachability. Branching VASSs are associated with a natural decision problem: *reachability* asks, given a BVASS $\mathcal{B}$, a root state q_r, and a leaf state q_ℓ, whether there exists a deduction tree with root label $(q_r, \bar{0})$ and every leaf labelled $(q_\ell, \bar{0})$; such a deduction tree is called a *reachability witness*. De Groote et al. [9] have shown that this problem is recursively equivalent to **MELL** provability, and it is currently unknown whether it is decidable—both problems are however known to be of non-elementary computational complexity [11].

Let us introduce some additional notation that will be handy in proofs. We write '$\mathcal{B}, T, q_\ell \triangleright q, \bar{v}$' if there exists a deduction tree of $\mathcal{B}$ with root label $(q, \bar{v})$ and leaves labelled by $(q_\ell, \bar{0})$, which uses each rule in $T \subseteq T_u \uplus T_s$ at least once. Such *root judgements* can be derived through the deduction system

$$\frac{}{\mathcal{B}, \emptyset, q_\ell \triangleright q_\ell, \bar{0}} \qquad \frac{\mathcal{B}, T, q_\ell \triangleright q_1, \bar{v} + \bar{e}_i}{\mathcal{B}, T \cup \{q \xrightarrow{\bar{e}_i} q_1\}, q_\ell \triangleright q, \bar{v}}$$

$$\frac{\mathcal{B}, T, q_\ell \triangleright q_1, \bar{v}}{\mathcal{B}, T \cup \{q \xrightarrow{-\bar{e}_i} q_1\}, q_\ell \triangleright q, \bar{v} + \bar{e}_i} \qquad \frac{\mathcal{B}, T_1, q_\ell \triangleright q_1, \bar{v}_1 \quad \mathcal{B}, T_2, q_\ell \triangleright q_2, \bar{v}_2}{\mathcal{B}, T_1 \cup T_2 \cup \{q \to q_1 + q_2\}, q_\ell \triangleright q, \bar{v}_1 + \bar{v}_2}$$

We write more simply '$\mathcal{B}, q_\ell \triangleright q, \bar{v}$' if there exists $T \subseteq T_u \uplus T_s$ such that $\mathcal{B}, T, q_\ell \triangleright q, \bar{v}$. With these notations, the reachability problem asks whether $\mathcal{B}, q_\ell \triangleright q_r, \bar{0}$.

3.2 Root Coverability

Our interest in this paper lies in a relaxation of the reachability problem, where we ask instead to *cover* the root: given as before $\langle \mathcal{B}, q_r, q_\ell \rangle$, we ask whether there exists a *coverability witness*, i.e. a deduction tree with root $(q_r, \bar{v})$ for some $\bar{v}$ in $\mathbb{N}^d$ and leaves $(q_\ell, \bar{0})$; in other words whether $\mathcal{B}, q_\ell \triangleright q_r, \bar{v}$ for some $\bar{v}$ in $\mathbb{N}^d$.

This problem was shown decidable by Verma and Goubault-Larrecq [23], and was later proven 2-ExpTime-complete by Demri et al. [6] in a slight variant called *branching vector addition systems* (BVAS):

Fact 1 (6, Thm. 8 and Thm. 21). *BVAS coverability is* 2-ExpTime-*complete.*

Branching VAS are not equipped with a state space Q. Their coverability problem is stated slightly differently, but is easy to reduce in both directions to

BVASS coverability. This would not be worth mentioning here if it were not for the following instrumental corollary of their proof, which exploits an encoding of d-dimensional BVASSs into $(d+6)$-dimensional BVASs:

Corollary 2. *Coverability in a BVASS $\mathcal{B} = \langle Q, d, T_u, T_s \rangle$ can be solved in deterministic time $2^{2^{O(n \cdot \log(n \cdot \log |Q|))}}$, where n denotes the size of the representation of $\langle d, T_u \rangle$.*

Corollary 2 entails that coverability remains in 2-EXPTIME for BVASSs with double exponential state space. This is an easy result, which we show in the full paper.

4 Upper Bound

In order to show a 2-EXPTIME upper bound for $\mathbf{R}_{\rightarrow}$ provability, we introduce as an intermediate decision problem the *expansive reachability* problem for BVASS (Sec. 4.1). Then, the first step of our proof in Sec. 4.2 takes us from the sequent calculus $L\mathbf{R}_{\rightarrow}$ to reachability in expansive BVASS. This is a simple construction that relies on the subformula property of $L\mathbf{R}_{\rightarrow}$, and is actually a particular case of a more general reduction shown in [11, Prop. 9]. The new technical result here is the second step: a reduction from expansive BVASS reachability to BVASS coverability, which is shown in Sec. 4.3. This new reduction also entails new upper bounds for provability in extensions of $L\mathbf{R}_{\rightarrow}$ studied in [11]; see Sec. 6.

4.1 Expansive Reachability

An *expansive BVASS* is a BVASS with an additional deduction rule:

$$\frac{q, \bar{\mathsf{v}} + \bar{\mathsf{e}}_i}{q, \bar{\mathsf{v}} + 2\bar{\mathsf{e}}_i}\ (\mathsf{expansion})$$

Note that expansions could be simulated by unary rules $q \xrightarrow{-\bar{\mathsf{e}}_i} q_i \xrightarrow{\bar{\mathsf{e}}_i} q'_i \xrightarrow{\bar{\mathsf{e}}_i} q$ for all q in Q and $0 < i \leq d$; we prefer to see them as new deduction rules.

This yields a new rule for root judgements, which we denote using '$\rhd_e$' to emphasise that we allow expansion rules:

$$\frac{\mathcal{B}, T, q_\ell \rhd_e q, \bar{\mathsf{v}} + 2\bar{\mathsf{e}}_i}{\mathcal{B}, T, q_\ell \rhd_e q, \bar{\mathsf{v}} + \bar{\mathsf{e}}_i}$$

The *expansive reachability* problem then asks, given an expansive BVASS $\mathcal{B}$ and two states q_r and q_ℓ, whether $\mathcal{B}, q_\ell \rhd_e q_r, \bar{0}$.

4.2 From $L\mathbf{R}_{\rightarrow}$ to Expansive Reachability

We prove here the following reduction:

Proposition 3. *There is a logarithmic space reduction from provability in $\mathbf{R}_{\rightarrow}$ to expansive reachability in ordinary BVASSs.*

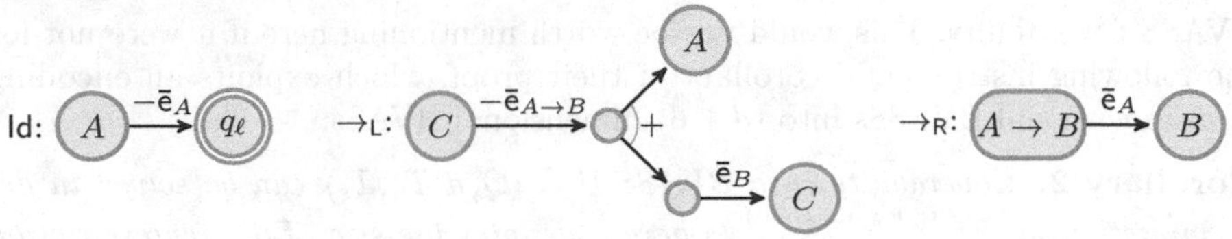

Fig. 1. The rules and intermediate states of $\mathcal{B}_F$

Let us consider an instance $\langle F \rangle$ of the provability problem for $\mathbf{R}_\to$ for an implicational formula F. The instance is positive if and only if we can find a proof for $\vdash F$ in $L\mathbf{R}_\to$. Thanks to the subformula property, we know that in such a proof, all the sequents $\Gamma \vdash A$ must use subfomulæ of F. That is, if we denote by S the set of subformulæ of F, then Γ is in $\mathbb{N}^S$ and A in S.

We construct from F an expansive BVASS $\mathcal{B}_F$ that implements proof search in $L\mathbf{R}_\to$ restricted to subformulæ of F. We define for this $\mathcal{B}_F \stackrel{\text{def}}{=} \langle Q_F, |S|, T_u, T_s \rangle$ where the state space Q_F includes S and a distinguished leaf state q_ℓ. It also includes some intermediate states as introduced in the translations of the rules of $L\mathbf{R}_\to$ into rules in $T_u \cup T_s$ depicted in Fig. 1. Note that (C) has no associated rule; it relies instead on expansions in $\mathcal{B}_F$. The full state space Q_F of $\mathcal{B}_F$, including intermediate states, is thus of size $O(|F|^2)$.

Let us write $\bar{v}_\Gamma$ for the vector in $\mathbb{N}^{|S|}$ associated with a multiset Γ in $\mathbb{N}^S$. The proof of Thm. 3 is a consequence of the following claim instantiated with $A = F$ and $\Gamma = \emptyset$:

Claim 3.1. For all Γ in $\mathbb{N}^S$ and A in S, $\Gamma \vdash A$ if and only if $\mathcal{B}_F, q_\ell \rhd_e A, \bar{v}_\Gamma$.

This results from a straightforward induction on the structure of proofs in $L\mathbf{R}_\to$ and expansive root judgements for $\mathcal{B}_F$; see the full paper for details.

4.3 From Expansive Reachability to Coverability

The second step of our proof that $\mathbf{R}_\to$ provability is in 2-ExpTime is then to reduce expansive reachability to coverability in BVASS. Our reduction incurs an exponential blow-up in the number of states, but thanks to Thm. 2, this still results in a 2-ExpTime algorithm:

Proposition 4. *There is a polynomial space reduction from BVASS expansive reachability to BVASS coverability.*

Topmost Increments. Consider an instance $\langle \mathcal{B}, q_r, q_\ell \rangle$ of the expansive reachability problem with $\mathcal{B} = \langle Q, d, T_u, T_s \rangle$. Because the root vector of an expansive reachability witness must be $\bar{0}$, we can identify along each branch of the witness and for each coordinate $0 < i \leq d$ the topmost (i.e. closest to the root) application of an (incr) rule—possibly no such increment ever occurs on some branches.

Assume without loss of generality that q_ℓ has no outgoing transition in $\mathcal{B}$. We construct a new BVASS $\mathcal{B}^\dagger = \langle Q^\dagger, d, T_u^\dagger, T_s\rangle$ with additional states q_ℓ^i and unary rules $q_\ell \xrightarrow{\bar{\mathrm{e}}_i} q_\ell^i \xrightarrow{-\bar{\mathrm{e}}_i} q_\ell$ for every $0 < i \leq d$. Then $\mathcal{B}, q_\ell \rhd_e q_r, \bar{0}$ if and only if $\mathcal{B}^\dagger, q_\ell \rhd_e q_r, \bar{0}$ (observe in particular that no expansion in q_ℓ^i can occur in an expansive reachability witness). Additionally, the new rules allow us to assume that there *is* a topmost increment for each branch and every coordinate of an expansive reachability witness of $\mathcal{B}^\dagger$.

Let $[d] \stackrel{\text{def}}{=} \{1, \ldots, d\}$. The root judgement relation can be refined as '$\rhd_e^s$' with a set $s \subseteq [d]$ of coordinates. The intended semantics for $i \in s$ is that there is at least one increment on coordinate i earlier on the path from the root in the expansive reachability witness. Formally, at the leaves

$$\frac{}{\mathcal{B}^\dagger, \emptyset, q_\ell \rhd_e^{[d]} q_\ell, \bar{0}}$$

since by assumption every coordinate must see an increase. Then, an increment is either topmost or not:

$$\frac{\mathcal{B}^\dagger, T, q_\ell \rhd_e^{s\uplus\{i\}} q_1, \bar{\mathrm{w}} + \bar{\mathrm{e}}_i}{\mathcal{B}^\dagger, T \cup \{q \xrightarrow{\bar{\mathrm{e}}_i} q_1\}, q_\ell \rhd_e^s q, \bar{\mathrm{w}}} \qquad \frac{\mathcal{B}^\dagger, T, q_\ell \rhd_e^{s\cup\{i\}} q_1, \bar{\mathrm{v}} + \bar{\mathrm{e}}_i}{\mathcal{B}^\dagger, T \cup \{q \xrightarrow{\bar{\mathrm{e}}_i} q_1\}, q_\ell \rhd_e^{s\cup\{i\}} q, \bar{\mathrm{v}}}$$

where $\bar{\mathrm{w}}(i) = 0$ and '$\uplus$' denotes disjoint union. Decrements and expansions are necessarily dominated by the topmost increment:

$$\frac{\mathcal{B}^\dagger, T, q_\ell \rhd_e^{s\cup\{i\}} q_1, \bar{\mathrm{v}}}{\mathcal{B}^\dagger, T \cup \{q \xrightarrow{-\bar{\mathrm{e}}_i} q_1\}, q_\ell \rhd_e^{s\cup\{i\}} q, \bar{\mathrm{v}} + \bar{\mathrm{e}}_i} \qquad \frac{\mathcal{B}^\dagger, T, q_\ell \rhd_e^{s\cup\{i\}} q, \bar{\mathrm{v}} + 2\bar{\mathrm{e}}_i}{\mathcal{B}^\dagger, T, q_\ell \rhd_e^{s\cup\{i\}} q, \bar{\mathrm{v}} + \bar{\mathrm{e}}_i}$$

Finally, the same topmost increments have been seen on both branches of a split:

$$\frac{\mathcal{B}^\dagger, T_1, q_\ell \rhd_e^s q_1, \bar{\mathrm{v}}_1 \quad \mathcal{B}^\dagger, T_2, q_\ell \rhd_e^s q_2, \bar{\mathrm{v}}_2}{\mathcal{B}^\dagger, T_1 \cup T_2 \cup \{q \to q_1 + q_2\}, q_\ell \rhd_e^s q, \bar{\mathrm{v}}_1 + \bar{\mathrm{v}}_2}$$

The refined root judgements verify

$$\mathcal{B}, q_\ell \rhd_e q_r, \bar{0} \text{ implies } \mathcal{B}^\dagger, q_\ell \rhd_e^\emptyset q_r, \bar{0}\ , \tag{1}$$

the converse implication being immediate by removing the 's' annotations.

Reduction to Coverability. We construct a BVASS $\mathcal{B}^\ddagger = \langle Q^\dagger \times 2^{[d]}, d, T_u^\ddagger, T_s^\ddagger\rangle$ and build a coverability instance $\langle \mathcal{B}^\ddagger, (q_r, \emptyset), (q_\ell, [d])\rangle$. The idea is to maintain a set $s \subseteq [d]$ as in the refined judgements $\rhd_e^s$; however since we cannot test to zero we will rely instead on nondeterminism. Let

$$\begin{aligned} T_u^\ddagger &\stackrel{\text{def}}{=} \{(q, s) \xrightarrow{\bar{\mathrm{e}}_i} (q_1, s \cup \{i\}) \mid q \xrightarrow{\bar{\mathrm{e}}_i} q_1 \in T_u^\dagger, s \subseteq [d]\} && (\text{incr}^\ddagger)\\ &\cup \{(q, s \cup \{i\}) \xrightarrow{-\bar{\mathrm{e}}_i} (q_1, s \cup \{i\}) \mid q \xrightarrow{-\bar{\mathrm{e}}_i} q_1 \in T_u^\dagger, s \subseteq [d]\}\ , && (\text{decr}^\ddagger)\\ T_s^\ddagger &\stackrel{\text{def}}{=} \{(q, s) \to (q_1, s) + (q_2, s) \mid q \to q_1 + q_2 \in T_s^\dagger, s \subseteq [d]\}\ . && (\text{split}^\ddagger)\end{aligned}$$

For $s \subseteq [d]$ and $\bar{\mathrm{v}}$ in $\mathbb{N}^d$, we define $s \cdot \bar{\mathrm{v}}$ for each $0 < i \leq d$ by

$$(s \cdot \bar{\mathrm{v}})(i) \stackrel{\text{def}}{=} \begin{cases} \bar{\mathrm{v}}(i) & \text{if } i \in s\ , \\ 0 & \text{otherwise.} \end{cases} \tag{2}$$

We show the following claims in the full paper:

Claim 4.1. If $\mathcal{B}^\dagger, q_\ell \rhd_e^s q, \bar{v}$, then there exists $\bar{v}' \geq \bar{v}$ such that $\mathcal{B}^\ddagger, (q_\ell, [d]) \rhd (q, s), \bar{v}'$.

Claim 4.2. If $\mathcal{B}^\ddagger, (q_\ell, [d]) \rhd (q, s), \bar{v}$, then $\mathcal{B}^\dagger, q_\ell \rhd_e q, s \cdot \bar{v}$.

Proof of Thm. 4. If $\mathcal{B}, q_\ell \rhd_e q_r, \bar{0}$, then by (1), $\mathcal{B}^\dagger, q_\ell \rhd_e^\emptyset q_r, \bar{0}$, thus by Claim 4.1, there exists $\bar{v}$ such that $\mathcal{B}^\ddagger, (q_\ell, [d]) \rhd (q_r, \emptyset), \bar{v}$, i.e. we can cover $(q_r, \emptyset)$ in $\mathcal{B}^\ddagger$.

Conversely, if $\mathcal{B}^\ddagger, (q_\ell, [d]) \rhd (q_r, \emptyset), \bar{v}$, then by Claim 4.2, $\mathcal{B}^\dagger, q_\ell \rhd_e q_r, \emptyset \cdot \bar{v}$ where $\emptyset \cdot \bar{v} = \bar{0}$. Therefore $\mathcal{B}, q_\ell \rhd_e q_r, \bar{0}$ in the original BVASS $\mathcal{B}$. □

Theorem 5. *Provability in* $\mathbf{R}_\rightarrow$ *is in* 2-ExpTime.

Proof. By Thm. 3 and Thm. 4, from a provability instance $\langle F \rangle$, we can reduce to a coverability instance $\langle \mathcal{B}_F^\ddagger, (q_r, \emptyset), (q_\ell, [|F|]) \rangle$ where $\mathcal{B}_F^\ddagger$ has dimension $|F|$ and a number of states in $2^{p(|F|)}$ for a polynomial p. By Thm. 2, this coverability instance can be solved in double exponential time in $|F|$. Note that the coverability check can be performed on-the-fly from F to avoid the explicit construction of $\mathcal{B}_F^\ddagger$. □

5 Lower Bound

In this section, we exhibit a reduction from BVASS coverability to $\mathbf{R}_\rightarrow$ provability, thereby showing its 2-ExpTime-hardness.

Previous reductions from counter machines to substructural logics in [12, 22, 11] actually reduce to provability in the logic extended with a *theory* encoding the rules of the system, which is then reduced to the basic logic. This last step relies in an essential way on the presence of exponential or additive connectives to 'dispose' of unused rules.

Having neither exponential nor additive connectives at our disposal, we introduce in Sec. 5.1 a *comprehensive* variant of the expansive reachability problem, where every rule should be employed at least once in the deduction. We further avoid the use of a theory and define in Sec. 5.2 a *focusing* calculus for $\mathbf{R}_\rightarrow$, from which the correctness of the reduction given in Sec. 5.3 will be facilitated.

5.1 Comprehensive Reachability

Given an expansive BVASS $\mathcal{B} = \langle Q, d, T_u, T_s \rangle$ and two states q_r and q_ℓ, the *comprehensive reachability* problem asks whether there exists a deduction tree of $\mathcal{B}$ with root label $(q_r, \bar{0})$ and leaves label $(q_\ell, \bar{0})$, such that every rule in $T_u \cup T_s$ is used at least once. Termed differently, it asks whether $\mathcal{B}, T_u \cup T_s, q_\ell \rhd_e q_r, \bar{0}$. We show that BVASS coverability can be reduced to comprehensive expansive reachability, hence by Thm. 1:

Proposition 6. *Comprehensive reachability in expansive ordinary BVASS is* 2-ExpTime*-hard.*

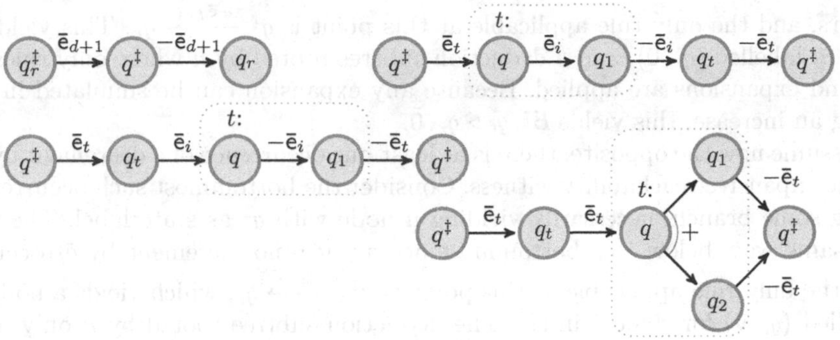

Fig. 2. The rules of $\mathcal{B}^\ddagger$ in the proof of Thm. 6, where t ranges over $T_u^\dagger \cup T_s$

Increasing Reachability. Let us consider an instance $\langle \mathcal{B}, q_r, q_\ell \rangle$ of the coverability problem in an ordinary BVASS $\mathcal{B} = \langle Q, d, T_u, T_s \rangle$. As a first step, we construct an ordinary BVASS $\mathcal{B}^\dagger \stackrel{\text{def}}{=} \langle Q, d, T_u^\dagger, T_s \rangle$ with additional *increases* $q \xrightarrow{\bar{e}_i} q$ for every q in Q and $0 < i \leq d$. We claim that coverability in $\mathcal{B}$ is equivalent to reachability in $\mathcal{B}^\dagger$:

Claim 6.1. There exists $\bar{v}$ in $\mathbb{N}^d$ such that $\mathcal{B}, q_\ell \triangleright q_r, \bar{v}$ iff $\mathcal{B}^\dagger, q_\ell \triangleright q_r, \bar{0}$.

Proof Sketch. Clearly, if $\mathcal{B}, q_\ell \triangleright q_r, \bar{v}$ for some $\bar{v}$ in $\mathbb{N}^d$, then $\mathcal{B}^\dagger, q_\ell \triangleright q_r, \bar{v}$, and using increases in q_r shows $\mathcal{B}^\dagger, q_\ell \triangleright q_r, \bar{0}$. Conversely, if there is a reachability witness for $\langle \mathcal{B}^\dagger, q_r, q_\ell \rangle$, then we can assume that increases $q \xrightarrow{\bar{e}_i} q$ occur as close to the root as possible. As increases occurring right below increments, decrements, or splits can be permuted locally to occur right above, such a reachability witness has all its increases at the root. The deduction tree below those increases is labelled $(q_r, \bar{v})$ for some $\bar{v}$ in $\mathbb{N}^d$ and is also a deduction tree of $\mathcal{B}$. $\square$

Comprehensive Root Rules. The second step of the reduction from BVASS coverability builds an ordinary BVASS $\mathcal{B}^\ddagger \stackrel{\text{def}}{=} \langle Q^\ddagger, d+1+|T_u^\dagger \cup T_s|, T_u^\ddagger, T_s \rangle$ where $Q^\ddagger \stackrel{\text{def}}{=} Q \uplus \{q^\ddagger, q_r^\ddagger\} \uplus \{q_t \mid t \in T_u^\dagger \cup T_s\}$. It features an additional set of unary 'root' rules—depicted in Fig. 2—designed to allow any rule in $T_u^\ddagger \cup T_s$ to be employed in a reachability witness.

The idea is to introduce a new state $q^\ddagger$ and a new coordinate for each rule t in $T_u^\dagger \cup T_s$. Starting from $q^\ddagger$, $\mathcal{B}^\ddagger$ can simulate any rule t from $T_u^\dagger \cup T_s$ by first incrementing by the corresponding unit vector $\bar{e}_t$, then applying the rule, and finally decrementing by $\bar{e}_t$ to return to $q^\ddagger$. This ensures that, if $\mathcal{B}^\ddagger, T, q_\ell \triangleright_e q^\ddagger, \bar{v}$ for some $\bar{v}$ in $\mathbb{N}^d$, then $\mathcal{B}^\ddagger, T', q_\ell \triangleright_e q^\ddagger, \bar{v}$ where $T_u^\dagger \cup T_s \subseteq T'$. The additional states and rules from $q_r^\ddagger$ and to q_r then show that:

Claim 6.2. If $\mathcal{B}^\dagger, T, q_\ell \triangleright q_r, \bar{0}$ for some $T \subseteq T_u^\dagger \cup T_s$, then $\mathcal{B}^\ddagger, T_u^\ddagger \cup T_s, q_\ell \triangleright_e q_r^\ddagger, \bar{0}$.

Conversely, assume that $\mathcal{B}^\ddagger, q_\ell \triangleright_e q_r^\ddagger, \bar{0}$. This entails $\mathcal{B}^\ddagger, q_\ell \triangleright_e q^\ddagger, \bar{e}_{d+1}$.

First assume that no decrement by $\bar{e}_t$ for any t in $T_u^\dagger \cup T_s$ is ever used in the corresponding expansive reachability witness. Then also no increment by $\bar{e}_t$

occurs, and the only rule applicable at this point is $q^\ddagger \xrightarrow{-\bar{\mathsf{e}}_{d+1}} q_r$. This yields a node n labelled $(q_r, \bar{0})$, and a deduction subtree rooted by n where only rules of $\mathcal{B}^\dagger$ and expansions are applied. Because any expansion can be simulated in $\mathcal{B}^\dagger$ using an increase, this yields $\mathcal{B}^\dagger, q_\ell \rhd q_r, \bar{0}$.

Assume now the opposite: there is at least one occurrence of a decrement by $\bar{\mathsf{e}}_t$ in the expansive reachability witness. Consider the bottommost such occurrence along some branch, necessarily yielding a node with $q^\ddagger$ as state label. Then in the same way, below this bottommost occurrence, no increment by $\bar{\mathsf{e}}_t$ occurs, and the only rule applicable at this point is $q^\ddagger \xrightarrow{-\bar{\mathsf{e}}_{d+1}} q_r$, which yields a node n labelled $(q_r, \bar{\mathsf{v}})$ for some $\bar{\mathsf{v}}$ in $\mathbb{N}^d$. The deduction subtree rooted by n only uses rules of $\mathcal{B}^\dagger$ and expansions, thus as in the previous case $\mathcal{B}^\dagger, q_\ell \rhd q_r, \bar{\mathsf{v}}$. Using increases in q_r then shows $\mathcal{B}^\dagger, q_\ell \rhd q_r, \bar{0}$. Therefore, in all cases:

Claim 6.3. If $\mathcal{B}^\ddagger, T_u^\ddagger \cup T_s, q_\ell \rhd_e q_r^\ddagger, \bar{0}$, then $\mathcal{B}^\dagger, q_\ell \rhd q_r, \bar{0}$.

By Claims 6.1, 6.2, and 6.3, $\mathcal{B}^\ddagger, T_u^\ddagger \cup T_s, q_\ell \rhd_e q_r^\ddagger, \bar{0}$ if and only if there exists $\bar{\mathsf{v}}$ in $\mathbb{N}^d$ such that $\mathcal{B}, q_\ell \rhd q_r, \bar{\mathsf{v}}$, thereby showing the correctness of our reduction.

5.2 Focusing Proofs in $\mathbf{R}_\to$

We enforce a particular proof policy in our simulation of BVASSs in $\mathbf{R}_\to$, which is inspired by the *focusing proof* techniques [2] employed to reduce non-determinism during proof search in sequent calculi. With only implication at our disposal, we find ourselves in a 'negative fragment', where focusing proofs have a very simple calculus $F\mathbf{R}_\to$. This is equivalent to restricting oneself to long normal forms in the associated λ-calculus.

A *focusing sequent* is of one of the two forms '$\Gamma, [A] \Vdash B$' or '$\Gamma \Vdash A$' where '$[A]$' is called a *focused* formula. We let $\Gamma, \Delta, \ldots$ denote as before multisets of implicational formulæ and A, B, C implicational formulæ. Here are the rules of the focusing calculus $F\mathbf{R}_\to$:

$$\frac{}{[a] \Vdash a}\,(\mathsf{atomic}) \qquad \frac{\Gamma, [A] \Vdash a}{\Gamma, A \Vdash a}\,(\mathsf{focus}) \qquad \frac{\Gamma, A, A \Vdash a}{\Gamma, A \Vdash a}\,(\mathsf{C}^f)$$

$$\frac{\Gamma \Vdash A \quad \Delta, [B] \Vdash a}{\Gamma, \Delta, [A \to B] \Vdash a}\,(\to_{\mathsf{L}}^f) \qquad \frac{\Gamma, A \Vdash B}{\Gamma \Vdash A \to B}\,(\to_{\mathsf{R}}^f)$$

Note that our focusing calculus $F\mathbf{R}_\to$ gives the priority to right implications $(\to_{\mathsf{R}}^f)$ over the left implications $(\to_{\mathsf{L}}^f)$, focus (focus) and contractions (C^f): the latter can only be applied to sequents with atomic succedents a in $\mathcal{A}$. A similar observation is that a focusing sequent $\Gamma, [a] \Vdash A$ is provable if and only if $A = a$ is atomic and $\Gamma = \emptyset$ is the empty multiset, since (atomic) is the only rule yielding a sequent with a focused atomic formula $[a]$.

Theorem 7 ($F\mathbf{R}_\to$ is sound and complete). *A sequent $\Gamma \vdash A$ is provable in $L\mathbf{R}_\to$ if and only if the focusing sequent $\Gamma \Vdash A$ is provable in $F\mathbf{R}_\to$.*

We prove Thm. 7 in the full paper, using the admissibility of a suitable cut rule in $F\mathbf{R}_\to$.

5.3 From Comprehensive Expansive Reachability to $\boldsymbol{FR}_{\rightarrow}$

Let us consider a comprehensive expansive reachability instance $\langle \mathcal{B}, q_r, q_\ell \rangle$ with $\mathcal{B} = \langle Q, d, T_u, T_s \rangle$. We are going to construct an implicational formula F such that $\vdash F$ if and only if $\mathcal{B}, T_u \cup T_s, q_\ell \rhd_e q_r, \bar{0}$.

We work for this on the set of atomic formulæ $Q \uplus \{e_i \mid 0 < i \leq d\}$, and associate to a root judgement $\mathcal{B}, T, q_\ell \rhd_e q, \bar{\mathsf{v}}$ a focusing sequent

$$q_\ell, \Delta_T, \Gamma_{\bar{\mathsf{v}}} \Vdash q \tag{3}$$

where Δ_T encodes the rules in $T \subseteq T_u \cup T_s$ and $\Gamma_{\bar{\mathsf{v}}}$ encodes $\bar{\mathsf{v}}$: let $T = \{t_1, \ldots, t_k\}$ and $\bar{\mathsf{v}} = c_1 \bar{\mathsf{e}}_1 + \cdots + c_d \bar{\mathsf{e}}_d$, then

$$\Delta_T \stackrel{\text{def}}{=} \ulcorner t_1 \urcorner, \ldots, \ulcorner t_k \urcorner \,, \tag{4}$$

$$\Gamma_{\bar{\mathsf{v}}} \stackrel{\text{def}}{=} e_1^{c_1}, \ldots, e_d^{c_d} \,, \tag{5}$$

where '$\ulcorner t \urcorner$' is the individual encoding of rule t and 'A^c' stands for c repetitions of the formula A. We use the following individual rule encodings:

$$\ulcorner q \xrightarrow{\bar{\mathsf{e}}_i} q_1 \urcorner \stackrel{\text{def}}{=} (e_i \rightarrow q_1) \rightarrow q \,, \tag{6}$$

$$\ulcorner q \xrightarrow{-\bar{\mathsf{e}}_i} q_1 \urcorner \stackrel{\text{def}}{=} q_1 \rightarrow (e_i \rightarrow q) \,, \tag{7}$$

$$\ulcorner q \rightarrow q_1 + q_2 \urcorner \stackrel{\text{def}}{=} q_1 \rightarrow (q_2 \rightarrow q) \,. \tag{8}$$

Then proof search in $\boldsymbol{FR}_{\rightarrow}$ is easily seen to implement deductions in $\mathcal{B}$:

Claim 8.1 (Completeness). If $\mathcal{B}, T, q_\ell \rhd_e q, \bar{\mathsf{v}}$, then $q_\ell, \Delta_T, \Gamma_{\bar{\mathsf{v}}} \Vdash q$.

Proof. We proceed by induction on the structure of the root judgement. For the base case, i.e. for $\mathcal{B}, \emptyset, q_\ell \rhd_e q_\ell, \bar{0}$, we have

$$\dfrac{\dfrac{}{[q_\ell] \Vdash q_\ell}\ (\mathsf{atomic})}{q_\ell \Vdash q_\ell}\ (\mathsf{focus})$$

as desired.

For the induction step, if the last applied rule is an increment $t = q \xrightarrow{\bar{\mathsf{e}}_i} q_1$ on a judgement $\mathcal{B}, T, q_\ell \rhd_e q_1, \bar{\mathsf{v}} + \bar{\mathsf{e}}_i$, then

$$\dfrac{\dfrac{\dfrac{\overset{\text{i.h.}}{q_\ell, \Delta_T, \Gamma_{\bar{\mathsf{v}}}, e_i \Vdash q_1}}{q_\ell, \Delta_T, \Gamma_{\bar{\mathsf{v}}} \Vdash e_i \rightarrow q_1}\ (\rightarrow^f_{\mathsf{R}}) \quad \dfrac{}{[q] \Vdash q}\ (\mathsf{atomic})}{q_\ell, \Delta_T, \Gamma_{\bar{\mathsf{v}}}, [(e_1 \rightarrow q_1) \rightarrow q] \Vdash q}\ (\rightarrow^f_{\mathsf{L}})}{q_\ell, \Delta_T, \Gamma_{\bar{\mathsf{v}}}, (e_1 \rightarrow q_1) \rightarrow q \Vdash q}\ (\mathsf{focus})$$

and an additional contraction (C^f) if $t \in T$ shows $q_\ell, \Delta_{T \cup \{t\}}, \Gamma_{\bar{\mathsf{v}}} \Vdash q$ as desired.

If the last applied rule is a decrement $t = q \xrightarrow{-\bar{\mathsf{e}}_i} q_1$ on a judgement $\mathcal{B}, T, q_\ell \rhd_e q_1, \bar{\mathsf{v}}$, then

$$\dfrac{\text{i.h.}\quad q_\ell, \Delta_T, \Gamma_{\bar{\mathrm{v}}} \Vdash q_1 \qquad \dfrac{\dfrac{\dfrac{}{[e_i] \Vdash e_i}\,(\text{atomic})}{e_i \Vdash e_i}\,(\text{focus}) \qquad \dfrac{}{[q] \Vdash q}\,(\text{atomic})}{e_i, [e_i \to q] \Vdash q}\,(\to_L^f)}{\dfrac{q_\ell, \Delta_T, \Gamma_{\bar{\mathrm{v}}}, e_i, [q_1 \to (e_i \to q)] \Vdash q}{q_\ell, \Delta_T, \Gamma_{\bar{\mathrm{v}}}, e_i, q_1 \to (e_i \to q) \Vdash q}\,(\text{focus})}\,(\to_L^f)$$

and a contraction (C^f) if $t \in T$ shows $q_\ell, \Delta_{T \cup \{t\}}, \Gamma_{\bar{\mathrm{v}}+\bar{\mathrm{e}}_i} \Vdash q$ as desired.

If the last applied rule is an expansion on a judgement $\mathcal{B}, T, q_\ell \rhd_e q, \bar{\mathrm{v}} + 2\bar{\mathrm{e}}_i$, then

$$\dfrac{\overset{\text{i.h.}}{q_\ell, \Delta_T, \Gamma_{\bar{\mathrm{v}}}, e_i, e_i \Vdash q}}{q_\ell, \Delta_T, \Gamma_{\bar{\mathrm{v}}}, e_i \Vdash q}\,(\mathsf{C}^f)$$

as desired.

Finally, if the last applied rule is a split $t = q \to q_1 + q_2$ on two judgements $\mathcal{B}, T_1, q_\ell \rhd_e q_1, \bar{\mathrm{v}}_1$ and $\mathcal{B}, T_2, q_\ell \rhd_e q_2, \bar{\mathrm{v}}_2$, then

$$\dfrac{\dfrac{\text{i.h.}\quad q_\ell, \Delta_{T_1}, \Gamma_{\bar{\mathrm{v}}_1} \Vdash q_1 \qquad \dfrac{\overset{\text{i.h.}}{q_\ell, \Delta_{T_2}, \Gamma_{\bar{\mathrm{v}}_2} \Vdash q_2} \qquad \dfrac{}{[q] \Vdash q}\,(\text{atomic})}{q_\ell, \Delta_{T_2}, \Gamma_{\bar{\mathrm{v}}_2}, [q_2 \to q] \Vdash q}\,(\to_L^f)}{q_\ell, q_\ell, \Delta_{T_1}, \Delta_{T_2}, \Gamma_{\bar{\mathrm{v}}_1}, \Gamma_{\bar{\mathrm{v}}_2}, [q_1 \to (q_2 \to q)] \Vdash q}\,(\to_L^f)}{q_\ell, \Delta_{T_1 \cup T_2 \cup \{t\}}, \Gamma_{\bar{\mathrm{v}}_1 + \bar{\mathrm{v}}_2} \Vdash q}\,(\mathsf{C}^f)$$

as desired. □

The interest of the focusing calculus $\boldsymbol{F}\mathbf{R}_\to$ is that, starting from a sequent $q_\ell, \Delta, \Gamma, [\ulcorner t \urcorner] \Vdash q$ where Δ is in $\mathbb{N}^{\Delta_{T_u \cup T_s}}$ and Γ in $\mathbb{N}^{\{e_i \mid 0 < i \leq d\}}$ and the focus is on the encoding of a rule t, there is *no choice* but to follow the proof trees shown in the proof of Claim 8.1. Given a multiset m in $\mathbb{N}^E$ for some set E, we write

$$\sigma(m) \stackrel{\text{def}}{=} \{e \in E \mid m(e) > 0\} \tag{9}$$

for the *support* of m.

Claim 8.2 (Soundness). Let Δ be in $\mathbb{N}^{\Delta_{T_u \cup T_s}}$, Γ in $\mathbb{N}^{\{e_i \mid 0 < i \leq d\}}$, q in Q, and $n > 0$. Then $q_\ell^n, \Delta, \Gamma \Vdash q$ implies $\mathcal{B}, \sigma(\Delta), q_\ell \rhd_e q, \bar{\mathrm{v}}_\Gamma$.

Proof. Note that $n = 0$ would yield an unprovable sequent. We proceed by induction on the structure of a proof tree for the focusing sequent. The only applicable rules in a proof search from $q_\ell^n, \Delta, \Gamma \Vdash q$ are (focus) and (C^f). In the latter case, we distinguish three cases depending on the contracted formula A:

- If $A = q_\ell$, i.e. if $q_\ell^{n+1}, \Delta, \Gamma \Vdash q$, then by induction hypothesis $\mathcal{B}, \sigma(\Delta), q_\ell \rhd_e q, \bar{\mathrm{v}}_\Gamma$ as desired.
- If $A = e_i$ in Γ, then by induction hypothesis $\mathcal{B}, \sigma(\Delta), q_\ell \rhd_e q, \bar{\mathrm{v}}_\Gamma + \bar{\mathrm{e}}_i$, and an expansion yields $\mathcal{B}, \sigma(\Delta), q_\ell \rhd_e q, \bar{\mathrm{v}}_\Gamma$ as desired.
- If $A = \ulcorner t \urcorner$ for some rule t in Δ, then the support $\sigma(\Delta)$ is not changed and by induction hypothesis $\mathcal{B}, \sigma(\Delta), q_\ell \rhd_e q, \bar{\mathrm{v}}_\Gamma$ as desired.

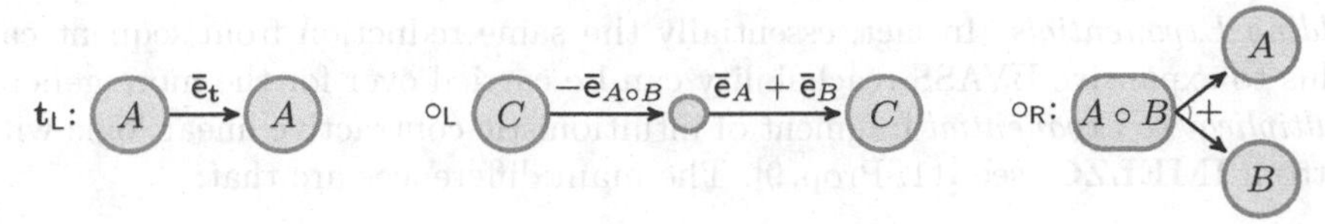

Fig. 3. The additional BVASS rules for $LR^{t}_{\to,\circ}$

In the former case, we also distinguish three cases depending on which formula A receives the focus:

- If $A = q_\ell$, necessarily Δ and Γ are empty and $q = q_\ell$, and indeed $\mathcal{B}, \emptyset, q_\ell \rhd_e q_\ell, \bar{0}$.
- If $A = e_i$ in Γ, then proof search fails since $q \neq e_i$.
- If $A = \ulcorner t \urcorner$ in Δ, then proof search needs to follow the proof trees used in the proof of Claim 8.1, and applying the induction hypothesis on the open leaves of these trees allows to conclude in each case. □

Theorem 8. *Provability in* $\mathbf{R}_\to$ *is* 2-ExpTime-*hard.*

Proof. We reduce from the comprehensive expansive reachability problem, which is 2-ExpTime-hard by Thm. 6. From an instance $\langle \mathcal{B}, q_r, q_\ell \rangle$ where $\mathcal{B} = \langle Q, d, T_u, T_s \rangle$, we construct a formula $F \stackrel{\text{def}}{=} q_\ell \to \varphi(T_u \cup T_s, q_r)$ defined by

$$\varphi(\emptyset, q_r) \stackrel{\text{def}}{=} q_r\ , \qquad \varphi(T \uplus \{t\}, q_r) \stackrel{\text{def}}{=} \ulcorner t \urcorner \to \varphi(T, q_r)\ . \tag{10}$$

By Thm. 7, $\vdash F$ if and only if $\Vdash F$. The latter holds if and only if $q_\ell, \Delta_{T_u \cup T_s} \Vdash q_r$ since we can only apply $(\to^f_{\mathsf{R}})$. Then this occurs if and only if $\mathcal{B}, T_u \cup T_s, q_\ell \rhd_e q_r, \bar{0}$ by Claim 8.1 and Claim 8.2. □

6 Extensions

Adding Multiplicatives. The sequent system $LR_\to$ for $\mathbf{R}_\to$ can be extended to accommodate further multiplicative connectives: the *fusion* connective $\circ$ (aka. 'co-tenability' in [1]) and the sentential constant $\mathbf{t}$:

$$\frac{\Gamma \vdash A}{\Gamma, \mathbf{t} \vdash A}\,(\mathsf{t_L}) \qquad \frac{}{\vdash \mathbf{t}}\,(\mathsf{t_R})$$

$$\frac{\Gamma, A, B \vdash C}{\Gamma, A \circ B \vdash C}\,(\circ_\mathsf{L}) \qquad \frac{\Gamma \vdash A \quad \Delta \vdash B}{\Gamma, \Delta \vdash A \circ B}\,(\circ_\mathsf{R})$$

Let us call $LR^{t}_{\to,\circ}$ the resulting sequent system. The BVASS $\mathcal{B}_F$ presented in Sec. 4.2 can be extended in a straightforward manner with the rules of Fig. 3 and by identifying q_ℓ with $\mathbf{t}$. Thanks to Thm. 4 and Thm. 2, this shows that provability in $LR^{t}_{\to,\circ}$ is in 2-ExpTime.

Note that the sequent system $LR^{t}_{\to,\circ}$ is the same as that of intuitionistic multiplicative contractive linear logic (**IMLLC**), where $\to$, $\circ$, and $\mathbf{t}$ are usually noted respectively $\multimap$, $\otimes$, and $\mathbf{1}$.

Adding Exponentials. In fact, essentially the same reduction from sequent calculus to expansive BVASS reachability can be carried over for the more general *multiplicative exponential* fragment of intuitionistic contractive linear logic with bottom **IMELZC**, see [11, Prop. 9]. The main differences are that:

1. The exponential connectives incur an exponential blow-up in the number of states of the constructed BVASS $\mathcal{B}_F$: its state space now contains $S \times 2^{S_!}$ where $S_!$ is the set of exponential subformulæ of F. The subsequent reduction to coverability in Sec. 4.3 then performs a product with $2^{S \setminus S_!}$ (contractions in exponential subformulæ being already handled), hence the resulting state space remains of size $2^{p(|F|)}$ for some polynomial p.
2. The exponential connectives also require an additional operation of *full zero test*: as shown in [11, Lem. 3], this operation can be eliminated at no cost in complexity.

By Thm. 8, Thm. 4 and Thm. 2, we conclude:

Theorem 9. *Provability in any logic between* $\mathbf{R}_{\rightarrow}$ *and* **IMELZC** *is* 2-ExpTime-*complete.*

This applies in particular to $\mathbf{R}^{\mathbf{t}}_{\rightarrow}$, **IMLLC**, and **IMELLC**. It seems likely that the non-intuitionistic variants **MLLC** and **MELLC**, i.e. multiplicative and multiplicative exponential contractive linear logic, are also 2-ExpTime-complete: the upper bound follows from the bound on **IMELZC**.

7 Concluding Remarks

Besides closing a longstanding open problem, the proof that $\mathbf{R}_{\rightarrow}$ is 2-ExpTime-complete paves the way for new investigations:

- In spite of the high worst-case complexity of BVASS coverability, Majumdar and Wang [13] have recently presented a practical algorithm with encouraging initial results. The reduction in Sec. 4 allows to transfer their techniques to $\mathbf{R}_{\rightarrow}$ provability, but might incur a worst-case exponential blow-up.
- Provability in the related implicational fragment $\mathbf{T}_{\rightarrow}$ of ticket entailment has recently been proven decidable independently by Padovani [16] and Bimbó and Dunn [3]. Although the complexity of this problem is currently unknown, the latter proof relies on provability in $L\mathbf{R}^{\mathbf{t}}_{\rightarrow}$, which we prove to be 2-ExpTime-complete in Sec. 6.

Acknowledgements. The author thanks David Baelde for his excellent suggestion of employing focusing proofs and helpful discussions around their uses.

References

1. Anderson, A.R., Belnap, Jr., N.D.: Entailment: The Logic of Relevance and Necessity, vol. I. Princeton University Press (1975)

2. Andreoli, J.M.: Logic programming with focusing proofs in linear logic. J. Logic Comput. 2(3), 297–347 (1992)
3. Bimbó, K., Dunn, J.M.: On the decidability of implicational ticket entailment. J. Symb. Log. 78(1), 214–236 (2013)
4. Church, A.: The weak theory of implication. Kontrolliertes Denken, Untersuchungen zum Logikkalkül und zur Logik der Einzelwissenschaften, pp. 22–37. Kommissions-Verlag Karl Alber, Munich (1951)
5. Curry, H.H., Feys, R.: Combinatory Logic. Studies in Logic and the Foundations of Mathematics, vol. 1, 22. North-Holland Publishing Company (1958)
6. Demri, S., Jurdziński, M., Lachish, O., Lazić, R.: The covering and boundedness problems for branching vector addition systems. J. Comput. Syst. Sci. 79(1), 23–38 (2012)
7. Dunn, J.M., Restall, G.: Relevance logic. In: Handbook of Philosophical Logic, pp. 1–128. Kluwer Academic Publishers (2002)
8. Figueira, D., Figueira, S., Schmitz, S., Schnoebelen, P.: Ackermannian and primitive-recursive bounds with Dickson's Lemma. In: LICS 2011, pp. 269–278. IEEE Computer Society (2011)
9. de Groote, P., Guillaume, B., Salvati, S.: Vector addition tree automata. In: LICS 2004, pp. 64–73. IEEE Computer Society (2004)
10. Kripke, S.A.: The problem of entailment. In: ASL 1959. J. Symb. Log., vol. 24, p. 324 (1959) (abstract)
11. Lazić, R., Schmitz, S.: Non-elementary complexities for branching VASS, MELL, and extensions. In: CSL-LICS 2014, ACM (to appear, 2014), arXiv:1401.6785 [cs.LO]
12. Lincoln, P., Mitchell, J., Scedrov, A., Shankar, N.: Decision problems for propositional linear logic. Ann. Pure App. Logic 56(1-3), 239–311 (1992)
13. Majumdar, R., Wang, Z.: Expand, enlarge, and check for branching vector addition systems. In: D'Argenio, P.R., Melgratti, H. (eds.) CONCUR 2013. LNCS, vol. 8052, pp. 152–166. Springer, Heidelberg (2013)
14. Mayr, E.W., Meyer, A.R.: The complexity of the word problems for commutative semigroups and polynomial ideals. Adv. Math. 46(3), 305–329 (1982)
15. Moh, S.K.: The deduction theorems and two new logical systems. Methodos 2, 56–75 (1950)
16. Padovani, V.: Ticket entailment is decidable. Math. Struct. Comput. Sci. 23(3), 568–607 (2013)
17. Riche, J., Meyer, R.K.: Kripke, Belnap, Urquhart and relevant decidability & complexity. In: Gottlob, G., Grandjean, E., Seyr, K. (eds.) CSL 1998. LNCS, vol. 1584, pp. 224–240. Springer, Heidelberg (1999)
18. Schmitz, S.: On the computational complexity of dominance links in grammatical formalisms. In: ACL 2010, pp. 514–524. ACL Press (2010)
19. Statman, R.: Intuitionistic propositional logic is polynomial-space complete. Theor. Comput. Sci. 9(1), 67–72 (1979)
20. Urquhart, A.: The undecidability of entailment and relevant implication. J. Symb. Log. 49(4), 1059–1073 (1984)
21. Urquhart, A.: The complexity of decision procedures in relevance logic. Truth or Consequences: Essays in honour of Nuel Belnap, pp. 61–76. Kluwer Academic Publishers (1990)
22. Urquhart, A.: The complexity of decision procedures in relevance logic II. J. Symb. Log. 64(4), 1774–1802 (1999)
23. Verma, K.N., Goubault-Larrecq, J.: Karp-Miller trees for a branching extension of VASS. Discrete Math. Theor. Comput. Sci. 7(1), 217–230 (2005)

Near Semi-rings and Lambda Calculus

Rick Statman

Carnegie Mellon University
Department of Mathematical Sciences
Pittsburgh, PA 15213
statman@cs.cmu.edu

Abstract. A connection between lambda calculus and the algebra of near semi-rings is discussed. Among the results is the following completeness theorem.

A first-order equation in the language of binary associative distributive algebras is true in all such algebras if and only if the interpretations of the first order terms as lambda terms beta-eta convert to one another. A similar result holds for equations containing free variables.

1 Near Semirings and Bad Algebra

A "near semi-ring" consists of a monoid $(S, +, 0)$ together with a semigroup $(S, *)$ satisfying the right distributive law $(x + y) * z = x * z + y * z$ and the left annihilator law $0 * x = 0$ [11] (for motivation see [13]). Examples are the non-negative integers (natural numbers), which is a near ring, and the ordinal numbers. An interesting case is when $(S, *)$ possesses an identity 1, and the monoid $(S, *, 1)$ replaces $(S, *)$. Now if $(S, *, 1)$ is any monoid then Cayley's representation sends the point $s : S$ to the function $f_s : S \to S$ defined by $f_s(t) = s * t$,with $f_1 =$ the identity function. This is sometimes called the regular representation, [12]. Now if + is defined pointwise $(f_r + f_s)(t) = f_r(t) + f_s(t)$ from $(S, +, 0)$, then the Cayley representation satisfies

$$f_0 = \text{the function identically } 0$$
$$f_r + f_s = f_{r+s}.$$

This is the representation theorem of Hoogewijs [7]. Thus if we begin with a pair of monoids we can use the Cayley representation of $(S, +, 0)$, say $(S', +', 0')$, where $+'$ is composition of functions $f : S' \to S'$, with $*$ lifted to some operation $*'$ on $S' \to S' \times S' \to S'$. Now we can apply the above Cayley representation of the result embedding it in $(S' \to S') \to (S' \to S')$ so that the operations become

final $+ =$ pointwise composition
final $* =$ composition.

There are, however, interesting examples with more than two binary operations. One example is the polynomial ring over a field together with substitution. A second is finite cardinal numbers together with inf, sup, + and $*$. In

G. Dowek (ed.): RTA-TLCA 2014, LNCS 8560, pp. 410–424, 2014.

each case "lower" operations distribute over "higher" ones. In the latter case, although each semigroup is a monoid the law of left annihilation fails. This suggests the following definition. A "binary associative distributive (bad) algebra of height n" consists of a set S together with $n+1$ binary operations $*_0, ..., *_n$ such that for each $i = 0, \ldots, n$, $(S, *_i)$ is a semigroup and the distributive laws $(r *_i s) *_j t = (r *_j t) *_i (s *_j t)$ hold when $j < i$. We say that the bad algebra is "monoidal" if there exist $n+1$ elements $I_0, ..., I_n : S$, such that for each $i = 0, \ldots, n$, $(S, *_i, I_i)$ is a monoid and when $j > i$ $I_j *_i x = I_j$.

For a given S, a type structure is a map $S(0) = S$, $S(m+1)$ is non- empty, and $S(m+1) \subset S(m) \to S(m)$. We can generalize Hoogewijs' representation theorem as follows.

Theorem 1. *Every bad algebra can be embedded into a type structure.*

Proof. The construction embeds a bad algebra of height n into $S(n+1)$ recursively by starting with $*_{n-1}$ and $*_n$ and working down the $*_i$ while up the $S(n-i+1)$. At the stage for $*_i$ the $*_j$ for $j > i$ are embedded pointwise and the $*_j$ for $j < i$ are embedded by definition. The construction clearly works for the monoidal case. End of proof.

2 Bad Algebra and Lambda Calculus

Some notation will be useful. We adopt for the most part the notation and terminology of [1].

$$
\begin{aligned}
B &:= \lambda xyz.\ x(yz) \\
K &:= \lambda xy.\ x \\
I &:= \lambda x.\ x \\
1 &:= \lambda xy.\ xy. \\
\sim &:= \text{beta} - \text{eta conversion} \\
\to_0 &:= \text{beta} - \text{eta reduction.}
\end{aligned}
$$

For the most part we shall use the colon ":" for membership and as an abbreviation for "has type", except when it is used in ":=" for syntactic identity and definitional equality.

Both Church and Curry observed that the combinators form a monoid under multiplication B, identity I, left zero KI, and beta-eta conversion. The same is true for addition $\lambda xyuv.\ xu(yuv)$, identity KI (the Church numeral 0), left zero K, and beta-eta conversion $\sim$. Since these satisfy the right distributive law

$$
\begin{aligned}
&(\lambda xyz.\ x(yz))((\lambda xyuv.\ xu(yuv))ab)\ c \\
&\sim (\lambda xyuv.\ xu(yuv))\ ((\lambda xyz.\ x(yz))ac)\ ((\lambda xyz.\ x(yz))bc)
\end{aligned}
$$

they form a near semi-ring.

Many years ago, see [6], I noticed a generalization of this near semi-ring structure to a hierarchy of monoids. Define

$$A_n := \lambda xyu_1 \dots u_n v.\ xu_1 \dots u_n(yu_1 \dots u_n v)$$

so $A_0 := B$ and A_1 is Church's addition. Then combinators form a monoid with multiplication A_n, identity $K^n I$, left zero K^{n+1} and beta-eta conversion. There are level $m+1$ integers $n \sim$

$$A_{m+1}(K^m I)(\dots(A_{m+1}(K^m I)(K^m I))\dots) \sim \lambda u_1 \dots u_m xy.\ x(\dots(xy)\dots),$$

which for $m = 1$ coincide with Church's finite ordinals. Again the right distributive law holds. More precisely we have

(associativity) $A_m(A_n xy) \sim A_0(A_n x)(A_n y)$ if $m = n$, (distributivity) $A_m(A_n xy) \sim A_{n+1}(A_m x)(A_m y)$ if $m < n$,

and in addition,

$$\begin{array}{lll} (i) & x \sim A_m x(K^m I) & \\ (ii) & A_m x \sim A_{m+1}(Kx)I & \\ (iii) & K(A_m xy) \sim A_{m+1}(Kx)(Ky) & \\ (iv) & K^n x \sim A_m(K^n x) & \text{if } m < n. \end{array}$$

Thus we have a monoidal bad algebra of height n.

The entire hierarchy of near semi-rings admits a very simple combinatory theory. If we endow the combinators I, K, and A_n for $n = 0, \dots$ with their weak beta reduction rules

$$\begin{array}{ll} I\,x & \to_1 x \\ K\,xy & \to_1 x \\ A_n xyu_1 ... u_n v & \to_1 xu_1 \dots u_n(yu_1 \dots u_n v) \end{array}$$

then termination can be proved. Indeed, one way to prove termination is to interpret the re-write arrow "$\to_1$" as strict descent in a well-ordered structure with a strictly monotone function. The following was observed in [9].

Let O be the set of countable ordinals and $V : O \times O \to O$ Veblen's function defined by

$V(0, b) = \omega^b$
$V(a, b) =$ the bth common fixed point of the functions $\lambda y V(x, y)$ for $x < a$.

Let $W = (O, V, 1)$. W is a structure, in the logician's sense, for any language with a single binary function symbol interpreted as V, and any number of individual constants interpreted as the (finite) ordinal 1. A proper combinator P with reduction rule

$$Px_1 \dots x_n \to_1 X$$

is said to terminate in W if for all $x_1, \dots, x_n$ in O we have

$$X < Px_1 \ldots x_n$$

Theorem 2. *[9]; The proper combinators which terminate in W are precisely those which convert to applicative combinations of I, K and the A_n.*

For our purposes we need only the weaker result that if the simpler structure $(\omega,\ \lambda xy.\ 2^x + y, 1)$ is used, then I, K, and all the A_n terminate. But we can add to this the reductions

$$\begin{array}{lll} \text{(associativity)} & A_m\ (A_m xy)z \to_2 A_m x(A_m yz) & \\ \text{(monoidalitry)} & A_m\ (K^m I)x \to_2 x & \\ & A_m\ x(K^m I) \to_2 x & \\ \text{(annihilation)} & A_m\ (K^n I)x \to_2 K^n I & \text{if } n > m \end{array}$$

which will be useful below. Our use of the symbol $\to_2$ is intended to also include $\to_1$ as our use of the symbol $\to_3$ below is intended to include $\to_2$. Since $\to_2$ obviously satisfies the weak diamond property, each $\to_2$ reduction terminates with a unique normal form. If we add

$$\text{(distributivity) } A_m(A_n xy)z \to_3 A_{n+1}(A_m xz)(A_m yz) \text{ if } m < n$$

then the weak diamond property holds but it is not so straightforward to verify Church-Rosser as before. We shall do this in the needed context by a roundabout argument below. The congruence generated by associativity, monoidalitry, annihilation and distributivity will be referred to as "bad conversion". Interestingly, when viewed as lambda terms there are non normalizable combinations; for example,

$$A_1 IK \twoheadrightarrow_0 \lambda xy.\ xx$$

and

$$(\lambda xy.\ xx)(\lambda xy.\ xx) \to_0 \lambda z.\ (\lambda xy.\ xx)(\lambda xy.\ xx \to_0 \ldots .$$

We believe that this can only happen if K is not in function position. We shall prove that this can only happen with K or an A_m not in function position.

3 Functional Completeness of Bad Algebra for Lambda Calculus

Simple types are built up from type variables $p, q, r, \ldots$ by $\to$. We shall adopt the Church typing discipline for terms requiring sub-terms to be explicitly typed. Nevertheless, it will be convenient to use Curry style notation

$$X : T$$

for X has type T. Fix a type variable s. We define the numerical types

Num(m) by
Num(0) $= s$
Num($m+1$) $=$ Num(m) $\to$ Num(m).

Our combinators have most general (principal) types

$$A_m : (r_1 \to (...(r_m \to (p \to q))...))$$

$$\to ((r_1 \to (...(r_m \to (r \to p))...)) \to (r_1 \to (...(r_m \to (r \to q))...)))$$

$K : p \to (q \to p)$
$I : p \to p.$

Now substituting Num(n) for p, q, r and Num($m - i + n + 1$) for r_i we obtain

A_m : Num($m + n + 1$) $\to$ Num($m + n + 2$)
K : Num(n) $\to$ Num($n + 1$)
I : Num($n + 1$).

If we suppose that M is an applicative combination of the A_m, K and I such that every occurrence of an A_m or K is in function position then M has a numerical type and is strongly beta-eta normalizable. Indeed, we may assume that M is in combinatory normal form, and by conversion (ii), that every occurrence of A_m has two arguments, and by conversion (iii), that every occurrence of K is in a sub-term of the form $K^m I$. Thus M is the interpretation of a term in the language of monoidal bad algebras under the generalized Hoogewijs representation. A term, not assumed to be in combinatory normal form, where every A_m has at least two arguments and each K occurs is in a sub-term $K^m I$ is said to be "equable". A conversion generated by the reduction relation $\to_3$ is said to be equable if every term in the conversion is equable.

It is easily seen that every term in an equable conversion can be simply typed with a numerical type using the typings above for A_m, K, and I

If M is equable let $M!$ result from M by replacing each sub-term $A_n XY$ by $\lambda u_1...u_n v.\ X u_1...u_n (Y u_1...u_n v)$, and each maximal $K^n I$ by $\lambda u_1...u_n v.\ v$. $M!$ is a beta reduct of M and $M!$ has the property

(modesty) If a free variable z occurs in a sub-term Z then Z has numerical type not larger than the type of z.

Modesty is preserved under beta reduction and eta expansion so it is shared by the long beta-eta normal form of $M!$. We now proceed to show the converse.

We say that X is "openly equable" if X is a combination of A^m, K, I and the variables $x_k, x_{k+1}, ..., x_n$, for $k <$ or $= n$, such that X : Num(k), x_i : Num(i) for $i = k, ..., n$, each occurrence of A_m has at least two arguments, and each occurrence of K is in a sub-term $K^m I$. If X is openly equable and X is in combinatory normal form then $X!$ is modest. If X is openly equable we inductively construct a term $[x_k]X$ satisfying the same properties as X with k replaced by $k+1$ such that $([x_k]X)x_k \twoheadrightarrow_1 X$. This is "bracket abstraction" for our context.

The basis case is $X := x_k$ in which case we let $[x_k]X := I$. We can also include in this step $X := K^m I$ in which case we set $[x_k]X := K^{m+1}I$. For the induction step we distinguish three cases with $l > 0$.

Case 1: $X := x_j X_1 ... X_l$. Then $x_j : \text{Num}(j), X_1 : \text{Num}(j-1)..., X(l) : \text{Num}(j-l) = \text{Num}(k)$, so by modesty x_k occurs only in X_l. In this case we can put $[x_k]X := A_0(x_j X_1 ... X_{l-1})([x_k]X(l))$.

Case 2: $X := A_m X_1 ... X_l$. This case is similar to Case 1 if $l > 2$. If $l = 2$ then $X_i : \text{Num}(k)$ for $i = 1, 2$ and we can put $[x_k]X := A_{m+1}([x_k]X_1)([x_k]X_2)$.

Case 3: $X := (K_m I)X_1 ... X_l [x_k]X := A_0((K_m I)X_1 ... X_{l-1})([x_k]X_l)$.

Theorem 3. *(functional completeness) If M is a modest simply typed term of numerical type then M beta-eta converts to an equable applicative combination of A_m, K, and I.*

Proof. By a previous remark it suffices to consider normal M. For modest normal X with variables $x_k, x_{k+1}, ..., x_n$, for $k <$ or $= n$, such that $X : \text{Num}(k)$ and $x_i : \text{Num}(i)$ for $i = k, ..., n$, we construct \$X, a combinatory normal applicative combination of A^m, K, I, and the variables $x_k, x_{k+1}, ..., x_n$ such that $X \twoheadrightarrow_1 \$X$, each occurrence of A_m has two arguments, and each occurrence of K is in a sub-term $K^m I$. We proceed by induction and we can write
$X := \lambda x_{k-1} ... x_{k-l}.\ x_i X_1 ... X_{i-k+l}$ where by assumption $x_j : \text{Num}(j)$, and $X_j : \text{Num}(i-j)$. If $l > 0$ the induction hypothesis applies to $\lambda x_{k-2} ... x_{k-l}.\ x_i X_1 ... X_{i-k+l}$ and we can apply $[x_{k-1}]$. Otherwise, x_k can only occur in X_{i-k} by modesty. Thus $X := YZ$ and the induction hypothesis applies to Y and Z. End of proof.

4 Logical Completeness of Lambda Calculus for Bad Algebra

Lemma 1. *If X is openly equable, $X!$ is modest and $X \rightarrow_1 Y$ then*

$$[x_k]X \twoheadrightarrow_3 [x_k]Y.$$

Proof. By induction on the length of X. The result follows immediately from the induction hypothesis unless the contracted redex is the head redex. If the contracted redex is the head redex we distinguish two cases

Case 1: $X = K^m I X_1 ... X_l$.
Subcase 1: $m > 0$. If $l > 1$ then $Y = K^{m-1} I X_2 ... X_l$ and $[x_k]X \rightarrow_1 [x_k]Y$. If $l = 1$ then $[x_k]X \rightarrow_2 [x_k]Y$

Subcase 2: $m = 0$. In case $l = 1$, $[x_k]X \to_2 [x_k]Y$. In case $l > 1$, by modesty x_k occurs only in X_l so $[x_k]X = A_0(IX_1...X_{l-1})([x_k]X_1) \to_1 [x_k]Y$, since X is (openly) equable.

Case 2: $X = A_m X_1 X_2 X_3 ... X_l$. This case is clear if $l > 3$ since X is modest. If $l = 3$ then

Subcase 1: $m = 0$. So $Y = X_1(X_2X_3)$, $[x_k]X = A_0(A_0X_1X_2)([x_k]X_3) \to_2$ by associativity $A_0X_1(A_0X_2([x_k]X_3)) = [x_k]Y$

Subcase 2: $m > 0$. Then $Y = A_{m-1}(X_1X_3)(X_2X_3)$. As before $[x_k]X = A_0(A_mX_1X_2)([x_k]X_3) \to_3$ by distributivity $= A_m(A_0X_1[x_k]X_3)(A_0X_2[x_k]X_3) = [x_k]Y$. End of proof.

Lemma 2. *If X is openly equable, $X!$ is modest and $X \to_3 Y$ then $[x_k]X) \twoheadrightarrow_3 [x_k]Y$.*

Proof. We have $[x_k](A_m(A_{m'}XY)Z =$ by modesty $(A_m(A_{m'}[x_k]X[x_k]Y)$ $[x_k]Z \to_3$ by distributivity $(A_{m'}(A_m[x_k]X[x_k]Z)(A_m[x_k]Y[x_k]Z) = [x_k]$ $(A_{m'}(A_mXZ)(A_mYZ)$ when $m' > m$. Also when $m' = m$, $[x_k](A_m(A_mXY)Z =$ by modesty $(A_m(A_m[x_k]X[x_k]Y)[x_k]Z \to_2$ by associativity $(A_m([x_k]$ $X(A_m[x_k]Y[x_k]Z) = [x_k](A_mX(A_mYZ)$. The lemma now follows from Lemma 1. End of proof.

Lemma 3. *If X is openly equable, $X!$ is modest, and Y is the beta normal form of X then $X \twoheadrightarrow_3 \$Y$ where $\$Y$ is as in the proof of Theorem 3.*

Proof. There is a standard reduction of X to Y. The proof is by induction on the length of the standard reduction with a subsidiary induction on the length of X. Here we note that if X is an application then either X begins with a variable or X begins with a combinatory redex and contracting the redex commutes with the ! operation, or X is combinatory head normal and $X!$ begins with λ. We consider these cases.

Case 1: X begins with a variable. Then the induction hypothesis applies to the arguments of the variable.

Case 2: X has a combinatory head redex. Then the induction hypothesis applies to the result of contracting the head redex.

Case 3: X is in combinatory head normal form, say $X := A_mUV$. Thus $X! = \lambda u_1...u_nv.\ Uu_1...u_n(Vu_1...u_nv)$. Now $Y = \lambda u_1...u_nv.\ Z$ and the induction hypothesis applies to $Uu_1...u_n(Vu_1...u_nv)$ and Z. Thus $Uu_1...u_n(Vu_1...u_nv) \twoheadrightarrow_3 \$Z$, so by Lemma 2

$[u_1, ..., u_n, v]Uu_1...u_n(Vu_1...u_nv) \twoheadrightarrow_3 [u_1, ..., u_n, v]\$Z$ hence

$X = [u_1, ..., u_n, v]Uu_1...u_n(Vu_1...u_nv)$
$\twoheadrightarrow_3 [u_1, ..., u_n, v]\$Z = \$Y$. End of proof.

Lemma 3 clearly extends to beta-eta normal forms since $[x](Xx) = A_0XI$.

Proposition 1. *(Church-Rosser theorem for $\twoheadrightarrow_3$) If M and N are equable and are linked by an equable conversion then there exists an equable P such that $M \twoheadrightarrow_3 P_3 \twoheadleftarrow N$. Moreover P can be taken to be in combinatory normal form.*

Proof. If M and N are linked by an equable conversion then they beta-eta convert. Thus there exists beta-eta normal Q such that M and N both beta- eta reduce to Q. By Lemma 3 $M \twoheadrightarrow_3 Q_3 \twoheadleftarrow N$. End of proof.

Theorem 4. *(logical completeness); If M and N are normal equable applicative combinations of A_m, K and I then M bad converts to N if and only if M beta-eta converts to N.*

We can state a corollary to Theorem 4 in informal terms as follows. A first-order equation in the language of monoidal bad algebras is true in all monoidal bad algebras if, and only if, the corresponding lambda terms beta-eta convert. A similar result holds for equations containing free variables. For this it suffices to extend our typing to intersection types. The first order free variables appear as parameters with types which are intersections of numerical types. We shall have a less trivial application of intersection types below.

5 The B, I Monoid

The lowest level, the monoid generated by B and I alone, turns out to be of independent interest. It is the "positive part" of Richard Thompson's group F, and it generates this group. In 1965, Thompson discovered this group in connection to his theorem characterizing the groups with solvable word problems. The group F was rediscovered by Peter Freyd and Alex Heller in 1993 in connection with homotopy retracts $x * x \equiv x$ which do not split $y * z \equiv I$ and $x \equiv z * y$. It will be convenient to write Bxy in infix "monoid" notation $x * y$. The following beta-eta conversions hold

Nx	$\sim x * ... * x * 1$	if N is the Church numeral for n and there are n occurrences of x
$B(x * y)$	$\sim$	$Bx * By$
$B * N$	$\sim$	$N * B$
$B(Bx) * B$	$\sim$	$B * Bx$
$K(x * y)$	$\sim$	$Bx * Ky$
Kx	$\sim$	$Bx * 0$
$B(Bx) * K$	$\sim$	$K * x$.

Note that the 3rd and 4th equation have B not in function position in contrast to the previous paragraphs. "Monomials" have the form

$$B(...(BA)...) \qquad d \text{ occurrences of } B$$

where A is one of the atoms B, K or a Church numeral N. d is the "degree" of the monomial and A is its "sort". A product of monomials is a "multinomial". We have the

Lemma 4. *Every applicative combination of B, K, I and Church numerals converts to a multinomial.*

Proof. Observe that if P is normal w.r.t. weak beta reductions and associativity then P can be reduced a product of monomials by the reductions

$BI \rightarrow_4 1$
$Nx \rightarrow_4 x * ... * x * 1$ if N is the Church numeral for n
and there are n occurrences of x
$B(x * y) \rightarrow_4 Bx * By$
$K(x * y) \rightarrow_4 Bx * Ky$
$Kx \rightarrow_4 Bx * 0$.

In a multinomial M, if for any two consecutive monomials
$B(...(BA')...) * B(.....(BA").....)$,
of degrees d' and d'', we have $|d'' - d'| < 2$, M is said to be "semi-regular". A maximal consecutive sub-product of monomials of the same degree is called a "block" of M. If M is semi-regular and all of its blocks have the same number of monomials we say that M is "regular".

Proposition 2. *Every multinomial converts to one which is regular.*

This can be seen by inserting factors $B(...(B1)...)$.

Church observed that the Church numeral for n applied to the Church numeral for m beta reduces to the Church numeral for m^n. Consequently, every applicative combination of Church numerals beta-eta strongly normalizes. We have already proved that every applicative combination of B, K, I, and Church numerals, where every occurrence of B and K is in function position, is strongly normalizable. We shall now show that the condition on K can be relaxed. Given a regular multinomial M we may write $M = M(m) * ... * M(1)$ where each $M(i)$ is a block. Put $s(i) =$ the number of monomials in $M(i)$ whose sort is B or K. We call the sequence $s(1)...s(m)$ the "statistics of M".

Theorem 5. *An applicative combination of B, K, and Church numerals where each occurrence of B is in function position is strongly normalizable.*

Proof. First we will show that if a $\rightarrow_1$ or $\rightarrow_4$ reduct of a combination is strongly beta-eta normalizable then so is the combination. We show this by showing that intersection types [4] lift from contracta to redexes. Lifting is straightforward for combinatory redexes. As to $\rightarrow_4$ consider the following beta-eta conv's. Note that in each, eta is only used in the reductions from right to left. We observe that in the reduction from left to right, if a given r.h.s. has an intersection type then its expansion has the same type. In addition, in the reduction from right to left, if a term has an intersection type then its reduct has the same type by the subject reduction theorem. This includes eta reductions. These remarks cover all $\rightarrow_4$ except the special case $(1x) \rightarrow_4 x$. For this case, observe that in application x will always be a closed term beginning with lambda, so $(1x)$ has the desired type by the normal form for intersection types.

(i) $B(Bxy)z \twoheadrightarrow_0 \lambda v.\ (Bxy)(zv) \twoheadrightarrow_0 \lambda v.\ x(y(zv))_0 \twoheadleftarrow Bx(Byz)$
(ii) $Kx \rightarrow_0 \lambda y.\ x\ {}_0\!\leftarrow \lambda yv.\ xv\ {}_0\!\twoheadleftarrow \lambda yv.\ x(KIyv)\ {}_0\!\twoheadleftarrow \lambda y.\ Bx(KIy)\ {}_0\!\leftarrow (Bx)(KI)$
(iii) $K(Bxy) \rightarrow_0 \lambda u(Bxy) \twoheadrightarrow_0 \lambda uv.\ x(yv)\ {}_0\!\twoheadleftarrow \lambda uv.\ x(yv)\ {}_0\!\twoheadleftarrow \lambda uv.\ x(Kyu)\ {}_0\!\twoheadleftarrow \lambda u.(Bx)(Kyu)_0 \twoheadleftarrow B(Bx)(Ky)$
(iv) $B(Bxy) \twoheadrightarrow_0 \lambda uv.\ (Bxy)(uv) \rightarrow_0 \lambda uv.\ x(y(uv))\ {}_0\!\twoheadleftarrow \lambda uv.\ x(Byuv)\ {}_0\!\twoheadleftarrow \lambda u.\ (Bx)(Byu)\ {}_0\!\twoheadleftarrow B(Bx)(By)$
(v) $Nx \rightarrow_0 \lambda y.\ x(\ldots(xy)\ldots)\ {}_0\!\twoheadleftarrow \ \ldots\ {}_0\!\twoheadleftarrow \lambda y.\ x((...(Bxx)...)y)\ {}_0\!\twoheadleftarrow Bx(\ldots(Bxx)\ldots)$. The cases for the remaining are straightforward.

Now suppose that M is a multinomial all of whose sorts are either K or Church numerals. We shall show that M has an intersection type. To simplify notation we shall assume that each Church numeral is 2. The cases of 0 and 1 are straightforward and the cases of $n > 2$ are treated exactly like the case of 2, except for much notational complication. Suppose that $M = M(m) * ... * M(1)$ where each $M(i)$ is a monomial of degree $d(i)$ with sort 2 or K. We construct a rooted tree with the following properties.

(0) The root is a line incident with only a single point; the "low point".
(1) The tree is binary branching; that is, each point has either 0 (a leaf), 1, or 2 ancestors. All points, except the low point, have a unique descendent.
(2) Each path from the root to a leaf has length $m + 1$, as measured by the number of points on the path.
(3) Each line of the tree is labeled with a simple type. This includes the root. The types associated with lines joining a point to its ancestors are "ancestor types", and the type associated with the line joining a point to its descendent (or, in the case of the low point, the type associated with the root) is the "descendent type".

The tree will be constructed and the types will be assigned in such a way that

(i) if $M(i)$ has a Church numeral sort then each point at depth i has exactly two ancestors and if $T", T'$ are the types associated with these two ancestors and T is the descendent type then

$$M(i) : (T" \wedge T') \rightarrow T$$

(ii) If $M(i)$ has K as its sort then each point at depth i has a unique ancestor and if T' is a type associated with this ancestor and T is the descendent type then

$$M(i) : T' \rightarrow T.$$

First we construct the tree without labels. This is completely determined by (i) and (ii). Next, we label each line with 1 or 2 simple types s.t.

(a) A line incident with a leaf, and the root, are labeled with a single simple type.

(b) All other lines are labeled with two simple types: +type and −type. The pair of types associated with a given line will have to be unified in order to achieve the final labelling. This preliminary labelling goes as follows.

We define the revised degree

$$d(i)' = \begin{cases} d(i) \text{ if the sort of } M(i) \text{ is } K \\ d(i)+1 \text{ is the sort of } M(i) \text{ is } 2 \end{cases}$$

and let $d = max\{d(i)'|i = 1,\ldots,m\}$. Set $e(i) = d - d(i) + s(1) + \ldots + s(i-1)$.

If the sort of $M(i)$ is 2 then we consider any point at depth i and select new type variables $r_1 \ldots r_{d(i)}, p_1, \ldots p_{e(i)}, p, q_1, \ldots, q_{e(i)}, q, r$. The left ancestor of this point gets −type
$r_1 \to (\ldots (r_{d(i)} \to ((q_1 \to (\ldots (q_{e(i)} \to q)\ldots)))) \ldots)$
and the right ancestor gets −type

$$r_1 \to (\ldots (r_{d(i)} \to ((q_1 \to (\ldots (q_{e(i)} \to q)\ldots)) \to (p_1 \to (\ldots (p_{e(i)} \to r)\ldots)))) \ldots)$$

and the descendent of this point gets +type

$$r_1 \to (\ldots (r_{d(i)} \to (p \to (p_1 \to (\ldots (p_{e(i)} \to r)\ldots)))\ldots).$$

If the sort of this point is K then we consider any point at depth i and select new type variables $r_1 \ldots r_{d(i)}, p, q_1, \ldots, q_{e(i)}, q$. The ancestor point of this point gets −type

$$r_1 \to (\ldots (r_{d(i)} \to (q_1 \to (\ldots (q_{e(i)} \to q)\ldots))) \ldots)$$

and the descendent of this point gets +type

$$r_1 \to (\ldots (r_{d(i)} \to (p \to (q_1 \to (\ldots (q_{e(i)} \to q)\ldots)))) \ldots).$$

Now if $M(i)$ has sort 2 then

$$r_1 \to (\ldots (r_{d(i)} \to (p \to (q_1 \to (\ldots (q_{e(i)} \to q)\ldots)))) \ldots) \wedge$$
$$r_1 \to (\ldots (r_{d(i)} \to ((q_1 \to (\ldots (q_{e(i)} \to q)\ldots)) \to (p_1 \to (\ldots (p_{e(i)} \to r)\ldots)))) \ldots)$$

which is Coppo and Dezani [4] type equivalent to

$$r_1 \to (\ldots (r_{d(i)} \to ((p \to (q_1 \to (\ldots (q_{e(i)} \to q)\ldots))) \wedge$$
$$((q_1 \to (\ldots (q_{e(i)} \to q)\ldots) \to (p_1 \to (\ldots (p_{e(i)} \to r)\ldots)))) \ldots)$$

so (i) and (ii) are clearly satisfied.

Finally there is a single substitution which simultaneously unifies all the pairs of +type and −type. This is defined directly by recursion on $m-$ the depth of the point noting that any such pair have the same number of components.

Now if we put the same variable x at each leaf with type Coppo and Dezani equivalent to the intersection of all the types at the leaves we see that Mx has and is strongly normalizable. End of proof.

6 Church's Theorem

If we endow the set of natural numbers with the discrete topology and the set of functions from natural numbers to natural numbers with the product topology then a term M could compute a uniformly continuous functional from functions to numbers if there exists a k such that for any sequence of Church numerals $N(1) \ldots N(k)$

$$MN(1) \ldots N(k)$$

converts to a Church numeral.

A consequence of our simple type assignment is that if M is an applicative combination of the A_m, K and I such that every occurrence of an A_m or K is in function position then M computes a uniformly continuous functional. The reason is that M has a numerical type, and so we can pick the types of $N(1), \ldots, N(k)$ so that the type of

$$MN(1) \ldots N(k)$$

is Num(2), and the only long beta-eta normal forms of type Num(2) are Church numerals. Similarly, if M is an applicative combination of B, K and Church numerals such that every Church numeral is in function position then M is strongly normalizable and M computes a uniformly continuous functional. This can be seen as follows. We have the following beta-eta conversions

$$
\begin{aligned}
B(x * y) &\sim Bx * By \\
K(x * y) &\sim Bx * Ky \\
K(Bx) &\sim B * Kx \\
B(Kx) &\sim K(Kx) \\
Kx * y &\sim Kx \\
K * Bx &\sim B(Bx) * K \\
K * B &\sim BB * K \\
K * K^m I &\sim K^m I \qquad \text{if } m > 0.
\end{aligned}
$$

The first four permit a multinomial form where each monomial has the form

(a) $B^m B$
(b) $B^m(KI)$
(c) K

The fifth conversion insures that if a monomial of the form (b) with $m = 0$ occurs then it is the rightmost monomial in the multinomial. Finally, the conversions six, seven, and eight, together with the first, insure that all K's occur rightmost left of such a monomial, and all B's rightmost left of them.

It follows that every monomial is beta-eta convertible to one built up from I by the operation

$$X \to_5 BX * B \ldots B * K \ldots K$$

possibly with KI appended. Now it is clear that

$$B \dots B * K \dots K * KI$$

has a simple type of the form $\mathrm{Num}(n_2) \rightarrow \mathrm{Num}(n_1)$. Thus by induction if X has a simple type

$$\mathrm{Num}(n_k) \rightarrow (\dots(\mathrm{Num}(n_2) \rightarrow \mathrm{Num}(n_1)\dots))$$

then

$$BX * B \dots B * K \dots * K * KI$$

has simple type

$$\mathrm{Num}(n'_{k+1}) \rightarrow (\dots(\mathrm{Num}(n'_2) \rightarrow \mathrm{Num}(n'_1)\dots).$$

Now we can proceed as before. The multinomial M is said to be non-decreasing (non-increasing) if the degrees of the blocks of M are non-decreasing (non-increasing) from right to left.

Examples:

(i) An applicative combination of B, K and Church numerals where each positive Church numeral is in function position converts to a non-decreasing multinomial.
(ii) An applicative combination of B, K, and Church numerals where each occurrence of B, and K is in function position converts to a multinomial all of whose sorts are Church numerals (statistics all 0's).
(iii) A regular,non-decreasing multinomial with no sort a positive integer, whose statistics $s(1) \dots s(m)$ satisfies

$$i > s(i) + \dots + s(1)$$

converts to a non-increasing multinomial.

A binary product $M'' * M'$ of regular monomials with the same number of blocks and statistics $s(1)'' \dots s(m)''$, $s(1)' \dots s(m)'$ respectively is called an "oscillator" if M'' is non-increasing, M' is non-decreasing, and $s(i)'' = s(m+1-i)'$. A multinomial M is said to be "harmonic" if it is a product of oscillators.

Examples(cont'd): the following kinds of multinomials convert to harmonic ones

(iv) Monomials where every sort is an integer.
(v) $M * M$, where M is regular,non-decreasing, no sort is a positive integer and the statistics $s(1) \dots s(m)$ satisfy $i > s(i) + \dots + s(1)$.

We have the following generalization of Church's theorem.

Theorem 6. *If M is harmonic then M is strongly normalizable and for sufficiently large k, for any sequence of Church numerals $N(1) \ldots N(k)$*

$$MN(1) \ldots N(k)$$

converts to a Church numeral.

Proof. Ommitted for lack of space.

7 The Game of Alonzo

We would like to know Theorem 5 for all applicative combinations of B, K, and Church numerals. We would like to know that every applicative combination of B, K, and Church numerals is strongly normalizable. Indeed, we would like to know the truth of the following conjecture.

Conjecture: If M is a product of monomials each of whose sorts is either B or the Church numeral 2, then for sufficiently large k,

$$M \underbrace{1 \ldots 1}_{k} \text{ conv.} 1 \quad .$$

Now M 1 can be calculated as follows:

$$\begin{array}{ll} BP(M\ 1) & \sim P * (M\ 1) \\ B(P * Q) & \sim BP * BQ \\ B\ 1 & \sim 1 \\ 2\ P & \sim P * P \end{array}$$

so we can formulate an equivalent conjecture about a game, which we call the "game of Alonzo". The game of Alonzo consists of n players and a dealer. The players and the dealer are seated around a circular table. Play takes place counter-clockwise around the table (right to left). Each player has a stack of 1,000$ chips and is dealt a single card, which is either an ace (2) or a duce (B), by the dealer. As play rotates counter-clockwise to a given player three rules apply; the "ante", the "cashout", and the "split".

(1) If the player has at least one chip then he ante's up by putting one 1,000$ chip in the center of the table. This is the "ante".
(2) If the player has no chips and his card is a duce then he must pay for and give one 1,000$ chip to each player to his left and to the right of the dealer. These chips are bought and payed for at a bank separate from the game. This is "cashing out". The player then leaves the game, unless he is the last player.
(3) If a player has no chips and his card is an ace then he gets to play a copy of each of the hands of the players to his left and to the right of the dealer. Each copy gets the corresponding number of chips and all can be played

independently. This is "splitting". If there are no such players to his left he leaves the game unless he is the last player. Play continues to rotate counter-clockwise through the dealer and no new cards are dealt. Play ends when there is only one player left, and he wins the pot.

Conjecture: Every game ends.

References

1. Barendregt, H.P.: The Lambda Calculus. North Holland (1984)
2. Birget, J.C.: Monoid generalizations of the Richard Thompson groups. Journal of Pure and Applied Algebra 213, 264–278 (2009)
3. Church, A.: The Calculi of Lambda Conversion. PUP (1941)
4. Coppo, M., Dezani, M.: A new type assignment for lambda terms. Archiv fur Math. Logik 19 (1978)
5. Curry, H.B.: Combinatory Logic, vol. 1. North Holland (1958)
6. Freyd, P., Heller, A.: Splitting homotopy idempotents II. Journal of Pure and Applied Algebra 89, 93–195 (1993)
7. Hoogewijs, J.: Semi-nearring embeddings. Med. Konink. Vlaamse Acad. Wetensch. Lett. Schone Kunst. Belgie Kl. Wetensch 32(2), 3–11 (1970)
8. Pottinger, G.: A type assignment for strongly normalizable lambda terms. Curry Festschrift (1980)
9. Statman, R.: Freyd's hierarchy of combinator monoids. In: LICS (1991)
10. Thompson, R.: Embeddings into finitely generated groups which preserve the word problem. In: Adian, Boone, Higman (eds.) The Word Problem. Studies in Logic and the Foundation of Mathematics, vol. 95, pp. 401–441. North Holland (1980)
11. Van Hoorn, W., Van Rootselaar, B.: Fundamental notions in the theory of semi-near rings. Comp. Math. 18, 65–78 (1967)
12. Artin, M.: Algebra. Prentice Hall (2011)
13. Internet, `http://www.algebra.uni-linz.ac.at/Nearrings`

All-Path Reachability Logic*

Andrei Ştefănescu[1], Ştefan Ciobâcă[2], Radu Mereuta[1,2],
Brandon M. Moore[1], Traian Florin Şerbănuţă[3], and Grigore Roşu[1,2]

[1] University of Illinois at Urbana-Champaign, USA
{stefane1,radum,bmmoore,grosu}@illinois.edu
[2] University "Alexandru Ioan Cuza", Romania
stefan.ciobaca@info.uaic.ro
[3] University of Bucharest, Romania
traian.serbanuta@fmi.unibuc.ro

Abstract. This paper presents a language-independent proof system for reachability properties of programs written in non-deterministic (e.g. concurrent) languages, referred to as *all-path reachability logic*. It derives partial-correctness properties with all-path semantics (a state satisfying a given precondition reaches states satisfying a given postcondition on all terminating execution paths). The proof system takes as axioms any unconditional operational semantics, and is sound (partially correct) and (relatively) complete, independent of the object language; the soundness has also been mechanized (Coq). This approach is implemented in a tool for semantics-based verification as part of the $\mathbb{K}$ framework.

1 Introduction

Operational semantics are easy to define and understand. Giving a language an operational semantics can be regarded as "implementing" a formal interpreter. Operational semantics require little formal training, scale up well and, being executable, can be tested. Thus, operational semantics are typically used as trusted reference models for the defined languages. Despite these advantages, operational semantics are rarely used directly for program verification (i.e. verifying properties of a given program, rather than performing meta-reasoning about a given language), because such proofs tend to be low-level and tedious, as they involve formalizing and working directly with the corresponding transition system. Hoare or dynamic logics allow higher level reasoning at the cost of (re)defining the language as a set of abstract proof rules, which are harder to understand and trust. The state-of-the-art in mechanical program verification is to develop and prove such language-specific proof systems sound w.r.t to a trusted operational semantics [1–3], but that needs to be done for each language separately.

Defining more semantics for the same language and proving the soundness of one semantics in terms of another are highly uneconomical tasks when real programming languages are concerned, often taking several years to complete. Ideally, we would like to have only one semantics for a language, together with a generic theory and a set of generic tools and techniques allowing us to get all the benefits of any other semantics

* Full version of this paper, with proofs, available at http://hdl.handle.net/2142/46296

G. Dowek (ed.): RTA-TLCA 2014, LNCS 8560, pp. 425–440, 2014.

without paying the price of defining other semantics. Recent work [4–7] shows this *is possible*, by proposing a *language-independent proof system* which derives program properties directly from an operational semantics, *at the same proof granularity and compositionality* as a language-specific axiomatic semantics. Specifically, it introduces *(one-path) reachability rules*, which generalize both operational semantics reduction rules and Hoare triples, and give a proof system which derives new reachability rules (program properties) from a set of given reachability rules (the language semantics).

However, the existing proof system has a major limitation: it only derives reachability rules with a *one-path* semantics, that is, it guarantees a program property holds on one but not necessarily all execution paths, which suffices for deterministic languages but not for non-deterministic (concurrent) languages. We here remove this limitation, proposing the first generic all-path reachability proof system for program verification.

Using *matching logic* [8] as a configuration specification formalism (Section 2), where a pattern φ specifies all program configurations that match it, we first introduce the novel notion of an *all-path reachability rule* $\varphi \Rightarrow^{\forall} \varphi'$ (Section 3), where φ and φ' are matching logic patterns. Rule $\varphi \Rightarrow^{\forall} \varphi'$ is valid iff any program configuration satisfying φ reaches, on any complete execution path, some configuration satisfying φ'. This subsumes partial-correctness in non-deterministic languages. We then present a proof system for deriving an all-path reachability rule $\varphi \Rightarrow^{\forall} \varphi'$ from a set $\mathcal{S}$ of semantics rules (Section 4). $\mathcal{S}$ consists of reduction rules $\varphi_l \Rightarrow^{\exists} \varphi_r$, where φ_l and φ_r are simple patterns as encountered in operational semantics (Section 6), which can be non-deterministic. The proof system derives more general sequents "$\mathcal{S}, \mathcal{A} \vdash_C \varphi \Rightarrow^{\forall} \varphi'$", with $\mathcal{A}$ and C two sets of reachability rules. Intuitively, $\mathcal{A}$'s rules (*axioms*) are already established valid, and thus can be immediately used. Those in C (*circularities*) are only claimed valid, and can be used only after taking execution steps based on the rules in $\mathcal{S}$ or $\mathcal{A}$. The most important proof rules are

Step :

$$\frac{\begin{array}{ll}\models \varphi \rightarrow \bigvee_{\varphi_l \Rightarrow^{\exists} \varphi_r \,\in\, \mathcal{S}} \exists FreeVars(\varphi_l)\, \varphi_l & \\ \models \exists c\, (\varphi[c/\square] \wedge \varphi_l[c/\square]) \wedge \varphi_r \rightarrow \varphi' & \text{for each } \varphi_l \Rightarrow^{\exists} \varphi_r \,\in\, \mathcal{S}\end{array}}{\mathcal{S}, \mathcal{A} \vdash_C \varphi \Rightarrow^{\forall} \varphi'}$$

Circularity :

$$\frac{\mathcal{S}, \mathcal{A} \vdash_{C \cup \{\varphi \Rightarrow^{\forall} \varphi'\}} \varphi \Rightarrow^{\forall} \varphi'}{\mathcal{S}, \mathcal{A} \vdash_C \varphi \Rightarrow^{\forall} \varphi'}$$

Step is the key proof rule which deals with non-determinism: it derives a sequent where φ reaches φ' in one step on all paths. The first premise ensures that any configuration satisfying φ has successors, the second that all successors satisfy φ' ($\square$ is the configuration placeholder). Circularity adds the current goal to C at any point in a proof, and generalizes language-independently the various language-specific axiomatic semantics invariant rules (this form was introduced in [4]).

We illustrate on examples how our proof system enables state exploration (similar to symbolic model-checking), and verification of program properties (Section 6). We show that our proof system is sound and relatively complete (Section 5). We describe our implementation of the proof system as past of the $\mathbb{K}$ framework [9] (Section 7).

Contributions. This paper makes the following specific contributions:

1. A language-independent proof system for deriving all-path reachability properties, with proofs of its soundness and relative completeness; the soundness result has also been mechanized in Coq, to serve as a foundation for certifiable verification.
2. An implementation of it as part of the $\mathbb{K}$ framework.

2 Matching Logic

Here we briefly recall matching logic [8], which is a logic designed for specifying and reasoning about arbitrary program and system configurations. A matching logic formula, called a *pattern*, is a first-order logic (FOL) formula with special predicates, called basic patterns. A *basic pattern* is a configuration term with variables. Intuitively, a pattern specifies both structural and logical constraints: a configuration satisfies the pattern iff it matches the structure (basic patterns) and satisfies the constraints.

Matching logic is parametric in a signature and a model of configurations, making it a prime candidate for expressing state properties in a language-independent verification framework. The configuration signature can be as simple as that of IMP (Fig. 3), or as complex as that of the C language [10] (with more than 70 semantic components).

We use basic concepts from multi-sorted first-order logic. Given a *signature* Σ which specifies the sorts and arities of the function symbols (constructors or operators) used in configurations, let $T_\Sigma(Var)$ denote the *free Σ-algebra* of terms with variables in *Var*. $T_{\Sigma,s}(Var)$ is the set of Σ-terms of sort s. A valuation $\rho: Var \to \mathcal{T}$ with $\mathcal{T}$ a Σ-algebra extends uniquely to a (homonymous) *Σ-algebra morphism* $\rho: T_\Sigma(Var) \to \mathcal{T}$. Many mathematical structures needed for language semantics have been defined as Σ-algebras, including: boolean algebras, natural/integer/rational numbers, lists, sets, bags (or multisets), maps (e.g., for states, heaps), trees, queues, stacks, etc.

Let us fix the following: (1) an algebraic signature Σ, associated to some desired configuration syntax, with a distinguished sort *Cfg*, (2) a sort-wise infinite set *Var* of variables, and (3) a Σ-algebra $\mathcal{T}$, the *configuration model*, which may but need not be a term algebra. As usual, $\mathcal{T}_{Cfg}$ denotes the elements of $\mathcal{T}$ of sort *Cfg*, called *configurations*.

Definition 1. *[8] A matching logic formula, or a* ***pattern****, is a first-order logic (FOL) formula which additionally allows terms in $T_{\Sigma,Cfg}(Var)$, called* ***basic patterns****, as predicates. A pattern is* ***structureless*** *if it contains no basic patterns.*

We define satisfaction $(\gamma,\rho) \models \varphi$ over configurations $\gamma \in \mathcal{T}_{Cfg}$, valuations $\rho: Var \to \mathcal{T}$ and patterns φ as follows (among the FOL constructs, we only show $\exists$):

$(\gamma,\rho) \models \exists X\, \varphi$ *iff* $(\gamma,\rho') \models \varphi$ *for some* $\rho' : Var \to \mathcal{T}$ *with* $\rho'(y) = \rho(y)$ *for all* $y \in Var \backslash X$

$(\gamma,\rho) \models \pi$ *iff* $\gamma = \rho(\pi)$ *where* $\pi \in T_{\Sigma,Cfg}(Var)$

We write $\models \varphi$ when $(\gamma,\rho) \models \varphi$ for all $\gamma \in \mathcal{T}_{Cfg}$ and all $\rho : Var \to \mathcal{T}$.

A basic pattern π is satisfied by all the configurations γ that *match* it; in $(\gamma,\rho) \models \pi$ the ρ can be thought of as the "witness" of the matching, and can be further constrained in a pattern. For instance, the pattern from Section 6

$$\langle \texttt{x:=x+1} \parallel \texttt{x:=x+1},\ \texttt{x} \mapsto n \rangle \wedge (n = 0 \vee n = 1)$$

is matched by the configurations with code "$\texttt{x:=x+1} \parallel \texttt{x:=x+1}$" and state mapping program variable `x` into integer n with n being either 0 or 1. We use *italic* for mathematical variables in *Var* and `typewriter` for program variables (program variables are represented in matching logic as constants of sort *PVar*, see Section 6).

Next, we recall how matching logic formulae can be translated into FOL formulae, so that its satisfaction becomes FOL satisfaction in the model of configurations, $\mathcal{T}$. Then, we can use conventional theorem provers or proof assistants for pattern reasoning.

Definition 2. *[8] Let* $\square$ *be a fresh Cfg variable. For a pattern* φ*, let* $\varphi^\square$ *be the FOL formula formed from* φ *by replacing basic patterns* $\pi \in \mathcal{T}_{\Sigma,Cfg}(Var)$ *with equalities* $\square = \pi$*. If* $\rho : Var \to \mathcal{T}$ *and* $\gamma \in \mathcal{T}_{Cfg}$ *then let the valuation* $\rho^\gamma : Var \cup \{\square\} \to \mathcal{T}$ *be such that* $\rho^\gamma(x) = \rho(x)$ *for* $x \in Var$ *and* $\rho^\gamma(\square) = \gamma$*.*

With the notation in Definition 2, $(\gamma, \rho) \models \varphi$ iff $\rho^\gamma \models \varphi^\square$, and $\models \varphi$ iff $\mathcal{T} \models \varphi^\square$. Thus, matching logic is a methodological fragment of the FOL theory of $\mathcal{T}$. We drop $\square$ from $\varphi^\square$ when it is clear in context that we mean the FOL formula instead of the matching logic pattern. It is often technically convenient to eliminate $\square$ from φ, by replacing $\square$ with a *Cfg* variable c and using $\varphi[c/\square]$ instead of φ. We use the FOL representation in the Step proof rule in Fig. 1, and to establish relative completeness in Section 5.

3 Specifying Reachability

In this section we define one-path and all-path reachability. We begin by recalling some matching logic reachability [6] notions that we need for specifying reachability.

Definition 3. *[6] A (one-path)* ***reachability rule*** *is a pair* $\varphi \Rightarrow^\exists \varphi'$*, where* φ *and* φ' *are patterns (which can have free variables). Rule* $\varphi \Rightarrow^\exists \varphi'$ *is* ***weakly well-defined*** *iff for any* $\gamma \in \mathcal{T}_{Cfg}$ *and* $\rho : Var \to \mathcal{T}$ *with* $(\gamma, \rho) \models \varphi$*, there exists* $\gamma' \in \mathcal{T}_{Cfg}$ *with* $(\gamma', \rho) \models \varphi'$*. A* ***reachability system*** *is a set of reachability rules. Reachability system* $\mathcal{S}$ *is* ***weakly well-defined*** *iff each rule is weakly well-defined.* $\mathcal{S}$ *induces a* ***transition system*** $(\mathcal{T}, \Rightarrow^{\mathcal{T}}_{\mathcal{S}})$ *on the configuration model:* $\gamma \Rightarrow^{\mathcal{T}}_{\mathcal{S}} \gamma'$ *for* $\gamma, \gamma' \in \mathcal{T}_{Cfg}$ *iff there is some rule* $\varphi \Rightarrow^\exists \varphi'$ *in* $\mathcal{S}$ *and some valuation* $\rho : Var \to \mathcal{T}$ *with* $(\gamma, \rho) \models \varphi$ *and* $(\gamma', \rho) \models \varphi'$*. A* $\Rightarrow^{\mathcal{T}}_{\mathcal{S}}$***-path*** *is a finite sequence* $\gamma_0 \Rightarrow^{\mathcal{T}}_{\mathcal{S}} \gamma_1 \Rightarrow^{\mathcal{T}}_{\mathcal{S}} \ldots \Rightarrow^{\mathcal{T}}_{\mathcal{S}} \gamma_n$ *with* $\gamma_0, \ldots, \gamma_n \in \mathcal{T}_{Cfg}$*. A* $\Rightarrow^{\mathcal{T}}_{\mathcal{S}}$*-path is* ***complete*** *iff it is not a strict prefix of any other* $\Rightarrow^{\mathcal{T}}_{\mathcal{S}}$*-path.*

We assume an operational semantics is a set of (unconditional) reduction rules "$l \Rightarrow^\exists r$ if b", where $l, r \in T_{\Sigma,Cfg}(Var)$ are program configurations with variables and $b \in T_{\Sigma,Bool}(Var)$ is a condition constraining the variables of l, r. Styles of operational semantics using only such (unconditional) rules include evaluation contexts [11], the chemical abstract machine [12] and $\mathbb{K}$ [9] (see Section 6 for an evaluation contexts semantics). Several large languages have been given semantics in such styles, including C [10] (about 1200 rules) and R5RS Scheme [13]. The reachability proof system works with any set of rules of this form, being agnostic to the particular style of semantics.

Such a rule "$l \Rightarrow^\exists r$ if b" states that a ground configuration γ which is an instance of l and satisfies the condition b reduces to an instance γ' of r. Matching logic can express terms with constraints: $l \wedge b$ is satisfied by exactly the γ above. Thus, we can regard such a semantics as a particular weakly well-defined reachability system $\mathcal{S}$ with rules of the form "$l \wedge b \Rightarrow^{\exists} r$". The weakly well-defined condition on $\mathcal{S}$ guarantees that if γ matches the left-hand-side of a rule in $\mathcal{S}$, then the respective rule induces an outgoing transition from γ. The transition system induced by $\mathcal{S}$ describes precisely the behavior of any program in any given state. In [4–6] we show that reachability rules capture one-path reachability properties and Hoare triples for deterministic languages.

Formally, let us fix an operational semantics given as a reachability system $\mathcal{S}$. Then, we can specify reachability in the transition system induced by $\mathcal{S}$

Definition 4. *An **all-path reachability rule** is a pair $\varphi \Rightarrow^{\forall} \varphi'$ of patterns φ and φ'.*

An all-path reachability rule $\varphi \Rightarrow^{\forall} \varphi'$ is satisfied, $\mathcal{S} \models \varphi \Rightarrow^{\forall} \varphi'$, iff for all complete $\Rightarrow^{\mathcal{T}}_{\mathcal{S}}$-paths τ starting with $\gamma \in \mathcal{T}_{Cfg}$ and for all $\rho : Var \to \mathcal{T}$ such that $(\gamma, \rho) \models \varphi$, there exists some $\gamma' \in \tau$ such that $(\gamma', \rho) \models \varphi'$.

A one-path reachability rule $\varphi \Rightarrow^{\exists} \varphi'$ is satisfied, $\mathcal{S} \models \varphi \Rightarrow^{\exists} \varphi'$, iff for all $\gamma \in \mathcal{T}_{Cfg}$ and $\rho : Var \to \mathcal{T}$ such that $(\gamma, \rho) \models \varphi$, there is either a $\Rightarrow^{\mathcal{T}}_{\mathcal{S}}$-path from γ to some γ' such that $(\gamma', \rho) \models \varphi'$, or there is a diverging execution $\gamma \Rightarrow^{\mathcal{T}}_{\mathcal{S}} \gamma_1 \Rightarrow^{\mathcal{T}}_{\mathcal{S}} \gamma_2 \Rightarrow^{\mathcal{T}}_{\mathcal{S}} \cdots$ from γ.

The racing increment example in Section 6 can be specified by

$$\langle \texttt{x:=x+1} \,||\, \texttt{x:=x+1},\ \texttt{x}\mapsto m \rangle \Rightarrow^{\forall} \exists n\, (\langle \texttt{skip},\ \texttt{x}\mapsto n \rangle \wedge (n = m +_{Int} 1 \vee n = m +_{Int} 2)$$

which states that every terminating execution reaches a state where execution of both threads is complete and the value of **x** has increased by 1 or 2 (this code has a race).

A Hoare triple describes the resulting state after execution finishes, so it corresponds to a reachability rule where the right side contains no remaining code. However, all-path reachability rules are strictly more expressive than Hoare triples, as they can also specify intermediate configurations (the code in the right-hand-side need not be empty) Reachability rules provide a unified representation for both language semantics and program specifications: $\varphi \Rightarrow^{\exists} \varphi'$ for semantics and $\varphi \Rightarrow^{\forall} \varphi'$ for all-path reachability specifications. Note that, like Hoare triples, reachability rules can only specify properties of complete paths (that is, terminating execution paths). One can use existing Hoare logic techniques to break reasoning about a non-terminating program into reasoning about its terminating components.

4 Reachability Proof System

Fig. 1 shows our novel proof system for all-path reachability. The target language is given as a weakly well-defined reachability system $\mathcal{S}$. The soundness result (Thm. 1) guarantees that $\mathcal{S} \models \varphi \Rightarrow^{\forall} \varphi'$ if $\mathcal{S} \vdash \varphi \Rightarrow^{\forall} \varphi'$ is derivable. Note that the proof system derives more general sequents of the form $\mathcal{S}, \mathcal{A} \vdash_{C} \varphi \Rightarrow^{\forall} \varphi'$, where $\mathcal{A}$ and C are sets of reachability rules. The rules in $\mathcal{A}$ are called *axioms* and rules in C are called *circularities*. If either $\mathcal{A}$ or C does not appear in a sequent, it means the respective set is empty: $\mathcal{S} \vdash_{C} \varphi \Rightarrow^{\forall} \varphi'$ is a shorthand for $\mathcal{S}, \emptyset \vdash_{C} \varphi \Rightarrow^{\forall} \varphi'$, and $\mathcal{S}, \mathcal{A} \vdash \varphi \Rightarrow^{\forall} \varphi'$ is a shorthand for $\mathcal{S}, \mathcal{A} \vdash_{\emptyset} \varphi \Rightarrow^{\forall} \varphi'$. Initially, both $\mathcal{A}$ and C are empty. Note that "$\to$" in Step and Consequence denotes implication.

Step :

$$\frac{\models \varphi \rightarrow \bigvee_{\varphi_l \Rightarrow^{\exists} \varphi_r \,\in\, \mathcal{S}} \exists FreeVars(\varphi_l)\, \varphi_l \qquad \models \exists c\, (\varphi[c/\square] \wedge \varphi_l[c/\square]) \wedge \varphi_r \rightarrow \varphi' \quad \text{for each } \varphi_l \Rightarrow^{\exists} \varphi_r \in \mathcal{S}}{\mathcal{S}, \mathcal{A} \vdash_C \varphi \Rightarrow^{\forall} \varphi'}$$

Axiom :

$$\frac{\varphi \Rightarrow^{\forall} \varphi' \in \mathcal{A}}{\mathcal{S}, \mathcal{A} \vdash_C \varphi \Rightarrow^{\forall} \varphi'}$$

Reflexivity :

$$\frac{\cdot}{\mathcal{S}, \mathcal{A} \vdash \varphi \Rightarrow^{\forall} \varphi}$$

Transitivity :

$$\frac{\mathcal{S}, \mathcal{A} \vdash_C \varphi_1 \Rightarrow^{\forall} \varphi_2 \qquad \mathcal{S}, \mathcal{A} \cup C \vdash \varphi_2 \Rightarrow^{\forall} \varphi_3}{\mathcal{S}, \mathcal{A} \vdash_C \varphi_1 \Rightarrow^{\forall} \varphi_3}$$

Case Analysis :

$$\frac{\mathcal{S}, \mathcal{A} \vdash_C \varphi_1 \Rightarrow^{\forall} \varphi \qquad \mathcal{S}, \mathcal{A} \vdash_C \varphi_2 \Rightarrow^{\forall} \varphi}{\mathcal{S}, \mathcal{A} \vdash_C \varphi_1 \vee \varphi_2 \Rightarrow^{\forall} \varphi}$$

Abstraction :

$$\frac{\mathcal{S}, \mathcal{A} \vdash_C \varphi \Rightarrow^{\forall} \varphi' \qquad X \cap FreeVars(\varphi') = \emptyset}{\mathcal{S}, \mathcal{A} \vdash_C \exists X\, \varphi \Rightarrow^{\forall} \varphi'}$$

Consequence :

$$\frac{\models \varphi_1 \rightarrow \varphi_1' \qquad \mathcal{S}, \mathcal{A} \vdash_C \varphi_1' \Rightarrow^{\forall} \varphi_2' \qquad \models \varphi_2' \rightarrow \varphi_2}{\mathcal{S}, \mathcal{A} \vdash_C \varphi_1 \Rightarrow^{\forall} \varphi_2}$$

Circularity :

$$\frac{\mathcal{S}, \mathcal{A} \vdash_{C \cup \{\varphi \Rightarrow^{\forall} \varphi'\}} \varphi \Rightarrow^{\forall} \varphi'}{\mathcal{S}, \mathcal{A} \vdash_C \varphi \Rightarrow^{\forall} \varphi'}$$

Fig. 1. Proof system for reachability. We make the standard assumption that the free variables of $\varphi_l \Rightarrow^{\exists} \varphi_r$ in the Step proof rule are fresh (e.g., disjoint from those of $\varphi \Rightarrow^{\forall} \varphi'$).

The intuition is that the reachability rules in $\mathcal{A}$ can be assumed valid, while those in C have been postulated but not yet justified. After making progress from φ (at least one derivation by Step or by Axiom with the rules in $\mathcal{A}$), the rules in C become (coinductively) valid (can be used in derivations by Axiom). During the proof, circularities can be added to C via Circularity, flushed into $\mathcal{A}$ by Transitivity, and used via Axiom. The desired semantics for sequent $\mathcal{S}, \mathcal{A} \vdash_C \varphi \Rightarrow^{\forall} \varphi'$ (read "$\mathcal{S}$ with axioms $\mathcal{A}$ and circularities C proves $\varphi \Rightarrow^{\forall} \varphi'$") is: $\varphi \Rightarrow^{\forall} \varphi'$ holds if the rules in $\mathcal{A}$ hold and those in C hold after taking at least on step from φ in the transition system $(\Rightarrow^{\mathcal{T}}_{\mathcal{S}}, \mathcal{T})$, and if $C \neq \emptyset$ then φ reaches φ' after at least one step on all complete paths. As a consequence of this definition, any rule $\varphi \Rightarrow^{\forall} \varphi'$ derived by Circularity has the property that φ reaches φ' after at lest one step, due to Circularity having a prerequisite $\mathcal{S}, \mathcal{A} \vdash_{C \cup \{\varphi \Rightarrow^{\forall} \varphi'\}} \varphi \Rightarrow^{\forall} \varphi'$ (with a non-empty set of circularities). We next discuss the proof rules.

Step derives a sequent where φ reaches φ' in one step on all paths. The first premise ensures any configuration matching φ matches the left-hand-side φ_l of some rule in $\mathcal{S}$ and thus, as $\mathcal{S}$ is weakly well-defined, can take a step. Formally, if $(\gamma, \rho) \models \varphi$, then there exists some rule $\varphi_l \Rightarrow^{\exists} \varphi_r \in \mathcal{S}$ and some valuation ρ' of the free variables of φ_l such that $(\gamma, \rho') \models \varphi_l$, and thus γ has at least one $\Rightarrow^{\mathcal{T}}_{\mathcal{S}}$-successor generated by the rule $\varphi_l \Rightarrow^{\exists} \varphi_r$. The second premise ensures that each $\Rightarrow^{\mathcal{T}}_{\mathcal{S}}$-successor of a configuration matching φ matches φ'. Formally, if $\gamma \Rightarrow^{\mathcal{T}}_{\mathcal{S}} \gamma'$ and γ matches φ then there is some rule $\varphi_l \Rightarrow^{\exists} \varphi_r \in \mathcal{S}$ and $\rho : Var \rightarrow \mathcal{T}$ such that $(\gamma, \rho) \models \varphi \wedge \varphi_l$ and $(\gamma', \rho) \models \varphi_r$; then the second part implies γ' matches φ'.

Designing a proof rule for deriving an execution step along all paths is non-trivial. For instance, one might expect Step to require as many premises as there are transitions going out of φ, as is the case for the examples presented later in this paper. However, that is not possible, as the number of successors of a configuration matching φ may be unbounded even if each matching configuration has a finite branching factor in the transition system. Step avoids this issue by requiring only one premise for each rule by which some configuration φ can take a step, even if that rule can be used to derive multiple transitions. To illustrate this situation, consider a language defined by $\mathcal{S} \equiv \{\langle n_1 \rangle \wedge n_1 >_{Int} n_2 \Rightarrow^{\exists} \langle n_2 \rangle\}$, with n_1 and n_2 non-negative integer variables. A configuration in this language is a singleton with a non-negative integer. Intuitively, a positive integer transits into a strictly smaller non-negative integer, in a non-deterministic way. The branching factor of a non-negative integer is its value. Then $\mathcal{S} \models \langle m \rangle \Rightarrow^{\forall} \langle 0 \rangle$. Deriving it reduces (by Circularity and other proof rules) to deriving $\langle m_1 \rangle \wedge m_1 >_{Int} 0 \Rightarrow^{\forall} \exists m_2\, (\langle m_2 \rangle \wedge m_1 >_{Int} m_2)$. The left-hand-side is matched by any positive integer, and thus its branching factor is infinity. Deriving this rule with Step requires only two premises, $\models (\langle m_1 \rangle \wedge m_1 >_{Int} 0) \rightarrow \exists n_1 n_2\, (\langle n_1 \rangle \wedge n_1 >_{Int} n_2)$ and $\models \exists c\, (c = \langle m_1 \rangle \wedge m_1 >_{Int} 0 \wedge c = \langle n_1 \rangle \wedge n_1 >_{Int} n_2) \wedge \langle n_2 \rangle \rightarrow \exists m_2\, (\langle m_2 \rangle \wedge m_1 >_{Int} m_2)$. A similar situation arises in real life for languages with thread pools of arbitrary size.

Axiom applies a trusted rule. Reflexivity and Transitivity capture the corresponding closure properties of the reachability relation. Reflexivity requires C to be empty to ensure that all-path rules derived with non-empty C take at least one step. Transitivity enables the circularities as axioms for the second premise, since if C is not empty, the first premise is guaranteed to take at least one step. Consequence, Case Analysis and Abstraction are adapted from Hoare logic. Ignoring circularities, these seven rules discussed so far constitute formal infrastructure for symbolic execution.

Circularity has a coinductive nature, allowing us to make new circularity claims. We typically make such claims for code with repetitive behaviors, such as loops, recursive functions, jumps, etc. If there is a derivation of the claim using itself as a circularity, then the claim holds. This would obviously be unsound if the new assumption was available immediately, but requiring progress (taking at least on step in the transition system $(\mathcal{T}, \Rightarrow^{\mathcal{T}}_{\mathcal{S}})$) before circularities can be used ensures that only diverging executions can correspond to endless invocation of a circularity.

One important aspect of concurrent program verification, which we do not address in this paper, is proof compositionality. Our focus here is limited to establishing a sound and complete language-independent proof system for all-path reachability rules, to serve as a foundation for further results and applications, and to discuss our current implementation of it. We only mention that we have already studied proof compositionality for earlier one-path variants of reachability logic [5], showing that there is a mechanical way to translate any Hoare logic proof derivation into a reachability proof of similar size and structure, but based entirely on the operational semantics of the language. The overall conclusion of our previous study, which we believe will carry over to all-path reachability, was that compositional reasoning can be achieved methodologically using our proof system, by proving and then using appropriate reachability rules as lemmas. However, note that this works only for theoretically well-behaved languages which enjoy a compositional semantics. For example, a language whose semantics assumes a bounded heap size, or which has constructs whose semantics involve the entire program, e.g., call/cc, will lack compositionality.

5 Soundness and Relative Completeness

Here we discuss the soundness and relative completeness of our proof system. Unlike the similar results for Hoare logics and dynamic logics, which are separately proved for each language taking into account the particularities of that language, we prove soundness and relative completeness once and for all languages.

Soundness states that a syntactically derivable sequent holds semantically. Because of the utmost importance of the result below, we have also mechanized its proof. Our complete Coq formalization can be found at `http://fsl.cs.illinois.edu/rl`.

Theorem 1 (Soundness). *If* $\mathcal{S} \vdash \varphi \Rightarrow^{\forall} \varphi'$ *then* $\mathcal{S} \models \varphi \Rightarrow^{\forall} \varphi'$.

Proof (sketch — complete details in [14]). Unfortunately, due to Circularity, a simple induction on the proof tree does not work. Instead, we prove a more general result (Lemma 1 below) allowing sequents with nonempty $\mathcal{A}$ and C, which requires stating semantic assumptions about the rules in $\mathcal{A}$ and C.

First we need to define a more general satisfaction relation than $\mathcal{S} \models \varphi \Rightarrow^{\forall} \varphi'$. Let $\delta \in \{+, *\}$ be a flag and let $n \in \mathbb{N}$ be a natural number. We define a new satisfaction relation $\mathcal{S} \models_n^{\delta} \varphi \Rightarrow^{\forall} \varphi'$ by restricting the paths in the definition of $\mathcal{S} \models \varphi \Rightarrow^{\forall} \varphi'$ to length at most n, and requiring progress (at least one step) when $\delta = +$.

Formally, we define $\mathcal{S} \models_n^{\delta} \varphi \Rightarrow^{\forall} \varphi'$ to hold iff for any complete path $\tau = \gamma_1 ... \gamma_k$ of length $k \leq n$ and for any ρ such that $(\gamma_1, \rho) \models \varphi$, there exists $i \in \{1, ..., k\}$ such that $(\gamma_i, \rho) \models \varphi'$. Additionally, when $\delta = +$, we require that $i \neq 1$ (i.e. γ makes progress). The indexing on n is required to prove the soundness of circularities. Now we can state the soundness lemma.

Lemma 1. *If* $\mathcal{S}, \mathcal{A} \vdash_C \varphi \Rightarrow^{\forall} \varphi'$ *and* $\mathcal{S} \models_n^{+} \mathcal{A}$ *and* $\mathcal{S} \models_{n-1}^{+} C$ *then* $\mathcal{S} \models_n^{*} \varphi \Rightarrow^{\forall} \varphi'$*, and furthermore, if* C *is nonempty then* $\mathcal{S} \models_n^{+} \varphi \Rightarrow^{\forall} \varphi'$.

Theorem 1 follows by showing that $\mathcal{S} \models \varphi \Rightarrow^{\forall} \varphi'$ iff $\mathcal{S} \models_n \varphi \Rightarrow^{\forall} \varphi'$ for all $n \in \mathbb{N}$.

Lemma 1 is proved by induction on the derivation (with each induction hypothesis universally quantified over n). Consequence, Case Analysis, and Abstraction are easy. Axiom may only be used in cases where $\mathcal{S} \models_n^{+} \mathcal{A}$ includes $\mathcal{S} \models_n^{+} \varphi \Rightarrow^{\forall} \varphi'$ (as $\mathcal{S}$ contains only one-path rules). Reflexivity may only be used when C is empty, and $\mathcal{S} \models^{*} \varphi \Rightarrow^{\forall} \varphi$ unconditionally. The premises of Step are pattern implications which imply that any configuration matching φ is not stuck in $\Rightarrow_{\mathcal{S}}^{\mathcal{T}}$, and all of its immediate successors satisfy φ'. This directly establishes that $\mathcal{S} \models^{+} \varphi \Rightarrow^{\forall} \varphi'$. Transitivity requires considering execution paths more carefully. If C is empty, then the proof is trivial. Otherwise the induction hypothesis gives that $\varphi_1 \Rightarrow^{\forall} \varphi_2$ holds with progress. Therefore, when proving $\varphi_2 \Rightarrow^{\forall} \varphi_3$, the circularities are enabled soundly. Circularity proceeds by an inner well-founded induction on n. The outer induction over the derivation gives an induction hypothesis showing the desired conclusion under the additional assumption that $\varphi \Rightarrow^{\forall} \varphi'$ holds for any m strictly less than n, which is exactly the induction hypothesis provided by the inner induction on n. □

We next show relative completeness: any valid all-path reachability property of any program in any language with an operational semantics given as a reachability system $\mathcal{S}$ is

$$step(c,\ c') \equiv \bigvee_{\mu \equiv \varphi_l \Rightarrow^{\exists} \varphi_r \in \mathcal{S}} \exists FreeVars(\mu)\ (\varphi_l[c/\square] \wedge \varphi_r[c'/\square])$$

$$coreach(\varphi) \equiv \forall n \forall c_0 ... c_n \left(\square = c_0 \rightarrow \bigwedge_{0 \le i < n} step(c_i,\ c_{i+1}) \rightarrow \neg \exists c_{n+1}\ step(c_n,\ c_{n+1}) \rightarrow \bigvee_{0 \le i \le n} \varphi[c_i/\square] \right)$$

Fig. 2. FOL encoding of one step transition relation and all-path reachability

derivable with the proof system in Fig. 1 from $\mathcal{S}$. As with Hoare and dynamic logics, "relative" means we assume an oracle capable of establishing validity in the first-order theory of the state, which here is the configuration model $\mathcal{T}$. An immediate consequence of relative completeness is that Circularity is sufficient to derive any repetitive behavior occurring in any program written in any language, and that Step is also sufficient to derive any non-deterministic behavior! We establish the relative completeness under the following assumptions: (1) $\mathcal{S}$ is finite; (2) the model $\mathcal{T}$ includes natural numbers with addition and multiplication; and (3) the set of configurations $\mathcal{T}_{Cfg}$ is countable (the model $\mathcal{T}$ includes some injective function $\alpha : \mathcal{T}_{Cfg} \rightarrow \mathbb{N}$). Assumption (1) ensures Step has a finite number of prerequisites. Assumption (2) is a standard assumption (also made by Hoare and dynamic logic completeness results) which allows the definition of Gödel's β predicate. Assumption (3) allows the encoding of a sequence of configurations into a sequence of natural numbers. We expect the operational semantics of any reasonable language to satisfy these conditions. Formally, we have the following

Theorem 2 (Relative Completeness). *If $\mathcal{S} \models \varphi \Rightarrow^{\forall} \varphi'$ then $\mathcal{S} \vdash \varphi \Rightarrow^{\forall} \varphi'$, for any semantics $\mathcal{S}$ satisfying the three assumptions above.*

Proof (sketch — complete details in [14]). Our proof relies on the fact that pattern reasoning in first-order matching logic reduces to FOL reasoning in the model $\mathcal{T}$. A key component of the proof is defining the *coreach*(φ) predicate in plain FOL. This predicate holds when every complete $\Rightarrow^{\mathcal{T}}_{\mathcal{S}}$-path τ starting at c includes some configuration satisfying φ. We express *coreach*(φ) using auxiliary predicate *step*(c, c') which encodes the one step transition relation ($\Rightarrow^{\mathcal{T}}_{\mathcal{S}}$). Fig. 2 shows both definitions. As it is, *coreach*(φ) is not a proper FOL formula, as it quantifies over a sequence of configurations. This is addressed using the injective function α to encode universal quantification over a sequence of configurations into universal quantification over a sequence of integers, which is in turn encoded into quantification over two integer variables using Gödel's β predicate (encoding shown in [14]).

Next, using the definition above we encode the semantic validity of an all-path reachability rule as FOL validity: $\mathcal{S} \models \varphi \Rightarrow^{\forall} \varphi'$ iff $\models \varphi \rightarrow coreach(\varphi')$. Therefore, the theorem follows by Consequence from the sequent $\mathcal{S} \vdash coreach(\varphi') \Rightarrow^{\forall} \varphi'$. We derive this sequent by using Circularity to add the rule to the set of circularities, then by using Step to derive one $\Rightarrow^{\mathcal{T}}_{\mathcal{S}}$-step, and then by using Transitivity and Axiom with the rule itself to derive the remaining $\Rightarrow^{\mathcal{T}}_{\mathcal{S}}$-steps (circularities can be used after Transitivity). The formal derivation uses all eight proof rules. □

IMP language syntax	IMP evaluation contexts syntax
$\mathit{PVar} ::=$ program variables	$\mathit{Context} ::= \blacksquare$
$\mathit{Exp} ::= \mathit{PVar} \mid \mathit{Int} \mid \mathit{Exp}\ \mathtt{op}\ \mathit{Exp}$	$\mid \langle \mathit{Context},\ \mathit{State} \rangle$
$\mathit{Stmt} ::= \mathtt{skip} \mid \mathit{PVar} := \mathit{Exp}$	$\mid \mathit{Context}\ \mathtt{op}\ \mathit{Exp} \mid \mathit{Int}\ \mathtt{op}\ \mathit{Context}$
$\mid \mathit{Stmt};\ \mathit{Stmt} \mid \mathit{Stmt} \parallel \mathit{Stmt}$	$\mid \mathit{PVar} := \mathit{Context} \mid \mathit{Context};\ \mathit{Stmt}$
$\mid \mathtt{if}(\mathit{Exp})\ \mathit{Stmt}\ \mathtt{else}\ \mathit{Stmt}$	$\mid \mathit{Context} \parallel \mathit{Stmt} \mid \mathit{Stmt} \parallel \mathit{Context}$
$\mid \mathtt{while}(\mathit{Exp})\ \mathit{Stmt}$	$\mid \mathtt{if}(\mathit{Context})\ \mathit{Stmt}\ \mathtt{else}\ \mathit{Stmt}$

IMP operational semantics			
lookup	$\langle C, \sigma\rangle[x] \Rightarrow^{\exists} \langle C, \sigma\rangle[\sigma(x)]$	**op**	$i_1\ \mathtt{op}\ i_2 \Rightarrow^{\exists} i_1\ op_{Int}\ i_2$
asgn	$\langle C, \sigma\rangle[x := i] \Rightarrow^{\exists} \langle C, \sigma[x \leftarrow i]\rangle[\mathtt{skip}]$	**seq**	$\mathtt{skip};\ s \Rightarrow^{\exists} s$
cond$_1$	$\mathtt{if}(i)\ s_1\ \mathtt{else}\ s_2 \Rightarrow^{\exists} s_1$ if $i \neq 0$	**cond**$_2$	$\mathtt{if}(0)\ s_1\ \mathtt{else}\ s_2 \Rightarrow^{\exists} s_2$
while	$\mathtt{while}(e)\ s \Rightarrow^{\exists} \mathtt{if}(e)\ s; \mathtt{while}(e)\ s\ \mathtt{else}\ \mathtt{skip}$	**finish**	$\mathtt{skip} \parallel \mathtt{skip} \Rightarrow^{\exists} \mathtt{skip}$

Fig. 3. IMP language syntax and operational semantics based on evaluation contexts

6 Verifying Programs

In this section we show a few examples of using our proof system to verify programs based on an operational semantics. In a nutshell, the proof system enables generic symbolic execution combined with circular reasoning. Symbolic execution is achieved by rewriting modulo domain reasoning.

First, we introduce a simple parallel imperative language, IMP. Fig. 3 shows its syntax and an operational semantics based on evaluation contexts [11] (we choose evaluation contexts for presentation purposes only). IMP has only integer expressions. When used as conditions of `if` and `while`, zero means false and any non-zero integer means true (like in C). Expressions are formed with integer constants, program variables, and conventional arithmetic constructs. Arithmetic operations are generically described as op. IMP statements are assignment, `if`, `while`, sequential composition and parallel composition. IMP has shared memory parallelism without explicit synchronization. The examples use the parallel construct only at the top-level of the programs. The second example shows how to achieve synchronization using the existing language constructs.

The program configurations of IMP are pairs $\langle \mathtt{code},\ \sigma\rangle$, where code is a program fragment and σ is a state term mapping program variables into integers. As usual, we assume appropriate definitions for the integer and map domains available, together with associated operations like arithmetic operations ($i_1\ op_{Int}\ i_2$, etc.) on the integers and lookup ($\sigma(x)$) and update ($\sigma[x \leftarrow i]$) on the maps. We also assume a context domain with a plugging operation ($C[t]$) that composes a context and term back into a configuration. A configuration context consists of a code context and a state. The definition in Fig. 3 consists of eight reduction rules between program configurations, which make use of first-order variables: x is a variable of sort *PVar*; e is a variable of sort *Exp*; s, s_1, s_2 are variables of sort *Stmt*; i, i_1, i_2 are variables of sort *Int*; σ is a variable of sort *State*; C is a variable of sort *Context*. A rule reduces a configuration by splitting it into a context and a redex, rewriting the redex and possibly the context, and then plugging the resulting term into the resulting context. As an abbreviation, a context is not mentioned if not used; e.g., the rule **op** is in full $\langle C, \sigma\rangle[i_1\ \mathtt{op}\ i_2] \Rightarrow^{\exists} \langle C, \sigma\rangle[i_1\ op_{Int}\ i_2]$. For example, configuration $\langle \mathbf{x} := (2+5) - 4,\ \sigma\rangle$ reduces to $\langle \mathbf{x} := 7 - 4,\ \sigma\rangle$ by applying the

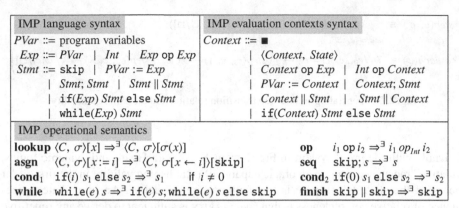

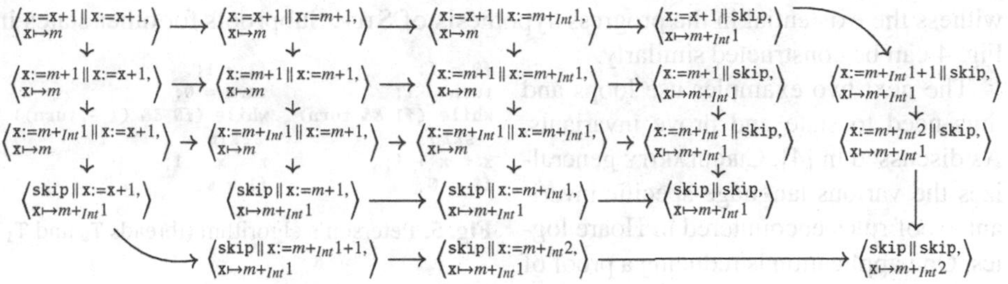

Fig. 4. State space of the racing increment example

op$_+$ rule with $C \equiv \mathbf{x} := \blacksquare - 4$, $\sigma \equiv \sigma$, $i_1 \equiv 2$ and $i_2 \equiv 5$. We can regard the operational semantics of IMP above as a set of reduction rules of the form "$l \Rightarrow^\exists r$ if b", where l and r are program configurations with variables constrained by boolean condition b. As discussed in Section 3, our proof system works with any rules of this form.

Next, we illustrate the proof system on a few examples. The first example shows that our proof system enables exhaustive state exploration, similar to symbolic model-checking but based on the operational semantics. Although humans prefer to avoid such explicit proofs and instead methodologically use abstraction or compositional reasoning whenever possible (and such methodologies are not excluded by our proof system), a complete proof system must nevertheless support them. The code $\mathbf{x} := \mathbf{x}+1 \parallel \mathbf{x} := \mathbf{x}+1$ exhibits a race on $\mathbf{x}$: the value of $\mathbf{x}$ increases by 1 when both reads happen before either write, and by 2 otherwise. The all-path rule that captures this behavior is

$$\langle \mathtt{x}{:=}\mathtt{x}{+}1 \parallel \mathtt{x}{:=}\mathtt{x}{+}1,\ \mathtt{x}{\mapsto} m\rangle \Rightarrow^\forall \exists n\, (\langle \mathtt{skip},\ \mathtt{x}{\mapsto} n\rangle \land (n = m +_{Int} 1 \lor n = m +_{Int} 2)$$

We show that the program has exactly these behaviors by deriving this rule in the proof system. Call the right-hand-side pattern G. The proof contains subproofs of $c \Rightarrow^\forall G$ for every reachable configuration c, tabulated in Fig. 4. The subproofs for c matching G use Reflexivity and Consequence, while the rest use Transitivity, Step, and Case Analysis to reduce to the proofs for the next configurations. For example, the proof fragment below shows how $\langle \mathbf{x} := m+1 \parallel \mathbf{x} := \mathbf{x}+1,\ \mathbf{x} \mapsto m\rangle \Rightarrow^\forall G$ reduces to $\langle \mathbf{x} := m +_{Int} 1 \parallel \mathbf{x} := \mathbf{x}+1,\ \mathbf{x} \mapsto m\rangle \Rightarrow^\forall G$ and $\langle \mathbf{x} := m+1 \parallel \mathbf{x} := m+1,\ \mathbf{x} \mapsto m\rangle \Rightarrow^\forall G$:

$$\dfrac{\text{Step}\ \dfrac{\cdots}{\left\langle {\mathtt{x}:=m+1 \parallel \mathtt{x}:=\mathtt{x}+1, \atop \mathtt{x}\mapsto m} \right\rangle \Rightarrow^\forall \left\langle {\mathtt{x}:=m+_{Int}1 \parallel \mathtt{x}:=\mathtt{x}+1, \atop \mathtt{x}\mapsto m} \right\rangle \lor \left\langle {\mathtt{x}:=m+1 \parallel \mathtt{x}:=m+1, \atop \mathtt{x}\mapsto m} \right\rangle} \qquad \dfrac{\dfrac{\cdots}{\left\langle {\mathtt{x}:=m+_{Int}1 \parallel \mathtt{x}:=\mathtt{x}+1, \atop \mathtt{x}\mapsto m} \right\rangle \Rightarrow^\forall G} \quad \dfrac{\cdots}{\left\langle {\mathtt{x}:=m+1 \parallel \mathtt{x}:=m+1, \atop \mathtt{x}\mapsto m} \right\rangle \Rightarrow^\forall G}}{\left\langle {\mathtt{x}:=m+_{Int}1 \parallel \mathtt{x}:=\mathtt{x}+1, \atop \mathtt{x}\mapsto m} \right\rangle \lor \left\langle {\mathtt{x}:=m+1 \parallel \mathtt{x}:=m+1, \atop \mathtt{x}\mapsto m} \right\rangle \Rightarrow^\forall G}\ \text{CA}}{\langle \mathbf{x} := m+1 \parallel \mathbf{x} := \mathbf{x}+1,\ \mathbf{x} \mapsto m\rangle \Rightarrow^\forall G}\ \text{Trans}$$

For the rule hypotheses of Step above, note that all rules but **lookup** and **op**$_+$ make the overlap condition $\exists c \left(\left\langle {\mathtt{x}:=m+1 \parallel \mathtt{x}:=\mathtt{x}+1, \atop \mathtt{x}\mapsto m} \right\rangle[c/\square] \land \varphi_l[c/\square]\right)$ unsatisfiable, and only one choice of free variables works for the **lookup** and **op**$_+$ rules. For **lookup**, φ_l is $\langle C, \sigma\rangle[x]$ and the overlap condition is only satisfiable if the logical variables C, σ and x are equal to $(\mathbf{x} := m+1 \parallel \mathbf{x} := \blacksquare + 1)$, $(\mathbf{x} \mapsto m)$, and $\mathbf{x}$, resp. Under this assignment, the pattern $\varphi_r = \langle C, \sigma\rangle[\sigma(x)]$ is equivalent to $\langle \mathbf{x} := m+1 \parallel \mathbf{x} := m+1,\ \mathbf{x} \mapsto m\rangle$, the right branch of the disjunction. The **op**$_+$ rule is handled similarly. The assignment for **lookup** can also witness the existential in the progress hypothesis of Step. Subproofs for other states in Fig. 4 can be constructed similarly.

The next two examples use loops and thus need to state and prove invariants. As discussed in [4], Circularity generalizes the various language-specific invariant proof rules encountered in Hoare logics. One application is reducing a proof of $\varphi \Rightarrow^{\forall} \varphi'$ to proving $\varphi_{\text{inv}} \Rightarrow^{\forall} \varphi_{\text{inv}} \vee \varphi'$ for some pattern invariant φ_{inv}. We first show $\models \varphi \rightarrow \varphi_{\text{inv}}$, and use Consequence to change the goal to $\varphi_{\text{inv}} \Rightarrow^{\forall} \varphi'$. This is claimed as a circularity, and then proved by transitivity with $\varphi_{\text{inv}} \vee \varphi'$. The second hypothesis $\{\varphi_{\text{inv}} \Rightarrow^{\forall} \varphi'\} \vdash \varphi_{\text{inv}} \vee \varphi' \Rightarrow^{\forall} \varphi'$ is proved by Case Analysis, Axiom, and Reflexivity.

```
f0 = 1;                      f1 = 1;
turn = 1;                    turn = 0;
while (f1 && turn)           while (f0 && (1 - turn))
  skip                         skip
x = x + 1;                   x = x + 1;
f0 = 0;                      f1 = 0;
```

Fig. 5. Peterson's algorithm (threads $\texttt{T}_0$ and $\texttt{T}_1$)

Next, we can use Peterson's algorithm for mutual exclusion to eliminate the race as shown in Fig. 5. The all-path rule $\varphi \Rightarrow^{\forall} \varphi'$ that captures the new behavior is

$$\begin{aligned}&\langle \texttt{T}_0 \,\|\, \texttt{T}_1,\ (\texttt{f0} \mapsto 0,\ \texttt{f1} \mapsto 0,\ \texttt{x} \mapsto N)\rangle\\ &\Rightarrow^{\forall} \exists t\, \langle \texttt{skip},\ (\texttt{f0} \mapsto 0,\ \texttt{f1} \mapsto 0,\ \texttt{x} \mapsto N +_{Int} 2,\ \texttt{turn} \mapsto t)\rangle\end{aligned}$$

Similarly to the unsynchronized example, the proof contains subproofs of $c \Rightarrow^{\forall} \varphi'$ for every configuration c reachable from φ. The main difference is that Circularity is used with each of these rules $c \Rightarrow^{\forall} \varphi'$ with one of the two threads of c in the `while` loop (these rules capture the invariants). Thus, when we reach a configuration c visited before, we use the rule added by Circularity to complete the proof.

The final example is the program SUM ≡ "`s := 0; LOOP`" (where LOOP stands for "`while (n>0) (s := s+n; n := n-1)`"), which computes in s the sum of the numbers from 1 up to n. The all-path reachability rule $\varphi \Rightarrow^{\forall} \varphi'$ capturing this behavior is

$$\langle \texttt{SUM},\ (\texttt{s} \mapsto s,\ \texttt{n} \mapsto n)\rangle \wedge n \geq_{Int} 0 \ \Rightarrow^{\forall}\ \langle \texttt{skip},\ (\texttt{s} \mapsto n *_{Int} (n +_{Int} 1)/_{Int} 2,\ \texttt{n} \mapsto 0)\rangle$$

We derive the above rule in our proof system by using Circularity with the invariant rule $\exists n' (\langle \texttt{LOOP},\ (\texttt{s} \mapsto (n -_{Int} n') *_{Int} (n +_{Int} n' +_{Int} 1)/_{Int} 2,\ \texttt{n} \mapsto n')\rangle \wedge n' \geq_{Int} 0) \Rightarrow^{\forall} \varphi'$. Previous work [4–7] presented a proof system able to derive similar rules, but which hold along *some* execution path, requiring a separate proof that the program is deterministic.

7 Implementation

Here we briefly discuss our prototype implementation of the proof system in Fig. 1 in $\mathbb{K}$ [9]. We choose $\mathbb{K}$ because it is a modular semantic language design framework, it is used for teaching programming languages at several universities, and there are several languages defined in it including C [10], PHP [15], Python, and Java. Due to space limitations, we do not present $\mathbb{K}$ here. We refer the reader to `http://kframework.org` for language definitions, a tutorial, and our prototype. As discussed in Section 3, we simply view a $\mathbb{K}$ semantics as a set of reachability rules of the form "$l \wedge b \Rightarrow^{\exists} r$".

The prototype is implemented in Java, and uses Z3 [16] for domain reasoning. It takes an operational semantics and uses it to perform concrete or symbolic execution. At its core, it performs *narrowing* of a conjunctive pattern with reachability rules between conjunctive patterns, where a conjunctive pattern is a pattern of the form $\exists X(\pi \wedge \psi)$, with X a set of variables, π a basic pattern (program configurations with variables), and ψ a structureless formula. Narrowing is necessary when a conjunctive pattern is too abstract to match the left-hand side of any rule, but is unifiable with the left-hand sides of some rules. For instance, consider the IMP code fragment "`if (`b`) then x = 1; else x = 0;`". This code does not match the left-hand sides of either of the two rules giving semantics to `if` (similar to $\mathbf{cond}_1$ and $\mathbf{cond}_2$ in Fig. 3), but it is unifiable with the left-hand sides of both rules. Intuitively, if we use the rules of the semantics, taking steps of rewriting on a ground configuration yields concrete execution, while taking steps of narrowing on a conjunctive pattern yields symbolic execution. In our practical evaluation, we found that conjunctive patterns tend to suffice to specify both the rules for operational semantics and program specifications.

For each step of narrowing, the $\mathbb{K}$ engine uses *unification modulo theories*. In our implementation, we distinguish a number of mathematical theories (e.g. booleans, integers, sequences, sets, maps, etc) which the underlying SMT solver can reason about. Specifically, when unifying a conjunctive pattern $\exists X(\pi \wedge \psi)$ with the left-hand side of a rule $\exists X_l(\pi_l \wedge \psi_l)$ (we assume $X \cap X_l = \emptyset$), the $\mathbb{K}$ engine begins with the syntactic unification of the basic patterns π and π_l. Upon encountering corresponding subterms (π' in π and π'_l in π_l) which are both terms of one of the theories above, it records an equality $\pi' = \pi'_l$ rather than decomposing the subterms further (if one is in a theory, and the other one is in a different theory or is not in any theory, the unification fails). If this stage of unification is successful, we end up with a conjunction ψ_u of constraints, some having a variable in one side and some with both sides in one of the theories. Satisfiability of $\exists X \cup X_l(\psi \wedge \psi_u \wedge \psi_l)$ is then checked by the SMT solver. If it is satisfiable, then narrowing takes a step from $\exists X(\pi \wedge \psi)$ to $\exists X \cup X_l \cup X_r(\pi_r \wedge \psi \wedge \psi_u \wedge \psi_l \wedge \psi_r)$, where $\exists X_r(\pi_r \wedge \psi_r)$ is the right-hand side of the rule. Intuitively, "collecting" the constraints $\psi_u \wedge \psi_l \wedge \psi_r$ is similar to collecting the path constraint in traditional symbolic execution (but is done in a language-generic manner). For instance, in the `if` case above, narrowing with the two semantics rules results in collecting the constraints $b =$ `true` and $b =$ `false`.

The $\mathbb{K}$ engine accepts a set of user provided rules to prove together, which capture the behavior of the code being verified. Typically, these rules specify the behavior of recursive functions and while loops. For each rule, the $\mathbb{K}$ engine searches starting from the left-hand side for formulae which imply the right-hand side, starting with $\mathcal{S}$ the semantics and C all the rules it attempts to prove. By a derived rule called *Set Circularity*, this suffices to show that each rule is valid. As an optimization, Axiom is given priority over Step (use specifications rather than stepping into the code).

Most work goes into implementing the Step proof rule, and in particular calculating how $\rho \models \exists c\ (\varphi[c/\square] \wedge \varphi_l[c/\square])$ can be satisfied. This holds when $\rho^\gamma \models \varphi$ and $\rho^\gamma \models \varphi_l$, which can be checked with unification modulo theories. To use Step in an automated way, the $\mathbb{K}$ tool constructs φ' for a given φ as a disjunction of $\varphi_r \wedge \psi_u \wedge \psi \wedge \psi_l$ over each rule $\varphi_l \Rightarrow^\exists \varphi_r \in \mathcal{S}$ and each way ψ_u of unifying φ with φ_l. As discussed in Section 4, in general this disjunction may not be finite, but it is sufficient for the examples that we considered. The Consequence proof rule also requires unification modulo theories, to check validity of the implication hypothesis $\models \varphi_1 \rightarrow \varphi'_1$. The main difference from Step is that the free variables of φ' become universality quantified when sending the query to the SMT solver. The implementation of the other proof rules is straight-forward.

8 Related Work

Using Hoare logic [17] to prove concurrent programs correct dates back to Owicki and Gries [18]. In the rely-guarantee method proposed by Jones [19] each thread relies on some properties being satisfied by the other threads, and in its turn, offers some guarantees on which the other threads can rely. O'Hearn [20] advances a Separation Hypothesis in the context of separation logic [21] to achieve compositionality: the state can be partitioned into separate portions for each process and relevant resources, respectively, satisfying certain invariants. More recent research focuses on improvements over both of the above methods and even combinations of them (e.g., [22–25]).

The satisfaction of all-path-reachability rules can also be understood intuitively in the context of temporal logics. Matching logic formulae can be thought of as state formulae, and reachability rules as temporal formulae. Assuming CTL* on finite traces, the semantics rule $\varphi \Rightarrow^{\exists} \varphi'$ can be expressed as $\varphi \rightarrow E \bigcirc \varphi'$, while an all-path reachability rule $\varphi \Rightarrow^{\forall} \varphi'$ can be expressed as $\varphi \rightarrow A\Diamond\varphi'$. However, unlike in CTL*, the φ and φ' formulae of reachability rules $\varphi \Rightarrow^{\exists} \varphi'$ or $\varphi \Rightarrow^{\forall} \varphi'$ share their free variables. Thus, existing proof systems for temporal logics (e.g., the CTL* one by Pnueli and Kesten) are not directly comparable with our approach.

Bae et al [26], Rocha and Meseguer [27], and Rocha et al [28] use narrowing to perform symbolic reachability analysis in a transition system associated to a unconditional rewrite theory for the purposes of verification. There are two main differences between their work and ours. First, they express state predicates in equational theories. Matching logic is more general, being first-order logic over a model of configurations T. Consequently, the Step proof rule takes these issues into account when considering the successors of a state. Second, they use rewrite systems for symbolic model checking. Our work is complementary, in the sense that we use the operational semantics for program verification, and check properties more similar to those in Hoare logic.

Language-independent proof systems. A first proof system is introduced in [6], while [5] presents a mechanical translation from Hoare logic proof derivations for IMP into derivations in the proof system. The Circularity proof rule is introduced in [4]. Finally, [7] supports operational semantics given with conditional rules, like small-step and big-step. All these previous results can only be applied to deterministic programs.

9 Conclusion and Future Work

This paper introduces a sound and (relatively) complete language-independent proof system which derives program properties holding along all execution paths (capturing partial correctness for non-deterministic programs), directly from an operational semantics. The proof system separates reasoning about deterministic language features (via the operational semantics) from reasoning about non-determinism (via the proof system). Thus, all we need in order to verify programs in a language is an operational semantics for the respective language.

We believe that existing techniques such as rely-guarantee and concurrent separation logic could be used in conjunction with our proof system to achieve semantically grounded and compositional verification.

Our approach handles operational semantics given with unconditional rules, like $\mathbb{K}$ framework, PLT-Redex, and CHAM, but it cannot handle operational semantics given with conditional rules, like big-step and small-step (rules with premises). Extending the presented results to work with conditional rules would boil down to extending the Step proof rule, which derives the fact that φ reaches φ' in one step along all execution paths. Such a extended Step would have as prerequisites whether the left-hand side of a semantics rule matches (like the existing Step) and additionally whether its premises hold. The second part would require an encoding of reachability in first-order logic, which is non-trivial and mostly likely would result in a first-order logic over a richer model than $\mathcal{T}$. The difficulty arises from the fact that Step must ensure all successors of φ are in φ'. Thus, this extension is left as future work.

Acknowledgements. We would like to thank the anonymous reviewers and the FSL members for their valuable feedback on an early version of this paper. The work presented in this paper was supported in part by the Boeing grant on "Formal Analysis Tools for Cyber Security" 2014-2015, the NSF grant CCF-1218605, the NSA grant H98230-10-C-0294, the DARPA HACMS program as SRI subcontract 19-000222, the Rockwell Collins contract 4504813093, and the (Romanian) SMIS-CSNR 602-12516 contract no. 161/15.06.2010.

References

1. Nipkow, T.: Winskel is (almost) right: Towards a mechanized semantics textbook. Formal Aspects of Computing 10, 171–186 (1998)
2. Jacobs, B.: Weakest pre-condition reasoning for Java programs with JML annotations. J. Logic and Algebraic Programming 58(1-2), 61–88 (2004)
3. Appel, A.W.: Verified software toolchain. In: Barthe, G. (ed.) ESOP 2011. LNCS, vol. 6602, pp. 1–17. Springer, Heidelberg (2011)
4. Roşu, G., Ştefănescu, A.: Checking reachability using matching logic. In: OOPSLA, pp. 555–574. ACM (2012)
5. Roşu, G., Ştefănescu, A.: From hoare logic to matching logic reachability. In: Giannakopoulou, D., Méry, D. (eds.) FM 2012. LNCS, vol. 7436, pp. 387–402. Springer, Heidelberg (2012)
6. Roşu, G., Ştefănescu, A.: Towards a unified theory of operational and axiomatic semantics. In: Czumaj, A., Mehlhorn, K., Pitts, A., Wattenhofer, R. (eds.) ICALP 2012, Part II. LNCS, vol. 7392, pp. 351–363. Springer, Heidelberg (2012)
7. Roşu, G., Ştefănescu, A., Ciobâcă, C., Moore, B.M.: One-path reachability logic. In: LICS 2013. IEEE (2013)
8. Roşu, G., Ellison, C., Schulte, W.: Matching logic: An alternative to hoare/Floyd logic. In: Johnson, M., Pavlovic, D. (eds.) AMAST 2010. LNCS, vol. 6486, pp. 142–162. Springer, Heidelberg (2011)

9. Roşu, G., Şerbănută, T.F.: An overview of the K semantic framework. J. Logic and Algebraic Programming 79(6), 397–434 (2010)
10. Ellison, C., Roşu, G.: An executable formal semantics of C with applications. In: POPL, pp. 533–544. ACM (2012)
11. Felleisen, M., Findler, R.B., Flatt, M.: Semantics Engineering with PLT Redex. MIT (2009)
12. Berry, G., Boudol, G.: The chemical abstract machine. Theoretical Computer Science 96(1), 217–248 (1992)
13. Matthews, J., Findler, R.B.: An operational semantics for Scheme. JFP 18(1), 47–86 (2008)
14. Ştefănescu, A., Ciobâcă, C., Moore, B.M., Şerbănuţă, T.F., Roşu, G.: Reachability Logic in K. Technical Report. University of Illinois (November 2013), `http://hdl.handle.net/2142/46296`
15. Filaretti, D., Maffeis, S.: An executable formal semantics of php. In: ECOOP. LNCS (to appear, 2014)
16. de Moura, L., Bjørner, N.S.: Z3: An efficient SMT solver. In: Ramakrishnan, C.R., Rehof, J. (eds.) TACAS 2008. LNCS, vol. 4963, pp. 337–340. Springer, Heidelberg (2008)
17. Hoare, C.A.R.: An axiomatic basis for computer programming. Communications of the ACM 12(10), 576–580 (1969)
18. Owicki, S.S., Gries, D.: Verifying properties of parallel programs: An axiomatic approach. Communications of the ACM 19(5), 279–285 (1976)
19. Jones, C.B.: Specification and design of (parallel) programs. In: Mason, R.E.A. (ed.) Information Processing 1983: World Congress Proceedings, pp. 321–332. Elsevier (1984)
20. O'Hearn, P.W.: Resources, concurrency, and local reasoning. Theoretical Computer Science 375(1-3), 271–307 (2007)
21. Reynolds, J.C.: Separation logic: A logic for shared mutable data structures. In: LICS, pp. 55–74. IEEE (2002)
22. Feng, X.: Local rely-guarantee reasoning. In: POPL, pp. 315–327. ACM (2009)
23. Vafeiadis, V., Parkinson, M.J.: A marriage of rely/guarantee and separation logic. In: Caires, L., Vasconcelos, V.T. (eds.) CONCUR 2007. LNCS, vol. 4703, pp. 256–271. Springer, Heidelberg (2007)
24. Reddy, U.S., Reynolds, J.C.: Syntactic control of interference for separation logic. In: POPL, pp. 323–336. ACM (2012)
25. Hayman, J.: Granularity and concurrent separation logic. In: Katoen, J.-P., König, B. (eds.) CONCUR 2011. LNCS, vol. 6901, pp. 219–234. Springer, Heidelberg (2011)
26. Bae, K., Escobar, S., Meseguer, J.: Abstract logical model checking of infinite-state systems using narrowing. In: RTA, pp. 81–96 (2013)
27. Rocha, C., Meseguer, J.: Proving safety properties of rewrite theories. In: Corradini, A., Klin, B., Cîrstea, C. (eds.) CALCO 2011. LNCS, vol. 6859, pp. 314–328. Springer, Heidelberg (2011)
28. Rocha, C., Meseguer, J., Muñoz, C.A.: Rewriting modulo smt and open system analysis. In: WRLA. LNCS (to appear, 2014)

Formalizing Monotone Algebras for Certification of Termination and Complexity Proofs*

Christian Sternagel and René Thiemann

Institute of Computer Science, University of Innsbruck, Austria
{christian.sternagel,rene.thiemann}@uibk.ac.at

Abstract. Monotone algebras are frequently used to generate reduction orders in automated termination and complexity proofs. To be able to certify these proofs, we formalized several kinds of interpretations in the proof assistant Isabelle/HOL. We report on our integration of matrix interpretations, arctic interpretations, and nonlinear polynomial interpretations over various domains, including the reals.

1 Introduction

Since the first termination competition[1] in 2004 it is of great interest whether a proof – that has been automatically generated by a termination or complexity tool – is indeed correct. The increasing complexity of generated proofs makes certification (i.e., checking correctness) more and more tedious for humans. Hence the interest in automated certification of termination and complexity proofs. This led to the general approach of using proof assistants for certification.

In this paper we present one of the key ingredients of our certifier, CeTA [34], namely the machinery for checking order constraints for (weakly) monotone algebras in the form of polynomial, matrix, and arctic interpretations. These constraints frequently arise in both termination and complexity proofs. For example, during the full run on the termination problem database in 2013, 3759 certifiable proofs have been generated. In 3170 of these proofs interpretations are used. Hence, they would not be certifiable by CeTA without the results of this paper.

In order to properly certify such proofs – and not just implement an independent but untrusted machinery for constraint checking – we take a two-phase approach. In the first phase, we prove general properties in Isabelle/HOL [25]. For example, we show that indeed all of the above interpretations are sound, i.e., they may be used for termination proofs.

In the second phase, we have to check concrete applications of interpretations on a concrete set of constraints. For example, at this point we need to ensure monotonicity of a given polynomial or to validate the growth rate of some matrix interpretation. To this end, we develop appropriate algorithms, prove them correct within Isabelle/HOL, and then invoke Isabelle's code generator [11] to obtain our certifier CeTA. The consequences of applying code generation are twofold:

* Supported by the Austrian Science Fund (FWF) projects P22767 and J3202.

[1] http://termination-portal.org/wiki/Termination_Competition

G. Dowek (ed.): RTA-TLCA 2014, LNCS 8560, pp. 441–455, 2014.

while we obtain a high execution speed, there is the additional requirement that all the algorithms that are used for certification have to be fully executable (in the sense of functional programming).

Contribution and Overview. After giving some preliminaries in Section 2, we present our main contributions. In Section 3 we start with a generic formalization of polynomial and matrix interpretations that may be instantiated by several carriers like the naturals, the rationals, and the reals; we also support recent monotonicity criteria that are not present in other certifiers. Our integration of arctic interpretations (Section 4) reveals how the theory on arctic naturals and arctic integers can be unified. Moreover, it shows how to support arctic interpretations which are monotone in presence of a fresh binary symbol, a novelty. We further report on how we achieve executability for an interesting subset of the real numbers (Section 5). At this point, we also have to develop algorithms for computing n-th roots of numbers in order to efficiently factor numbers. Afterwards, we present our work on certifying complexity proofs (Section 6): as far as we know CeTA is the first certifier which supports complexity proofs at all. We finally conclude in Section 7.

All of the proofs that are presented (or omitted) in the following have been made available in the archive of formal proofs [29,30,33] or in IsaFoR,[2] an Isabelle/HOL formalization of rewriting. We further provide example termination and complexity proofs[3] which show applications of the various kinds of interpretations and can all be certified by CeTA.

2 Preliminaries

We assume familiarity with term rewriting (see, e.g., Baader and Nipkow [1]) but briefly recall notions that are used in the following. Terms are defined inductively: a *term* is either a variable x or is constructed by applying a function symbol f from the *signature* $\mathcal{F}$ to a list of argument terms $f(t_1, \ldots, t_n)$.

A pair of terms (s, t) is sometimes considered a *(rewrite) rule*, then we write $s \to t$. A set $\mathcal{R}$ of rules is called a *term rewrite system* (TRS for short). In contrast to many authors, we do not assume any a priori restrictions on rules of TRSs (the most frequent ones being that the left-hand side of a rule is not a variable and that rules do not introduce fresh variables on their right-hand side; both or either of the previous conditions are sometimes referred to as *variable condition* in the literature). Whenever there are restrictions, we mention them explicitly. TRSs induce a *rewrite relation* by closing their rules under contexts and substitutions. More precisely the rewrite relation of $\mathcal{R}$, denoted by $\to_\mathcal{R}$, is defined inductively by $s \to_\mathcal{R} t$ whenever there are a rule $\ell \to r \in \mathcal{R}$, a context C, and a substitution σ such that $s = C[\ell\sigma]$ and $t = C[r\sigma]$.

[2] `http://cl-informatik.uibk.ac.at/software/ceta`
[3] `http://cl-informatik.uibk.ac.at/software/ceta/experiments/interpretations`

A *semiring* is a structure $(A, +, \cdot, 0, 1)$ such that $(A, +, 0)$ is a commutative monoid with neutral element 0 and $(A, \cdot, 1)$ is a monoid with neutral element 1. Moreover, $\cdot$ distributes over $+$, $0 \neq 1$, and $0 \cdot x = x \cdot 0 = 0$ for all $x \in A$.

An *($\mathcal{F}$-)algebra* $\mathcal{A}$ is a carrier set A equipped with an interpretation function $f_\mathcal{A} : A^n \to A$ for every n-ary $f \in \mathcal{F}$. We call an algebra $\mathcal{A}$ *monotone* w.r.t. a binary relation $>$ when all interpretation functions are monotone, i.e., $f_\mathcal{A}(\ldots, a, \ldots) > f_\mathcal{A}(\ldots, b, \ldots)$ whenever $a > b$. A *well-founded monotone algebra* is a monotone algebra $(\mathcal{A}, >)$ such that $>$ is well-founded. For any algebra $\mathcal{A}$, terms can be interpreted w.r.t. an assignment α, written $[t]_\alpha$. Then, $s >_\mathcal{A} t$ denotes $[s]_\alpha > [t]_\alpha$ for all α.

A binary relation $\to$ is *terminating* (or *well-founded*) if there are no infinite derivations $a_1 \to a_2 \to a_3 \to \cdots$. Given two binary relations $\to_\alpha$, $\to_\beta$ we write $\to_\alpha/\to_\beta$ to abbreviate $\to_\beta^* \cdot \to_\alpha \cdot \to_\beta^*$, i.e., the rewrite relation of $\to_\alpha$ *relative to* $\to_\beta$. Termination of $\to_\alpha/\to_\beta$ is also called relative termination of $\to_\alpha$ w.r.t. $\to_\beta$.

We call a pair of two orders on terms $(\succ, \succeq)$ a *reduction pair* whenever it satisfies the following requirements: $\succ$ is well-founded, $\succeq$ and $\succ$ are compatible (i.e., $\succeq \cdot \succ \subseteq \succ$) and stable (i.e., closed under substitutions), and $\succeq$ is monotone (i.e., closed under contexts). If in addition $\succ$ is monotone, we call $(\succ, \succeq)$ a *monotone* reduction pair. Reduction pairs are employed for termination proofs inside the dependency pair framework, monotone reduction pairs for direct termination and complexity proofs.

3 Polynomial and Matrix Interpretations

Two widely used approaches to synthesize reduction pairs are polynomial interpretations (Lankford [18]) and matrix interpretations (Endrullis et al. [8]).

To support polynomial interpretations within CeTA, we formalized nonlinear multivariate polynomials [30] within Isabelle/HOL. Since similar tasks have already been conducted CoLoR [3] and Coccinelle [6] (using the approach of CiME [7]), we just shortly mention two distinguishing features of our work.

A formalization of polynomial orders has already been described by Blanqui and Koprowski [3]. Whereas their formalization fixes the carrier to $\mathbb{N}$, our polynomial orders are parametric in the carrier, cf. theory `Poly_Order`. Hence, we can treat polynomial orders over $\mathbb{N}$, $\mathbb{Q}$, and $\mathbb{R}$ within the same framework by just instantiating the carrier to the respective ordered semiring. Here, for both $\mathbb{Q}$ and $\mathbb{R}$ we use *δ-orders* as the strict order to achieve well-foundedness – as described by Lucas [20]: $x >_\delta y := x - y \geq \delta \wedge y \geq 0$ where δ is some fixed positive number. Notice that each carrier may have its own specialties, e.g., x^2 is monotone over $\mathbb{N}$, but not monotone for δ-orders if $\delta < 1$. This required the addition of properties in the parametric setting. For example, we added a Boolean parameter *power-mono* which describes whether polynomials like x^k with $k > 1$ are strictly monotone; it is always satisfied for $\mathbb{N}$ but demands $\delta \geq 1$ for $\mathbb{Q}$ and $\mathbb{R}$.

As far as we know we provide the first formalization of the improved monotonicity criteria for polynomials over $\mathbb{N}$ of Neurauter et al. [24] that can ensure monotonicity of polynomials like $2x^2 - x$ which are not monotone over $\mathbb{Q}$ and $\mathbb{R}$. We made them available in the archive of formal proofs [30], theory `NZM`.

To support matrix interpretations, we basically follow the ideas of Courtieu et al. [7], i.e., we integrate matrix interpretations as linear polynomial interpretations where the carrier consists of matrices. To this end, we first developed a list-based and executable formalization of matrices [29] within Isabelle/HOL and afterwards connected it in the theory `Linear_Poly_Order` to obtain matrix interpretations in IsaFoR. Again, one of the distinguishing features of our work is the parametric carrier – for example in [7] the carrier is fixed to $\mathbb{N}$. Note that the demand for other carriers like $\mathbb{Q}$ and $\mathbb{R}$ was clearly shown by Neurauter and Middeldorp [23]: matrix interpretations over $\mathbb{R}$ are strictly more powerful than those over $\mathbb{Q}$ which in turn are strictly more powerful than those over $\mathbb{N}$.

Having developed the abstract results on these interpretations, it was easy to integrate executable criteria within CeTA that check applications of polynomial or matrix interpretations within concrete termination proofs – if the carrier consists of (matrices over) rational or natural numbers. However, more work had to be done for the reals. Before we discuss these problems in Section 5, we consider another kind of semiring in the next section.

4 Arctic Interpretations

The semirings $(\mathbb{A}_{\mathbb{N}}, \max, +, -\infty, 0)$ and $(\mathbb{A}_{\mathbb{Z}}, \max, +, -\infty, 0)$ are called *arctic semiring* and *arctic semiring below zero*, respectively. Here, $\mathbb{A}_A$ denotes the extension of A by the element $-\infty$, i.e., $A \cup \{-\infty\}$, $\max\{x, -\infty\} = x$, and $x + -\infty = -\infty + x = -\infty$ for all x. Waldmann and Koprowski [16] first used these semirings in the well-founded monotone algebra setting.

In the following we unify and extend (see Sternagel and Thiemann [28] for an earlier account) the arctic interpretations introduced by Waldmann and Koprowski. To do so, we first introduce the notion of an ordered arctic semiring.

Definition 1. *Let $(A, +, \cdot, 0, 1)$ be a semiring. Then an* ordered arctic semiring, *denoted by $(A, +, \cdot, 0, 1, >, \geq, \mathsf{pos})$, satisfies the additional requirements:*

- *$\geq$ is reflexive and transitive; $> \cdot \geq \subseteq >$ and $\geq \cdot > \subseteq >$*
- *$1 \geq 0$; $\neg\mathsf{pos}(0)$; $\mathsf{pos}(1)$; $x > 0$; $x \geq 0$; and $x = 0$ whenever $0 > x$*
- *$+$ is left-monotone w.r.t. $\geq$, i.e., $x + z \geq y + z$ whenever $x \geq y$*
- *$+$ is monotone w.r.t. $>$, i.e., $w + x > y + z$ whenever $w > y$ and $x > z$*
- *$\cdot$ is left- and right-monotone w.r.t. $\geq$ and left-monotone w.r.t. $>$*
- *$\mathsf{pos}(x + y)$ whenever $\mathsf{pos}(x)$; and $\mathsf{pos}(x \cdot y)$ whenever $\mathsf{pos}(x)$ and $\mathsf{pos}(y)$*
- *$\{(x, y) \mid x > y \wedge \mathsf{pos}(y)\}$ is well-founded*

Interpretation into an ordered arctic semiring yields a reduction pair.

Theorem 2. *Let $\mathcal{A}$ be an algebra over an ordered arctic semiring with interpretations $f_{\mathcal{A}}(x_1, \ldots, x_n) = f_0 + f_1 \cdot x_1 + \cdots + f_n \cdot x_n$ such that $\mathsf{pos}(f_i)$ for some $0 \leq i \leq n$. Then $(\succ_{\mathcal{A}}, \succeq_{\mathcal{A}})$ is a reduction pair.*

Examples for ordered arctic semirings are given in the following:

Example 3. The arctic semiring, the arctic semiring below zero, and the arctic rational semiring $(\mathbb{A}_\mathbb{Q}, \max, +, -\infty, 0)$ are ordered arctic semirings for $x > y := (y = -\infty \vee (x \neq -\infty \wedge x >_A y))$ (where $>_A$ is $>_\mathbb{N}$ and $>_\mathbb{Z}$ for naturals and integers, respectively; and $x >_\delta y$ for some $\delta > 0$ for the rationals), $x \geq y := (y = -\infty \vee (x \neq -\infty \wedge x \geq_{\mathbb{N}/\mathbb{Z}/\mathbb{Q}} y))$, and $\mathsf{pos}(x) := (x \neq -\infty \wedge x \geq_{\mathbb{N}/\mathbb{Z}/\mathbb{Q}} 0)$.

Note that the ordered arctic semiring over $\mathbb{A}_\mathbb{Q}$ together with Theorem 2, unifies and extends Theorems 12 and 14 of Waldmann and Koprowski [16]. The main advantage of our approach is that we only require interpretations to have at least one positive f_i (instead of always requiring the constant part f_0 to be positive). Although our result is slightly more general, we could completely reuse the original proof structure of [16] to formalize Theorem 2.

Waldmann and Koprowski also showed that for string rewriting (i.e, terms over a signature of function symbols that are at most unary) arctic interpretations are monotone and thus may be used for rule removal on standard termination problems. In order to apply this technique in the dependency pair framework together with *usable rules* we also need $\mathcal{C}_\mathcal{E}$-compatibility, i.e., the rules of the TRS $\mathcal{C}_\mathcal{E} = \{\mathsf{c}(x,y) \to x, \mathsf{c}(x,y) \to y\}$ must be oriented where c is some fresh symbol. But by considering $\mathcal{C}_\mathcal{E}$ we leave the domain of string rewriting.

Nevertheless, we want to obtain $\mathcal{C}_\mathcal{E}$-compatibility for monotone arctic interpretations. As an application consider the technique of Giesl et al. [9, Thm. 28] which allows us to remove all non-usable rules and all strictly oriented rules from a dependency pair problem, provided that the dependency pairs and rules are weakly oriented, the rules in $\mathcal{C}_\mathcal{E}$ are strictly oriented, and $\succ$ is monotone. To this end we first need signature extensions for relative termination (see also [31]).

Theorem 4 (Signature Extensions Preserve Relative Termination). *Let $\mathcal{R}$ and $\mathcal{S}$ be TRSs over a common signature $\mathcal{F}$. Moreover, suppose that no right-hand side of a rule in $\mathcal{S}$ introduces fresh variables. Then $\to_\mathcal{R}/\to_\mathcal{S}$ terminates for terms over arbitrary extensions of $\mathcal{F}$, whenever it does so for terms over $\mathcal{F}$.*

The above statement is not true when $\mathcal{S}$ violates the variable condition.

Example 5. Consider $\mathcal{R} = \{\mathsf{a} \to \mathsf{b}\}$ relative to $\mathcal{S} = \{\mathsf{c} \to x\}$. Over the common signature $\mathcal{F} = \{\mathsf{a}/0, \mathsf{b}/0, \mathsf{c}/0\}$ we have relative termination. However, extending $\mathcal{F}$ by $\{\mathsf{f}/2\}$ yields the infinite derivation where $C = \mathsf{f}(\mathsf{b}, \square)$.

$$s = \mathsf{f}(\mathsf{a},\mathsf{c}) \to_\mathcal{R} C[\mathsf{c}] \to_\mathcal{S}^* C[s] \to_\mathcal{R} C[C[\mathsf{c}]] \to_\mathcal{S}^* \cdots$$

Finally, we have to show that for monotone $\succ$ we get $\mathcal{C}_\mathcal{E}$-compatibility.

Lemma 6. *Consider a reduction pair $(\succ, \succsim)$ and TRSs $\mathcal{R}$, $\mathcal{S}$ over a common signature of at most unary function symbols $\mathcal{F}$ such that no rule of $\mathcal{S}$ introduces fresh variables. Moreover, let $\mathcal{R} \subseteq \succ$ and $\mathcal{S} \subseteq \succsim$. Then monotonicity of $\succ$ implies termination of $\to_{\mathcal{C}_\mathcal{E} \cup \mathcal{R}}/\to_{\mathcal{C}_\mathcal{E} \cup \mathcal{S}}$.*

Proof. By Theorem 4 together with monotonicity of $\succ$, we obtain termination of $\to_\mathcal{R}/\to_\mathcal{S}$ for arbitrary extensions of $\mathcal{F}$. Consider a lexicographic path order where all symbols are equal in precedence in combination with an argument filter that projects unary function symbols to their argument and keeps all other symbols unchanged. Then, this combination yields a monotone reduction pair $(>, \geq)$. Moreover, $\mathcal{R}$ and $\mathcal{S}$ are compatible with $\geq$ (since all terms are collapsed to a single variable or constant). Since also $\mathcal{C}_\mathcal{E} \subseteq >$ we obtain relative termination of $\rightarrow_{\mathcal{C}_\mathcal{E}}$ w.r.t. $\rightarrow_{\mathcal{R}\cup\mathcal{S}}$ and thus termination of $\rightarrow_{\mathcal{C}_\mathcal{E}\cup\mathcal{R}}/\rightarrow_{\mathcal{C}_\mathcal{E}\cup\mathcal{S}}$. □

While the above lemma does not quite yield $\mathcal{C}_\mathcal{E}$-compatibility, it can be used to show that from every reduction pair $(\succ, \succsim)$ that satisfies the above conditions we obtain a corresponding $\mathcal{C}_\mathcal{E}$-compatible reduction pair $(\succ', \succsim')$. More specifically, take $\succ' = (\rightarrow_{\mathcal{C}_\mathcal{E}\cup\mathcal{R}}/\rightarrow_{\mathcal{C}_\mathcal{E}\cup\mathcal{S}})^+$ and $\succsim' = \rightarrow^*_{\mathcal{C}_\mathcal{E}\cup\mathcal{S}}$.

Now if we start from a monotone reduction pair $(\succ, \succsim)$, and a set of rules $\mathcal{P}$ over an at most unary signature and take $\mathcal{R} = \mathcal{P} \cap \succ$ and $\mathcal{S} = \mathcal{P} \cap \succsim$, then the resulting reduction pair $(\succ', \succsim')$ is $\mathcal{C}_\mathcal{E}$-compatible, monotone, and orients all rules of $\mathcal{P}$ that were also oriented by the original reduction pair.

5 Interpretations over the Reals

Whereas all basic operations on $\mathbb{Q}$ are executable, this is not the case for $\mathbb{R}$. To solve this problem, automated tools only work on a subset of the real numbers [21,36,37]. For example, in the setting of Zankl and Middeldorp [36] numbers may be chosen from $\mathbb{Q}[\sqrt{2}]$, the field extension of $\mathbb{Q}$ by $\sqrt{2}$. All these numbers are of the form $p + q\sqrt{2}$ where p and q range over $\mathbb{Q}$. In [37], Zankl et al. allow even more generic forms, e.g., where $\sqrt{2}$ may be replaced by $\sqrt{b}$ for a fixed natural number b with $\sqrt{b} \notin \mathbb{Q}$, i.e., we consider $\mathbb{Q}[\sqrt{b}]$.

Fixing the base b, all numbers in $\mathbb{Q}[\sqrt{b}]$ can be represented by pairs (p, q) (encoding $p + q\sqrt{b}$), where all ordered semiring operations are again executable. For example, $(p_1, q_1) \cdot (p_2, q_2) := (p_1 p_2 + b q_1 q_2, p_1 q_2 + p_2 q_1)$ and $(p, q) > 0 := (p \geq 0 \land q > 0) \lor (p > 0 \land q \geq 0) \lor (p \geq 0 \land q < 0 \land p^2 > bq^2) \lor (p \leq 0 \land q > 0 \land p^2 < bq^2)$ where the definition of $(p, q) > 0$ can be used to decide the comparison $p_1 + q_1\sqrt{b} > p_2 + q_2\sqrt{b}$ by choosing $p = p_1 - p_2$ and $q = q_1 - q_2$ [37, Def. 10].

A larger subset of the real numbers, namely the algebraic real numbers, has been formalized by Cohen [5]. However, since this formalization has been conducted using Coq, it cannot easily be integrated into our Isabelle/HOL development. Moreover, as far as we know, all real numbers which are currently generated by automated termination tools are contained in $\mathbb{Q}[\sqrt{b}]$ for some fixed b which can be chosen in the configuration of the tool. Hence, for certification it suffices to formalize this subset of the real numbers. Although a full integration of real algebraic numbers in Isabelle might be welcome, we pursued the more lightweight approach which was sufficient to support $\mathbb{Q}[\sqrt{b}]$.

We consider some alternatives for representing $\mathbb{Q}[\sqrt{b}]$ in Isabelle/HOL. The first alternative is to fix some b and create a new type $pair_b$ consisting of pairs of rational numbers. Then addition, multiplication, comparison, etc. are defined as above, and we have to prove that this new type forms an ordered semiring (one that is completely independent from the real numbers). The disadvantage of this approach is that for each b we have to define a new type. As we can only define finitely many types within a finite Isabelle formalization, our certifier will be limited to a fixed number of choices for b.

For higher flexibility, we can alternatively create a type *triples* that additionally takes the parameter b as third component. The problem in this approach is to give a total definition for all operations, e.g., what should be the result of $(1,1,2)\cdot(1,1,3)$, i.e., how can we represent the number $(1+1\cdot\sqrt{2})\cdot(1+1\cdot\sqrt{3}) = 1+\sqrt{2}+\sqrt{3}+\sqrt{6}$ as a triple $p+q\sqrt{b}$ for suitable values of $p,q \in \mathbb{Q}$ and $b \in \mathbb{N}$.

A third possibility would be to not create a type at all, but use locales [2] and explicit domains. We do not go into the details here, but just mention that this approach is currently not applicable, since other parts of the formalization – like the theories on nonlinear polynomial interpretations – utilize the type class for semirings, and do not support the more flexible locales.

Our final solution is to not define a new type to form the semirings $\mathbb{Q}[\sqrt{b}]$, but to perform data refinement [10] instead, i.e., provide an implementation type for the reals. This has the following advantages: For a start, the Isabelle distribution already contains the result that $\mathbb{R}$ is an ordered semiring. Thus all the properties of the reals can be used when formalizing monotone algebras over the reals. Moreover, our implementation can be partial, e.g., we do not have to support the multiplication of arbitrary numbers like $(1+\sqrt{2})\cdot(1+\sqrt{3})$. Finally, as soon as a better implementation is available, we can just replace the current one by the new one, and do not have to change the theories which show that the reals can be used to generate monotone algebras.

5.1 A First Implementation of $\mathbb{R}$ via Triples (p, q, b)

In the following we implement the reals by the type *mini-alg* containing all triples $(p,q,b) \in \mathbb{Q}\times\mathbb{Q}\times\mathbb{N}$ that satisfy the invariant $q = 0 \vee \sqrt{b} \notin \mathbb{Q}$. Such a quotient type is easily created and accessed via the lifting and transfer package [15].

For this data refinement, we first have to declare how *mini-alg* is mapped into the reals. This is done by a function *real-of*: *mini-alg* $\to \mathbb{R}$, defined as:

$$\mathit{real\text{-}of}\,(p,q,b) = p + q\sqrt{b}$$

Next, we tell the code generator that *real-of* should be seen as the constructor for real numbers, i.e., from now on we consider the reals as being defined by the datatype definition ***datatype*** $\mathbb{R} =$ *real-of mini-alg* where *real-of* is the unique constructor which takes a triple as input. Afterwards, the desired operations on reals must be implemented via lemmas on this new "constructor" *real-of*. E.g., the unary minus operation is implemented by proving the following lemma:

$$-\mathit{real\text{-}of}\,(p,q,b) = \mathit{real\text{-}of}\,(-p,-q,b)$$

Often, we only implement partial operations, e.g., for some binary operations we require triples with compatible bases. For example, addition is defined by the lemma

$real\text{-}of\ (p_1, q_1, b_1) + real\text{-}of\ (p_2, q_2, b_2) =$
  $\mathit{if}\ compatible\ (p_1, q_1, b_1)\ (p_2, q_2, b_2)$
  $\mathit{then}\ (\mathit{if}\ q_1 = 0\ \mathit{then}\ real\text{-}of\ (p_1 + p_2, q_2, b_2)\ \mathit{else}\ real\text{-}of\ (p_1 + p_2, q_1 + q_2, b_1))$
  $\mathit{else}\ abort\ (\lambda\_.\ real\text{-}of\ (p_1, q_1, b_1) + real\text{-}of\ (p_2, q_2, b_2))\ ()$

where *compatible* $(p_1, q_1, b_1)\ (p_1, q_2, b_2)$ is defined as $q_1 = 0 \vee q_2 = 0 \vee b_1 = b_2$, and *abort* $f\ x = f\ x$. That is, two triples are compatible iff one of them encodes a rational number, or the bases are identical. The equation for *abort* allows us to prove the above lemma, but is not used to generate code, since this would lead to nontermination in case of incompatible triples. Instead, the code generator issues an appropriate error in the target language at this point. This trick was already described by Lochbihler [19].

Above we defined several operations on reals like addition, multiplication, greater-than, and a mapping from $\mathbb{Q}$ into $\mathbb{R}$. Some of the binary operations are partial and require compatible triples as input. However, the above mentioned operations do not make use of the invariant of *mini-alg*. The invariant is required for operations like equality and inverse. For example, for the multiplicative inverse of a triple (p, q, b) we use the triple $(p/d, -(q/d), b)$ where the divisor is $d = p^2 - bq^2$. To ensure that $d \neq 0$ whenever *real-of* $(p, q, b) \neq 0$ we need the invariant that $\sqrt{b}$ is irrational. Similarly, also for equality – which is defined as $p_1 = p_2 \wedge q_1 = q_2 \wedge (q_1 = 0 \vee b_1 = b_2)$ for compatible triples (p_1, q_1, b_1) and (p_2, q_2, b_2) – we require the invariant that $\sqrt{b}$ is irrational. Otherwise, the above implementation of equality would return false for the inputs $(0, 1, 4)$ and $(2, 0, 4)$, but *real-of* $(0, 1, 4) = 0 + 1 \cdot \sqrt{4} = 2 = 2 + 0 \cdot \sqrt{4} =$ *real-of* $(2, 0, 4)$.

So far, we defined all required field operations and comparisons, each of which is implemented by a constant number of operations on rational numbers. However, we are lacking a way to really construct irrational numbers. To this end we provide a partial implementation of the square root function that is only defined for input triples encoding rational numbers. The definition for nonnegative rational numbers with numerator n and denominator d is

$$sqrt\ (\frac{n}{d}, 0, b) = \mathit{if}\ \sqrt{nd} \in \mathbb{Z}\ \mathit{then}\ (\frac{\sqrt{nd}}{d}, 0, 0)\ \mathit{else}\ (0, \frac{1}{d}, nd) \tag{1}$$

where the case-analysis is solely performed to satisfy the invariant of triples of type *mini-alg*. In (1) we make use of a square root function on integers which can decide for a given integer i whether $\sqrt{i} \in \mathbb{Z}$ or not. If so, it also returns the resulting number, cf. Thiemann [32, Thm. 14]. We modified this square root function such that it can additionally compute $\lfloor\sqrt{i}\rfloor$ and $\lceil\sqrt{i}\rceil$. In this way, we are able to implement $\lfloor\cdot\rfloor$ and $\lceil\cdot\rceil$ on triples of type *mini-alg*.

In total, our implementation provides the following operations on reals: $+$, $-$, $\times$, $\cdot^{-1}$, $>$, $\geq$, $=$, $\lfloor\cdot\rfloor$, $\lceil\cdot\rceil$, and $\sqrt{\cdot}$. Only the last three require the computation of square roots. All binary operations succeed if their operands are compatible (which is always the case if a fixed base b is chosen), and only the last operation is restricted to rational numbers as input. This implementation supports all operations that we require for monotone algebras except for one.

5.2 A Second Implementation of $\mathbb{R}$ via Triples (p, q, b)

First note that CeTA not only accepts or rejects a termination proof, but also provides a detailed error message in case of rejection. To this end, we have to print numbers occurring in interpretations, i.e., we need a function *show* : $\mathbb{R} \rightarrow$ *string*.

An easy solution might be to postulate such a function and provide an implementation via the axiom: *show* (*real-of* (p, q, b)) = the string "p + q * sqrt(b)".

One might argue that adding this axiom is not really relevant, as it is only used for error messages. However, adding it immediately introduces an inconsistency in the logic:

$$\begin{aligned} \text{"4 + 1 * sqrt(18)"} &= \mathit{show}\ (\mathit{real\text{-}of}\ (4,1,18)) &&= \mathit{show}\ (4+1\cdot\sqrt{18}) \\ = \mathit{show}\ (4+3\cdot\sqrt{2}) &= \mathit{show}\ (\mathit{real\text{-}of}\ (4,3,2)) &&= \text{"4 + 3 * sqrt(2)"} \end{aligned}$$

That is, the wrong fact that the first and the last string are identical is derivable.

As a consequence, we want to avoid this inconsistent axiom which stems from the fact that *real-of* is not injective, e.g., the number $\sqrt{18}$ can be represented by both $(0, 3, 2)$ and $(0, 1, 18)$. To this end, we define a new type of triples, *mini-alg-unique*. It is similar to *mini-alg* but adds another invariant: every triple (p, q, b) must satisfy $q = 0 \land b = 0 \lor q \neq 0 \land \mathit{prime\text{-}product}\ b$, where *prime-product* b demands that b is a product of distinct primes. For example, 2 and $6 = 2 \cdot 3$ are prime products, but $18 = 2 \cdot 3 \cdot 3$ is not, since 3 occurs twice.

In the remainder of this section we assume that we perform data refinement of $\mathbb{R}$ by implementing it via triples of type *mini-alg-unique*. While most of the algorithms work as for *mini-alg*, we list the most important differences.

The main advantage of *mini-alg-unique* is that *real-of* is now injective. As a result, equality of reals can easily be implemented as equality of triples without checking for compatibility. For example, since $(1, 2, 3) \neq (2, 2, 2)$ we conclude $1 + 2 \cdot \sqrt{3} \neq 2 + 2 \cdot \sqrt{2}$. This also allows us to define a total function for comparisons which is implemented via $\lfloor\cdot\rfloor$: if the numbers are equal, then the result is determined, and otherwise we multiply both numbers iteratively by 1024 until there is a difference after applying $\lfloor\cdot\rfloor$. For example, the algorithm shows $1+2\cdot\sqrt{3} < 2+2\cdot\sqrt{2}$ since $\lfloor 1024\cdot(1+2\cdot\sqrt{3})\rfloor = 4571 < 4944 = \lfloor 1024\cdot(2+2\cdot\sqrt{2})\rfloor$.

As *real-of* is injective, it is now also possible to define *show* on the reals, and later on implement it for triples of type *mini-alg-unique*. To this end, assume we have already defined a function *mau-show* which pretty prints triples t as strings. The specification of *show* in the logic is

$$\mathit{show}\ x = (\mathsf{if}\ \exists t.\ x = \mathit{real\text{-}of}\ t\ \mathsf{then}\ \mathit{mau\text{-}show}\ (\mathsf{THE}\ t.\ x = \mathit{real\text{-}of}\ t)\ \mathsf{else}\ \text{"nothing"})$$

while its implementation is given by the lemma: *show* (*real-of* t) = *mau-show* t, where THE $t.\ P\ t$ results in the unique t satisfying $P\ t$, if such a t exists, and is undefined, otherwise. In the definition of *show*, existence of t is established before calling *mau-show* and uniqueness follows from the injectivity of *real-of*.

The last algorithm that requires an adaptation is the implementation of *sqrt*. The definition from (1) is not suitable any more, as $\sqrt{nd} \notin \mathbb{Z}$ does not guarantee nd to be a prime product. Therefore, a preprocessing is required which factors every natural number m (like nd) into $m = s^2 \cdot p$ where $s, p \in \mathbb{N}$ and either p is a prime product or p is 1. In the latter case, $\sqrt{m}$ is the natural number s, and in the former case the triple $(0, s, p)$ represents the number $\sqrt{m}$ and also satisfies the invariant of *mini-alg-unique*.

For the factorization algorithm, we do not fully decompose m into prime factors – which would roughly require $\sqrt{m}$ iterations – but use the following algorithm, requiring only $\sqrt[3]{m}$ iterations. First check whether $\sqrt{m}$ is irrational. If not, then we are done by returning $(\sqrt{m}, 1)$. Otherwise, we check whether m has a factor between 2 and $\lfloor \sqrt[3]{m} \rfloor$. If we detect such a factor p, then we store p, and continue to search factors of m/p with a new upper bound of $\lfloor \sqrt[3]{m/p} \rfloor$. If there is no such factor, then we conclude that m is a prime product as follows: assume that m is not a prime product, i.e., $m = p \cdot p \cdot q$ for some prime p and natural number q. Then $q \neq 1$ since otherwise $\sqrt{m} = p \in \mathbb{N}$ is not irrational. Hence, m has both p and q as factors. But since we tested that m does not divide any of the numbers up to $\lfloor \sqrt[3]{m} \rfloor$, we know that both p and q are larger than $\sqrt[3]{m}$. Hence, $m = p \cdot p \cdot q > (\sqrt[3]{m})^3 = m$, a contradiction.

Note that, for implementing the factorization algorithm, we not only need the square root algorithm of [32, Sect. 6], but also require an algorithm to compute $\lfloor \sqrt[3]{m} \rfloor$. To this end, we extended the work of [32] to arbitrary n-th roots, i.e., we can check $\sqrt[n]{p} \in \mathbb{Q}$ and compute $\lfloor \sqrt[n]{p} \rfloor$ for every $n \in \mathbb{N}$ and $p \in \mathbb{Q}$. Here, Porter's [26] formalization of Cauchy's mean theorem was extremely helpful to show soundness of our n-th root algorithm. The algorithm itself uses a variant of Newton iteration to compute precise roots, which uses integer divisions where one usually works on rational or floating point numbers.

5.3 Summary

We performed data refinement to implement the subset $\mathbb{Q}[\sqrt{b}]$ of the reals as triples (p, q, b) representing $p + q \cdot \sqrt{b}$. The first implementation has the advantage of being more efficient, but several operations are only supported partially, where in binary operations the same basis $\sqrt{b}$ must be present. The second implementation is less partial and even allows to define a show function on reals, at the cost of having to perform a factorization of prime products, which we implemented via an algorithm with $\sqrt[3]{n}$ iterations. To this end, we also formalized a generic n-th root algorithm. This part of the formalization has been made available in the archive of formal proofs [33].

Using this formalization, we are able to certify each application of monotone algebras over the reals within termination proofs generated by $\mathsf{T_TT_2}$ [17].

6 Complexity Proofs

Monotone algebras are not only a useful tool for termination analysis, but also for complexity analysis. In the following, we first introduce basic notions regarding complexity – including a modularity result of Zankl and Korp [35] – and afterwards provide details on complexity results for matrix interpretations and polynomial interpretations which have been integrated into IsaFoR and CeTA.

6.1 Complexity and Modularity

To measure the complexity of a TRS $\mathcal{R}$, we use the following notions and ideas of [4,12,14]. The *derivation height* of a term w.r.t. some binary relation $\to$ on terms is defined as $\mathsf{dh}_\to(t) = \max\{n \mid \exists s.\, t \to^n s\}$ and measures the height of the derivation tree of t. The *derivational complexity* of $\to$ is given by $\mathsf{dc}_\to(n) = \max\{\mathsf{dh}_\to(t) \mid |t| \leq n\}$. That is, the maximal length of derivations with starting terms of size n is bounded by $\mathsf{dc}_\to(n)$. While in $\mathsf{dc}_\to(n)$ all starting terms are considered, allowing for terms like $\mathsf{double}^n(1)$, there is the alternative notion of *runtime complexity*, where starting terms are restricted to basic terms. In detail, $\mathsf{rc}_{\to,\mathcal{C}}(n) = \max\{\mathsf{dh}_\to(t) \mid |t| \leq n, t \in \mathcal{B}_\mathcal{C}\}$ where $\mathcal{B}_\mathcal{C}$ denotes the set of *basic terms*, i.e., terms of the form $f(c_1,\ldots,c_n)$, with all the c_i only built over symbols from $\mathcal{C}$. Here $\mathcal{C}$ typically is the set of constructors of the TRS of interest.

Note that all of $\mathsf{dh}_\to$, $\mathsf{dc}_\to$, and $\mathsf{rc}_{\to,\mathcal{C}}$ are only well-defined if max is applied to a finite set. However, this is not necessarily the case, as on the one hand $\to$ may be infinitely branching or nonterminating, and on the other hand, there might be infinitely many terms of size n if the signature is infinite. In order to avoid having to worry about these side-conditions within our formalization, we instead define the following function

$$\textit{deriv-bound-rel}_\to(SE, f) = (\forall n\, t.\, (t \in SE(n) \implies \nexists s.\, t \to^{f(n)+1} s))$$

checking whether a given function f is an upper bound for the complexity. Here, *SE* describes the set of starting elements depending on a natural number n, usually limiting the size of elements. We can easily model runtime and derivational complexity: $\textit{deriv-bound-rel}_\to(\lambda n.\, \{t \mid |t| \leq n\}, f)$ and $\textit{deriv-bound-rel}_\to(\lambda n.\, \{t \mid |t| \leq n, t \in \mathcal{B}_\mathcal{C}\}, f)$ express that $\mathsf{dc}_\to$ and $\mathsf{rc}_{\to,\mathcal{C}}$ are bounded by f, respectively.

The above definitions are contained in the theory `Complexity`, which also contains the first formalization of the modularity result by Zankl and Korp [35, Thm. 4.4]. Here, we stay in an abstract setting where $\to_1$, $\to_2$, and $\to_3$ are arbitrary binary relations (not necessarily ranging over terms), cf. the theorem *deriv-bound-relto-class-union* that is part of IsaFoR.

Theorem 7. *Let $\to_i$ be binary relations (with $i \in \{1,2,3\}$). Moreover, let $\textit{deriv-bound-rel}_{\to_1/(\to_2\cup\to_3)}(SE, g_1)$ and $\textit{deriv-bound-rel}_{\to_2/(\to_1\cup\to_3)}(SE, g_2)$ for two functions $g_1, g_2 \in \mathcal{O}(f)$. Then there is a function $g \in \mathcal{O}(f)$ such that $\textit{deriv-bound-rel}_{(\to_1\cup\to_2)/\to_3}(SE, g)$.*

6.2 Complexity via Monotone Interpretations

In order to ensure complexity bounds via some monotone algebra $(\mathcal{A}, >)$ with carrier A, one first needs a function of type $A \to \mathbb{N}$ which bounds the number of decreases. To be more precise, in the generic setting for semirings within IsaFoR we require a function *bound* such that for each $a \in A$ there are no $a_1, a_2, \ldots$ such that $a > a_1 > \ldots > a_{bound(a)+1}$, and moreover *bound* has to grow linearly in its argument, cf. `Complexity_Carrier` for further details.

We defined various valid *bound* functions for the different kind of carriers. For example, we have chosen $bound_\mathbb{N}(n) = n$ for the naturals, $bound_\delta(x) = \lceil x/\delta \rceil$

for δ-orders on the rationals and reals, and $bound_{mat(A)}(m) = bound_A(||m||)$ for matrices over carrier A, where $||\cdot||$ denotes the linear norm of a matrix.

Obviously, whenever each reduction step $t \to s$ within a derivation corresponds to a decrease $[t]_\alpha > [s]_\alpha$ then $\mathsf{dh}_{\to}(t) \leq bound([t]_\alpha)$ for every assignment α and term t. Thus, $\mathsf{dh}_{\to_\mathcal{R}/\to_\mathcal{S}}(t) \leq bound([t]_\alpha)$ whenever $\ell >_\mathcal{A} r$ for each $\ell \to r \in \mathcal{R}$ and $\ell \geq_\mathcal{A} r$ for each $\ell \to r \in \mathcal{S}$. At this point in the formalization we fix α to be the zero-assignment α_0 where $\alpha_0(x) = 0$ for all x.

Since *bound* has to grow linearly in its argument, to get asymptotic bounds it suffices to estimate $[t]_{\alpha_0}$ for each $t \in SE(n)$, depending on n.

For *polynomial interpretations* we formalized the criterion of strongly linear interpretations of Hofbauer [13].

Theorem 8. *Let $\mathcal{F}$ be a subset of the signature. Whenever $f_\mathcal{A}(x_1, \ldots, x_n) = c_f + \sum_{i=1}^{n} x_i$ for each $f \in \mathcal{F}$ then*

- *$[t]_{\alpha_0}$ is linearly bounded in $|t|$ whenever $\mathcal{F}$ is the full signature.*
- *$[t]_{\alpha_0}$ is bounded by $\mathcal{O}(|t|^d)$ whenever $\mathcal{F} = \mathcal{C}$, $t \in \mathcal{B}_\mathcal{C}$, and d is the largest degree of a polynomial within the interpretation.*

The two alternatives have been formalized in `Poly_Order`, where the first one (*linear-bound*) is used for derivational complexity and the second one (*degree-bound*) for runtime complexity. Further note that the above theorem can be combined with several ordered semirings, so that currently CeTA can check complexity proofs involving polynomial interpretations over $\mathbb{N}$, $\mathbb{Q}$, and $\mathbb{R}$.

Furthermore, we also support complexity proofs via matrix interpretations. To be more precise, we provide the first formalization of the criterion of Moser, Schnabl, and Waldmann [22] that for upper triangular matrix interpretations we get an upper bound of $[t]_{\alpha_0} \in \mathcal{O}(|t|^d)$ where d is the dimension of the matrix.

To this end, we have first proven [22, Lem. 5] that $||m^n|| \in \mathcal{O}(n^{d-1})$ is satisfied for an upper triangular matrix m of dimension d. This fact has been made available as *upper-triangular-mat-pow-value* in `Matrix_Comparison` within the archive of formal proofs [29]. Here, we want to stress that the formalization has been much more verbose than the paper: in [22] the proof is two lines long, whereas the formalization takes 300 lines. However, this is not surprising since the two lines have been expanded to a more detailed paper proof (one full page) in Schnabl's PhD thesis [27], and even this proof contains a "straightforward" inner induction which is not spelled out.

In `Matrix_Comparison` we also prove that the linear norm is sub-multiplicative, i.e., $||m_1 \times m_2|| \leq ||m_1|| \cdot ||m_2||$, a property that is required to achieve [22, Thm. 6], but is not mentioned in the paper.

Again, all of our results have been proven in a generic way for several semirings, which includes the semirings on $\mathbb{N}$, $\mathbb{Q}$, and $\mathbb{R}$. In this way, we generalized [22, Thm. 6] which was only proven for the natural numbers. Especially the proof that the linear norm is sub-multiplicative required the development of a completely new proof: at least three mathematical textbooks contain the same incomparable statement, which states the property for a whole class of norms, but only for the reals. However, we required the property only for the linear norm, but for matrices of type $A^{n \times m}$ where A is generic. Therefore, the proofs within the textbooks – which all use a limit construction on the reals – could not be formalized. Instead, we performed an inductive proof over the shared dimension of the matrices, cf. *linear-norm-submultiplicative* for more details.[4]

7 Conclusion

We presented an overview of our Isabelle/HOL formalization of interpretations over various carriers, which is part of the formalized library IsaFoR and employed in the fully verified certifier CeTA. The kinds of interpretations we support are linear polynomial interpretations, which also allow for matrix interpretations, and nonlinear polynomial interpretations. As we have shown above, supported carriers range from natural numbers, over integers and rational numbers, to real numbers, as well as corresponding arctic carriers. This unifies and extends previous work. Since CeTA needs to certify given proofs containing explicit numbers and interpretation functions we also had to take care that our formalization supports executable algorithms for all required operations (like addition, multiplication, various comparisons, the square root function, etc.). For real numbers this is not a trivial task. Our solution was to perform data refinement to a subset of the reals that suffices for our purposes. Finally we presented our formalization of complexity related results. In contrast to typical formulations in the literature, we only provide upper bounds, but in return do not have to care about well-definedness issues that would arise otherwise.

Acknowledgments. We are grateful to Bertram Felgenhauer for pointing us to Cauchy's mean theorem when proving soundness of our root algorithm.

The authors are listed in alphabetical order regardless of individual contributions or seniority.

References

1. Baader, F., Nipkow, T.: Term Rewriting and All That. Cambridge University Press (1998)
2. Ballarin, C.: Locales: A module system for mathematical theories. J. Autom. Reasoning 52(2), 123–153 (2014), doi:10.1007/s10817-013-9284-7
3. Blanqui, F., Koprowski, A.: CoLoR: a Coq library on well-founded rewrite relations and its application to the automated verification of termination certificates. Math. Struct. Comp. Sci. 21(4), 827–859 (2011), doi:10.1017/S0960129511000120

[4] In the formalization there is a precondition that all elements have to be nonnegative. This is due to the fact that the linear norm is defined as the sum of all values in the matrix, without taking absolute values. This was done on purpose, so that we do not even have to require the existence of an absolute value function.

4. Cichon, A., Lescanne, P.: Polynomial interpretations and the complexity of algorithms. In: Kapur, D. (ed.) CADE 1992. LNCS, vol. 607, pp. 139–147. Springer, Heidelberg (1992), doi:10.1007/3-540-55602-8_161
5. Cohen, C.: Construction of real algebraic numbers in COQ. In: Beringer, L., Felty, A. (eds.) ITP 2012. LNCS, vol. 7406, pp. 67–82. Springer, Heidelberg (2012), doi:10.1007/978-3-642-32347-8_6
6. Contejean, E., Courtieu, P., Forest, J., Pons, O., Urbain, X.: Automated certified proofs with CIME3. In: Proc. 22nd RTA. LIPIcs, vol. 10, pp. 21–30. Schloss Dagstuhl (2011), doi:10.4230/LIPIcs.RTA.2011.21
7. Courtieu, P., Gbedo, G., Pons, O.: Improved matrix interpretation. In: van Leeuwen, J., Muscholl, A., Peleg, D., Pokorný, J., Rumpe, B. (eds.) SOFSEM 2010. LNCS, vol. 5901, pp. 283–295. Springer, Heidelberg (2010), doi:10.1007/978-3-642-11266-9_24
8. Endrullis, J., Waldmann, J., Zantema, H.: Matrix interpretations for proving termination of term rewriting. J. Autom. Reasoning 40(2-3), 195–220 (2008), doi:10.1007/s10817-007-9087-9
9. Giesl, J., Thiemann, R., Schneider-Kamp, P.: The dependency pair framework: Combining techniques for automated termination proofs. In: Baader, F., Voronkov, A. (eds.) LPAR 2004. LNCS (LNAI), vol. 3452, pp. 301–331. Springer, Heidelberg (2005), doi:10.1007/978-3-540-32275-7_21
10. Haftmann, F., Krauss, A., Kunčar, O., Nipkow, T.: Data refinement in Isabelle/HOL. In: Blazy, S., Paulin-Mohring, C., Pichardie, D. (eds.) ITP 2013. LNCS, vol. 7998, pp. 100–115. Springer, Heidelberg (2013), doi:10.1007/978-3-642-39634-2_10
11. Haftmann, F., Nipkow, T.: Code generation via higher-order rewrite systems. In: Blume, M., Kobayashi, N., Vidal, G. (eds.) FLOPS 2010. LNCS, vol. 6009, pp. 103–117. Springer, Heidelberg (2010), doi:10.1007/978-3-642-12251-4_9
12. Hirokawa, N., Moser, G.: Automated complexity analysis based on the dependency pair method. In: Armando, A., Baumgartner, P., Dowek, G. (eds.) IJCAR 2008. LNCS (LNAI), vol. 5195, pp. 364–379. Springer, Heidelberg (2008), doi:10.1007/978-3-540-71070-7_32
13. Hofbauer, D.: Termination Proofs and Derivation Lengths in Term Rewriting Systems. Dissertation, Technische Universität Berlin, Germany (1991), Available as Technical Report 92-46, TU Berlin (1992)
14. Hofbauer, D., Lautemann, C.: Termination proofs and the length of derivations. In: Dershowitz, N. (ed.) RTA 1989. LNCS, vol. 355, pp. 167–177. Springer, Heidelberg (1989), doi:10.1007/3-540-51081-8_107
15. Huffman, B., Kunčar, O.: Lifting and transfer: A modular design for quotients in isabelle/HOL. In: Gonthier, G., Norrish, M. (eds.) CPP 2013. LNCS, vol. 8307, pp. 131–146. Springer, Heidelberg (2013), doi:10.1007/978-3-319-03545-1_9
16. Koprowski, A., Waldmann, J.: Arctic termination ... below zero. In: Voronkov, A. (ed.) RTA 2008. LNCS, vol. 5117, pp. 202–216. Springer, Heidelberg (2008), doi:10.1007/978-3-540-70590-1_14
17. Korp, M., Sternagel, C., Zankl, H., Middeldorp, A.: Tyrolean Termination Tool 2. In: Treinen, R. (ed.) RTA 2009. LNCS, vol. 5595, pp. 295–304. Springer, Heidelberg (2009), doi:10.1007/978-3-642-02348-4_21
18. Lankford, D.: On proving term rewriting systems are Noetherian. Technical Report MTP-3, Louisiana Technical University, Ruston, LA, USA (1979)
19. Lochbihler, A.: Light-weight containers for Isabelle: Efficient, extensible, nestable. In: Blazy, S., Paulin-Mohring, C., Pichardie, D. (eds.) ITP 2013. LNCS, vol. 7998, pp. 116–132. Springer, Heidelberg (2013), doi:10.1007/978-3-642-39634-2_11

20. Lucas, S.: On the relative power of polynomials with real, rational, and integer coefficients in proofs of termination of rewriting. Appl. Algebr. Eng. Comm. 17(1), 49–73 (2006), doi:10.1007/s00200-005-0189-5
21. Lucas, S.: Practical use of polynomials over the reals in proofs of termination. In: Proc. 9th PPDP, pp. 39–50. ACM (2007), doi:10.1145/1273920.1273927
22. Moser, G., Schnabl, A., Waldmann, J.: Complexity analysis of term rewriting based on matrix and context dependent interpretations. In: Proc. 28th FSTTCS. LIPIcs, vol. 2, pp. 304–315. Schloss Dagstuhl (2008), doi:10.4230/LIPIcs.FSTTCS.2008.1762
23. Neurauter, F., Middeldorp, A.: On the domain and dimension hierarchy of matrix interpretations. In: Bjørner, N., Voronkov, A. (eds.) LPAR-18 2012. LNCS, vol. 7180, pp. 320–334. Springer, Heidelberg (2012), doi:10.1007/978-3-642-28717-6_25
24. Neurauter, F., Middeldorp, A., Zankl, H.: Monotonicity criteria for polynomial interpretations over the naturals. In: Giesl, J., Hähnle, R. (eds.) IJCAR 2010. LNCS, vol. 6173, pp. 502–517. Springer, Heidelberg (2010), doi:10.1007/978-3-642-14203-1_42
25. Nipkow, T., Paulson, L.C., Wenzel, M.T.: Isabelle/HOL. LNCS, vol. 2283. Springer, Heidelberg (2002), doi:10.1007/3-540-45949-9
26. Porter, B.: Cauchy's mean theorem and the Cauchy-Schwarz inequality. Archive of Formal Proofs (March 2006), `http://afp.sf.net/entries/Cauchy.shtml`
27. Schnabl, A.: Derivational Complexity Analysis Revisited. PhD thesis, University of Innsbruck, Austria (2011)
28. Sternagel, C., Thiemann, R.: Certification extends termination techniques. In: Proc. 11th WST (2010), arXiv:1208.1594
29. Sternagel, C., Thiemann, R.: Executable matrix operations on matrices of arbitrary dimensions. Archive of Formal Proofs (June 2010), `http://afp.sf.net/entries/Matrix.shtml`
30. Sternagel, C., Thiemann, R.: Executable multivariate polynomials. Archive of Formal Proofs (August 2010), `http://afp.sf.net/entries/Polynomials.shtml`
31. Sternagel, C., Thiemann, R.: Signature extensions preserve termination. In: Dawar, A., Veith, H. (eds.) CSL 2010. LNCS, vol. 6247, pp. 514–528. Springer, Heidelberg (2010), doi:10.1007/978-3-642-15205-4_39
32. Thiemann, R.: Formalizing bounded increase. In: Blazy, S., Paulin-Mohring, C., Pichardie, D. (eds.) ITP 2013. LNCS, vol. 7998, pp. 245–260. Springer, Heidelberg (2013), doi:10.1007/978-3-642-39634-2_19
33. Thiemann, R.: Implementing field extensions of the form $\mathbb{Q}[\sqrt{b}]$. Archive of Formal Proofs (February 2014), `http://afp.sf.net/entries/Real_Impl.shtml`
34. Thiemann, R., Sternagel, C.: Certification of termination proofs using CeTA. In: *Proc. 22nd TPHOLs*. LNCS, vol. 5674, pp. 452–468. Springer, Heidelberg (2009), doi:10.1007/978-3-642-03359-9_31
35. Zankl, H., Korp, M.: Modular complexity analysis via relative complexity. In: Proc. 21st RTA. LIPIcs, vol. 6, pp. 385–400. Schloss Dagstuhl (2010), doi:10.4230/LIPIcs.RTA.2010.385
36. Zankl, H., Middeldorp, A.: Satisfiability of non-linear (Ir)rational arithmetic. In: Clarke, E.M., Voronkov, A. (eds.) LPAR-16 2010. LNCS, vol. 6355, pp. 481–500. Springer, Heidelberg (2010), doi:10.1007/978-3-642-17511-4_27
37. Zankl, H., Thiemann, R., Middeldorp, A.: Satisfiability of non-linear arithmetic over algebraic numbers. In: Proc. SCSS. RISC-Linz Technical Report, vol. 10-10, pp. 19–24 (2010)

Conditional Confluence (System Description)*

Thomas Sternagel and Aart Middeldorp

University of Innsbruck, Innsbruck, Austria
{thomas.sternagel,aart.middeldorp}@uibk.ac.at

Abstract. This paper describes the *Con*ditional *Con*fluence tool, a fully automatic confluence checker for first-order conditional term rewrite systems. The tool implements various confluence criteria that have been proposed in the literature. A simple technique is presented to test conditional critical pairs for infeasibility, which makes conditional confluence criteria more useful. Detailed experimental data is presented.

Keywords: conditional term rewriting, confluence, automation.

1 Introduction

Confluence of term rewrite systems (TRSs) is an undecidable property. Nevertheless there are a number of tools [1, 10, 17] available to check for confluence of TRSs. For *conditional* TRSs (CTRSs) checking confluence is even harder and to date there was no automatic support. The Conditional Confluence tool—ConCon—aims to change this picture. The tool implements three different confluence criteria for oriented CTRSs that have been reported in the literature [2, 8, 16]. A simple technique for infeasibility of conditional critical pairs based on the tcap function is presented to (mildly) enhance the applicability of two of the confluence criteria.

The remainder of this paper is structured as follows. In Section 2 we sum up some basic facts about (conditional) rewriting the reader should be familiar with and we recall two transformations that are used to test for effective termination and confluence. The three implemented confluence criteria are described in Section 3. Section 4 is about infeasibility and contains a larger example. The tool is described in Section 5. A number of experiments have been conducted with ConCon. They are presented in Section 6. The paper concludes with some remarks on implementation issues, thoughts on extensions, and future work in Section 7.

2 Preliminaries

We assume knowledge of the basic notions regarding CTRSs (cf. [3, 15]). Let $\mathcal{R}$ be a CTRS. Let $\ell_1 \to r_1 \Leftarrow c_1$ and $\ell_2 \to r_2 \Leftarrow c_2$ be variants of rewrite rules

* The research described in this paper is supported by FWF (Austrian Science Fund) projects P22467 and I963.

G. Dowek (ed.): RTA-TLCA 2014, LNCS 8560, pp. 456–465, 2014.

of $\mathcal{R}$ without common variables and let $p \in \mathcal{P}os_{\mathcal{F}}(\ell_2)$ such that ℓ_1 and $\ell_2|_p$ are unifiable. Let σ be a most general unifier of ℓ_1 and $\ell_2|_p$. If $\ell_1 \to r_1 \Leftarrow c_1$ and $\ell_2 \to r_2 \Leftarrow c_2$ are not variants or $p \neq \epsilon$ then the conditional equation $\ell_2\sigma[r_1\sigma]_p \approx \ell_2\sigma \Leftarrow c_1\sigma, c_2\sigma$ is called a *conditional critical pair* of $\mathcal{R}$. A conditional critical pair $s \approx t \Leftarrow c$ of a CTRS $\mathcal{R}$ is *joinable* if $s\sigma \downarrow_{\mathcal{R}} t\sigma$ for every substitution σ that satisfies c. Since in this paper we are concerned with *oriented* CTRSs, the latter means that $u\sigma \to_{\mathcal{R}}^* v\sigma$ for every equation $u \approx v$ in c. We say that $s \approx t \Leftarrow c$ is *infeasible* if there exists no substitution σ that satisfies c. The TRS obtained from a CTRS $\mathcal{R}$ by dropping the conditional parts of the rewrite rules is denoted by $\mathcal{R}_{\mathsf{u}}$. We say that $\mathcal{R}$ is *normal* if every right-hand side of every condition in every rule is a ground normal form with respect to $\mathcal{R}_{\mathsf{u}}$. Rewrite rules $\ell \to r \Leftarrow c$ of CTRSs are classified according to the distribution of variables among ℓ, r, and c, as follows:

type	requirement	type	requirement
1	$\mathcal{V}\mathsf{ar}(r) \cup \mathcal{V}\mathsf{ar}(c) \subseteq \mathcal{V}\mathsf{ar}(\ell)$	3	$\mathcal{V}\mathsf{ar}(r) \subseteq \mathcal{V}\mathsf{ar}(\ell) \cup \mathcal{V}\mathsf{ar}(c)$
2	$\mathcal{V}\mathsf{ar}(r) \subseteq \mathcal{V}\mathsf{ar}(\ell)$	4	no restrictions

An n-CTRS contains only rules of type n. So a 1-CTRS contains no extra variables, a 2-CTRS may only contain extra variables in the conditions, and a 3-CTRS may also have extra variables in the right-hand sides provided these occur in the corresponding conditional part. The set of variables occurring in a sequence of terms $t_1, \ldots, t_n$ is denoted by $\mathcal{V}\mathsf{ar}(t_1, \ldots, t_n)$. Likewise the function $\mathsf{var}(t_1, \ldots, t_n)$ returns the elements of $\mathcal{V}\mathsf{ar}(t_1, \ldots, t_n)$ in an arbitrary but fixed order. An oriented CTRS $\mathcal{R}$ is called *deterministic* if for every rule $\ell \to r \Leftarrow s_1 \approx t_1, \ldots, s_n \approx t_n$ in $\mathcal{R}$ we have $\mathcal{V}\mathsf{ar}(s_i) \subseteq \mathcal{V}\mathsf{ar}(\ell, t_1, \ldots, t_{i-1})$ with $1 \leqslant i \leqslant n$.

An oriented CTRS $\mathcal{R}$ is *quasi-decreasing* if there exists a well-founded order $>$ with the subterm property that extends $\to_{\mathcal{R}}$ such that $\ell\sigma > s_i\sigma$ for all $\ell \to r \Leftarrow s_1 \approx t_1, \ldots, s_n \approx t_n \in \mathcal{R}$, $1 \leqslant i \leqslant n$, and substitutions σ with $s_j\sigma \to_{\mathcal{R}}^* t_j\sigma$ for $1 \leqslant j < i$. Quasi-decreasingness ensures termination and, for finite CTRSs, computability of the rewrite relation. We recall two transformations from deterministic 3-CTRSs to TRSs that can be used to show quasi-decreasingness.

Unravelings were first introduced in [13]. Unravelings split conditional rules into several unconditional rules and the conditions are encoded using new function symbols. Originally they were used to study the correspondence between properties of CTRSs and TRSs as well as modularity of CTRSs. The unraveling defined below goes back to [12]. We use the formulation in [15, p. 212]. It simulates the conditional rules from a CTRS $\mathcal{R}$ by a sequence of applications of rules from the TRS $\mathrm{U}(\mathcal{R})$, in effect verifying the conditions from left to right until all the conditions are satisfied and the last rule yielding the original right-hand side may be applied.

Definition 1. *Every deterministic 3-CTRS $\mathcal{R}$ is mapped to the TRS $\mathbb{U}(\mathcal{R})$ obtained from $\mathcal{R}$ by replacing every conditional rule $\rho\colon \ell \to r \Leftarrow s_1 \approx t_1, \ldots, s_n \approx t_n$ with $n \geqslant 1$ in $\mathcal{R}$ with*

$$\begin{aligned} \ell &\to U_\rho^1(s_1, \mathsf{var}(\ell)) \\ U_\rho^1(t_1, \mathsf{var}(\ell)) &\to U_\rho^2(s_2, \mathsf{var}(\ell, t_1)) \\ &\cdots \\ U_\rho^n(t_n, \mathsf{var}(\ell, t_1, \ldots, t_{n-1})) &\to r \end{aligned}$$

where U_ρ^i are fresh function symbols.

In our implementation we use the variant of $\mathbb{U}$ sketched in [15, Example 7.2.49] and formalized in [8, Definition 6]. In this variant certain U-symbols originating from different rewrite rules are shared, in order to reduce the number of critical pairs and thereby increasing the chances of obtaining a confluent TRS.

The second transformation, introduced in [2], from deterministic 3-CTRSs to TRSs does not use any additional symbols and it does not aim to simulate rewriting in the CTRS. Hence its use is limited to show quasi-decreasingness.

Definition 2. *Every deterministic 3-CTRS $\mathcal{R}$ is mapped to the TRS $\mathbb{V}(\mathcal{R})$ obtained from $\mathcal{R}$ by replacing every conditional rule $\ell \to r \Leftarrow s_1 \approx t_1, \ldots, s_n \approx t_n$ with $n \geqslant 1$ in $\mathcal{R}$ with*

$$\ell \to s_1\sigma_0 \qquad \cdots \qquad \ell \to s_n\sigma_{n-1} \qquad \ell \to r\sigma_n$$

for the substitutions $\sigma_0, \ldots, \sigma_n$ inductively defined as follows:

$$\sigma_i = \begin{cases} \varepsilon & \text{if } i = 0 \\ \sigma_{i-1} \cup \{x \mapsto s_i\sigma_{i-1} \mid x \in \mathcal{V}\mathsf{ar}(t_i) \setminus \mathcal{V}\mathsf{ar}(\ell, t_1, \ldots, t_{i-1})\} & \text{if } 0 < i \leqslant n \end{cases}$$

The following lemma shows how the transformations are used to obtain quasi-decreasingness. The first condition is from [15, p. 214] while the second one is a combination of [15, Lemma 7.2.6] and [15, Proposition 7.2.68]. It is unknown whether the first condition is implied by the second (cf. [15, p. 229]).

Lemma 3. *A deterministic 3-CTRS $\mathcal{R}$ is quasi-decreasing if $\mathbb{U}(\mathcal{R})$ is terminating or $\mathbb{V}(\mathcal{R})$ is simply terminating.* □

3 Three Confluence Criteria

Our tool implements three known confluence criteria [2, 8, 16]. The first criterion is from Avenhaus and Loría-Sáenz [2, Theorem 4.1]. Its applicability is restricted to quasi-decreasing and *strongly irreducible* deterministic 3-CTRSs. A term t is called *strongly irreducible* if $t\sigma$ is a normal form for every normalized[1]

[1] A normalized substitution maps variables to normal forms.

substitution σ. We say that $\mathcal{R}$ is strongly irreducible if the right-hand side of every condition in every conditional rewrite rule is strongly irreducible. Strong irreducibility is undecidable. In our tool we use the following decidable approximation [2]: no non-variable subterm of a right-hand side of a condition unifies with the left-hand side of a rule (after renaming).

Theorem A. *A quasi-decreasing strongly irreducible deterministic 3-CTRS $\mathcal{R}$ is confluent if and only if all critical pairs of $\mathcal{R}$ are joinable.* □

The second confluence criterion is from Suzuki *et al.* [16, Section 7]. It does not impose any termination assumption, but forbids (feasible) critical pairs and requires the properties defined below, which are obviously computable. A CTRS $\mathcal{R}$ is *right-stable* if every rewrite rule $\ell \to r \Leftarrow s_1 \approx t_1, \dots, s_n \approx t_n$ in $\mathcal{R}$ satisfies $\mathcal{V}\mathsf{ar}(\ell, s_1, \dots, s_i, t_1, \dots, t_{i-1}) \cap \mathcal{V}\mathsf{ar}(t_i) = \varnothing$ and t_i is either a linear constructor term or a ground $\mathcal{R}_\mathsf{u}$-normal form, for all $1 \leqslant i \leqslant n$. An oriented CTRS $\mathcal{R}$ is *properly oriented* if for every rewrite rule $\ell \to r \Leftarrow c$ with $\mathcal{V}\mathsf{ar}(r) \nsubseteq \mathcal{V}\mathsf{ar}(\ell)$ in $\mathcal{R}$ the conditional part c can be written as $s_1 \approx t_1, \dots, s_m \approx t_m, s'_1 \approx t'_1, \dots, s'_n \approx t'_n$ such that the following two conditions are satisfied: $\mathcal{V}\mathsf{ar}(s_i) \subseteq \mathcal{V}\mathsf{ar}(\ell, t_1, \dots, t_{i-1})$ for all $1 \leqslant i \leqslant m$ and $\mathcal{V}\mathsf{ar}(r) \cap \mathcal{V}\mathsf{ar}(s'_i, t'_i) \subseteq \mathcal{V}\mathsf{ar}(\ell, t_1, \dots, t_m)$ for all $1 \leqslant i \leqslant n$.

Theorem B. *Almost orthogonal properly oriented right-stable 3-CTRSs are confluent.* □

The third criterion is a recent result by Gmeiner *et al.* [8, Theorem 9]. It employs the notion of *weak left-linearity*, which is satisfied for a deterministic CTRS $\mathcal{R}$ if $x \notin \mathcal{V}\mathsf{ar}(r, s_1, \dots, s_n)$ for every rule $\ell \to r \Leftarrow s_1 \approx t_1, \dots, s_n \approx t_n$ in $\mathcal{R}$ and variable x that appears more than once in $\ell, t_1, \dots, t_n$.

Theorem C. *A weakly left-linear deterministic CTRS $\mathcal{R}$ is confluent if $\mathbb{U}(\mathcal{R})$ is confluent.* □

Here the modified version of the unraveling $\mathbb{U}$ described after Definition 1 is used. The three criteria are pairwise incompatible, as shown in the following examples.

Example 4. The oriented 1-CTRS $\mathcal{R}$ consisting of the following four rules

$$\mathsf{a} \to \mathsf{b} \qquad \mathsf{b} \to \mathsf{a} \qquad \mathsf{f}(x,x) \to \mathsf{a} \qquad \mathsf{g}(x) \to \mathsf{a} \Leftarrow \mathsf{g}(x) \approx \mathsf{b}$$

is weakly left-linear and deterministic and its unraveling $\mathbb{U}(\mathcal{R})$

$$\mathsf{a} \to \mathsf{b} \qquad \mathsf{b} \to \mathsf{a} \qquad \mathsf{f}(x,x) \to \mathsf{a} \qquad \mathsf{g}(x) \to \mathsf{U}(\mathsf{g}(x), x) \qquad \mathsf{U}(\mathsf{b}) \to \mathsf{a}$$

is confluent, hence $\mathcal{R}$ is confluent by Theorem C. Since $\mathcal{R}$ is neither left-linear nor strongly irreducible, Theorems A and B are not applicable.

Example 5. The normal 2-CTRS consisting of the rule

$$\mathsf{h}(x) \to \mathsf{g}(x) \Leftarrow \mathsf{f}(x,y) \approx \mathsf{b}$$

is orthogonal, properly oriented, and right-stable and therefore confluent by Theorem B but not deterministic so Theorems A and C do not apply. Hence this example contradicts the claim in [8] that Theorem B is a corollary of Theorem C.

Example 6. The normal 1-CTRS consisting of the rule

$$\mathsf{f}(x,x) \to \mathsf{a} \Leftarrow \mathsf{g}(x) \approx \mathsf{b}$$

is quasi-decreasing, strongly irreducible, and non-overlapping, and thus confluent by Theorem A. Since the system is not (weakly) left-linear, Theorems B and C do not apply.

4 Infeasibility

The applicability of Theorems A and B strongly depends on the presence of critical pairs. Many natural examples employ rules which only yield a couple of critical pairs which are in fact all infeasible.

Infeasibility is undecidable in general. Two sufficient conditions are described in the literature: [9, Appendix A] and [2, Definition 4.4]. The former method can only be used in very special cases (left-linear constructor-based join systems without extra variables and using "strict" semantics). The use of the latter method is restricted to quasi-decreasing strongly irreducible deterministic 3-CTRSs (like in Theorem A) and is described below.

Given a critical pair $s \approx t \Leftarrow c$, the conditions in c are transformed into a TRS $\mathcal{C} = \{\overline{u} \to \overline{v} \mid u \approx v \in c\}$ where $\overline{t}$ is the result of replacing every $x \in \mathcal{V}\mathsf{ar}(c)$ occurring in t by a fresh constant c_x. If there is a left-hand side u in c such that $\overline{u}_2 \;{}_{\mathcal{R}\cup\mathcal{C}}^{*}\!\!\leftarrow \overline{u} \to_{\mathcal{R}\cup\mathcal{C}}^{*} \overline{u}_1$ and u_1, u_2 are strongly irreducible and not unifiable then $s \approx t \Leftarrow c$ is infeasible. The same method can also be used as sufficient condition for joinability [2]: $s \approx t \Leftarrow c$ is joinable if $\overline{s} \to_{\mathcal{R}\cup\mathcal{C}}^{*} \cdot {}_{\mathcal{R}\cup\mathcal{C}}^{*}\!\!\leftarrow \overline{t}$.

Our new technique for infeasibility is based on the tcap function, which was introduced to obtain a better approximation of dependency graphs [7] and later used as a sufficient check for non-confluence for TRSs [17]. It is defined as follows. If t is a variable then $\mathsf{tcap}(t)$ is a fresh variable and if $t = f(t_1,\ldots,t_n)$ then we let $u = f(\mathsf{tcap}(t_1),\ldots,\mathsf{tcap}(t_n))$ and define $\mathsf{tcap}(t)$ to be u if u does not unify with the left-hand side of a rule in $\mathcal{R}$, and a fresh variable otherwise.

Lemma 7. *Let $\mathcal{R}$ be an oriented CTRS. A conditional critical pair $s \approx t \Leftarrow c$ of $\mathcal{R}$ is infeasible if there exists an equation $u \approx v \in c$ such that $\mathsf{tcap}(u)$ does not unify with v.* □

We conclude this section with an example illustrating that Theorem A benefits from the new infeasibility criterion of Lemma 7.

Example 8. Consider the following CTRS $\mathcal{R}_{\mathsf{min}}$ from [11]:

$$\begin{array}{rlll}
0 < \mathsf{s}(x) \to \mathsf{true} & \mathsf{min}(x : \mathsf{nil}) \to x & & (1)\\
x < 0 \to \mathsf{false} & \mathsf{min}(x : xs) \to x & \Leftarrow x < \mathsf{min}(xs) \approx \mathsf{true} & (2)\\
\mathsf{s}(x) < \mathsf{s}(y) \to x < y & \mathsf{min}(x : xs) \to \mathsf{min}(xs) & \Leftarrow x < \mathsf{min}(xs) \approx \mathsf{false} & (3)\\
 & \mathsf{min}(x : xs) \to \mathsf{min}(xs) & \Leftarrow \mathsf{min}(xs) \approx x & (4)
\end{array}$$

To check whether Theorem C is able to show confluence of $\mathcal{R}_{\mathsf{min}}$ we look at the result of the optimized version of the unraveling $\mathbb{U}$ from Definition 1 on rules (2) to (4):

$$\begin{aligned} \mathsf{min}(x:xs) &\to \mathsf{U}_1(x < \mathsf{min}(xs), x, xs) & \mathsf{min}(x:xs) &\to \mathsf{U}_2(\mathsf{min}(xs), x, xs) \\ \mathsf{U}_1(\mathsf{true}, x, xs) &\to x & \mathsf{U}_2(x, x, xs) &\to \mathsf{min}(xs) \\ \mathsf{U}_1(\mathsf{false}, x, xs) &\to \mathsf{min}(xs) \end{aligned}$$

There is a peak $\mathsf{U}_1(x < \mathsf{min}(xs), x, xs) \leftarrow \mathsf{min}(x : xs) \to \mathsf{U}_2(\mathsf{min}(xs), x, xs)$ between different normal forms of $\mathbb{U}(\mathcal{R}_{\mathsf{min}})$ and hence $\mathbb{U}(\mathcal{R}_{\mathsf{min}})$ is non-confluent. So Theorem C cannot show confluence of $\mathcal{R}_{\mathsf{min}}$. Theorem B does not apply here because $\mathcal{R}_{\mathsf{min}}$ is not right-stable. For Theorem A we compute critical pairs. There are twelve but for symmetry reasons we only have to consider six of them:

$$\begin{aligned} x &\approx x & &\Leftarrow x < \mathsf{min}(\mathsf{nil}) \approx \mathsf{true} && (1,2) \\ \mathsf{min}(\mathsf{nil}) &\approx x & &\Leftarrow x < \mathsf{min}(\mathsf{nil}) \approx \mathsf{false} && (1,3) \\ \mathsf{min}(\mathsf{nil}) &\approx x & &\Leftarrow \mathsf{min}(\mathsf{nil}) \approx x && (1,4) \\ \mathsf{min}(xs) &\approx x & &\Leftarrow x < \mathsf{min}(xs) \approx \mathsf{true},\ x < \mathsf{min}(xs) \approx \mathsf{false} && (2,3) \\ \mathsf{min}(xs) &\approx x & &\Leftarrow x < \mathsf{min}(xs) \approx \mathsf{true},\ \mathsf{min}(xs) \approx x && (2,4) \\ \mathsf{min}(xs) &\approx \mathsf{min}(xs) & &\Leftarrow x < \mathsf{min}(xs) \approx \mathsf{false},\ \mathsf{min}(xs) \approx x && (3,4) \end{aligned}$$

The pairs (1,2) and (3,4) are trivial. The terms $\mathsf{tcap}(x < \mathsf{min}(\mathsf{nil})) = x' < \mathsf{min}(\mathsf{nil})$ and false are not unifiable, hence (1,3) is infeasible by Lemma 7. The critical pair (2,3) can be shown to be infeasible by the method described at the top of page 460 (as well as by Lemma 7) and (1,4) and (2,4) can be shown to be joinable by the same method. So Theorem A applies and we conclude that $\mathcal{R}_{\mathsf{min}}$ is confluent.

5 Design and Implementation

ConCon is written in Scala 2.10, an object-functional programming language. Scala compiles to Java byte code and therefore is easily portable to different platforms. ConCon is available under the LGPL license and may be downloaded from:

`http://cl-informatik.uibk.ac.at/software/concon/`

In order to use the full power of ConCon one needs to have some termination checker understanding the TPDB[2] format and some confluence checker understanding the same format installed on one's system. One may have to adjust the paths and flags of these programs in the file `concon.ini`, which should reside in the same directory as the `concon` executable. For input we support the XML[3] format as well as a modified version of the TRS format of the TPDB.[4]

[2] `http://www.lri.fr/~marche/tpdb/format.html`

[3] `http://www.termination-portal.org/wiki/XTC_Format_Specification`

[4] `http://termination-portal.org/wiki/TPDB`

The modification concerns a new declaration CONDITIONTYPE, which may be set to SEMI-EQUATIONAL, JOIN, or ORIENTED. Although for now ConCon works on oriented CTRSs we designed the CONDITIONTYPE to anticipate future developments. In the conditional part of the rules we only allow == as relation, since the exact interpretation is inferred from the CONDITIONTYPE declaration:

```
(CONDITIONTYPE ORIENTED)
(VAR x)
(RULES
  not(x) -> false | x == true
  not(x) -> true  | x == false
)
```

This modified TRS format is closer to the newer XML version and makes it very easy to interpret, say, a given join CTRS as an oriented CTRS (by just modifying the CONDITIONTYPE).

ConCon is operated through a command line interface described below. In addition to the command line version there is also an easy to use web interface available on the ConCon website.

Usage. Just starting the tool without any options or input file as follows

```
java -jar concon_2.10-1.1.0.0.min.jar
```

will output a short usage description. We will abbreviate this command by ./concon in the following. The flag --conf may be used to configure the employed confluence criteria. The flag takes a list of criteria which are tried in the given order. If a method is successful the rest of the list is skipped. By default ConCon uses all the available confluence criteria in the following order:[5]

U Check whether the input system is unconditional, if so give it to an external unconditional confluence checker.
B Try Theorem B.
C Try Theorem C using an external unconditional confluence checker.
A Try Theorem A using an external termination checker.

One may always add a timeout at the end of ConCon's parameter list. The default timeout is 60 seconds. When calling ConCon with an input file like

```
./concon 292.trs
```

it will just try to apply all confluence criteria in sequence with the default timeout as explained above. The first line of the output will be one of YES, NO, or MAYBE, followed by the input system, and finally a textual description of how ConCon did conclude the given answer. One may use -a, -s, and -p to prevent output of the answer, the input system, and the textual description, respectively.

If one is only interested in the critical pairs of the system and which of them can be shown to be infeasible, one may use the following call

```
./concon -c 292.trs
```

[5] Theorem B does not need calls to external programs and in our experiments Theorem C produced an answer faster than Theorem A on average.

the -c causes ConCon to print all overlaps and the associated (conditional) critical pairs of the system, and indicates whether they could be shown infeasible.

In order to check the input system for quasi-decreasingness the flag -q may be used. In addition one may use the option --ter together with one of the strings u or v to restrict the transformation to use for the termination check. This method gives the transformed unconditional system to an external termination checker.

The flag -t may be used to tell ConCon to just apply a transformation and output the result. The flag takes a string parameter specifying which transformation to use. The available options are u, uopt and v standing for the transformations of Definition 1, its modified version, and Definition 2, respectively.

Many syntactic criteria for CTRSs, like proper-orientedness or weak left-linearity, are tedious to check by hand. Other properties of interest, like quasi-decreasingness, are undecidable. Executing the call

```
./concon -l 292.trs
```

results in a list of properties of the input CTRS.

6 Experiments

We have collected a number of examples from the literature. Our collection currently consists of 129 CTRSs, including the 3 new examples in Section 3, which we extracted from 32 different sources. Of these 129 CTRSs, 101 are presented as oriented CTRSs in the literature. The corresponding files in the modified TPDB format can be downloaded from the ConCon website. Additionally they have been added to the confluence problems database.[6] This collection should also be of interest for the termination competition[7] since the CTRS category of TPDB contains a mere 7 examples.

The experiments we describe here were carried out on a 64bit GNU/Linux machine with an Intel® Core™ i7-3520M processor clocked at 2.90GHz and 8GB of RAM using the tool parallel.[8] The kernel version is 3.14.1-1-ARCH. The version of Java on this machine is 1.7.0_55. We had to increase the stack size to 20MB using the JVM flag -Xss20M to prevent stack overflows caused by parsing deep terms like in the file 313.trs. The following external tools were used in the experiments:

- CSI, version 0.4 (call: csi - trs 30)[9]
- $\mathsf{T_TT_2}$, version 1.16 (call: ttt2 - trs 30)[10]

First we checked confluence of the given systems. The timeout was set to one minute.

Figure 1a gives an overview of how many systems could be shown to be confluent by which of the three theorems. More details for 6 of the CTRSs are listed

[6] http://coco.nue.riec.tohoku.ac.jp/problems/
[7] http://termcomp.uibk.ac.at/
[8] http://www.gnu.org/s/parallel
[9] http://cl-informatik.uibk.ac.at/software/csi/
[10] http://cl-informatik.uibk.ac.at/software/ttt2/

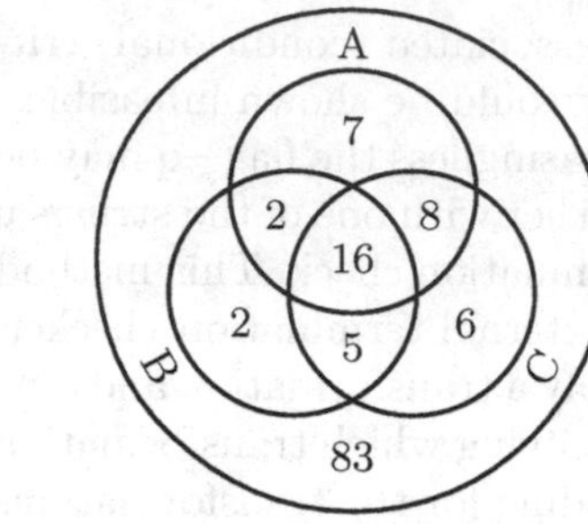

(a) 129 CTRSs

		theorem			CCPs			
CTRS	source	A	B	C	#	=	⋈	↓
`264.trs`	6	0.7	–	–	0	–	–	–
`286.trs`	4	–	–	2.1	0	–	–	–
`287.trs`	5	–	0.5	–	0	–	–	–
`292.trs`	8	0.8	–	×	12	4	6	2
`324.trs`	[14]	–	–	×	4	0	0	0
`336.trs`	[16]	–	0.6	–	2	0	2	0

(b) 6 selected CTRSs

Fig. 1. Confluence results

in Figure 1b. The columns 'A' to 'C' list the time in seconds in case of success, '–' if the theorem was not applicable because of the syntactic preconditions, and '×' if the method will never be able to show confluence for the system. The first 3 CTRSs can only be shown to be confluent by one of the theorems. The next one (`292.trs`) is Example 8. Example `324.trs` cannot be shown confluent by any of the implemented methods. As can be seen in Figure 1a, this actually holds for the majority of the 129 CTRSs. Example `336.trs` requires the method of Lemma 7 to show its conditional critical pairs infeasible, afterwards Theorem B is applicable. In the last four columns we list for each of the 6 CTRSs the number of conditional critical pairs, and whether they are trivial (column '='), or could be shown to be infeasible (column '⋈') or joinable (column '↓').

The ConCon website contains more detailed experimental results, including a comparison of the two sufficient conditions in Lemma 3 for quasi-decreasingness.

7 Concluding Remarks

We presented a tool which implements three different methods to show confluence of oriented CTRSs. A simple sufficient criterion for infeasibility increased the applicability of two of the methods. This is clearly a first step and our experiments show that there is lots of room for improvements:

- Several of the systems in our test bed are in fact non-confluent, so methods to prove non-confluence of CTRSs are in demand.
- Many of the systems have infeasible critical pairs but our current criterion for infeasibility is not powerful enough to show this. Our own investigations show that progress here is hard to achieve. For oriented CTRSs, infeasibility is a reachability problem and techniques based on tree automata completion are a natural candidate for investigation.
- Furthermore, so far we have no good support for conditional rewriting, which is needed to check joinability of (feasible) critical pairs. From early investigations by Ganzinger [6], Zhang and Rémy [18, 19], Avenhaus and Loría-Sáenz [2], and others in we know that implementing conditional rewriting is a highly complex problem.

Finally, confluence methods for semi-equational [4] and in particular join CTRSs [5, 16] are well-investigated and should be implemented.

References

1. Aoto, T., Yoshida, J., Toyama, Y.: Proving confluence of term rewriting systems automatically. In: Treinen, R. (ed.) RTA 2009. LNCS, vol. 5595, pp. 93–102. Springer, Heidelberg (2009)
2. Avenhaus, J., Loría-Sáenz, C.: On conditional rewrite systems with extra variables and deterministic logic programs. In: Pfenning, F. (ed.) LPAR 1994. LNCS, vol. 822, pp. 215–229. Springer, Heidelberg (1994)
3. Baader, F., Nipkow, T.: Term Rewriting and All That. CUP (1998)
4. Bergstra, J.A., Klop, J.W.: Conditional rewrite rules: Confluence and termination. Journal of Computer and System Sciences 32(3), 323–362 (1986)
5. Dershowitz, N., Okada, M., Sivakumar, G.: Confluence of conditional rewrite systems. In: Kaplan, S., Jouannaud, J.-P. (eds.) CTRS 1987. LNCS, vol. 308, pp. 31–44. Springer, Heidelberg (1988)
6. Ganzinger, H.: A completion procedure for conditional equations. Journal of Symbolic Computation 11(1/2), 51–81 (1991)
7. Giesl, J., Thiemann, R., Schneider-Kamp, P.: Proving and disproving termination of higher-order functions. In: Gramlich, B. (ed.) FroCos 2005. LNCS (LNAI), vol. 3717, pp. 216–231. Springer, Heidelberg (2005)
8. Gmeiner, K., Nishida, N., Gramlich, B.: Proving confluence of conditional term rewriting systems via unravelings. In: Proc. 2nd IWC, pp. 35–39 (2013)
9. González-Moreno, J.C., Hortalá-González, M.T., Rodríguez-Artalejo, M.: Denotational versus declarative semantics for functional programming. In: Kleine Büning, H., Jäger, G., Börger, E., Richter, M.M. (eds.) CSL 1991. LNCS, vol. 626, pp. 134–148. Springer, Heidelberg (1992)
10. Hirokawa, N., Klein, D.: Saigawa: A confluence tool. In: Proc. 1st IWC, p. 49 (2012)
11. Kounalis, E., Rusinowitch, M.: A proof system for conditional algebraic specifications. In: Okada, M., Kaplan, S. (eds.) CTRS 1990. LNCS, vol. 516, pp. 51–63. Springer, Heidelberg (1991)
12. Marchiori, M.: On deterministic conditional rewriting. Computation Structures Group Memo, vol. 405. MIT Laboratory for Computer Science (1987)
13. Marchiori, M.: Unravelings and ultra-properties. In: Hanus, M., Rodríguez-Artalejo, M. (eds.) ALP 1996. LNCS, vol. 1139, pp. 107–121. Springer, Heidelberg (1996)
14. Nishida, N., Sakai, M., Sakabe, T.: Soundness of unravelings for conditional term rewriting systems via ultra-properties related to linearity. Logical Methods in Computer Science 8, 1–49 (2012)
15. Ohlebusch, E.: Advanced Topics in Term Rewriting. Springer (2002)
16. Suzuki, T., Middeldorp, A., Ida, T.: Level-confluence of conditional rewrite systems with extra variables in right-hand sides. In: Hsiang, J. (ed.) RTA 1995. LNCS, vol. 914, pp. 179–193. Springer, Heidelberg (1995)
17. Zankl, H., Felgenhauer, B., Middeldorp, A.: CSI – A confluence tool. In: Bjørner, N., Sofronie-Stokkermans, V. (eds.) CADE 2011. LNCS, vol. 6803, pp. 499–505. Springer, Heidelberg (2011)
18. Zhang, H.: Implementing contextual rewriting. In: Rusinowitch, M., Remy, J.-L. (eds.) CTRS 1992. LNCS, vol. 656, pp. 363–377. Springer, Heidelberg (1993)
19. Zhang, H., Remy, J.-L.: Contextual rewriting. In: Jouannaud, J.-P. (ed.) RTA 1985. LNCS, vol. 202, pp. 46–62. Springer, Heidelberg (1985)

Nagoya Termination Tool*

Akihisa Yamada[1], Keiichirou Kusakari[2], and Toshiki Sakabe[1]

[1] Graduate School of Information Science, Nagoya University, Japan
[2] Faculty of Engineering, Gifu University, Japan

Abstract. This paper describes the implementation and techniques of the Nagoya Termination Tool, a termination prover for term rewrite systems. The main features of the tool are: the first implementation of the weighted path order which subsumes most of the existing reduction pairs, and the efficiency due to the strong cooperation with external SMT solvers. We present some new ideas that contribute to the efficiency and power of the tool.

1 Introduction

Proving termination of term rewrite systems (TRSs) has been an active field of research. In this paper, we describe the *Nagoya Termination Tool* (NaTT), a termination prover for TRS, which is available at

`http://www.trs.cm.is.nagoya-u.ac.jp/NaTT/`

NaTT is powerful and fast; its power comes from the novel implementation of the *weighted path order (WPO)* [24, 26] that subsumes most of the existing reduction pairs, and its efficiency comes from the strong cooperation with state-of-the-art *satisfiability modulo theory (SMT)* solvers. In principle, any solver that complies with the SMT-LIB Standard[1] version 2.0 can be incorporated as a back-end into NaTT.

In the next section, we recall the dependency pair framework that NaTT is based on, and present existing techniques that are implemented in NaTT. Section 3 describes the implementation of WPO and demonstrates how to obtain other existing techniques as instances of WPO. Some techniques on cooperating with SMT solvers are presented in Section 4. After giving some design details in Section 5, we assess the tool by its results in the *termination competition*[2] in Section 6. Then we conclude in Section 7. Due to page limit, some experimental results are found in the full version of this paper [25].

2 The Dependency Pair Framework

The overall procedure of NaTT is illustrated in Figure 1. NaTT is based on the *dependency pair framework (DP framework)* [1,9,10], a very successful technique

* This work was supported by JSPS KAKENHI #24500012.
[1] `http://www.smtlib.org/`
[2] `http://termination-portal.org/wiki/Termination_Competition`

G. Dowek (ed.): RTA-TLCA 2014, LNCS 8560, pp. 466–475, 2014.

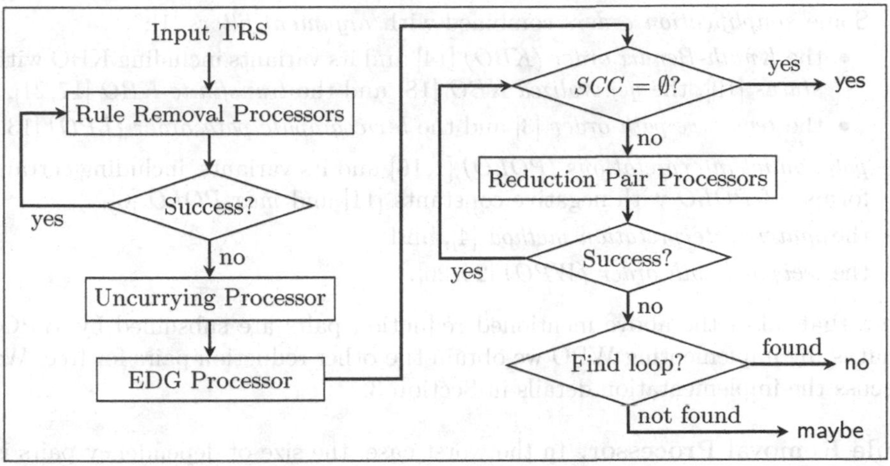

Fig. 1. Flowchart of NaTT

for proving termination of TRSs which is implemented in almost all the modern termination provers for TRSs. In the DP framework, dependencies between function calls defined in a TRS $\mathcal{R}$ is expressed by the set $\mathsf{DP}(\mathcal{R})$ of *dependency pairs*. If a function f is defined by a rule

$$f(s_1, \dots, s_n) \to C[g(t_1, \dots, t_m)] \in \mathcal{R}$$

where g is also defined in $\mathcal{R}$, then this dependency is described by the following dependency pair:

$$f^\sharp(s_1, \dots, s_n) \to g^\sharp(t_1, \dots, t_m) \in \mathsf{DP}(\mathcal{R})$$

The DP framework (dis)proves termination of $\mathcal{R}$ by simplifying and decomposing *DP problems* $\langle \mathcal{P}, \mathcal{R} \rangle$, where initially $\mathcal{P} = \mathsf{DP}(\mathcal{R})$. To this end, many *DP processors* have been proposed. NaTT implements the following DP processors:

Dependency Graph Processor. This processor decomposes a DP problem $\langle \mathcal{P}, \mathcal{R} \rangle$ into $\langle \mathcal{P}_1, \mathcal{R} \rangle \dots \langle \mathcal{P}_n, \mathcal{R} \rangle$ where $\mathcal{P}_1, \dots, \mathcal{P}_n$ are the *strongly connected components (SCCs)* of the *dependency graph* [7,10]. Since the dependency graph is not computable in general, several approximations called *estimated dependency graphs (EDGs)* have been proposed. NaTT implements the EDG proposed in [8].

Reduction Pair Processor. This processor forms the core of NaTT. A *reduction pair* is a pair $\langle \succsim, \succ \rangle$ of orders s.t. $\succ$ is *compatible* with $\succsim$ (i.e., $\succsim \cdot \succ \cdot \succsim \subseteq \succ$), both of $\succsim$ and $\succ$ are stable under substitution, $\succsim$ is monotone and $\succ$ is well-founded. From a DP problem $\langle \mathcal{P}, \mathcal{R} \rangle$, if all the involved rules are weakly decreasing (i.e., $\mathcal{P} \cup \mathcal{R} \subseteq \succsim$), strictly decreasing rules in $\mathcal{P}$ (w.r.t. $\succ$) can be removed. A great number of techniques for obtaining reduction pairs have been proposed so far. NaTT supports the following ones:

- Some *simplification orders* combined with *argument filters* [1]:
  - the *Knuth-Bendix order (KBO)* [14] and its variants including KBO with *status* [19], the *generalized KBO* [18] and the *transfinite KBO* [17,21],
  - the *recursive path order* [3] and the *lexicographic path order (LPO)* [13],
- *polynomial interpretations (POLO)* [1,16] and its variants, including certain forms[3] of *POLO* with negative constants [11] and *max-POLO* [6],
- the *matrix interpretation method* [4], and
- the *weighted path order (WPO)* [24,26].

Note that all of the above mentioned reduction pairs are subsumed by WPO. That is, by implementing WPO we obtain the other reduction pairs for free. We discuss the implementation details in Section 3.

Rule Removal Processor. In the worst case, the size of dependency pairs is quadratic in the size of the input TRS $\mathcal{R}$. Hence it is preferable to reduce the size of $\mathcal{R}$ before computing dependency pairs. To this end NaTT applies the *rule removal processor* [7]. If all rules in $\mathcal{R}$ are weakly decreasing w.r.t. a *monotone* reduction pair, then the processor removes strictly decreasing rules from $\mathcal{R}$. The required monotonicity of a reduction pair is obtained by choosing appropriate parameters for the implementation of WPO described above.

Uncurrying Processor. Use of uncurrying for proving termination is proposed for *applicative* rewrite systems in [12]. The uncurrying implemented in NaTT is similar to the generalized version proposed in [20], in the sense that it does not assume *application symbols* to be binary. A symbol f is considered as an application symbol if all the following conditions hold:

- f is defined and has positive arity,
- a subterm of the form $f(x, \dots)$ does not occur in any left-hand-sides of $\mathcal{R}$,
- a subterm of the form $f(g(\dots), \dots)$ occurs in some right-hand-side of $\mathcal{R}$.

If such application symbols are found, then $\mathcal{R}$ is uncurried w.r.t. the uncurrying TRS $\mathcal{U}$ that consists of the following rules:[4]

$$f(f^l g(x_1, \dots, x_m), y_1, \dots, y_n) \to f^{l+1} g(x_1, \dots, x_m, y_1, \dots, y_n)$$

for every $g \neq f$ and l less than the *applicative arity*[5] of g, where $f^0 g$ denotes g and $f^{l+1} g$ is a new function symbol of arity $m + n$.

3 The Weighted Path Order

As we mentioned in the introduction, NaTT implements only WPO for obtaining reduction pairs. WPO is parameterized by (1) a *weight algebra* which specifies

---

[3] Here, negative values are allowed only for the constant part.

[4] The notation is derived from the *freezing* technique [22].

[5] Applicative arities are taken so that η-*saturation* is not needed.

how weights are computed, (2) a *precedence* on function symbols, and (3) a *status function* which specifies how arguments are compared. In the following sections, we present some options which NaTT provides for specifying search spaces for these parameters.

3.1 Templates for Weight Algebras

One of the most important tasks in proving termination by WPO is finding an appropriate weight algebra. In order to reduce the task to an SMT problem, NaTT considers *template algebras* over integers. Currently the following template algebras are implemented:

- The algebra $\mathcal{P}ol$ indicates that weights of terms are computed by a linear polynomial. Interpretations are in the following shape:

$$f_{\mathcal{P}ol}(x_1, \ldots, x_n) = w_f + \sum_{i=1}^{n} c_{f,i} \cdot x_i \qquad (1)$$

  where the *template variables* w_f and $c_{f,1}, \ldots, c_{f,n}$ should be decided by an external SMT solver.
- The algebra $\mathcal{M}ax$ indicates that weights are computed using the max operator. A symbol f with arity ≥ 1 is interpreted in the following shape:

$$f_{\mathcal{M}ax}(x_1, \ldots, x_n) = \max_{i=1}^{n}(p_{f,i} + c_{f,i} \cdot x_i) \qquad (2)$$

  where $p_{f,1}, \ldots, p_{f,n}$ are template variables. For constant symbols, interpretations of the shape (1) are used. Since the operator max is not usually supported by SMT solvers, these interpretations are encoded as quantifier-free formulas using the technique presented in [24].
- The algebra $\mathcal{MP}ol$ combines both forms of interpretations described above. Since it is inefficient to consider all combinations of these interpretations, $\mathcal{MP}ol$ decides the shape of interpretations according to the following intuition: If a constraint such as $f(x) > g(x, x)$ appears, then g is interpreted as $g_{\mathcal{M}ax}$, because the imposed constraint $c_{f,1} \geq c_{g,1} \wedge c_{f,1} \geq c_{g,2}$ is easier than $c_{f,1} \geq c_{g,1} + c_{g,2}$, which would be imposed by the interpretation $g_{\mathcal{P}ol}$.

The template variables introduced above are partitioned into two groups: template variables $w_f, p_{f,1}, \ldots, p_{f,n}$ are grouped in the *constant part*, and template variables $c_{f,1}, \ldots, c_{f,n}$ are in the *coefficient part*. For efficiency, it is important to properly restrict the range of these variables.

3.2 Classes of Precedences

NaTT offers "quasi" and "strict" precedences, as well as an option to disable them (i.e., all symbols are considered to have the same precedence). For reduction pairs using precedences, we recommend quasi-precedences which are chosen by default, as the encoding follows the technique of [27] that naturally encodes quasi-precedences.

Table 1. Parameters for some monotone reduction pairs

Technique	template	coefficient	constant	precedence	status
Linear POLO	$\mathcal{Pol}$	$\mathbb{Z}_+$	$\mathbb{N}$	no	empty
LPO	$\mathcal{Max}$	$\{1\}$	$\{0\}$	yes	total
KBO[6]	$\mathcal{Pol}$	$\{1\}$	$\mathbb{N}$	yes	total
Transfinite KBO[6]	$\mathcal{Pol}$	$\mathbb{Z}_+$	$\mathbb{N}$	yes	total

3.3 Classes of Status Functions

NaTT offers three classes of *status functions*: "total", "partial" and "empty" ones. The standard notions of *status functions* are total ones that were introduced to admit *permutation* of arguments when comparing them lexicographically from left to right (cf. [19]). Such a comparison appears in many well-known reduction pairs; famous examples are LPO and KBO. By combining the idea of argument filters, status functions have recently been generalized to *partial* ones, that do not only *permute* but may also drop some arguments [23]. A partial status is beneficial for KBO, and even more significant when combined with WPO [26]. The extreme case of a partial status is the "empty" status, that drops all arguments and so no comparison of arguments will be performed. This option corresponds to the nature of interpretation methods, e.g. POLO, if precedences are also disabled.

3.4 Obtaining Well-Known Reduction Pairs

Although most of the existing reduction pairs are subsumed by WPO, some of them are still useful for improving efficiency, due to the restricted search space and simplified SMT encoding. We list parameters that correspond to some known reduction pairs in Tables 1 and 2. Note here that the effects of non-collapsing argument filters are simulated by allowing 0-coefficients in the weight algebra. Thus NaTT has a dedicated implementation only for *collapsing* argument filters, and implementations of usable rules for interpretation methods and path orders are smoothly unified.

4 Cooperation with SMT Solvers

NaTT is designed to work with any SMT-LIB 2.0 compliant solvers that support at least `QF_LIA` logic, for which various efficient solvers exist.[7] NaTT extensively uses SMT encoding techniques for finding appropriate reduction pairs; the conditions of reduction pair processors are encoded into the following SMT constraint:

$$\bigwedge_{l \to r \in \mathcal{R}} [\![l \succsim r]\!] \wedge \bigwedge_{s \to t \in \mathcal{P}} [\![s \succsim t]\!] \wedge \bigvee_{s \to t \in \mathcal{P}} [\![s \succ t]\!] \qquad (3)$$

[6] Further constraints for *admissibility* are imposed.

[7] Cf. the Satisfiability Modulo Theories Competition, `http://smtcomp.org/`

Table 2. Parameters for some (non-monotone) reduction pairs

Technique	template	coefficient	constant	precedence	status
Linear POLO	$\mathcal{P}ol$	$\mathbb{N}$	$\mathbb{N}$	no	empty
Max-POLO	$\mathcal{MP}ol$	$\mathbb{N}$	$\mathbb{Z}$	no	empty
LPO + argument filter	$\mathcal{M}ax$	$\{0,1\}$	$\{0\}$	yes	total
KBO + argument filter	$\mathcal{P}ol$	$\{0,1\}$	$\mathbb{N}$	yes	total
Matrix interpretations	$\mathcal{P}ol$	$\mathbb{N}^{d\times d}$	$\mathbb{N}^d$	no	empty
WPO($\mathcal{MS}um$)	$\mathcal{MP}ol$	$\{0,1\}$	$\mathbb{N}$	yes	partial

where each $[\![l \mathrel{_{(}\succsim_{)}} r]\!]$ is an SMT formula that represents the condition $l \mathrel{_{(}\succsim_{)}} r$. In the remainder of this section, we present two techniques for handling such constraints that contribute to the efficiency of NaTT.

4.1 Use of Interactive Features of SMT Solvers

In a typical run of termination verification, constraints of the form (3) are generated and solved many times, and each encoding sometimes involves thousands of lines of SMT queries with a number of template and auxiliary variables. Hence runtime spent for the SMT solver forms a large part of the overall runtime of the tool execution. NaTT tries to reduce the runtime by using *interactive* features of SMT solvers,[8] which are specified in SMT-LIB 2.0.

For each technique of reduction pairs, the encoded formula of the constraint $\bigwedge_{l\to r\in\mathcal{R}}[\![l \succsim r]\!]$ need not be changed during a run, as far as $\mathcal{R}$ is not modified.[9] Hence, when a reduction pair processor is applied for the first time, the back-end SMT solver is initialized according to the following pseudo-script:

(assert ($\bigwedge_{l\to r\in\mathcal{R}}(u_{l\to r} \Rightarrow [\![l \succsim r]\!])$))
(push)

where $u_{l\to r}$ is a boolean variable denoting whether the rule $l \to r$ is usable or not. When the processor is applied to an SCC $\mathcal{P}$, the following script is used:

(assert ($\bigwedge_{s\to t\in\mathcal{P}}[\![s \succsim t]\!] \wedge \bigvee_{s\to t\in\mathcal{P}}[\![s \succ t]\!]$))
(check-sat)

Then, if a solution is found by the SMT solver, NaTT analyzes the solution using the `get-value` command. After this analysis, the command

```
(pop)
```

is issued to clear the constraints due to $\mathcal{P}$ and go back to the context saved by the `(push)` command. In order to derive the best performance of the solver,

```
(reset)
```

[8] NaTT is not the first tool to use the interactive features of SMT solvers; e.g., Boogie makes use of these features in its Houdini implementation [15].

[9] Although rules in $\mathcal{R}$ may be removed by considering *usable rules*, the formula still need not be changed, since it can be simulated by negating a propositional variable that represents whether the rule is usable or not.

is also issued in case sufficiently many rules become unusable (e.g., 1/3 of the rules in $\mathcal{R}$) from $\mathcal{P}$. All these commands, `push`, `pop` and `reset` are expected to be available in SMT-LIB 2.0 compliant solvers.

4.2 Use of Linear Arithmetic

Note that expressions of the form (1) or (2) are nonlinear, due to the coefficients $c_{f,1}, \ldots, c_{f,n}$. However, not many SMT solvers support *nonlinear* arithmetic, and even if they do, they are much less scalable than they are for linear arithmetic. Hence, we consider reducing the formulas to linear ones by restricting the range of $c_{f,1}, \ldots, c_{f,n}$ e.g. to $\{0, 1\}$. Although the idea is inspired by [2], NaTT uses a more straightforward reduction using `ite` (*if-then-else*) expressions. Each coefficient $c_{f,i}$ is replaced by the expression `(ite` $b_{f,i}$ `1 0)` where $b_{f,i}$ is a propositional variable, and then multiplications are reduced according to the rule:

$$\texttt{(* (ite } e_1\ e_2\ e_3\texttt{)}\ e_4\texttt{)} \to \texttt{(ite } e_1\ \texttt{(* } e_2\ e_4\texttt{) (* } e_3\ e_4\texttt{))}$$

It is easy to see that this reduction terminates and linearizes expressions of the form (1) or (2). It is also possible to avoid an explosion of the size of formulas by introducing a auxiliary variable for the duplicated expression e_4.

Example 1. Consider the constraint $\mathsf{f}(\mathsf{f}(\mathsf{a})) > \mathsf{b}$ interpreted in the algebra $\mathcal{P}ol$, and suppose that the range of $c_{\mathsf{f},1}$ is restricted to $\{1, 2\}$. The interpretation of the term $\mathsf{f}(\mathsf{f}(\mathsf{a}))$ is reduced as follows (written as S-expressions):

$$\begin{aligned} [\![\mathsf{f}(\mathsf{f}(\mathsf{a}))]\!] &= \texttt{(+ } w_{\mathsf{f}}\ \texttt{(* (ite } b_{\mathsf{f},1}\ \texttt{2 1)}\ [\![\mathsf{f}(\mathsf{a})]\!]\texttt{))} \\ &\to \texttt{(+ } w_{\mathsf{f}}\ \texttt{(ite } b_{\mathsf{f},1}\ \texttt{(* 2 } [\![\mathsf{f}(\mathsf{a})]\!]\texttt{)}\ [\![\mathsf{f}(\mathsf{a})]\!]\texttt{))} \end{aligned}$$

Similarly, for $\mathsf{f}(\mathsf{a})$ we obtain

$$[\![\mathsf{f}(\mathsf{a})]\!] \to \texttt{(+ } w_{\mathsf{f}}\ \texttt{(ite } b_{\mathsf{f},1}\ \texttt{(* 2 } w_{\mathsf{a}}\texttt{)}\ w_{\mathsf{a}}\texttt{))}$$

Now, the constraint $[\![\mathsf{f}(\mathsf{f}(\mathsf{a})) > \mathsf{b}]\!]$ is expressed by the following script:

```
(define-fun v (+ w_f (ite b_f,1 (* 2 w_a) w_a)))
(assert (> (+ w_f (ite b_f,1 (* 2 v) v) w_b)))
```

In contrast to SAT encoding techniques [4–6], we do not have to care about the bit-width for the constant part and intermediate results. It is also possible to indicate that NaTT should keep formulas nonlinear, and solve them using SMT solvers that support `QF_NIA` logic. Our experiments on TPDB[10] problems, however, suggests that use of nonlinear SMT solving is impractical for our purpose.

[10] The Termination Problem Data Base, `http://termination-portal.org/wiki/TPDB`

5 Design

The source code of NaTT consists of about 6000 lines of code written in OCaml.[11] About 23% is consumed by interfacing SMT solvers, where some optimizations for encodings are also implemented. Another 17% is for parsing command-lines and TRS files. The most important part of the source code is the 40% devoted to the implementation of WPO, the unified reduction pair processor. Each of the other processors implemented consumes less than 3%. For computing SCCs, the third-party library `ocamlgraph`[12] is used.

5.1 Command Line Interface

The command line of NaTT has the following syntax:

```
./NaTT [FILE] [OPTION]... [PROCESSOR]...
```

To execute NaTT, an SMT-LIB 2.0 compliant solver must be installed. By default, `z3` version 4.0 or later[13] is supposed to be installed in the path. Users can specify other solvers by the `--smt "COMMAND"` option, where the solver invoked by `COMMAND` should process SMT-LIB 2.0 scripts given on the standard input.

The TRS whose termination should be verified is read from either the specified `FILE` or the standard input.[14] Each `PROCESSOR` is either an order (e.g. `POLO`, `KBO`, `WPO`, *etc.*, possibly followed by options), or a name of other processors (`UNCURRY`, `EDG`, or `LOOP`). Orders preceding the `EDG` processor should be monotone reduction pairs and applied as rule removal processors before computing the dependency pairs. Orders following the `EDG` processor are applied as reduction pair processors to each SCC in the EDG. A list of available `OPTION`s and `PROCESSOR`s can be obtained via `NaTT --help`.

5.2 The Default Strategy

In case no `PROCESSOR` is specified, the following default strategy will be applied:

- As a rule removal processor, POLO with coefficients in $\{1, 2\}$ and constants in $\mathbb{N}$ is applied.
- Then the uncurrying processor is applied.
- The following reduction pair processors are applied (in this order):
  1. POLO with coefficients in $\{0, 1\}$ and constants in $\mathbb{N}$,
  2. algebra $\mathcal{M}ax$ with coefficients in $\{0, 1\}$ and constants in $\mathbb{N}$,
  3. LPO with quasi-precedence, status and argument filter,
  4. algebra $\mathcal{MP}ol$ with coefficients in $\{0, 1\}$ and constants in $\mathbb{Z}$,
  5. WPO with quasi-precedence, partial status, algebra $\mathcal{MP}ol$, coefficients in $\{0, 1\}$ and constants in $\mathbb{N}$,
  6. matrix interpretations with $\{0, 1\}^{2\times 2}$ matrices and $\mathbb{N}^2$ vectors.
- If all the above processors fail, then a (naive) loop detection is performed.

---

[11] `http://caml.inria.fr/`

[12] `http://ocamlgraph.lri.fr/`

[13] `http://z3.codeplex.com/`

[14] The format is found at `https://www.lri.fr/~marche/tpdb/format.html`

6 Assessment

Many tools have been developed for proving termination of TRSs, and the international termination competition has been held annually for a decade. NaTT participated in the *TRS Standard* category of the full-run 2013, where the other participants are versions of: AProVE,[15] $\mathsf{T_TT_2}$,[16] MU-TERM,[17] and WANDA.[18] Using the default strategy described in Section 5.2, NaTT (dis)proves termination of 982 TRSs out of 1463 TRSs, and comes next to (the two versions of) AProVE, the constant champion of the category. It should be noticed that NaTT proved termination of 34 TRSs out of the 159 whose termination could not be proved by any other tool. NaTT is notably faster than the other competitors; it consumed only 21% of the time compared to AProVE, the second fastest. We expect that we can further improve efficiency by optimizing to multi-core architecture; currently, NaTT runs in almost single thread. NaTT also participated in the *SRS Standard* category. However, the result is not as good as it is for TRSs. This is due to the fact that the default strategy of Section 5.2 is designed only for non-unary signatures. It should be improved by choosing a strategy depending on the shape of input TRSs.

7 Conclusion

We described the implementation and techniques of the termination tool NaTT. The novel implementation of the weighted path order is described in detail, and some techniques for cooperating SMT solvers are presented. Together with these efforts, NaTT is one of the most efficient and strongest tools for proving termination of TRSs.

References

1. Arts, T., Giesl, J.: Termination of term rewriting using dependency pairs. TCS 236(1-2), 133–178 (2000)
2. Borralleras, C., Lucas, S., Navarro-Marset, R., Rodríguez-Carbonell, E., Rubio, A.: Solving non-linear polynomial arithmetic via SAT modulo linear arithmetic. In: Schmidt, R.A. (ed.) CADE-22. LNCS, vol. 5663, pp. 294–305. Springer, Heidelberg (2009)
3. Dershowitz, N.: Orderings for term-rewriting systems. TCS 17(3), 279–301 (1982)
4. Endrullis, J., Waldmann, J., Zantema, H.: Matrix interpretations for proving termination of term rewriting. JAR 40(2-3), 195–220 (2008)
5. Fuhs, C., Giesl, J., Middeldorp, A., Schneider-Kamp, P., Thiemann, R., Zankl, H.: SAT solving for termination analysis with polynomial interpretations. In: Marques-Silva, J., Sakallah, K.A. (eds.) SAT 2007. LNCS, vol. 4501, pp. 340–354. Springer, Heidelberg (2007)

---
[15] `http://aprove.informatik.rwth-aachen.de/`
[16] `http://cl-informatik.uibk.ac.at/software/ttt2/`
[17] `http://zenon.dsic.upv.es/muterm/`
[18] `http://wandahot.sourceforge.net/`

6. Fuhs, C., Giesl, J., Middeldorp, A., Schneider-Kamp, P., Thiemann, R., Zankl, H.: Maximal termination. In: Voronkov, A. (ed.) RTA 2008. LNCS, vol. 5117, pp. 110–125. Springer, Heidelberg (2008)
7. Giesl, J., Thiemann, R., Schneider-Kamp, P.: The dependency pair framework: Combining techniques for automated termination proofs. In: Baader, F., Voronkov, A. (eds.) LPAR 2004. LNCS (LNAI), vol. 3452, pp. 301–331. Springer, Heidelberg (2005)
8. Giesl, J., Thiemann, R., Schneider-Kamp, P.: Proving and disproving termination of higher-order functions. In: Gramlich, B. (ed.) FroCos 2005. LNCS (LNAI), vol. 3717, pp. 216–231. Springer, Heidelberg (2005)
9. Giesl, J., Thiemann, R., Schneider-Kamp, P., Falke, S.: Mechanizing and improving dependency pairs. JAR 37(3), 155–203 (2006)
10. Hirokawa, N., Middeldorp, A.: Dependency pairs revisited. In: van Oostrom, V. (ed.) RTA 2004. LNCS, vol. 3091, pp. 249–268. Springer, Heidelberg (2004)
11. Hirokawa, N., Middeldorp, A.: Polynomial interpretations with negative coefficients. In: Buchberger, B., Campbell, J. (eds.) AISC 2004. LNCS (LNAI), vol. 3249, pp. 185–198. Springer, Heidelberg (2004)
12. Hirokawa, N., Middeldorp, A., Zankl, H.: Uncurrying for termination and complexity. JAR 50(3), 279–315 (2013)
13. Kamin, S., Lévy, J.J.: Two generalizations of the recursive path ordering (1980) (unpublished note)
14. Knuth, D., Bendix, P.: Simple word problems in universal algebras. In: Computational Problems in Abstract Algebra, pp. 263–297. Pergamon Press, New York (1970)
15. Lal, A., Qadeer, S., Lahiri, S.K.: A solver for reachability modulo theories. In: Madhusudan, P., Seshia, S.A. (eds.) CAV 2012. LNCS, vol. 7358, pp. 427–443. Springer, Heidelberg (2012)
16. Lankford, D.: On proving term rewrite systems are Noetherian. Tech. Rep. MTP-3, Louisiana Technical University (1979)
17. Ludwig, M., Waldmann, U.: An extension of the knuth-bendix ordering with LPO-like properties. In: Dershowitz, N., Voronkov, A. (eds.) LPAR 2007. LNCS (LNAI), vol. 4790, pp. 348–362. Springer, Heidelberg (2007)
18. Middeldorp, A., Zantema, H.: Simple termination of rewrite systems. TCS 175(1), 127–158 (1997)
19. Steinbach, J.: Extensions and comparison of simplification orders. In: Dershowitz, N. (ed.) RTA 1989. LNCS, vol. 355, pp. 434–448. Springer, Heidelberg (1989)
20. Sternagel, C., Thiemann, R.: Generalized and formalized uncurrying. In: Tinelli, C., Sofronie-Stokkermans, V. (eds.) FroCoS 2011. LNCS, vol. 6989, pp. 243–258. Springer, Heidelberg (2011)
21. Winkler, S., Zankl, H., Middeldorp, A.: Ordinals and knuth-bendix orders. In: Bjørner, N., Voronkov, A. (eds.) LPAR-18 2012. LNCS, vol. 7180, pp. 420–434. Springer, Heidelberg (2012)
22. Xi, H.: Towards automated termination proofs through freezing. In: Nipkow, T. (ed.) RTA 1998. LNCS, vol. 1379, pp. 271–285. Springer, Heidelberg (1998)
23. Yamada, A., Kusakari, K., Sakabe, T.: Partial status for KBO. In: Proceedings WST 2013, pp. 74–78 (2013)
24. Yamada, A., Kusakari, K., Sakabe, T.: Unifying the Knuth-Bendix, recursive path and polynomial orders. In: Proceedings PPDP[15], pp. 181–192 (2013)
25. Yamada, A., Kusakari, K., Sakabe, T.: Nagoya Termination Tool. CoRR abs/1404.6626 (2014)
26. Yamada, A., Kusakari, K., Sakabe, T.: A unified order for termination proving. CoRR abs/1404.6245 (2014) (submitted to SCP)
27. Zankl, H., Hirokawa, N., Middeldorp, A.: KBO orientability. JAR 43(2), 173–201 (2009)

Termination of Cycle Rewriting

Hans Zantema[1,2], Barbara König[3], and H.J. Sander Bruggink[3]

[1] TU Eindhoven, Department of Computer Science, P.O. Box 513, 5600 MB Eindhoven, The Netherlands
h.zantema@tue.nl

[2] Radboud University Nijmegen, Institute for Computing and Information Sciences, P.O. Box 9010, 6500 GL Nijmegen, The Netherlands

[3] Duisburg Essen University, Abteilung Informatik und Angewandte Kognitionswissenschaft, 47048 Duisburg, Germany
{barbara_koenig,sander.bruggink}@uni-due.de

Abstract. String rewriting can not only be applied on strings, but also on cycles and even on general graphs. In this paper we investigate termination of string rewriting applied on cycles, shortly denoted as cycle rewriting, which is a strictly stronger requirement than termination on strings. Most techniques for proving termination of string rewriting fail for proving termination of cycle rewriting, but match bounds and some variants of matrix interpretations can be applied. Further we show how any terminating string rewriting system can be transformed to a terminating cycle rewriting system, preserving derivational complexity.

1 Introduction

A string rewriting system (SRS) consists of a set of rules $\ell \to r$ where ℓ and r are strings, that is, elements of Σ^* for some alphabet Σ. String rewriting means that for such a rule $\ell \to r$ an occurrence of ℓ is replaced by r. In the standard interpretation this only works on strings: a string of the shape $u\ell v$ is replaced by urv. However, it is natural also to apply this on a cycle, that is, a string in which the start point is connected to its end point. For instance, for an SRS R containing the rule $ab \to cba$ we want to allow the cycle rewrite step

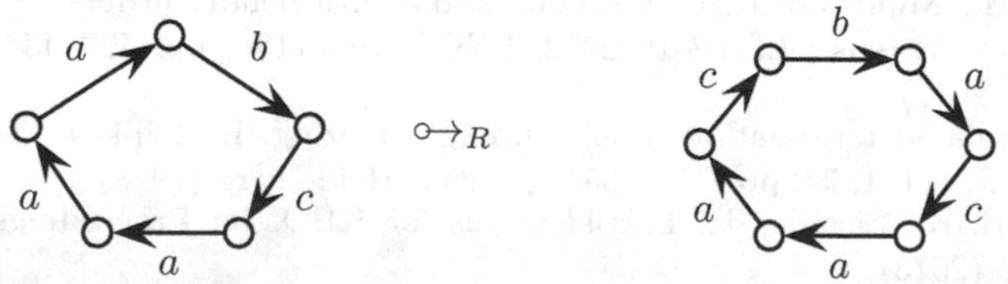

We use the notation $\looparrowright_R$ for a cycle rewrite step. In this paper we investigate termination of cycle rewriting. It is easy to see that termination of cycle rewriting implies termination of string rewriting. However, the other way around does not hold: the single rewrite rule $ab \to ba$ is terminating in the setting of string rewriting, but not in the setting of cycle rewriting, since the cycle ab of length 2 rewrites to the cycle ba which is equal to the cycle ab, so this rewriting can go on forever.

G. Dowek (ed.): RTA-TLCA 2014, LNCS 8560, pp. 476–490, 2014.

Cycle rewriting can be seen as a special instance of graph transformation [2]. In a separate paper [1] we investigate how the techniques of this paper extend to the general setting of graph transformation. In particular there we show that for a graph transformation system in which all rules are string rewrite rules, termination on all cycles coincides with termination on all graphs. So from the perspective of termination of graph transformation it is more natural to consider string rewriting to be applied on cycles rather than on strings, justifying an investigation of cycle rewriting as a separate topic. In describing communication protocols with message passing on a ring structure, the steps are essentially cycle rewrite steps, again a motivation for investigating cycle rewriting.

Standard techniques for proving termination of term rewriting and string rewriting like the recursive path order, dependency pairs and polynomial interpretation fail to prove termination of cycle rewriting as they all easily prove termination of the single rule $ab \to ba$ which is not terminating in the cycle setting. They all exploit the term structure by which every string has a begin and an end, while a cycle has not. Nevertheless, for a few other powerful techniques, in particular match bounds, arctic matrices and tropical matrices, we show that these can be applied to prove termination of cycle rewriting. It turns out that the techniques of arctic and tropical matrices can be interpreted by weighting in type graphs: tropical matrices correspond to the requirement that every morphism of a left hand side to the type graph admits a morphism of the corresponding right hand side to the type graph of a lower weight, while arctic matrices correspond to the requirement that every morphism of a right hand side to the type graph admits a morphism of the corresponding left hand side to the type graph of a higher weight. Further match bound proofs can be seen as a particular case of tropical matrix proofs. We developed an implementation automatically finding proofs based on a combination of these techniques.

We investigate derivational complexity of cycle rewrite systems for which termination is proved by these techniques. Arctic matrices and match bounds have been exploited before for proving bounds on derivational complexity of term rewriting and string rewriting: single applications of these techniques yield linear bounds, while combined application may yield higher bounds. In this paper we give similar results in the setting of cycle rewriting. In particular we give examples of length preserving systems for which termination can be proved by combining the above mentioned techniques, while any polynomial can be achieved as a lower bound for derivational complexity. For non-length-preserving systems we show that exponential derivational complexity can be reached.

We investigate a particular shape of SRSs for which we show that termination of string rewriting and cycle rewriting coincide; it is characterized by *end symbols* that only occur as the last symbol of a left hand side or right side of a rule. We show how any SRS can be transformed to an SRS of this special shape, preserving termination and derivational complexity. As a consequence, termination of cycle rewriting is undecidable, and for every computable function an SRS R exists for which cycle rewriting is terminating and the derivational complexity exceeds the computable function.

The paper is organized as follows. Section 2 presents basic definitions and observations related to cycle rewriting. Section 3 presents the techniques for proving termination of cycle rewriting. In Section 4 derivational complexity of these techniques is investigated. Section 5 introduces the special format with end symbols and presents the corresponding theory. In Section 6 our implementation is discussed. We conclude in Section 7.

2 Cycle Rewriting

We consider cycles over an alphabet Σ which are essentially strings over Σ in which the leftmost element is connected to the rightmost element. We represent cycles by strings, where for all strings u, v the string uv represents the same cycle as vu. More precisely, for any alphabet Σ we define the set $\mathsf{Cycle}(\Sigma)$ of cycles over Σ by

$$\mathsf{Cycle}(\Sigma) \;=\; \Sigma^*/\sim$$

where $\sim$ is the equivalence relation on Σ^* defined by

$$u \sim v \iff \exists u_1, u_2 \in \Sigma^* : u = u_1 u_2 \wedge v = u_2 u_1.$$

It is straightforward to check that indeed $\sim$ is an equivalence relation. The cycle represented by a string u, i.e., the equivalence class of u w.r.t $\sim$, is denoted by $[u]$.

As usual we define a string rewrite system (SRS) over Σ to be a subset R of $\Sigma^* \times \Sigma^*$. Elements (ℓ, r) of an SRS are called (string rewrite) rules and are usually written as $\ell \to r$, where ℓ is called the left hand side (lhs) and r the right hand side (rhs) of the rule. As usual, the string rewrite relation $\to_R$ on Σ^* is defined by $u \to_R v \iff \exists x, y \in \Sigma^*, \ell \to r \in R : u = x\ell y \wedge v = xry$.

For an SRS R over Σ we define the corresponding cycle rewrite relation $\circ\!\!\to_R$ on $\mathsf{Cycle}(\Sigma)$ as follows:

$$[u] \circ\!\!\to_R [v] \iff \exists x \in \Sigma^*, \ell \to r \in R : \ell x \sim u \wedge rx \sim v.$$

Equivalently, one can state $[u] \circ\!\!\to_R [v] \iff \exists u' \in [u], v' \in [v] : u' \to_R v'$.

The main goal of this paper is to study $\circ\!\!\to_R$, in particular how to prove termination, that is, does not allow an infinite reduction.

Lemma 1. *Let R be an SRS over an alphabet Σ for which the relation $\circ\!\!\to_R$ on $\mathsf{Cycle}(\Sigma)$ is terminating. Then the string rewrite relation $\to_R$ on Σ^* is terminating too.*

Proof. If $u \to_R v$ for $u, v \in \Sigma^*$ then $[u] \circ\!\!\to_R [v]$. Hence an infinite $\to_R$-reduction transforms to an infinite $\circ\!\!\to_R$-reduction, proving the lemma. □

The converse of Lemma 1 does not hold: the SRS consisting of the single rule $ab \to ba$ is clearly terminating, but since $[ab] = [ba]$ the corresponding cycle rewrite relation $\circ\!\!\to$ is not terminating.

So termination of $\circ\!\!\to_R$ is a stronger requirement than termination of $\to_R$.

A natural question to ask is how confluence of $\circ\!\!\to_R$ is related to confluence of $\to_R$. It turns out that none of the possible implications holds. As a first example consider the SRS over $\{a,b\}$ consisting of the two rules $ab \to ba,\ ab \to b$. Since the string ab rewrites to both ba and b, both being normal forms, the relation $\to_R$ is not confluent for this SRS. However, with respect to $\circ\!\!\to_R$ every string containing n b-s rewrites to b^n while every string containing no b-s is a normal form, hence $\circ\!\!\to_R$ is confluent. Conversely consider the SRS consisting of the two rules $ab \to aa,\ ba \to bb$. Straightforward critical pair analysis shows that $\to_R$ is locally confluent; since it is terminating (proved e.g. using dependency pairs) it is confluent too. However, $\circ\!\!\to_R$ is neither confluent since ab admits two normal forms aa and bb, nor terminating since $aab \sim aba \to_R abb \to_R aab$.

Also with respect to weak normalization the relations $\circ\!\!\to_R$ and $\to_R$ are incomparable: for the single rule $ab \to ba$ the relation $\to_R$ is weakly normalizing, while $\circ\!\!\to_R$ is not. Conversely, for the SRS consisting of the two rules $ab \to ab,\ ba \to a$ the relation $\circ\!\!\to_R$ is weakly normalizing, while $\to_R$ is not.

3 Termination by Type Graphs

From now on we concentrate on developing techniques to prove termination of $\circ\!\!\to_R$. We start by a most basic technique exploiting decreasing weights. A weight function $W : \Sigma \to \mathbb{N}$ is extended to a weight function $W : \Sigma^* \to \mathbb{N}$ by defining inductively $W(\epsilon) = 0$ and $W(ax) = W(a) + W(x)$ for $a \in \Sigma, x \in \Sigma^*$: the weight of a string is simply the sum of the weights of its elements.

Lemma 2. *Let R be an SRS over Σ and let $W : \Sigma \to \mathbb{N}$ satisfy*

- *$W(\ell) \geq W(r)$ for all $\ell \to r \in R$, and*
- *$\circ\!\!\to_{R'}$ is terminating for $R' = \{\ell \to r \in R \mid W(\ell) = W(r)\}$.*

Then $\circ\!\!\to_R$ is terminating.

Proof. We prove that $W(u) = W(v)$ for all u, v satisfying $[u] \circ\!\!\to_{R'} [v]$ and $W(u) > W(v)$ for all u, v satisfying $[u] \circ\!\!\to_{R \setminus R'} [v]$. Then termination of R follows from termination of R' and well-foundedness of $>$. So let $[u] \circ\!\!\to_{R\setminus R'} [v]$, then we can write $u = u_1u_2, v = v_1v_2, u_2u_1 = \ell x, v_2v_1 = rx$ for some $\ell \to r \in R \setminus R'$. Then

$$\begin{aligned} W(u) &= W(u_1u_2) = W(u_1) + W(u_2) = W(u_2u_1) \\ &= W(\ell x) = W(\ell) + W(x) \\ &> W(r) + W(x) = W(rx) \\ &= W(v_2v_1) = W(v_2) + W(v_1) = W(v_1v_2) = W(v). \end{aligned}$$

For the remaining case $[u] \circ\!\!\to_{R'} [v]$ we obtain exactly the same derivation with '$>$' replaced by '$=$', hence concluding $W(u) = W(v)$ which we had to prove. □

In simple applications of Lemma 2 we have $W(\ell) > W(r)$ for all $\ell \to r \in R$, by which R' is empty and hence $\circ\!\!\to_R$ is trivially terminating. Then the only thing to be done for proving termination of $\circ\!\!\to_R$ is choosing a $W(a) \in \mathbb{N}$ for every $a \in \Sigma$ such that $W(\ell) > W(r)$ for all $\ell \to r \in R$.

Example 3. For the SRS R consisting of the four rules

$$aa \to bc,\ bb \to cd,\ cc \to ddd,\ ddd \to ac$$

the relation $\circ\!\!\to_R$ is terminating due to Lemma 2 by choosing $W(a) = 30, W(b) = 27, W(c) = 32$ and $W(d) = 21$, for which it is checked that $W(\ell) > W(r)$ for all four rules $\ell \to r$. These numbers are the smallest possible ones.

Next we give a generalization of Lemma 2 inspired by the notion of a *type graph* as it appears in graph transformation systems [2] and coinciding with the approach of tropical and arctic matrix interpretations [4,7]. Also the proofs could be given in the setting of matrices over semirings as we discuss later.

We define a *type graph* (V, E, W) over a signature Σ to be a directed graph in which the edges are labeled by symbols from Σ, and have a weight $W(e) \in \mathbb{N}$, that is, $E \subseteq V \times \Sigma \times V$ and $W : E \to \mathbb{N}$.

For $u = a_1 a_2 \cdots a_n \in \Sigma^+$ and $p, q \in V$ we define a *u-path from p to q in a type graph* (V, E, W) to be a sequence $(p_1, a_1, q_1)(p_2, a_2, q_2) \cdots (p_n, a_n, q_n)$ of edges in E such that $p_1 = p$, $q_n = q$ and $q_i = p_{i+1}$ for $i = 1, \ldots, n-1$. The *weight* $W(u)$ of such a u-path is defined to be $\sum_{i=1}^{n} W(p_i, a_i, q_i)$. In case $p = q$, the u-path is called a *u-cycle*.

We distinguish two kinds of criteria to conclude $\circ\!\!\to_R$ termination from morphisms from paths to type graphs: tropical and arctic. Tropical means that every path morphism of a left hand side to the type graph admits a morphism of the corresponding right hand side to the type graph of a lower weight, while arctic means that every path morphism of a right hand side to the type graph admits a morphism of the corresponding left hand side to the type graph of a higher weight. This terminology is inspired by the corresponding terminology for matrix interpretations.

Theorem 4. *Let $R' \subseteq R$ be SRSs over Σ. Let (V, E, W) be a type graph over Σ. Assume*

- *there exists $p \in V$ such that $(p, a, p) \in E$ for all $a \in \Sigma$, and*
- *$\circ\!\!\to_{R'}$ is terminating, and*
- *either*
  - *(tropical) for every $\ell \to r \in R$, $p, q \in V$ and for every ℓ-path from p to q in (V, E, W) having weight w there is an r-path from p to q in (V, E, W) having weight w' with $w \geq w'$, and $w > w'$ if $\ell \to r \notin R'$, or*
  - *(arctic) for every $\ell \to r \in R$, $p, q \in V$ and for every r-path from p to q in (V, E, W) having weight w there is an ℓ-path from p to q in (V, E, W) having weight w' with $w' \geq w$, and $w' > w$ if $\ell \to r \notin R'$.*

Then $\circ\!\!\to_R$ is terminating, and any $\circ\!\!\to_R$ reduction of a cycle $[u]$ contains at most $|u| \cdot w$ steps with respect to $R \setminus R'$ for $w = \max_{e \in E} W(e)$.

Proof. Assume $[u_0] \circ\!\!\to_R [u_1] \circ\!\!\to_R [u_2] \circ\!\!\to_R \cdots \circ\!\!\to_R [u_n]$ contains more than $|u_0| \cdot w$ steps with respect to $R \setminus R'$ for $w = \max_{e \in E} W(e)$. We will derive a contradiction; this proves the theorem as termination immediately follows.

For the tropical case choose any u_0-cycle; this exists due to the first assumption of the theorem. Let w_0 be the weight of this cycle.

Next, for $i = 1, 2, 3, \ldots$ we can choose a u_i-cycle by replacing the ℓ-path being the part of the u_{i-1}-cycle by the corresponding r-path as indicated in the tropical condition of the theorem, where $\ell \to r$ is the rule applied in $[u_{i-1}] \circ\!\!\to_R [u_i]$. Let w_i be the weight of this new cycle; if $\ell \to r \in R \setminus R'$ then $w_i < w_{i-1}$, otherwise $w_i \leq w_{i-1}$. As by the assumption there are more than $|u_0| \cdot w$ steps with '<', and $w_n \geq 0$, we conclude $w_0 > |u_0| \cdot w$. This contradicts the definition of $w = \max_{e \in E} W(e)$.

For the arctic case choose any u_n-cycle; this exists due to the first assumption of the theorem. Let w_n be the weight of this cycle.

Next, for $i = n-1, n-2, n-3, \ldots$ we can choose a u_i-cycle by replacing the r-path being the part of the u_{i+1}-cycle by the corresponding r-path as indicated in the arctic condition of the theorem, where $\ell \to r$ is the rule applied in $[u_i] \circ\!\!\to_R [u_{i+1}]$. Let w_i be the weight of this new cycle; if $\ell \to r \in R \setminus R'$ then $w_i < w_{i-1}$, otherwise $w_i \leq w_{i-1}$. The rest of the argument is as before. □

In case the type graph consists of a single node, every path in this type graph consists of a sequence of edges of this node to itself. In this case the conditions for the tropical case and the termination conclusion of Theorem 4 coincide with the conditions and the conclusion of Lemma 2. Hence indeed we can state that Theorem 4 is a generalization of Lemma 2. The next example shows that it is a strict generalization.

Example 5. For the SRS R consisting of the single rule $aa \to aba$ Lemma 2 does not apply, since $2W(a) > 2W(a) + W(b)$ has no solutions in the natural numbers. Instead we define a type graph consisting of two nodes 1 and 2, and four edges $(1, a, 1), (1, b, 1), (1, a, 2), (2, b, 1)$, of which $(1, a, 1)$ has weight 1 and all others have weight 0, as indicated in the picture.

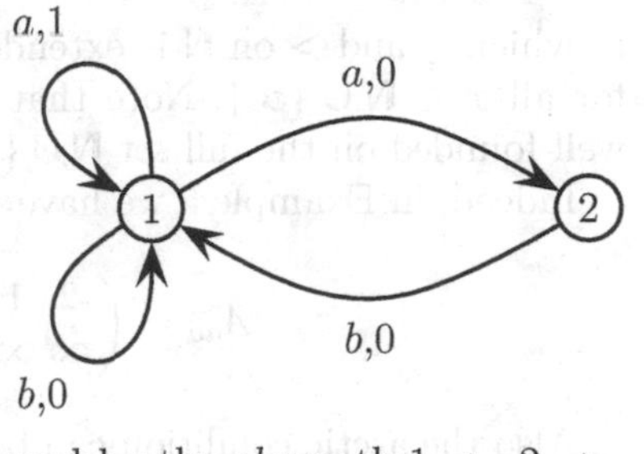

Now there are exactly two aa-paths:

- $1 \to_a 1 \to_a 1$ of weight 2 that may be replaced by the aba-path $1 \to_a 2 \to_b 1 \to_a 1$ of weight 1, and
- $1 \to_a 1 \to_a 2$ of weight 1 that may be replaced by the aba-path $1 \to_a 2 \to_b 1 \to_a 2$ of weight 0.

So all conditions of the tropical version of Theorem 4 are satisfied in choosing R' to be empty, proving that $\circ\!\!\to_R$ is terminating.

Type graphs can be represented by matrices in the following natural way. Number the nodes of the type graphs from 1 to n. For every $a \in \Sigma$ let A_a be the matrix such that $A_a(i, j) = w$ if and only if an edge (i, a, j) exists of weight w, while $A_a(i,j) = \infty$ if no such edge exist. So for example, the type graph of Example 5 is represented by the following matrices:

$$A_a = \begin{pmatrix} 1 & 0 \\ \infty & \infty \end{pmatrix}, \quad A_b = \begin{pmatrix} 0 & \infty \\ 0 & \infty \end{pmatrix}.$$

Now we consider the semiring $(\mathbb{N} \cup \{\infty\}, \min, +)$, that is, the semiring consisting of $\mathbb{N} \cup \{\infty\}$ on which the binary operator min acts as the semiring addition and the normal addition acts as the semiring multiplications. Here on $\mathbb{N}$ the operators min and $+$ act as usual, while it is extended to $\mathbb{N} \cup \{\infty\}$ by defining

$$\min(\infty, x) = \min(x, \infty) = x \text{ and } \infty + x = x + \infty = \infty$$

for all $x \in \mathbb{N} \cup \{\infty\}$. Now ∞ acts as the semiring zero and 0 acts as the semiring unit. This semiring is called the *tropical semiring* after its study by the Brazilian mathematician Imre Simon [8]. Now it is easily checked that path concatenation corresponds to matrix multiplication over this semiring, more precisely, if A_u is defined by $A_u(i,j)$ to be the lowest weight of a u-path from i to j, and ∞ if no such path exists, then $A_{uv} = A_u \times A_v$, where $\times$ is matrix multiplication with respect to this semiring. For instance, in the above example we have $A_{ab} = \begin{pmatrix} 0 & \infty \\ \infty & \infty \end{pmatrix}$. In this notation the tropical condition can be reformulated to $A_\ell \geq A_r$ for all $\ell \to r \in R$ and $A_\ell > A_r$ for all $\ell \to r$ not in R'. Here $\geq$ and $>$ on matrices are defined by

$$A \geq B \iff \forall i,j : A(i,j) \geq B(i,j), \quad A > B \iff \forall i,j : A(i,j) > B(i,j),$$

in which $\geq$ and $>$ on $\mathbb{N}$ is extended to $\mathbb{N} \cup \{\infty\}$ by defining $\infty \geq x$ and $\infty > x$ for all $x \in \mathbb{N} \cup \{\infty\}$. Note that this also yields $\infty > \infty$, by which $>$ is not well-founded on the full set $\mathbb{N} \cup \{\infty\}$.

Indeed, in Example 5 we have

$$A_{aa} = \begin{pmatrix} 2 & 1 \\ \infty & \infty \end{pmatrix} > A_{aba} = \begin{pmatrix} 1 & 0 \\ \infty & \infty \end{pmatrix}.$$

Also the arctic condition can be described by a matrix condition over a semiring: the arctic semiring $(\mathbb{N} \cup \{-\infty\}, \max, +)$, so similar to the tropical semiring, but now with max as the semiring addition, having $-\infty$ as its zero, in which $-\infty$ is less than all other elements. In this notation the arctic condition can be reformulated to $A_\ell \geq A_r$ for all $\ell \to r \in R$ and $A_\ell > A_r$ for all $\ell \to r$ not in R'. These arctic matrix interpretations have been studied in [7], being a modification of matrix interpretations [4]. There the termination proofs for term rewriting (including string rewriting) are based on monotone algebras. As pointed out by an anonymous referee the monotone algebra approach can be adjusted for our cycle setting by taking the *trace* of a matrix, that is, the sum of the diagonal, as its interpretation; compatibility with the cycle interpretation then follows from the well-known property that AB and BA have the same traces for all matrices

A, B, for any commutative semiring. In this way also matrix interpretations over the semiring $(\mathbb{N}, +, \times)$ can be applied for proving termination of cycle rewriting.

This section is concluded by showing how the method of match-bounds for proving termination of string rewriting can be seen as a special instance of tropical type graphs, and therefore also proves termination of cycle rewriting. Here we refer to the basic version of match-bounds which is also used for proving linear derivational complexity of string rewriting, and not the version based on forward closures which is more powerful for proving termination of string rewriting. Surprisingly, this basic theorem of match-bounds uses exactly the same data structure of a type graph: a directed graph in which every edge is labeled by symbols from Σ and has a natural number assigned to it. Where in type graphs this natural number serves as a *weight*, denoted by W, in match-bounds it serves as a *height* and is denoted by H.

Theorem 6. *Let R be SRSs over Σ. Let (V, E, H) be a type graph over Σ. Assume*

- *there exists $p \in V$ such that $(p, a, p) \in E$ for all $a \in \Sigma$, and*
- *for every $\ell \to r \in R$, $p, q \in V$ and for every ℓ-path from p to q in (V, E, H) there is an r-path from p to q in (V, E, H) such that the height of every edge in this r-path is $1 + m$, where m is the smallest height of an edge in the ℓ-path.*

Then $\circ\!\!\to_R$ is terminating.

If the conclusion of Theorem 6 is weakened to termination of $\to_R$, it coincides with the basic version of the match-bound theorem for string rewriting from [5,10,3].

Proof. For proving Theorem 6 we apply the tropical case of Theorem 4, in which we define $W(u, a, v) = s^{h-H(u,a,v)}$, where s is a number higher than the length of the longest right hand side of R, and h is the highest value of H occurring in the type graph (V, E, H). We choose $R' = \emptyset$. Then all conditions of Theorem 4 hold, where the tropical condition for a rule $\ell \to r$ and an ℓ-path from u to v with smallest height m follows from

$$W(\ell) \geq s^{h-m} > |r| \cdot s^{h-(m+1)} = W(r),$$

in which the r path from u to v is chosen according to the second condition of Theorem 6. So $\circ\!\!\to_R$ is terminating according to Theorem 4. □

The typical use of match-bounds is that one tries to construct a corresponding type graph by completion: start by a single node with an a-loop of height 0 for every $a \in \Sigma$, and complete it by continuously investigating all ℓ-paths in the graph and add a corresponding r-path if it does not yet exist. If this process ends, termination has been proved. Note that the second condition of Theorem 6 may be weakened: the existence an r-path with total weight less than the weight of the ℓ-path is sufficient, also if not all edges in the r-path have height exactly $m + 1$.

4 Derivational Complexity

In term rewriting and string rewriting *derivational complexity* of a terminating rewrite system is defined to be the longest reduction length expressed in the size of the initial term. For cycle rewriting we do exactly the same: for an SRS R over Σ we define

$$\mathsf{dc}_R(n) \;=\; \max\{k \mid \exists t, s \in \Sigma^* : |t| \leq n \wedge t \circ\!\!\to_R{}^k s\}.$$

Here $\circ\!\!\to_R{}^k$ means the composition of k $\circ\!\!\to_R$-steps. An SRS R is said to have linear (quadratic, cubic, ...) derivational complexity with respect to cycle rewriting if $\mathsf{dc}_R(n) = \Theta(n)$ $(\Theta(n^2), \Theta(n^3), \ldots)$.

Our first theorem on derivational complexity states that combined application of Lemma 2 only proves termination of systems with linear derivational complexity.

Theorem 7. *Let* $n \geq 1$. *Let* $R = \bigcup_{i=1}^n R_i$ *for which for every* $k = 1, \ldots, n$ *termination of* $\bigcup_{i=1}^k R_i$ *is proved by Lemma 2 by choosing* $R' = \emptyset$ *for* $k = 1$ *and* $R' = \bigcup_{i=1}^{k-1} R_i$ *for* $k > 1$. *Then* $\mathsf{dc}_R(n) = O(n)$.

Proof. We apply induction on n. If $n = 1$ then Lemma 2 is applied with $R' = \emptyset$, meaning that $W(\ell) > W(r)$ for all $\ell \to r \in R$, from which we obtain $W(t) > W(t')$ for every $t \circ\!\!\to_R t'$. If C is the highest value of $W(a)$ for $a \in \Sigma$, then from $t \circ\!\!\to_R{}^k s$ we conclude $k \leq W(t) \leq C|t|$, proving $\mathsf{dc}_R(n) = O(n)$.

For $n > 1$ the termination proof of $R_1 \cup R_2$ is given by weights W_1 and W_2 satisfying $W_1(\ell) > W_1(r)$ for $\ell \to r \in R_1$, $W_2(\ell) = W_2(r)$ for $\ell \to r \in R_1$, and $W_2(\ell) > W_2(r)$ for $\ell \to r \in R_2$. Choose $C \in \mathbb{N}$ such that $C > W_1(\ell) - W_1(r)$ for all $\ell \to r \in R_2$. Define $W(a) = CW_2(a) + W_1(a)$ for $a \in \Sigma$, then combining the above properties yields $W(\ell) > W(r)$ for all $\ell \to r \in R_1 \cup R_2$. So termination of $R_1 \cup R_2$ is proved by Lemma 2 by choosing $R' = \emptyset$ and the new weight W. Now the theorem follows by applying the induction hypothesis on $R = \bigcup_{i=1}^{n-1} R'_i$ in which $R'_1 = R_1 \cup R_2$ and $R'_k = R_{k+1}$ for $k = 2, \ldots, n-1$. □

If termination of $\circ\!\!\to_R$ is proved by applying Theorem 4 for $R' = \emptyset$, then $\mathsf{dc}_R(n) = O(n)$ immediately follows from Theorem 4. In particular, this holds for proofs by match-bounds via Theorem 6, as Theorem 6 was proved by applying Theorem 4 for $R' = \emptyset$. However, by combined application of Theorem 4 much longer derivation lengths can be achieved: even by combining a single application of Theorem 4 with a single application of Lemma 2 exponential derivation lengths can be obtained, as is shown in the following example.

Example 8. Let the SRS R consist of the four rules

$$aL \to Lbb,\; Rb \to aR,\; BL \to R,\; RB \to LB.$$

One easily shows that Ba^kLB rewrites to $a^{2k}LB$ for every $k \geq 0$, so B^kaLB rewrites to $a^{2^k}LB$. As the increase of size is exponential in the size of the original string, it has at least exponential derivation length: $\mathsf{dc}_R(n) = \Omega(2^n)$. Note that the reduction does not exploit the cycle structure, so the exponential derivation length both holds for cycle rewriting and string rewriting.

For proving termination of $\circ\!\!\to_R$ we apply Lemma 2 by choosing $W(B) = 2$, $W(R) = 1$ and $W(a) = W(b) = W(L) = 0$. As $W(\ell) > W(r)$ for the last two rules $\ell \to r$, and $W(\ell) = W(r)$ for the first two rules, it remains to prove termination for the first two rules. For doing so we apply the tropical case of Theorem 4 for $R' = \emptyset$ and the following type graph:

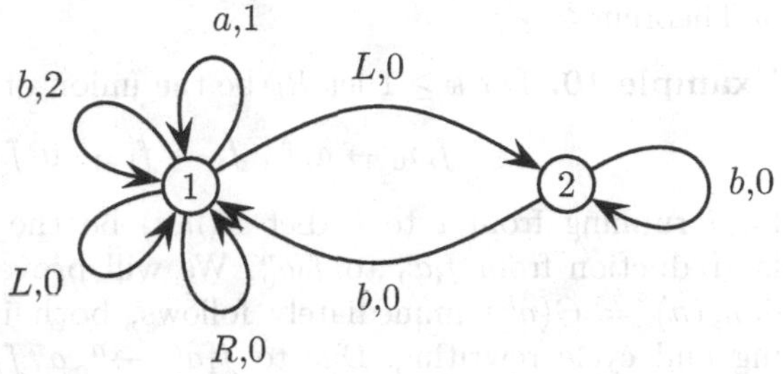

It contains the following paths labeled by left hand sides:

- $1 \to_a 1 \to_L 1$ of weight 1 to be replaced by $1 \to_L 2 \to_b 2 \to_b 1$ of weight 0,
- $1 \to_a 1 \to_L 2$ of weight 1 to be replaced by $1 \to_L 2 \to_b 2 \to_b 2$ of weight 0,
- $1 \to_R 1 \to_b 1$ of weight 2 to be replaced by $1 \to_a 1 \to_R 1$ of weight 1,

by which all requirements of Theorem 4 hold and termination of $\circ\!\!\to_R$ can be concluded.

In a first view the following observations look quite contradictory to the observation that R has exponential derivation length. The first two rules have linear derivation lengths since we found a proof by Theorem 4 in which $R' = \emptyset$, and by the application of Lemma 2 we concluded that the number of applications of the other two rules is linear in the size of the original string. But the example clearly shows what is going on: between consecutive applications of the third and fourth rule the length of the string is doubled, and after doubling a linear number of times the length has increased to exponential size, after which the first two rules can be applied an exponential number of times.

The following theorem states that the situation is quite different if the lengths of the strings do not increase. An SRS R is called *non-length-increasing* if $|r| \leq |\ell|$ for all $\ell \to r \in R$.

Theorem 9. *Let R be a non-length-increasing SRS and let $R' \subseteq R$ satisfy all requirements of Theorem 4, and assume $\mathsf{dc}_{R'}(n) = O(f(n))$ for some function f. Then $\mathsf{dc}_R(n) = O(nf(n))$.*

Proof. Take an arbitrary $\circ\!\!\to_R$ reduction $[u_0] \circ\!\!\to_R [u_1] \circ\!\!\to_R \cdots \circ\!\!\to_R [u_k]$ of length k starting with a string u_0 for which $|u_0| = n$. Since R is non-length-increasing we obtain $|u_i| \leq n$ for all $i = 0, \ldots, k$. From the proof of Theorem 4 we conclude that the total number of $\circ\!\!\to_{R \setminus R'}$-steps in this reduction is at most Cn for some constant C. From $|u_i| \leq n$ for all $i = 0, \ldots, k$ and $\mathsf{dc}_{R'}(n) = O(f(n))$ we conclude that the maximal number of consecutive $\circ\!\!\to_{R'}$-steps in this reduction is at most $Df(n)$ for some constant D. Combining these observations yields $k \leq Cn + (1 + Cn)Df(n)$, from which we conclude $k = O(nf(n))$. □

So as a consequence, we conclude that if for a non-length-increasing SRS R termination of $\circ\!\!\to_R$ is proved only by consecutive application of Theorem 4, then $\mathsf{dc}_R(n) = O(n^k)$ for k being the number of consecutive applications: R has polynomial derivational complexity. Next we give an example showing that every polynomial derivational complexity can be achieved by non-length-increasing systems for which termination can be proved by repeated application of Theorem 4.

Example 10. For $k \geq 1$ let R_k be the union of

$$f_i a_0 \to a_i f_i,\ f_i \to f_{i-1},\ a_i f_0 \to f_{i-1} a_0,$$

for i running from 1 to k. Let $F(n,i)$ be the number of steps of a particular reduction from $f_i a_0^n$ to $f_0 a_0^n$. We will prove $F(n,k) = \Theta(n^k)$, from which $\mathsf{dc}_{R_k}(n) = \Theta(n^k)$ immediately follows, both in the setting of string rewriting and cycle rewriting. Due to $f_1 a_0^n \to^n a_1^n f_1 \to a_1^n f_0 \to^n f_0 a_0^n$ we obtain $F(n,1) = 2n+1$. For $i > 1$ we consider the reduction

$$\begin{aligned} f_i a_0^n \to^n a_i^n f_i \to^i a_i^n f_0 &\to a_i^{n-1} f_{i-1} a_0 \to^{F(1,i-1)} a_i^{n-1} f_0 a_0 \\ &\to a_i^{n-2} f_{i-1} a_0^2 \to^{F(2,i-1)} a_i^{n-2} f_0 a_0^2 \\ &\to a_i^{n-3} f_{i-1} a_0^3 \to^{F(3,i-1)} a_i^{n-3} f_0 a_0^3 \\ &\cdots \\ &\to f_{i-1} a_0^n \to^{F(n,i-1)} f_0 a_0^n, \end{aligned}$$

yielding $F(n,i) \geq \sum_{j=1}^{n} F(j,i-1)$. Now one proves $F(n,i) > (1/i!)n^i$ by induction on i, using the well-known property $\sum_{j=1}^{n} j^{i-1} \geq (1/i) * n^i$, concluding the proof.

Next we prove termination of $\circ\!\!\to_{R_k}$ by repeated application of Theorem 4 and its special instance Lemma 2. First remove $f_k \to f_{k-1}$ by counting f_k, that is, choose $W(f_k) = 1$ and $W(a) = 0$ for all $a \neq f_k$ in Lemma 2. Next apply Theorem 4 by choosing the following type graph, where both a_i and f_i in the left stand for $k+1$ copies, for i running from 0 to k. One easily checks that both $f_k a_0$-paths can be replaced by an $a_k f_k$-path of lower weight, while for all other rules $\ell \to r$ all ℓ-paths can be replaced by an r-path of the same weight. So the rule $f_k a_0 \to a_k f_k$ may be removed. Next remove $a_k f_0 \to f_{k-1} a_0$ by counting a_k. The remaining system now is R_{k-1}, on which the argument is repeated until all rules have been removed.

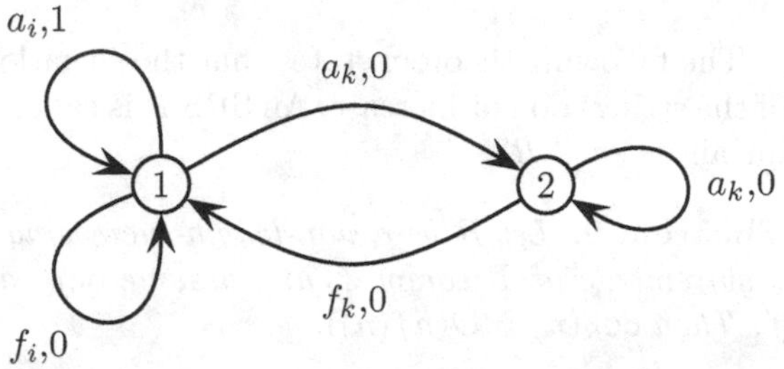

5 End Symbols

In this section we show that for SRSs R of a particular shape termination of $\to_R$ and termination of $\circ\!\!\to_R$ coincide. The special shape is characterized by an *end symbol*: a symbol E that only occurs as the last element of left hand sides and right hand sides of rules. We show that any SRS can be transformed to an SRS of this special shape, preserving termination and derivational complexity. As a consequence, termination of $\circ\!\!\rightarrow_R$ is undecidable, and for every computable function F an SRS R exists such that $\mathsf{dc}_R(n) = \Omega(F(n))$.

Lemma 11. *Let R be an SRS over Σ and let $\mathsf{E} \in \Sigma$. Let $R' \subseteq R$ consist of the rules of R in which E does not occur. Assume*

1. *every rule of $R \setminus R'$ is of the shape $u\mathsf{E} \to v\mathsf{E}$ for $u, v \in (\Sigma \setminus \{\mathsf{E}\})^*$,*
2. *$\circ\!\!\rightarrow_{R'}$ is terminating, and*
3. *$\rightarrow_R$ is terminating.*

Then $\circ\!\!\rightarrow_R$ is terminating.

Proof. Assume $\circ\!\!\rightarrow_R$ admits an infinite reduction $[u_1] \circ\!\!\rightarrow_R [u_2] \circ\!\!\rightarrow_R [u_3] \circ\!\!\rightarrow_R \cdots$. Then due to assumptions 1 and 2 it contains steps with respect to $R \setminus R'$, so the symbol E occurs in u_1, so there is a string of the shape $v_1\mathsf{E} \in [u_1]$. From $[u_1] \circ\!\!\rightarrow_R [u_2]$ we conclude that $v_1\mathsf{E} = u'u''$ where $u''u' = \ell x$ for some rule $\ell \to r$ in R and some $x \in \Sigma^*$ and $rx \in [u_2]$. If $u'' = \epsilon$ we may also choose $u' = \epsilon$ and $u'' = \ell x$, so in all cases we may assume that u'' is non-empty and ends in E.

As $u''u' = \ell x$, and ℓ contains no E other than in its last position by assumption 1, from u'' ending in E we conclude that $u'' = \ell y$ for some y, and $x = yu'$. If y is non-empty it ends in E, if $y = \epsilon$ then ℓ ends in E, and then by assumption 1 also r ends in E. In all cases ry ends in E. Write $u'ry = v_2\mathsf{E}$, then $rx = ryu' \simeq u'ry = v_2\mathsf{E}$, so $v_2\mathsf{E} \in [u_2]$ since $rx \in [u_2]$.

Summarizing, from $v_1\mathsf{E} \in [u_1]$ we constructed $v_2\mathsf{E} \in [u_2]$ such that

$$v_1\mathsf{E} = u'u'' = u'\ell y \to_R u'ry = v_2\mathsf{E} \in [u_2].$$

Repeating this construction yields the infinite reduction $v_1\mathsf{E} \to_R v_2\mathsf{E} \to_R v_3\mathsf{E} \to_R \cdots$, contradicting assumption 3. □

Next we define a transformation ϕ on SRSs such that an SRS R is terminating if and only if $\circ\!\!\rightarrow_{\phi(R)}$ is terminating, exploiting Lemma 11. Moreover, this transformation preserves reduction lengths. For a signature Σ we define $\Sigma' = \{f' \mid f \in \Sigma\}$ in which for every $f \in \Sigma$ the symbol f' is fresh. Apart from these f's we introduce three more fresh symbols $\mathsf{L}, \mathsf{R}, \mathsf{E}$, of which E will act as the end symbol from Lemma 11. Now for an SRS R over Σ the SRS $\phi(R)$ over $\Sigma \cup \Sigma' \cup \{\mathsf{L}, \mathsf{R}, \mathsf{E}\}$ is defined to consist of the rules

$$\begin{array}{lll} (a) & \mathsf{RE} \to \mathsf{LE} & \\ (b) & f\mathsf{L} \to \mathsf{L}f' & \text{for all } f \in \Sigma \\ (c) & \mathsf{R}f' \to f\mathsf{R} & \text{for all } f \in \Sigma \\ (d) & \ell\mathsf{L} \to r\mathsf{R} & \text{for all } \ell \to r \in R \end{array}$$

Theorem 12. *In the above setting, for any SRS R over Σ the following three properties are equivalent:*

1. $\to_R$ *is terminating,*
2. $\to_{\phi(R)}$ *is terminating,*
3. $\circ\!\!\to_{\phi(R)}$ *is terminating.*

Moreover, if a string $u \in \Sigma^*$ *admits a* $\to_R$ *reduction of* n *steps for some* n*, then* $u\mathsf{RE}$ *admits a* $\to_{\phi(R)}$ *reduction of* n *steps.*

For proving this theorem we first need a lemma.

Lemma 13. *Let* BC *consist of the rules of type* (*b*) *and* (*c*) *above. Then* $\circ\!\!\to_{BC}$ *is terminating.*

Proof. Similar to Example 8 we apply Theorem 4 using the following type graph, in which f stands for all $f \in \Sigma$ and f' stands for all f' for which $f \in \Sigma$. Now by choosing $R = BC$ and $R' = \emptyset$ all requirements of Theorem 4 hold, hence proving that $\circ\!\!\to_{BC}$ is terminating. □

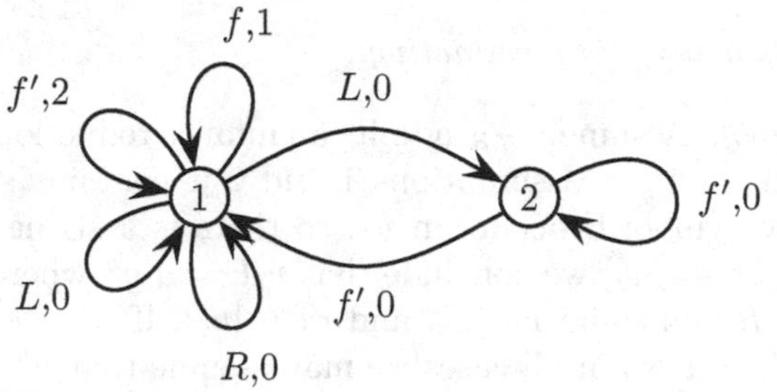

Now we arrive at the proof of Theorem 12.

Proof. 1 ⇒ 2:
Assume $\to_R$ is terminating and $\to_{\phi(R)}$ admits an infinite reduction. If this infinite $\to_{\phi(R)}$ reduction contains only finitely many (*d*) steps, then there is an infinite $\to_{\phi(R)}$ reduction that only consists of (*a*), (*b*), (*c*) steps. Then by counting R symbols there is also an infinite $\to_{\phi(R)}$ reduction only consisting of (*b*), (*c*) steps, contradicting Lemmas 1, 13. So the infinite $\to_{\phi(R)}$ reduction contains infinitely many (*d*) steps. In this reduction remove every L symbol and every R symbol, and replace every symbol f' by f, for every $f \in \Sigma$. Then every (*a*), (*b*), (*c*) step is replaced by an equality and every (*d*) steps is replaced by an R step, yielding an infinite $\to_R$ reduction, contradiction.

2 ⇒ 1:
An infinite $\to_R$ reduction is transformed to an infinite $\to_{\phi(R)}$ by putting RE behind and the following observation, also proving the 'moreover' remark in the theorem:
if $u \to_R v$ then $u\mathsf{RE} \to^+_{\phi(R)} v\mathsf{RE}$.
This is shown as follows: write $u = x\ell y$ and $v = xry$ for $\ell \to r \in R$, then

$$u\mathsf{RE} = x\ell y\mathsf{RE} \to_{(a)} x\ell y\mathsf{LE} \to^*_{(b)} x\ell\mathsf{L}y'\mathsf{E} \to_{(d)} xr\mathsf{R}y'\mathsf{E} \to^*_{(c)} xry\mathsf{RE} = v\mathsf{RE}.$$

3 ⇒ 2: Immediate from Lemma 1.

2 ⇒ 3:
We apply Lemma 11 on the SRS $\phi(R)$. First observe that $\phi(R)$ satisfies condition 1 by construction. So it remains to prove condition 2: the rules of type (*b*), (*c*), (*d*) are terminating. By counting L symbols it suffices to prove that the rules of type (*b*), (*c*) are terminating, which follows from Lemma 13. □

Theorem 14. *Termination of* $\hookrightarrow_R$ *is an undecidable property, and for every computable function F an SRS R exists such that* $\mathsf{dc}_R(n) = \Omega(F(n))$.

Proof. The main result from [6] states that every Turing machine M can be transformed to an SRS R_M such that $\to_{R_M}$ is terminating if and only if M halts on every input, proving that termination of string rewriting is undecidable. Applying Theorem 12 we obtain that every Turing machine M can be transformed to an SRS $\phi(R_M)$ such that $\hookrightarrow_{\phi(R_M)}$ is terminating if and only if M halts on every input. So also termination of $\hookrightarrow$ is undecidable.

For the second claim take a uniformly halting Turing machine M such that for every n there is a configuration of size $O(n)$ admitting $\Omega(F(n))$ transitions before halting; a Turing machine computing $F(n)$ satisfies this property. Now due to Theorem 12 $\hookrightarrow_{\phi(R_M)}$ is terminating and satisfies $\mathsf{dc}_{\phi(R_M)}(n) = \Omega(F(n))$. □

6 Implementation

We implemented all techniques presented in this paper in our tool TORPAcyc. More precisely, for a given SRS the tool tries to prove termination stepwise by tropical and arctic type graphs for increasing graph size running from 1 to 3. As soon as a rule is found yielding a strict decrease while all other rules yield a weeks decrease, this rule is removed and the process continues with the rest. For graph size 1 this corresponds to simple weight arguments, so this is by what the procedure always starts. Apart from this also the match-bound method is applied, in which the corresponding type graph is constructed by completion, typically yielding graphs of up to hundreds of nodes.

For match-bounds the implementation from the tool TORPA has been reused, as presented in [10]. For searching for type graphs the real work is done by the external SMT solver `Yices`: the requirements are expressed in an SMT formula similar to other implementations of arctic and tropical matrix interpretations, and whenever `Yices` finds a satisfying assignment, the corresponding type graph is constructed from this satisfying assignment and presented in matrix notation.

A zip file containing the source code, a Linux executable, a parameter file, the external tool `Yices` and several examples can be downloaded from `http://www.win.tue.nl/~hzantema/torcyc.zip`.

The parameter file `param` contains several parameters which may be edited. For instance, if you want to find a type graph proof with weights as small as possible, you may stepwise decrease the maximal value for type graph weights.

7 Conclusions

Syntactically cycle rewriting is the same as string rewriting, but the semantics is different: as in cycle rewriting the start of the string is connected to the end, more rewrite steps are possible, and the notion of termination of cycle rewriting is strictly stronger than termination of string rewriting. Techniques for proving termination of string rewriting based on monotone algebras and dependency pairs strongly exploit the structure of strings having a start and an end, and do not serve for modifications proving termination of cycle rewriting. In this paper we showed that tropical matrix interpretations, arctic matrix interpretations and match-bounds apply for proving termination of cycle rewriting. These techniques are the same that are used for proving linear derivational complexity for term and string rewriting in [9]. An anonymous referee pointed out that also matrix interpretations over natural numbers with the usual addition and multiplication can be applied for proving termination of cycle rewriting, even for quadratic derivational complexity.

Our result for match-bounds follows from observing that match-bounds are a special case of tropical matrix interpretations.

Apart from only proving termination we also investigated derivational complexity of our techniques, and developed transformations by which we could show that termination of cycle rewriting is undecidable and every computable derivational complexity can be reached.

References

1. Bruggink, H.J.S., König, B., Zantema, H.: Termination analysis for graph transformation systems (submitted, 2014), `http://www.ti.inf.uni-due.de/fileadmin/public/bruggink/gtst.pdf`
2. Corradini, A., Montanari, U., Rossi, F.: Graph processes. Fundamenta Informaticae 26(3/4), 241–265 (1996)
3. Endrullis, J., Hofbauer, D., Waldmann, J.: Decomposing Terminating Rewrite Relations. In: Proc. Workshop on Termination (WST 2006), pp. 39–43 (2006)
4. Endrullis, J., Waldmann, J., Zantema, H.: Matrix interpretations for proving termination of term rewriting. Journal of Automated Reasoning 40, 195–220 (2008)
5. Geser, A., Hofbauer, D., Waldmann, J.: Match-bounded string rewriting systems. Applicable Algebra in Engineering, Communication and Computing 15(3-4), 149–171 (2004)
6. Huet, G., Lankford, D.S.: On the uniform halting problem for term rewriting systems. Rapport Laboria 283. INRIA (1978)
7. Koprowski, A., Waldmann, J.: Arctic termination ... below zero. In: Voronkov, A. (ed.) RTA 2008. LNCS, vol. 5117, pp. 202–216. Springer, Heidelberg (2008)
8. Simon, I.: Recognizable sets with multiplicities in the tropical semiring. In: Koubek, V., Janiga, L., Chytil, M.P. (eds.) MFCS 1988. LNCS, vol. 324, pp. 107–120. Springer, Heidelberg (1988)
9. Zankl, H., Korp, M.: Modular complexity analysis via relative complexity. In: Proceedings of the 21st Conference on Rewriting Techniques and Applications (RTA). Lipics, pp. 385–400 (2010)
10. Zantema, H.: Termination of string rewriting proved automatically. Journal of Automated Reasoning 34, 105–139 (2004)

Author Index